Pacific International Maritime Conference (imc 2019)

Sydney, Australia
8-10 October 2019

ISBN: 978-1-7138-0871-8

Printed from e-media with permission by:

Curran Associates, Inc.
57 Morehouse Lane
Red Hook, NY 12571

Some format issues inherent in the e-media version may also appear in this print version.

Printed with permission by Curran Associates, Inc. (2020)

For permission requests, please contact The Royal Institution of Naval Architects
at the address below.

The Royal Institution of Naval Architects
8-9 Northumberland Street
London, WC2N 5DA
United Kingdom

Phone: 020 7235 4622
Fax: 020 7259 5912

publications@rina.org.uk

Additional copies of this publication are available from:

Curran Associates, Inc.
57 Morehouse Lane
Red Hook, NY 12571 USA
Phone: 845-758-0400
Fax: 845-758-2633
Email: curran@proceedings.com
Web: www.proceedings.com

TABLE OF CONTENTS

SESSION 16

SESSION 17

SESSION 18

SESSION 19

SESSION 20

OPPORTUNITIES FOR ENHANCING DEFENCE CAPABILITY GENERATION:
THE FOURTH INDUSTRIAL REVOLUTION

Josephine Delore, Lockheed Martin Australia, josephine.r.delore@lmco.com

James Keane, RAN, james.keane@defence.gov.au

Matthew Paes, RAN, matthew.paes@fleet.defence.gov.au

Michael Van Den Berg, RAN, michael.vandenberg@fleet.defence.gov.au

John Wilbraham, RAN, john.wilbraham@defence.gov.au

Martin Renilson, Royal Institution of Naval Architects, martin@renilson-marine.com

ABSTRACT

The current and future threat environment in which the Royal Australian Navy (RAN) will operate is becoming increasingly congested, lethal and asymmetric. To ensure that the Australian Defence Force 'keeps pace' with the industrial and commercial sectors while remaining ahead of its adversaries, there needs to be a constantly evolving, agile capability program that leverages from the fourth industrial revolution, Industry 4.0.

Defence industry has historically driven a significant number of the cutting-edge technologies that are used every day - from canned foods through to the internet. Ever more advanced technologies are becoming prevalent in areas of operations and regions of interest. The tempo at which these threat systems can emerge is a significant challenge to RAN's ability to acquire, certify, commission and employ capabilities that effectively counter the threats before they can evolve. The acquisition system of the future must be agile and minimally impacted by unnecessary complexity, inconsistent application of policy, or industry engagement limitations.

This paper presents a tiered framework for capability acquisition that matches the exponential growth of technological capacity. Defence 4.0 is a paradigm shift in the way capability is employed and generated. This framework has been modelled around the OODA loop and is comprised by the following four pillars: 1) Observe – Technology Intelligence Horizon Scanning, 2) Orient – Test and Evaluation, 3) Decide – Rapid Acquisition Process, and 4) Act – Deployment of Solution.

1. INTRODUCTION

Australian Department of Defence (Defence) have laid out broad innovation and reform plans for an Australian ship building capability to 2040, in the Naval Shipbuilding Plan, 2017 [1], and the 2016 White Paper [2]. The plan commits to maximising Australian industry involvement in naval shipbuilding in Australia and throughout the capability life cycle for new naval capabilities. There are tangible programs to bring together industry, academia and Defence research through initiatives such as the Centre for Defence Industry Capability (CDIC), the Defence Innovation Hub (DIH) and the Next Generation Technologies Fund (NGTF).

The Naval Shipbuilding Plan and initiatives are part of a broader objective and recognition that older styles of undertaking capability development and acquisition are not up to the task of keeping Defence capabilities effective against the rapid technological advances being adopted by adversaries.

The fourth industrial revolution, referred to as Industry 4.0, is a term that originated in Germany and defines the fundamental shift to cyber-physical systems [3]. Exciting new technological developments from Industry 4.0 are emerging such as:

1. Advanced automation and robotics;
2. Artificial intelligence (AI) and machine learning;
3. Additive manufacturing such as 3D printing; and
4. The Internet of Things (IoT).

The great promise of Industry 4.0 can only be realised if Defence has an agile capability acquisition system that works in concert with industry and academia. This paper will examine a broad view of Defence capability acquisition with a focus on the Naval context, and the application of Industry 4.0, which will be referred to as Defence 4.0.

Defence 4.0 is a paradigm shift in how capability is employed and generated. The framework has been modelled around the Observe-Orient-Decide-Act (OODA) loop and is comprised of the following four pillars:
1) Observe – Technology Intelligence Horizon Scanning,
2) Orient – Test and Evaluation,
3) Decide – Rapid Acquisition Process, and
4) Act – Deployment of Solution.

The loop is designed to illustrate how the capability phases flow into and loop back to create an organic model, as shown in Figure 1.

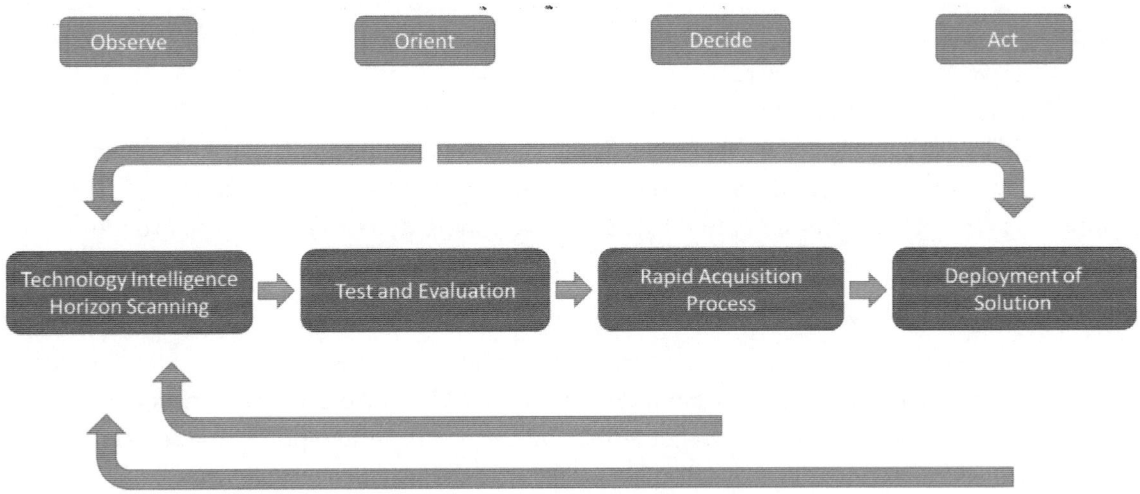

Figure 1: Defence 4.0 Capability Loop.

2. OBSERVE – TECHNOLOGY INTELLIGENCE HORIZON SCANNING

The purpose of horizon scanning is to identify and predict important global trends. Technology intelligence is a component of horizon scanning that focuses on identifying technology that can disrupt future growth. This section will explore technology intelligence horizon scanning activities being conducted by the Defence community (with a focus on RAN) and provide recommendations on how RAN can strengthen their technology intelligence.

2.1. Horizon Scanning in Defence

Defence uses several mechanisms to assist with horizon scanning, as illustrated in Figure 2. As part of Vice Chief Defence Force (VCDF) Group's Force Design Process, VCDF methodically analyses the current force structure for gaps and opportunities [4]. Defence also relies heavily on DST Group for advice. DST identified the task of scanning the environment to understand emerging threats and opportunities as one of its key roles [5]. To capture ideas from industry and academia, initiatives such as the DIH and NGTF have been established. These programs help Australia identify potentially disruptive technologies, enabling these areas to be proactively evaluated.

On an international level, Defence participates in The Technical Cooperation Program (TTCP). TTCP is an alliance between defence scientists from Australia, Canada, New Zealand, United Kingdom and the United States (Five Eyes Alliance). It is Australia's strongest tie to defence-related research being conducted in partner nations, encouraging idea sharing and open platforms. TTCP recognises that member nations do not have the individual resources to completely cover every area of defence research and that all partners can enhance their national defence at reduced cost through collaboration [6, 7].

These principles need to be applied in Defence's relationship with industry and academia. The private sector's research investment into areas such as "Big data" (managing large datasets) and AI can be leveraged by RAN. The DIH and NGTF are initiatives of the 2016 Defence Industry Policy Statement [8] that provide funding to companies and organisations to develop technologies that have been identified as being important to Defence. Another initiative was the Emerging Disruptive Technology Assessment Symposium (EDTAS) [9]. EDTAS are a series of presentations and workshops that have been run since 2015 and are co-hosted by DST, academia and industry. They explore the potential for technology to disrupt areas of interest to DST and provide guidance on priority areas for the NGTF. These programs are a step in the right direction, but more can be done to engage the broader Defence community enabling collaborative research.

As advanced threats are rapidly manifesting, RAN's ability to develop and bring into service new capabilities will be reliant on an increased capability for horizon scanning. Plan Pelorus 2022 outlines the direction of RAN from Chief of Navy's Senior Advisory Committee and aims to prepare RAN for future engagements. The Centre for Innovation (CFI) is designed for the RAN workforce to trial and develop ideas [10]. The CFI has also hosted Launch Pad events, which have successfully engaged industry and given them a platform to showcase to RAN [11]. The Navy Fleet Air Arm (FAA) hosts Shark Tanks at HMAS Albatross as part of FAA Innovation Program. The program gives RAN aviation personnel a platform to present ideas to improve capability [12].

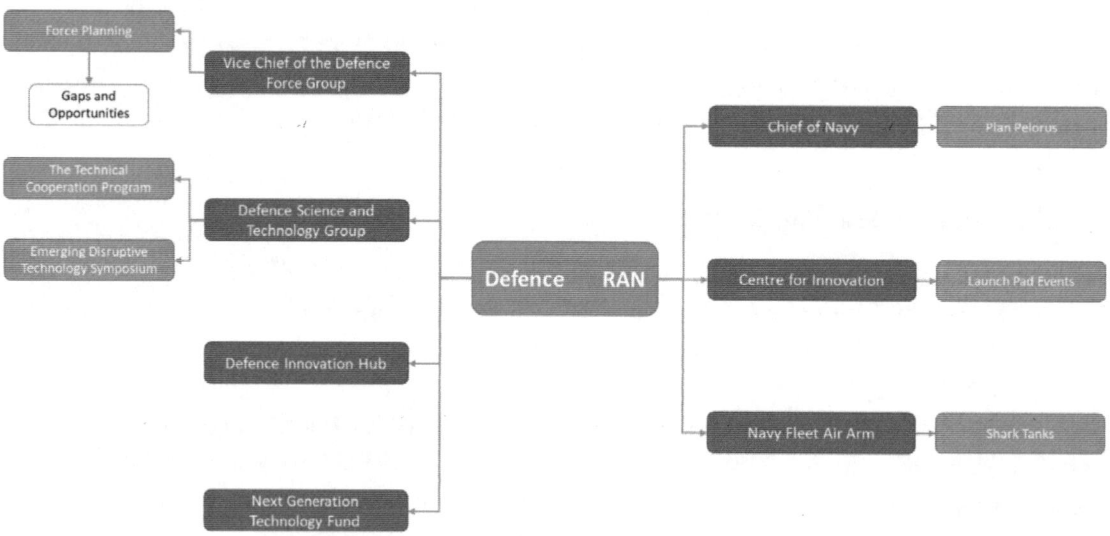

Figure 2: Defence and RAN Organisations for Innovation.

2.2. Harnessing Industry 4.0 to Improve Technology Intelligence

Horizon scanning is conducted in industry to drive innovation and help maintain a competitive edge. An important component of this is Internal Research and Development (IRAD) and research labs. Investment into research areas is guided by technology forecasting but also supports the prototyping of innovative concepts that can lead to disruptive technology. RAN encourages internal innovation, but the opportunity exists to refine and institutionalise efforts to engage with, and hence leverage, industry research. This is particularly true of international Primes who can seek to provide access to large (particularly United States) IRAD programs.

Defence benefits from engaging academia, which provides the ability to monitor the fundamental research conducted by universities (blue skies research) and recognise when the idea is mature enough to invest in. However, the pace of technology evolution is such that rather than trying to structure defence expectations around 20 to 30-year predictions, emphasis should be placed on the 10-year time horizon. Using technology readiness levels (TRLs) Defence should be aiming to support university research once it reaches TRL 4 or TRL 5 – laboratory breadboard or laboratory testing of integrated/semi-integrated system [13].

2.3. Recommendations

The following recommendations aim to develop a truly collaborative Defence-Industry relationship and maintain a competitive edge in the technology intelligence field:

- Establish a Naval Innovation Day – inspired by Army Innovation Day [14]. Yearly calls for proposals could be made, inviting industry and academia to showcase R&D activities that would enhance RAN capability. This event could focus on RAN high-priority interest areas. The aim of this day will be forging closer links with industry, encouraging/celebrating new ideas and a means to identify new concepts worthy of further investment;
- Structure within Defence how technology assessment is applied to university research. Encourage dialogue between universities by sending out Defence personnel to universities with the intent of creating an "Open Day" where universities can present research directly to Defence. Another initiative are joint research labs between Defence and universities. An example of this is the Jericho Smart Sensing Lab [15], which is a collaboration between the Royal Australian Air Force (RAAF) and the University of Sydney;
- Maintain and increase active involvement in the DIH and single service initiatives – reducing turnaround times; and
- Ensure involvement as stakeholder in the DST led NGTF programs.

3. ORIENT – TEST AND EVALUATION

To continually improve how Defence develops requirements and acquires capability, the technology options identified in horizon scanning need to be processed, analysed and synthesised into viable solutions that align with the high priority areas of Defence and RAN. The acquisition process relies heavily on the Test and Evaluation process (T&E). T&E informs decision makers of risk areas in cost, schedule and capability. It provides insight into how these risks can be managed or eliminated [16]. The following phases of T&E, shown in Figure 3, run alongside the capability life cycle:

- Preview T&E – supports the processing and assessment of options;
- Developmental T&E – supports design and development of system during acquisition;
- Acceptance T&E – verifies that the system is compliant to requirements; and
- Operational T&E – validates that system capability satisfies mission requirements.

This section will cover the preview and developmental T&E that inform how Defence sort and select options. The acceptance and operational T&E will be covered in Section 5 – Act, which will examine the verification and validation process (V&V).

Figure 3: Capability life cycle showing how the T&E phases align with the acquisition milestones [16].

3.1. Current Test and Evaluation Process

The Australian Defence Test and Evaluation Office (ADTEO) was established in 2007 to be the central authority for Defence T&E [16]. ADTEO provides support for Defence during the entire acquisition process, from helping to refine options during First Pass government approval through to Final Operational Capability.

One of the roles of the Force Integration Division of the VCDF Group is to oversee the Joint Test and Evaluation. This maintains that T&E is consistent across Defence. The Defence Test and Evaluation Roadmap 2008 [17] emphasised the need for centralised T&E as the current organisations within Defence could not cover all phases of T&E over the capability life cycle. The 2012 Senate Inquiry into Defence Procurement Procedures [16] recommended improved T&E capability to better identify and control risks pre-First Pass.

The Royal Australian Navy Test and Evaluation Authority (RANTEA) and the Aircraft Maintenance and Flight Trials Unit (AMAFTU) conduct the T&E activities within RAN. In 2015 the Australian National Audit Office (ANAO) released a report investigating Defence T&E [16]. One of the case studies used in the report was the Landing Helicopter Dock (LHD), HMAS Canberra, where the report highlighted that due to workforce vacancies capability risk had been transferred to RAN. The report recommended that T&E training is strengthened and becomes more centralised so that Defence can have a complete overview of the organisation's T&E capability.

3.2. Integrating Industry 4.0 with Test and Evaluation

As acquisitions of complex systems increase, T&E tasks will become more critical. The 2008 Roadmap [17] outlined the important role of industry in conducting T&E and providing Defence with independent test facilities. It highlighted that obtaining information from industry was an issue that impeded T&E. The 2018 Department of Defense (DoD) Digital Engineering Strategy [18] recommends that one option to increase industry engagement is by integrating virtual environments to allow easier transfer of information such as digital design and lifecycle management embodied by the digital twin concept. A digital twin is a dynamic and functioning virtual representation of a physical system. This helps to inform decision making by increasing design transparency and mitigates risks around systems complying to standards. RMIT University recently proposed the Virtual Design, Optimisation and Testing framework (VDOT), which is a system using AI that will be able to process the increasing amount of data available to Defence [19].

The technological developments in AI and digital design that can optimise T&E will not be able to be realised without more investment in a T&E workforce that has consistent training across Defence.

Recommendations

The following recommendations would help Defence improve agility and risk management in the T&E process:

- Feedback mechanisms from T&E must take priority in Defence 4.0 system engineering frameworks;
- A strong, centralized T&E force that is capable of surging to meet project needs;
- Industry should be given more transparency into the T&E process to build trust and open pathways to data sharing; and
- Incorporating end-to-end digital environments.

4. DECIDE – RAPID ACQUISTION PROCESS

The Observe phase of the Defence 4.0 cycle gives data-driven insights into the decisions that are needed. However, these observations only drive probabilistic recommendations. A decision must still be made, and the capability acquisition, and sustainment gears must begin turning in order to align a new capability with Defence's existing systems and assets.

4.1. Current Acquisition Process

Traditional acquisition processes and cycles are not prepared for the demands of timeliness and quality that will be needed by Defence 4.0. Contemporary efforts to modernise the acquisition cycle through Capability Acquisition and Sustainment Group (CASG) have trialled different project administrative strategies, such as the DDG alliance. CASG current goals include implementing First Principles Review (FPR) recommendations, to align strategic partnerships with industry, and to improve relationships with owners (Government) and customers (Capability Managers).

Implementing the Smart Buyer process is an example of modernising the acquisition phase of the capability life cycle. Yet modern acquisition strategies are still aimed at supporting technology available circa 2020. The acquisition process must continue to evolve in order to be able to integrate modular, sophisticated technologies into existing and future systems.

4.2. Rapid Acquisition Processes Under Defence 4.0

The Australian Standard for Defence Contracting (ASDEFCON) suite of contracting templates are used to provide a standardised set of documents of procurement [20]. Under Defence 4.0, rapid acquisition processes will need to be adaptable. In the same manner that the Observe phase recommends which capability to acquire; suitable acquisition contracts must be matched to that capability. Future contracts can adapt to the technology or system being acquired based upon 1) its TRL [13], 2) trust in the supplier (based off past performance and other metrics such as the company scorecard), and 3) capability priority and urgency.

The commitment to the fourth wave of deterrence research, that has emerged from the mid-1990s, demonstrates the scale and scope of the spectrum of future technologies that will need to be integrated [21]. Specialised technologies will require increasingly niche skillsets to acquire and implement. It is imperative for Defence 4.0 as a whole that this is achieved, so that capability managers will have trust in their system when Acting (deploying solutions). ASDEFCON will be automatically matched to the technology under consideration, critically reducing time and resources required. Standardised T&E techniques and measures will become a critical framework providing system feedback.

Matching an acquisition with a suitable contract structure should be the perfect 'task' for existing machine learning technologies. A key risk however will remain data quality, in that any recommender system is only as good as its inputs, and it will only deliver outputs asked of it. The proposal herein is certainly not intended to be a complete solution, it is intended to assist human decision makers in order for them to make better decisions in a timelier manner. Notwithstanding, this concept will likely prove to become a critical component of capability acquisition. This will be driven by necessity where complex software refresh cycles must be integrated into hardware and software lifecycles with vastly different requirements.

An emerging disruptive technology to acquisition capability is blockchain. Blockchain defines digital records (blocks) that are stored in a public database (chain). In 2018 it was announced that AUCloud will partner with Quant Network to develop a blockchain operating system for the Australian Government [22]. This system will be used by Defence and has the potential to significantly improve the efficiency of the acquisition process by allowing smart contracts between industry and Defence. A smart contract would be committed to the blockchain and be executed automatically by the computer. This allows for the construction of contracts from standard templates by AI.

A shake-up to the traditional acquisition model employed by the United States Intelligence community is the use of the independent venture capital firm, In-Q-Tel. The firm is backed by the Central Intelligence Agency (CIA), and has disrupted how the United States Intelligence community conduct their scanning, partnering and acquisition process. The Australian Government has formed a partnership with In-Q-Tel to allow Australian agencies access to emerging technologies acquired by In-Q-Tel [22]. The NGTF is an example of Defence scanning and acquiring emerging technology, however this has focused on mid to large scale programs. The DIH provides an avenue for start-ups to submit proposals to receive Defence funding whereas the In-Q-Tel model proactively seeks out start-ups that have technology that can be ready for use within 6 to 36 months [23].

4.3. Recommendations

Defence has taken the first steps toward mastering the 21st century acquisition environment, however, adoption and evolution of Defence 4.0 rapid acquisition processes is critical. Some techniques have been proposed, however new innovations should be continually be assessed. The following recommendations would add value to CASG in the short-medium term:

- Immediate focus on ability to acquire software and robotics systems of increasing autonomous sophistication;

- Put T&E at the core of the capability life cycle;
- Continually incorporate new information from the Observe phase to maximise the likelihood that adequate resources are available to acquire and implement new capabilities;
- Launch an Australian version of In-Q-Tel; and
- Implement blockchain-based contracts and use AI to automate their construction from ASDEFCON templates, reducing cost and time of acquisition.

5. ACT – DEPLOYMENT OF SOLUTION

The final phase, once the course of action has been decided, is the rapid deployment of the solution. Prospective threats are changing ever more rapidly, and the combination of longer lead times to develop new equipment, coupled with the increased rate of change of threats, is a significant concern for military planners. Rapid deployment relies on understanding the situation and workforce management – as vehicles and equipment become more complex the user group needs to be selected, trained and retained so that deployed systems have maintained support.

5.1. Current Transition from Acquisition to Operational

The maritime industry is not immune from the increasing length of time required to develop new equipment. There are big challenges to bringing new systems online after the acquisition phase. Currently Defence needs to integrate new technology into legacy systems, interconnect networks and collaborate with contractors. Defence is respected as a leader in providing consistent structure to certification, V&V and intellectual property (IP) strategy but as start-ups, new industry partners and academia are increasingly involved in Defence projects more can be done to streamline these processes and build strong relationships between industry and Defence.

Defence strategy recognises that innovative technologies require significant effort and funds to develop, and there needs to be trust that industry IP will be protected at the end of the acquisition cycle. In 2016 an IP strategy was released [24] to encourage industry to participate in the DIH. Coordination of cross-disciplinary environments across Defence, academia and industry will streamline each stage of a project life cycle and create an intuitive feedback loop back into generating new capability.

RAN has initiated a large investment in innovation, this includes many unit level initiatives to embrace Industry 4.0 such as additive manufacture trials [10]. Additive manufacture allows RAN to replace parts and experiment with solutions while at sea. This is done in workshops modelled off the CFI mentioned in Section 2.1.

5.2. Rapid Deployment using Defence 4.0

Current Defence capability generation within the Australian maritime domain could be arguably described as reactive. Data collected through the Observe phase informs the acquisition cycle on where there are gaps and opportunities, but there is also an immense amount of data being collected from intelligence, surveillance and reconnaissance (ISR) that can inform decisions at the deployment stage. A large step that is needed in processing ISR data is the standardisation of data formats, communication architectures and enhanced, automated meta-data tagging (e.g. automatic generation of semantic labels for entities in satellite imagery). This will not only be useful in the identification and evaluation of threats but in the sustainment and optimisation of systems to combat the threats. An example of this is the Global Fishing Watch [26] which connects data from Automatic Identification Systems (AIS), satellite imagery and IoT to determine a vessel's purpose. Primarily this has been used in the identification of illegal fishing vessels, one of the RAN's busiest constabulary tasks. The data gathered using these techniques can then also inform horizon scanning and acquisition.

Operational data can also be paired with game theory to provide modern tactics or be used in simulation software/equipment and serious games for training the Defence workforce. Serious games or gamified training creates an engaging and motivating learning experience. Simulation software and equipment such as augmented reality provide new ways to easily process large amounts of information through visualisation and enable collaboration between remote sites.

Open source data, standards and architectures enhances collaboration between industry, academia and Defence. This allows rapid responses to emerging problems and threats leveraging extant data sources and simulation capability. Free and open source systems are often supported and reviewed by larger communities than proprietary systems, increasing the likelihood that defects and vulnerabilities are detected earlier. The challenges of IP and security policy also apply here and would need to be closely managed by the data manager who is the single source of truth. The importance of open systems is supported by the memorandum signed by the Secretaries of the United States military, which stated that a Modular Open Systems Approach (MOSA) is imperative to mission success [27].

V&V and the sustainment of operational assets would both be helped by using a digital environment, as introduced in Section 3.2. Digital environments allow the creation of a digital twin, which is a digital representation of the physical system. Digital twins are continually updated with the physical system's performance and maintenance data throughout its life cycle. This is currently utilised in monitoring physical health of systems such as aircraft engines and Formula 1 cars [25] for accurate engineering fatigue and maintenance requirements and would offer high value to RAN for materiel seaworthiness.

5.3. Recommendations

The following recommendations would help the rapid deployment of an acquisition:

- Increased emphasis and high-level support for standardisation of data formats and architectures;
- Defence policy supporting collaboration with industry is critical to stimulate the industry investment into the disruptive technologies;
- Defence resources are allocated to include mentoring startups and new industry partners through the certification and V&V phase;
- V&V is supported by digital twins; and
- Defence workforce is supported by gamified training.

6. CONCLUSION

The motivation for designing Defence 4.0 around the OODA loop was to demonstrate how each phase of capability development informs the next but also has the potential to improve those preceding it. The recommendations put forward in the four phases: Observe – Technology Intelligence Horizon Scanning, Orient – Test and Evaluation, Decide – Rapid Acquisition Process, and Act – Deployment of Solution, could have significant impact on how Defence generates capability by leveraging the technological developments of Industry 4.0.

REFERENCES

1. Department of Defence. (2017). *Naval Shipbuilding Plan*. Retrieved from http://www.defence.gov.au/NavalShipbuilding/Docs/NavalShipbuildingPlan.pdf
2. Department of Defence. (2016). *2016 Defence White Paper*. Retrieved from http://www.defence.gov.au/WhitePaper/Docs/2016-Defence-White-Paper.pdf
3. RMIT University. (2019). What is industry 4.0? Retrieved from https://www.rmit.edu.au/industry/develop-your-workforce/tailored-workforce-solutions/c4de/industry-40
4. Blain, J. (2017). *Force Design Presentation to the Defence Innovation Hub Industry Update*, presented at Force Design, Canberra, 2017. Retrieved from https://www.business.gov.au/-/media/Business/CDIC/Documents/Industry-update-event/BRIG-Jason-Blain-Force-Design-PDF.pdf?la=en&hash=43C98C4CBFEF9780057C85130873D5BF5452DAA9
5. Defence Science and Technology. (2018). *DST Strategic Plan 2013-2018*. Retrieved from https://www.dst.defence.gov.au/sites/default/files/publications/documents/DST_Strategy_Completion_Report.pdf
6. Defence Science and Technology. (2019). The Technical Cooperation Program. Retrieved from https://www.dst.defence.gov.au/partnership/technical-cooperation-program
7. The Technical Cooperation Program. (2018). The Technical Cooperation Program: Overview. Retrieved from https://www.acq.osd.mil/ttcp/overview/
8. Department of Defence. (2016). *2016 Defence Industry Policy Statement*. Retrieved from http://www.defence.gov.au/WhitePaper/Docs/2016-Defence-Industry-Policy-Statement.pdf
9. Defence Science and Technology. (2019). Emerging Disruptive Technology Assessment Symposium (EDTAS). Retrieved from https://www.dst.defence.gov.au/events/emerging-disruptive-technology-assessment-symposium-edtas
10. Willee, B. (2018, November 5). Centre for Innovation continues to push boundaries with new technology. *Navy Daily.* Retrieved from http://news.navy.gov.au/en/Nov2018/Fleet/4908/Centre-for-Innovation-continues-to-push-boundaries-with-new-technology.htm#.XTep_OgzYuV
11. Kemp, F. (2016, October 23). Droning on about innovation. *Navy Daily.* Retrieved from http://news.navy.gov.au/en/Oct2016/Fleet/3293/Droning-on-about-innovation.htm#.XTlH2egzYuV

12. McMaugh, D. (2018, September 14). Innovation pays off for Fleet Air Arm. *Navy Daily.* Retrieved from http://news.navy.gov.au/en/Sep2018/Fleet/4819/Innovation-pays-off-for-Fleet-Air-Arm.htm#.XSKiFegzYuV
13. Defence Science and Technology. (2019). Technology Readiness Levels Definitions and Descriptions. Retrieved from https://www.dst.defence.gov.au/sites/default/files/basic_pages/documents/TRL%20Explanations_1.pdf
14. Australian Army. (2019, June 13). *Call for submissions for Army Innovation Day 2019* [Press Release]. Retrieved from https://www.army.gov.au/media-room/media-releases/call-for-submissions-for-army-innovation-day-2019
15. Royal Australian Air Force. (2019). Jericho Smart Sensing Lab. Retrieved from https://www.airforce.gov.au/our-mission/plan-jericho/jericho-smart-sensing-lab
16. Australian National Audit Office. (2015, November 24). *Test and Evaluation of Major Defence Equipment Acquisitions*. Retrieved from https://www.anao.gov.au/work/performance-audit/test-and-evaluation-major-defence-equipment-acquisitions-0
17. Department of Defence. (2008). *Defence Test and Evaluation Roadmap 2008*. Retrieved from http://www.defence.gov.au/publications/docs/Defence_Test_Evaluation_Roadmap_2008.pdf
18. Department of Defense. (2018). *Digital Engineering Strategy 2018*. Retrieved from https://www.acq.osd.mil/se/docs/2018-DES.pdf
19. RMIT University. (2019, February 25). *Industry 4.0 comes to Defence* [Press Release]. Retrieved from https://www.rmit.edu.au/news/all-news/2019/feb/vdot-defence-system
20. Capability Acquisition and Sustainment Group. (2019). ASDEFCON Suite of Tendering and Contracting Templates. Retrieved from http://www.defence.gov.au/casg/DoingBusiness/ProcurementDefence/ContractingWithDefence/PoliciesGuidelinesTemplates/ContractingTemplates/asdefcon.aspx
21. Knopf, J. W. (2010). The fourth wave in deterrence research. *Contemporary Security Policy*, *31*(1), 1-33.
22. Grubb, B. (2019, June 11). The CIA's investment fund is stalking Australian tech startups and has opened a local office. *The Sydney Morning Herald*. Retrieved from https://www.smh.com.au/business/banking-and-finance/the-cia-s-investment-fund-is-stalking-australian-tech-startups-and-has-opened-a-local-office-20190610-p51w3r.html
23. In-Q-Tel. (2019). Insights & Access. Retrieved from https://www.iqt.org/
24. Department of Defence. (2016). *Defence Innovation Hub Intellectual Property Strategy*. Retrieved from https://www.dst.defence.gov.au/sites/default/files/basic_pages/documents/Innovation_Hub_IP_Strategy.pdf
25. Egan, M. (2015, October 4). *'Digital Twin' Technology Changed Formula 1 and Online Ads. Planes, Trains and Power Are Next.* Retrieved from https://www.ge.com/reports/digital-twin-technology-changed-formula-1-online-ads-planes-trains-power-next/
26. Google. (2019). Oceans of data: tracking illegal fishing over 1.4 billion square miles. Retrieved from https://sustainability.google/projects/fishing-watch/

27. Spencer, R. V., Esper, M. T., Wilson, H. (2019). *Memorandum for Service Acquisition Executives and Program Executive Officers.* Retrieved from Defense Standardization Program website: https://www.dsp.dla.mil/Portals/26/Documents/PolicyAndGuidance/Memo-Modular_Open_Systems_Approach.pdf?ver=2019-01-18-122921-933

Developing Australia's Independent Surface Combatant Design Capability – Challenges and Options

Levi Catton CPEng MIEAust, Gibbs & Cox Australia (levi.catton@gibbscox.com.au)
Dr David Shackleton AO, Gibbs & Cox Australia (david.shackleton@gibbscox.com.au)

ABSTRACT

The Government's 2017 Naval Shipbuilding Plan indicates that over the next decade or more, Australia intends to develop significant new naval industrial capabilities. In particular, the plan states a policy requirement for... *"future Australian Governments being able to plan and execute: the design, construction and sustainment of future fleets of major surface combatants ... in Australia."* This objective has subsequently been emphasised through various ministerial statements.

Nations that maintain a warship design capability do so for strategic national security and military-economic reasons. The ability to design a competitive warship takes multiple decades to develop and must be proactively managed. A variety of measures and significant investments in skilling and technology are necessary to achieve a credible capability. In the context of this policy objective, this paper addresses some important questions relating to how the RAN will be supported in its exploration of the future of the Australian surface combatant fleet. These include:

a. Can Australia reasonably achieve an independent ability to develop globally competitive warship designs?
b. What are the benefits of such a capability?
c. How can such an ability be developed?
d. How can Australia's naval sector gain the skills and experience needed to establish, execute and assure such an undertaking?

These questions are central to how an independent sovereign warship development capability, and the operational capability of the RAN, will be developed in a manner which responds to long-term strategic uncertainty.

INTRODUCTION

In 2017 Government released the Naval Shipbuilding Plan (NSP), which outlined a structured approach to developing a sustainable naval sector (Commonwealth, 2017). Government's plans for investment in the naval sector indicate a then-year approx. $0.5 trillion enterprise over acquisition and sustainment of the currently planned naval programs. This context establishes, for the first time in Australia, conditions in which the pursuit of a sovereign, sustainable, end-to-end surface combatant design capability is a credible goal.

The Hunter Class Frigate Program (HCFP) represents a major portion of the investment under which the objective of sovereign combatant design is intended to be developed. In June 2018, Defence Minister Marise Payne said *"by the conclusion of the frigate build, ASC Shipbuilding will be a strategic national asset capable of independently designing, developing and leading the construction of complex, large naval warships"* (Commonwealth, 2018a). Under the

SEA5000-1 Head Contract, Government has formed an agreement with ASC Shipbuilding to develop a design capability to an extent that will be defined later in the program.

The Defence Industry Capability Plan states that *"... a motivated, innovative, cost-competitive and sustainable Australian industrial base, underpinned by experienced ship designers and builders who translate these attributes to Australian industry"* is a *"key enabler"* for delivering Australia's maritime and ASW capabilities (Commonwealth, 2018b).

The Government's objective of sovereign combatant design presents a range of significant opportunities, challenges, and necessities for the Australian naval, marine, technology, manufacturing and education sectors. This paper outlines some of those opportunities and challenges, and provides some concepts around how a plan for a truly sovereign combatant design capability might be developed.

MAINTAINING OPTION SPACE

At the cornerstone of a design capability is the Navy's place as the end user, with CASG as its acquisition agent, and the need to maintain a clearly understood option space enabling strong decision making throughout the capability development process. To achieve this, Australia needs the ability to explore the relationships between user need, system capability, industry capability and investment. Design is the activity which mediates these key program considerations. Without the ability to conduct advanced early design and cost analysis, the ability to respond to the decision space across these considerations is limited, and acquisition decisions are forced into accepting compromises.

As the program progresses, Government will seek evidence that the large investment it is making in the naval sector will be maximised for the benefit of Australia's naval capability, industry, knowledge capital, and strategic independence. In future assessment, reporting and decision making, it is important that Navy, CASG and Government have access to advice that is not linked to the commercial exigencies of a specific program or entity.

CAN AUSTRALIA SUSTAIN A CREDIBLE WARSHIP DESIGN CAPABILITY?

Figure 1 indicates the relationship of GDP to Defence budget for naval designing nations, not including the US (c. USD$649 Billion 2018 spend) and China (c. USD$250 Billion 2018 spend). Annex A provides further details related to the GDP and defence spend of naval designing nations. Figure 1 indicates that, above a basic threshold, maintenance of a naval design capability is not strongly correlated to GDP or total defence budget, with credible frigate design capabilities being maintained across two orders of magnitude of defence spending between $4 Billion and $650 Billion. Figure 1 shows that in terms of real defence spending, Australia is placed centrally amongst middle powers, with resourcing comparable to some important naval design centres including the Netherlands, Spain, Italy and South Korea.

By comparison with the case of Spain, we can conclude that independent combatant design, including integration of key US systems such as AEGIS, can reasonably be sustained within current Australian naval spending. In this comparison, the higher cost of operation in Australia is offset by the 50% greater defence budget applied by Australia.

Ultimately, the sustainment of a design capability depends not on defence spending level, but on how well the domestic naval sector is managed in terms of consistency and design cycle times. For instance, South Korea, Turkey, Spain and Japan have all taken a progressive system program approach to their naval development programs, which provides the consistency of design scope to sustain and evolve their naval design capabilities.

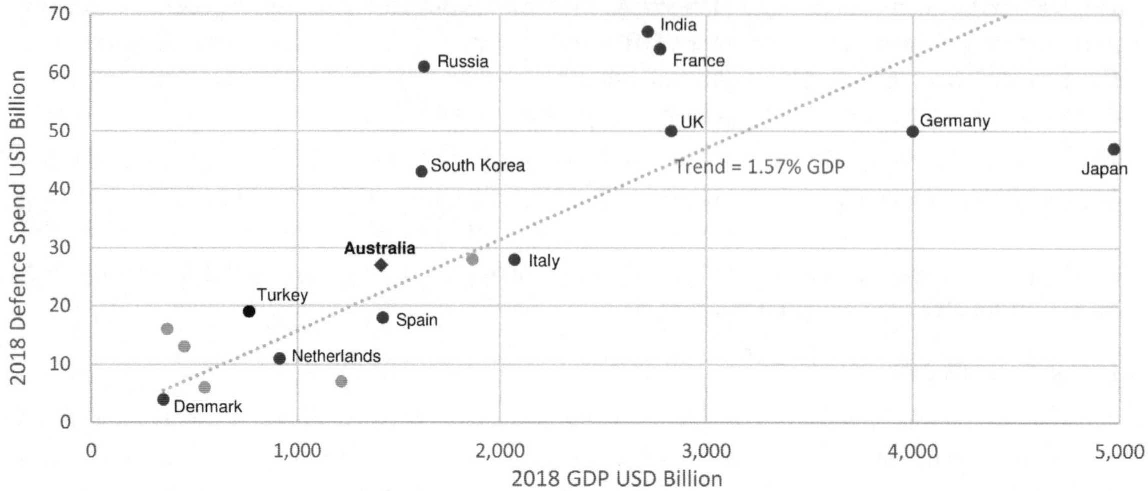

Figure 1 – Relationship of GDP to Defence budget, for nations with a credible naval combatant design capability, not including US and China. *Nations with a high-level combatant design capability are labelled against a black point, while nations with a lower-level design capability are shown as grey points. See also Annex A.*

Although most seafaring European nations have held a naval design capability over the long term, the Spanish capability was substantially re-vitalised based on US design technology transfer during the 1980's. The South Korean capability was also substantially established based on US design technology transfer during the 1990's. In both of these cases, the capability was established or re-established based on a long-term strategic objective of the national government to achieve an independent modern naval design capability, and in both cases the effort has been substantially successful. Nearly every country that has maintained or established a naval design capability has realised some export opportunities; with some countries, such as Spain, France and Germany, establishing significant industrial capacity and revenue on the basis of naval exports.

The broader Australian naval sector is unlikely to be cost competitive in the foreseeable future due to the cost of operation in Australia, but can achieve a competitive advantage through technology and capability leadership.

WHY PURSUE INDEPENDENT SOVEREIGN COMBATANT DESIGN?

The benefits of sovereign naval design fall into four broad areas. These are: strategic independence; technological and economic benefits; capability alignment; and sustainment benefits. These four areas of benefits are mutually reinforcing.

Addressing Strategic Uncertainty

Sovereign design provides Australia with an increased level of self-sufficiency in naval capability generation and sustainment, although combat system development is likely to continue to rely on foreign, in particular US, programs and participation. Removing reliance on foreign support for the provision of platform design markedly improves optionality for

Government and increases the long-term security of Australia's capacity to generate and sustain a competitive naval force.

The development pathways in combat systems are complex and demanding, and becoming more so with the advent of hypersonics, lasers, etc. These systems all have to be integrated in a survivable platform with a long service life, during which time every warfare technology we know and use now will change multiple times. This requires a level of capacity in the platform system to support future warfare systems development. Design provides Navy with the ability to define and realise the types and amounts of growth capacity that is needed to address the capability uncertainty that Navy faces in the longer term.

Related to this issue is the need to build naval ships in smaller design batches, allowing for a more progressive development of capability through the build program. Australia's capability margins are narrow, and require regular technology updates to sustain. Smaller build batches with larger capability development steps between design updates allows major technology changes to be incorporated in a timely way, providing a framework to consistently maintain Australia's capability margins.

Realising Technological and Economic Benefits

The economic impact of Australian industry content in a future Australian-designed naval ship can be expected to be greater than on previous and current programs, due to the large proportion of equipment and parts from foreign supply chains under those programs. Designing Australia's future combatant fleet in Australia will provide significantly greater opportunities for Australian industry to participate in the supply chain. Furthermore, a local design will be optimised for sustainment using the Australian supply chain and industry resources. This would be a ground-breaking step forward in the Australian naval sector, which has historically subsisted in the context of substantial foreign involvement in the acquisition and sustainment supply chain.

By providing a clearer understanding of the trade-offs between user need, system capability, industry capability and investment, an early design capability will help Government and Defence to set priorities and give guidance to other parts of the naval sector about where they should invest and develop.

A local design capability supports local development of naval systems. Combatant design demands a broad range of specialist technical skills, and represents a significant technological capacity, which has spin-off effects on participating and adjacent sectors. Australia already has some world-leading naval technology development programs and such development will be strengthened by the existence of a system program approach to platform design, which will align with and help support current combat system development programs. This will ultimately lead to more competitive capability outcomes.

Alignment to Australian End-User Needs

An Australian design is able to comprehensively address the end user capability requirements within the available investment envelope. Cole (2017) addressed the question of the relative value presented by an off-the-shelf approach to naval acquisition. While a developmental acquisition approach typically requires a higher up-front cost and schedule investment, it also delivers long-term offsets in total cost of ownership through better alignment to the support system, and increased value through better alignment of the capability to the end user need. For these reasons, full design development generally provides a stronger response to longer-term interests. An off-the-shelf approach is clearly indicated in cases where a schedule imperative is driving the acquisition, and the typically longer development cycle of a more developmental design is impractical, such as in the current case of the HCFP.

An Australian design capability supports Defence's role as an informed customer. Sovereign design will enable the Navy to make well-informed choices and compromises on what is required, within the level of resourcing Government is prepared to provide.

Alignment to Australian Sustainment Context

Some of the most significant advantages of sovereign design come in sustainment. Naval ship designs are generally designed to be sustained using the industrial resources of the parent country. By fully aligning a design with the Australian supply chain and the local sustainment ecology, significant savings in cost of ownership can be achieved. A smaller batch approach with more regular design updates also provides for opportunities to improve the supportability characteristics of the design by incorporating support lessons from early hulls into follow-on design updates.

TOWARDS AN INDEPENDENT SOVEREIGN CAPABILITY

A Short-Term Need for a Long-Term Strategy

The long-term horizon associated with naval development requires steps to be taken sooner than later to have confidence that they will reach maturation at the appropriate time. There are a range of examples past and present where failure to adopt a long-term view has prevented the most satisfactory solution from being undertaken. Although HCFP will be delivered over a long period of time, the ability to design a competitive warship also takes multiple decades to develop and requires early and proactive management to provide a reasonable assurance of success. Having established a plan, it requires long-term commitment to develop the capability.

The Critical Link Between Capability Requirements, Cost and Design

Costing is a critical and complementary part of design. The capability analysis and decision support aspects of early design are not meaningful without reliable cost information, and later-stage design activities must be conducted in the context of ongoing affordability assurance. The link between design and costing activities is critical to program cost control and must be foundational to the design approach. This is link is significantly stronger in a responsive local design capability.

Related to this is the need to manage affordability against user need, and control the cost and scope of requirements. A common cause of program problems is the growth of requirements and cost past an authorised baseline. The solution to this problem is the use of smaller build batches with larger capability development steps between design updates. This allows capability and requirements to be developed more progressively, with requirements, cost and design baselines aggressively protected after they are approved. This approach provides improved cost assurance, and protects cost performance in later stage design and production activities.

International Partnering Considerations

BAE Systems has previously outlined an approach to designing a future Australian surface combatant via collaboration with a future UK program to replace the Royal Navy's Type 45

(Stewart, 2019). This approach inherently limits the level of sovereign, independent warship design capability that BAE Systems can be expected to pursue in Australia, because building such a capability would create a competitor to the sovereign warship design capability of the UK and impact the future export opportunities of the UK Naval Ships business. The proposed partnering arrangement falls short of the truly sovereign approach sought by Government.

The vast majority of attempted international collaborations in warship design have failed when the necessary compromises on requirements, participation and supply chain could not be achieved. Of the many attempts, there is one known successfully executed multilateral design program in the history of NATO surface combatant design; the French / Italian Horizon frigate – which itself went through a major program split, with the UK retiring to develop T45. It is instructive to note that one of the key reasons that the UK retired from the Horizon program was because of their insistence that UK primes take the lead in the development effort on Horizon.

Finally, previous experiences such as Collins indicate that major program issues are hard to export back to international contractors, and ultimately are most likely to be resolved by the sovereign nation that owns the problem.

Approach Alternatives

There are a few typical high-level approaches to distributing scope across a naval ship design activity. Three approaches relevant to the current Australian context are suggested below. Other possible options do not adequately take account of the current agreement in place with ASC Shipbuilding for the development of a design capability, or are fundamentally variations of the options indicated.

Approach 1 – Extend current ASC Shipbuilding intentions beyond development of a design modification / design support capability, to develop an end-to-end sovereign design capability, independent of other programs. Examples of this model include the UK Type 45 and Type 26 programs.

Approach 2 – Establish an independent design partner to provide concept, preliminary and functional design capabilities, which collectively with the design capability to be developed in ASC Shipbuilding, provides a sovereign design capability. An analogue of this model is the Canadian Arctic Offshore Patrol Ship (AOPS) project, although in that case not all of the participating designers were Canadian.

Approach 3 – Establish a national collaborative design capability, incorporating existing and foreshadowed elements of design capability in Australia including the future ASC Shipbuilding design capability, and developing the additional capabilities necessary through design partners (such as but not limited to Gibbs & Cox) in a teaming arrangement. This model is comparable to the approach to combat systems development through CMPS, or the Naval Design Partnering teams in the UK.

A COLLABORATIVE MULTI-PARTICIPANT APPROACH

A credible Australian design capability must involve a range of participants. We envisage the formation of a naval design centre based on collaborative involvement of key qualified government and industry participants, including Navy Capability, Navy Engineering, CASG, DSTG, ASC Shipbuilding, and specialist design partners.

An Australian Naval Design Centre would be established based on a commercial management model, possibly under the management of a lead contractor, and reporting into CASG, probably under FAS Ships. Personnel from the various participants would be assigned, posted, or contracted into the organisation as applicable, to support the required labour profile.

The mission of the Australian Naval Design Centre (ANDC) would be to provide naval surface ship design solutions to Defence through the collaboration of Defence, industry and academia.

The ANDC would be funded mainly by the CASG project offices for whom the ANDC is providing design services. It's likely that some Defence and ADF positions within the ANDC could be supported directly by secondment from participating branches and groups, such as HNC, HNE, CASG and DSTG. Commercial participants such as ASC Shipbuilding and other specialist design partners would be funded through appropriately structured strategic contracts with CASG, managed under FAS Ships. The ANDC would likely have two physical centres, in Canberra and Adelaide, with whole-ship design work likely experiencing a transition from Canberra-centric in requirements and concept design development, to Adelaide-centric from functional design through to productionisation.

IP management would be a key issue requiring early attention. It is expected that foreground IP developed by the ANDC would be owned by the Commonwealth. However, there are complicating factors associated with how commercial participants bring background IP to their participation in the design work. In relation to the Commonwealth this can be handled through the development of appropriate licensing agreements, however, there will be concerns around distribution of proprietary background IP across participants. This will need further examination. A reasonable level of compromise might be required from commercial participants in relation to background IP, in order to enable them to participate efficiently in the ANDC. This could represent an important test of an entity's willingness to bring the necessary collaborative behaviours to the work.

Significant benefits are available through the implementation of a collaborative approach, and these are outlined in Table 1.

Key Benefit	**Characteristics supporting realisation of benefit**
Best practices applied	• Various participants with diverse experiences bring their key strengths to the enterprise • Establishes an internal marketplace of design and engineering practices and solutions, from which the best practices and solutions are more likely to arise • Collaboration of ASC Shipbuilding and other design partners across design phases drives design for production interests into early design, and maintains functional design intent through detail design • Brings Systems Engineering processes that are aligned with the Commonwealth's approach to acquisition
Increased protections and optionality for Defence	• The Commonwealth is not locked in to a single supplier, and retains enduring flexibility in how scope is allocated across participants • Provides an inherent challenge function between parties that helps regulate inappropriate behaviours, and reduces risk of any one party's commercial motivations negatively influencing the program • Reduces the risk of any one party holding too much freedom to interpret the Commonwealth's requirements in a manner most beneficial to itself • Substantially frees design work from commercial influence of the construction scope, but objectively considers production cost impacts

Key Benefit	Characteristics supporting realisation of benefit
Broadens impact of technological and economic benefits	• Builds on existing sovereign capability by integrating and building up existing niche contributors in the Australian naval engineering community • Distribution of the design capability across a range of organizations makes the technological benefits more impactful across the sector
Best alignment with policy direction	• Reflects Defence and Navy Industry policy by strengthening collaboration • Mirrors the approach to sovereign combat systems development capability • Incorporating existing Commonwealth capabilities such as DSTG leverages existing investments

Table 1 – Some benefits of a collaborative multi-participant approach to naval design

Parallels with UK MOD's Naval Design Partnering Teams

The ANDC concept is comparable to the UK MOD's Submarine and Surface Ship Naval Design Partnering (NDP) teams, which are Defence and Industry consortia. These constructs arose due to a shortage of qualified and experienced personnel within the MOD. In response, NDP teams were formed to support the MOD in retaining its informed customer status and maintaining sovereign capabilities. The NDP teams have involved parallel participation from multiple first and second tier naval sector suppliers.

The teams have been involved in activities such as pre-concept and concept phase whole boat/ship design and technical studies; developing the embodiment of new technology into existing platforms; supporting the MOD in undertaking through-life technology management studies; technical support to Technology Demonstration Programmes (TDPs); and dealing with emerging technical issues in current platforms (UK Ministry of Defence, 2016).

The NDP construct shows that a participative approach can be constructed which provides Defence with the necessary expertise and support to fulfil its objectives, by drawing together the expertise which exists across competitor parties within the national naval sector.

A System Program Approach to Platform Design and Integration

The HCFP CS development effort is being directed by the Combat Management and Payload Systems (CMPS) branch of CASG. The recent formation of CMPS represents the introduction of a system program approach to naval combat systems in Australia. This will continue to evolve towards an integrated warfare systems approach as sovereign capability develops.

The need for a collaborative approach was recognised in structuring the development program for the HCF CS, acknowledging that the scope of functions was likely beyond the extant capabilities of any one business, and would have created unhealthy competition in the marketplace. Therefore, for CS development, lead participants deliberately took a partnering approach, to build on existing strengths in the sector, share risk, and establish sovereign capability across a range of government and commercial entities.

Having established a system program approach to Combat Systems development, Australia should consider the implementation of a system program approach to platform design and ship integration. Nearly all naval designing countries that have established or re-established their design capability in the post WWII era have done so via a system program approach, which sets out a progressive development of increasingly sophisticated combatants. Some specific examples include:

- The South Korean KDX program, consisting of 3 KDX-I frigates, 6 KDX-II frigates, 6 KDX-III AEGIS Destroyers, and a planned 6 KDX-IIA AEGIS frigates
- The Turkish MILGEM program, consisting of 4 corvettes, followed by 4 frigates, then a planned further 4 hulls to a more advanced frigate design
- The Spanish F100 program, which produced three hulls to the F100 design, made progressive design updates to hulls F104 and F105, then commenced development of the significantly updated and modernized F110 design, with a planned 5 hulls
- The Japanese Kongo-Atago-Maya class progression (4, 2, and 2 hulls respectively), which demonstrates a consistent development pathway for Japan's AEGIS destroyer fleet

Each of these programs commenced with an introductory combatant design, and made step changes in capability and technology across relatively small build batches, typically representing around a third of the combatant fleet size. In the context of Australia's planned combatant fleet of 12 hulls, such an approach would indicate significant design updates at every 3-4 hulls. This aligns with the proposed build of approximately 3-hull batches in the HCFP. Under the system program approach to platform design taken in comparable naval designing nations, each of these design updates would be used to drive an increasing level of ship capability and Australian design capability, with the full design capability being realised and matured through the design of an AEGIS DDG-equivalent ship for the Hobart class replacement program.

RESOURCING AND SKILLING THE DESIGN ENTERPRISE

Developing the necessary qualified and experienced personnel is the central challenge to Australia's development of a competitive naval enterprise. In response to this challenge, Government recently released the Naval Shipbuilding Strategic Workforce Discussion Paper (Commonwealth, 2019).

In development of a design capability, the time horizons are long. Full maturity is only achieved across an entire class lifecycle, by incorporating the design, supply chain, operational and support experiences from the previous class. Patience is needed in this strategic endeavour.

Australia is an unforgiving environment when criticism can be levelled at shortcomings in major government programs. Key to preventing such circumstances is to have expert staff on the government team who understand the implications of taking decisions now, and on outcomes which do need to emerge over a decade hence. This means that APS and Naval officers must be appropriately capable and experienced, and the necessary advisory staff established to provide such expertise.

Australia has made long-term commitments to some major US combat system elements. Therefore, it is in Australia's interest to further develop our technical engagement with these key US combat system elements and the US naval sector generally. Additionally, development of the workforce necessary to sustain a competitive naval sector in Australia will be enhanced by deliberate engagement and cooperation with the naval sectors in both the US and UK. One avenue that could support this is to progress the 4 EYES Defence free trade zone.

SUSTAINING THE DESIGN ENTERPRISE

One of the next steps in developing Australia's design capability is to design the work profile. As stated earlier, the sustainment of a design capability depends but on how well domestic naval work is managed in terms of consistency and design cycle times. Sovereign design can be developed and sustained by correctly phasing naval ship design tasking over the next 25 years, in the same way that Australia's naval production capacity will be sustained by appropriately phasing the ship construction scope under the NSP. There is sufficient design work available in the current naval investment program to achieve this, if it is structured appropriately. Figure 2 indicates the scope of design work available to sustain combatant design. Some guiding constraints arise through the contractual arrangements established for current programs.

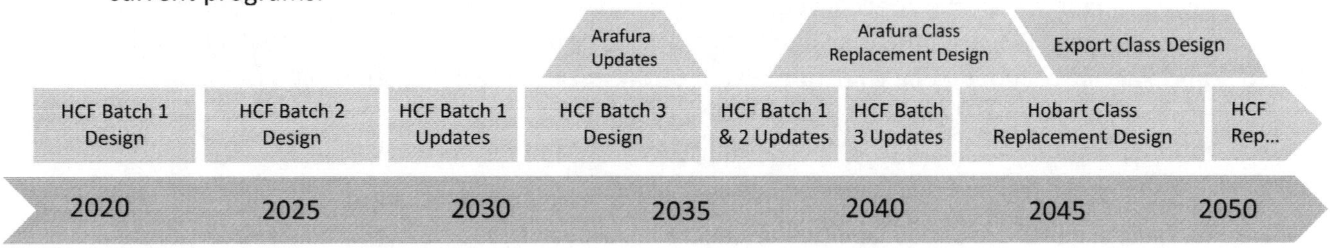

Figure 2 - Notional timeline of future naval surface ship design scope

AREAS FOR FURTHER DISCUSSION

This paper outlines a broad case and presents some basic concepts for consideration. There are a range of important questions that need to be examined as the suitability of a collaborative design capability is assessed. These questions include:

- Who are the wider stakeholders and how do they need to be engaged?
- What are the necessary elements of a sovereign warship design capability?
- What elements currently exist in Australia?
- What elements will be provided through the HCFP Head Contract?
- What elements need to be supplemented?
- Who are the participants, and what are their roles across the design enterprise?
- How are these participants best integrated through contracts and agreements?
- What is the appropriate management structure to maintain a strong focus on cost effective and timely delivery of design products, while appropriately integrating Defence and Industry personnel across a range of participants?
- What are the key risks that exist, and what backstops have to be established?
- How do we define the roadmap to a mature sovereign design capability and how will progress be measured along the way?

CONCLUSIONS

A long-term continuous naval ship production program in Australia allows, for the first time, credible pursuit of a combatant design capability. Establishing an independent combatant design capability is a Government policy, and an important response to increasing strategic uncertainty. It is a long-term objective which will require patient development.

Most other middle power nations which rely on seaborne trade for economic security have maintained, established or are establishing a naval design capability. Compared with other nations, including high labour cost nations, Australia's current naval spending is sufficient to sustain an advanced combatant design capability. Sustainment of naval design is enabled by two key conditions: firstly a willingness to accept some program risk in order to reap the benefits of a sovereign capability, and secondly an appropriate phasing of design work scope.

Most naval designing nations, including other high labour cost nations, have realised some level of export success, including both ship designs, technology transfer, and built ships. The international OPV and frigate market will continue to grow over the next 2-3 decades. Australia will be able to compete in this market in the future through technology and capability leadership.

The Government has stated an objective for Australia to develop a sovereign, independent design capability, but the specific approach to achieve this objective has not yet been clearly defined. There is a stated expectation that BAE Systems will develop this capability in its subsidiary ASC Shipbuilding over the course of the Hunter Class Frigate program. However, previous statements about their preferred approach to future warship design in Australia, and potentially competing commercial interests, are somewhat in conflict with this goal.

A national collaborative design capability may be the most beneficial approach to establishing Australia's naval design capability. An Australian Naval Design Centre would incorporate existing and foreshadowed elements of design capability in Australia, and develop the additional capabilities necessary through design partners in a teaming arrangement. Key benefits of this approach include the application of best practices, increased protections and optionality for Defence, and broader impact of technological and economic benefits across the naval sector. A comparable approach has been demonstrated in the UK naval sector.

Australia should consider the implementation of a progressive system program approach to platform design and ship integration, as has been done by most other nations that have recently established or re-established a naval design capability, and as has been done for combat system development in Australia.

Development of the workforce capabilities necessary to grow and sustain a competitive naval design capability and broader naval sector in Australia will be greatly enhanced by strong links and cooperation with the well-established and experienced naval sectors in the US and UK.

This paper has discussed some concepts around the establishment of Australia's naval design capability. In further assessing the suitability of a collaborative design capability, there are some key questions that need to be examined by stakeholders. Defence has an opportunity to lead a more open and inclusive debate about what Australia's future naval design capability looks like.

Australia faces a major strategic opportunity to develop a truly sovereign naval design capability, and realise greater strategic independence; broad technological and economic benefits; an increasingly competitive naval capability; and improvement of the sustainment system and cost of ownership. We are now at the inflexion point between planning and implementing Australia's future naval enterprise, and the course we establish for sovereign design in the next few years will substantially define the extent to which we reap these benefits over the next 50 years.

REFERENCES

Cole, J. (2017). First and Second Order Technical Implications of an Off-The-Shelf Approach to Naval Ship Acquisition. *Pacific 2017 International Maritime Conference.*

Commonwealth. (2017). *Naval Shipbuilding Plan.* Retrieved from Department of Defence - National Naval Shipbuilding Enterprise: http://www.defence.gov.au/NavalShipbuilding/Plan/

Commonwealth. (2018a). *The Hunter class - defending Australia and securing our shipbuilding sovereignty*. Retrieved from Department of Defence - Ministers: https://www.minister.defence.gov.au/minister/christopher-pyne/media-releases/joint-media-release-prime-minister-minister-defence-2

Commonwealth. (2018b). *Defence Industry Capability Plan.* Retrieved from Department of Defence - Strategic Policy & Intelligence: http://www.defence.gov.au/SPI/Industry/CapabilityPlan/

Commonwealth. (2019). *Release of Naval Shipbuilding Strategic Workforce Discussion Paper.* Retrieved from Department of Defence - Ministers: https://www.minister.defence.gov.au/minister/cpyne/media-releases/release-naval-shipbuilding-strategic-workforce-discussion-paper

Stewart, N. (2019). *Continuous Naval Shipbuilding: prosperity from a sovereign industry*. Retrieved from BAES Australia Newsroom: https://www.baesystems.com/en-aus/article/nigel-stewart-at-the-adm-congress

UK Ministry of Defence. (2016). *Enduring Submarine Naval Design Partnering (NDP) Industry Day.* Retrieved from Defence Contracts Online: https://www.contracts.mod.uk/do-mod-announcements/enduring-submarine-naval-design-partnering-ndp-industry-day/

ANNEX A

Comparison of GDP and Defence Budgets of Naval Designing Nations

Table A1 provides a list of countries that maintain a credible independent naval design capability, ranked by 2018 Gross Domestic Product (GDP). Capability is indicated at two broad levels; the higher level (noted by a green indicator) being the ability to design an internationally competitive multi-role heavy frigate / destroyer sized combatant, and the lower level (noted by an amber indicator) being the ability to design a regionally competitive corvette / light frigate sized combatant. The table also gives the respective 2018 defence budgets for context, and some representative examples of programs that demonstrate the design capability.

Country	GDP Rank	2018 GDP USD BB	2018 Defence Spend USD BB	G20 Member	Combatant Design Capability	Representative Programs	Exports
United States	1	20,494	649	Yes	✔	DDG1000, DDG51, LCS	Yes
China	2	13,407	About 250	Yes	✔	Type 052D, Type 054A, Type 055	Yes
Japan	3	4,972	47	Yes	✔	Atago, Maya	No
Germany	4	4,000	50	Yes	✔	F124, F125	Yes
United Kingdom	5	2,829	50	Yes	✔	Type 45, Type 26, Type 31	Yes
France	6	2,775	64	Yes	✔	Horizon, Aquitaine	Yes
India	7	2,717	67	Yes	✔	Shivalik, Kolkata, Visakhapatnam	Yes
Italy	8	2,072	28	Yes	✔	Horizon, Bergamini	Yes
Brazil	9	1,868	28	Yes	●	Barroso (Corvette)	No
Russia	11	1,631	61	Yes	✔	Admiral Gorshkov	Yes
South Korea	12	1,619	43	Yes	✔	Daegu, KDX-III Sejong the Great	Yes
Spain	13	1,426	18	Via EU	✔	F100, F110	Yes
Australia	14	1,418	27	Yes	⊗	N/A	N/A
Mexico	15	1,223	7	Yes	●	Durango (Corvette)	No
Netherlands	17	913	11	Via EU	✔	De Zeven Provinciën, Holland	Yes
Turkey	19	766	19	Yes	✔	Istanbul, Ada	Bid
Sweden	22	551	6	Via EU	●	Visby (Corvette)	No
Iran	27	452	13	No	●	Moudge (Light Frigate)	No?
Israel	32	370	16	No	●	Sa'ar 5-6 (Corv., shared designs)	Yes
Denmark	36	351	4	Via EU	✔	Iver Huitfeldt, Absalon	Bid

Table A1 – List of countries with a credible naval design capability, ranked by 2018 GDP

ANNEX B

Summary Characteristics of Representative Levels of Design Capability

Design Capability	Relative risk and technological benefit	Significant Outcomes
No design capability - build to print		• Reduced programmatic risk in design activity • International supply chain
Develop production design drawings	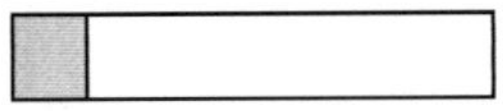	• Productionisation capability • Aligns production documentation with build yard administrative processes
Develop production design	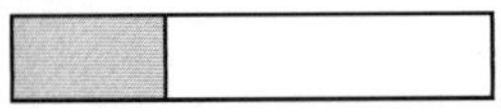	• As above, plus: • Detailed design capability • Increased alignment of production design with build yard production processes • Substantially international supply chain • Some opportunity for Australian supplier inclusion for minor parts and stock materials
Design modifications & updates	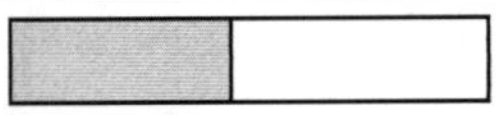	• As above, plus: • Limited functional design capability
Early design and cost analysis	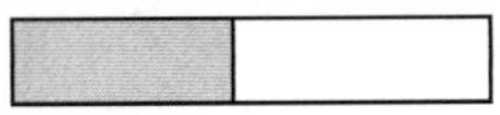	• Requirements analysis and decision support capability • Concept design capability • Preliminary design capability • Cost analysis capability
Participate in multilateral project to evolve design	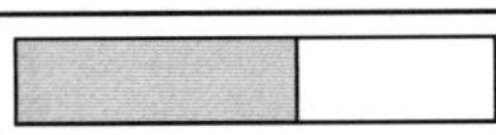	• Detailed design and productionisation capabilities • Limited concept, preliminary and functional design capabilities; the extent depends on cooperation structure and differences in partner programs • Opportunity to negotiate inclusion of Australian suppliers in the bilateral design, but Australian suppliers would compete with partner nation suppliers • Cost-sharing benefits arising from collaboration • Challenges and constraints arising from design, programmatic and supply chain collaboration
Independent major design evolution	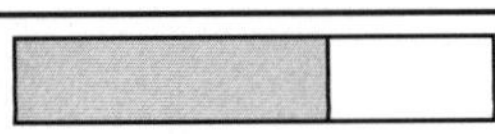	• Detailed design and productionisation capabilities • Substantial concept, preliminary and functional design capabilities; extent depends on design change scope • Opportunity to include additional Australian suppliers at the cost of design change
Sovereign naval ship design capability	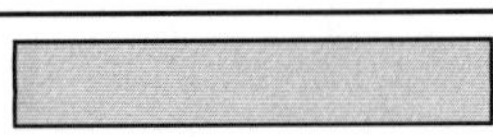	• Establishes end-to-end design capability • Maximum opportunity for Australian suppliers • Maximum alignment to RAN user need • Higher cost and technical risk profile • Maximum technological benefits in supporting sectors • Sovereign IP and ownership of export opportunities

Table B1 – Relative risk/benefit and significant outcomes associated with a notional set of tiered design capability levels

Why Smart Ship Technology is the Future of the Australian High Speed Craft Industry

Max van Someren, Austal, max.van-someren@austal.com; Robert Jackson, Austal, Robert.jackson@austal.com

ABSTRACT

Since the early 1990's the Australian maritime industry has led the world-wide market for large, high-speed, light-weight aluminium craft. However international competition in this sector is significant and growing.

Emerging smart ship technologies offer a once in a generation opportunity for the Australian high speed craft (HSC) industry to continue to build on its successful track record of innovation. Data can help us to build, operate and maintain more efficient, greener and more comfortable ships. The industry needs to move to producing ships that are not only fast but also *smart*.

High speed craft require a smart ship solution tailored to their unique design characteristics and operating conditions. We will describe the journey that Austal and its partners has been on over the last two years to develop MARINELINK-Smart, a smart ship system designed specifically to meet the needs of high speed craft customers.

We will explain how techniques such as optimisation of trim, throttle, route and other ship settings can add up to significant fuel savings and motions improvements for both naval and commercial ship operators.

Finally, we will provide case studies from at-sea installations of MARINELINK-Smart to quantify the financial and operational benefits possible from the application of smart ship technology to high speed ships.

CHALLENGES FACING THE AUSTRALIAN HIGH SPEED CRAFT INDUSTRY

Since the early 1990's the Australian maritime industry has led the world-wide market for large, high-speed, light-weight aluminium craft. Figure 1 shows Australian leadership of the market for HSC over 60 m in length in most years since 1990.

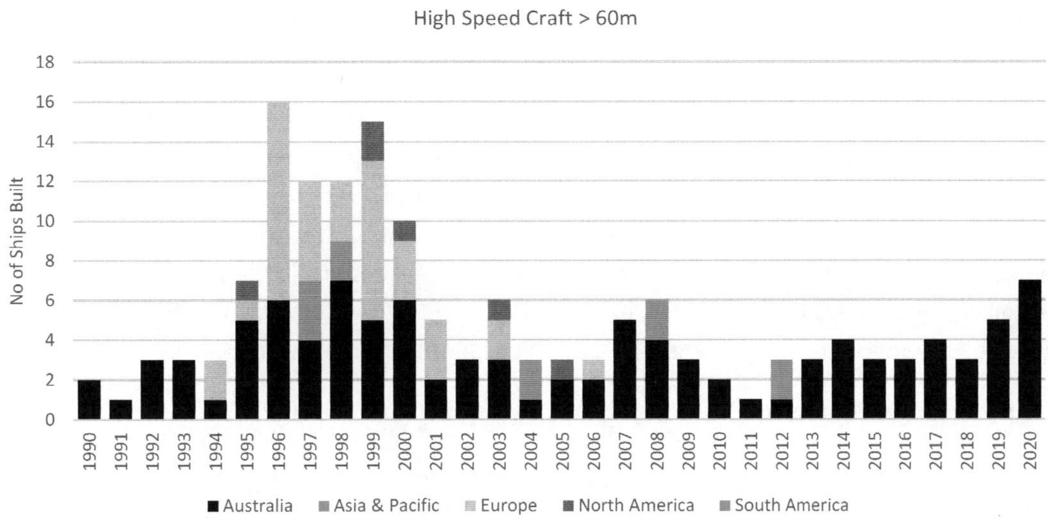

Figure 1 - Annual Large HSC Designs By Country 1990 – 2020 (Commercial & Naval) [1]

The demand for HSC capable of carrying cars and trucks as well as foot passengers led to an increase in the size of HSC in the late 1980's and early 1990's. Australian industry pushed the boundaries of technology, competing to offer customers ever larger, faster and more capable ships. As ferry customers built up confidence in the product and its ability to win revenue from slower speed ships the market for large HSC flourished. This led to the rise of the first competition outside Australia in the mid 1990's, primarily from Europe. Several shipyards produced a range of large HSC designs. However most new entrants to the market found the challenges of successfully welding large aluminium marine structures difficult to overcome and by the end of the decade the majority of large HSC were once again being built in Australia.

The proven expertise of the Australian large HSC industry led to continued export success for Australian ships, designers and equipment suppliers over the next twenty years. Perhaps the most significant example of this was the success Austal achieved in selling Australian-designed ships to the Littoral Combat Ship and Expeditionary Fast Transport programs for the US Navy.

The market for large HSC is now stronger than at any time since the early 1990s. Reasons for this include commercial HSC operators ordering ships to replace the ageing HSC fleet from the previous building boom. However to maintain the Australian HSC industry's leading position as the market continues to grow will require continual innovation to stay ahead of the competition. What options are available to the Australian HSC industry to build on its previous successes?

THE INTERNET OF THINGS AT SEA AS AN OPPORTUNITY TO DIFFERENTIATE

There are a variety of ways in which the Australian HSC industry might differentiate its products. The solution advocated in this paper is that the industry complement more traditional forms of R&D investment (e.g. hydrodynamics) with emerging digital sensor, communications and computing technology to build a differentiated proposition. Now commonly referred to as the Internet of Things, or IoT, this collection of technologies is both a challenge and an opportunity to every manufacturing industry (shipbuilding or otherwise).

Competitors can also design and build ships which integrate off the shelf IoT technologies to gather real-time data on ship operations. But to turn this in to enhanced operational efficiency, reduced fuel bills and improved customer comfort has so far not proven (in Austal's experience at least) to be a simple process. The Australian maritime industry, which has more experience of HSC design and construction than anywhere else in the world, is well-suited to overcome this challenge. In doing so it can built a long-term, competitive advantage which differentiates an Australian designed HSC from the global competition.

WHAT VALUE CAN DATA PROVIDE HSC CUSTOMERS?

The majority of large HSC operating today are commercial ferries carrying either passengers or passengers and vehicles. Operators of passenger ships tend to measure three specific forms of value when considering the performance of their ship:

1. **Efficiency.** Reducing fuel consumption to transport a given load will improve the profitability of the ship operator's business, all other factors being equal. HSC offer passengers faster journeys but generally at additional operational cost per passenger mile. This means that improved efficiency often offers relatively larger profit improvements for operators of HSC compared to ships operating at slower speeds. Improved efficiency is also associated with reductions in greenhouse gas emissions which is becoming increasingly important to operators.

2. **Comfort.** Passengers have choices on whether and how they travel. If they have an uncomfortable journey they may not decide to travel on the ship a second time which will impact a ship operator's revenue. The impact can be magnified if the negative experience is shared with others through social or conventional media. Certain HSC routes have been cancelled because (rightly or wrongly) they developed a poor reputation for motions and hence comfort. More positively a comfortable on-board environment is likely to lead to increased on-board passenger spend.

3. **Reliability.** A ship which is not available to perform its revenue earning function is of little use to the ship operator. As well as the direct impact of lost revenue the operator may suffer the indirect consequences of reputational impact, particularly if passengers have no other travel options. Reduced reliability also impacts planned and unplanned maintenance costs even if it does not lead to the total unavailability of the ship.

Different HSC operators may place different priorities on each of the three value types above. For instance a ship operator who competes directly with airlines may place particular emphasis on the customer experience even if this is creates additional operational expense. In contrast another operator may have an operating licence which is dependent on achieving a certain reliability of service, and so may see maintaining this as paramount.

IoT technologies offer various different possibilities for creating or enhancing each of these types of value. In generic terms IoT is about creating relevant data that was not previously available and then converting this in to actionable information for the operator.

Actionable information can be delivered to different users and on different timescales. For instance advice may be provided to the captain of the ship on adjustments that they can make to reach a more efficient running trim, and thus save fuel. The advice can be provided in a general form which is always broadly applicable (e.g. "always aim for a level static trim when loading"), in a journey specific form for a particular journey ("for the current loading and forecast weather conditions the optimal static trim is 0.2 degrees") or as advice which adapts to the moment-to-moment changing conditions ("reduce the trim tabs by 10 degrees following a 90 degree change of heading").

In another example use case IoT enables a much larger range of equipment condition data to be collected. Cloud computing power can harness these large datasets to forecast availability and drive predictive, rather than periodic or reactive, maintenance strategies. In this case the user of the data is more likely to be the on-shore asset manager, operating on a timescale of weeks, months or even years.

But what if we move away from the paradigm of providing advice to existing human users and automate the actions recommended by IoT systems? This takes us in to the realm of autonomous operations. Whilst significant focus (and maritime industry investment) in recent years has been put in to the development of systems for navigational autonomy, increasing autonomy at the ship system level should also be examined for its potential to deliver enhanced value to ship operators.

A prime candidate for this is the adaption of ride control system settings, since many aspects of the operation of these systems are already highly autonomous. A 'smart' (i.e. IoT enabled) ship has a far wider range of sensor and off-board information sources (such as external weather model data) accessible to inform the changes to a ride control system's settings. The smart ship can make a much more sophisticated assessment of the adaptions to the ride control settings necessary as the environmental conditions around the ship change.

The objective in developing Australian HSC in to smart ships is not to identify one specific adaptation to the ship which will revolutionise its value to an operator. Instead it is to put in place a hardware and software platform which allows for the systematic gathering, analysis and action upon as much relevant data as possible. This can then be used and re-used to generate a broad variety of different forms of information, advice or autonomous actions. Together these features will make the ship a highly appealing product. Furthermore, the in-built connectivity makes the ship able to continually update and improve (both hardware and software) as IoT technology continues to develop and evolve.

Figure 2 summarises some of the ways in which Smart Ships can significantly improve HSC customer value.

Figure 2 – Ways in which Smart Ships can Improve HSC Customer Value

AUSTAL'S JOURNEY IN PIONEERING IOT AT SEA FOR HIGH SPEED CRAFT

As IoT came to prominence in other adjacent industries (for example aerospace, power generation and rail) Austal launched an internal initiative to identify the implications for the future of HSC. This led to a broad review of both the potential customer value possible and the ability of emerging technology to deliver against this.

The company decided to look outside of industry to academia to see if their skills could help build a roadmap for the implementation of IoT to HSC. Austal partnered with Curtin University for a week long Design Thinking 'sprint' where academics, Austal engineers and customer representatives were taken through a structured workshop to tease out the main features of the proposed system. The final output of this process was a series of software mock ups which became the 'first draft' of the IoT system proposed.

The final result was the creation of the Austal MARINELINK-Smart platform. MARINELINK-Smart is a combination of new on-board sensors, ship data gathering hardware and software, cloud-computing storage and analysis, and live advice delivered to both ship crews and shore-side operational staff. The MARINELINK-Smart platform is summarised in Figure 3.

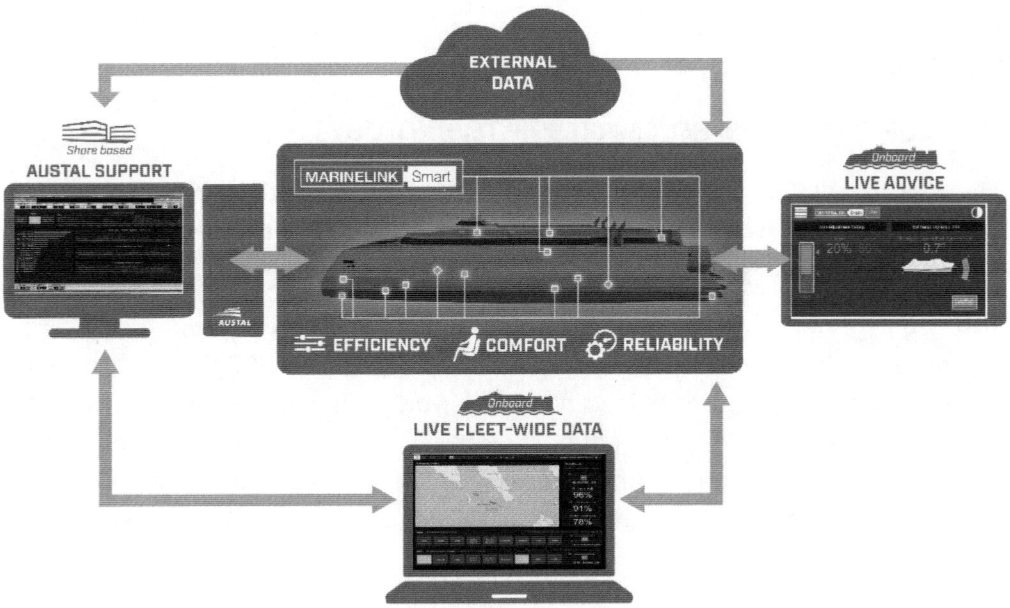

Figure 3 – Austal's MARINELINK-Smart IoT platform

Austal realised that they would need to deploy MARINELINK-Smart and prove its value quickly if they were to succeed in convincing customers of the value inherent in the platform. To achieve this the company formed two important partnerships with operators. The first was with Rottnest Express, a fast ferry operator local to Austal's Henderson headquarters. The ships operated by Rottnest Express are not built by Austal. However the company was able to trial various sensors for understanding ship motions and sea state on Rottnest Express vessels. Many important lessons on how to build a robust IoT system at sea were learned as a result of this first trial.

The first full deployment of the MARINELINK-Smart system, providing crew advice as well as gathering data, would require a vessel capable of gathering a much broader dataset. This would be easiest and fastest to prototype on an Austal-built vessel.

Fred Olsen SA is Austal's largest commercial customer. They are also one of the world's largest and most sophisticated HSC operators, currently operating a fleet of 6 large HSC. Therefore they were a natural choice to partner with in the development of MARINELINK-Smart. Fred Olsen SA currently operate three Austal-built ships whilst two more trimarans are currently under construction by Austal.

A customer partnership means the company can ensure the development of MARINELINK-Smart remains focussed at all times on the needs of the customer, who can provide rapid feedback on the utility (or not) of the new features of the system as they are deployed.

The connection to academia has also continued in to the development of the system. The MARINELINK-Smart development team consists of carefully chosen software developers, control engineers, data analysts and naval architects who work on campus at the University of Western Australia. Being on-campus enables the team to access leading academic developments in relevant disciplines such as machine learning, sensor technologies and oceanography.

EXAMPLES OF VALUE FROM HSC SMART SHIP TECHNOLOGY

Austal's MARINELINK-Smart is now in-service gathering HSC operational data across a number of ships and operators. As the data set available grows so too does the evidence of the value that the system is able to deliver to the ship's crew.

The first use case that MARINELINK-Smart was applied to is the provision of dynamic trim setting advice to HSC crew. Dynamic trim can be adjusted on an HSC through altering static trim (either by moving vessel payload if possible, or filling and emptying ballast tanks if available) or by adjusting the setting of a trim setting device such as a trim tab or transom mounted interceptor. An interceptor is a vertical plate parallel to the transom which creates a 'wedge' of water in front of it when deployed. It has a similar lift and drag effect to a trim tab.

In one example catamaran installation MARINELINK-Smart measured speed, fuel consumption, static trim, interceptor setting and a variety of other factors over a 4-month period to build a baseline model of the ship's fuel consumption performance. This model was then used to determine the optimal interceptor setting, and hence dynamic trim angle, to minimise fuel consumption in different loading and weather conditions. The model indicates that fuel savings of up to 4% may be possible through applying the optimum trim device setting for specific journeys.

Other forms of customer value have become clearer as Austal has built up experience with customers on the use of the MARINELINK-Smart data platform in-service. For example customers are provided with a dashboard which provides access to the data gathered by the system. This led to a number of conversations with customers on how this data could be used to summarise and track fleet performance against KPIs such as fuel consumption, estimated passenger sickness and on-time arrival. An example of a customer dashboard developed to meet this customer requirement is shown in Figure 4.

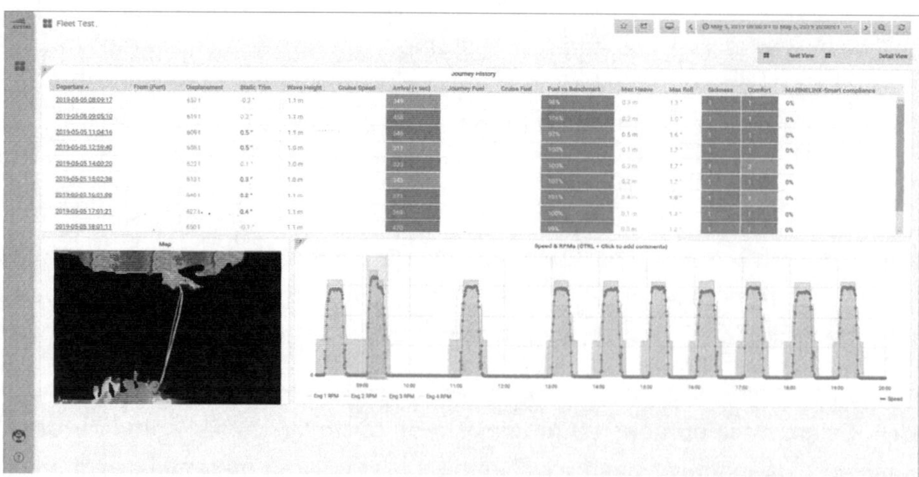

Figure 4 – Fleet Dashboards to Track Ferry Customer Journey KPIs

IMPLICATIONS FOR THE DEFENCE MARKET

Austal has a long history of pioneering technology for commercial customers and then offering this to the Defence market. This was most notable in the development of the trimaran hull form, which was first used on the 127m Benchijigua Express (launched 2004) before being incorporated in to the US Navy Independence-class Littoral Combat Ship.

In this example the end users were different but many of the things they valued were similar. HSC as ferries offer customers a faster passage, whilst the USN was interested in speed to deploy ships and equipment. Comfort is important for commercial customers to ensure their passengers will continue to choose to travel with them, whilst for the USN operational effectiveness of crew was an important concern [2].

There are similar analogies we can draw for the current smart ship developments. The fuel saving features of the system (such as trim and throttle optimisation) will also deliver enhanced range and endurance for a Defence vessel. Improvements to comfort from adjustments to ride control system settings will improve habitability, crew effectiveness and ultimately crew retention. Improved reliability, delivered by providing the necessary infrastructure on-board for comprehensive and continual condition-monitoring of equipment, will improve ship availability.

Austal has recognised this and so is working with the Royal Australian Navy to trial the deployment of the MARINELINK-Smart data gathering system to a RAN-operated patrol boat. In this first stage the system will provide a proof of concept of the secure transfer of ship data to shore, as a basis for building the value-adding services described above. In parallel Austal is working with development partner SMEC on an Enterprise Asset Management System (EAMS) to provide fully digital maintenance management, enhanced asset oversight and actionable predictions of equipment future state. MARINELINK-Smart will also provide the secure interface to the ship's data for this system. This will ensure a consistent flow of information and understanding of asset condition between the ship and shore operational teams.

It is important to also recognise the differences between commercial and defence customers. All customers desire strong security and strict access controls to their operational data. However defence customers will tend to have their own specific handling requirements for data, and probably also a hierarchy of data security requirements. This may mean that some data items have to be segregated from the majority of the ships data, and enhanced security procedures applied to this (ship location as reported by GPS being a prime example). Defence customers may also have additional questions on where there data resides if it is stored in the cloud, or may insist on shore storage on their own servers exclusively.

These challenges can be significant but they must be balanced against the potential benefits of IoT to navies. Ultimately a Navy's goal is to deploy a superior capability at an affordable cost. If an adversary is prepared to harness emerging IoT and other technologies to achieve a capability advantage then the investigation of the benefits at least is well worth doing.

WHERE NEXT FOR HIGH SPEED, SMART SHIPS?

Predicting the future direction of technology development is notoriously fraught. The final chapter of the 1993 reference book 'Worldwide High Speed Ferries' [3], entitled the 'The Future', got many things right. The author predicting the growth of vessel sizes and the expansion of the large, car and truck carrying variants of HSC.

Less accurate were forecasts of the uptake of Surface Effect Ships (hybrids of hovercraft and catamarans) by operators (see Figure 5). Surface effect ships (SES) proved unable to deliver comfort and efficiency to operators and so in the end had little impact on the HSC industry, despite the initial excitement around the potential of the technology. Smart ship technology is certainly currently subject to a lot of hype. In the end will it go the way of the SES, or will it be a game-changer as the light-weight catamaran before it proved to be?

Figure 5 – 1990 SEC Designed Surface Effect Ship Concept (image courtesy SEC) [3]

There are reasons to believe that the impact of IoT technology on the HSC industry will be significant and sustained. Unlike novel hull forms or even lightweight diesel engines the technologies of IoT are applicable to almost every industry. Therefore large investments are being made in the technology, and the costs of deploying IoT are reducing as a result. The relatively low (and ever reducing) costs mean that the benefits have to be relatively modest to justify the investment.

However to have a revolutionary impact on the industry IoT will need to offer more than simply a standard suite of incremental efficiency and comfort improvements at an affordable price.

Austal foresee potentially the most significant impact of IoT as being its use in combination with other emerging technologies, particularly propulsion technologies. Making efficient use of hybrid and battery technologies will depend on a much tighter integration of operational and propulsion system management information.

The high energy density of conventional fossil fuels has meant the amount of stored energy in the fuel tanks has been of limited operational concern beyond ensuring the ship has sufficient range to make the next bunkering opportunity. But the equivalent weight of electric energy storage makes the viability of electric propulsion much more dependent on an efficient use of the energy stored on-board. The optimal balance between vessel schedule, charging time and range is therefore closely coupled with ship loading, environmental conditions and propulsive efficiency. To operate efficiently and safely low or zero-emission ships will increasingly rely on highly accurate forecasting of their energy usage, which will in turn be dependent on sophisticated performance models which bring together engineering analysis and read-world ship data.

Efficiency will also be measured at the fleet rather than individual ship level for electric ships and extend to a consideration of the through-life costs incurred through operations. For instance the depth of battery discharge can have a significant impact on battery life. Therefore decision-making at the fleet level (i.e. choosing to switch ships between routes to minimise through-life battery costs) will require an aggregation of data at the fleet level. This is simply not possible without a shore data repository and IoT platform. To assist operators on preparing for this future Austal offers not only an IoT solution for new Austal ships but also a standalone IoT data collector which can be retrofitted to other ships in an operator's fleet.

CONCLUSIONS

The question ship operators should ask themselves is whether they can afford *not* to invest in IoT for their fleet. Every journey operated where data is not gathered is a potential missed opportunity to build a data set from which future efficiency, comfort and reliability improvements can be derived. Austal believes that the costs of IoT are now sufficiently low that every HSC Austal builds should be gathering this data set for their customers from the moment it enters service. Therefore every new Austal large HSC has MARINELINK-Smart installed. This also includes an Ethernet smart-sensor network throughout the vessel, which enables new sensors and equipment to be easily installed and integrated with the MARINELINK-Smart platform as these become available. This 'future proofs' Austal HSC for our customers, since the IoT devices available now are likely to very different to those available to our customers in 10 or 20 years' time.

Austal's objective in investing in IoT technology is to differentiate its ships in an increasingly competitive global market. The company recognises that this cannot be achieved in isolation and that a much more compelling proposition can be built by working with customers, academia and other industry to collaboratively build the IoT platforms of the future, for the benefit of the Australian HSC industry as a whole.

REFERENCES

1. ShipPax Database. URL: https://www.shippax.com/en/shippax-database.aspx (accessed November 2018)
2. McCauley, Michael & Pierce, E & Matsangas, Panagiotis. (2007). The High Speed Navy: Vessel Motion Influences on Human Performance. Naval Engineers Journal - NAV ENG J. 119.
3. Hynds, Paul. (1992). Worldwide High Speed Ferries. London: Conway Maritime Press Ltd.

Seaway Loads Applied to a Frigate by the Smoothed Particle Hydrodynamics Technique

Bruce Cartwright, Lex Mulcahy, and Rob Melchers, The University of Newcastle
Bruce.Cartwright@newcastle.edu.au, Lex.Mulcahy@newcastle.edu.au, and Rob.Melchers@newcastle.edu.au
Teresa Magoga and Seref Aksu, Defence Science Technology Group
Teresa.Magoga@dst.defence.gov.au, Seref.Aksu@dst.defence.gov.au
Damian McGuckin and Martin Renilson, Pacific Engineering Systems International
damianm@esi.com.au, martinr@esi.com.au

ABSTRACT

Reduction of maintenance and life-cycle management costs of ships and offshore structures remains a major concern for both naval and commercial operators. Accordingly, the development and utilisation of improved loads and degradation modelling and structural assessment tools are required. This includes the incorporation of non-linearities due to the fluid-structure interaction (FSI), because non-linear loads can considerably contribute to the long-term stresses and structural damage. Smoothed Particle Hydrodynamics (SPH), a mesh-free particle method, can provide a cost-effective and robust solution to determining the non-linear responses of ships in long crested waves.

This paper presents a summary of developments within the Structural Integrity of Maritime Platforms (SIoMP) project. The paper describes the implementation of SPH, coupling between SPH and Finite Element (FE) analysis, and the extraction of loads and stresses from a model of a naval ship. A key aspect of the analysis is the exploitation of a technique to develop regular waves at constant amplitude over many wavelengths. Test cases are used to study the model sensitivity to various numerical parameters, and the results computed using the SPH-FE method are compared to those computed using a strip theory method. Based on the results of the sensitivity analysis appropriate domain boundaries (depth and width), and SPH particle diameter are recommended.

GLOSSARY

2D	Two-dimensional
3D	Three-dimensional
CFD	Computational fluid dynamics
FE	Finite element
FSI	Fluid structure interaction
H	Wave height (m)
L_{WL}	Waterline length (m)
p	Pitch response (rad)
p'	Non-dimensionalised pitch
SIoMP	Structural Integrity of Maritime Platforms
SPH	Smoothed particle hydrodynamics
v	Ship speed (kn)
VPS	Virtual Performance Solution
y	Heave response (m)
y'	Non-dimensionalised heave

INTRODUCTION

Robust and efficient numerical methods are increasingly being developed and exploited to analyse the hydro-elastic responses of ships to ocean waves [1]. The incorporation of non-linearities due to the fluid-structure interaction (FSI), and extreme seas, is important because non-linear loads can have a considerable influence on the long-term stresses and structural damage when compared to accounting for the linear wave response alone [2, 3].

There is a need to reduce maintenance and life-cycle management costs and to have the ability to conduct accurate service life extension review naval, and other, ships [9]. These needs may be met through the development and use of improved loads and structural assessment tools.

Different methodologies to predict the external loads and structural responses of a ship in a seaway are available (Figure 1) [4]. These have varying complexity and computational expense. Time-domain analysis like FSI, is considered more accurate but considerably more computationally expensive than frequency-domain and quasi-static procedures. In general, computational constraints have made time-domain analysis of ships for long-term structural assessment not feasible [5, 6]. However, recent hardware advances may result in time-domain analysis being more commonly used in the maritime industry [5, 7].

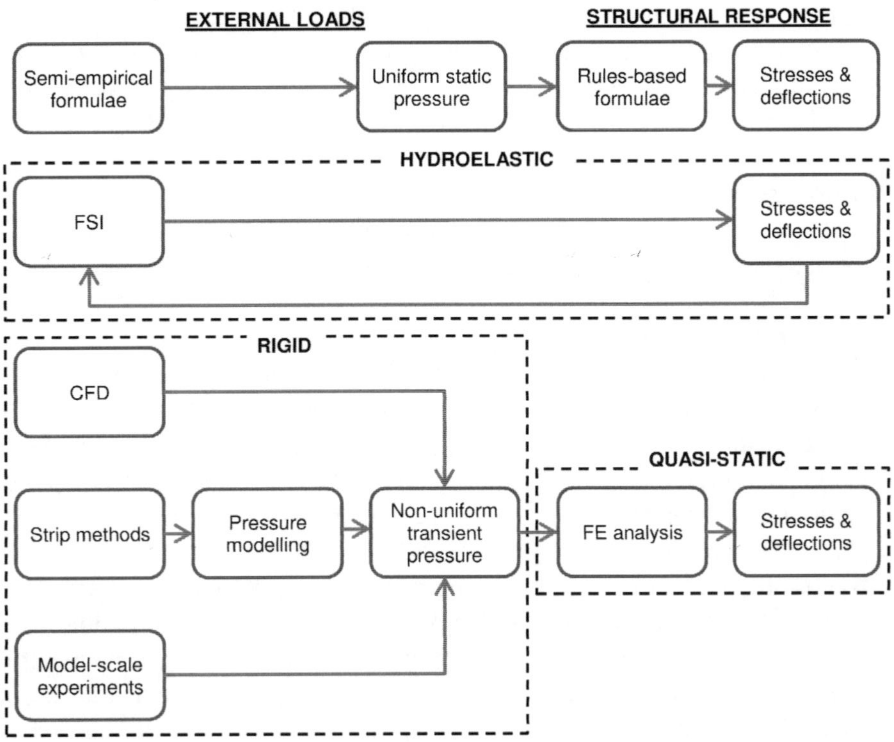

Figure 1: Schematic of different approaches for load and response assessment [4]

One type of hydro-elastic analysis is Smoothed Particle Hydrodynamics (SPH), a mesh-free particle method based on Lagrangian formulation. SPH can provide a cost-effective and robust solution to determining the non-linear responses of ships in long crested waves and of novel hullforms [7, 8].

This paper presents a summary of developments within the Structural Integrity of Maritime Platforms (SIoMP) project [10]. The aim of the SIoMP project is to improve the understanding of the effect of structural age-related degradation, due to long-term loads and corrosion, of naval and other ship and offshore structures. The paper describes the implementation of SPH, coupling between SPH and Finite Element (FE) analysis, the model set-up, and the extraction of loads and stresses. This is followed by test cases to study the model sensitivity to various numerical parameters, and presentation of a comparison between the results generated by the fully coupled SPH-FE methodology and a strip theory method. A key aspect of the analysis is the exploitation of a technique to develop regular waves at constant amplitude over many wavelengths [8].

METHODOLOGY

Smoothed Particle Hydrodynamics (SPH) and Implementation

Smoothed Particle Hydrodynamics (SPH) is a mesh-free particle based method that is particularly well-suited to modelling fluids and gases [11]. It was originally developed for modelling the development of galaxies and stars in astrophysics [12, 13]. Liu [11] provides a comprehensive review of the history and development of the commonly applied SPH formulations.

The mesh-free nature of SPH allows the investigation of complex fluid flows such as breaking waves, the parting of waves by the bow of the ship, green water on deck, and ingress of water into the ship through openings. These features are particularly difficult to handle efficiently in the grid-based schemes of conventional computational fluid dynamics (CFD).

The present paper uses the SPH implementation within the commercial software suite Virtual Performance Solution (VPS) [14]. VPS is a general purpose, multi-physics analysis tool, having a comprehensive material library that includes linear and non-linear material models.

Fluid Structure Interaction (FSI)

FSI is the analysis of the interaction and response of a fluid and a structure. In VPS, the SPH solver handles the fluid behaviour and the FE analysis solver handles the structure. These two solvers are integrated and are resolved simultaneously. The solution involves an explicit time-marching scheme. The time-step is chosen automatically by the software and is typically in the order of a fraction of a millisecond or so for metal structures.

To prevent structural elements from penetrating one another, VPS uses penalty-force sliding contact algorithms [15] along with an enhanced contact algorithm for the handling of discrete SPH particles interfacing with structural shell elements [16]. The contact interface between SPH particles and shell elements defines the distance that a SPH particle is allowed to approach a shell element (relative to the centre of the particle). It also defines any damping that may be applied to the contact, and the sliding friction that may exist. The project's experience to date has shown that a friction value of zero works well. The SPH and FE analysis solvers interact at each time-step through the contact interface. There is implied continuous and complete interaction of the fluid and the structure at every time step.

Modelling of Waves

The approach to model waves is to define a simulation domain, which is a portion of a large body of water. Deep-water waves are developed to be analogous to wind-generated waves that occur on the free surface of oceans. The edges of the simulation domain are established to respond in a way that mimics the larger ocean beyond the edges of the domain. One of the challenges of this simulation is determining the appropriate domain dimensions such that the finite domain is representative of the ocean environment.

This paper considers the case of a ship travelling into long-crested head seas. The simulation domain is a finite number of wavelengths in length, with upstream and downstream ends, depth, and width (Figure 2) [17].

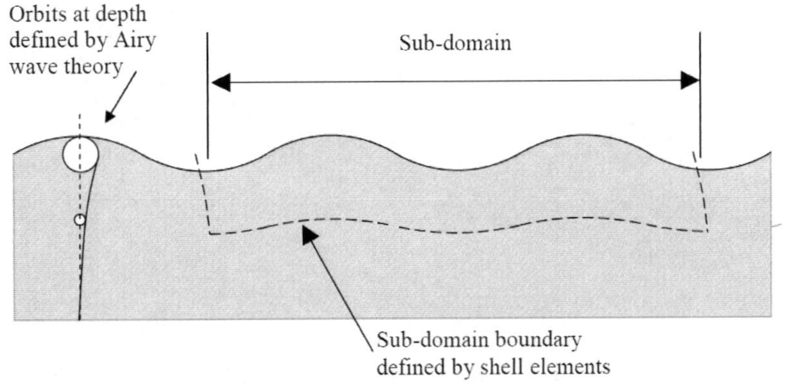

Figure 2: Concept of a subdomain for boundary definitions of moving floor [17]

All vertical edges of the simulation domain are defined by periodic boundary conditions. A periodic boundary condition is a numerical definition that can model an infinite length of fluid by passing fluid behaviour from one edge of the domain to the other continuously. For example, for a wave propagating through the domain, as the wave approaches the downstream end, the particles encounter the periodic boundary condition. This transfers their interactions to the upstream end of the domain. The wave will effectively be reinstated at the upstream end without any loss. Hence, the wave can propagate continuously through the domain.

The depth of the simulation domain is less than half the wavelength of interest. Thus, according to deep wave theory, the surface wave should create oscillatory motions of the water at that depth. The deep-water wave is simulated by assigning circular movements, equivalent to the orbital motion of the water, at the bottom of the domain. As a result, deep-water waves are simulated in a finite depth domain. When the motion of the bottom of the domain is in phase, and of the same frequency as the surface wave, the surface wave will propagate to mimic the deep water condition [18].

Modelling of Ship

The ship model used in the SIoMP project is a full-scale structural FE model of a frigate constructed of high-tensile steel. The global model comprises shell, beam and bar elements.

To enable the ship to establish an equilibrium in the water, the ship model commences the simulation at zero forward speed out of the water. Gravity is applied gradually to all elements and nodal masses of the ship, causing the ship to sink into the water to achieve equilibrium. Damping is applied to the SPH particles to prevent excessive movement during this period. For simulation at non-zero speed, the ship is accelerated to the desired speed.

For the case of head seas, the ship is restrained from sway and yaw by fixing a few nodes on the centreline at the bow and stern of the ship in lateral displacement. The ship is free to move in pitch, heave, and roll. Forward speed is fixed by a velocity boundary condition that acts on a group of nodes of the ship, close to where the propeller thrust would be developed on the ship.

Figure 3 presents a still shot of the FE model within the SPH domain during a simulation in regular waves.

Figure 3: Example still shot of FE model of ship within SPH domain during a simulation

To extract the hydrodynamic loads, the hull is divided into athwartships strips between each frame with a unique contact interface defined for each of these strips with the water. The sum of these contact interfaces governs the response of the ship to the water. In addition, the contact force on each strip is stored, allowing post-processing to reveal the variation of forces along the hull with time. In still water at zero forward speed when the ship has come to equilibrium, the longitudinal distribution of the contact forces acting between each frame of the ship coincides with the distribution of buoyancy along the length of the ship. As waves develop and the ship moves relative to them, the forces developed on the hull induce the ship motions and responses.

The effect of added mass acting on the ship is not calculated as an 'added' component. Instead the effect of added mass is developed in the same way the real ship experiences the effect of added mass. That is, the SPH particles in close proximity to the ship are influenced by the presence of the ship, and are thus accelerated or decelerated to allow the ship to occupy that volume of water.

TEST CASES

To demonstrate the application of the SPH-FE methodology, test cases are presented. This includes a study of the effect of the model sensitivity to various numerical parameters, and a comparison between the results generated by the fully coupled SPH-FE methodology and a strip theory method.

A naval ship, with waterline length (L_{WL}) of approximately 110 m, and length to beam ratio of approximately eight, is studied. The mass distribution is for the full scantling departure load condition, with mass of the structure plus the application of lumped masses to selected nodes to represent internal machinery, fluids and stores.

The numerical domain should conform to existing rules and guidelines to avoid the known issues with behaviours as blockage factor and wall (domain) interference [19]. The test cases, defined by the SPH particle diameter, domain width, and domain depth, are given in Table 1.

Table 1: Test matrix, all runs at ship speed (v) of 9 kn and in head seas

Run #	SPH particle diameter (m)	Domain width (m)	Domain depth (m)
1	1.2	50	10
2	1.2	50	20
3	1.2	50	25
4	1.2	50	30
5	0.9	50	20
6	1.5	50	20
7	1.8	50	20
8	1.2	25	20
9	1.2	75	20
10	1.2	100	20

For each test case, the ship heave response (y) and pitch response (p) about the centre of gravity are extracted. The non-dimensionalised heave response (y') and pitch response (p'), calculated using Equations 1 and 2 [19] respectively, are presented in the paper, where H is the wave height and λ is the wave length.

$$y' = \frac{y}{H} \quad \text{Equation 1}$$

$$p' = \frac{p}{\frac{2\pi}{\lambda}\times H} \quad \text{Equation 2}$$

Tools used in Validation

The results of the SPH simulation are validated against the results generated using traditional approaches:

- The quasi-static approach for the calculation of the still water and wave-induced shear forces and bending moments.
- The two-dimensional (2D) and three-dimensional (3D) approaches for calculating ship motions and responses when the ship is operating in a seaway.

The implementations of these approaches in the commercially available naval architectural tool MAESTRO 11.6.1 [20] are utilised in the present study.

Using the quasi-static assessment approach, only the hydrostatic and inertia force are considered. The centre of gravity and the centre of buoyancy of the model of a ship align vertically in the global coordinate system so that the ship is in equilibrium in heave, pitch, and roll. This allows the load distribution to be checked, and the closure of the bending moment distribution and shear force distribution is a means to validate equilibrium.

In the case of 2D strip theory, it is assumed that the body is rigid, motions are small, ship-hull sections are wall-sided, and the body has no effect on the waves. These assumptions permit the division of the submerged part of a ship into a finite number of strips. Then, 2D hydrodynamic coefficients for added mass can be computed for each strip and then summed over the length of the body to yield the 3D coefficients [21].

RESULTS AND DISCUSSION

Effect of Domain Depth

Figure 4a and b show the non-dimensionalised heave and pitch responses of the ship with varying domain depth, for a domain width of 50 m and SPH particle diameter of 1.2 m (test runs #1 to #4). Whilst the pitch varies minimally, the heave varies 30% over the studied domain depth range.

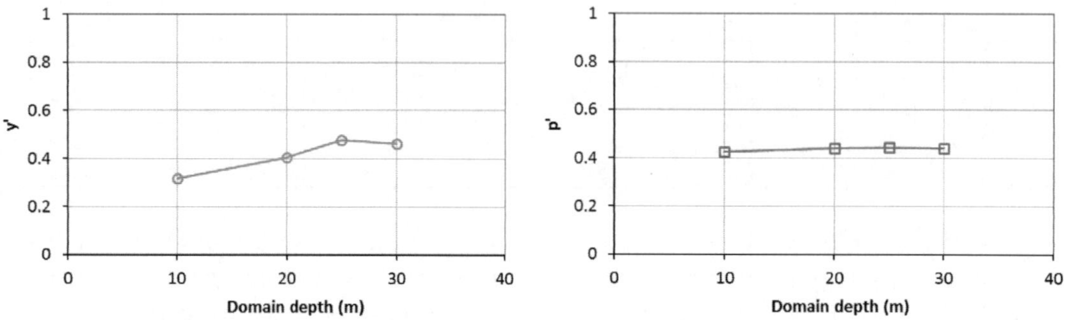

Figure 4: Domain depth versus a) y', and b) p' (test runs #1 to #4)

Effect of Domain Width

Figure 5a and b show the non-dimensionalised heave and pitch responses of the ship with varying domain width, for a constant domain depth of 20 m and SPH particle diameter of 1.2 m (test runs #2, #5, #6 and #7).

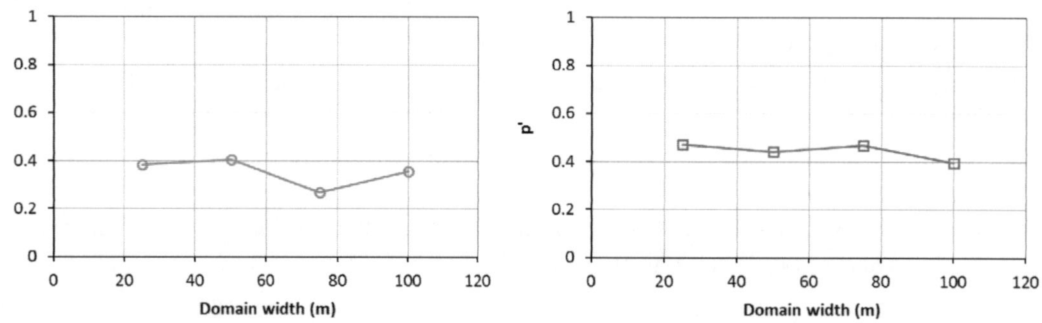

Figure 5: Domain width versus a) y', and b) p' (test runs #2, #8, #9 and #10)

Similar to that observed above, the pitch varies minimally with domain width. However, the heave response at a domain width of 75 m is relatively small. ITTC 7.5-02-07-02.1 [19] provides recommendations for dimensions to avoid tank (that is, domain) wall interference, based on running conditions. For the conditions considered, the minimum recommended ratio of domain width to model length is 1.0. Only the widest domain considered (100 m) meets this condition. This indicates that the ship response is likely to be influenced by wall reflections for the narrower domains.

Effect of SPH Particle Diameter

Figure 6a and b display the non-dimensionalised heave and pitch responses of the ship with varying SPH particle diameter width for a constant domain depth and width of 20 m and 50 m, respectively (test runs #2, #8, #9 and #10). Both the heave and pitch responses tend to be smaller when larger SPH particle diameters are used. That is, for the range of particle diameters considered, convergence (relatively small change in the motions with decreasing particle diameter) was not reached. This suggests that the SPH particle diameter should be as small as practical.

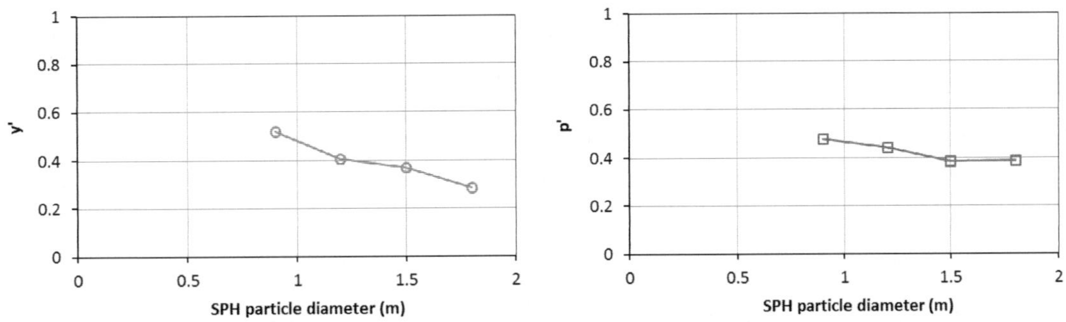

Figure 6: SPH particle diameter versus a) y', and b) p' (test runs #2, #5, #6 and #7)

Particle diameter has other influences. Larger particles have more mass, and so the interaction of each particle with the ship results in larger impact forces. This has the effect of making the interaction between the hull and the water noisy, requiring some form of filtering or smoothing. Smaller particles approach the continuum of water, and have a noticeably smoother force signal. Figure 7 demonstrates this effect where the contact force with time between the water particles and one athwartships strip of the ship for 0.9 m and 1.8 m diameter SPH particles. The 0.9 m diameter particles exhibit a less noisy force, and the ship response is smaller when larger SPH particles are used. The larger impacts from particles can also excite anomalous structural vibrations in the ship.

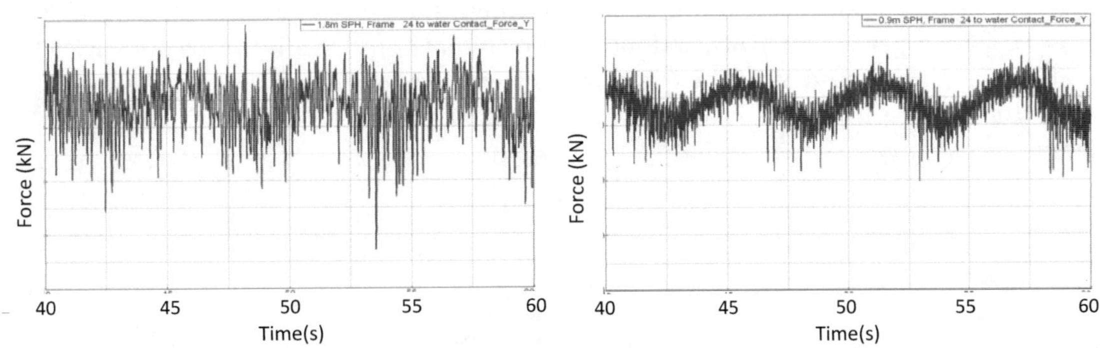

Figure 7: Time history of contact force between hull and water for a) SPH particle diameter of 1.8 m (test case #7), and (b) SPH particle diameter of 1.8 m (test case #5)

Still Water Vertical Bending Moment

VPS enables forces and moments to be extracted at sections defined within the model. There were 10 equi-spaced sections defined over a 55 m section in the centre region of the ship to resolve these forces and moments, as shown in Figure 8.

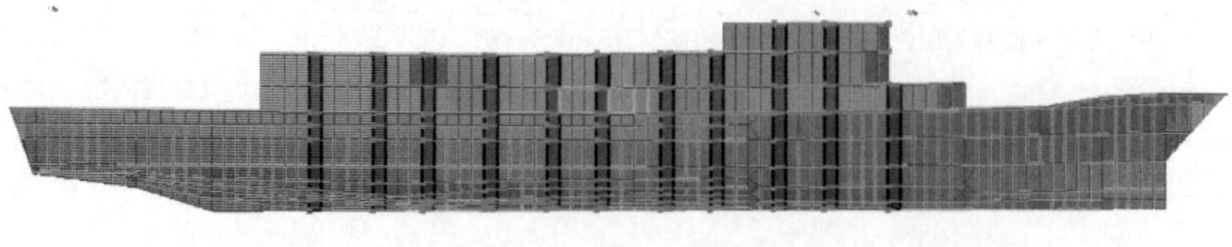

Figure 8: Location of sections for loads extraction

A comparison of the still water bending moments generated using MAESTRO and VPS is given in Figure 9. The maximum bending moment generated using VPS is approximately 11% greater than that predicted using MAESTRO. This difference is considered acceptable, as it is likely to be due to minor differences in the mass distributions between the two models. This aspect is being investigated further.

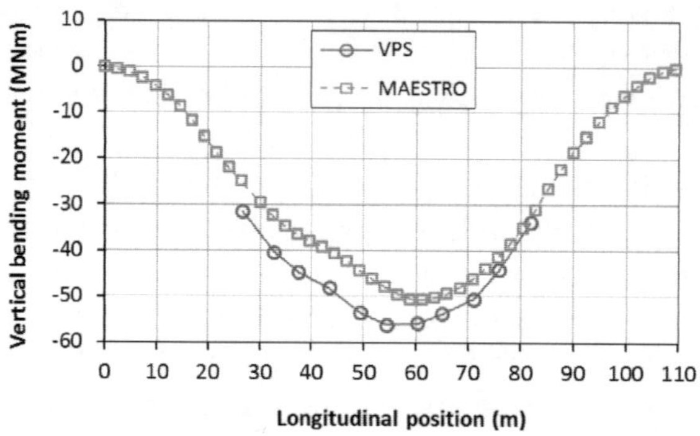

Figure 9: Still water vertical bending moment distributions calculated using VPS and MAESTRO

Vertical Bending Moment Distribution at Forward Speed in Waves

Figure 10 shows a comparison of the bending moment distribution calculated using 2D strip theory, implemented in MAESTRO, and the bending moment distribution calculated using VPS for λ/L_{wl} of 1.0, ship speed of 9 kn, wave height of 1 m, and head seas. The maximum bending moment generated using VPS is approximately 10% less than that predicted using 2D strip theory.

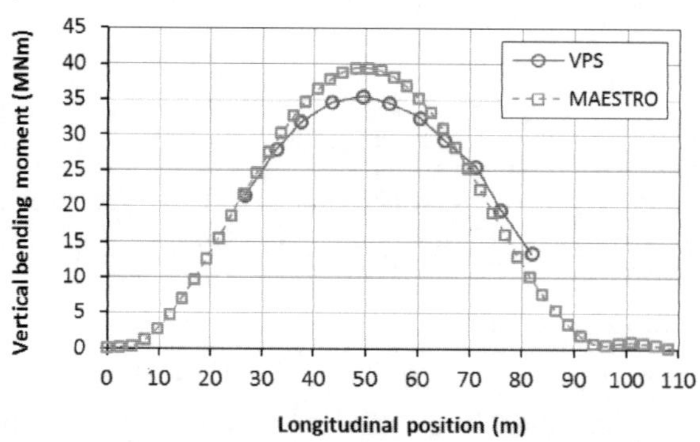

Figure 10: Vertical bending moment distributions calculated using VPS and 2D strip theory (MAESTRO), for λ/L_{wl} = 1.0, v = 9 kn, H = 1 m, and head seas

DISCUSSION

Based on the results of the sensitivity analysis, for the size of the ship, ship speed, and wave length considered, the following domain parameters are considered acceptable in terms of the accuracy achieved and computational effort required:

- Domain depth of 20 m.
- Domain width of 50 m (although best practice for experiments is 100 m, based on the result in Figure 5 the preliminary recommendation is that a domain width of 50 m provides reasonable results).
- SPH particle diameter of 1.2 m.

Based on the limited comparisons between the vertical bending moments distributions of the ship in the still water and at forward speed in regular waves, efforts to validate the proposed SPH-FE methodology are promising. Further work includes expansion of this study to other speeds, frequencies, and ship sizes.

The computer used was a 24 core Intel(R) Xeon(R) CPU E5-2650 v4 @ 2.20 GHz. The FE ship consisted of approximately 6,500 shell elements, and 15,000 bars and beams. In one example, with 552,630 SPH particles of 1.2 m diameter, filling a domain of 654 x 50 x 20 m, the simulation of 105 s of real time took 114 hours of wall clock computational time. Each SPH particle effectively requires the same computational effort, irrespective of particle diameter. Halving the diameter of the SPH particles for a specific domain size results in an 8-fold increase in the number of particles, and hence an 8-fold increase in computational effort. This becomes an important consideration in the pursuit of an economical solution.

CONCLUDING REMARKS

Based on the results of the sensitivity analysis, a domain depth of 20 m, width of 50 m, and SPH particle diameter of 1.2 m are considered acceptable in terms of the accuracy achieved and computational effort required. Further work includes expansion of this study to other speeds, frequencies, and ship sizes.

The next phase of the work in the SloMP project is to model the degradation of the ships' structure according to ship survey reports, and compare the stresses developed in the ship in critical areas to the as-built condition. It is proposed that this approach will be effective in helping to provide more informed decisions about future or immediate repair work.

ACKNOWLEDGEMENTS

The authors acknowledge the financial and in-kind support provided by the Australian Research Council, Defence Science and Technology Group, and Pacific ESI.

REFERENCES

[1] RAFIEE, A. & THIAGARAJAN, K. 2009. An SPH projection method for simulating fluid-hypoelastic structure interaction. *Computer Methods in Applied Mechanics and Engineering,* 198**,** 2785-2795.

[2] DRUMMEN, I., STORHAUG, G. & MOAN, T. 2008. Experimental and numerical investigation of fatigue damage due to wave-induced vibrations in a containership in head seas. *Journal of Marine Science and Technology,* 13**,** 428-445.

[3] KWON, K., FRANGOPOL, D. & KIM, S. 2013. Fatigue performance assessment and service life prediction of high-speed ship structures based on probabilistic lifetime sea loads. *Structure and Infrastructure Engineering,* 9, 102-115.
[4] RAZOLA, M., ROSEN, A. & GARME, K. 2014. Experimental evaluation of slamming pressure models used in structural design of high-speed craft. *International Shipbuilding Progress,* 61.
[5] MOHAMMADI, S. F., GALGOUL, N. S., STAROSSEK, U. & VIDEIRO, P. M. 2016. An efficient time domain fatigue analysis and its comparison to spectral fatigue assessment for an offshore jacket structure. *Marine Structures,* 49, 97-115.
[6] TUITMAN, J. 2010. *Hydro-elastic response of ship structures to slamming induced whipping.* Doctor of Philosophy, Technishe Universiteit Delft.
[7] ZHANG, A. M., SUN, P.-N., MING, F.-R. & COLAGROSSI, A. 2017. Smoothed particle hydrodynamics and its applications in fluid-structure interactions. *Journal of Hydrodynamics,* 29, 187-216.
[8] GROENENBOOM, P., CARTWRIGHT, B., MCGUCKIN, D., AMOIGNON, O., METTICHI, M. Z., GARGOURI, Y. & KAMOULAKOS, A. 2019. Numerical studies and industrial applications of the hybrid SPH-FE method. *Computers & Fluids,* 184, 40-63.
[9] HESS, P., AKSU, S., BLAKE, J., BOOTE, D., CARIDIS, P., EGOROV, A., FJELDSTAD, A., HOOGELAND, M., MURAYAMA, H., ANDERSON, M. R. & TAMMER, M. 2015. Structural Longevity. *Proceedings of the 19th International Ship and Offshore Structures Congress.* Cascais, Portugal.
[10] CARTWRIGHT, B., MELCHERS, R., AKSU, S., MAGOGA, T., SONG, C., MCGUCKIN, D. & RENILSON, M. 2017. Structural Integrity of Maritime Platforms. *Pacific 2017 International Maritime Conference.* Sydney, Australia.
[11] LIU, G., 2002. *Mesh Free Methods: Moving Beyond the Finite Element Method*, CRC Press, Boca Raton..
[12] LUCY, L., 1977, A numerical approach to the testing of the fission hypothesis, *The Astronomical Journal*, 82(12), 1013-1024.
[13] GINGOLD, R. & MONAGHAN, J. 1977, Smoothed particle hydrodynamics - Theory and application to non-spherical stars, *Monthly Notices of the Royal Astronomical Society*, 181, 375-389.
[14] 2018. VPS, Virtual Performance Solution User Manual, ESI Group, Paris, France.
[15] BOURAGO, N. & KUKUDZHANOV, V. 2005, A review of contact algorithms. *Mechanics of Solids*, 1, 45-87.
[16] LOBOVSKY, L. & GRONENBOOM, P. 2009. Remarks on FSI simulation using SPH, *4th SPHERIC Workshop*, Nantes, France, 378-383.
[17] CARTWRIGHT, B. 2012. *The Study of Ship Motions in Regular Waves using a Mesh-Free Numerical Method*, Master of Philosophy, University of Tasmania.
[18] CARTWRIGHT, B., McGUCKIN, D., TURNER, T., CANNON, S. 2006. The Modelling of Landing Craft Motions Inside a Flooded Well Deck using Smoothed Particle Hydrodynamics, *Pacific 2006 International Maritime Conference*, Sydney, Australia.
[19] ITTC 7.5-02-07-02.1. 2017. Recommended Procedures and Guidelines, Testing and Extrapolation Methods Loads and Responses, Sea Keeping, Sea Keeping Experiments. 28th Internationation Towing Tank Conference, Wuxi, China
[20] 2018. MAESTRO 11.6.1. Optimum Structural Design, Inc., United States.
[21] FALTINSEN, O. 1990. *Sea Loads on Ships and Offshore Structures*, Cambridge, United Kingdom, Cambridge University Press.

Real time structural loads monitoring for a large high-speed wave-piercing catamaran using numerical simulation and linear regression

Islam Almallah [a*], Jason Lavroff [a], Damien S Holloway [a], Babak Shabani [a], Michael R Davis [a]

[a] School of Engineering, University of Tasmania, Private Bag 65, Hobart 7001, Australia.

* Corresponding author: islam.almallah@utas.edu.au

ABSTRACT

High-speed wave-piercing catamarans encounter several kinds of global loads; some are similar to those for traditional monohull sea going ships, such as longitudinal bending moment (LBM), shear, torsional and dynamic wave slamming loads. However, due to the twin-hull configuration of the catamaran, additional types of global load are introduced, such as pitch connecting moment (PCM) and transverse bending moment (TBM). This paper investigates the structural health monitoring (SHM) of a high-speed wave-piercing catamaran through prediction of global loads from real-time processing of signals from a large distributed network of strain and acceleration sensors. Finite element method (FEM) and computational fluid dynamic (CFD) analyses at full scale are deployed to simulate catamaran response to loading cases and sea waves. A transformation equation based on a linear regression of the FEA results is applied to measured signals to convert strain signals to global loads. Global loads are also estimated using rigid body dynamics alongside computational fluid dynamics (CFD) simulation of HSV2. Motion and strain data collected from sea trials runs of a 98m high-speed wave-piercing catamaran (HSV-2) are used to confirm the proposed method. The sea trials runs of HSV2 Incat catamaran have been carried out in 2004 in a wide range of sea states, ship speed, wave heights, wave periods and sea wave headings. Longitudinal bending moment (LBM) is estimated for headseas run at 20 knots using two methods based on CFD and FEM simulation.

Figure 1 A Photograph of Incat HSV-2 Incat High-speed Wave-piercing catamaran [hull 061] during sea trials [1]

KEYWORDS: Global wave loads, finite element method (FEM), computational fluid dynamic (CFD), Wave-piercing catamaran, Full-scale, High-speed vessel.

NOMENCLATURE:

M_y	Longitudinal bending moment, sagging positive (N.m)	f_z	Vertical component of hydrodynamic force per unit length on the segment (positive up) (N/m)
m_x	forward segment mass (kg)	M_h	Hydrodynamic moment on the segment, evaluated at aft end of segment (N.m)
$\ddot{h}$	Heave acceleration at centre of gravity (*CG*), positive up (m/s^2)	I_y	Mass moment of inertia of the segment (kg.m^2)
$\ddot{p}$	Pitch angular acceleration at CG, positive bow down (rad/s^2)	ε_x	Strain
x_G	Distance of segment *CG'* forward of ship's CG (m)	z	Vertical height of neutral axis at the midship section (m)
x'_G	Distance from aft end of segment to the segment's *CG'*	I	Second moment of area of section about x-axis (m^4)
L	Length of vessel (m)	E	Young's modulus (Pa)

1. INTRODUCTION

High-speed catamarans encounter various types of structural loads because of their hull shape and sea wave conditions and thus it is important to predict the vessel structural performance during the design stage. While longitudinal and transverse bending moments and shear forces acting on a catamaran hull can be important in all headings depending on speed and wave height, pitch connecting and torsional moments and splitting forces have their highest effects in oblique seas.

When strain gauge records are available from hull structural monitoring, the data can be used with comprehensive finite-element analyses (FEA) to estimate global loads, [2] and [3]. For example, in the study by Jensen et al. (2001) [4] fibre optic Bragg strain gauges were attached at a cross section near amidships to measure the maximum strain responses for global loads, and strain/load relations were used to measure the global load response continuously and in real-time during tests. The internal ship loads were obtained from strain measurements through data processing using either the method of distortion modes or the conversion matrix [5]. Kefal and Oterkus (2016) [6] also studied the inverse Finite Element Method (iFEM) for real-time reconstruction of full-field structural displacements and stresses in plate and shell structures that are instrumented with strain sensors.

In relation to wave piercing catamarans, the global and slamming loads have been studied previously through experimental tests of an elastic segmented model. The magnitude and location of the dynamic wave slam force and slam induced hull bending moments resulting from dynamic sea wave slamming were investigated by Lavroff et al. (2013) [7] and Shabani et al. (2019) [8]. Global and slam loads of a 112 m Incat wave piercing catamaran (WPC) design were presented [9]. The computed global loads were compared with DNV rule-based loads and empirically derived loads based on full-scale measurements undertaken on similar types of vessels.

On the other hand, computational fluid dynamics (CFD) is considered to be a good tool to numerically study WPCs motions and loads if it is validated through towing-tank and full-scale tests, [10-12]. Centre bow slam loads of wave piercing catamarans can further be determined using computational fluid dynamics (CFD) simulation and compared with model tests for validation [13] whilst slam-induced bending of wave-piercing catamarans in head seas has also been investigated through fluid–structure interaction CFD simulations as reported by McVicar et al. (2018) [14]. Based on using Reynolds-Averaged Navier-Stokes Equations (RANSE), further CFD simulations were done to identify transient slam loads on a 112 m catamaran based on its hydroelastic motion response [15].

In the present study, the global longitudinal bending moment is estimated for a high-speed wave-piercing catamaran. Data records collected in sea trials of Incat catamaran HSV-2 Swift (Figure 1) are studied for this purpose. Results collected from FEA were used as input to the transformation matrix. This is achieved using finite element models provided by the Incat design office, Revolution Design. Load cases applied in the FEA were based on the Det Norske Veritas DNV (2011) [16] code rules for design loads.

2. METHOD

Catamaran global loads include longitudinal bending moment, transverse bending moment, pitch connecting moment, shear and splitting forces. The value of these loads depend on relative sea wave heading angle, wave characteristics and ship speed. The longitudinal bending moment (LBM) is one of the main global loads occurring at all heading angles and is investigated in the current work presented. The chart in Figure 2 represents the work process conducted in this study. Catamaran global loads are estimated by applying the load transfer matrix to strain gauge records from sea trials runs. This transformation is based on linear regression of DNV load cases applied in several FEA analyses. Catamaran global loads are also estimated through CFD simulation. Wave characteristics are determined directly from sea trial runs and applied to the CFD run.

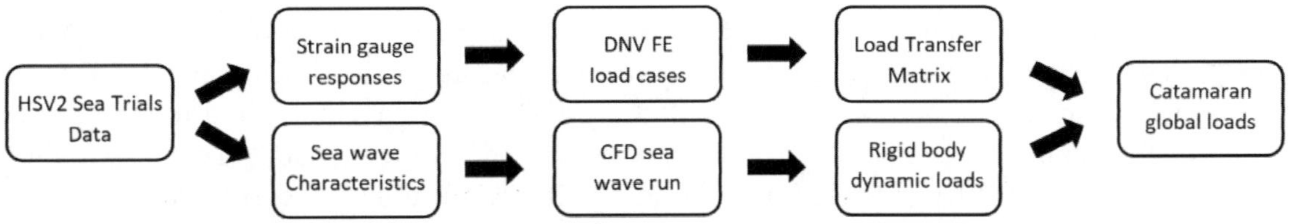

Figure 2 Process of work based on the development of load prediction using FE transformation matrix and CFD analyses

2.1 Rigid body dynamics

Catamaran internal global loads were estimated through motion and inertia responses using sea trials full-scale strain and motion data [17]. The vertical longitudinal bending moment is estimated using equation (1) which is derived by summing forces and moments about the selected section and eliminating the shear force V_x. Positive bending moments correspond to sagging moment and negative to hogging. Figure 3 shows the main forces and moments essential for calculation of internal bending moment at a specified section at distance x measured from the forward end of the catamaran.

$$M_y = M_h - m_x\, x'_G \left(g + \ddot{h} - \ddot{p}\, x_G \right) - I_y\, \ddot{p} \qquad (1)$$

The hydrodynamic moment term, M_h, is the moment calculated about the aft end of the forward segment of f_z, the vertical component of the hydrodynamic force acting on the segment (clockwise positive in Figure 3).

2.2 Structural model

Global loads can be estimated through strain responses at specific determined positions within the WPC hull. Strain to load conversion is a well-known method to define global bending moment. The strain signal record at strain gauge G5 is used for estimating the longitudinal bending moment (LBM). Strain gauge G5, located on the port keel at frame 26 (details are given in Figure 5 of Section 3), was the most responsive to longitudinal bending moment based on finite element (FE) runs that were undertaken by the design office Revolution Design prior to the sea trials.

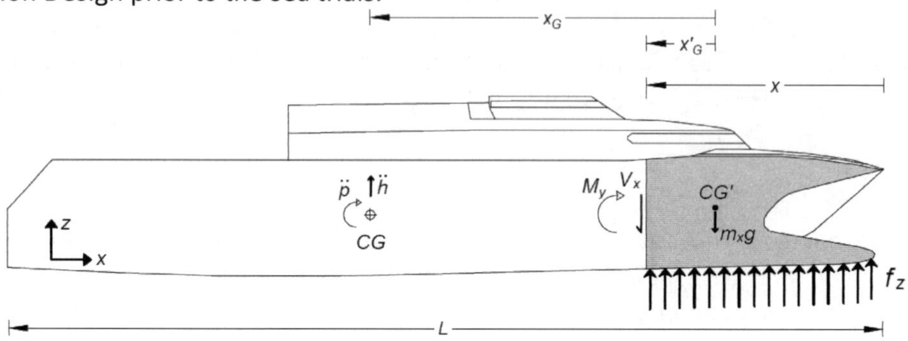

Figure 3 Rigid body dynamic internal bending load of HSV-2 high-speed catamaran

A conversion formulation is required in order to convert from strain to global bending moment. Several finite-element runs were performed to estimate the relation between strain response to applied bending load, [18]. The distribution of nodal forces in the FEA model is aimed at replicating the same method of loads application as set by the DNV code rules [16]. Figure 4 shows forces and exaggerated deflection for the longitudinal bending moment case.

Based on the principle of beam analysis there is a one-to-one correspondence between strain and global load, as presented in equation (2). Strain values depend on the applied load value and material and sectional properties *E*, *I* and *z*, which has a constant value for the same location. Consequently, the constant relation *C* between the applied load and corresponding strain is determined through these finite element (FE) cases at the location of strain gauge. It is important to note that this concept would not be valid in the event of slamming and whipping of the catamaran structure. A slamming load is a dynamic impulse load that occurs in a relatively small time period.

$$\varepsilon_x = \frac{M_y \times z}{I \times E} = C\, M_y \qquad (2)$$

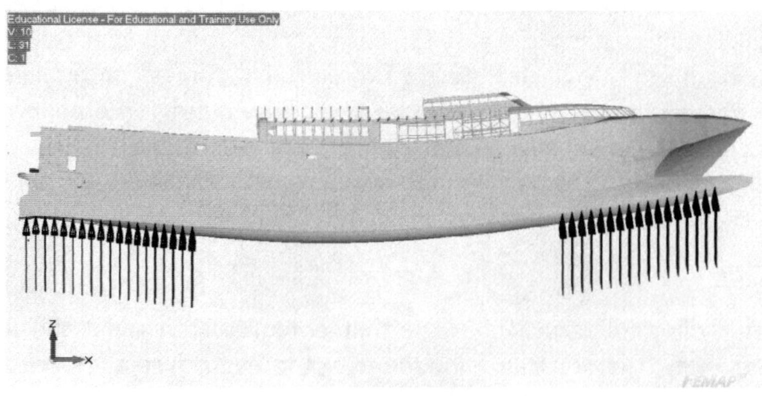

Figure 4 FE Longitudinal bending FE case [18]

3. HSV-2 SWIFT SEA TRIALS PROGRAMME

Incat catamaran HSV-2 Swift was extensively tested through a large number of sea trial runs by the US Navy and the Naval Surface Warfare Centre Carderock Division (NSWCCD) to assess motion and structural responses of the vessel over a wide range of wave heights, wave periods, wave headings and ship speeds [19]. The sea trials consisted of 22 octagons, each undertaken in a specific sea state, wave height and period, vessel speed and ride control setting. Each octagon consisted of five runs at different heading angles (the three remaining sides of the octagon were assumed to be equivalent by symmetry).

HSV-2 Swift catamaran was equipped with a large number of sensors to record all motions and structural responses. A TSK wave radar was placed on the bow of the forward deck to measure relative wave height during the sea trials. 53 strain gauges were installed on the catamaran to measure structural responses; these gauges were divided into groups, each group relating mainly to a particular global load. Each strain gauge was positioned at the appropriate location to pick up a specific dominant loading response of the vessel. Strain gauge G5 was located at frame number 26, in the X - axis direction at approximately 31.2 m from the stern at the keel level to pick up primarily the longitudinal bending response of the ship, whether for sagging or hogging. Figure 5 shows the locations of strain gauge G5 and the TSK wave radar.

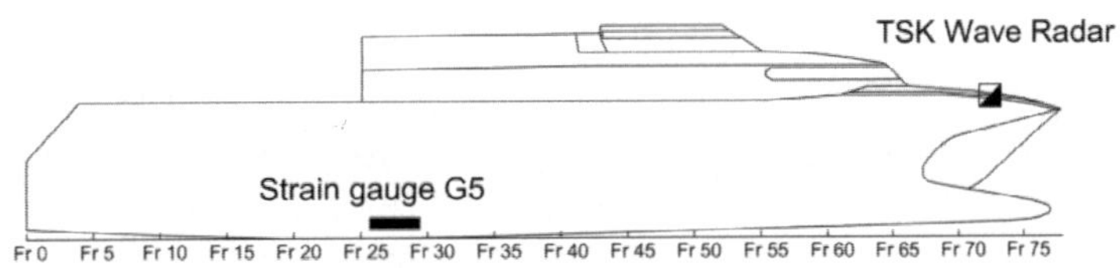

Figure 5 The location of strain gauge G5 and TSK wave height meter on HSV-2 Swift during sea trials

One run is selected from sea trials for the present analysis for prediction of the global load. It has been conducted in head seas at a speed of 20 knots for 20 minutes with significant wave height of 2.59 m in low sea state 5. A time record of 30 seconds was selected from the 20 minute run to represent wave loading on the HSV-2 without slamming events and as close as possible to a regular sea wave. This is because slamming events involve shock load and subsequent whipping response, and these effects are not accounted for in the present quasi-static finite element analysis. Figure 6 shows time records of wave height, pitch angle, heave acceleration and strain gauge G5.

4. CFD SIMULATION

Computational fluid dynamics (CFD) analysis was undertaken at full-scale to simulate the sea trial run 163 in order to investigate prediction of the global longitudinal bending moment. As mentioned above, the time record selected from run 163 has a wave profile close to a regular wave. Wave characteristics are identified from the 30-sec record. The wave height is set to be 1.7 m and the encounter wave period is taken to be 3.75 seconds peak to peak in this CFD simulation to closely match the data in Figure 6. Since the sea trials were in quasi-regular but random waves no attempt was made to match the phase of the incoming wave with that in the sea trials results.

The flow type of sea water waves used in this simulation is VOF inviscid fluid flow as it provides accurate results in comparatively less running time in full scale simulations. Separate CFD runs were developed to investigate the best flow type that should be used in the current simulation. Viscous laminar, viscous K-Omega turbulent and inviscid VOF fluid flows of sea water were used and compared. Inviscid flow is considered to be most efficient for simulating the full-scale WPC, as it gave comparable accuracy to viscous K-Omega turbulent flow but for significantly less running time. This is because the oscillatory nature of motion in waves is dominated by inertia effects over viscosity effects of water flow [10 and 12]. The overset mesh technique is used in CFD to simulate motions of the catamaran. Figure 7 shows pitch angle motion responses of these three fluid flow simulation types during the initial transient phase as steady-state is approached. It is noted that the amplitudes of pitch motion have less than 10% variation between the three simulations. However, as expected, the viscous turbulent and inviscid solutions are in closer agreement than the viscous laminar solution. This strongly supports the use of the inviscid simulation in the present work since the Reynolds number for the full-scale ship will be very high and the flow will be fully turbulent.

The mesh is refined near the hull and gradually increased towards the domain boundaries [14]. Figure 8 shows the mesh for the fluid surrounding the demihulls in the midship area. The smallest size of fluid mesh element close to hull is set to be 0.3125 m and the largest size of the mesh is 5 m.

Wave damping is applied near the outlet CFD boundary to prevent any wave reflection from the outlet surface back to the model. The domain block size is 4 times the length of the catamaran, 10 times the breadth, and the water depth is 15 times the catamaran draught. These domain dimensions assure that the developed motions of the catamaran in waves is free from any significant reflections that might occur at the boundaries.

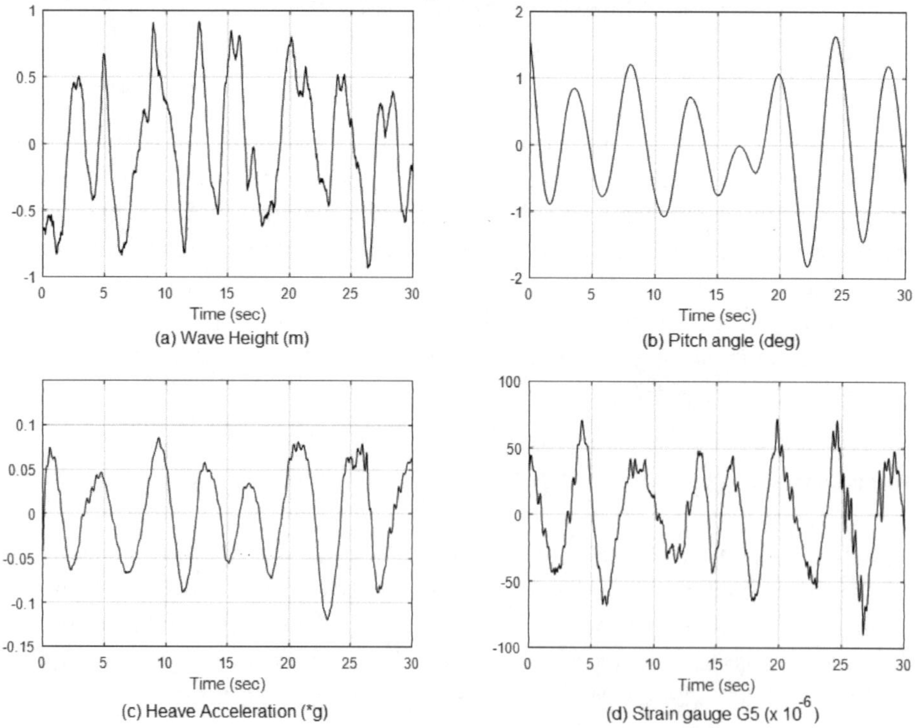

Figure 6 Sample wave and motion records of run 163 at a speed of 20 knot in head seas with wave height of 1.7m (Data Source : HSV 2 sea trials data [19])

The CFD simulation is set to run for 60 seconds and after the stabilization of motion responses and after the flow has reached steady state, a 30 seconds period is extracted for validation. A screenshot of the CFD model during bow entry in a wave is shown in Figure 9. This simulation is undertaken using High-performance computing (HPC) with 22 cores at 2.1 GHz and 6 GB RAM. About 48 hours was required for CPU time to run this simulation.

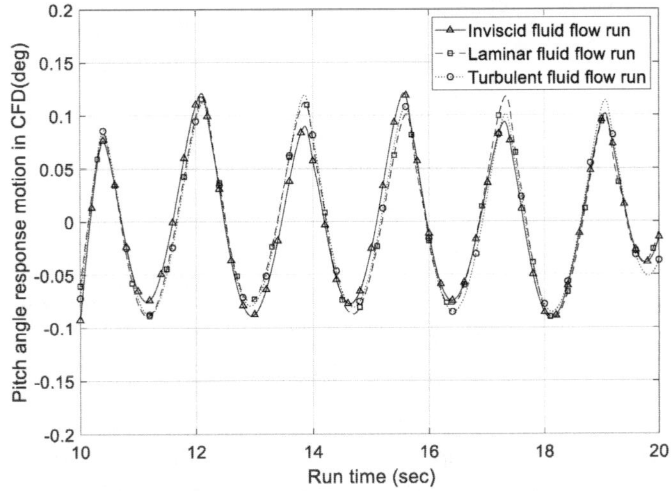

Figure 7 Pitch motion responses in CFD simulations undertaken with different fluid flow models in 2.44 m wave height of regular wave sea at 20 knots of forward speed

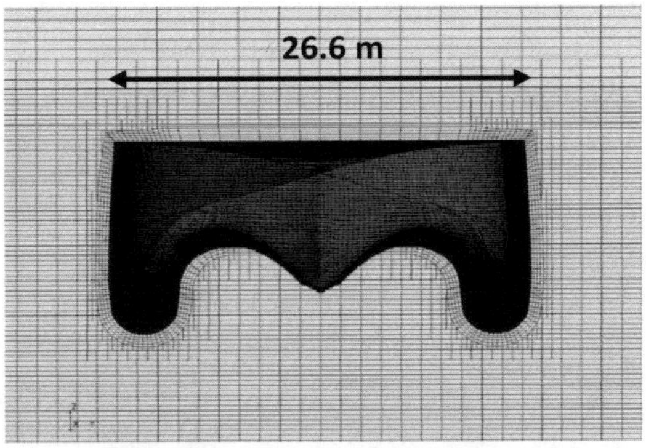

Figure 8 CFD Model of HSV-2 with mesh

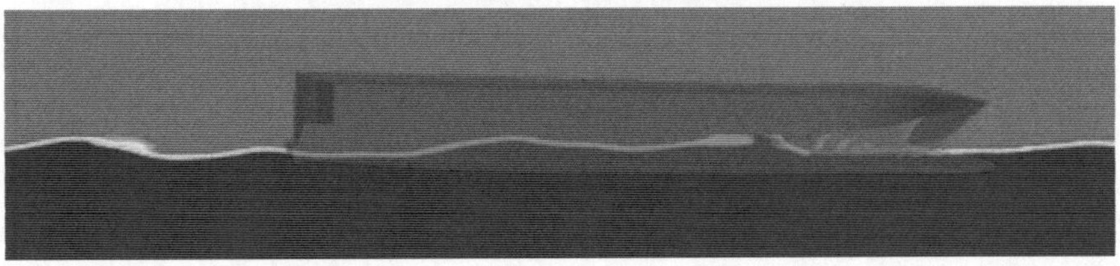

Figure 9 Wave and motion scene of HSV-2 in regular wave CFD simulation at 20 knots forward speed

Motion responses are compared between CFD simulation and sea trial records, as shown in figure 10. It is observed that there is agreement between the amplitudes of pitch angle and heave acceleration with those from sea trial run. However, a phase shift is also seen between motion records. As explained above, the CFD simulation assumes a regular wave with specific wave period, whereas the sea trial record has some small irregularities in the wave period and amplitude due to the random nature of sea waves as expected. In the CFD simulation it is difficult to set the same wave profile as in the sea trials runs, so it has been approximated to a degree where comparisons and validation between simulation and real sea trial run are attained.

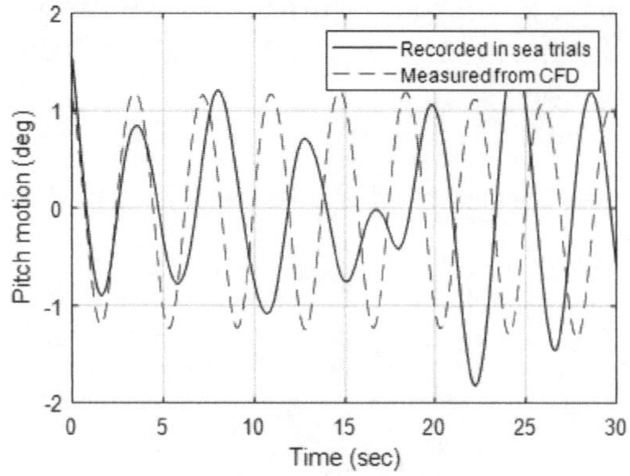

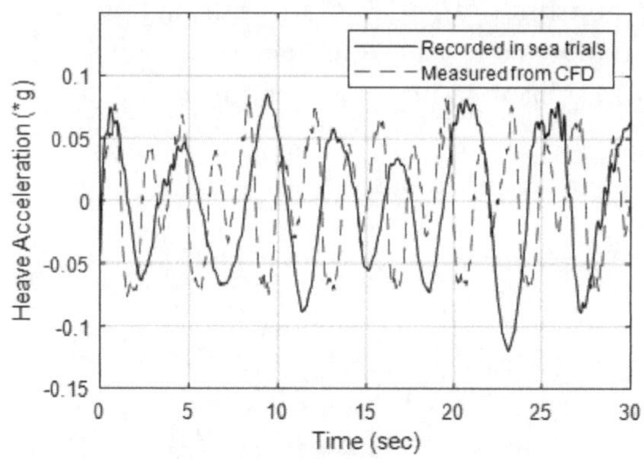

Figure 10 Motion of sea trial run compared with inviscid CFD simulation run undertaken at 20 knots of forward speed and 1.7m wave height.

Longitudinal bending moment (LBM) is estimated through a CFD simulation and rigid body dynamics formulation as discussed in section 2.1. This LBM is calculated at the sectional location of frame 26. It is compared with the longitudinal bending moment that is estimated based on strain gauge G5 record, through the strain-load conversion matrix. The still-water bending moment is subtracted from both load predictions (in the case of the sea trials by subtracting the mean from the strain gauges since it is impossible to quantify a true zero as there will always be some stress in the hull when the gauges are zeroed). Thus, the dynamic component of the load is investigated. Figure 11 compares the two longitudinal bending moment predictions. Although there is again a phase shift between the two load predictions, the load amplitudes agree well: longitudinal bending moment (LBM) values range between peaks of +/- 32 MN.m in sagging and hogging repetitively.

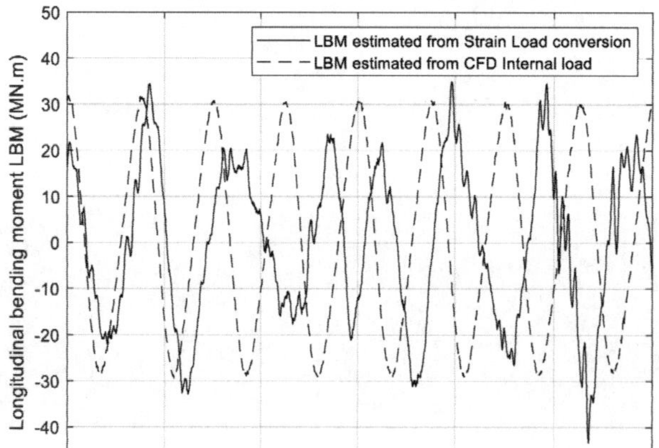

Figure 11 Predicted value of Longitudinal bending moment (LBM) of run 163 records based on CFD compared with FE conversion load at 20 knots of forward speed and 1.7m wave height.

The strong magnitude correlation between the motion and load responses of sea trial run records with the simulated CFD run implies that the applied methods in this study are reliable for prediction of global loads, despite minor deviation of the computed results due to the approximation of the random wave sea trial conditions with a near equivalent regular wave in the CFD simulation. It is recommended to apply this method to investigate further types of internal global loads of WPCs, such as torsional moments, transverse bending and splitting forces, and in different sailing conditions. This will assist in prediction of structural integrity at a future time and allows operators to respond to sea conditions to improve safety and increase the operating range. Consequently, the future design of new high-speed wave-piercing catamaran will be based on better understanding of motions and dynamic structural responses.

5. CONCLUSIONS

Sea trial records have been used to estimate the global longitudinal bending moment on HSV-2 high-speed wave-piercing catamaran. Two different methods are applied to calculate the longitudinal bending moment. Computational fluid dynamics (CFD) is combined with basic rigid body dynamics to estimate the internal bending moment. Then, strain records collected from a sea trial run are used with a finite element (FE) based matrix to convert to longitudinal bending moment.

The finite element method (FEM) using quasi-static loading is found to be sufficient for determining the trend of strain gauge responses observed in the sea trials runs in absence of slamming [18]. Furthermore, the conversion of strain data collected to corresponding global loads is achievable using linear regression of finite element analyses FEA. However, in order to estimate wave slamming loads, a transient finite element analysis is recommended of course.

Inviscid computational fluid dynamics (CFD) combined with a rigid body dynamics formulation is found to be a reliable method to study motion and loads associated with high-speed catamarans. CFD simulation has been validated on a broad basis with motion responses undertaken in full-scale sea trials tests, and the additional computational cost of a turbulent simulation was shown not to be warranted for this type of analysis. Longitudinal bending moment (LBM) estimated from FE and CFD based approaches showed good agreement of load value prediction. Loads within CFD simulations are estimated from accumulative addition of fluid element forces acting on the WPC hull. It is important to note that the CFD based approach is also effective especially when considering wave slamming loads occurring at higher range of speed [14].

The CFD simulation approach adopted in this study is beneficial for load estimation purposes for future high-speed catamarans during early design stages. The future research work of this project will be to identify further types of global loads, such as transverse bending moment, using the adopted methods. Both regular and irregular sea wave profile will be considered, and as well, headings other than headseas should be investigated to generate torsional loads of larger magnitude important for ship design.

6. ACKNOWLEDGEMENTS

This work has been supported by INCAT Tasmania Pty. Ltd., Revolution Design Pty. Ltd., Australian Research Council and University of Tasmania. Naval Surface Warfare Center, Carderock Division (NSWCCD) is also acknowledged for providing access to data collected from sea trials on HSV-2 Swift.

7. REFERENCES

1. Incat Australia Pty Ltd. [www.incat.com.au]. [Accessed : 01/11/2018].
2. Panciroli R, Biscarini C, Giovannozzi A, Maggiorana P, Jannelli E (2015) . Structural health monitoring through fiber bragg grating strain sensing. AIP Conference Proceedings. Rhodes, Greece: AIP Publishing.
3. Temarel P, Bai W, Bruns A, Derbanne Q, Dessi D, Dhavalikar S (2016). Prediction of wave-induced loads on ships: Progress and challenges. Ocean Engineering. 119:274-308.
4. Jensen A, Taby J, Pran K, Sagvolden G, Wang G (2001). Measurement of global loads on a full-scale SES vessel using networks of fiber optic sensors. Journal of ship research. 45(3):205-15.
5. Bigot F, Derbanne Q, Baudin E (2013). A review of strains to internal loads conversion methods in full scale measurements. Proc of PRADS2013.
6. Kefal A, Oterkus E (2016). Displacement and stress monitoring of a Panamax containership using inverse finite element method. Ocean Engineering. 119:16-29.
7. Lavroff J, Davis MR, Holloway DS, Thomas G (2013). Wave slamming loads on wave-piercer catamarans operating at high-speed determined by hydro-elastic segmented model experiments. Marine Structures. 33:120-42.
8. Shabani B, Lavroff J, Davis MR, Holloway DS, Thomas GA (2019). Slam loads and pressures acting on high-speed wave-piercing catamarans in regular waves. Marine Structures. 66:136-53.
9. Davidson G, Roberts T, Thomas G (2006). Global and slam loads for a large wavepiercing catamaran design. Australian Journal of Mechanical Engineering. 3(2):155-64.
10. Castiglione T, Stern F, Bova S, Kandasamy M (2011). Numerical investigation of the seakeeping behavior of a catamaran advancing in regular head waves. Ocean Engineering.38(16):1806-22.
11. Mancini S, Begovic E, Day AH, Incecik A (2018). Verification and validation of numerical modelling of DTMB 5415 roll decay. Ocean Engineering. 162:209-23.
12. Broglia R, Zaghi S, Campana E, Dogan T, Sadat-Hosseini H, Stern F (2019). Assessment of Computational Fluid Dynamics Capabilities for the Prediction of Three-Dimensional Separated Flows: The DELFT 372 Catamaran in Static Drift Conditions. Journal of Fluids Engineering. 141(9):091105.
13. Lavroff J, Davis M, Holloway D, Thomas G, McVicar J (2017). Wave impact loads on wave-piercing catamarans. Ocean Engineering. 131:263-71.
14. McVicar J, Lavroff J, Davis MR, Thomas G (2018). Fluid–structure interaction simulation of slam-induced bending in large high-speed wave-piercing catamarans. Journal of Fluids and Structures. 82:35-58.
15. McVicar J, Lavroff J, Davis MR, Davidson G (2016). Transient slam load estimation by RANSE simulation and by dynamic modeling of a hydroelastic segmented model. The 30th Symposium on Naval Hydrodynamics.
16. DNV. DET NORSKE VERITAS (2011). Rules for High Speed, Light Craft and Naval Surface Craft.
17. Holloway D, Davis M, Thomas G (2003). Rigid body dynamic hull bending moments, shear forces and PCM in fast catamarans. The 7th International Conference on Fast Sea Transportation.
18. Almallah I, Lavroff J, Holloway D, Davis M (2019). HIGH-SPEED WAVE-PIERCING CATAMARAN GLOBAL LOADS DETERMINED BY FEA AND SEA TRIALS. International Journal of Maritime Engineering. 161(A2):139-54.
19. Brady TF, Donnelly MJ, Griggs DB (2004). HSV-2 Swift Instrumentation and Technical Trials Plan. Naval Surface Warfare Center, Cardrock Division, Report NO: NSWCCD-65-TR-2004/18.

PACIFIC 2019 Maritime Exposition

Keynote speech

Category:

Author: Ian Irving, Chief Executive Naval Shipbuilding Institute

Ian.Irving@navalshipbuildingcollege.com.au

Title: Creating a skilled, sovereign workforce for ships and submarines

Overview:

The Naval Shipbuilding College (the College) is an Australian Government initiative managed by Naval Shipbuilding Institute (Australia) Pty Ltd, a joint venture between KBR and Huntington Ingalls Industries (HII).

The College has been established to ensure a skilled and accessible, sovereign naval shipbuilding workforce is available to deliver and maintain the Australian Government's $90 billion continuous naval shipbuilding enterprise over the next 50 years.

Since the official opening in November 2018, the undertaking in uncharted waters involved implementing the operational model against the back-drop of a recovering mining industry, the establishment of a new space industry, a robust aviation industry and an already established and growing defence industry has presented unique challenges.

The College has developed innovative partnerships with education and training providers across Australia. Vocational and tertiary training methods and courses are being assessed and adapted for naval shipbuilding requirements, identified through a whole of industry approach to taxonomy priorities and future needs.

Review and establishment of effective workforce retention strategies and techniques have been undertaken to encourage the skilled workforce to remain in Australia, to ensure the sustainment of the naval shipbuilding industry, and risk mitigate salaries from steep increases.

The result is an on-track workforce attraction, training and retention model which will deliver the largest skilled workforce in Australia's history.

A healthy and competitive Australian naval shipbuilding enterprise will strengthen a maritime industry by creating increased import and export opportunities as well as a highly skilled regional workforce.

The scale and scope of Australia's $90 billion Shipbuilding Enterprise engineering project is unprecedented in Australia's industrial history, with the cutting-edge upgrade to the Royal

Australian Navy's fleet of ships and submarines having all the hallmarks of a nation-building endeavour.

This represents the largest single engineering project that this country has ever embarked upon, a 40 year plus continuous shipbuilding enterprise. But it's much broader than just the naval projects themselves and is indeed a national endeavour that requires skills and expertise established at a sovereign level, not seen in this country before.

With much discussion about the availability and capability of the required workforce as one of the major challenges for this enterprise - this is where the Naval Shipbuilding College comes in.

Operated by the Naval Shipbuilding Institute - a partnership between KBR and America's largest naval shipbuilding company Huntington Ingalls Industries - the Naval Shipbuilding College is based at both the Osborne Naval Shipyard in South Australia and at South Metropolitan TAFE's Naval Base campus in Western Australia.

The Naval Shipbuilding College is partnering with Defence, industry and a national network of training and education providers. Its purpose is to play an integral role in developing the Australian workforce required to ensure the successful delivery of the critical naval capabilities that support our national security.

The activity of the College is now gaining real momentum and is poised to transform Australia's naval shipbuilding workforce of the future.

The ambition of the Australian Government's $90 billion engineering project is unprecedented in Australia's industrial history, and to succeed it will require the unrivalled development of a sovereign shipbuilding and sustainment capability that reaches into every state and territory, underpinned by a 'new level' of collaboration amongst industry to identify, attract, train and retain a shipbuilding workforce.

Global prime shipbuilding companies through to small, bespoke local industry suppliers will all play a critical role in contributing to the realisation of a truly national and sustainable Enterprise. It is estimated that more than 15,000 qualified and non-qualified workers will be required directly and indirectly in the first ten years of the Enterprise by naval shipbuilders and the associated supply chain and sustainment industry. Thousands of jobs will be secured for Australians over the coming decades and there are significant opportunities for local businesses to prosper too, as the supply chain engagement fans out across the whole country. Thousands of small to medium companies within the supply chain and sustainment categories are relying on industry to seize this opportunity.

The strengthening of Australia's naval shipbuilding industrial network provides a unique opportunity for career development and upskilling across a multitude of job categories. Workers will have the opportunity to gain training and qualifications across many high-end skills and professions that will give them a strong foundation for a rewarding and meaningful career. As workers become increasingly skilled, their employment options will widen and many will pursue promotional pathways and new experiences.

New career opportunities will soon spawn throughout Australia, offering jobs across a diversity of disciplines, from fabrication roles in the construction yards, through to engineering roles in the design bureaus, and leadership opportunities within management teams. They will be jobs of the future supported by state-of-the-art digital tools embedded throughout the design and construction processes, as well as during integration across the supply chain.

Indeed, the College is already well advanced in partnering with training and education providers across Australia to ensure there is a sustainable pipeline of skilled and job-ready workers available. Today's high school, TAFE and university students will form a critical element of Australia's future naval shipbuilding construction, supply chain, and sustainment workforce, coupled with experienced personnel sourced from across Australia's industrial sectors. However, large parts of this workforce are still three to four years away from graduating and the major companies' current needs are for workers who are skilled and experienced to lead and establish the core of their design and construction teams.

The Mission and Vision of the College

The Mission of the Naval Shipbuilding College is - To Lead the creation and sustainment of the enterprise that identifies, attracts and skills multi-generational sovereign capability for naval shipbuilding.

The Vision that the College has is that - all stakeholders know that Australia has the shipbuilding capability to meet the Naval Shipbuilding Plan

The College has been established to serve a number of stakeholder groups.

a. Firstly, the Commonwealth – the Department of Defence. To lead the creation and sustainment of the Enterprise that identifies, attracts and skills multi-generational sovereign capability for naval shipbuilding for Australia - from school students to experienced trades and tertiary graduates and ensures that the workforce is available with sufficient skills and capacity to meet Australia's Naval Shipbuilding Plan - including construction, supply chain, sustainment, and ship repair activities across the nation.

b. Secondly – Industry. Partnering with industry to create a national skills base. Ensuring that the industry workforce is available in the required locations in the required capacity when needed. Ensuring that this workforce has the required skills, knowledge and experience required to enter the shipbuilding sector. Reducing industry's cost of sourcing its required labour force, and assisting with the overall management of sector workforce evolution (up and down) over time.

c. Thirdly – the education and training sector. Connecting the education and training sector to where they can contribute to the national Naval Shipbuilding Enterprise. Ensuring that suitable education and training offerings are made available to meet the workforce training needs, and ensuring that the quality of the skilling offerings is endorsed and maintained at the requisite level.

d. And finally - the future workforce. Increasing the overall pool of workforce candidates through national industry sector promotion. Promoting the opportunities available for careers in Australia's Naval Shipbuilding Enterprise. Connecting potential future employees with the sector opportunities and industry companies.

The Naval Shipbuilding College Model

The College operational model is a 'hub and spoke.' The College is the Hub, headquartered at Osborne, South Australia and with an office in Henderson Western Australia, adjacent to the two major shipbuilding construction yards, and with potentially a third location in NSW to serve the Garden Island ship sustainment precinct. This is the Hub. The spokes are Australia's training provider network. The College has engagement with the national training provider network across the vocational and tertiary sectors in every state and territory of Australia – these are the spokes.

Naval Shipbuilding College Activities:

On behalf of Government and its industry stakeholders the College conducts activities across the complete spectrum of workforce management and development for the naval shipbuilding sector. From workforce demand and supply modelling, through workforce attraction and community outreach and engagement, to coordination of skilling and training initiatives to build job ready candidates to enter shipbuilding industry roles.

The College is at the very core of the national continuous shipbuilding enterprise – ensuring that the workforce is there for success. Establishing a comprehensive set of workforce development services across the sector that will underpin the establishment of Australia's Naval Shipbuilding Enterprise.

Workforce Management

In conjunction with industry the College has documented a common Industry Job Taxonomy – for common job categorisation across the sector, as the foundation of the workforce management function. The College is conducting labour market supply analyses across the sector for the primary job roles, and is collating and analysing future projected workload demand from industry. Identification of the priority and critical roles for industry has been assessed over time, and engagement is being driven across the whole supply chain to fully understand the complete workload demands, particularly those not within the perimeters of the construction yards themselves. This information is used as the basis of analysis for the available supply within the market and from training providers, in order to identify the requirements for market interventions. Specific job supply strategies are developed for all of the priority roles, and current and projected shortages, and appropriate interventions, identified to assure supply is planned and enacted.

Talent Acquisition:

To help meet this demand the College launched a national Workforce Register in July, 2018. First time entrants, people transitioning from adjacent industries or those wishing to upskill within the industry can use the Workforce Register to be connected with appropriate training and education providers and potential employers across Australia. Candidate Endorsement Consultants provide tailored training, education and career path guidance to people seeking to acquire the skills and experience required by future employers. More than 1500 people have already registered interest to work on some of the most technologically advanced projects on the planet and thousands more are expected to follow as the College extends its engagement across the country, to elevate naval shipbuilding as a career of choice.

Skilling, Training and Development services

The College has established a national network of training providers across the vocational and tertiary levels which now spans all states and territories. The College is working with these providers to identify gaps within current training and education offerings to ensure that they are suitable for the skilling requirements of the Enterprise. The College is developing and overseeing gap closure plans, and where needed identifying specific non-accredited training modules and electives required to be available to meet the specific skilling and competence requirements of the naval shipbuilding roles. On behalf of industry the College also provides the training and education entities with a formal College course endorsement, once the gap closure plans have been implemented and industry is satisfied that the training offerings will create job suitable candidates. The College also manages the 'Community of Excellence' of these endorsed providers through the duration of the National Shipbuilding Plan.

Communications and Engagement and Community Outreach

One of the key roles and responsibilities of the College on behalf of the sector is to generate awareness within the broader community of the available careers within the Enterprise, and to attract the future workforce candidates to these new roles. The College has a growing strategic outreach campaign that spans the nation, aimed at connecting with and attracting the future workforce, from high school pupils through TAFE and university students, to existing experienced trades and professionals already within industry. This campaign will leverage existing education and employment programs at the federal and state levels where possible. The College will be conducting its own national events in each state and territory on a yearly basis leveraging existing education and engagement programs at headline events. The College is promoting the sector through secondary schools and tertiary institutions for industry attraction, and developing the website/social media to connect with future generations. The College is also active within Australia's professional communities of practice to promote the sector opportunities and raise the profile of shipbuilding careers.

Conclusion:

The College is operating at the centre of Australia's Naval Shipbuilding Enterprise - partnering with industry and the education and training sector, to ensure that the workforce is available with sufficient skills and capacity to meet Australia's Naval Shipbuilding Plan.

The College is quickly building its capacity to provide workforce solutions for the Enterprise and is ready to remain an integral part of Australia's defence industry for generations to come. It is partnering with Defence, industry and a national network of training and education providers to play an integral role in developing the Australian workforce required to ensure our nation successfully delivers the critical naval capabilities that support our national security.

The collaboration and cooperation seen across the major global shipbuilding companies involved in the Enterprise is unprecedented. All are committed to pulling in the same direction to collaboratively grow the talent pool, through establishing and developing new workforce pipelines to ensure the success of the Enterprise. It's a very exciting time to be part of establishing Australia's Naval Shipbuilding Enterprise.

Forecasting Australian Shipbuilding Skill Demand

Hugh McCann, KPMG, hmccann1@kpmg.com.au
Joshua Rodgers, KPMG, joshuarodgers@kpmg.com.au
Mike Kalms, KPMG, mkalms@kpmg.com.au

INTRODUCTION

KPMG works with a number of clients, including members of the National Shipbuilding College, to understand the challenges associated with the National Shipbuilding Program (NSP) and the implications for the planned construction program. This paper draws on examples of this work, including scenario analyses conducted using KPMG's Shipbuilding Enterprise Workforce Model[1]. Insights into workplace relations considerations are also used, which influence the leavers the Commonwealth may use to help achieve the workforce required.

The Naval Shipbuilding Strategic Workforce Discussion Paper (NSWDP) (National Naval Shipbuilding Office, 2019) acknowledges a number of the risks associated with the workforce supply for the NSP, and the Commonwealth's role in addressing some of these by identifying levers to address shortages of supply. However it did not address the Commonwealth's role in creating or smoothing the demand side of the equation. Key drivers for this demand are:

- size and structure of the overall fleet
- the planned in-service life of different vessel classes including upgrade paths and capability assurance programs (CAPs)
- plan for sequential vessel development and delivery (see Figure 1)

These elements represents considerable levers within the Government's control to achieve a sustainable sovereign shipbuilding industry. The modelling presented in this paper identifies the boom and bust cycle will likely continue without further changes to the NSP program and the longer term plans for the force structure (Birkler & Et al., 2015).

The remainder of this paper will explore the context of the challenge, provide updated estimates for the scale of the workforce demand and how this compares with the 2015 RAND analysis and 2019 NSWDP, and offers a high level overview of the key levers available to balance supply and demand and further smooth the predicted boom and bust cycle.

---

[1]Analysis conducted in June 2018 prior to the announcement of BAE Systems Type 26 as the winner of the SEA5000 competitive evaluation process.

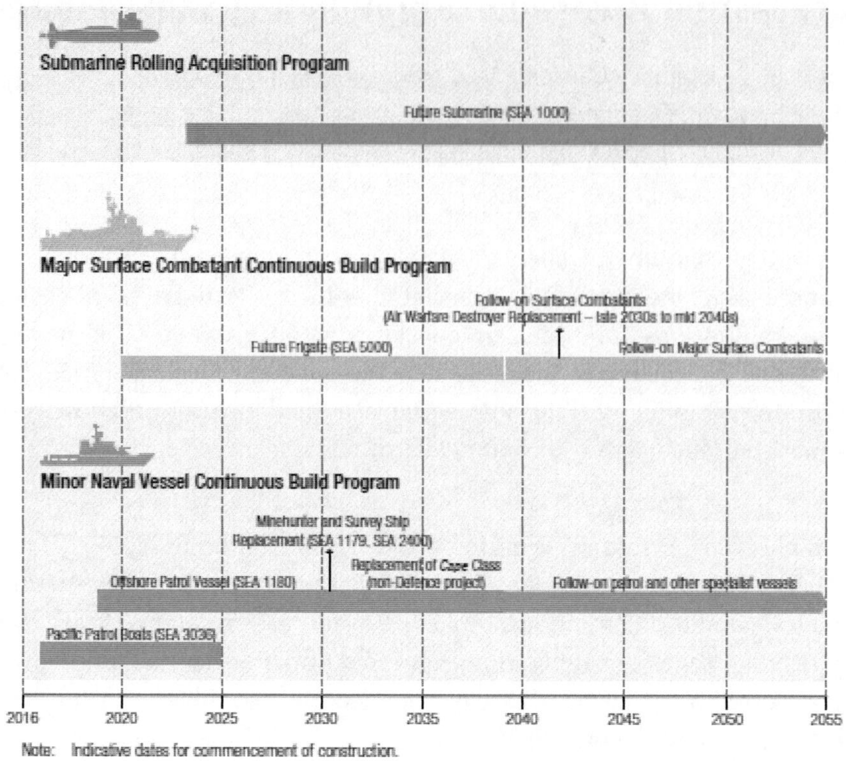

Figure 1: Naval Shipbuilding Continuous Build Programs (Commonwealth of Australia, 2017)

CONTEXT OF THE CHALLENGE AND DRIVERS WHICH HAVE RESULTED IN THE CURRENT STATE

There are a number of factors that have contributed to the current state and challenges faced by Australia in implementing the NSP.

The length of service of Royal Australian Navy (RAN) vessels has been increasing since Federation, but has remained relatively constant since the 1970s with a median in-service duration of 34 years for major surface combatants (Figure 2 and Figure 3). Lengthening service life for ships is a consistent global trend, but for a relatively small naval fleet such as Australia's, this makes continuous demand for Naval Shipbuilding difficult.

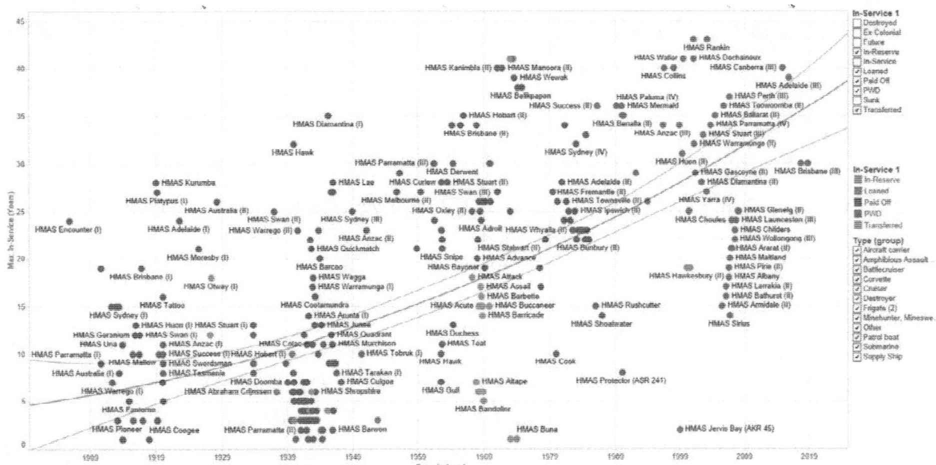

Figure 2: In-Service Duration of Australian Warships Scatter plot against commissioning date[2]

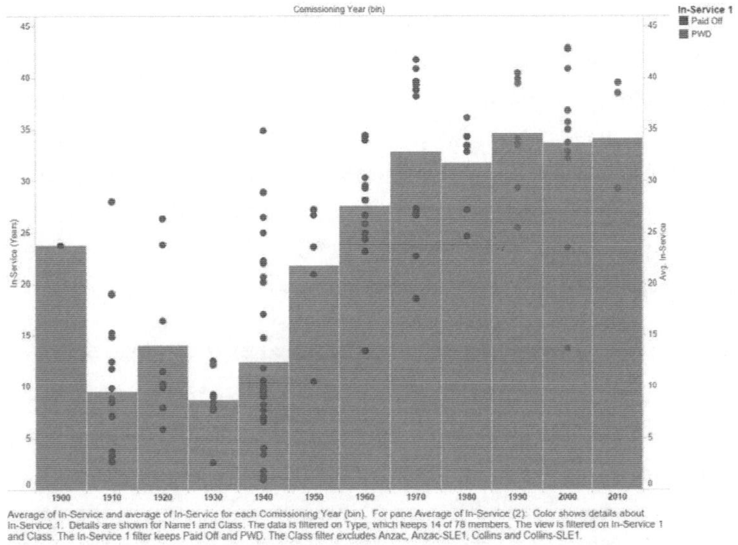

Figure 3: In-Service Duration of Australian Major Surface Combatants and Other Large Vessels (Inc. Submarines) Bar chart binned by commissioning year[3]

The trend of increasing length of service or Life of Type (LOT) is not restricted to the maritime (sea) domain. It is common across Air and Land domains as well. This is a function of the various technology cycles of complex weapon systems that see the structure (hull / airframe) last 30+ years, but combat systems upgraded several times within that period as part of Capability Assurance Programs (CAPs). This in turn is influenced by the capital cost of the

[2] Shown are the commissioning date for RAN Ships against their in-service years (including indicative Planned Withdrawal Dates (PWDs) for the current fleet) data sourced primarily from (Royal Australian Navy, 2019) with Service Life Extensions and PWDs inferred from NSP press releases from CoA and Primes over the period 2016-2019.

[3] This figure excludes Patrol boats and other minor vessels. Outliers to this trend include a number of second hand amphibious ships, where the original (USN) commissioning dates were used.

initial acquisition proportional to the Total Lifecycle Cost (TOTEX)[4,5], and by investments in long term planning to achieve the right balance of spend on new acquisition projects compared to "upgrades".

Most major surface combatants experience a mid-life upgrade, but more recently Service Life Extension (SLE) or Life of Type Extension (LOTEX) programs have also been common. The RAN also has a history of purchasing second hand vessels, in particular Logistic Support and Amphibious Ships (Sea Lift).

Analyses by the RAND Corporation (Birkler & Et al., 2015), commissioned by the Commonwealth, sought to inform strategies to address the looming "Valley of Death" in shipbuilding activity and the associated loss of specialist skills due to the forecast end of the Air Warfare Destroyer (AWD) construction program and the start of Future Frigate construction. Issues of accelerated ageing due to changes of operational usage of the Patrol Boat fleet were also considered for their impact on demand for new Offshore Patrol Vessels (OPVs). Subsequent program decisions, such the draw out of the AWD build, have made marked improvements; however, the forecast "Valley of Death" has not been averted. Whilst the OPV program will mitigate the required maximum growth rate for the skills and workforce required to meet SEA5000, it also adds to the peak demand. RAND (Birkler & Et al., 2015) analyses did not consider the submarine construction workforce requirements, nor those of the sustainment enterprise and its expansion post 2049 when force structure increases. The timing of SEA1000 (Figure 1), concurrent with SEA1180 and SEA5000 further exacerbates the tension in supply vs. demand for key skill types and resources.

QUANTIFYING THE NATIONAL SHIPBUILDING WORKFORCE DEMAND

Building on the previous analyses of RAND (Birkler & Et al., 2015) and others is important to measure the impact of recent program decisions, test assumptions, and move to a unified approach suitable for informing the long-lead-time decisions required for the enterprise to respond. Whilst the range, type and efficacy of response options is a topic of significant interest, the focus of this paper is to offer quantitative estimates of the demand for specific scenarios and related program assumptions.

A variety of supply and demand scenarios were used to forecast likely slippage in the shipbuilding program due to gaps and growth constraints in the Australian industrial base. They were also used to forecast the peaks and troughs in workforce capacity requirements based on fixed schedules. A high level description of the methodology is provided below.

Methodology

The activity was conducted in five distinct stages:

1 **Data Collection** included detailed review of previous modelling in the literature and of published Defence investment plans; and expert elicitation surveys to generate different

[4] ~50% for Major Surface Combatants and ~70% for Combat Aircraft. (Mislick & Nussbaum, 2015)

[5] "For example, the labour required to build an AWD or assemble an LHD is about three to seven times the labour required for large modernization projects or deep maintenance projects. Moreover, the labour required to build a new ship could be 30 to 70 times more than the labour required for routine maintenance." (Birkler & Et al., 2015)

shipbuilding production scenarios, constraints and risks for the building of each type of naval platform.

2. Confirm **Model Architecture** to accept inputs on workforce structure and fleet plans and perform sequential simulations to determine the scale and type of specific skills required for sustainment, construction and design activities for each vessel.
3. **Exercise the Model** to dynamically grow and contract the workforce to meet demand on a quarterly basis, or alter the production schedule based on a constrained workforce and growth rate.
4. **Verify and Validate** modelling results by comparison with the results of previous analyses with more constrained scope, including effective recreation of the previous analyses of the RAND Corporation conducted in 2015 (Birkler & Et al., 2015) see Figure 4.
5. **Analysis and Presentation** of results in an interactive analytics tool (Tableau) for simple exploration of all scenarios.

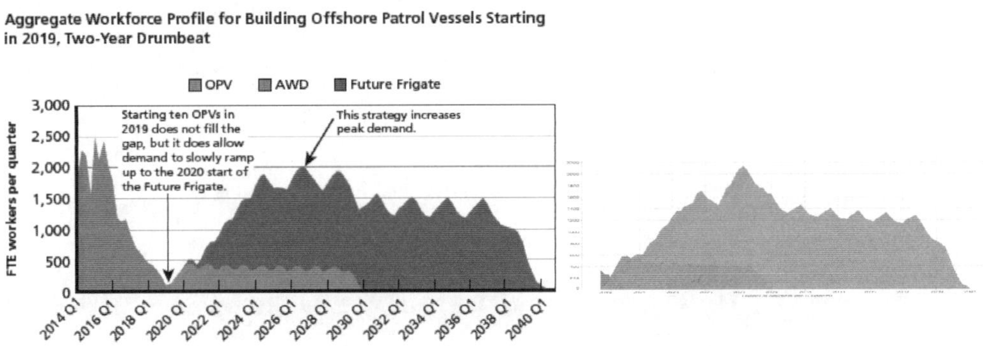

Figure 4: The model (Right) was verified against the RAND model results (Birkler & Et al., 2015)(Left) from Appendix D: Exploring the Option of Producing Offshore Patrol Vessels (Scenario D.6) (Birkler & Et al., 2015)

BASELINE SCENARIO RESULTS

The workforce required to meet the overall production plan can be estimated in several ways. The key differentiators are in consideration of both constrained and unconstrained labour pools i.e. unconstrained assumes the workforce can grow to satisfy both skill demand and schedule. For an unconstrained labour pool the workforce required to deliver the desired acquisition scenario is shown Figure 5. Trade specialisations are separated into categories and sub-categories depicted in Figure 6**Error! Reference source not found.**.

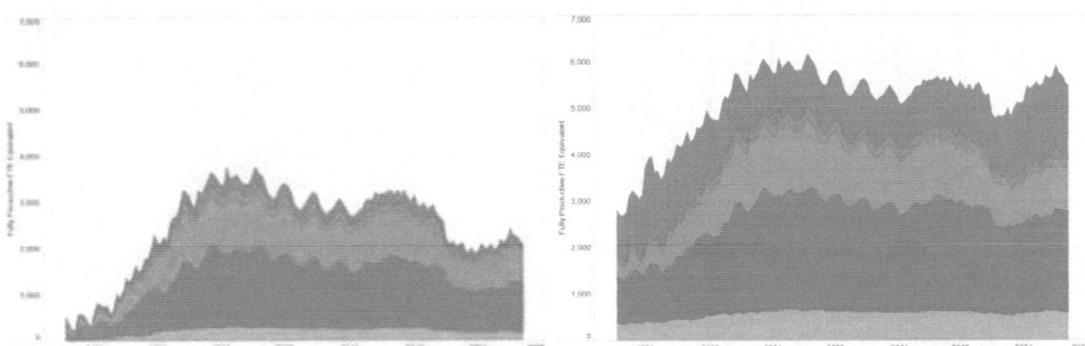

Figure 5: Shipbuilding workforce forecast by Shipbuilding Skill Sub-Category (Trade Type), unconstrained baseline scenario (Left - Surface Ships Only, Right – Submarine Design and Construction Included), both scenarios include follow on builds for AWD and OPVs only.

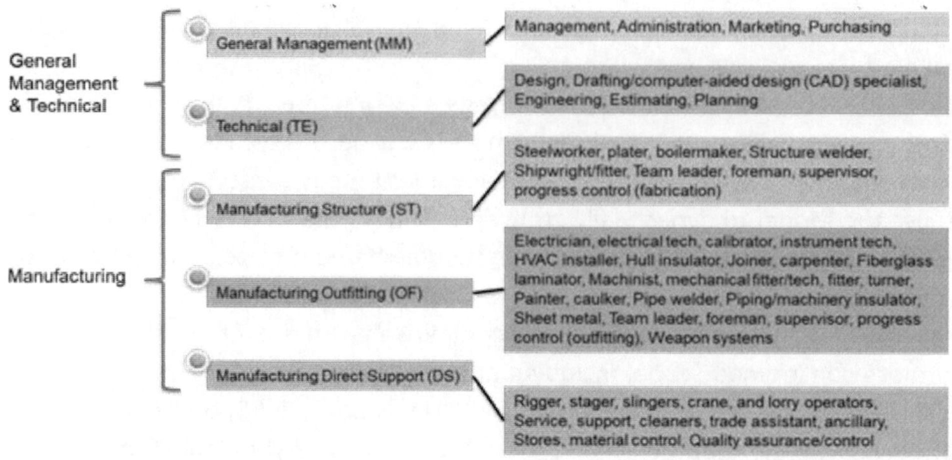

Figure 6: Shipbuilding workforce trade descriptions and identifiers consistent with RAND 2015 (Birkler & Et al., 2015) showing skill categories, specific skills, and skill competencies.

Figure 7 illustrates the demand by class and hull (A), trade type and experience level (B), and geography and shipyard (B).

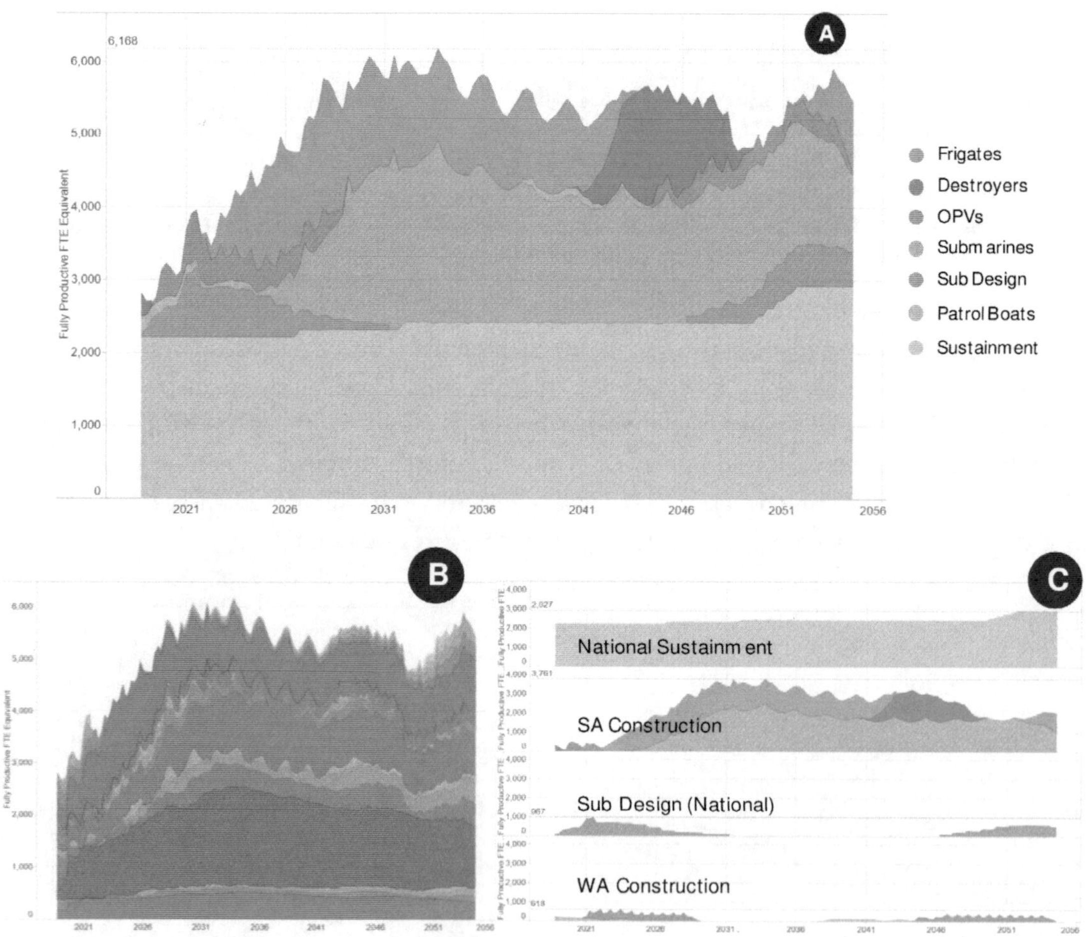

Figure 7: Shipbuilding workforce estimate example outputs for the New Baseline. A: Overall Demand by Program, B: Demand by Skill Type and Experience Level, C: Demand by Program and Location.

Figure 8 shows how trade types break down further to specific skills and skill competencies which enable direct translation of the workforce estimates to associated demand for vocational, graduate and post graduate education and training. This is a fundamental next step required of the Shipbuilding enterprise, and must be taken in collaboration with the education sector to quantify and plan for demand nationally and regionally in coordination with the other levers available (see *OTHER CONSIDERATONS* section of this paper).

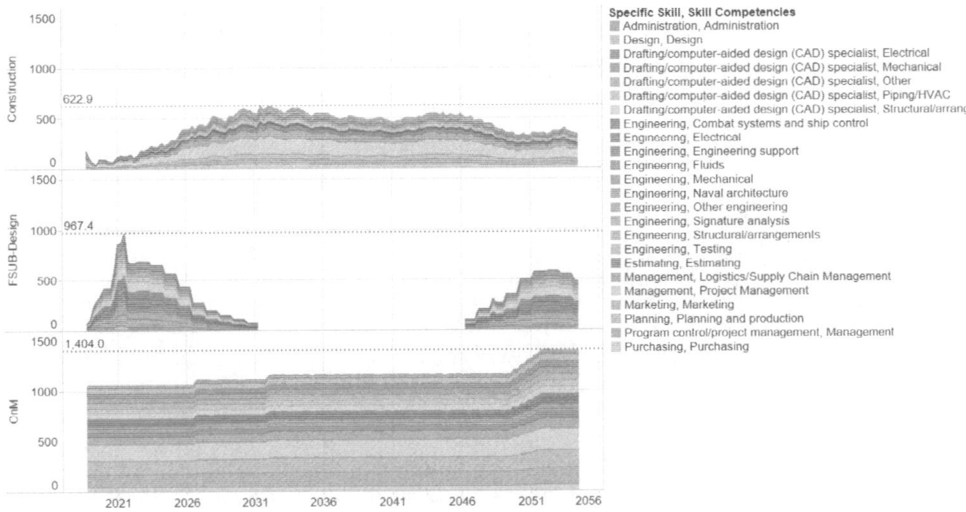

Figure 8: Shipbuilding workforce estimate, showing demand for specific skill competencies for the New Baseline by location and skill competency

COMPARISON WITH THE NATIONAL SHIPBUILDING WORKFORCE DISCUSSION PAPER

Comparison between the analysis presented in this paper and the NSWDP is challenging due to the differences in fidelity of the respective approaches. The analysis in this paper shows time based variations over a 35 year period and were generated for various scenarios, which provide limited comparison opportunities with the single point estimates in the NSWDP.

The scenario selected for comparison against the NSWDP modelling results took into consideration the latest announcements regarding Anzac class Service Life Extension (SLE) and Hunter class Frigate construction plans and is shown for selected data points in Table 1.

Two major points of difference between the two estimates are clear. The NSWDP does not include workforce requirements for design activities for the Attack class submarine (which this analysis forecasts to peak at over 600 FTE in the early 2020s), nor is the sustainment workforce considered. In this analysis the sustainment workforce is forecast to remain relatively consistent at around 2200 FTE until 2049 where the introduction of additional submarines expands the force structure and the associated sustainment resources required.

This analysis estimates the fully productive FTE and the forecast number of actual personnel required to deliver that productivity. The latter can be substantially higher based on the workforce experience mix and skill specific productivity levels; and can be heavily influenced by industrial relations policies related to staff employment continuity in periods of slow down. Table 1 shows the peak direct jobs estimated from the analysis presented in this paper as well as the quarter the results approximate the workforce peak estimated by the NSWDP.

Table 1: Comparison with NSWDP total workforce estimates (2018-2050). In parentheses is the quantity and quarter where the results described in this paper approximate the workforce peak estimated by the NSWDP.

Program	NSWDP Direct jobs	Analyses presented in this paper, completed in 2018 Direct jobs Fully Productive[6]	Direct jobs Personnel[7]
Guardian class pacific patrol boats	200	225 Peak Q4 2018	249 Peak Q4 2018
Arafura class patrol vessels	400	453 Peak Q1 2020 (SA) 560 Peak Q1 2022 (WA)[8]	734 Peak Q1 2020 (SA) 665 Peak Q1 2022 (WA)[8]
Hunter class frigates[9]	1,500	1832 Peak Q2 2028 (1539 Occurs Q4 2025)	1864 Peak Q2 2028 (1675 Occurs Q4 2025)
Attack class submarines construction	1,100	2475 Peak Q3 2031 (1146 Occurs Q2 2028)	2777 Peak Q3 2031 (1509 Occurs Q2 2028)
Attack class submarine design[10]		967 Absolute Peak Q3 2021[11] 682.6 Long Term Max Q1 2023	1311 Absolute Peak Q3 2021[11] 694 Long Term Max Q1 2023
Overall Peak	3,200	3752 Peak Q2 2030 (3439 Occurs Q1 2028)	4108 Peak Q2 2030 (3816 Occurs Q1 2028)

There is an unquantified risk of increased competition for specialist technical skills. This is specifically between the Hunter class and Attack class design activities in the years 2019-2026, which if not effectively managed could significantly delay one or both programs. The modelling presented in this paper included a higher design (technical trade-hours) requirement in the first of class for what was still Future Frigate at the time of analysis, however in relation to an Australianised Military off the Shelf (MOTS) acquisition of the type-26 frigate this was likely an underestimate. The model already showed tension between the technical trade requirement for this program and the Attack submarine design, generating a backlog of submarine design work that resulted in a rapidly growing workforce, oversized to meet the longer term requirements. If this is constrained to a more sustainable ceiling there would be even greater delay implications for the start of submarine construction.

In summary, the major points of difference to the NSWDP analysis are:

- Under the baseline assumptions, with a relatively unconstrained model, it is predicted that the proposed delivery of the 2017 NSP will be delayed.
- With Attack class design resources included, the NSWDP underestimates shipbuilding workforce growth to 2030 by approximately 17%.
- However, with revision of the 2017 NSP schedule to incorporate proposed service life extension of the Anzac Class Frigates, under the same conditions, the NSP can be delivered on time (i.e. to a New Baseline). This assumes relatively unconstrained workforce growth during the peak period 2019 to 2030 which requires significant coordination.

[6] Baseline scenario, Fully Productive FTE Equivalent.
[7] Baseline scenario, Number of Personnel (FTE).
[8] This number occurs after the Guardian construction completed, and assumes that of the total 560 jobs, 335 are new and 225 would be met from the Guardian program.
[9] Body of effort excludes Detailed Design (It was assumed that this was a relatively off the shelf acquisition).
[10] This number is quite high, but mitigates the absence of detailed Frigate design.
[11] Heavily influenced by demand from submarine design in the initial 5 years

WORKFORCE SUPPLY SCENARIOS, CONSTRAINTS AND POTENTIAL IMPACTS

Translating Forecasts into Reality

This paper presents an unconstrained scenario, however there are practical limits to the overall workforce size, the local workforce size which can be productive around a fixed number of vessels, and the minimum duration of each production activity. For example a single shipyard can typically build two large ships simultaneously: one alongside being outfitted, the other under construction. Similarly, submarine construction cannot start in earnest if design has not been completed.

The concurrent build restrictions or "slot constraints" in combination with the estimated demand profiles (which include resource requirements and build durations) for shipbuilding programs, and assumptions around drumbeats for construction and learning curves, means slippage in the 2017 NSP production schedule results even with relatively unconstrained workforce growth rates.

These schedule impacts are summarised as follows:

- Under the baseline assumptions the 2017 Naval Shipbuilding Plan is forecast to be delivered late.
- However, if this schedule is revised from the 2017 NSP to incorporate recent plans to extend the service life of the Anzac Class Frigates, a revised production schedule can be delivered with significantly reduced risk of delays (New Baseline). To achieve the fully productive FTE required to deliver the Technical Engineering and General Management disciplines, peak quarterly growth rates of 22% and 10.3% respectively are required. Growth in Technical Engineering peaks relatively early due to the demands of submarine design, whilst General Management peaks at the height of construction (see Figure 9).
- Comparison of the New Baseline against different levels of constraint for the 2017 NSP acquisition and construction schedule is shown in Table 2. This highlights that for the 2017 NSP, practical limits to the productive workforce size at both national and local levels results in schedule slippage for the production program. In the constrained cases, where workforce variation rates are lower, greater impacts are observed.

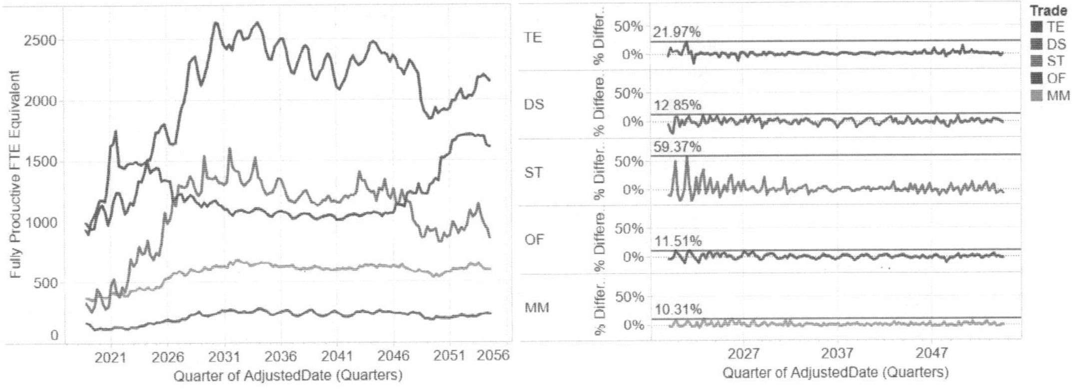

Figure 9: Forecast fully productive FTE and quarterly growth rates for each skill category

Table 2: Forecast production impacts from minimal practical constraints to a relatively unstrained workforce model. Without an ANZAC SLE, practical constraints impact the future submarine and future frigate most significantly.

Program	New Baseline Unconstrained	2017 NSP Unconstrained	2017 NSP Constrained Growth for General Management & Technical
Guardian class pacific patrol boats	16th Hull onwards delayed 2-3 Quarters due to prioritisation of OPVs	14th Hull onwards delayed 4 Quarters due to prioritisation of OPVs	No Change from 2017 NSP
Arafura class patrol vessels (OPVs)	2nd of class (SA) 2 Quarters delayed, WA On time	2nd of class (SA) 3 Quarters delayed, WA On time	No Change from 2017 NSP
Hunter class frigates	2nd of class onwards 1 Quarter delayed	2nd of class 1 Quarter delayed	2nd of class onwards 2 Quarters delayed
Attack class submarines construction	2nd of class 1 Quarter delayed, 12th of class 9 Quarters	2nd of class 8 Quarters delayed, 12th of class 16 Quarters	1st of class 12 Quarters delayed, by the 8th of class the program is 16 Quarters delayed

In Figure 10, shown in blue is the 2017 NSP production scenario (unconstrained rate of labour pool growth). Applying a constraint on the maximum growth rate of Technical Engineering and General Management trade types only of 15% per quarter results in further schedule delays of:

- 0.5 year schedule slip on the delivery of the 2nd-9th Future Frigates
- 4 year schedule slip on the Submarine program.

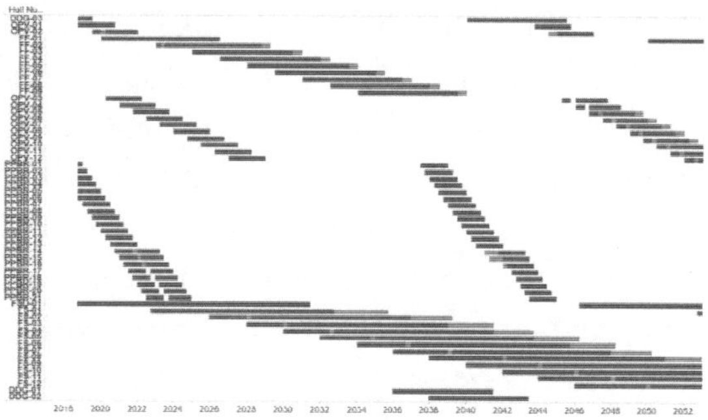

Figure 10: Forecast 2017 NSP production scenario (blue) and delayed scenario (orange) due to limits on rates of workforce growth

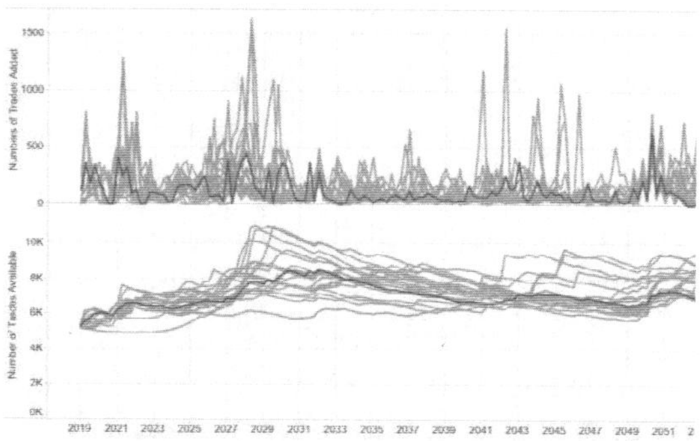

Figure 11: Trades person demand showing the baseline scenario (blue) and several other scenarios (grey) which vary considerably in growth requirements at different times (due to inefficient schedule overlaps which result). These estimates are for total number of personnel and include maintaining a sustainment workforce of approx. 2200 FTE.

Table 3: Estimated peak resource demand and the compound annual growth rate from 2018

		Peak		
Skill Type	**2018 Q4**	**Personnel**	**Quarter**	**CAGR[12]**
Technical Engineering	932	1749	2021 Q3	26%
Manufacturing Direct Support	171	275	2031 Q3	4%
Manufacturing Structure	333	1603	2031 Q3	13%
Manufacturing Outfitting	993	2635	2030 Q2	9%
General Management	374	683	2032 Q2	5%
Overall workforce	2802	6073	2031 Q3	6%

The required growth rates for specific skills and trades (including experienced personnel) is quite high and consistent (though does experience localised fluctuations) (see

[12] Compound Annual Growth Rate (CAGR) between Q4 2018 and peak

Table 3). This has implications for the domestic education sector, and in particular the pipeline of vocational training in relevant courses to balance supply and demand.

The baseline scenario shown in Figure 11 reflects the impact of the ANZAC Class SLE. The New Baseline avoids several peaks of activity associated with more aggressive production schedules. These points of concentration reflect the sensitivity of the overall system to variations in the production program and therefore the benefits of harnessing dynamic models during the decision making process.

OTHER CONSIDERATIONS

Measures to increase workforce productivity and potentially reduce the peaks in skill demand should be considered, along with increased support for workplace relations negotiations. Shipbuilding must also learn from other sectors and position itself as the industry of choice for longer term sustainability. The following workforce planning considerations were not covered in this paper but are presented here to highlight key considerations and control levers which should be factored into coordinated planning across the shipbuilding enterprise:

- Understanding of and understudying in critical roles
- Structural sustainability of trade professions
- Growing skills not qualifications
- Consideration of technical capability in the Australian Public Service
- Establishing shipbuilding as an Industry of Choice for new workforce entrants
- Workforce management at an industry level; including critical factors such as productivity, flexibility and leveraging lessons from other sectors to encourage diversity and retention
- Forecasts of competing workforce demand from other sectors with similar skill requirements

CONCLUSIONS

A national and coordinated approach needs to be taken to quantify resource demands, understand the critical roles and skill areas, and assist the strategic planning of industry, Government and education institutions in order for successful delivery of the National Shipbuilding Plan. The Naval Shipbuilding College provides a significant opportunity to deliver and continuous improve such an approach.

In addition to these measures, clear lead indicators of a mobilised and growing shipbuilding sector need to be established and tracked. An enterprise approach requires resource estimates to be continually updated; and effective coordination mechanisms put in place across the NSP programs, and between government and industry, to introduce appropriate incentives and support mechanisms and minimise delays which could introduce or exacerbate resource boom and bust cycles.

REFERENCES

ANAO. (2018). *Naval Construction Programs—Mobilisation (Auditor-General Report No. 39 of 2017–2018).* Australian National Audit Office.

ANAO. (2019). *ANZAC class Frigates — Sustainment.* Australian National Audit Office.

Birkler, J., & Et al. (2015). *Australia's Naval Shipbuilding Enterprise Preparing for the 21st Century (RR1093).* Santa Monica: RAND Corporation.

Commonwealth of Australia. (2017). *Naval Shipbuilding Plan .*

Mislick, G. K., & Nussbaum, D. A. (2015). *Cost Estimation: Methods and Tools, First Edition.* Wiley.

National Naval Shipbuilding Office. (2019). *Naval Shipbuilding Strategic Workforce Discussion Paper.*

Royal Australian Navy. (2019). *Availalbe Ship Histories*. Retrieved from http://www.navy.gov.au/fleet/ships-boats-craft/available-ship-histories

APPENDIX A – KEY PROGRAM AND MODELLING ASSUMPTIONS

The following provides key assumptions used modelling the National Shipbuilding Plan construction schedule.

1) The size of the future fleet is fixed:
   a) 12 OPVs: 2 in SA to be built by ASC at Osborne, and 10 in WA to be built by Civmech at Henderson.
   b) 9 Frigates to be built at Osborne South. 12 Submarines to be built at Osborne North.
2) The construction of the future fleet will not employ module consolidation:
   a) This assumption should be challenged through stakeholder engagement. The impact of changes to this assumption would potentially be less man-hours, or shorter duration builds.
   b) This also impacts the number of hulls that can be in construction concurrently. Note that while this is typically two for large ships, Naval Group has indicated that a modular approach to Attack class construction was intended, hence this model allows up to 4 future submarines in construction at once.
   c) A similar argument could be made for the use of advanced manufacturing such as digital shipyards with higher levels of automated manufacturing. Whilst standard learning curve rates have been incorporated into the modelling, no consideration has been included for such disruptive elements which would likely result in major reductions in labour required in particular for structural, management and technical trade's types.
   d) For the purposes of modelling workforce, we will look at the constraints associated with build location and labour availability.
   e) The size of the workforce required to achieve the acquisition/production plan is an output of the model. Further input from shipbuilding and industry experts is required to define final inputs (i.e. acquisition/production plan).
3) 100% of all construction occurs in Australia.
4) As of July 2018, the Osborne South and North, and Civmech's new shipbuilding facilities have yet to be built or are still under construction. This may introduce further risk of delayed production start dates and should be considered as part of industry engagement.
5) Combat Systems Integration (along with its risks) is included as part of construction.
6) The aim of the Commonwealth's acquisition schedule is to roughly align with the planned withdrawal dates (PWDs) of the current fleet:
   a) This may not be achievable resulting in capability gaps or requiring life extensions
   b) This may require a larger peak workforce that does not achieve the desired stable workforce, one of the design intents of the NSP is a Continues Shipbuilding program and Rolling Submarine Acquisition
   c) This analyses considered the Future Force Structure with the original PWDs for the Collins class Submarine and with a Service Life Extension (SLE) program
   d) The analyses considered the Future Force Structure with the an assumed in service life of approximately 30 years for Anzac class Frigates, which factored in a minor Service Life Extension (SLE), about half that referenced in the recent ANAO report (ANAO, 2019)

   e) Multiple scenarios where run with combinations of 1.5 and 2 year drumbeats for Frigate construction and 25 and 30 year in-service lives, delayed starts to various programs, different man-hours per hull (larger more complex builds – size variations), different learning curve rates
   f) Further information from industry is required to determine the likely capability gaps or SLE requirements to replace current baseline assumptions.
7) At the time of this analysis, announcement of BAE Systems Type 26 as the winner of the SEA5000 competitive evaluation process had not occurred.
   a) Size of vessel and estimated man-hours were not known (Baseline assumed 5M man-hours for future frigate, based on an AWD sized vessel), Hunter class vessels appear to have a larger full load displacement than the AWD.
   b) Design requirement and prototyping was not factored in, instead:
   c) Start date for Hunter class was modelled as 2020 instead of 2022 (prototyping to start in 2020), but with a longer build for the first of class, but without a proportionally different trade distribution build profile.
   d) Drumbeat and Anzac SLE were also not known at the time. It became clear though through modelling of various scenarios that an Anzac class SLE would most likely be needed. Also that a drum beat of 1.5 years and LOT of 30 years meant that the continuous shipbuilding program would run out of ships in the mid 2040s.

A Framework for Assessing the Impact of Integrating New Equipment onto Naval Vessels

Lily Webster[1], Brett A. Morris[2] and Dylan M. Dwyer[2]
1. Navy Technical Bureau, Navy Engineering Division
2. Maritime Division, Defence Science and Technology Group
brett.morris@dst.defence.gov.au

ABSTRACT

A nation's Defence Force has to adapt to changing geopolitical, technological and economic challenges. This particularly impacts capability systems with relatively long lifecycles such as naval ships. In Australia, government policy has meant the Royal Australian Navy (RAN) currently uses the off-the-shelf (OTS) acquisition strategy. If the RAN is to remain adaptable while using an OTS strategy, approaches to evaluate the capacity of OTS designs to integrate new equipment at various stages in the lifecycle are required. By understanding the impact of integrating new equipment onto a naval ship design, the RAN can manage the risk of not being able to respond to changing strategic and technological circumstances.

Recent research in the Maritime Division of the Defence Science and Technology (DST) Group has focused on developing methods to assess the impacts of integrating new equipment onto OTS naval ship designs. These assessments can be made throughout the capability lifecycle, including tender evaluation, when establishing potential batch-build strategies and when considering ship subsystem upgrades. To support these assessments, an Integration Impact Assessment framework has been constructed. The framework allows an assessment of the capacity of a ship design to integrate a range of equipment configurations and investigate the implications on the ship's mission performance.

This paper focuses on the use of the Integration Impact Assessment framework to assess the impact on ship designs of integrating new systems for additional military roles. The paper covers a pilot study undertaken jointly by the Naval Technical Bureau and DST to assess Multi-Role Vessel designs for their ability to accommodate new equipment in support of forecasted future roles of maritime constabulary, rapid environmental assessment and mine counter measures.

INTRODUCTION

The OTS acquisition strategy may result in the need for the Royal Australian Navy (RAN) to perform design changes to align the OTS vessel design with RAN specific requirements. Additionally, vessels typically have long service-lives and must be periodically upgraded to meet evolving RAN requirements. Consequently, there is likely to be a need for the RAN to integrate new systems and equipment during acquisition, and update or upgrade this equipment at different stages throughout the vessel service-life. The need to integrate new equipment can arise from changing geopolitical, technological and economic challenges that necessitate changes in Australian naval operational requirements [1]. The need to integrate new equipment can also arise from the maturation and uptake of suitable technologies, a current example of which are uninhabited systems [2]. Ideally, RAN vessel designs will inherently have a degree of flexibility, or adaptability built in through a range of appropriate margins. Throughout the vessel's lifecycle, these margins will be consumed as new or upgraded systems and equipment are integrated onto the vessel. Evaluating and understanding these margins and therefore the flexibility of the base platform will assist in identifying and managing capability risks and assist the RAN in making evidence-based design decisions [1].

The current Australian Chief of Navy, VADM Noonan, noted at the Defence and Industry Conference in 2018 that *"The future is not static and the context and needs of the future will continue to evolve. With it, we will be required to continually assess our capability"* [3]. In order to continually assess and manage the margins available on RAN vessels so they can adapt to these evolving needs, a methodical framework and associated tool is required. This paper covers the development and use of the Integration Impact Assessment (IIA) Framework and associated Tool that was developed to satisfy this requirement[1].

The IIA Framework provides the ability to undertake an assessment of various equipment configurations (ECs) to ascertain what integration impacts there will be on a naval vessel's margins and functional capability [4]. Integration of new equipment onto a vessel for which it was not originally designed may occur at any time in its lifecycle, including during the initial build process, major dockings or in batches of vessels being constructed using a continuous build program. This means that the IIA Framework and Tool could potentially be used to support decision-making during [4]:

1. Tender Evaluation: both tendered vessel performance and projected whole of life performance.
2. Modification of a current vessel to achieve evolving capability requirements.
3. Mid-life upgrade evaluation: what equipment will have the biggest capability increase with the smallest impact?
4. Development of batching strategies for classes of vessels in a continuous build program.
5. Life of Type Extension (LOTE) studies.
6. Exploration of Multi-Role Vessel concepts.

The IIA Framework has been developed to be flexible in application; the margins assessed could include consumable physical parameters such as space and weight, as well as measurable Key Performance Parameters (KPPs), such as stability compliance, endurance, and sprint speed. Additional analytical approaches such as the Ship Performance Modelling and Simulation Framework proposed by Dwyer and Morris [5] could be utilised to generate these KPPs, allowing the impact on more complex KPPs such as 'Percent Time Operable at Sea State 5' to be calculated and considered.

---

[1] Initial development of the Integration Impact Assessment Tool was conducted by Frazer-Nash Consultancy, under the direction of the Defence Science Technology (DST) Group, to assess the impacts of integrating various technologies and subsystems on board the base platform for the SEA2400 project (Survey Vessel Hydrographic & Oceanographic). Subsequent development was conducted in a collaboration between the Navy Engineering Branch and DST.

This paper introduces the IIA Framework and covers a test implementation in a spreadsheet-based tool. The test implementation investigates the integration of a range of equipment configurations onto a Multi-Role Vessel (MRV) base platform. The paper concludes with some observations on the utility of the tool.

THE INTEGRATION IMPACT ASSESSMENT FRAMEWORK AND TOOL

Due to the nature of the activities that lead up to the integration impact assessments, the steps required to execute the IIA Framework could conceivably be implemented in any particular order the user deems fit. However, a general approach to the steps involved in the IIA Framework is depicted in Figure 1.

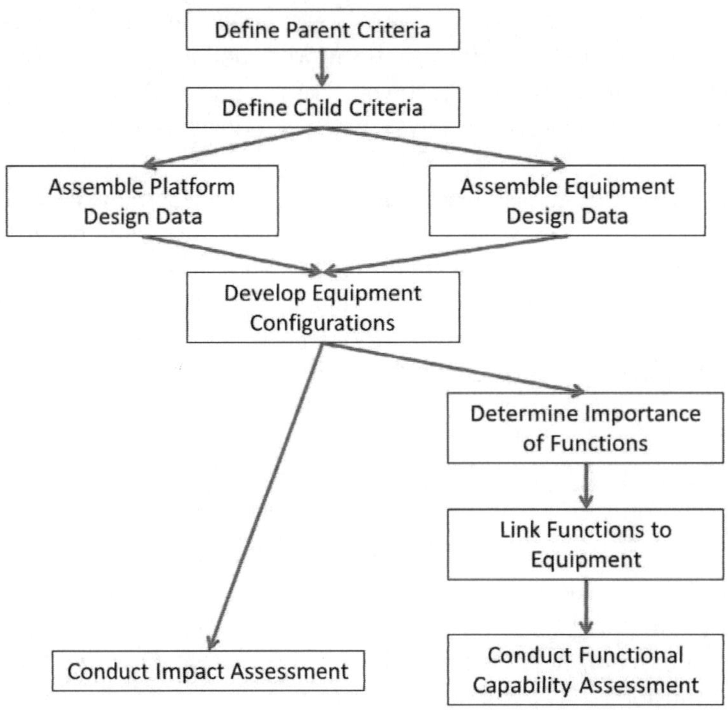

Figure 1: Overview of steps involved in the Integration Impact Assessment Framework.

Impact Assessment

The integration impact assessment component of the IIA Framework provides a qualitative assessment of the impact of integrating new equipment on a vessel's margins. The parent and child criteria are the metrics of the IIA Framework. They should represent the remaining vessel margins, over and above the original design, which the newly integrated equipment will consume.

Five common parent margin categories that are important for current and future naval vessels were identified by Schank *et al.* [2], they are; power, cooling, support for personnel, space and bandwidth margins, as well as the well-established weight and stability margins [2]. A summary of the expected impacts on these margins from emerging naval technology trends according to Schank *et al.* [2] is shown in Figure 2.

	Power	Cooling	Personnel	Space	Bandwidth
Unmanned systems	Little change	No change	Increase	Unclear	Increase
Electromagnetic weapons	Increase	Increase	Little change	Increase	No change
Long-range targeting	Increase	Increase	Little change	Increase	No change
Increasing Networking	Increase	Increase	More technical	Unclear	Increase

Figure 2: Effect of some key technology trends on vessel margins [2].

The final step of the Impact Assessment in the IIA Tool involves a comparison between the margins available on a vessel and those consumed by integrating an equipment configuration, to calculate an impact score for each criterion. An overall impact assessment score is calculated using a Multi-Criteria Decision Making (MCDM) method that combines the scores from all of the parent and child criteria. While this can be useful, testing of the IIA Framework has highlighted that the overall score should never be used without investigating all individual parent and child criteria to check whether individual criteria, or margins on the vessel are over consumed. These aspects are covered in more detail in Morris *et al.* [4], as well as in the test implementation covered later in this paper.

Functional Capability Assessment

The functional capability assessment provides an estimate of the functional capability provided by an equipment configuration. The estimate is based on a weighted summation of the number of required functions that an equipment configuration can perform. The functions that a vessel is required to perform must be defined and their importance estimated as shown in Figure 1. Once this has been done, the functions need to be linked to the equipment that could perform them. For the vessel to achieve a functional capability contribution of a function, all pieces of equipment required to perform the function must be present in an equipment configuration; e.g. if an equipment configuration contained a davit without a seaboat, it would not be able to perform boarding party functions. The functions can be defined in a top-down manner based on the content typically provided in an Operational Concept Document.

The functional capability assessment component of the IIA Framework should only be considered with reference to the impact assessment. This is to ensure that functional capability assessments are conducted with an understanding of the equipment configurations that can be feasibly integrated onto a vessel.

Further details on the IIA Framework and Tool can be found in Morris *et al.* [4]. This paper focuses on a test implementation of the IIA Framework using the current version of the IIA Tool to investigate its utility.

TEST IMPLEMENTATION – IMPACT OF INTEGRATING EQUIPMENT ONTO MULTI-ROLE VESSELS

This test implementation involves the use of the IIA Framework and Tool to explore Multi-Role Vessel (MRV) concepts. The MRV concept in this case is a vessel that is to perform the military roles of Rapid Environment Assessment (REA), Mine Countermeasures (MCM) and Maritime Constabulary (MC). It is assumed that the vessel will spend 60 % of its time at sea undertaking MC operations, 30 % of its time at sea undertaking REA and 10 % performing MCM activities. In the test implementation, the IIA Framework is applied to three assessments:

1. Assessing the impact of including various Equipment Configurations (ECs) associated with the onto three candidate MRV designs.
2. A comparison of the three candidate MRV designs.
3. An assessment of the functional capability achieved by the ECs.

Background

The MRV concept has been gaining momentum in recent years, with many Navies developing MRV programs to combat the increasing number of tasks a naval vessel is required to perform [6]. An increasing amount of time is spent on military operations other than war such as; humanitarian relief after natural disasters, border protection, piracy interdiction, search and rescue. The single-mission navy vessel can be seen as an expensive and rigid solution, which requires expensive mid-life upgrades to keep up with competing technology advances. The modular concept of reconfigurable payloads has the potential to reduce the cost of mid-life upgrades and allows the vessel to be reconfigured to perform different roles as required [2, 7]. The interoperability of the same base platform and systems potentially reduces training requirements and cost of ownership while increasing operational efficiency [7, 8].

The roles of REA, MCM and MC were identified in an initial capability description for the SEA1180 MRV project in 2011. It was planned to replace the existing RAN Patrol, Hydrographic and MCM force, comprised of 26 vessels across four classes, with a single, multirole class [8]. This concept was not realised for SEA1180 at the time, however this concept has been used as the basis for the test implementation.

The test implementation uses the IIA Framework to investigate the impact of integrating various subsystems onto three MRV concepts and the associated functional capability provided with each EC. It is worth noting that the impact of placing additional equipment on the vessel on the centre of gravity has not been assessed in this test implementation and it is assumed that the vessel stability and structural strength will not limit the Equipment Configurations (ECs). In a full implementation of the IIA Framework, the impact on the vessel centre of gravity could be included as a criterion.

Define Integration Impact Criteria

The five parent criteria defined for this test implementation are: Space – non tank, Space – tank, Weight, Heating Ventilation Air-conditioning and Cooling (HVAC) and Power. The criteria and weightings were developed by DST researchers acting as defence stakeholders. Child level integration criteria were developed for each parent criteria. For each criterion, the margin available on a vessel is typically applied as the upper threshold, and a value of zero applied as the lower threshold. The equipment configuration's value for that criterion is then marked as a percentage of these thresholds. For the test implementation, all criteria were assigned an increasing value curve. This means that a lower integration impact score results in a lower utilisation of the available margins. As the vessels used in the test implementation are concepts, i.e. in the early design stage, a lower utilisation of margins is preferable as reserve margins may be required for future upgrades and remediation when the vessel is in service.

Assemble Platform and Equipment Characteristics

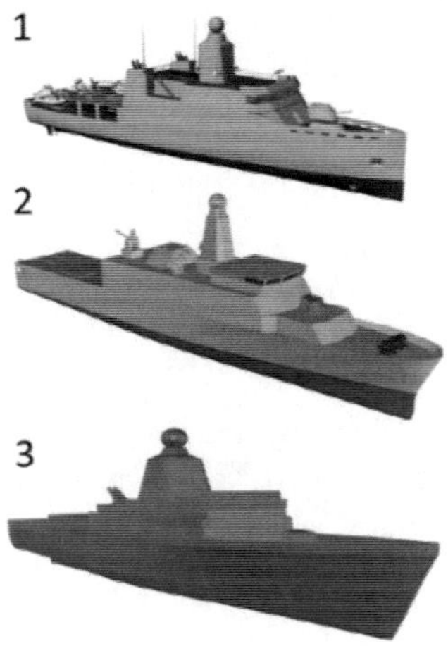

Figure 3: Candidate Vessel Designs

The three vessels used in this test implementation were final year projects by students at the Australian Maritime College, developed under direction from DST and are depicted in Figure 3.

These vessels were designed as multi-role vessels, with slightly varying requirements and particulars. Computer Aided Drawing (CAD) models, general arrangements and ship specification documents were supplied with each vessel and have been used to assemble the platform design data required to populate the IIA criteria thresholds.

The Equipment required for the roles was developed using a top-down mapping of the equipment required to achieve the operational functions described in Morris [9] in combination with engineering knowledge. The data required to populate the criteria values for the equipment was then sourced and the functional capability that the equipment could achieve were recorded using a traceability matrix, an example of which is provided in Reference [4].

Equipment Configurations

Twenty-two ECs were developed using the open literature sources and engineering knowledge. Three levels of EC were developed for each vessel role, representing a minimum, medium and maximum performance level. Combinations of these role-specific ECs were also generated, to identify any potential integration impact savings by utilising the same equipment across multiple roles. The hydrographic equipment configurations were developed with the assistance of data from Hayes and Webb [10]. The three equipment configuration levels generated were formed as:

1. Minimum performance level: shipborne sensors and a small specialist crew.
2. Medium performance level: minimum performance level equipment + towed sensors and a medium specialist crew.
3. Maximum performance level: Minimum + medium performance level equipment + off-board, UUV and UAV sensors with a maximum specialist crew.

As an example, the MCM role ECs are shown in Table 1.

Table 1: Mine Counter Measure Equipment Configuration levels

MCM Minimum	MCM Medium	MCM Max
UUV (Double Eagle Mk II)	MCM Min Plus	MCM Min + MCM Med Plus
Three MCM Crew	UAV (Camcopter s-100)	Two MCM Crew (7 total)
Small Deployable Survey Kit	UUV (Seafox Mine Disposure)	Clearance Divers (6)
Deployable Meteorological Kit	Two MCM Crew (5 total)	Clearance Diver RHIB
LARS A-frame (Seaview) Max1.5t	LARS Crane (SAAB Seaeye)	Transportable recompression Chamber
Ceremonial: Flags etc.	RHIB MCM (J3)	Davit (Clearance Diver RHIB)
Towing Gear	Davit (RHIB PRH 100 AP)	
Mooring Gear		

Link Functions to Equipment

The functional capability assessment within the IIA Framework is based on the number of functions the equipment within the EC can perform, and does not involve the candidate vessel; it is merely a weighted measure of the functions that can be performed by an EC. Hence, it should not be used in isolation from the Impact Assessment.

For the MRV test implementation of the IIA Tool, A previous study [9] into the functions required to be performed by an MRV was utilised. Morris' [9] list of 96 parent functions and 135 children functions has been refined, with functions that are performed by the base platform such as Transit and Seakeeping not being considered in this functional capability assessment. The remaining functions were segmented by role and assigned an importance, or weight based on the total 'time spent in role' defined in the vessel requirements: 60 % for MC, 30 % REA and 10 % MCM.

The importance or weight of a function is calculated by dividing the percentage of time the MRV will spend in the role by the number of functions within it. Furthermore, if a function contributes to a capability more than once, the importance was multiplied by the number of times that function occurred within the role. An example of this is the shallow water function within the MC role. Shallow water functions were present in many of the missions identified for MC role such as Littoral warfighting, Border protection and Counter terrorism, which resulted in its high contribution to capability. These functions are listed in Tables 2, 3 and 4 by each role respectively.

Table 2: Functions and their contribution to MC Functional Capability

Role	Top-Level Function	Second-Level Function	Contribution (%)
MC	Border Protection Functions	Boarding Party Functions	6.92%
	Launch, Recover and Stow	Boats	6.92%
		UAV	2.31%
		UUV	2.31%
		USV	2.31%
	Sensor Functions	Above Water Sensors	4.62%
		Underwater Sensors	2.31%
		Visual Surveillance and Detection	2.31%
	Shallow Water Functions	Shallow Water Functions	13.85%
	Counter Piracy Functions	Sensor Functions	4.62%
		Special Forces Functions	2.31%
	Counter Terrorism Functions	Littoral Warfighting Functions	4.62%
		Offshore Warfighting Functions	4.62%

Table 3 Functions and their contribution to REA Functional Capability

Role	Top-Level Function	Second-Level Function	Contribution (%)
REA	Bathymetric Mapping Functions	Shallow Water Functions	4.0%
		Sensor Functions	2.0%
		Deep Bathymetric Mapping	2.0%
		LRaSF.1-Boats	4.0%
		LRaSF.2-UUV	4.0%
		LRaSF.3-UAV	4.0%
		LRaSF.4-USV	4.0%
	Sensor Functions	Underwater Sensors	2.0%
		Above Water Sensors	2.0%
	Subsearch Functions	Hydrographic Survey Functions	2.0%

Table 4 Functions and their contribution to MCM Functional Capability

Role	Top-Level Function	Second-Level Function	Contribution (%)
MCM	Clearance Diver Functions	LRaSF.1-Boats	0.71%
	Ceremonial Functions	Ceremonial Functions	0.71%
	Minehunting Functions	LRaSF.1-Boats	1.43%
		LRaSF.3-UAV	1.43%
		LRaSF.4-USV	1.43%
		LRaSF.6-UUV	1.43%
	Sensor Functions	SF.1.2-Underwater Sensors	1.43%
	Minehunting Functions	ShF- Shallow Water Functions	0.71%
	Sensor Functions	SF.1.1-Above Water Functions	0.71%

Impact of Integrating Equipment Configurations onto Vessels

The impact assessment results for integrating the 22 ECs onto each of the three candidate MRV designs is summarised in Table 5. This provides only a high-level overview of the overall impact assessment that is calculated as a weighted value using an MCDM technique. An impact result of 100 % represents a vessel margin being completely consumed. An impact result greater than 100 % means the EC cannot be feasibly integrated within the vessel's margins and indicates some equipment may need to be removed or design changes made to the base platform.

As previously discussed, it is vital that the overall impact assessment criteria are not viewed in isolation from the individual parent and child integration impact criteria. For example, the child criteria results of EC10 provide a useful demonstration of this as the overall integration impact results for vessels one and two are less than 100 % (red cells highlighted in Table 5). However, when viewing the individual child-level integration impact criteria results for EC10 shown in Figure 3 (also see the blue highlighted cells in Table 5), it can be seen that six of the criteria are greater than 100 % for Vessel 1 and five of them are greater than 100 % for Vessels 2.

Table 5: Results Summary of the Integration Impact Assessment for the MRV Test Implementation [4].

ID	Equipment Configuration Description	Vessel 1		Vessel 2		Vessel 3	
		Impact (%)	No. Criteria > 100%	Impact (%)	No. Criteria > 100%	Impact (%)	No. Criteria > 100%
EC1	Base (nothing)	0	0	0	0	0	0
EC2	Rapid Environmental Assessment (REA) Minimum	40	2	32	2	74	4
EC3	REA Medium	59	3	51	4	105	5
EC4	REA Maximum	74	7	62	5	154	7
EC5	Mine Countermeasures (MCM) Minimum	7	0	9	2	25	3
EC6	MCM Medium	38	1	33	3	73	4
EC7	MCM Maximum	56	3	46	5	95	5
EC8	Constabulary (MC) Minimum	51	4	38	4	77	5
EC9	MC Medium	57	5	42	5	82	5
EC10	MC Maximum	76	6	57	5	128	5
EC11	REA + MCM Minimum	45	2	40	4	84	6
EC12	REA + MCM Medium	95	6	81	5	161	7
EC13	REA + MCM Maximum	120	9	101	9	224	9
EC14	REA + MC Minimum	100	7	77	6	154	6
EC15	REA + MC Medium	125	9	99	9	190	7
EC16	REA + MC Maximum	146	10	115	9	245	10
EC17	MC + MCM Minimum	68	5	54	7	106	6
EC18	MC + MCM Medium	93	7	73	7	139	6
EC19	MC + MCM Maximum	118	9	91	8	186	7
EC20	REA + MC + MCM Minimum	106	8	84	8	164	8
EC21	REA + MC + MCM Medium	149	9	120	9	227	10
EC22	REA + MC + MCM Max	170	13	137	10	278	12
Totals			125		126		137

These individual criterion results are useful as they provide suggestions for vessel design changes. For example, from Figure 3 it can be seen that for the child-level integration impact criteria 1.5, Machinery Space, EC10 has consumed only 31 % of the machinery space available on Vessel 2. This means there could be potential to install more auxiliary machinery on Vessel 2, thereby resolving some of the over allocated criteria such as HVAC (criteria 4.1) and power generation (criteria 5.1).

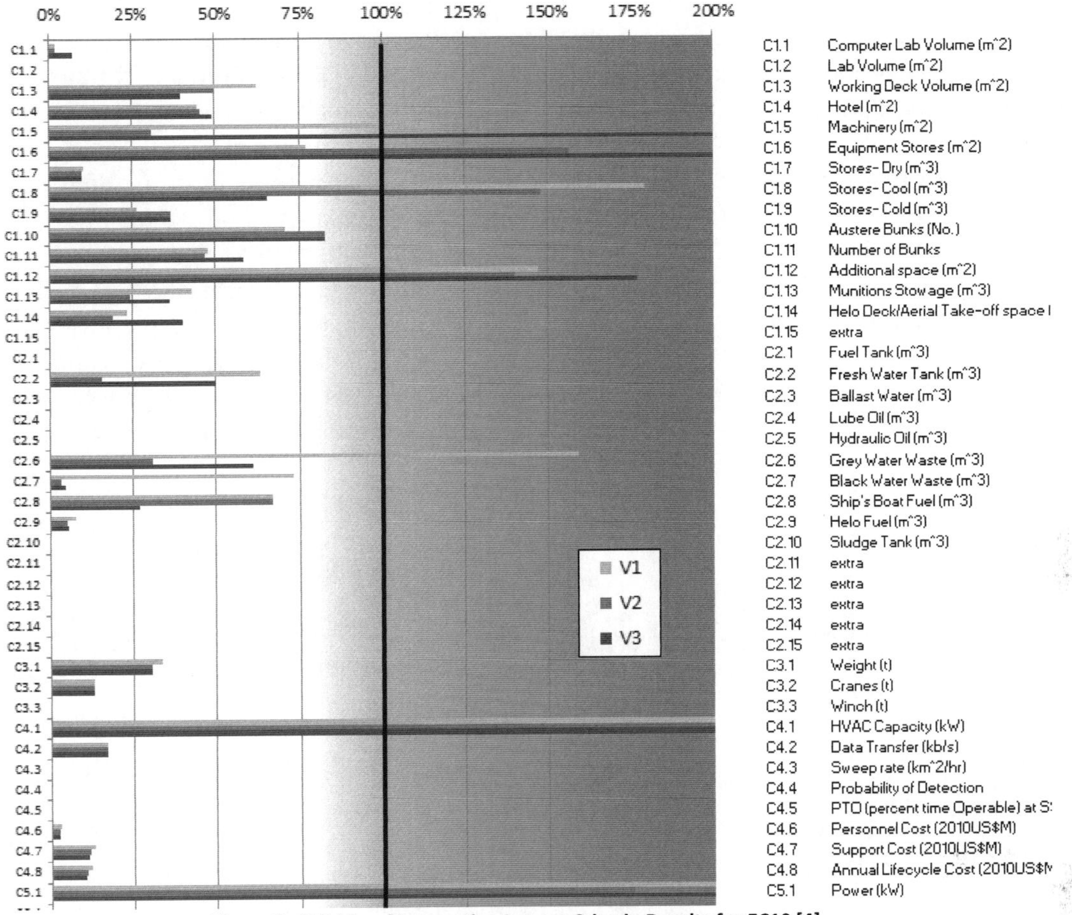

Figure 3: Child-level Integration Impact Criteria Results for EC10 [4].

Evaluating Vessels

A simple way of comparing the three vessels investigated for this test implementation is to sum the total number of over allocated (i.e. over 100 %) integration impact criteria for all of the ECs. In this case, by adding up the number of criteria greater than 100 % for each vessel as shown in the final row in Table 5, it can be seen that Vessel 1 has a total of 125 over allocated criteria across all ECs, Vessel 2 has 126 and Vessel 3 has 137. Again, this does not provide a full understanding of which vessel is preferable, as this will be dependent on the EC that is of most value to the capability stakeholders. Once this is known, the vessel evaluation will need to be conducted in light of all integration impact criteria.

Evaluating Functional Capability

The functional capability assessment component of the IIA Framework is based on the ECs and is independent of the vessels being evaluated. It is therefore important to remain cognisant of which ECs can be feasibly integrated onto any vessel under evaluation when conducting the functional capability assessment. Using the weightings of the importance of the functions for the functional capability given in Table 1, the relative capability of each of the ECs were calculated with the results shown in Table 6.

To achieve a functional capability score in the IIA Tool, all equipment required to perform the function must be present in the EC. One piece of equipment may be included in multiple functions. As a number of equipment pieces are typically required for each function, and these may not be coincident with a single EC, an EC may have a low capability score due to not including one specific piece of equipment; e.g. a RHIB, which contributes to the capability 'Shallow Water Functions', but only if there is a davit to 'Launch, Recover and Stow Boats'. Conversely, items such as RHIBs are included in several functions and the inclusion of a RHIB in an EC will typically increase the capability of that EC. Such items of equipment, when included in an EC can increase overall functional capability with neutral impact, as their impact is only counted once, but their contribution to capability may span across several functions. This can be seen in EC17, where the individual capability scores of MC Min (EC8) = 0.461 and MCM Min (EC5) = 0.19. However, due to the combination of equipment in these two ECs, an overall capability for EC17 is 0.664 because the equipment in the EC can also be used to perform some of HS role functions. This demonstrates the potential benefits of the multi-role vessel and shared equipment and the ability of the tool to highlight these benefits.

Table 6: Summary of the relative functional capability provided by each EC [4].

EC Number	EC	Functional Capability			
		Overall	Role 1	Role 2	Role 3
			MC	HS	MCM
EC1	Base (nothing)	0.000			
EC2	REA Minimum	0.190	0.023	0.160	0.007
EC3	REA Medium	0.266	0.023	0.200	0.043
EC4	REA Maximum	0.403	0.046	0.300	0.057
EC5	MCM Minimum	0.190	0.046	0.080	0.064
EC6	MCM Medium	0.398	0.139	0.160	0.100
EC7	MCM Maximum	0.398	0.139	0.160	0.100
EC8	MC Minimum	0.461	0.392	0.040	0.029
EC9	MC Medium	0.654	0.531	0.080	0.043
EC10	MC Maximum	0.654	0.531	0.080	0.043
EC11	REA + MCM Minimum	0.374	0.069	0.240	0.064
EC12	REA + MCM Medium	0.472	0.092	0.280	0.100
EC13	REA + MCM Maximum	0.700	0.300	0.300	0.100
EC14	REA + MC Minimum	0.728	0.485	0.200	0.043
EC15	REA + MC Medium	0.952	0.600	0.280	0.071
EC16	REA + MC Maximum	0.966	0.600	0.280	0.086
EC17	MC+ MCM Minimum	0.664	0.439	0.140	0.086
EC18	MC + MCM Medium	0.897	0.577	0.220	0.100
EC19	MC + MCM Maximum	0.897	0.577	0.220	0.100
EC20	REA + MC + MCM Minimum	0.857	0.531	0.240	0.086
EC21	REA + MC + MCM Medium	0.980	0.600	0.280	0.100
EC22	REA + MC + MCM Maximum	1.000	0.600	0.300	0.100

By comparing the functional capabilities for each EC in Table 6 with their impact in Table 5, an indication of the ECs that can be potentially integrated onto the vessels and achieve the most functionality can be identified. From the perspective of functional capability in Table 6, ideally, ECs 20, 21 and 22 would be integrated onto the MRV. However, from Table 5 it can be seen that the overall impacts for these ECs are greater than 100 % except for on instance. The number of individual criterion with an integration impact over 100 % is also significant, which indicates it is unlikely to be possible to integrate these ECs on these vessels.

CONCLUSIONS

The Integration Impact Assessment (IIA) Framework and associated spreadsheet-based Tool has been developed to build understanding of the impact of integrating new equipment onto a naval vessel. The IIA Framework and Tool builds this understanding by investigating the various vessel margins, such as space, weight, power and HVAC, that will be consumed by integrating an equipment configuration (EC). The IIA Framework and Tool can be used at various stages in a vessel's lifecycle including; tender evaluation, mid-life upgrades, life-of-type extension studies, and to assist with batching strategy development in continuous build programs.

The test implementation provided several key insights into the utility of the IIA Framework and Tool. Gathering and entering the design data of the vessels and equipment to be assessed within the IIA Tool can be time consuming. However, the ability to reuse parent and associated child-level criteria in subsequent assessments is a strength of the IIA Tool as it means this process does not need to be repeated. Furthermore, equipment and ECs can be reused, thereby significantly reducing the effort involved in conducting follow-on assessments. Hence, for a particular platform type the useability/efficiency of the IIA Tool improves with each subsequent study. This would be advantageous for a class of ship, which may be investigated for its capacity to integrate new equipment at various stages in its lifecycle. The test implementation also demonstrated the ability of the IIA Tool to highlight when vessel margins were being over utilised and underutilised by an EC. Investigating the impact of ECs on child criteria demonstrated how potential design changes could be identified in order to potentially increase critical margins.

Other insights provided by the test implementation include the need for engagement with appropriate Subject Matter Experts when developing ECs and the importance of the functions the vessel is required to perform. The IIA Tool itself requires an 'intelligent' user so that the data used in the impact and functional capability assessments is appropriate. The results of any assessment based on incorrect or inappropriate data is likely to yield untrustworthy results.

A final key conclusion, which was clearly demonstrated during the test implementation, is that the individual child and parent-level integration impact criteria need to be investigated as part of any integration impact assessment. This is because the rolled up, overall weighted values of the integration impact calculated using the Multi-Criteria Decision Making method can potentially hide key impacts and other insights. These insights can be used to identify potential equipment trade-offs and design changes. Overall, the IIA Framework and associated Tool has been demonstrated to provide useful insights for investigations into the impact of integrating new equipment onto a platform.

ACKNOWLEDGEMENTS
Funding to implement the idea of the IIA Tool was provided by the SEA2400 project and Frazer-Nash Consultancy undertook the initial development of the tool. The authors acknowledge the support Karl Slater provided during the initial phase of IIA Framework development. Andrew Tynan also acted as a sounding board during ideas development. The authors would also like to acknowledge the Australian Maritime College final year design project students for the vessel designs used in the test implementation presented in this paper.

REFERENCES

1. Dwyer, D.M., Magoga, T., and Morris, B.A., *Fatigue life as a factor in assessing warship design flexibility to support batch-building programs*, in *MARSTRUCT 2019*. 2019: Dubrovnik, Croatia.
2. Schank, J.F., Savitz, S., Munson, K., Perkinson, B., McGee, J., and Sollinger, J.M., *Designing Adaptable Ships: Modularity and Flexibility in Future Ship Designs*. 2016, RAND Corporation: Santa Monica, CA.
3. Noonan, M. *Chief of Navy Speeches: Address to the Defence and Industry Conference*. 2018 [cited 2019 20/5/2019]; Available from: http://www.navy.gov.au/media-room/publications/chief-navy-speeches-address-defence-and-industry-conference.
4. Morris, B.A., Webster, L., and Dwyer, D.M., *An Impact Assessment Framework for the Integration of Equipment onto Platforms*, Defence, Editor. 2019, Defence Science and Technology Group Technical Report: Fishermans Bend, VIC.
5. Dwyer, D. and Morris, B.A., *A Ship Performance Modelling and Simulation Framework to Support Requirements Setting*, in *Pacific 2017 International Maritime Conference*. 2017: Sydney, Australia.
6. Evans, G. *The rise of the multirole vessel (MRV)*. 2012 [cited 2019 20/5/2019]; Available from: https://www.naval-technology.com/features/featurethe-rise-of-the-multi-role-vessel/.
7. Largiader, C.A., *Modular Platform Based Surface Ship Design*, in *Ocean Engineering*. 2001, MIT: MIT.
8. Harvey, J. and Griggs, R.J. *Project SEA 1180 - Offshore Combatant Vessel Initial Capability Description*. 2011 [cited 2019 20/5/2019]; Available from: http://www.aph.gov.au/DocumentStore.ashx?id=bf32c3ce-39e7-4463-835a-0ff69404cc7a.
9. Morris, B., *A Process to Establish the Common Functions Performed by a Multi-Role Vessel*, Defence, Editor. 2010, DSTO Technical Report: Fishermans Bend.
10. Hayes, R. and Webb, S., *Technology Roadmap for SEA2400 Phase 1*. 2017.

More Efficient and Flexible Procurement Assessment with a Multi Criteria Decision Analysis Framework

Robert Hayes, Frazer-Nash Consultancy, r.hayes@fncaustralia.com.au

ABSTRACT

With the increasing complexity of capability requirements, available equipment, and integration considerations, assessing and optimizing vessels has become more onerous and time-consuming to perform using traditional methods. To improve the efficiency, flexibility, and auditability of tender evaluations, a desktop tool based on multi-criteria decision analysis was built for a procurement team to automate and visualize their decision-making process.

This paper examines the underlying framework that supports the tool. It introduces the underlying mathematics, provides an outline of the processes and steps involved, and explains what outputs the framework provides to the end user. The paper discusses the merits of using the framework, along with the pitfalls and challenges of using the tool. Finally, the paper presents suggested circumstances in which the tool may or may not be appropriate.

INTRODUCTION

In a procurement program there are a complex set of interrelated factors to consider, and an end-goal which is to maximise capability, minimise risk and provide value-for-money. In a more practical sense this means balancing the three competing drivers (figure 1):

- Capability – the ability of the platform to achieve the mission system;
- Plausibility – the ability of the platform to be viable from an engineering and physical constraint perspective; and
- Affordability – the ability of the platform to be economically viable.

Therefore, a program needs to seek out the solutions which offer a balanced compromise and support these trade-off decisions with quantifiable data.

In most defence procurement programs there are likely to be several tendered platform designs. Each of these may be offered in different configurations or fit-out states, resulting in a large number of potential procurement options to be assessed. Whilst all options must meet the essential requirements in order to be further considered, it is common to specify desirable requirements – those which the war fighter considers would improve the capability or useability of the system. These desirable characteristics hence form some of the criteria against which the options can be assessed.

Performing trade-off assessments of all these options manually is a very time intensive process, which is prone to error and-or bias. It requires a great deal of discipline and documentation to ensure processes are transparent, traceable and fair. Furthermore, when priorities change mid-program or designs are altered, the updated information needs to pass back through the decision-making process to ensure the best decision is still being made. Finally, with the ultimate decision-making resting with stakeholders not involved in the detail of the assessments, presenting them with information summaries that are clear and defensible is paramount.

The method and framework set out in this paper are drawn from work to develop an Integration Impact Assessment tool for DST & Navy. This tool was developed under the specific direction of DST with funding provided by Defence from a current RAN ship procurement project. We would like to acknowledge the contribution by both the RAN and DST in funding and supporting the development of this tool.

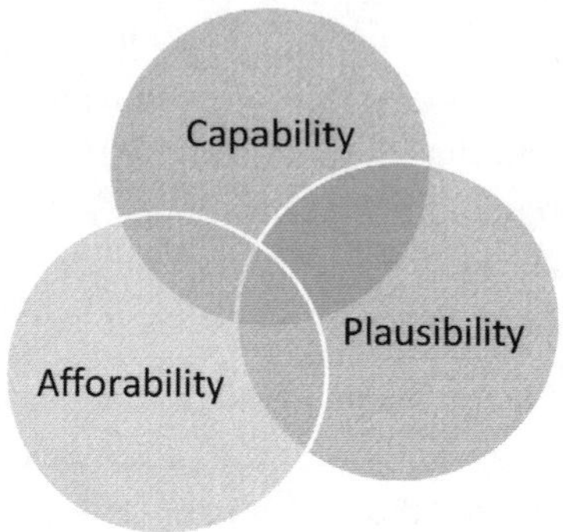

Figure 1: Three competing drivers in a procurement program

THE FRAMEWORK

The Context and Need

This conundrum calls for an efficient and flexible process. Therefore, a procurement program should look to adopt a structured assessment method that is traceable, flexible, scalable, auditable, and repeatable. One such method is to use a decision-making method such as multi-criteria decision analysis (MCDA), backed up by rigorous systems engineering methods. This paper focuses on the framework as a methodology rather than the implementation (e.g. a software tool). Successful implementation depends on the needs and complexities of the procurement program, and any software developed should be bespoke or heavily customised for the end-user to yield maximum benefit.

Developing a framework that can deal with a complex problem in a simple manner, ultimately requires bounds and assumptions to be made. To simplify the problem, we separate the framework into two halves; the first linking capability to the equipment, and the second assessing the impact of integrating that equipment on the platform.

Defining Capability and Relating it to Equipment

Capability of a system can be considered the result of nine fundamental inputs to capability (FICs), namely: personnel, organisation, collective training, major systems, supplies, facilities and training areas, and support. However, when evaluating tendered equipment designs we cannot control any factor but the equipment. Therefore, in terms of maximising our platform's capability, we must assume at the initial stage that equipment is proportional to capability. To determine how our equipment meets our platform's needs, an enterprise architecture such as DODAF should be used (figure 2). The functions of each operational activity can then be mapped to the equipment that enables these functions. This is not a linear mapping and there will be a mixture of one-to-one mapping, one-to-many, and many-to-one of functions to equipment (figure 3).

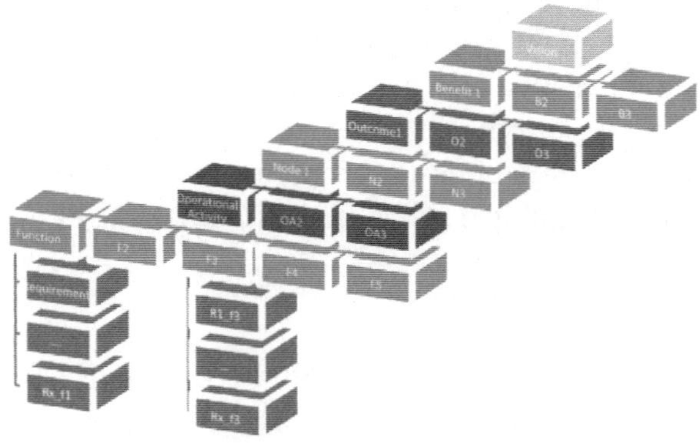

Figure 2: Indicative enterprise architecture

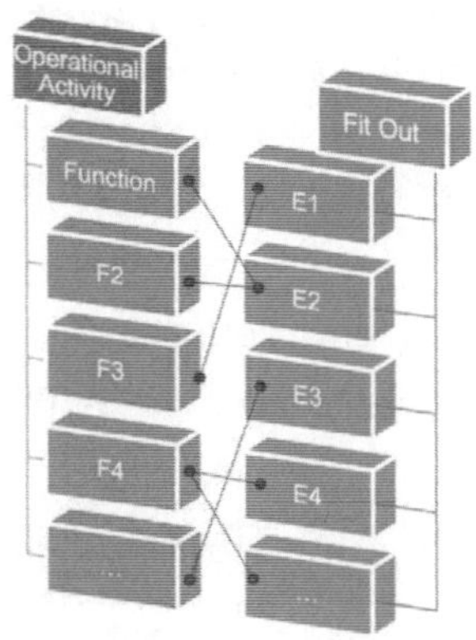

Figure 3: Illustration of mapping equipment to functions

Through these relationships, the functions can be assigned a contribution value (e.g. percentage) towards total platform capability. Since, the equipment is also mapped to these functions, we now have a relationship between equipment and capability. Henceforth, choosing a platform fit-out containing some of this equipment will implicitly indicate which functions have been met and contribution towards capability of that fit-out.

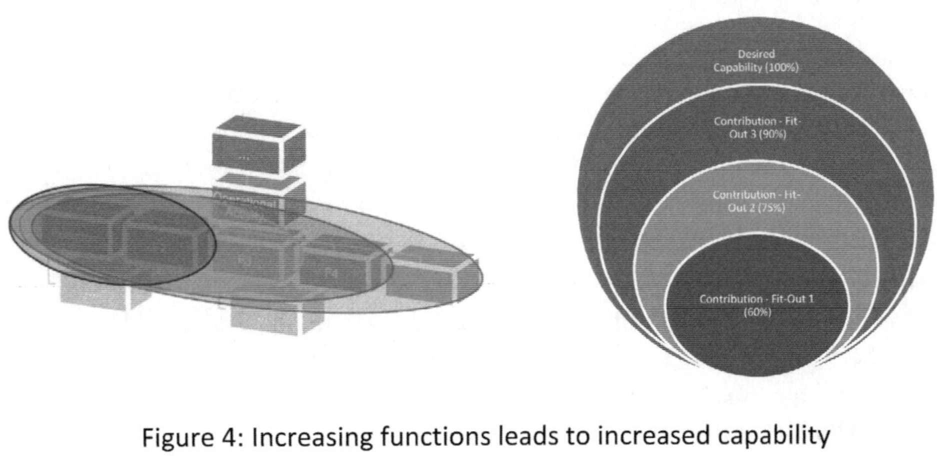

Figure 4: Increasing functions leads to increased capability

To take this further and allow for a multitude of assessments, a procurement program may choose to develop several equipment fit-outs, each representing more equipment. Hence, each fit-out achieves more functions and has an increased capability contribution. For example, you may have three fit-outs, "Fit-out 1, "Fit-out 2", and "Fit-out 3" that approximately provide 60%, 75%, 90% of desired capability respectively (figure 4). These fit-outs can then be assessed against multiple platforms using MCDA to determine the impact of integration as covered in the next section.

Introduction to Multi-Criteria Decision Analysis

Multi Criteria Decision Analysis (MCDA) is a quantitative tool which helps to aggregate a stakeholder's preferences against conflicting criteria to find the alternative with the highest overall score when all criteria are considered.

MCDA is comprised of four stages; first is defining the criteria, second is establishing the weighting of each criterion, third is scoring each of a number of discrete options against each criterion, and the fourth stage is calculating the weighted criterion function. The preferred option is provided by the highest value after summation of all weighted criteria functions. This can be automated by a spreadsheet, allowing multiple concurrent calculations to be performed easily.

We use MCDA in this framework to assess the integration difficulty of a number of fit-out options (defined in the previous section) against a number of proposed platforms. These combinations can be set up in a matrix as in table 1.

	Platform Design A	*Platform Design B*	*Platform Design C*
Fit-Out 1	*Option A-1*	*Option B-1*	*Option C-1*
Fit-Out 2	*Option A-2*	*Option B-2*	*Option C-2*
Fit-Out 3	*Option A-3*	*Option B-3*	*Option C-3*

Table 1:Matrix of options for MCDA

In each case we will assess the option against the defined criteria, clustering the criteria in a hierarchy as described below. We will use the product of the weighting and the evaluated score to determine the result for each tier (i.e., the sub-criteria). The results from each tier will be aggregated to provide the evaluated score for the next tier level, until the overall score is determined for all options.

Step 1: Define Criteria

We start by defining the criteria against which the candidate solutions will be assessed. In general, these criteria should consider a plethora of physical and integrated logistics support (ILS) requirements of the platform. Having accurate and representative criteria is crucial to a successful assessment, and input from stakeholders and subject matter experts should be solicited in developing the criteria.

It is recommended that the criteria and sub-criteria are clustered in a structured hierarchy, referred to as a value tree (see Multi-Criteria Analysis: a Manual, 2009, for further details). Using this method, similar criteria are grouped at the same level, and similar groups combined at the subsequent levels up the tree (figure 5). A criterion's position in the overall value tree can be determined, and competing or duplicate criterion can be identified, and re-defined.

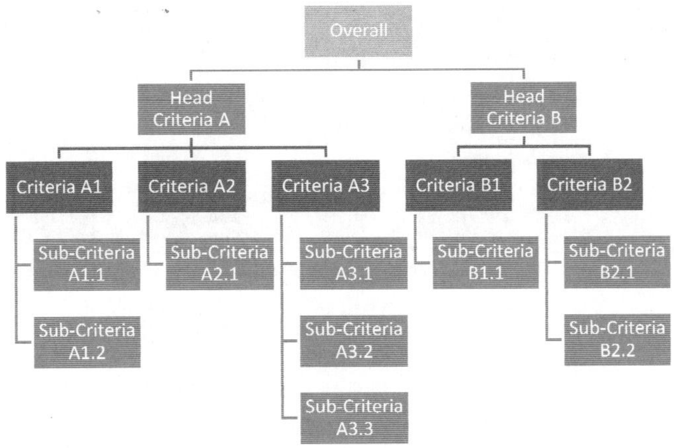

Figure 5: Value tree for determining criteria

Before finalising the choice of criteria, the provisional set needs to be assessed against a range of qualities. If the criteria do not adhere to all of these, it may be necessary to separate the criteria into multiple assessments. These qualities and their justification are:

- Completeness – have all necessary criteria been included? This will be an iterative affair, with multiple passes required to ensure that no major criterion is overlooked, and all the criteria are adequately defined.
- Redundancy – are all criteria necessary? Unimportant or superfluous criteria should be excluded to reduce the complexity of the assessment.
- Operationality – can each option be judged against each criterion? The assessment may be objective (e.g. weight or length) or subjective (e.g. vulnerability) and MCDA can accommodate both simultaneously but the criterion must be defined clearly enough to be assessed.
- Mutual independence of preferences – are scores associated with one criterion able to be assigned without knowledge of the scores of other criteria? This may not always be possible, in which case it may be necessary to combine criterion into a new preference independent criterion.
- Double counting – do any criteria have the same definition? However, just because two criteria appear similar does not necessarily mean there is double counting, likewise, in other instances doubling counting may not be immediately evident. An objective basis cannot be used to determine whether there is double counting, but fortunately checking for independent of preference will generally reveal double counting.
- Size – is there an excessive number of criteria? The structure should not be larger than necessary, and simple evaluation to eliminate criteria which may lead to inconsistencies should be performed early.
- Impacts occurring over time – are criteria affected by future events? MCDA does not have any ability to deal with time-dependent variables. Any criteria that varies with time must be approximated by another method (e.g. a non-linear value function) or modelled externally.

An example of a simple criteria set is in figure 6.

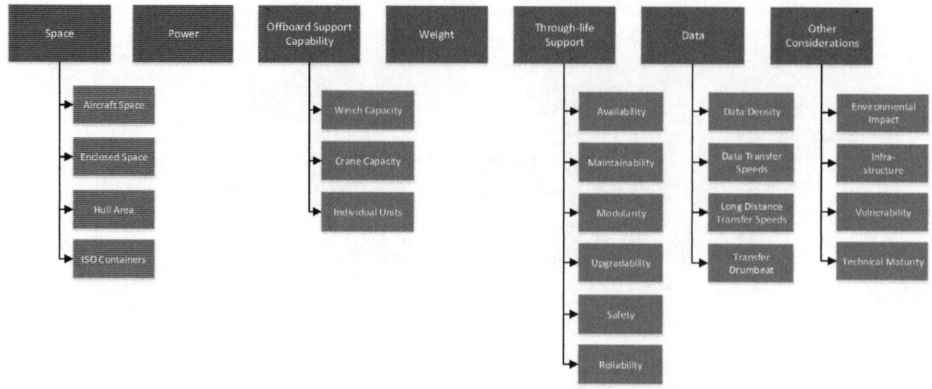

Figure 6: Exemplar criteria for a vessel procurement program

For a particular assessment, we need to populate each sub-criterion with data. In the evaluation (step 4), the data for the platform will be compared against the aggregate data from the fit-out. The criterion and whether it is a physical constraint, a variable value, or a non-physical concept, will drive what mathematical operation is used to ascertain this aggregate value (e.g. summation, averaging, maxima, minima, or otherwise).

For simplicity sake, if we consider a basic physical constraint such as Enclosed Space, we would take the sum of the individual package volume for each piece of equipment in this particular *Fit-Out* and use that define our upper and lower bounds for that criterion. This is because the package volume is the driving requirement for this criterion, that is to say what we as engineers are evaluating when assessing this criterion. Say for example our required value of all equipment is 100m^3, therefore, we can consider the lower bound to be 70% of this value (e.g. 70m^3), as not all the equipment is required to be on the platform at the same time. We can consider the upper bound to 120m^3 as this provides a 20% future upgrade margin. The bounds at this stage can be refined if after an initial pass it is determined they are too onerous or lax. From this example can be seen that having an awareness of what is being assessed is important to ensure the assessment reflects reality,

Step 2: Establishing the Weights of Each Criteria

We need to assign a weighting to each criterion such that when evaluated, the relative merit and preference of this criterion is considered. This also ensures that more important criteria have a stronger influence on the overall assessment than less important criteria.

There are a number of methods possible for eliciting weighting and scoring. This framework utilises a composite approach, with multi-attribute utility functions for scoring, and pair-wise comparisons from the Analytical Hierarchy Process (AHP) to derive weights. AHP is an indirect weighting technique, developed by Thomas R Saaty (Saaty 1987) and uses a series of questions to determine the relative importance of the criteria and hence their weighting. In this framework, the normalisation modifications proposed by Belton and Gear, 1983 and accepted by Saaty, 1994 (Ideal AHP) was implemented to prevent against logical inconsistencies arising from the original method.

It may seem counterintuitive to use pair-wise weighting, as if we are able to non-linearly determine the relationships of the value function (step 3), we may know enough about the function to use a geometrical weighting system. However, this implementation has been used as it enables a first pass to be established with a linear mapping. Further refinement can then be made as more information and understanding is established during the process.

AHP is effective as it does not ask stakeholders to rank criteria, instead asking a series of questions which compel stakeholders to compare two criteria directly to determine which one is more important, and the scale of that importance. This method can be successfully presented to both technical and non-technical stakeholders and if required use fuzzy terms such as strongly/weakly agree/disagree. Furthermore, due its nature of using matrices it is simple to implement and compute using a spreadsheet.

AHP presents a series of questions in turn such as in table 2, and for each question the stakeholder is to indicate their preference one way or the other (or if they are equal) on a scale between 1 and 9, where: 1 = equal; 3 = weakly more important; 5 = strongly more important; 7 = very strongly more important; 9 = absolutely more important.

If there are criteria, there are questions. This practically limits technique to around 20 objectives (190 questions), but more realistically it should be used for between 3 and 8 criteria (i.e. 3 and 28 questions).

Factor	Factor Weighting Score More Important Than	Equal	Less Important Than	Factor	
Criteria 1	9 8 7 6 5 4 3 2	1	2 3 4 5 6 7 8 9	Criteria 2	Q1
Criteria 2	9 8 7 6 5 4 3 2	1	2 3 4 5 6 7 8 9	Criteria 3	Q2
Criteria 3	9 8 7 6 5 4 3 2	1	2 3 4 5 6 7 8 9	Criteria 4	Q3
Criteria 4	9 8 7 6 5 4 3 2	1	2 3 4 5 6 7 8 9	Criteria 5	Q4
Criteria 5	9 8 7 6 5 4 3 2	1	2 3 4 5 6 7 8 9	Criteria 6	Q5
Criteria 6	9 8 7 6 5 4 3 2	1	2 3 4 5 6 7 8 9	Criteria 7	Q6
Criteria 7	9 8 7 6 5 4 3 2	1	2 3 4 5 6 7 8 9	Criteria 8	Q7
Criteria 8	9 8 7 6 5 4 3 2	1	2 3 4 5 6 7 8 9	Criteria 9	Q8
Criteria 9	9 8 7 6 5 4 3 2	1	2 3 4 5 6 7 8 9	Criteria 10	Q9
Criteria 10	9 8 7 6 5 4 3 2	1	2 3 4 5 6 7 8 9	Criteria 11	Q10
Criteria 11	9 8 7 6 5 4 3 2	1	2 3 4 5 6 7 8 9	Criteria 12	Q11

Table 2: AHP question format

Step 3: Defining the Value Function for Each Criteria

A value function must be defined for all criteria, this function will include the minimum acceptable value for each criterion, the desirable (or maximum) value for each criterion and a curve approximating the relative value of improving from the minimum value to the desirable/maximum value.

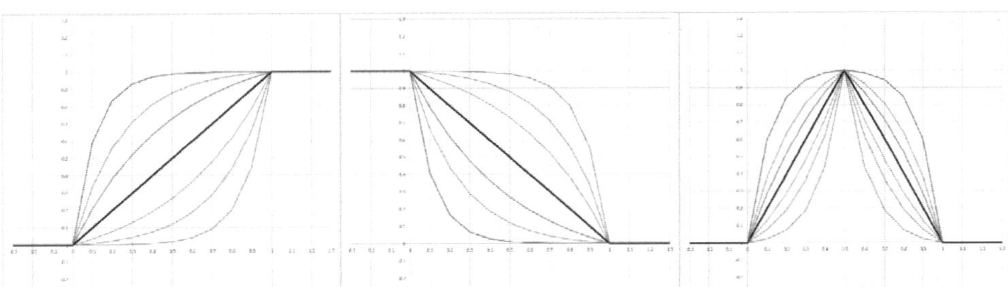

Figure 7: Example curves (from L to R): Increasing, Decreasing, Midpoint with stepwise functions applied

The value curves control how the evaluated value is mapped between the upper and lower limits using an exponential formula controlled by the α-value. For an increasing curve, an α-value of 0 is a linear mapping, a positive value creates a logarithmic curve of increasing steepness, and a negative value creates an exponential curve of increasing steepness. For a decreasing curve, it is the inverse. For midpoint curves, a positive value provides a shallower transition into the midpoint. Furthermore, by applying stepwise functions how values which fall outside the minimum/maximum are treated can be calculated (figure 7 shows such example curves).

These functions are highly flexible but conversely with this flexible, discipline needs to be applied. Experience and knowledge of how the function behaves for the particular criterion is required otherwise the assessment validity may be jeopardised. If in doubt, a basic first pass should always be a linear mapping.

Returning to example from part 1 – Equipment Volume (figure 8); experience may say that adopting an increasing curve with a medium positive α-value (e.g. 5) is most appropriate. This allows the value to rise towards 1.0 quickly, ensure that values that come close to the required value score well, yet leaves there is some headroom for upgradability to be scored.

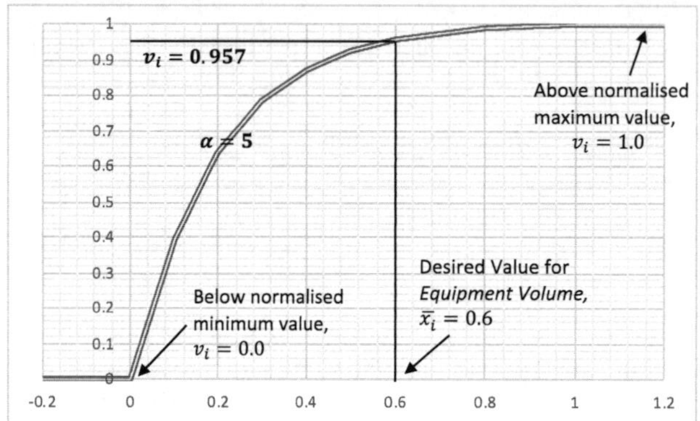

Figure 8: Exemplar value curve for equipment volume example from part 1

An alternative to using the composite approach is to apply REMBRANT (see Lootsma, 1992 & Olsen 1996) or MACBETH (Bana e Costa and Vansnick, 1997 & Bana e Costa et al., 1999) techniques, and more information about these methods can be found at their respective sources or in Multi-Criteria Analysis: a Manual, 2009.

Step 4: Calculate the Criteria Functions

For the last step of MCDA, and penultimate step of the framework we evaluate all the options concurrently. For each option we are determining the impact of integrating of that fit-out on that tendered platform design, against the criteria determined in step 1. The data for this is drawn from two sources; the aggregated equipment data (by means of the fit-outs– touched on in part 1), and the platform data. This forms part of the criteria function, which is for each option is the summation across all assessed criteria, of the product of the value function (from the value curve output) and the weighting.

Looking at table 3, we see the four sub-criteria under the space criteria in figure 6.

Since we are considering impact, we want to minimise this (hence a lower impact). To arrive at this, we invert the MCDA score. The option with the lowest overall impact is best option (e.g. Option A-1 in this example).

Considering the value tree (step 1), by the overall impact in the example above we refer to just one criteria or head criteria. The other criteria are assessed simultaneously, with their own weightings and data, and all these assessments can all be aggregated and used to generate the integration impact for the overall assessment. In this instance, the *total score* generated at the previous level becomes the *value function* at the next level up, and the weightings are calculated for that level. Essentially, this principle of stacking has no limit, but realistically two or three layers would be sufficient for most applications (e.g., sub criteria, criteria, aggregated assessments).

Criteria	Desired Value	Lower Bound	Upper Bound	Alpha	Weights	
Aircraft Space	25	15 (60%)	50 (200%)	5	0.412	
Enclosed Space	30	20 (67%)	41 (135%)	3	0.049	
Hull Area	130	98 (75%)	150 (75%)	2	0.248	
ISO Containers	5	5 (100%)	10 (200%)	0	0.29	
	Option A-1			Option B-1		
Criteria	Tendered Value	Value Function	Wt Value Funct	Tendered Value	Value Function	Wt Value Funct
Aircraft Space	20	0.514	0.212	25	0.766	0.162
Enclosed Space	30	0.809	0.040	20	0.003	0.000
Hull Area	145	0.970	0.241	100	0.106	0.026
ISO Containers	8	0.600	0.174	10	1.000	0.174
Total Score			0.666			0.362
Impact			33%			64%

Table 3: Exemplar results table from MCDA

This concludes the MCDA part of the framework. The final step of the tool is bridging the capability-to-equipment relationship and the equipment-to-platform integration impact assessment. Therefore, making the equipment the independent variable and capability and integration impact the dependent variables.

Bringing It All Together

The tool that was developed allowed the team to generate their assessments, and for each option have the raw impact assessment values and capability contribution values. However, the tool did not explicitly define how the information and results were summarised. This was because the team did not know at that stage how they wanted to present the information, and wanted to focus on the assessments first and foremost.

In terms of taking this framework forward, we can propose how the information can be presented. The options are either to use charts or tables. The example below gives an indication of a table format. In this example we can see the matrix of options presented, together with their cost in terms of the baseline (e.g. lowest cost option).

	Platform Design A *2.5x Baseline Cost*	*Platform Design B* *1.3x Baseline Cost*	*Platform Design C* *Lowest Cost (Baseline)*
Fit-Out 1 *3.0x Baseline Cost*	*Plausibility 80%* *5.5x Baseline Cost* *90% Capability*	*Plausibility 50%* *4.3x Baseline Cost* *90% Capability*	*Plausibility 30%* *3.0x Baseline* *90% Capability*
Fit-Out 2 *1.4x Baseline Cost*	*Plausibility 90%* *2.9x Baseline Cost* *75% Capability*	*Plausibility 70%* *2.7x Baseline Cost* *75% Capability*	*Plausibility 50%* *1.4x Baseline* *75% Capability*
Fit-Out 3 *Lowest Cost (Baseline)*	*Plausibility 100%* *2.5x Baseline Cost* *60% Capability*	*Plausibility 90%* *1.3x Baseline Cost* *60% Capability*	*Plausibility 70%* *Baseline Cost* *60% Capability*

Table 4: Indicative summary output from framework

By colour coding this matrix correspondingly, we can quickly visualise which options are the best, which are acceptable (with compromise) and which should be discounted.

Together with a knowledge of the cost of the platform and fit-out option leads to setting up a decision-making matrix such as table 4 (figures provided do not relate to previous worked example).

This table, although simple, provides an easily comparable and digestible overview of the options, which can be used by non-technical stakeholders to present and defend decisions.

DISCUSSION

Merits of the Frameworks

MCDA is an agreement process. It allows people to make decisions, informed by analysis. The power of the MCDA method in particular is it models a natural way of thinking, and then if the outcome is counterintuitive or subject to disagreement between stakeholders, sensitivity analysis can explore why and assist in reaching consensus.

In terms of the context of a procurement team it allows the team to compare several options to understand what capability is being achieved, the impact on the platform of achieving that capability, and the cost of that capability, and present this information cohesively and logically, increasing efficiency. Through sensitivity analysis and iteration, the team can understand and quantify trade-offs being made and make an informed decision on whether they are acceptable.

The framework encourages transparency and traceability by documenting criteria, value curves, and AHP judgement scores for each iteration and assessment. Finally, it provides flexibility by allowing a multitude of parameters to be adjusted to suit different assessments and aspects being considered.

Pitfalls of The Framework

It is important to consider that while the framework helps automating decision making, it does not replace a skilled team familiar with what they are procuring. The most important aspect is being mindful and realistic in the scope and limits of the framework.

The experience of procurement program team is key to driving the framework, as the framework requires well defined and accurate criteria and data to be inputted. If used by an inexperienced team there is a risk that the flexibility leads to mistakes being made in the data definition or over complication of scenarios. These in turn will lead to an unusable output and loss of time.

The final risk of the framework is either by intention or not can lead to a team trying to design by numbers rather than design by reality. It must be appreciated that while the tool may encourage combining the best elements, real-world constraints may prohibit this.

Appropriateness of The Framework

The underlying mathematics powering the decision-making MCDA, AHP, is well understood, but subject to significant criticism and debate. It relies on a well-founded and logical scenario to be set up in the assessment. Therefore, it is recommended that the framework be used by a user that is familiar with both the topic being assessed and MCDA. Furthermore, setting up scenarios is a time-consuming process with a steep learning process, so a commitment to using the framework should be made before embarking on the journey. If the assessment is relatively simple, other less complex methods may be more appropriate to get to an answer quicker.

The framework is dependent on the quality of the data input. Therefore, a review of the results should be undertaken by the user to ensure they are reasonable and plausible. The framework should be used to support the decision-making process rather than being relied upon as providing a definitive answer.

Ranking and weighting of criteria is crucial to a good assessment. Before undertaking any complex weightings, a first pass of weighting is recommended to be conducted separately (outside the tool) using another ranking method (e.g. Rank Order Centroid).

Any weighting that is used should consider stakeholder input and experience, and be verified and agreed amongst stakeholders to ensure overall happiness of ranking and order.

The weighting should be respective of the analysis being undertaken (i.e. the priorities should change depending on the characteristics of the platform or system). Additionally, the sensitivities of the input data should be considered, and ideally sensitivity analysis done outside the framework first.

CONCLUSION

In summary we have seen a framework presented that brings together well proven and understood concepts in a matter that is tailored to assessing options in a procurement program. To that end it serves three primary objectives:

- Compare an equipment fit-out against multiple platforms;
- Compare multiple equipment fits-outs against a platform;
- Map equipment to capability – determine what equipment is required.

The tool serves two further secondary purposes:

- Understand the outcome of changing the contribution of functions to capability; and
- Understand the sensitivity of a design to change.

By adopting this framework into a bespoke software tool, the procurement program can increase their efficiency and effectiveness of tender evaluation. This leads to a final outcome where trade-offs are clear, well presented, and platform limitations have been fully considered along the way. Ultimately leading to a successful procurement with no surprises.

ACKNOWLEDGEMENTS

We would like to acknowledge the technical contribution and direction of Karl Slater, Brett Morris, and Dylan Dwyer of DST in the development of this framework (and the resulting Integration Impact Assessment Tool). We would like to further acknowledge Defence in providing funding for the development of the framework and tool from a current RAN ship procurement project.

BIBLIOGRAPHY

1. Multiattribute Value Analysis (MVA), Dr Quoc Do and Prof Stephen Cook, University of South Australia.
2. The Engineering Design of Systems: Models and Methods, Dennis M. Buede, Wiley Publishing, 2000
3. The Analytical Hierarchy Process – What it is and how it is used, R.W. Saaty, Math Modelling, Vol 9, No 3-5, pp 161-176, 1987 https://core.ac.uk/download/pdf/82000104.pdf
4. Multi-criteria analysis: a manual, Department for Communities and Local Government: London, 2009
5. Lootsma, F.A. (1992) The REMBRANDT System for Multi-criteria Decision Analysis via Pairwise Comparisons or Direct Rating, Report 92–05, Faculty of Technical Mathematics and Informatics, Delft University of Technology, The Netherlands
6. Olson, D. (1996) Decision Aids for Selection Problems, Springer Verlag, New York
7. Belton, V,. and Gear, T. (1983) 'On a short-coming in Saaty's method of analytic hierarchies', Omega, 11, pp.228–30
8. Saaty, T.L. (1994). Fundamentals of Decision Making and Priority Theory with the AHP. RWS Publications, Pittsburgh, PA, U.S.A.

Frustration and Task Deviation Causes Accidents at Sea

Asanka Rajapakse[1], Gholam Reza Emad[1], Margareta Lützhöft[2], and Michelle Grech[3]
[1]Australian Maritime College, University of Tasmania, Launceston, Australia
[2]Maritime Studies Dept., Western Norway University of Applied Sciences, Haugesund, Norway
[3]Ship Operations, Australian Maritime Safety Authority, Canberra, Australia

Asanka.Rajapakse@utas.edu.au
Reza.Emad@utas.edu.au
Margareta.Holtensdotter.Lutzhoft@hvl.no
Michelle.Grech@amsa.gov.au

ABSTRACT

The surge in the number of maritime accidents in the last decades is the greatest challenge for the maritime industry to address. The majority of investigations on these accidents blame seafarers' poor performance as the main contributing factor. The international maritime authorities attempts to address the issue by adopting and updating maritime safety regulations and conventions has had limited success. Although accident investigation reports show that seafarers' task deviation is one of the major causal elements however, there are not many research projects that address this phenomenon. The finding of our qualitative research in task deviation in the maritime domain, discovered that seafarers' frustration is a key contributing factor which causes task deviations in maritime accidents.

INTRODUCTION

Maritime accidents involving commercial ships are continually on the rise [1,2] so as their average costs [3,4]. Accident investigations' reports show that the task deviations by seafarers is the major contributing factor in most accidents at sea [5]. Although, the world maritime regulatory body, International Maritime Organisation (IMO), has frequently adopted and amended maritime safety regulations to minimise accidents at sea due to task deviations by seafarers, the effort has achieved only partial success [6,7]. Research has shown that in devising contemporary shipping standards it is perceived that shipping is a linear operations which may be one of the reasons for this partial success [8].

Shipping is a complex activity that consisted of independent and interconnected elements which forms non-linear operations [9]. The humans conducting these tasks may add to this complexity due to attributes such as frustration [10,11]. Therefore, 'work as done' (WAD) on ships by seafarers (employees) could be unaligned with 'work as imagined' (WAI) by shipping operators (managers) who provide seafarers with procedures and checklists to comply with shipping standards [12]. A nonconformity towards the prescribed series of actions in a procedure is considered as 'task deviation' [5,13]. To study what are the underlaying factors that cause task deviation on board ships and how can it be prevented we designed a research project. The earlier results indicated there are several elements involved in the everyday activates of seafarers that facilitate their deviation from their main task. One the findings showed that frustration plays an important role among others in task deviation on board ships.

Using a socio-cultural approach, the aim of this paper is to investigate how frustration among seafarers may lead to discrepancies between WAI and WAD which have the potential to lead to consequential outcomes. Most marine accident investigations do not capture human element contributions to the accidents sufficiently to implement countermeasures [5, 14,15,16,17,18,19,20,21]. Further, this research utilises a novel approach to studying the task deviation using a socio-cultural theoretical framework to better understand how the inadequate alignment between WAI and WAD leads to frustration among seafarers.

Research shows the importance of socio-cultural attributes of workplaces and the way they influence safe practices of workers who deviate from work procedures provided by managers [22,23]. This study utilised Cultural Historical Activity Theory (CHAT) which is also called Activity Theory to examine this complex issue of seafarers' frustration. Using CHAT enabled us to understand the multi-mediational processes [24] in play in everyday activities on commercial ships, thereby provided insights to factors causing seafarers' frustration leading to accidents.

BACKGROUND

Studies show that frustration cause headaches, forgetfulness, lack of concentration, lack of attention to detail, confusion which may results in errors in decision-making, lack of confidence, risk taking and carelessness among others because of which people may deviate from their tasks leading to consequential outcomes [25,26,27,28,29,30,31,32,33,34]. Additionally, frustration is a source of distraction [35]. Thus, conducting tasks under the influence of frustration constitutes an impairment [36].

While extensive research on frustration is carried out in industries such as transportation, research on this topic in the maritime domain remains partial with limited discussion on effect of frustration on WAI and WAD. Lack of such research could be the reason for maritime industry's view of mitigating task deviations with more procedures, tighter training programs and technology [37,38,39]. Since this approach has not achieved the expected success [6], comprehension of seafarers' complex social and cultural context at sea in terms of frustration and how it causes task deviations through differences in WAI and WAD may have the potential to enhance maritime safety.

Any operational task is designed and executed at two levels in any organisation [40]. The design level is the blunt end (shipping operator) and the execution level is the sharp end (seafarers). WAI is what the shipping operators believe should happen to achieve a work task safely at sea. WAD is how the sharp end (the seafarers) conduct the work in the actual context [41]. The concept of WAI and WAD can be used in investigating factors causing discrepancies when work as done on the front line at sea could differ from the procedures work as imagined by those who design guidelines and checklists [42].

Due to many issues at the time of work, persons conducting work could deviate from procedures provided by shipping operators to complete the task despite the need to submit the signed checklist at the termination of the operation [43]. Thus, a signed checklist does not necessarily mean that the procedure has been properly followed. Although, task deviation does not always lead to accidents however, an adverse outcome is inevitable if the situation is not recovered from, before it is too late [5,44].

Although task deviation due to frustration has been addressed in literature in different perspectives such as sociobiological [45,46] and means of retaliation [47], yet it remains an elusive topic of study. In the context of this research, we utilised the Cultural historical activity theory (CHAT) to comprehend why frustration takes place at sea. CHAT provides a proper framework for analysing socio-cultural relations between seafarers and their work environment [48,49].

CHAT is based on the premise that a collection of sociocultural components containing complex mediational structures which, collectively, influence the behaviour of humans who are motivated to achieve a common goal [5,50,51]. Thus, the principal unit of analysis in CHAT is an activity system [50]. An activity system is a system of human doing. These elements within a single activity system are *subject, object, outcome, community, rules, division of labour* and *tools* as illustrated in figure 1 [5,49,50]. As depicted in figure 1, in any activity a subject uses tools to work on an object for obtaining a desired outcome. In the process the subject which is part of the division of labour in the community of work needs to follow certain rules to achieve the goal of the activity.

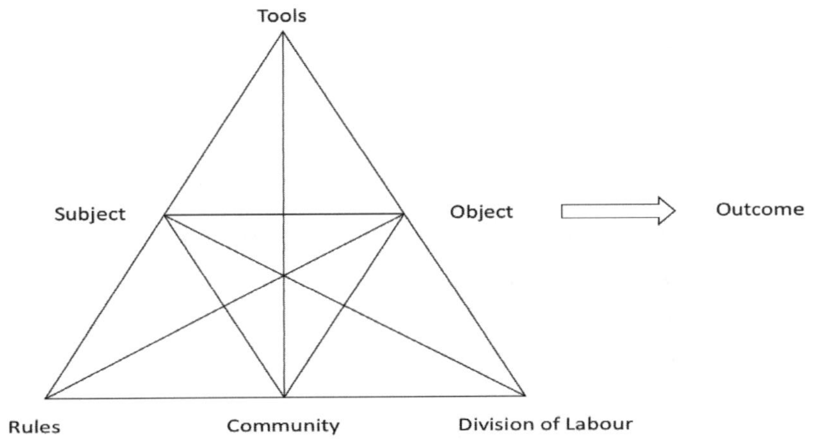

Figure 1. Activity System (Engeström, 1987, p. 78)

An activity system has internal and external tensions called *contradictions.* They cause changes within the activity system over time [52]. For example, seafarers need to follow rules and regulations (e.g. procedures and checklists) to conduct tasks. However, following inappropriate or badly designed rules and regulations may cause frustration which do not allow seafarers to perform their jobs safely. This creates a contradiction within the activity system. By examining these contradictions in the seafarers' activity system, this study is an attempt to establish why seafarers are frustrated at sea [5].

RESEARCH METHOD

This qualitative project utilised CHAT to examine contradictions between WAI and WAD on board commercial ships with the aim of exploring why seafarer frustration takes place. This study perceives shipboard tasks as part of the seafarers' activity system (figure 2).

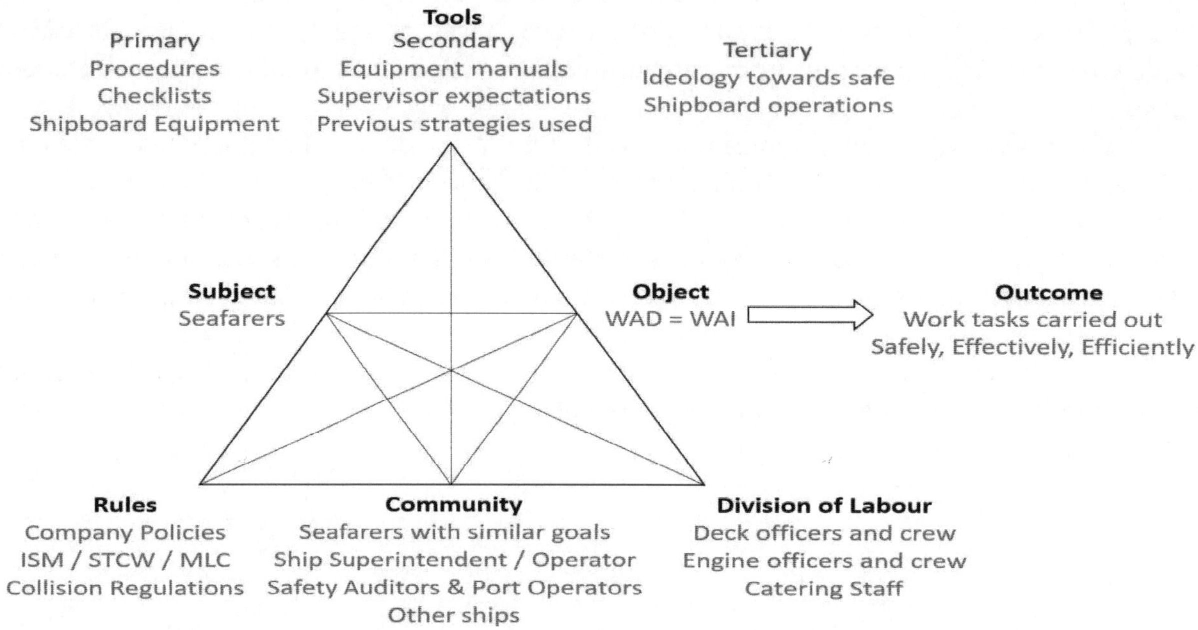

Figure 2. Seafarers' activity system [5]

This approach had three primary benefits for this study. Firstly, it facilitated comprehension of real-world data obtained from research participants. Secondly, it assisted in providing a structure for management of interpretations of research data gathered. Thirdly, this approach enabled identification of contradictions present in the seafarers' activity system leading to task deviations due to frustration.

Fourth generation evaluation was used to maintain trustworthiness of the study, where trustworthiness was measured in terms of credibility, transferability, dependability and confirmability [53].

Data Collection

Unstructured one-on-one interviews was the strategy used for data collection in this research. While using interviews led to the capture of reliable perspectives from respondents [48], they also provided supplementary useful data in terms of research participant's tone of voice, body gestures, pauses and mannerisms which added to the richness of data [54,55]. Thereby, using the qualitative approach resulted in accumulation of thorough accounts of seafarer behaviour and views within the maritime domain [55,56,57,58,59,60].

Data from sixty-one interviews were used for the study. They represented senior and junior officers with ratings from deck, engine and catering departments. Out of the interviewees one was a female. The average age of the interviewees was 47 years, while average sea service was 14 years. Interviewees represented Australia and eight countries in Asia, Europe, and West Africa. Research participants possessed work experience on different types of commercial ships such as tankers, containers, bulk carriers, general cargo ships, reefer ships, and offshore support vessels.

The interviewer requested the research participants to describe circumstances, situations and sequence of events which caused seafarer frustration leading to accidents due to task deviation. If interviewees were not directly involved in these accidents, they were witnesses. Accidents interviewees narrated were related to capsizing, collisions, groundings, fire, flooding, oil spills, and personal injuries.

The interviews lasted between thirty minutes to three hours. In total sequence of events and circumstances leading to seventy-four accidents were ascertained.

Data Analysis

Initial step was to conduct theoretical thematic analysis using Braun & Clarke's [61] 6-step framework to identify patterns within the gathered qualitative data [61,62]. Initial and major codes emerged through theoretical saturation of data [63,64,65,66] by constantly conceptualising patterns until no new attributes of patterns emerged [67,68].

Next step was to identify sociocultural contexts that caused frustration in seafarers at sea. This was done by using results from the first step in conjunction with the seafarers' activity system as presented in Figure 2. Analysis of dialectical relationships between the major codes using CHAT as a lens assisted in a wider understanding of sociocultural contexts causing frustration at sea. In order to achieve this, relationship codes were assigned between each component in the seafarers' activity system such as subject to subject, subject to object, subject to rules etc using the major codes established prior.

The third and the final step was to analyse contradictions within the seafarers' activity system using these relationship codes. The output of this step was identification of mediators in the seafarers' activity system that clashed with the object of conducting work at sea safely (WAI) due to seafarers engaging in tasks under the influence of frustration (WAD), thereby generating themes and categories.

FINDINGS

In this study, four categories emerged that contradict the object of getting WAD (by seafarers) aligned with WAI (by ship operators) as shown in Table 1. This study shows that seafarers' frustration at work may influence their decision making and behaviour to deviate from the prescribed procedures [69] leading to consequential outcomes. Most interviewees stated that accident investigations do not capture these areas of human element.

Table 1. Categories and themes

Categories and their themes			
Physical hardship to conduct task	**Loss or isolation from loved ones**	**Working with undesirable people**	**Severed Expectations**
Lack of physical fitness	Going through divorce or break up with partner	Unfair treatment	Postponed or cancelled shore leave due to work
Too tired after shore leave	Missing important functions at home	Perceived unfair appraisal reports	Delays to promised promotions / raises
Illness while at work	Single parents worried about kids at home	One or more team members disagree on execution method of the job	Lack of human resources on board to match work load
Adverse weather conditions	Someone ill in the family	Language barriers	Perceived unnecessary paper work
Fear of reprimand or negative appraisal if found resting	Death in the family	Perceived damage to reputation	Delays to jobs because of paperwork / permit renewal
Favouritism – physically strenuous tasks not assigned equally	Slow or no Wi-Fi onboard making it difficult to communicate with family	Working with colleagues who are not familiar with job tasks	Perceived financial losses
Simultaneous tasks		Supervisor blaming / insulting his own team with other supervisors	Unexpected pilot errors
Low quality food			Unexpected manoeuvring actions from other ships
Inadequate equipment to conduct jobs			Pressure from ship operator to continue operating vessel in spite of breakdowns
Audits and Inspections			Delayed sign off

Physical hardship to conduct task

Mandatory international regulations enforced by maritime authorities *(rules)* require medical examinations of all seafarers to assess their fitness for work at sea. However, some tasks which is asked to be undertaken by seafarers needs what is beyond their estimated physical fitness capacity which cause task deviation due to frustration. This situation is aggravated when seafarers fall ill or are suffering from seasickness. Older seafarers *(subject)* may also be vulnerable to become frustrated on board due to lack of fitness. Adverse weather conditions are also an issue for fitness. A seafarer *(subject)* (R8) explained how he narrowly escaped certain death due to frustration causing task deviation:

> *That time the temperature in the pump room (oil tanker ship) must have been over 60 degrees. It was so hot and I was drenched in sweat. I was feeling angry for no reason and couldn't think clearly. Rather than going up two stair cases to reach the (main) staircase (to access the bottom deck), I decided to jump across to a nearby hanging tank top (short cut) because I was exhausted. I did and held on to a tank top railing which came off. I was so lucky that my right foot landed in a clear area otherwise I would have fallen at least 25m to the bottom of the pump room.*

When seafarers undertake physically difficult work tasks, they can demonstrate anger and in some instances aggression due to frustration because of contradictions within the component of subject in the seafarers' activity system. For example, seafarers taking shore leave in ports sometimes results in insufficient rest before they take over duties again on board which frustrates them. Additionally, many research participants explained how the extra work hours required for safety audits at ports *(rules)* lead to violation of rest hours assigned to seafarers as per STCW convention. This frustrates them leading to task deviation causing discrepancies between WAI and WAD.

Another cause of frustration among seafarers leading to task deviation was the unavailability of correct tools and equipment resulting in more physical exertion to complete tasks. Further, fear of negative appraisal by supervisors may lead to seafarers continuing to work in fatigued conditions causing frustration and may lead to task deviation.

Engaging in simultaneous tasks *(rules)* [5] and favouritism among crew *(division of labour)* also may results in frustration. Also, low quality food on board *(tools)* can results in frustration.

This study shows that physical hardship to carry out tasks on board conflicts within the components of *subject*, *rules, tools* and *division of labour* in the seafarers' activity system by contradicting the object of conducting 'WAD aligned with WAI'.

Loss or isolation from loved ones

Seafarers most of their time at work are within the confines of a ship [70]. This research supports previous research findings showing seafarers consider their tenures on board as isolated for their family [71]. Interviewees spoke of how frustration due to this isolation leads to accidents [72]. Frustrations are heightened on important days such as birthdays and anniversaries. Also, being at sea, seafarers are likely to miss out on passing away of loved ones and most of the time unable to attend funerals. For example, a senior officer (R52) explained:

> *I received a message from my wife about the sudden passing away of my mother. We were sailing towards New-foundland and there was simply not even time to get to her funeral. The company asked me 3 times whether I need to go home from Canada but there was no point going home after 3 days of the funeral. I made the hard decision to remain on board because I was in a money critical time as well. It was torture to be onboard swearing at myself for not*

being with her. Without my own knowledge I had become a hazard in the engine room and the chief engineer asked me to stay in my cabin for some time until I could get my head right.

Seafarers (*subject*) are also distracted because of frustration when they hear that their loved ones are going through illnesses, which may lead to accidents by creating a contradiction in the seafarers' activity system. However, accident investigations do not capture these human element attributes, but seafarers are mostly blamed for not adhering to company's procedures or checklists.

Research participants also showed that the seafarers on board are distracted by frustration during periods of divorce or relationship troubles. Out of frustration, they at times display aggressive behaviour and engaged in risky decision making leading to accidents. Similarly, another source of frustration for seafarers are poor communication facilities on board [73,74] leading to task deviation which may result in accidents.

Our analysis show that loss or isolation from loved ones causes frustration among seafarers. Contradictions were observed in the *subject* component of the seafarers' activity system causing direct discrepancies to the system's object of WAD aligned with WAI where seafarers deviate from their work tasks with the potential to cause accidents.

Working with undesirable people

Working with undesirable people can cause frustration among seafarers which may lead to task deviation with undesirable consequential events. Unfair treatment is a reason why people are perceived to be undesirable. We found that although it is common in the workplace [75,76,77] it remains unreported most of the time. However, in some cases seafarers displayed aggression against their supervisors *(division of labour)* due to perceived unfair treatment [78]. A master (R32) explained:

Easiest way we can make a ship unsafe is by mistreating people. It is normal to be mistreated onboard by other shippies or department heads. Most of them try to deal with it alone motivated by their need to earn money. It has always happened and will happen in future. Makes men angry and helpless. Having said that, you find a swing of the wrist here and there. I feel that this is one of the most overlooked safety concerns on ships. An angry seaman is a dangerous seaman, sometimes a dead one.

Interviewees spoke of how seafarers are subjected to unfair treatment by shipping operators which frustrated them. They also showed that when seafarers "speak-up" to point out safety concerns, they are sometimes blamed and considered "trouble makers" affecting their chances of re-employment. Further, perceived racism is prevalent on commercial ships and research participants of how sometimes other seafarers didn't even respond when they greeted them. Severe abuse either verbal or physical by senior officers is very common as well leading to frustration. Nevertheless, these circumstances almost are never reported.

When seafarers consider their reputation is tarnished among colleagues due to being blamed from supervisors, rumours, and even issues at home *(subject / community / division of labour)*, they experience frustration and may engage in task deviation. Additionally, seafarers are frustrated when their supervisors *(division of labour)* spoke negatively about their team or individually to others on board. Similarly, perceived unfair negative appraisal reports prepared by supervisors is a severe cause of frustration which leads to task deviation.

Moreover, seafarers *(subject)* experience frustration when they do not agree with procedures arranged to be followed by the team members and supervisors *(community)*. Research participants stated how they experience frustration when their safety concerns as a team member are dismissed by their team or their supervisor. They do not desire to work in such teams. Seafarers are also frustrated when other work team members are not familiar with the equipment and job tasks having to compensate for their inadequacies as well. Furthermore, interviews spoke of how difficult it was to work with seafarers from some countries due to language barriers. These circumstances causes frustration which may lead to task deviation among seafarers potentially causing accidents.

Thus, this study shows that working with undesirable people leads to seafarer frustration. Contradictions within the elements of *subject, community and division of labour* in the seafarers' activity system cause conflicts to the system's object of WAD aligned with WAI. In that way, seafarers deviate from their tasks which may lead to accidents at sea.

Severed Expectations

This research confirms that frustration occur due to severed expectations leading to task deviations as shown by previous research work [79,80,81]. Seafarers have expectations from their work environment and towards their future. If expectations are cut short by unanticipated circumstances, they experience frustration. Seafarers can experience disappointment on board which may lead to task deviation causing accidents frequently. For example, an electrical engineer (R10) explained:

> *The 2nd Engineer postponed the fitter's shore leave to help me with the alignment of the lub oil pump motor. He had a red hot argument with the 2nd engineer because 2nd had promised him shore leave this trip. He might have missed it last time. He had that disappointed red face the whole time he was helping me. This is was the reason he put his hand in the wrong place when I was aligning the motor.*

Delays to salary raise and promotion is another reason seafarers *(subject)* are frustrated. Shipping operators *(community)* quote various reasons as excuses. On some occasions, salaries were cut down.

Seafarers don't expect shipping companies to operate ships in coercion towards seafarers, when critical safety machinery was broken down. However, this study shows that this is sometimes the case even with highly reputed shipping companies which frustrates seafarers leading to accidents.

Furthermore, unforeseen administrative work commonly referred to as paperwork on board frustrates seafarers. It is frustrating when job tasks are unexpectedly postponed because of paperwork such as permit validations at shift changes *(rules)*. Seafarers also believe that many administrative tasks on board are unnecessary and it distracts seafarers causing frustration. Paperwork frustration is a cause for seafarers taking short cuts, and sometimes neglecting job tasks altogether leading to task deviation, and thereby to potential accidents.

Many research participants mentioned that accidents may also happen at sea due to disappointments causing task deviation in terms of lack of human resources to carry out the actual workload on board. However, these shipping operators are in compliance with international manning regulations *(rules)*.

Interviewees further explained that when watch-keepers on the wheelhouse observed unexpected manoeuvres of other ships violating international collision regulations *(community)*, they experienced frustration. This may lead to aggressive risk-taking manoeuvres on the watch-keepers' part which could result in close quarters situations and collisions. Also, we found another source of frustration in the wheelhouse is unexpected pilot errors *(community)* leading to task deviation which may lead to accidents.

Thus, severed expectations *(rules / community)* cause discrepancies between WAD and WAI at sea resulting in contradiction of safe operation of ships *(object)* by seafarers *(subject)* in the seafarers' activity system which may lead to maritime accidents.

DISCUSSION

This research shows the socio-cultural attributes of seafarers' work practices in terms of conducting WAD aligned with WAI at sea can criticality compromise safety at sea. By analysing the findings of this study, it can be perceived that in the current dominated practice in shipping domain the elements in the seafarers' activity system clearly contradicted with the objective of aligned WAD and WAI causing task deviation due to seafarer frustration. These findings may assist the maritime industry to enhance safety at sea.

The findings of this study are in line with findings from previous research in other industries pertaining to issues with frustration in the work place resulting in unsafe practices [25,26,27,28,29,30,31,32,33,34], and a source of distraction and impairment [35, 36]. This research is the first of its kind carried out in the maritime industry as per the authors' knowledge. Therefore, the findings may assist maritime governing bodies to conduct maritime accident investigations with more emphasis toward the role of human element in accidents [5]. However, since research participants involved in this research were from only nine countries, further research consisting of seafarers from other shipping nations may shed more light towards the findings of this research. Another limitation of this study is that feedback from the respondents mirrored their individual perceptions. Therefore, additional research in terms of frustration leading to task deviation causing discrepancies between WAD and WAI may assist the maritime industry to align WAD with WAI to mitigate accidents at sea.

This research acknowledges that aligning WAD with WAI at sea to perform safe operations is a complex accomplishment [12]. This complexity is intensified by physical hardships experienced by seafarers to conduct tasks, loss or isolation from loved ones, working with undesirable people and severed seafarer expectations at sea.

These concerns could be some of the causes leading to accidents where the human element is adjudged the main cause, even after the implementation of countless rules and regulations to prevent them. As accident investigators don't capture frustration as a contributing reason for marine accidents, addressing these socio-cultural and contextual concerns could be a step forward in mitigation of task deviation of seafarers leading to accidents at sea.

REFERENCES

1. EMSA. (2018). Annual overview of marine casualties and incidents 2018. European Maritime Safety Agency. Lisbon, Portugal.
2. EMSA. (2017). Annual overview of marine casualties and incidents 2017. European Maritime Safety Agency. Lisbon, Portugal.
3. IUMI. (2017). IUMI reports continued increase in frequency of major vessel casualties. International Union of Marine Insurance. Hamburg.
4. EMSA. (2016). Annual overview of marine casualties and incidents 2016. European Maritime Safety Agency. Lisbon, Portugal.
5. Rajapakse, A., Emad, G. R., Lützhöft, M. and Grech, M. (2019). A study on time constraints and task deviations at sea leading to accidents – a cultural-historical perspective. Maritime Policy & Management, DOI: 10.1080/03088839.2019.1574407
6. Batalden, B. and Sydnes, A. K. (2014). Maritime safety and the ISM code: a study of investigated casualties and incidents. WMU Journal of Maritime Affairs, 13, 3-25.
7. Knudsen, O. F. and B. Hassler. (2011). IMO legislation and its implementation: Accident risk, vessel deficiencies and national administrative practices. Marine Policy 35: 201–207.
8. Oltedal, H. A. and M. Lützhöft. (2018). Managing Maritime Safety. Oxon: Routledge.
9. Caschili, S. and F. R. Medda. (2012). A review of the maritime container shipping industry as a complex system. Interdisciplinary Description of Complex Systems 10: 1-15.
10. Dingus, T. A., Guo, F., Lee, S., Antin, J. F., Perez, M., Buchanan-King, M., and Hankey, J. (2016). Driver crash risk factors and prevalence evaluation using naturalistic driving data. Proceedings of the National Academy of Sciences, 113(10), 2636–2641.
11. Strayer, D. L., Turrill, J., Coleman, J. R., Ortiz, E. V., and Cooper, J. M. (2014). Measuring cognitive distraction in the automobile II: Assessing in-vehicle voice-based interactive technologies. Washington, DC: AAA Foundation for Traffic Safety.
12. Hollnagel, E. (2013). A Resilience Engineering Approach to Assess Major Accident Risks. In: Albrechtsen, E. and D. Besnard. (eds.) Oil and Gas, Technology and humans assessing the Human Factors of Technological change. Trondheim: Ashgate.
13. Leveson, G. N. (2004). A New Accident Model for Engineering Safer Systems. Safety Science, 42: 237-270.
14. de Vries, L. (2017). Work as done? Understanding the practice of sociotechnical work in the maritime domain. Journal of Cognitive Engineering and Decision Making 11(3): 270-295, DOI: 10.1177/1555343417707664.

15. Praetorius, G., E. Hollnagel, and J. Dahlman. (2015). Modelling Vessel Traffic Service to understand resilience in everyday operations. Reliability Engineering and System Safety 141: 10-21.
16. Praetorius, G. and E. Hollnagel. (2014). Control and resilience within the maritime traffic management domain. Journal of Cognitive Engineering and Decision Making Dec 2014 8: 303-317, DOI:10.1177/1555343414560022.
17. Van Westrenen, F. (2014). Modelling arrival control in a vessel traffic management system. Cognition, Technology & Work 16(4): 501-508.
18. Praetorius, G. and M. Lundh. (2013). "Under Dangerous Conditions" : Safety Construction and Safety-Related Work Onboard of Merchant Vessels. Proceedings, 5th Symposium on Resilience Engineering: Managing Trade-offs : 24th-27th June 2013 at Soesterberg, The Netherlands. 61-66.
19. Bergström, J., Dahlström, N., Van-Winsen, R., Lützhöft, M., Nyce, J., and Dekker, S. (2009). Rule and role retreat: an empirical study of procedures and resilience. Journal of Maritime Research, 6(3), 41-58.
20. Lützhöft. M., B. Sherwood Jones, J. V. Earthy and C. Bergquist. (2006). Making safety by tying the knot: examining resilience in shipping. Proceedings of The 2nd Symposium on Resilience Engineering. Juan-les-Pins, France, November 8–10.
21. Morel, G. and C. Chauvin. (2006). A socio-technical approach of risk management applied to collisions involving fishing vessels. Safety Science, 44(7): 599-619.
22. Waring, J. and E. Rowley. (2011). A Socio-cultural Perspective on Patient Safety. Farnham Surrey, UK: Ashgate.
23. Woods, M. (2010). Cultural safety and the socioethical nurse. Nursing ethics 17: 715-25. 10.1177/0969733010379296.
24. Engeström, Y. (1999). Expansive visibilization at work: an activity theoretical perspective. Computer Supported Co-operative Work 8: 63-93.
25. Arkes H. R., Dawes, R. M., and Christensen, C. (1988) Factors influencing the use of a decision rule in aprobabilistic task. In: Dowie J, Elstein AS (eds) Professional judgment: a reader in clinical decisionmaking. Cambridge University Press, New York.
26. NTSB. (1994). Aircraft Accident Report, Controlled Collision With Terrain, Express II Airlines, Inc./Northwest Airlink Flight 5719, Jetstream BA-3100, N334PX, Hibbing, Minnesota, December 1, 1993 (PDF). National Transportation Safety Board.
27. Gasper, K. and Clore, G. L. (1998). The persistent use of negative affect by anxious individuals to estimate risk. J Pers Soc Psychol 74:1350–1363.
28. Lowenstein, G. and Lerner, J. (2003). The role of affect in decision-making. In: Davidson R, Scherer K,Goldsmith H (eds) Handbook of affective science. Oxford University Press, New York.
29. Fessler, D. M. T., Pillsworth, E. G., and Flamson, T. J. (2004). Angry men and disgusted Women: an evolutionaryapproach to the influence of emotions on risk-taking. Organ Behav Hum Decis Process 95:107–123.
30. Lerner, J. S. and Tiedens, L. Z. (2006) Portrait of the angry decision maker: how appraisal tendencies shapeanger's influence on cognition. J Behav Decis Making 19:115–137.

31. Mesken, J., Hagenzieker, M. P., Rothengatter, T., and de Waard, D. (2007). Frequency, determinants, and consequences of different drivers' emotions: An on the-road study using self-reports, (observed) behaviour, and physiology. Transportation Research Part F: Traffic Psychology and Behavior, 10(6), 458–475. https://doi.org/10.1016/j.trf.2007.05.001.
32. Beisswingert, B. M., Zhang, K., Goetz, T., Fang, P., and Fischbacher, U. (2015). The effects of subjective loss of control on risk-taking behavior: The mediating role of anger. Frontiers in Psychology, 6. https://doi.org/10.3389/fpsyg.2015.00774.
33. Brandenburg, S., Oehl, M., and Seigies, K. (2017). German taxi drivers' experiences and expressions of driving anger: Are the driving anger scale and the driving anger expression inventory valid measures? Traffic Injury Prevention, 18(8), 807–812. https://doi.org/10.1080/15389588.2017.1316493.
34. Ge, Y., Zhang, Q., Zhao, W., Zhang, K., and Qu, W. (2017). Effects of trait anger, driving anger, and driving experience on dangerous driving behavior: A moderated mediation analysis. Aggressive Behavior, 43(6), 544–552. https://doi.org/10.1002/ab.21712.
35. Nesbit, S. M., and Conger, C. C. (2011). Evaluation of cognitive responses to anger-provoking driving situations using the Articulated Thoughts during Simulated Situations procedure, Transportation Research Part F: Traffic Psychology and Behaviour, Volume 14, Issue 1, Pages 54-65, ISSN 1369-8478, https://doi.org/10.1016/j.trf.2010.09.006.
36. Dahlen, E. R. and White, R. P. (2006). The Big Five factors, sensation seeking, and driving anger in the prediction of unsafe driving. Personality and Individual Differences. 41:903–915.
37. Dekker, S. (2018). The Field Guide to understanding human error. London - Routledge.
38. Emad, G., and Roth, W.M. (2008). Contradictions in practices of training for and assessment of competency: A case study from the maritime domain. Education and Training, 50, 260–272.
39. Emad, G. and Roth, W. M. (2009). Policy as boundary object: A new way to look at educational policy design and implementation. Vocations and Learning, 2, 19–35.
40. Hollnagel, E. (2014). Safety 1 & Safety 2 The past and future of safety management. Farnham, Surrey, England: Ashgate.
41. Hollnagel, E. (2014). Why WAI is different from WAD. Presented at the third resilience health care net meeting, Middlefart, Denmark, 12–14 August 2014.
42. Hollnagel, E. (2012). FRAM: the functional resonance analysis method. Modelling complex socio-technical systems. Surrey, UK: Ashgate Publishing Limited.
43. Dekker, S. (2006). Chronicling the Emergence of Confused Consensus. In: HOLLNAGEL, E. & LEVESON, G. N. (eds.) Resilience Engineering : Concepts and Precepts, pp. 77–92. Aldershot: Taylor & Francis Ltd.
44. Kontogiannis, T. (1999). User Strategies in Recovering from Errors in Man-Machine Systems. Safety Science 32: 49-68.
45. Freud, S. (1930). Civilization and its discontents. London: Hogarth Press.
46. Lorenz, K. (1974). Civilized man's eight deadly sins. New York: Harcourt, Brace, and Jovanovich.
47. Berkowitz, L. (1993). Aggression: its causes, consequences, and control. New York: McGraw-Hill.

48. Devane, B. and K. Squire. (2012). Activity theory in the learning technologies. In D. Jonassen, & S. Land (Eds.) Theoretical foundations of learning environments, 242-268 New York, NY: Routledge.
49. Engeström, Y. (1987). Learning by expanding. An activity-theoretical approach to developmental research. Helsinki, Finland, Orienta-Konsultit Oy.
50. Sannino, A. and Engeström, Y. (2018). Cultural-historical activity theory: founding insights and new challenges. Cultural-Historical Psychology, 14(3), 43-56.
51. Cole, M., Engeström, Y., and Vasquez, O. (1997). Mind, Culture and Activity. Cambridge: Cambridge University Press.
52. Engeström, Y. and R. Miettinen. (1999). Introduction. In: Engeström, Y., R. Miettinen and R. Punamaki. (eds.) Perspectives on activity theory, pp 1-16. New York, NY: Cambridge University Press.
53. Guba, E. G. and Y. S. Lincoln. (1989). Fourth generation evaluation. Newbury Park, CA and London: Sage Publications.
54. Drew, P. (2009). Conversation analysis. In Jonathan A. Smith (Ed.), Qualitative psychology: A practical guide to research methods (pp. 133–159). Los Angeles: Sage.
55. Yin, R. K. (2011). Qualitative Research from start to finish. New York: Guilford Press.
56. Silverman, D. (2010). Qualitative research. London: Sage.
57. Cohen, L., Manion, L., & Morison, K. (2007). Research Methods in Education. (6th ed.). London: Routledge.
58. Berg, B. L. (2007). Qualitative research methods for the social sciences. London: Pearson.
59. Rubin, H. J. and Rubin, I. S. (2005). Qualitative interviewing: The art of hearing data (2nd ed.). Thousand Oaks, CA: Sage.
60. Brown, J. D. (2005). Research methods for applied linguistics. In: A. Davies and C. Elder,
61. Braun, V. and V. Clarke. (2006). Using thematic analysis in psychology. Qualitative Research in Psychology 3: 77-101.
62. Javadi, M. and Zarea, M. (2016). Understanding Thematic Analysis and its Pitfalls. Journal Of Client Care, 1 (1) , 33-39.
63. Bertaux, D. (1981). From the life-history approach to the transformation of sociological practice. In Biography and society: The life history approach in the social sciences, ed. by D. Bertaux, 29–45. London: Sage.
64. Kuzel, A. (1992). Sampling in qualitative inquiry. In Doing qualitative research, ed. B. Crabtree and W. Miller, 31–44. Newbury Park, CA: Sage
65. Creswell, J. W. (1994). Research Design: Qualitative and Quantitative Approaches. Thousand Oaks. CA: Sage.
66. Ryan, G. W. and Bernard, H. R. (2000). Data management and analysis methods. In Norman K. Denzin & Yvonna S. Lincoln (Eds.), Handbook of qualitative research (pp.769-802). Thousand Oaks: Sage.
67. Glaser, B. G. (2001). The grounded theory perspective: Conceptualization contrasted with description. Mill Valley, CA: Sociology Press.
68. Charmaz, K. (2006). Constructing grounded theory: A practical guide through qualitative analysis. London: Sage.
69. Lerner, J. and Keltner D. (2000). Beyond valence: toward a model of emotion-specific influences on judgement and choice. Cogn Emot 14:473–493.

70. Oldenberg, M. and Jensen, H.J. (2012), "Merchant seafaring: A changing and hazardous occupation", Occupational *and Environmental Medicine*, Vol.69, No.6, pp.132-148.
71. Dimitrova, D. N. and Blanpain, R. (2010). Seafarers' rights in the globalized maritime industry. Alphen aan den Rijn: Kluwer Law International.
72. Thomas, M. and Bailey, N. (2009). Out of time: Work, temporal synchrony and families. *Sociology*, Vol.43, No.4, pp.613-630.
73. Thomas, M. and Bailey, N. (2006). Square Pegs in Round Holes? Leave Periods and Role Displacement in UK-Based Seafaring Families. Work, Employment and Society 20(1): 129-149. doi:10.1177/0950017006061277.
74. Thomas, M., Sampson, H. and Zhao, M. (2003). Finding A Balance: Companies, Seafarers and Family Life. Maritime Policy and Management 30(1): 59-76. doi:10.1080/0308883032000051630
75. Baron, R. A. (1994). The physical environment of work settings; Effects on task performance, interpersonal relations, and job satisfaction. In B. M. Staw & L. L. Cummings (Eds.), Research in organizational behavior (Vol. 16, pp. 1-46). Greenwich, CT; JAI Press.
76. Kinney, J. A. (1995). Violence at work: How to make your company safer for employees and customers. Englewood Cliffs, NJ: Prentice Hall.
77. Mantell, M. R. (1994). Ticking bombs: Defusing violence in the workplace. Burr Ridge, IL: Irwin.
78. Neuman, J. H. and Baron, R. A. (1997). Aggression in the workplace. In R. Giacalone & J. Greenberg (Eds.), Antisocial behavior in organizations (pp. 37-67). Thousand Oaks, CA: Sage.
79. Hennessy, D. A. and Wiesenthal, D. L. (1999). Traffic congestion, driver stress, and driver aggression. Aggressive Behavior, 25(6), 409–423.
80. Fraine, G., Smith, S. G., Zinkiewicz, L., Chapman, R., and Sheehan, M. (2007). At home on the road? Can drivers' relationships with their cars be associated with territoriality? Journal of Environmental Psychology, 27(3), 204–214.
81. Hennessy, D. A. (2008). The impact of commuter stress on workplace aggression. Journal of Applied Social Psychology, 38(9), 2315–2335.

Assessing the determinants of safety culture in the maritime industry

Daniela Andrei
Curtin University
daniela.andrei@curtin.edu.au

Michelle R Grech
Australian Maritime Safety Authority
michelle.grech@amsa.gov.au

Mark Griffin
Curtin University
mark.griffin@curtin.edu.au

Andrew Neal
University of Queensland
a.neal@psy.uq.edu.au

ABSTRACT

The study approach, findings and the way forward on an Australian Linkage Council funded collaborative research project between the University of Queensland, University of Western Australia and the Australian Maritime Safety Authority, focusing on safety culture will be presented. The aim of the study was to identify the factors that have the strongest impact on safety on board international vessels operating in Australian waters, in order to provide recommendations on how to improve safety in the maritime industry. More than 1,000 seafarers from 197 ships comprising 23 flag States were surveyed. The survey assessed safety culture, work demands, fatigue, mental health, and well-being and safety performance. The results are being used to develop a set of recommendations regarding the best ways to improve safety on board ships. The recommendations centre on a) improving the quality of work procedures; b) introduction of effective fatigue management systems; and c) improving the quality of work design and organisational support. The findings of this study have been presented at a range of industry forums, briefings, and at the International Maritime Organization.

INTRODUCTION

Safety Climate, Safety Culture, Behaviour and Outcomes

The term "safety culture" refers to the way that an organisation manages safety, and reflects the core beliefs and attitudes that guide behaviour and decision-making (Casey, Griffin, Flatau Harrison & Neal, 2017). There are two broad elements of safety culture. The first are the policies, practices and procedures that the organisation has for managing safety. This first element is sometimes referred to as "safety climate" within the academic literature (Griffin & Neal, 2000; Neal, Griffin, & Hart, 2000). The second are the values, priorities, norms and motives held by people in the organisation. These two elements reflect the distinction between safety culture as something that an organisation *has* (i.e., policies, practices & procedures) and safety climate as something that an organisation *is* (i.e., people with a shared set of values and beliefs). Measures of safety climate and culture are highly correlated, and are not distinguishable for practical purposes (Casey, et al., 2017). In this report, we use the term "safety culture" rather than "safety climate", because it is a broader term, and is more widely recognized within the maritime industry.

In a recent review, Lützhöft, Grech & Porathe. (2011) identified safety culture as a critical risk factor for the maritime industry. They argued that whilst most accidents at sea are caused by human error, these errors are attributable to conditions created by the organisation. Specifically, safety-related policies and practices relating to communication, commitment, trust, incident reporting, risk management and training play an important role in shaping behaviour, which can either directly or indirectly affect safety. According to Lützhöft et al. (2011) maritime safety culture is a concern, because shipping operators are under significant cost pressures. While there is anecdotal evidence suggesting that maritime safety culture is a critical risk factor, research on maritime safety culture is limited and fragmented. Recent work indicates that the relationship that exists between safety culture, safety behaviour and safety outcomes observed in other high-risk industries may also exist within the maritime industry. While this evidence is encouraging more is needed to clarify the role of safety culture in the maritime industry, particularly in relation to seafarers aboard vessels operating in Australian waters. The remoteness of the work environment also creates other concerns with crew having limited social contact and may be isolated for long periods of time with little support, all of which can reduce performance, health and well-being (Oliver, Cheyne, Tomas, & Cox, 2002). Hence this study also included an evaluation of seafarer mental health and wellbeing in the context of safety culture.

The aim of the study was to identify the factors that have the strongest impact on safety on board international vessels operating in Australian waters, in order to provide recommendations on how to improve safety in the maritime industry. This paper presents a selection of the findings with a full report available upon request.

RESEARCH METHOD

Measures

For the purpose of this study the *Developmental Safety Culture Survey* (DSCS) developed by the research team was used to measure different levels of safety culture development. The DSCS distinguishes between different levels of safety culture development based on existing theory (Lawrie, Parker, & Hudson, 2006, Reason, 1997) and was validated during the pilot study. Twelve items were developed to tap into the "Systems and Processes" and "People" aspects of safety culture. An overview of the measurement model is presented in figure 1. The aspects measured in the "Systems and Processes" section included:

- Safety policies and procedures,
- Safety training,
- Communication,
- Role definitions,
- Reporting systems, and
- Operational schedules.

The "People" aspects included:

- Safety Values,

- Norms, and
- Motives.

Based on the existing literature three specific descriptors were developed to reflect:
1. dysfunctional/reactive safety culture;
2. compliance oriented culture, and
3. participative/generative safety culture.

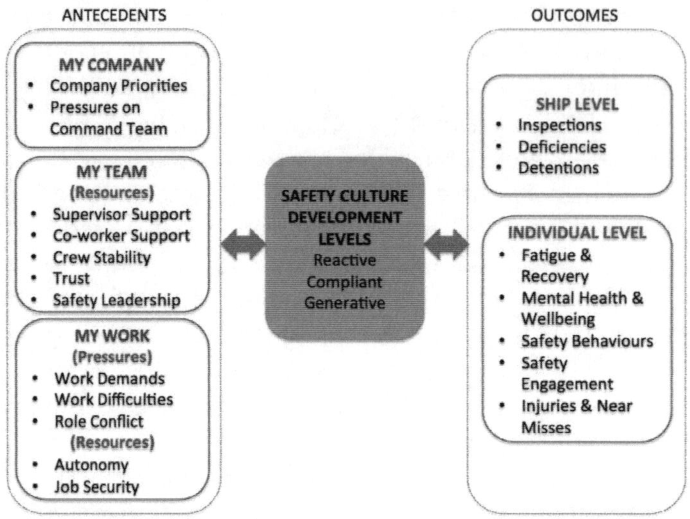

Figure 1: Overview of the overall measurement approach and variables included in this study.

Participants responded on a 5-point Likert scale, where 1 received descriptors that reflected a reactive safety culture and 5 represented descriptors that reflected a participative/generative safety culture. Participants indicated the value that best reflects the way each aspect was being managed on their ship. The survey also included a series of individual and work demographic questions. Examples of individual level factors included fatigue and recovery with participants quality of sleep also assessed by asking whether they experienced sleep problems onboard the ship. Beyond the general measures of compliance, the quality of safety behaviours were analysed by looking at two types of positive compliance behaviours: *adaptive compliance* and *deep compliance*; and two types of negative safety behaviours: *surface compliance* and *non-compliance*.

Procedure

Participation in the study was voluntary and anonymity was guaranteed. The option of electronic and paper based surveys was provided. Partnerships with various organisations with direct contact to ships and seafarers were developed to increase survey reach. These included training providers, AMSA inspectors, pilots and seafarer welfare centres.

Participants

The final sample consisted of 1026 seafarers. 164 participants completed the command team survey and 862 participants filled in the survey for the rest of the crew. The difference in surveys were related to *work pressure*, *safety behaviours* and *safety engagement* questions. For the Command Team, the *work pressure* measures were supplemented with an extra measure of *role conflict* arising from their critical position as mediators between company and crew. The safety behaviours and safety engagement measures had different referents in which the Command Team were asked to provide an overall assessment of their crew behaviours using the same items, while the crew were required to only report on their own work behaviours and safety engagement.

97.9% of the participants were male with an average age of 34.7 years (*SD*=10.4 years). The age range for 57.8% of the participants was between 18 to 37 years. Participants were mostly experienced seafarers, with an average overall tenure at sea of around 10 years (*M*=9.76, *SD*=8.78 years at sea). Most participants worked long contracts – in the region of 9 months to 1 year, especially evident for the officers and ratings. Most participants reported 4 months or less onboard the ship, with very few having been onboard for more than 9 months. The sample was represented by more than 16 nationalities with most of the participants coming from the Philippines. Participants were also asked to report how many different nationalities were on board their ship. On average, participants indicated that there were about 4 different nationalities on board the ship they were working on.

Ship level data

All responses from seafarers on the same ship, identified by its IMO number, were averaged to obtain an overall score for the ship. 195 distinct ships were identified across the sample. The ships were then categorised into the following ship types (Figure 2).

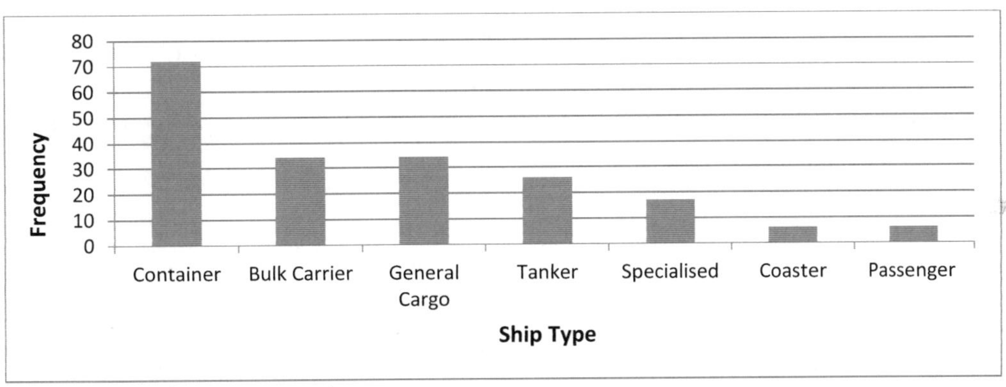

Figure 2. Type of ships surveyed.

The breakdown of Flag States represented in this study was fairly consistent with the flag state population of vessels coming into Australian ports during the same year this survey was conducted. Panama (*N*=30) was the most frequently represented, followed by Singapore (*N*=27), Hong Kong (*N*=22), Liberia (*N*=20), Malta (*N*=13), Marshall Islands (*N*=12), Australia (*N*=11) and Bahamas (*N*=10).

Data analysis and reporting

Using SPSS statistical analyses tool the data were analysed at two distinct levels. First, an analysis was carried out at the individual level, taking into consideration the main differences and associations between responses offered by individual seafarers. The, data were analysed at the ship level by aggregating all individual responses from the same ship. Cross-level interactions were also investigated in order to identify the effects of broader (ship level) factors on individual outcomes. The results highlight the strengths and weaknesses of safety culture, as well as its possible antecedents and consequences within the sample.

Due to increased pressures and uncertainty in the industry and the possible increased relevance of priorities communicated by companies, a multi-level analysis was also performed to investigate more closely the way perceptions of company priorities and operational uncertainty at the command team level might explain safety and well-being outcomes for the rest of the crew onboard the ship. The main interest was on the interplay between priority on safety and costs, but operational uncertainty was also added to the model. An overview of the predictors used in this analysis is presented in Figure 3.

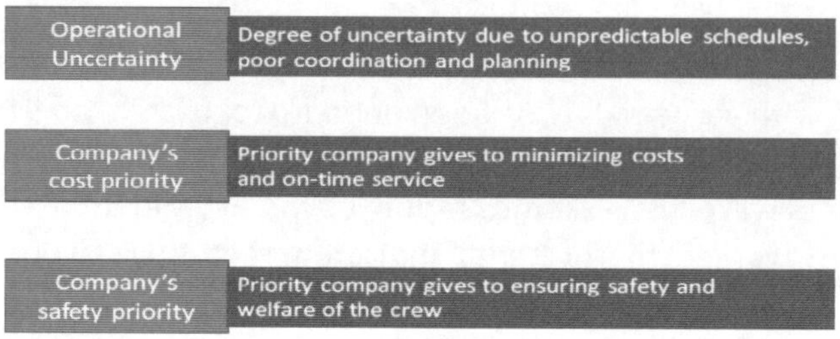

Figure 3. Overview of main predictors used in the multi-level analysis on priorities, operational uncertainty and their effects on safety and wellbeing.

At the ship level of the multi-level model, priorities and operational uncertainty were included as perceived by members of the command team. The reasoning was twofold. First, company priorities are usually communicated to seafarers by the command team onboard the ship which inform their decisions and management on the crew. Second, from a methodological perspective, using two different sources for the different data: the command team for priorities and operational uncertainty, and the rest of the crew for wellbeing and safety outcomes ensures more robust results.

FINDINGS and DISCUSSION

Safety Culture and its antecedents

Figure 4 shows the results of how participants evaluated safety culture across ships. Aspects that were most positively evaluated were those related to seafarers' perceived personal responsibility towards safety: *responsibilities and motives*. Although the overall findings show that safety culture was evaluated positively, there were still a number of cases for which safety culture was reported within the reactive–compliance based spectrum such as the *Planning and Scheduling* dimension. In addition, although formal mandatory requirements such as the International Safety Management Code are expected to have a positive impact on the evaluation of perception of systems and processes, it is also important to understand how these formal systems have an impact on safety behaviour and wellbeing of seafarers.

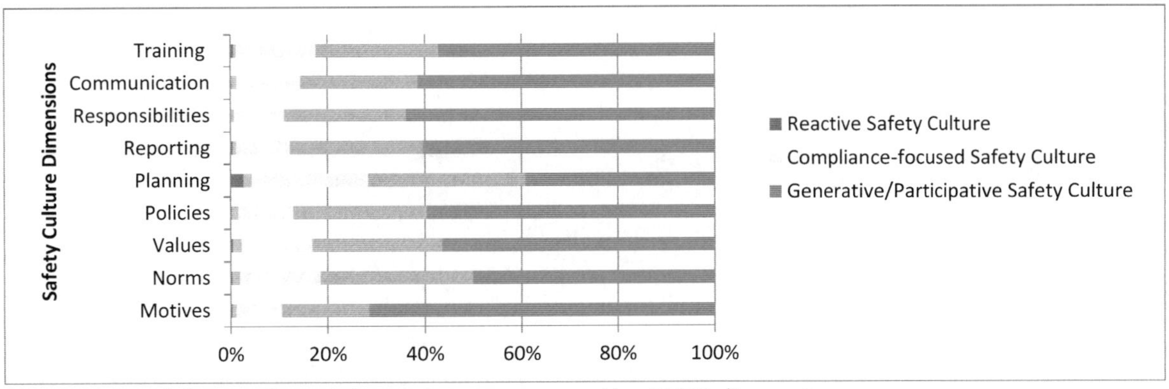

Figure 4 Breakdown of participants' responses on the SCDS dimensions.

Figure 5 presents an overview of how company priorities are perceived. Overall, seafarers perceive that companies place a great importance on preventing damage to the ship and cargo, as well as on the safety of the crew. However, about 20% of seafarers perceive that the company they work for places little or moderate importance on their welfare.

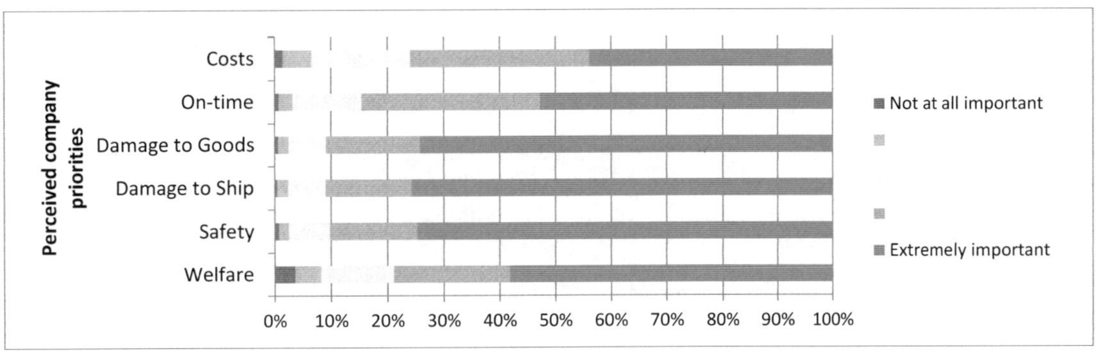

Figure 5. Breakdown of participants' responses on perceived company priorities items.

Work Demands

Participants reported working a average of 61.28 hours per week, with a standard deviation of 13.06 hours. The results indicate that a high proportion (almost 30%) of the participants are working long hours, exceeding 69 hours/week. Long working hours appear to be also coupled with increased work pressures and demands. More than 20% of participants reported that their working hours are unpredictable. Similarly, approximately 40% of participants reported working under time pressure, and about half of them reported experiencing high demands for vigilance at least sometimes in their work.

Work Difficulties

Descriptive data on the three types of shipboard conditions (*work difficulties*) that might affect safety culture and safety outcomes: *physical conditions*, *technology and resources*, and *operational uncertainty* were analysed. Two categories of *physical conditions* were measured: external (weather, visibility and ship motion) and internal conditions (see Figure 6). Approximately 40% of participants reported that bad weather often caused difficulties in performing their work. Additionally, more than 20% of participants reported that poor visibility and ship motion often created difficulties for them in performing their work. Results for internal physical working conditions were similar (Figure 6), with loud noise and cramped workspaces being reported as a source of frequent disturbance by a high proportion of participants.

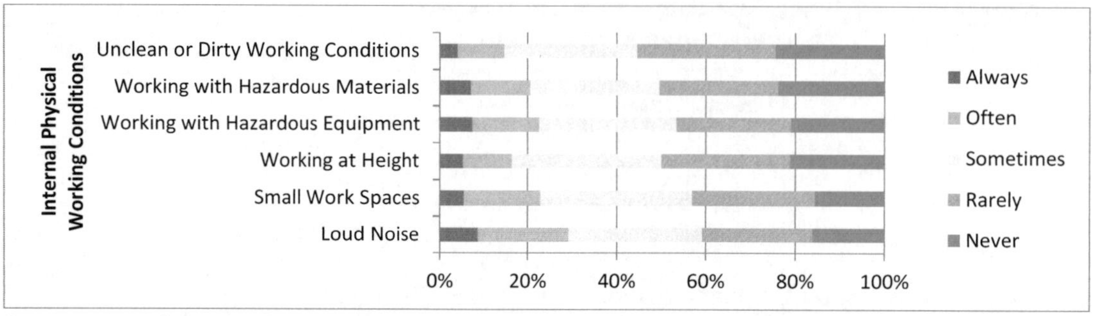

Figure 6. Breakdown of participants' responses evaluating how often internal physical conditions are creating difficulties for them in their work.

Scores for difficulties related to technology and resources were relatively homogenous. However, more participants (around 20%) reported that not having the needed supplies, and maintenance problems often created difficulties in performing work (Figure 7).

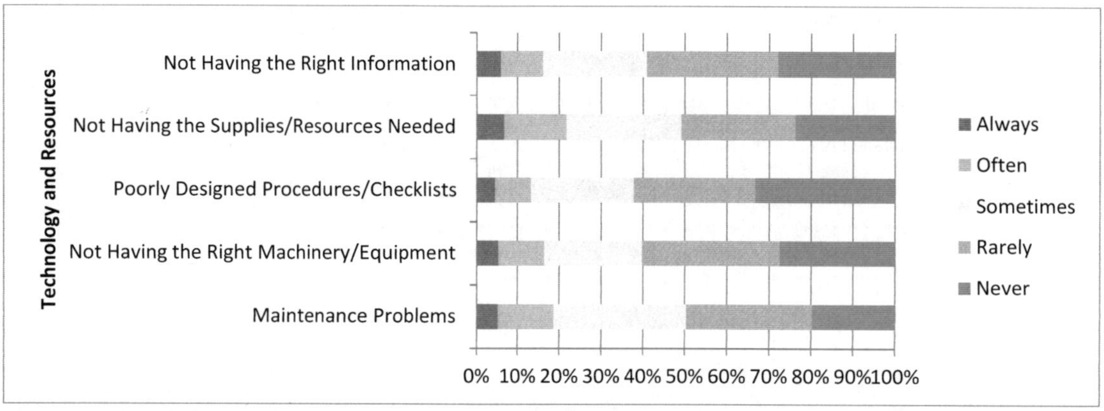

Figure 7. Breakdown of participants' responses evaluating how often conditions related to available technology and resources are creating difficulties for them in their work.

Approximately 40% of the sample reported difficulties related to operational uncertainty at least sometimes in their work. Scores are relatively homogenous across the factors measured, but frequent changes to schedule and manifest as well as disruptions or delays appear to be more common forms of difficulties (Figure 8).

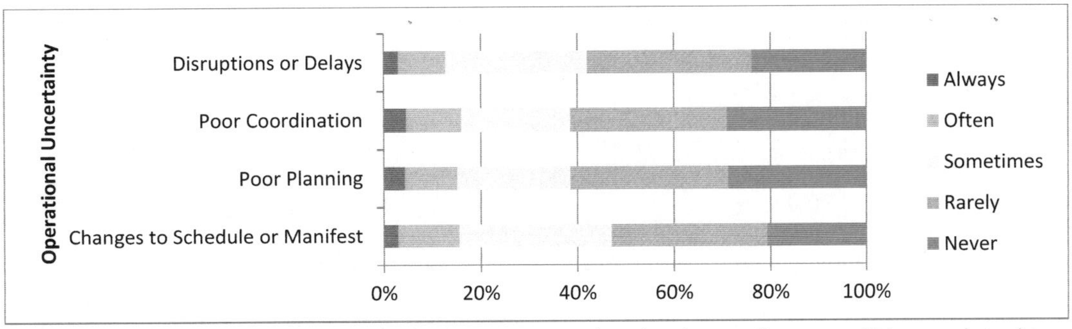

Figure 8. Breakdown of participants' responses evaluating how often conditions related to operational uncertainty are creating difficulties for them in their work.

Safety Leadership

Four different aspects of safety leadership were measured: *leverage*, *energise*, *adapt* and *defend*. Overall, all aspects of safety leadership received positive evaluations, with over 80% of participants agreeing that their supervisors exhibit all four of the surveyed safety leadership behaviours. It is indicated that the way leaders reflect and communicate safety goals, represents another type of work resource which plays an important role in health and safety outcomes.

Outcomes - Fatigue

Approximately 12% of the participants experienced sleep problems, while close to half of the participants reported no sleep-related difficulties. A similar pattern is observed in the participants' fatigue data (Figure 9). Approximately half of the participants reported experiencing low levels of fatigue, while close to 20% of the participants reported experiencing increased or high levels of fatigue, more notably, chronic fatigue.

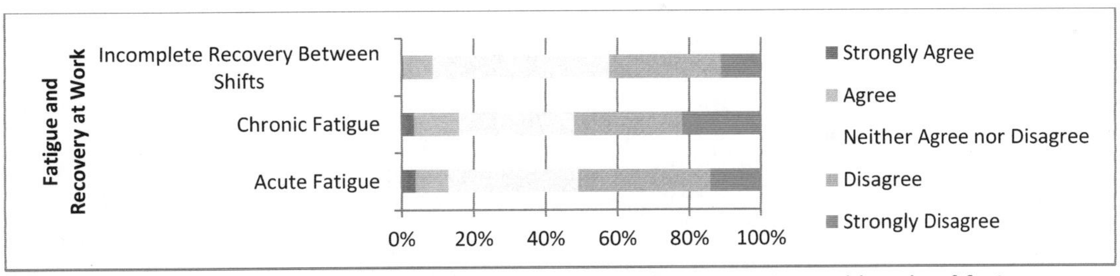

Figure 9. Breakdown of participants' responses regarding experienced levels of fatigue at work.

Outcomes - Mental Health and wellbeing

Figure 10 presents an overview of mental health and wellbeing. Three aspects of wellbeing were measured: *hedonic*, *psychological* and *social* wellbeing. Almost 40% of the participating seafarers reported experiencing negative symptoms at least sometimes, and around 10% of them reported more frequent levels of mental health (frequent symptoms of mental ill health – depression and anxiety). In terms of overall wellbeing, responses were more positive. However, the lowest percentages of wellbeing were found for social wellbeing. Not surprisingly, social wellbeing is the aspect of wellbeing that is more likely to be impacted by the working arrangements in the maritime industry which can be linked to social isolation which is supported by the literature (Iversen, 2012).

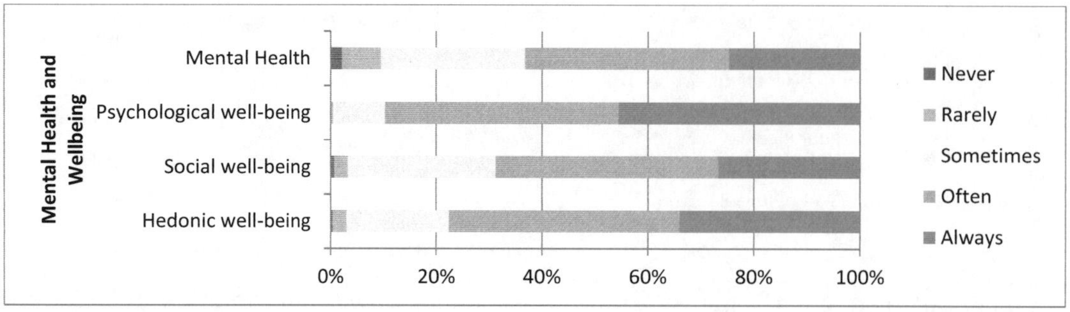

Figure 10. Breakdown of participants' responses regarding mental health and wellbeing.

Outcomes - Overall Safety Behaviours

Overall *safety behaviours* were measured in terms of *safety task performance*, *safety participation* and *safety innovation*. High levels of these behaviours were reported, especially for safety task performance. The positive results for safety compliance do not necessarily reflect mature levels of participative/generative safety on board the participating ships. Participative/generative safety cultures are usually associated with less emphasis on overall compliance (safety task performance) and more safety participation and innovation. While safety participation and innovation levels were relatively high in this sample, levels of safety task performance reported were even higher indicating a strong emphasis on compliance.

Outcomes - Types of Safety Compliance

Figure 11 presents the results for positive compliance behaviours. The results suggest a high level of positive compliance. Most of the participants (approximately 80%) reported trying their best to apply the correct procedures to the task (deep compliance) and being adaptive, such as drawing on knowledge and experience to come up with a solution to complete the task safely when circumstances make existing procedures not appropriate (adaptive compliance).

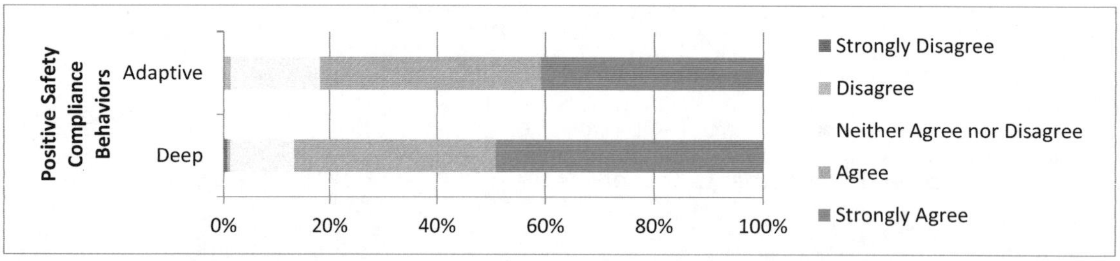

Figure 11. Breakdown of self-reported positive safety compliance behaviours.

However, when negative safety compliance behaviours are taken into account (Figure 12), the results indicate that non-compliance, and especially surface compliance, are also manifested by participants. Notably, more than 40% of participants reported that they sometimes just "tick the boxes" without paying too much attention to the actual procedures; and almost 20% reported some level of non-compliant behaviours (e.g. skip the procedures to get the work done). The results for positive and negative safety behaviours might appear contradictory at first glance. However, there are potential explanations for this pattern of findings. In particular, there are multiple procedures in place on any vessel, and seafarers might comply with some but not others. Even when overall compliance is positive, there might be situations of non-compliance or surface compliance that have the potential to put safety at risk.

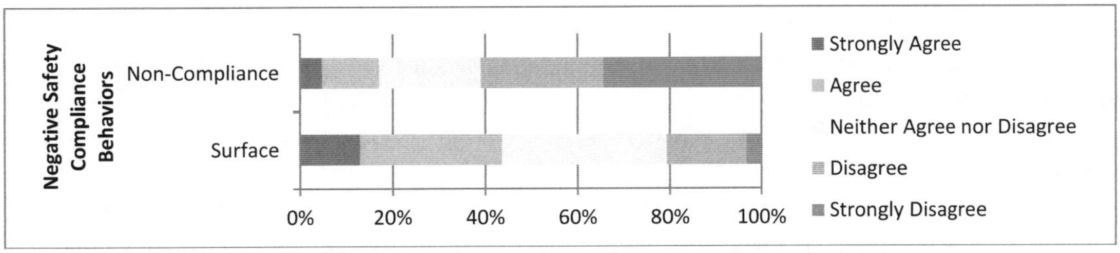

Figure 12. Breakdown of self-reported negative safety compliance behaviours.

Multi-level analysis

Perceived operational uncertainty, company priorities, and safety in relation to wellbeing and safety compliance outcomes.

The results (Figure 13) indicate that a priority on safety perceived at the command team level is not related to either wellbeing or safety compliance at the crew level. However, operational uncertainty and especially a company's priority on costs translate into negative outcomes for seafarers' wellbeing and safety compliance. These results converge toward the conclusion that prioritising costs and increased operational uncertainty might damage both safety and wellbeing, and a sole focus on safety would not be sufficient to counteract these effects.

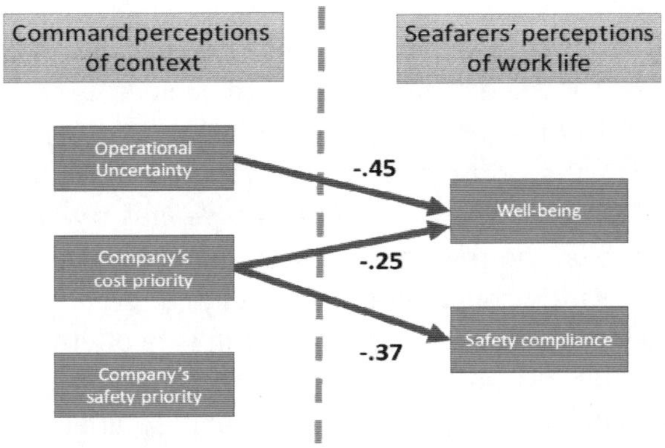

Figure 13. Overview of the multi-level analysis of the effects of perceptions of the overall context at the command team level on safety and wellbeing outcomes for the rest of the crew.

CONCLUSIONS

Approximately 40% of this study's sample indicated that they experienced difficulties in performing their tasks due to factors related to technology and resources, such as "poorly designed procedures/checklists" and "not having the right information". Similarly, conversations between seafarers and researchers during data collection revealed that a frequent complaint by seafarers was that there were too many procedures and many were too complicated for effective use. To encourage positive safety behaviours (e.g. deep compliance to safety rules and procedures), seafarers must have the necessary safety knowledge and motivation to perform their task safely, and this is determined, partly, by the degree of clarity and quality of the work procedures.

More than 20% of participants reported working more than 69 hours per week and that working hours were unpredictable. Approximately 12% of the participants reported experiencing sleep problems and 20% agreed that they experience some level of chronic and similarly 20% indicated experiencing acute fatigue. Further analyses revealed that chronic fatigue leads to reduced levels of psychological wellbeing that may impact on the overall functioning of employees.

RECOMMENDATIONS

A number of recommendations were developed based on information gathered from this study. The purpose is to propose research-based practices designed to manage the implications associated with this study's findings. This report focuses on two of these recommendations.

1. Work and Procedures:

The important role of the quality of work procedures in predicting compliance is evidenced in studies that show that procedures that were perceived as vague, inappropriate, poorly written or difficult to access were more likely to result in poor compliance (Dahl, Fenstad & Kongsvik, 2014). Hence, a focus on developing and ensuring high-quality work rules and procedures that are easily understood and are perceived as valid by those to whom they are addressed is critical.

2. Fatigue Management

Organisations need to develop fatigue management interventions that continuously monitor and manage fatigue risks to prevent fatigue-related incidents or impaired psychological wellbeing. Managing the risk of fatigue requires a combination of intervention strategies with some being more effective than others. The International Maritime Organization's (IMO) Maritime Safety Committee approved the revised *Guidelines on Fatigue* (MSC 1598) at its 100th session. Led by Australia, the revision resulted in a more useable guidance document. Central to these guidelines is the concept of a risk-based approach to fatigue management. This includes the approach that since fatigue affects the safe operation of the vessel, fatigue management should logically be an integral part of safety management systems. The Australian Maritime Safety Authority has developed useable guidelines to support fatigue risk management implementation in the maritime domain based on these guidelines.

REFERENCES

Dahl, O., Fenstad, J., & Kongsvik, T. (2014). Antecedents of safety-compliant behaviour on offshore service vessels: A multi-factorial approach. *Maritime Policy & Management, 41*(1), 20-41.

Grech, M., Horberry, T., & Koester, T. (2008). *Human factors in the maritime domain*. CRC Press.

Griffin, M. A., & Neal, A. (2000). Perceptions of safety at work: A framework for linking safety climate to safety performance, knowledge, and motivation. *Journal of Occupational Health Psychology, 5*, 347–358. doi:10.1037fl1076-8998.5.3.347

Griffin, M. A., & Hu, X. (2013). How leaders differentially motivate safety compliance and safety participation: The role of monitoring, inspiring, and learning. *Safety Science, 60*, 196-202.

Iverson, R. T. B. (2012). The mental health of seafarers. *International Maritime Health* 2012;*63(2)*:78-89

Lawrie, M., Parker, D., & Hudson, P. (2006). Investigating employee perceptions of a framework of safety culture maturity. *Safety Science*, *44*(3), 259-276.

Lützhöft, M., Grech, M. R., & Porathe, T. (2011). Information environment, fatigue, and culture in the maritime domain. *Reviews of Human Factors and Ergonomics*, *7*(1), 280-322.

Neal, A., Griffin, M. A., & Hart, P. M. (2000). The impact of organizational climate on safety climate and individual behavior. *Safety Science*, *34*(1), 99-109.

Reason, J. (1997). *Managing the risks of organizational accidents.* Aldershot: Ashgate.

Roberts, S. E., & Marlow, P. B. (2005). Traumatic work related mortality among seafarers employed in British merchant shipping, 1976–2002. *Occupational and environmental medicine*, *62*(3), 172-180.

A human centred approach to optimise human performance in complex marine environments and habitable spaces.

Dr Peter Schumacher, University of South Australia, peter.schumacher@unisa.edu.au
Prof. Siobhan Banks, University of South Australia, siobhan.banks@unisa.edu.au

This project undertaken under DST Group Research Agreement: 6961

ABSTRACT

The design, size and layout of submarine habitable spaces impacts people's ability to work, rest, sleep, engage in exercise and recreation. In addition, attracting and retaining talented people is important for the future submarine force and living conditions and habitability are factors that impact retention. In addition, managing crew fatigue and subsequently increasing operational effectiveness, is a significant challenge to ensure optimum decision making under challenging circumstances; part of this is ensuring the environment is designed to meet user needs.

In order to develop better habitable spaces a human-centred design (HCD) development and assurance methodology, including the application of anthropometric data, was developed for the Australian Defence Science and Technology Group (DSTG). The first application of the method was the design of sleeping berths, galleys and messes for submarines. An HCD approach will develop environments that can help reduce mental and physiological fatigue enhancing submariner cognitive performance and decision making as well as improving recruitment and retention. While HCD is an established discipline it is underutilised in shipbuilding and the early implementation of HCD in projects will provide improved outcomes for users.

PURPOSE AND BENEFITS

The capacity for operators to assimilate and understand the data provided by complex systems is severely tested in 24/7, operational environments. In addition, high workload, stress, and extreme fatigue impair cognitive performance and increase errors during operationally critical tasks. The huge personal, financial and tactical cost of high workload, stress and fatigue while on deployment must be addressed if we are to perform better and faster. The new operational edge comes, not only from technology, but from optimised operators that can make superior decisions in an information dense and complex undersea environment. As part of this optimising the environment for rest and relaxation is an important fatigue countermeasure.

The need to recruit and retain personnel for future submarine crews presents challenges and shipboard quality of life is one of several non-work factors that impact retention (Schwerin, Kline, Olmsted & Wilcove, 2006). A survey of US Navy personnel found that unsatisfactory living conditions, such as lack of privacy and personal space, limited room in the cabin areas and berths as well as poor showers, head spaces and mattresses have a negative effect on performance and crew retention (Wilcove & Schwerin, 2008). The design of habitable spaces requires further research, including the development of mock ups and simulations, to better understand the relative importance of different features (Wilcove & Schwerin, 2008).

Finally, poor design decisions result in poor crew satisfaction and expensive refits. A review of the psychological and physiological impact of the submarine environment, specifically looking at the US Virginia Class SSN, found that poor design of the mess and berth had a negative impact on the quality of life. This was evidenced by a larger number of psychologically based waivers and disqualifications among SSN crews compared to SSBN and these were exacerbated on the Virginia class due to deficiency in the design of the habitation areas (Shobe et al., 2003).

INTRODUCTION

The myth that fatigue and stress can be overcome by adequate motivation has influenced military operations for decades (Shay 1998). However, this attitude is changing and the need to improve habitability on board submarines in order to improve crew performance is well established. In a military operating environment there are many stressors in addition to actual warfighting, ranging from reduced sleep, shift-work, social pressure, environmental conditions and operational environments that frequently involve long work hours and less than optimal sleep environments (Miller, Matsangas & Shattuck, 2008). A lack of sleep results in impaired cognitive performance (Banks & Dinges, 2007; Nebes et al., 2009). Habitability cannot be compromised to other elements without impacting crew performance (OPNAVINST, 1996) and '...improvements in cabin design could have significant effects on crew fatigue and overall effectives.' (Ponton & Parera, 2015) In this way habitability could be used as a fatigue countermeasure, protecting against deteriorations in cognitive performance and increased errors.

HCD is an element of ergonomic processes and it is described in standards such as ISO/TS 18152 Ergonomics of Human-System Interaction, however, despite the recommendations of the standards, HCD practices are not used in the early stages of ship building design and development (Genez et al., 2018). What is unique in this case is the implementation of HCD in the early stages of the project to develop an understanding of the user needs and an HCD focused assurance processes.

APPROACH

The HCD approach draws on expertise from industrial design, anthropometry and bio-behavioural psychology. The process has been developed to design habitable spaces using a range of human-centred design and co-design methods to investigate the user needs, wants, activities, habits, physical requirements and psycho/social needs. The application of anthropometry to ensure human fit includes physical models and digital-human-mannequins (DHMs) within CAD and VR and using the human simulation software, JACK™, to evaluate designs. The Royal Australian Navy Submarine Force was involved throughout the project as part of the development of the design exemplars.

HCD is a framework for product development that involves the end users perspective in all steps of the design process. HCD practice emphasises user engagement methods such as observation and interviews as well as participatory design, co-design and contextual design techniques for investigating and describing the interactions between people and their environments (Beyer, Holtzblatt, 1997).

What people do while on watch and at work while onboard is governed by predetermined procedures and processes that are used to inform the design of the work environment.

However, this does not account for everything people do while working or what people do when not on watch or at work and are using their own time for their own purposes. These day-to-day activities on submarines are not well understood by non-submariners.

Submarine habitable environments are unique due to the extreme space restrictions, challenging watch keeping schedules and complicated technical context. As a result the design team, who have probably never experienced these extreme living conditions, cannot draw on their own personal experiences to understand how these environments are used. Even the simplest day-to-day activities on submarines are very different to 'top-side' activities, for example, taking a shower or having a meal. The HCD methods investigate these activities so that the submariner's actions and needs are recorded and made explicit so that design team can develop better environments that meet the submariner's requirements. Without this understanding the design team risks making uninformed and poor decisions resulting in sub-optimal living environments that impact of the submariner's quality of life.

A second element of the project was the application of anthropometry to ensure the best physical fit of the environment for the range of body shapes and sizes in the user population. This utilised the Anthropometry Survey of the Royal Australian Navy (ASRAN) (Ponton et al., 2019) data gathered specifically to support Navy ship building in Australia. The design for human fit involved the application of standards, anthropometric data, the Maritime Physical Accommodation Guidance for the Royal Australian Navy (MPAGRAN) (Ponton, 2019) and the use of DHMs based on the ASRAN data.

The DHMs were made in SolidWorks™ so that the design team could include dimensionally accurate human figures from the ASRAN data in the CAD model at all stages of the design development process. The DHMs were also used in VR to understand what spaces would be like when populated, for example to understand how the mess looks when full of people having a meal or to evaluate sightlines to display screens for briefings.

This development process includes empirical testing of the design proposals. A full-sized prototype of the junior mess was evaluated with 25 people to understand how people used the seating and tables meals and sleep trials are underway to evaluate the consequences of different orientations of the berths in the boat of the berths and the effectiveness of the tapered sleeping berth configuration.

THE HCD PROCESS

The HCD process used for this project is based on the 'Design Double Diamond' developed by the UK Design Council (Design Council, 2007), Figure 1, and the principles from ISO 9241-210:2010 Ergonomics of Human-System Interaction -- Part 210: Human-centred design for interactive systems (International Standards Organisation, 2010). This process is used for a variety of projects including user interface, user experience, service design as well as product design. Note, ISO 9241-210:2010 is part of the standard for Human Factors Integration ISO/TS 18152, the human systems model in ISO/TS 18152 conforms to and extends ISO 9241-210. This project focused on HCD methods and practices and so references ISO 9241-210 in preference to the broader ISO/TS 18152.

There are several ways the HCD process is described, they share similar characteristics, stages and structures but use different formats and representations to emphasize different aspects of the process. The Double Diamond was used for this project as it provides a simple and well-established format. It has four stages, Develop (divergent research), Define (convergent clarification), Develop (divergent ideation) and finally Define (convergent implementation). The use of HCD for ship design is being developed whereby user engagement, field research and collaborative practices are being used to develop bridges and ships engine rooms (Gernez, et al., 2018).

The Double Diamond can be mapped onto the system engineering lifecycle V-model used for managing the overall development process, The Discover and Define stages are analogous to 'developing the concept of operations' and 'systems level requirements' from the V-model and Develop and Deliver correspond to the 'sub-systems requirements' and 'detailed design' from the V-model. The Double Diamond process operates in conjunction with the main V-model process but independently to suit the flexible approach required for user engagement.

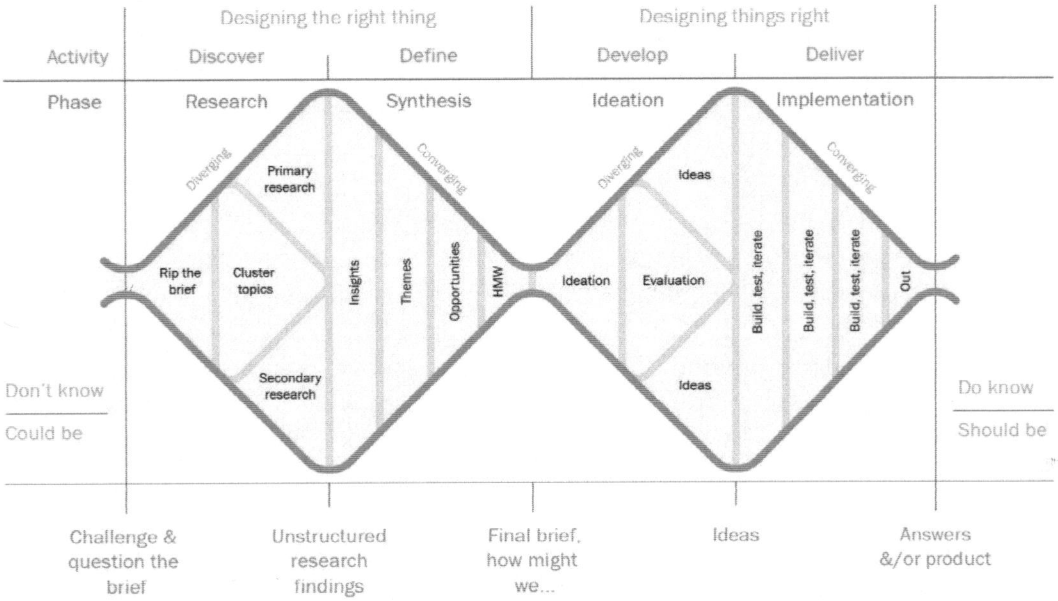

Figure 1 - The Design Double Diamond.

These HCD principles from ISO 9241-210:2010 are:

- The design is based on an explicit understanding of users, tasks and environments. Submarines are unusual environments that need to be investigated to ensure the design team understands the context and the user needs.
- Users are involved throughout the design and development process. Submariners are the best source of information about the activities and challenges that come with life on board submarines.
- User feedback is a critical source of information. Designs were evaluated with users and their feedback was used as the basis for improvements to the design.
- The process is iterative. Iteration involves repeating design cycles until the design satisfies the users' needs, iteration enables the progressive elimination of uncertainty and risk.
- The design addressees the whole user experience. HCD addresses ease of use as well as social, psychological, fatigue and emotional aspects.

- The design team includes multidisciplinary skills and perspectives. A strong HCD team includes a diverse range of skills and knowledge including industrial design, human factors, engineering, anthropometry and psychology.

The different stages of the HCD process used different methods. The first two stages, Discover and Define, dealt with understanding and documenting the user's needs. The user needs were then used as the basis of the design development in the Develop and Deliver stages. The following sections describe each stage and how they were implemented for the submarine habitable space exemplar designs.

Discover (Divergent Research)

This is the stage where the design team conducted research to understand the users and the context. A core method at this stage were workshops with submariners to understand their goals, activities, issues with the current platform and their requirements for future vessels. Other methods used at this stage included site visits, a literature and standards review, a technology search and an investigation of similar environments such as space habitation, aircraft and submarines used by other navies. This stage produced a large amount of qualitative information that was analysed and synthesised in the next stage.

Define (Convergent Clarification)

This stage focused on synthesizing and bringing structure to the findings of the Discovery stage in order to develop criteria, use scenarios and ergonomic and technology hard-points. It drew together the wide array of qualitative and dimensional information and presented it in a format to provide focus for the design development and evaluation stages. The data qualitative was analysed using methods from the social sciences and anthropology, such as affinity diagramming, to reveal patterns and insights. Diagrams were created to show the ergonomic hard-points, work-flows, spatial relations between components, functions and user needs.

Develop (Divergent Ideation)

This stage created several design proposals based on the hard points, scenarios and criteria from the Discover and Define stages, the designs were evaluated with users and developed iteratively. CAD models and physical prototypes were made to elicit feedback and responses from users. The human fit was continually evaluated using the DHMs in CAD and with the use of physical prototypes.

Submariners were consulted in workshops where illustrations, VR and physical models were used to communicate the proposals. The user workshops were held regularly as part of an iterative process to ensure the design proposal met the user needs.

The user workshops were summarised in reports that described the design development process, the workshop activities and the user feedback. These reports provided the evidence base for the final design proposal.

Define (Convergent Implementation)

This stage refined the design based on the input of the submariners. The human fit was evaluated with the DHM simulation software Jack™. The design proposal was documented and described explaining the design with reference to the user needs, criteria, use scenarios, anthropometry and technical constraints.

EXEMPLAR PROJECTS

Several design exemplars were developed for this project, the sleeping berth and cabin, the junior and senior sailors mess, wardroom and galley. These areas were identified by the DSTG as problematic on the Collins Class and requiring attention for future submarines (Ponton & Parera, 2015). The spaces were developed in isolation from the general arrangement.

User Engagement Activities

At the core of the HCD process was user engagement at all stages of the project, from the initial 'Discovery' stage through to the evaluation of design proposals in the 'Develop' stage. The following tables, Tables 1 and 2, show the user workshops conducted over the course of the project. These workshops involved 8 to 10 submariners which gave the group diversity in ranks, roles and experience. The participants were different for each workshop, except for the two iterations of the mess and galley development that used the same participants.

In addition, there were numerous informal conversations, boat tours and demonstrations with submariners that provided background information and enabled the design team to gather feedback on workshop outcomes and design proposals. The iterative nature of the workshops enabled the progressive elimination of uncertainty and risk and confidence the outcome would meet the users needs.

A trial of the junior mess prototype was conducted using 25 non-submariners to evaluate how people sit and eat meals in the space and to evaluate how the design supported other activities including recreation, damage control and triage.

Table 1 - Sleeping berth and cabin - user engagement activities

Stage / Date	Discovery: 29th – 31st May 2017	Develop: 20th – 21st February 2018
Duration	2 ½ days (6 hours / day)	2 days (6 hours / day)
Participants	8	10
Activities	Guided boat tour Focus group Co-design development with full scale physical models Affinity diagramming	Review of criteria Review of berth options Review of storage requirements Berth access and egress evaluation Evaluation of the cabin in VR

Table 2 - Mess and galley - user engagement activities

Stage / Date	Discovery: 7th – 8th June 2018	Develop Iteration 1: 28th -29th November 2018	Develop Iteration 2: 5th – 6th March 2019	Evaluation: 15th April 2019
Duration	2 days (6 hours / day)	2 days (6 hours / day)	2 days (6 hours / day)	1 day (6 hours / day)
Participants	8 submariners including 2 cooks	8 submariners including 2 cooks	8 submariners including 2 cooks	25 participants (no submariners)
Activities	Workshop on use scenarios Co-design development with full scale physical models	Review of design proposals, scenarios and criteria, with VR and illustrations	Review of full-sized physical mock-ups	Simulation of junior mess use scenarios in a full-sized mock up

RESULTS AND DISCUSSION

The outcome of this project was an HCD process and methods integrated with anthropometric design and assessment tools tailored to the task of developing habitable spaces on submarines. This process was used to develop a set of design concepts for the sleeping berths / cabin and the mess and galley for future submarines in order to demonstrate the HCD methods and to develop criteria and use scenarios for design assurance.

By taking a human centred approach, engaging with users at all project stages, applying anthropometry and implementing psychological and physiological considerations into the design the submariner's opportunity for effective rest is potentially improved resulting in better cognitive performance and decision making. This has implications for reducing fatigue, improving crew endurance and enhancing cognitive performance and decision making, giving operators a combat edge.

Working closely with submariners at all stages of the process provided confidence that the design is fit for purpose as well as appealing and comfortable which has implications for recruitment and retention.

Using DHMs and the anthropometric data at all stages of the process provides assurance that the final design will be sized to fit the range of body sizes in the RAN population.

The use of physical prototyping was vital for the development of the designs and for the user engagement. VR was also an important tool for design development and user engagement as it enabled designers and users to experience a variety of design proposals early in the process and provide feedback before physical prototypes were constructed.

Sleeping Berths and Cabins

For the purposes of this exercise a longitudinally orientated berth on the mid-deck in a cabin of six was used as the basis for the design. The berth and cabin, Figure 2, includes the following features:

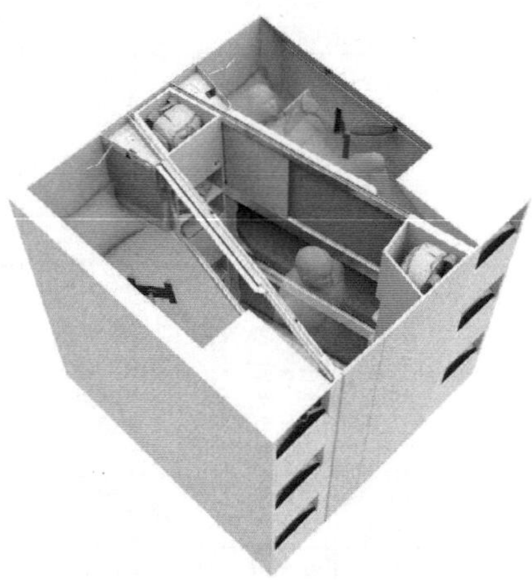

Figure 2 - The sleeping berth and cabin proposal showing the tapered berth and cabin space.

- A tapered berth and cabin to provides more room around the upper body in the berth and more space in the cabin. The berth is wider across the shoulders where people's bodies are at their widest and where most activity occurs, and narrower at the feet where less space is required. This uses the space more efficiently and makes the volumes appear larger. The tapered configuration was met with high approval from the submariners.
- A sliding panel at the head end of the berth was included to block out sound and light. A curtain was used for the remainder of the enclosure at the knees and feet for ease of egress, to reduce the perception of confinement and to allow for air movement.

- A variety of lighting options were developed to suit the various use scenarios. For example, the cabin has a bright light for use when in port while people are moving in and stowing their belongings and there is a dim red light for use when underway so that people can see what is in the cabin without disturbing others who are asleep.
- The lighting has different colours to match the occupant's circadian rhythm. The lights in the berth are set at 2000-3000^{0}K (red/orange) in the evening to assist with the transition to sleep and they set are at 6000^{0}K (daylight) in the morning to assist with the transition from sleep to work.
- A range of air-conditioning outlets allow the users tailor the air location and flow according to their preferences.
- Keeping bath towels and boots dry is an important factor for comfort, hygiene and odour reduction. The proposal includes a ventilated bay where towels and boots can be dried and stored.
- Lockable storage space is provided in the cabin for uniforms and larger items, in addition there is storage inside the berths for personal items with a power outlet for charging devices. The storage includes loops for tying down items so they don't shift and rattle as the boat moves.

Mess

The mess, Figure 3, includes the following features:

Figure 3 - The junior sailor's mess.

- An arrangement that allows for a range of use scenarios including meals, recreation, briefing, damage control and triage.
- The tables are arranged to reduce the amount of people that are 'sat-in' and unable to leave the table without asking others to move. The tables can be modified to make longer tables so people can be laid down for first aid and medical care.
- Three TV screens are arranged so everyone has a good line of sight for presentations and entertainment.
- A sideboard includes a drinks station, fridge and bin with space for the entertainment equipment and media.
- The banquette seating around the periphery allows for flexible seating arrangements as well as providing for a good amount of storage underneath the seats. The peripheral seating minimises the number of people with their back to the group so that people can see other people's faces in order to support social cohesion.

Galley

The galley, Figure 4, includes the following features:

- The space for appliances is configured for maximum flexibility and so appliances can be changed out as technology changes.

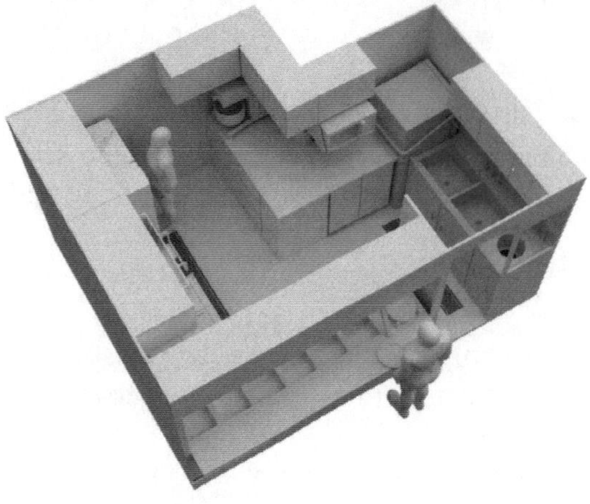

Figure 4 - The galley proposal showing the area designated for the dishwashing tasks.

- The arrangement accommodates the dishwashing, food preparation, cooking and serving activities in different zones within the galley for effective workflows and to avoid cross contamination of raw and cooked food.
- A servery for six standard 1/1 gastronorm trays is included so that a range of meal options can be provided.
- Storage is provided for utensils, standard gastronorm cooking trays, cleaning chemicals and food staples.
- Dishwashing takes place in a small zone separate from the food preparation zones. There is a linear workflow starting where the plates are deposited, scraped, then washed before being returned to storage.

ASSUNARNCE PROCESS

The assurance process developed as part of the project has been used to evaluate alternate design proposals using these scenarios, criteria and design exemplars as the basis of the evaluation. This provided the framework for comparing different design arrangements, defining the trade space and describing the advantages and trade-offs of various proposals. This outcome was a detailed, user centred evaluation that provided the client with a rich and detailed analysis inform the decision-making process.

FUTURE WORK

The work conducted to date demonstrates how the HCD process and methods can be implemented for the design of submarine habitable spaces. Elements of the designs will be tested empirically in an experimental environment to quantify cognitive and physiological benefits and the impact on fatigue and subsequently operational effectiveness. The relationship between improved habitability and crew recruitment and retention can be explored comparing the design proposals to the exiting accommodation on the Collins Class submarines.

CONCLUSIONS

The project resulted in the application of the HCD process for submarine habitable spaces that focuses on close engagement with users throughout the process, applied anthropometry and the consideration of the psychological and physiological dynamics. The HCD process is described in the existing ergonomics standards, what is novel is implementing it in an integrated way early in the process where it can have the greatest impact.

The outcomes of this project are innovative submarine habitable space design proposals and criteria and use scenarios for use in design assurance. It uses creative methods and a rigorous, transparent approach that has be applied to the mess/galley, sleeping berths and control room. This approach can also be applied to other areas such as the exercise areas, ablutions, and for maintenance access and usability.

REFERENCES

Banks, S. (2007). Behavioral and physiological consequences of sleep restriction. *Journal of clinical sleep medicine*, *3*(05), 519-528. Beyer, H., & Holtzblatt, K. (1997). *Contextual design: defining customer-centered systems*. Elsevier.

Council, D. (2007). Eleven lessons: Managing design in eleven global companies-desk research report. *Design Council*.

Chief of Naval Operations. (1996). OPNAVINST 9640.1A, Shipboard Habitability Program. Retrieved from http://neds.daps.dla.mil/Directives/table34.html

Gernez, E., Nordby, K., Seim, Ø., Brett, P. O., & Hauge, R. (2018). Human-centered, collaborative, field-driven design—a case study. In Marine Design XIII, Volume 1 (pp. 291-305). CRC Press.

International Standards Organisation. (2010) Ergonomics of human-system interaction -- Part 210: Human-centred design for interactive systems (ISO 9241-210:2010). Retrieved from http://iso.org

International Standards Organisation. (2010) Ergonomics of human-system interaction – Specification for the process assessment of human-system issues (ISO/TR 18152). Retrieved from http://iso.org

Miller, N. L., Matsangas, P., & Shattuck, L. G. (2008). Fatigue and its effect on performance in military environments. *Performance under stress*, *2007*, 231-49.

Nebes, R. D., Buysse, D. J., Halligan, E. M., Houck, P. R., & Monk, T. H. (2009). Self-reported sleep quality predicts poor cognitive performance in healthy older adults. *The Journals of Gerontology: Series B*, *64*(2), 180-187.

Ponton, K. (2019). A Revised *Maritime Physical Accommodation Guidance for the Royal Australian Navy.* DSTG.

Ponton, K., & Parera, D. (2015). *The Submarine Habitability Assessment Questionnaire SUB-HAQ: A Survey of RAN Submariners.* DSTG.

Ponton, K., Gamble, D., Coleman, J., Furnell, A., Sockshell, S., Fletcher K. (2019) *Revised Anthropometry Guidance for the Royal Australian Navy.* DSTG.

Schwerin, M. J., Kline, T. L., Olmsted, M. G., & Wilcove, G. L. (2006). Validation of a work/non-work life model of quality of life and retention among Navy personnel. Paper presented at the Center for Naval Analysis Workforce Research Conference, Falls Church, VA.

Shay, J. (1998). Ethical standing for commander self-care: the need for sleep. *Parameters*, *28*(2), 93.

Wilcove, G. L., & Schwerin, M. J. (2008). Shipboard habitability in the US Navy. *Military psychology*, 20(2), 115-133.

Shobe, K., Bing, M., Duplessis, C. L., Dyche, J., Fothergill, D., Horn, W., . . . Plott, C. (2003). *Psychological, Physiological, And Medical Impact Of The Submarine Environment On Submariners With Application To Virginia Class Submarines - Technical Report #TR-1229.*

Improving ship safety by integrating risk management into the SMS framework and tracking new key performance indicators

Stuart Williams

Maritime Safety Research Centre (MSRC)

University of Strathclyde

Stuart.Williams@Strath.ac.uk

ABSTRACT

Recent well publicized cruise and ferry ship accidents have been a wake-up call to the ship operating industry. These events raise the question of better integration of risk and safety, and using more advanced risk and safety key performance indicators by ship owners. This paper presents the current findings of research underway at the University of Strathclyde's Maritime Safety Research Centre (MSRC) from the author's PhD program. The paper summarizes the development of the current International Maritime Organization (IMO) approach to implementing a Safety Management System (SMS), looks at rail and airline industry's approaches to SMSs and then proposes a new framework for maritime implementation. The focus is on integrating a more holistic (Enterprise) risk management process into the SMS and generating a new set of performance indicators. SMS modelling is accomplished via application of Dr. Nancy Leveson's (MIT) System-Theoretic Process Analysis (STPA) on cruise and ferry operators. The preliminary results from two ship operators are presented. The current state of the art in maritime performance indicators is reviewed. The author proposes an extension of Pareto's 80/20 Rule and Parmenter's 10/80/10 Rule to generate a new set of generic key performance indicators that can be implemented and tracked by ship operators to guide the improvement in safety gained over time.

OVERVIEW

Over the past two and a half years the author has been part of the newly established Maritime Safety Research Centre (MSRC), part of the Naval Architecture Department at the University of Strathclyde. The initial approach was to look at a fairly broad spectrum of cruise and ferry operators and see how mature their holistic or Enterprise Risk Management (ERM) approaches were, and determine whether the companies with a more mature approach to risk management exhibited a corresponding improvement in their safety records. While conceptually this was straightforward, after gathering information from approximately 10 companies, it became clear that their focus was on meeting the requirements in the International Safety Management (ISM) Code and not employing holistic risk.

The ISM Code became mandatory for tankers, bulk carriers and passenger ship operators in 1998. (1) All other ship types had to meet the Code by 2002. Figure 1 summarizes the headings of the various sections in the ISM Code, which has had only fairly minor revisions since its development by the Maritime Safety Committee in the late 1980's and early 1990's.

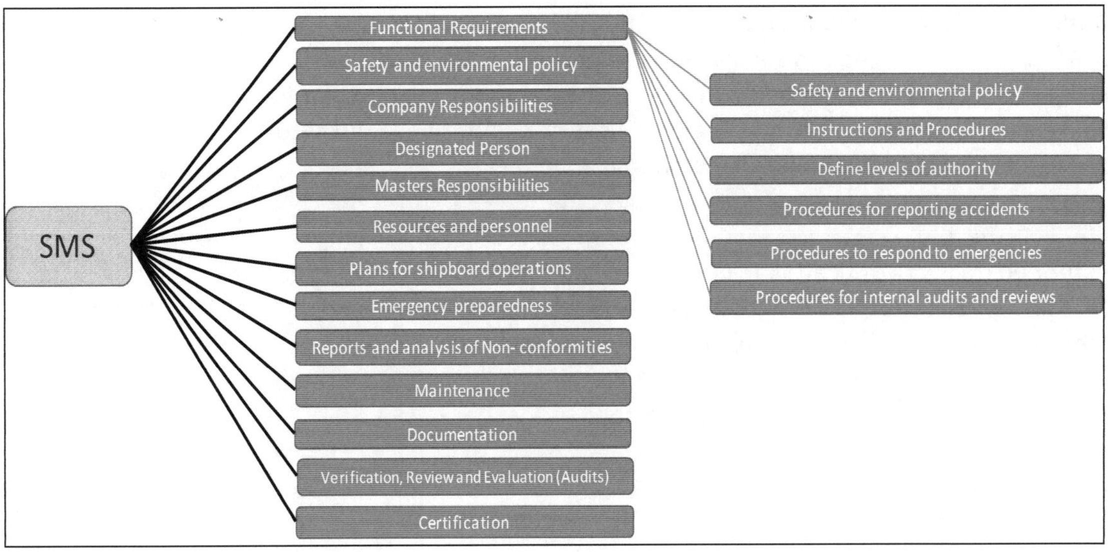

Fig. 1: ISM Code SMS required

Risk management is only mentioned in the Objective Section of the ISM Code, and is not integrated into the requirements set for the SMS implementation.

With this realization, the author's research evolved by studying the implementation of Safety Management Systems (SMSs) in other safety critical transportation industries. Rail and airline approaches were reviewed. In Europe, the European Union (EU) started requiring railroads to implement the use of SMS in 2004. (2) The International Civil Aviation Organization (ICAO), part of the United Nations, began requiring the use of a SMS in 2006. (3) Both the EU railroad requirements and the ICAO airline requirements benefited from the pioneering work done by the IMO. Both include a much more defined integration of hazard and risk procedures into their SMSs. After a review of the required components of the rail and airlines, an extension of the current maritime SMS was developed and used as the template for modelling. This template then served as the framework for assessing the first two ferry and cruise ship company's SMS implementations.

In order to study the functioning of each ship operator's SMS, research was conducted to determine what modelling technique should be implemented to study the performance of the existing systems. Since 2004, Professor Nancy Leveson at the Massachusetts Institute of Technology (MIT) has developed the System-Theoretic Accident Model and Process (STAMP) (4). Basically, this process looks at safety as an emergent property of the complex social-technical control system that ship operations represent.

STAMP assessments via the System Theoretic-Process Analysis (STPA) of each of the cruise and ferry companies were conducted to determine how effective the management barriers, trying to prevent hazardous conditions from accruing, are operating. The controllers, their control actions and the corresponding unsafe control actions are modelled based on each company's SMS implementation. An example of a controller would be the shore side safety managers providing safety alerts and guidance on safety audits to the ships in their control.

Besides defining the requirements that the SMS must address, the ultimate outcome of the STPA process is to generate a set of indicators for each company. These indicators provide insight into measuring the performance of the company in general and the SMS in particular. By analysing several companies a generic set of performance indicators can be created that are useful for any ship operating company.

The concept of Key Performance Indicators was assessed. Recent work by David Parmenter extended the work of the Italian economist Pareto's 80-20 Rule, by creating his 10/80/10 Rule. (5) Parmenter's argument is that the set of indicators should have 10% as Key Performance Indicators (KPIs) and 10% as Key Results Indicators (KRIs). The KRIs have a financial focus. For safety-critical transportation industries like ships, airlines and railroads, the author proposes the creation of key risk and safety indicators.

METHODOLOGY

A review of maritime SMS implementation

The ISM Code provides an outline of what ship operating company's Safety Management System (SMS) must address. (1) Figure 1 contains the outline of the information required. It can be noticed that the maritime implementation of the requirements for a SMS is not organized by functional areas, and certain key elements like hazard analysis and risk management are not directly addressed. General guidance on risk is provided in the Objectives section of the ISM Code. In Section 1.2.2, the following guidance is provided:

> *"Safety management objectives of the company should, inter alia;*
>
> *(1) Provide for safe practices in ship operations and safe working environment;*
> *(2) Assess all identified risks to its ship, personnel, and the environment, and establish appropriate safeguards; and*
> *(3) Continuously improve safety management skills of personnel ashore and aboard ships."*

Part (2) underlined above is the only specific mention of risk management in the regulation, which therefore gives a very broad avenue for interpretation by each ship operator.

Review of the airline and railroad SMS implementation

Figure 2 contains European Union (EU) guidance for implementing safety on rail systems. To see how one country interpreted this guidance, the United Kingdom's Network Rail system's approach to creating its SMS was reviewed.(6) The UK used the EU guidance, with their implementation having an almost identical framework for its SMS. The UK did create a small change by adding a set of Common Safety Methods (CSMs), which provides more specific guidance on risk and monitoring. One of the required areas is risk evaluation and the implementation of risk controls. As reported by the Enterprise Risk Manager (ERM) at Network Rail, Network Rail has a very mature ERM approach and seeking to achieve the highest level (7).

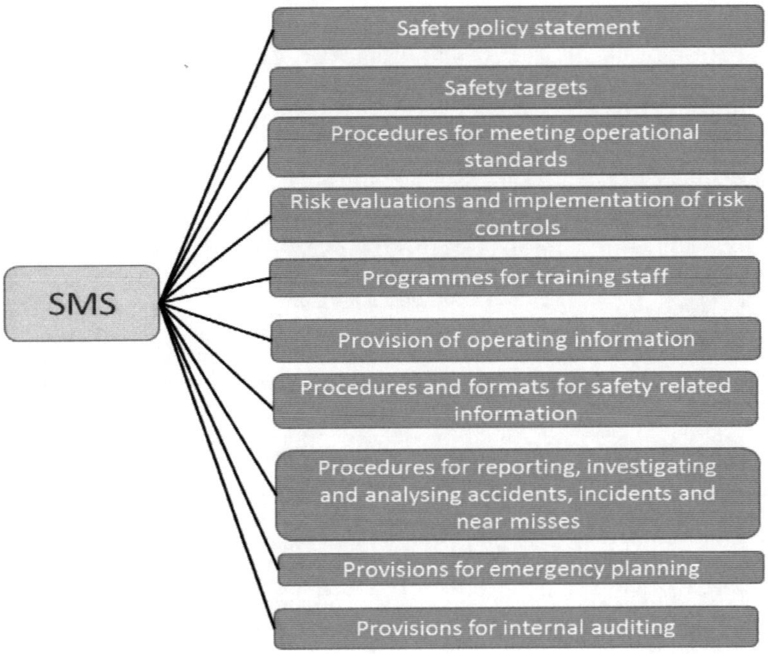

Fig. 2: European Union Railroad safety guidance

In a similar manner, the airline SMS guidance was reviewed. The International Civil Aviation Organization (ICAO) created the first requirement for airline use of a SMS in 2006, eight years later then the maritime implementation requiring the use of a SMS. ICAO's initial implementation benefited from the maritime work, but was much more comprehensive in its guidance. The first issue of the Safety Management Manual (1) provided general guidance in each area with a recommended 10 Steps to follow to implement a SMS. It was not until the 2nd edition in 2009 that the characteristic outline shown in Figure 3 emerged, with the four building blocks for creating a SMS implemented. (8) The second edition was a major re-write, focusing on how to implement and track the usage of airline SMSs. The manual was further improved with updates in 2013 and 2018.

The strength of the guidance in Figure 3 is that it organizes the various functions required for a strong safety management system. Each airline must establish clear safety polices, establish a visible integrated risk approach, measure how well the safety system is working and then train and communicate safety and risk information throughout the company. These high-level functional categories provide a critical and firm foundation to guide the long-term implementation of airline SMSs.

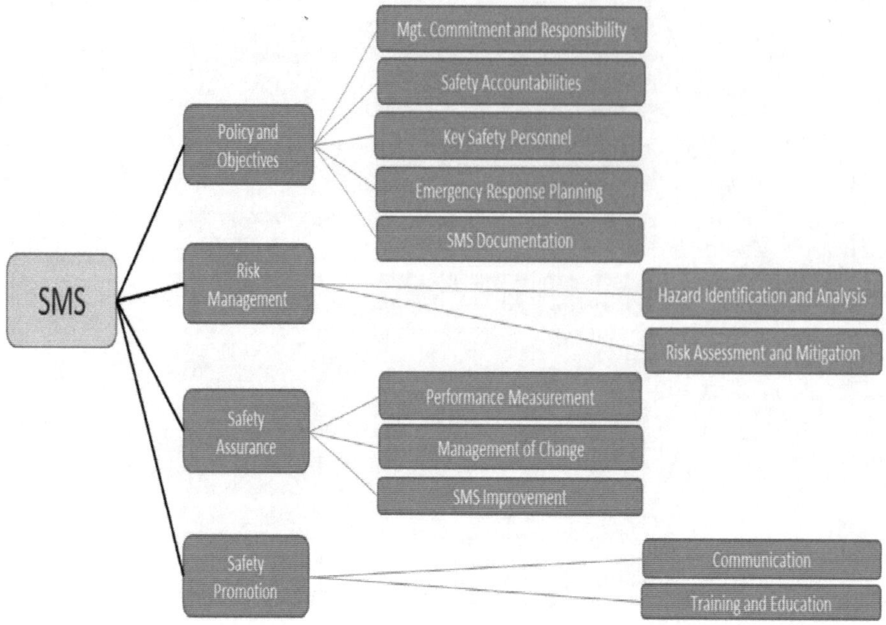

Fig. 3: ICAO SMS

Using the ICAO framework as a guide, a proposed generic SMS for ship operations was developed (Figure 4) which integrates the requirements contained in the marine ISM Code with the functional areas from the airline industry. A crucial aspect is the integration of risk management, hazard analysis and safety assurance as part of the requirements that the SMS must contain.

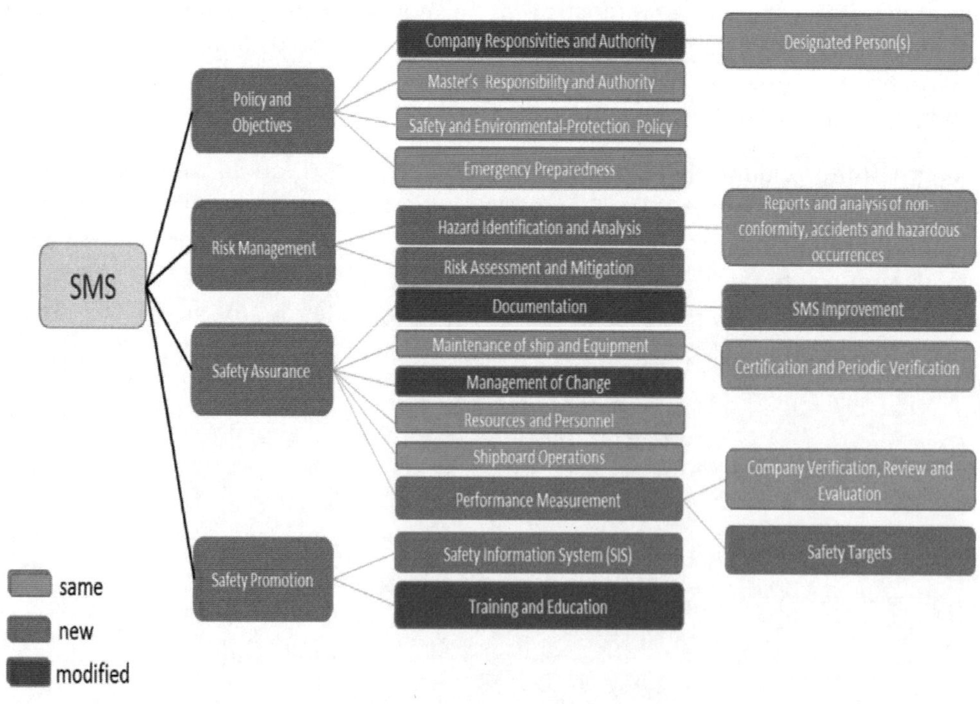

Fig. 4: Proposed Generic SMS for Ship Operators

System-Theoretic Process Analysis (STPA)

The System-Theoretic Process Analysis (STPA) applied is a systems tool, based on STAMP that can be used to analyse organizational structures, like safety management. (9) The process identifies loss scenarios that might cause potential accidents, then generates a set of requirements for safe operations and ultimately a corresponding set of performance indicators. The recent past accident, incident and near miss statistics are then analysed to establish baseline performance. To date, one ferry operator and one cruise ship operator have been analysed.

In STAMP the controllers are the various individuals throughout the structure that implement various safety barriers. For example, at a high level the Chief Executive Officer is responsible for providing the high-level Safety Policy that the company should follow. The Safety Policy is a control action. At a lower level in the organization the shore based ship management team the safety manager insures prompt release of Safety Alerts, another control action.

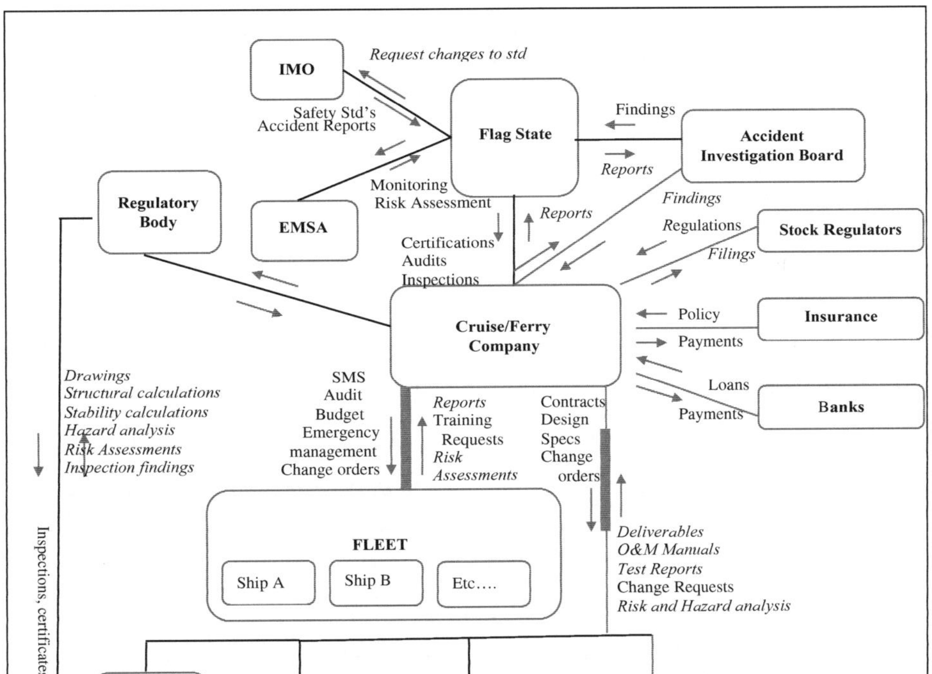

Figure 5: STPA Multi-Layer Control Structure

An example of an Unsafe Control Action (UCA) would be the safety manager failing to issue a Safety Alert, or issuing it late so that a similar problem occurs on another ship in the fleet.

The first step in the process was to create an overarching control diagram that places the ship operator within the entire system that ultimately influences the safe operation of their fleet of ships. Figure 5 captures this environment on one of the companies studied.

In this diagram the Stock Regulators, for publically traded companies, require that the company identify the risks to the company's performance. These risks provide insight in to the health of the company on a yearly basis and should be generated as part of the risk process being used at each company.

In the case of this company, the framework showed missing reports, or feedback documentation that was created to just pass an audit, but did not provide strong control of risks or hazards. The areas highlighted in Red were uncovered as a result of the STPA implementation.

The diagram models the controllers, the controlled processes and the control flows between them (as control actions and feedback). The safety and risk control actions and feedback typically manifest in the form of documents such as standards, manuals, reports, etc. Each of these Control Actions (CAs) creates a set of Unsafe Control Actions (UCAs). The STPA model then analyses these UCAs to determine the loss scenarios, system requirements and corresponding system indicators. STPA assess the functioning of the SMS and generates a set of requirements and indicators that the system should meet and produce, if it is working in accordance with regulations and the company's management approaches. This diagram represents a compilation of the barriers created to prevent unsafe conditions from developing. Since the amount of data generated is fairly large, an Access data base was created to track the information.

Key Performance Indicators (KPIs)

There is a rather large body of research and fair number of commercial products that address the development and use of indicators to try and track performance and improve the safe operation of ships. A search of the literature identified the following providers of systems based on developing indicators to track:

- Baltic and International Maritime Council (BIMCO) (10)
- Tanker Management Self-Assessment (TMSA) (11)
- American Bureau of shipping) as well as other commercial vendors have products that provide a set of KPIs for ship operators to implement (12)

Each of these are based on collecting data on operations and then using some process to filter that data to create meaningful information to manage ship operations. The underlying issue that each of these systems has to grapple with is that each of these processes can generate a large number of KPIs. The term "Key" becomes diminished when hundreds of "key performance indicators" are created and tracked.

ANALYSIS

STPA results

Table 1 summarizes the first two company's STPA findings. With the exception of the number of Controllers, there are similarities exhibited for the number of Control Actions and corresponding Unsafe Control Actions. This probably reflects the fact that each of the company's SMS implementations is patterned after the required number of items that must be included based on the ISM Code.

To understand the type of information generated by the STPA process, Table 2, an excerpt from the STPA Access database, gives an example of the shore based audit team executing an audit of one of the ships in the fleet. Working step wise down through the table shows that even if the Control Action is accomplished, there can be one or more Unsafe Control Actions created. In this case UCA C6-8-1-8 notes that if the audit is conducted, the analysis could be a superficial "check the box" approach which fails to uncover hazards or repeat findings. Based on this a loss scenario was created with a corresponding requirement that when audits are performed they must be of high quality uncovering new hazards and repeat findings.

Table 1 STPA Findings

Function	# Company A	# Company B
Controllers	26	37
Control Actions (CAs)	103	101
Unsafe Control Actions (UCAs)	358	314
Loss scenarios (LSs)	358	314
Safety System Requirements (SSRs)	358	314
Safety System Indicators (SSIs)	274	275

The final step in the STPA process is to generate an indicator that tracks this particular loss scenario. In this case, the number of repeat findings is the indicator that gives a very good assessment of the performance of the complete audit process. In all, as Table 1 shows, there were slightly more than 100 control actions that were generated for each company. The total number of safety system indicators was 274 and 275 for the two companies respectfully. Table 2 has the loss scenario, system requirement and indicator for one Unsafe Control Action. Due to the large number of these factors, each controller was asked to prioritize each loss scenario, system requirement and indicator on a scale of 1 to 3, with 1 being very important, two less and three lesser still. Using this process 61 level 1 indicators were created for Company A and 101 for Company B. The use of subject matter experts helped to focus which factors are most important. This is currently an area of research, especially in the analysing the output of STPA assessments of airline safety management systems. There is recent work trying to address this issue by Karanikas & Chatzimichalidou (13). Their paper proposes the use of continuous values for the behaviour of system components along with a weighing of each component relative to its hierarchical level in the organization

The prioritization process used here is just a first attempt to determine which of the UCAs and their corresponding factors are most important. Additional work is needed in this area.

Table 2 Steps in the STPA process

Control Actions (CA) By Controller – Fleet Operations		
Locator	Control Action Name	Description
C6-8	Audit	Support of audit analysis

Step 1

Unsafe Control Action (UCA)	
Unsafe Control Action Locator	Unsafe Control Action Description
C6-8-1	The audit is conducted, but is a superficial check the box approach that fails to uncover hazards, or repeat findings. [SSH5]

Step 2

List Loss Scenario				
List20	Locator	Priority	Loss Scenario Name	Description
C6-8-1	C6-8-1-LS1	1	Audit	Audit is superficial and therefore fails to uncover existing hazards or repeat findings.

Step 3

List Safety System Requirement				
Combo20	Locator	Priority	Safety System Requirement Name	Description
C6-8-1	C6-8-1-SSR1	1	Audit	Quality of audits shall be monitored periodically to insure that the breadth and depth of the audit coverage consistently uncovers new and repeat findings.

Step 4

System Indicator					
System Indicator Locator	Controller Name	Priority	System Indicator Name	Description	Unsafe Control Action Locator
C6-8-1-SI1	Audit Team	1	Audit	Number of repeat findings each review cycle shall be tracked and reviewed monthly.	C6-8-1

Step 5

Stu's (4x5)x80 Rule

The Italian economist Vilfredo Pareto in 1897 hypothesized that there is an 80/20 rule that roughly 80% of an effect comes from 20% of the causes. In his case, he was looking at the fact that 80% of the land in Italy was owned by 20% of the people. Richard Koch (15) extended this approach to the broader concept that worthwhile results come from a small minority of the effort. Using this thought, David Parmenter postulated (5) that of the range of performance indicators, roughly 80% should be tracked and reported lower in the organization, while another 20% are key and should be reported to senior management and the board of directors. For a traditional corporation (non-safety critical) he recommended using 10% Key Results Indicators and 10% for Key Performance Indicators, therefore his 10-80-10 rule. A majority of the result indicators are financial.

Table 3 shows the four key areas that need to be tracked: safety, risk, performance and financial. The initial work at the first two companies in the financial area focused on the safety related aspects of financial indicators. One example was to track the size of the budgets for safety training. Each company had different ideas as to tracking financial indicators and due to just two data points, more work is needed in this area.

Appendix A contains a summary of the characteristics, frequency of measurement and number of measures for each area. In each of the key areas up to a maximum of five parameters could be developed. From the STPA models of the initial two companies researched, this table shows an initial set of generic key metrics developed by combining overlapping results from both companies. These appear to provide real insight into the operation of the company's ships. Over time the number of key indicators in each area will grow and change, but the idea would be to keep each key area to five or less parameters. A much larger number of lower level indicators can be tracked in each category.

Table 3 Example of Stu's (4x5)x80 Rule

a	Metric
Key Safety Indicator (KSI)	Crew retention rate reported weekly
	Number of safety meetings held onboard reported weekly
Key Risk Indicator (KRI)	Percentage of company covered by holistic (enterprise) risk management, reported quarterly.
	Number of new high risks in the risk register reported weekly.
	Total number of incomplete risk assessments should be reported monthly
Key Performance Indicator (KPI)	Percentage of preventative maintenance items completed versus planned reported monthly
	Number of environmental violations reported weekly
	Number of repeat audit findings reported monthly

Key Financial Indicator (KFI)	Passenger advanced booking trend reported daily
	Average monthly vessel cost change per ship, reported monthly
	Percentage change in safety budgets for ship training, reported monthly

Each of these generic indicators captures in one parameter the functioning of a whole set of procedures or processes. For example, the audit indicator tracks the overall performance of the audit system by capturing the number of repeat items found during follow-on audits. If this number starts to increase, it is a very telling flag that the health of the whole audit process is starting to deteriorate.

CONCLUSIONS

Use of STAMP/STPA provides a structured framework to generate a set of indicators that assess how well a current ship operator's SMS is functioning. Using STPA on the first two company's safety and risk management systems provided a system's approach to generating a detailed set of loss scenarios, system requirements and indicators for each companies to use. By utilizing these requirements and indicators, that are linked to known hazards, the effectiveness of the current set of policies and safety management procedures can be tracked over time. Clearly the number of company's modelled needs to be increased validate the benefits of better integration of risk.

The author proposes the creation of a limited number of key indicators, Stu's (4x5)x80 Rule to sharpen the focus on the truly critical key indicators for a safety critical industry like ship operations. While a large number of indicators can be tracked, the key indicators for safety, risk, performance and financial are the ones that will provide the necessary information for the various control layers within the ship operating company to use to improve safety.

WAY FORWARD

This paper raises the prospect of incorporating more aggressive risk and hazard analysis in to the safety management approach used by ship operators. The author spent several weeks as an Intern at the IMO. During this period the 100th Maritime Safety Committee (MSC) meeting took place. The concepts describe here were presented to the Maritime Safety Division and Legal Affairs Divisions at the IMO. A paper is in preparation by one of the member nations to recommend starting work on an update to the ISM Code and will be submitted for consideration at MSC 102. This paper recommends starting work on assessing what changes are needed to the ISM Code. Assuming a working group is established a wide range of changes will be assessed. The recently approved International Standards Organization (ISO) standard on Occupational Health and Safety (14) should provide an excellent source of guidance for the MSC to use in the update of the ISM Code.

In order to assess the impact of better integration of risk management into ship operations, the path forward is to implement the recommended changes at a larger number of ship operators. The first two companies studied are starting the implementation of more aggressive risk and hazard analysis. The results of these implementations will be instrumental in generating the data needed to support the next evolution in maritime safety management.

REFERENCES

(1) International Maritime Organization [IMO]. (2018). International Safety Management (ISM) Code, (5th ed.,p.16), ISBN: 978-92-801-1696-0.

(2) L 220/16, Official Journal of the European Union, 21.16.04, Directive 2004/49/EC

(3) International Civil Aviation Organization [ICAO] Safety Management Manual. (2018). 4th Edition DOC 9859-AN/474), Retrieved from https://rise.articulate.com/share/v5Sm_0DJQvKI51ZQb6HJmBy7bOrhQfTE#/lessons/_5KUx8747VkDIz_znAbNxw_pCdPBQguw.

(4) Leveson, N. (2004). A New Accident Model for Engineering Safer Systems, *Safety Science, Vol. 42, No. 4.* April 2004. Pp. 237-270

(5) Parmenter, D. (2015). Key Performance Indicators, 3rd Edition, ISBN 978-1-119-01984-8

(6) Office of Rail and Road (ORR) The Railways and Other Guided Transport Systems (Safety) Regulations 2006 (as amended) A Guide to ROGS, April 2018

(7) Hunter-Jones, H., (2016). ERM as a Change Initiative at Network Rail, *Enterprise Risk*, Autumn 2016, p.10-13. Retrieved from https://enterpriseriskmag.com/enterprise-risk-magazine/

(8) International Civil Aviation Organization [ICAO] Safety Management Manual. (2018). 4th Edition
(DOC 9859-AN/474), Retrieved from https://rise.articulate.com/share/v5Sm_0DJQvKI51ZQb6HJmBy7bOrhQfTE#/lessons/_5KUx8747VkDIz_znAbNxw_pCdPBQguw

(9) Leveson, N. & Thomas, J. (2018). *System Theroetic Process and Analysis Handbook*, p.133.

(10) BIMCO, (2018). The Shipping KPI Standard, Released 1st January 2018

(11) Tanker Management Self-Assessment (TMSA) Version 3. (2019) Retrieved from https://OCIMF.org

(12) ABS launches new software for vessel efficiency. (2016, Jun 07). *Progressive Digital Media Transportation (Incl Airports, Roadways, Railways, Shipping, Automotive & Logistics) News.* Retrieved from https://search.proquest.com/docview/1794545159?accountid=14116

(13) Karanikas, N. & Chatzimichaidou, M. (2018). The COSYCO Concept: an indicator for Comparing System Configurations, *International Cross-industy Safety Conference (ICSC) 2017*, (*ADV 1*(1) p. 154-174), DOI: 10.5117/adv2018.1.011 Kara.

(14) Nisonger, T. (2008) The "80/20 Rule" and Core Journals, The Serials Librarian, 55:1-2, 62-84, DOI: 10.1080/03615260801970774

(15) International Standards Organization [ISO] Occupational Health and Safety Management Standard, 45001, March 2018, https://www.iso.org/iso-45001-occupational-health-and-safety.html

Appendix A: Summary of indicators

Types of Performance Indicators (PIs)	**Characteristics**	**Frequency of Measurement**	**Number of Measures**
Key Performance Indicators (KPIs) tell staff and management what to do to increase performance across the whole organization	Tells how well the various parts of the organization are performing	Monthly, Quarterly	Up to 5
Performance Indicators (PIs) are targeted measures that track preventative maintenance, environmental releases etc…		24/7, daily, Weekly, Monthly, Quarterly	No more than 20
Key Risk Indicators (KRIs) are based on an enterprise risk process that captures areas that may cause significant problems to the viability of the organization	Risk indicators guide where to focus mitigation efforts to reduce risk	Monthly, Quarterly	Up to 5
Risk Indicators (RIs) are more focused on potential areas that will impact operational safety, environmental issues or financial performance.		Weekly	No more than 20
Key Safety Indicators (KSIs) give a summary of the collective efforts of the various teams in meeting the goal of delivering safe operations	Safety indicators guide where potential unsafe actions may occur	Monthly, Quarterly	Up to 5
Safety Indicators (SIs) provide a sense of how the safety culture is performing		Weekly	No more than 20
Key Financial Indicators (KFIs) give an overview on the company's financial performance	Financial measures provide guidance on the overall performance of the company	Weekly, Monthly, Quarterly	Up to 5
Financial Indicators (FIs) are working level metrics that help guide day to day decision making		Daily, Weekly	No more than 20

NOVEL PLATFORM FOR LITTORAL RESEACH VESSELS

K. Goh, Knud E Hansen & **J. Chappell**, Australian Institute Of Marine Science

1 INTRODUCTION

Research vessels are normally based on fishing and offshore vessels owing to the similar type of handling operations required on the aft of the vessel. While a displacement monohull has proven to be an adequate platform for this purpose, they are challenged with low hydrodynamic efficiency, poor seakeeping and limited deck space, especially on smaller vessels.

In this paper the authors investigate the use of alternative hull forms to meet the challenging operational environment and the benefits they can provide over a standard monohull. Renewable and hybrid energy systems have also been applied resulting in a design that has better performance, comfort, functionality and flexibility at a significantly lower operating cost.

Also described is how 3D modelling and ShipSpace™ Virtual Reality collaboration tools have been used in the concept design phase.

2 BACKGROUND

2.1 CURRENT RESEARCH VESSELS

Research vessels have typically been mono-hulls often based on fishing trawlers or offshore supply vessels. These vessels have similar requirements to conduct handling operations on the aft working deck. Such hulls have proved sufficiently capable that few variations from this type have been explored.

Designers at Knud E Hansen Naval Architects have been challenged to develop a vessel concept that would significantly improve on the vessels currently in operation with the Australian Institute Of Marine Science (AIMS). The principle particulars of the two largest vessels the RV Cape Ferguson and RV Solander are presented below.

	RV Cape Ferguson	RV Solander
Design Length	23.9m	34.9m
Moulded beam	7.8m	8.6m
Hull depth	3.7m	4.2m
Design draught	2.5m	2.8m
Displacement	212 tons	415 tons
Deadweight	45 tons	100 tons
Gross tonnage	198 GT	383 GT
Complement	14 pers.	18 pers.

Table 1 - Current AIMS fleet principle particulars

Figure 1 - RV Solander (AIMS)

2.2 AIMS REQUIREMENTS

The AIMS brief was to improve both the capability of the future vessel while also reducing the vessel's operational costs and improving its 'green' profile. After conducting a survey from AIMS scientific stakeholders, capability was defined.

Although there was no specific requirement for the future vessel size, the AIMS stakeholder requirements clearly indicated that the vessel's dimensions were based around the larger 35 meter *RV Solander*. Other aspects of the stakeholder requirements, such as the aft deck space and number of tenders for example, also indicated a need for the largest vessel possible.

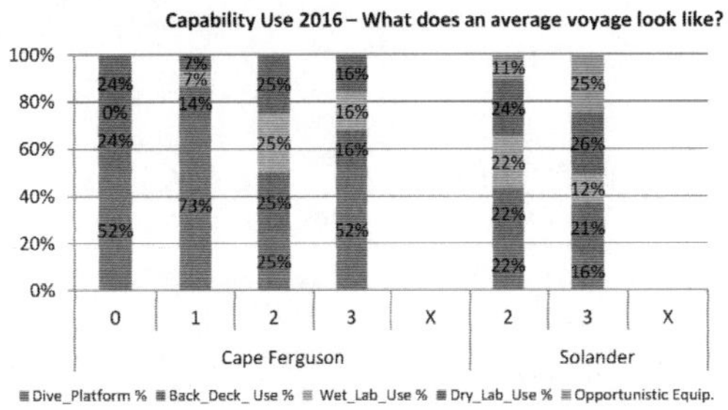

Figure 2 – Current AIMS fleet operations

The stakeholder requirement for a length of 35 meter is interpreted as the future vessel's maximum size, since a length of 35 meter equates to an Equivalent Gross Tonnage of 500 GT for certificate of competency according to the NSCV and subsequent crewing requirements.

Modularised science fitout was a high priority since the variation of research work being undertaken was increasing. For example, a containerised laboratory for seismic survey or an ROV. Modularising the science equipment would allow reduced preparation time in port and improved vessel utilisation. The vessel should also be able to operate UAVs and deploy larger and heavier mooring than the current vessels.

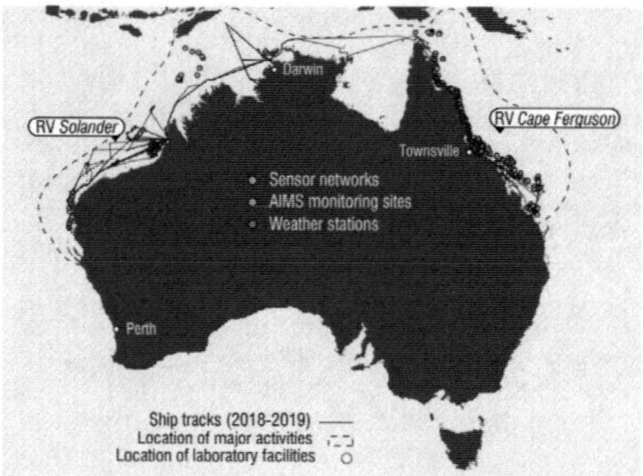

Figure 3 – Current AIMS fleet track 2018-2019

Low Underwater Radiated Noise (URN) to reduce the effect of the vessel on the sensitive aquatic environment was also important as was improved performance of vessels acoustic sensors.

Other important improvements for the future vessel included:

- Safer boarding for RHIB tender boats
- Better arrangement to enable science equipment, ship provisions and bunkers to be loaded simultaneously
- Improved crew habitability

The following aspects where considered for reducing operational costs:

- Crewing
- Fuel consumption
- Maintenance requirements

2.3 CONCEPT STUDY

To better understand and benchmark the potential benefits of any novel hull platform for a research vessel, three concept designs were developed. The first is a like-for-like design which uses a conventional mono-hull and mechanical propulsion system similar to the existing AIMS vessel. The second design is similar to the like-for-like design but with green technologies applied. The third design is the so-called X-Factor vessel.

The focus of the X-Factor design is both a step change increase in capability over the like-for-like designs and also a more visually interesting design that differentiates it from other vessels and helps attract public attention to AIMS science activities and mission.

3 HULL PLATFORMS

A brief review of common hull forms and their suitability for a littoral research vessel was undertaken.

3.1 MONO-HULLS

3.1 (a) Displacement Monohull

A displacement monohull, such as the existing RV Solander and the baseline designs are a conventional vessel monohull that largely meets the requirements of an oceanographic research vessel. They offer a good balance between load carrying requirements, seakeeping and low construction costs. The main limitation of small displacement monohulls is low cruising speeds, which are limited hydrodynamically by wave resistance. For a 35m length vessel the maximum speed is typically about 12-13 kns. The base concept design, being a like-for-like fleet replacement or addition shall be designed as a displacement monohull.

3.1 (b) Dynamic Monohull

A dynamic monohull is a hull form that provides dynamic lift at higher speeds and enables vessels to achieve much higher speeds than conventional displacement monohulls restricted by wave resistance. A higher maximum speed would allow quicker transit to areas of scientific interest and enable higher scientific utilisation of the vessel.

Dynamic monohulls are much more sensitive to weight and are usually constructed from aluminium or composites at considerably higher cost. The other penalty of a dynamic monohull is poorer seakeeping. It is well understood that the seakeeping characteristics of dynamic monohulls are not compatible with oceanographic research requirements. Susceptibility to surface waves creates high acceleration forces and motions that limit the usefulness of dynamic monohull vessels as a work platform.

3.2 MULTI-HULLS

3.2 (a) Catamaran

A conventional catamaran configuration can provide large stable platforms for oceanographic research in clam waters, however they can have higher accelerations in even moderate sea states with high degrees of pitch and roll coupling. They enable higher cruise speeds as each hull can have a much higher slenderness (length to beam) ratio without compromising stability. The trade-off is higher construction cost since two hulls need to be constructed and using more expensive materials if weight saving for dynamic lifting is also required.

The tunnel structure joining the hulls must be sufficiently high to prevent wave slap and slamming which can be jarring and uncomfortable or even dangerous motions. Anchoring arrangement of catamaran must be carefully considered since the vessel will swing if a single anchor deployed from only of the hulls.

3.2 (b) Trimaran

Trimaran hulls have been increasingly explored where good seakeeping and large deck areas are required. Although the costs might seem to be higher than a catamaran with three hulls required, the usual implementation is more akin to a stabilised monohull where only the centre hull is fitted out with equipment and facilities and the outer 'amah' hulls are small and essentially empty. Similar to catamarans, the height of the tunnel connecting the hulls must be high enough to avoid wave slamming.

Figure 4 – Trimaran ferry (Austal Ships)

A trimaran hull form generally has higher resistance at low speeds due to the increased wetted area and frictional losses of the three hulls when compared to monohulls. At higher speeds the more slender trimaran hulls have lower wave resistance and powering requirements compared to a monohull with similar requirements. However, interference waves form between the hulls that can increase or reduce resistance depending on the vessel speed.

3.2 (c) Small Waterplane Hulls

For smaller vessels operating in rough conditions, sea state 4 to 5, surface waves acting on the water plane area of the hull are primarily responsible for creating rolling and pitching vessel motions. Hulls designed to have Small Waterplane Areas (SWA) are called SWATH or SWASH designs depending on whether they are twin or single hulls. Small waterplane hulls area characterised by torpedo shaped underwater hulls which provide most of the buoyancy with the over water part of the vessel supported by narrow pylon structures which may be continuous or divided into sections.

Figure 5 – SWATH (Platypus Marine Inc.)

It is very common to incorporate some aspect of reduced waterplane area in catamaran or trimaran hulls to improve seakeeping leading to the term semi-SWA designs. Semi-SWA hulls designed for high speed usually only have the forward hulls so shaped, since this is sufficient to reduce pitching motions when the vessel in moving into waves at speed. For an oceanographic vessel a true small waterplane hull design is preferred since the vessel will likely be moving slowly during much of the time, with waves approaching from any direction.

Traditionally, the main drawback of small waterplane hull designs have been the limited space in the underwater hulls to install propulsion machinery and the engine exhaust and ventilation systems. With advancement of diesel-electric propulsion systems this is no longer an issue, since space consuming engine rooms can be arranged topside and only the compact electric motors are needed to be housed in the torpedo hulls. Construction from aluminium or composites is also required to reduce weight and the size of the torpedo hulls to minimise underwater resistance.

Another characteristic of small waterplane hull designs is that they have limited deadweight capability, since a small change in vessel displacement makes a large change in draught. It is common for SWA hull vessels have elaborate ballast systems to adjust for shifting loads and active stabiliser fins to control motion in waves. Some SWATH vessels are designed to operate as conventional catamarans when in transit and then ballast down to become a SWATH when reduced motions in waves is required.

Characteristics / Hull type	Low speed power	High speed power	Motion – low waves	Motion – med. waves	Sea-keeping	Dead-weight	Internal volume	Deck space	Cost
Displacement Monohull	+	-	+/-	+/-	+/-	+	+	-	+
Dynamic Monohull	+	+	-	-	-	+/-	+	-	+
Catamaran	-	+	+/-	-	+/-	+/-	+/-	+	+/-
Trimaran	-	+/-	+	+/-	+/-	+/-	+/-	+	+/-
SWATH	-	-	+	+	+	-	-	+	-

Table 2- Hull form characteristics pros and cons

4 GREEN TECHNOLGIES

4.1 ALTERNATIVE FUELS

Alternative fuels to diesel such as LNG, CNG, LPG and hydrogen were studied, however the availability of both suitable engines and bunkering facilities has meant that conventional diesel engines have been specified.

Fuel-cells were also investigated. These devices combines hydrogen and oxygen to make electricity with water and heat being the only waste products. Despite being a promising green technology, the production, transport and storage of hydrogen fuel remains both limited and inefficient and have not been investigated further.

Regardless, the concept designs presented have the flexibility to accept other fuel types and fuel-cells as these technologies mature and supporting infrastructure becomes available.

4.2 SHAFT GENERATOR / MOTOR

Shaft generators are coupled to the main propulsion engines and power the ship service electrical load more efficiently than using separate diesel generators. The larger diesel propulsion engines have higher thermal efficiency than the smaller generator engines. Furthermore, shaft generators generally allow main propulsion engines to run closer to their optimum efficiency. Shaft generators also reduce engine running hours and therefore maintenance costs on the diesel generator engines.

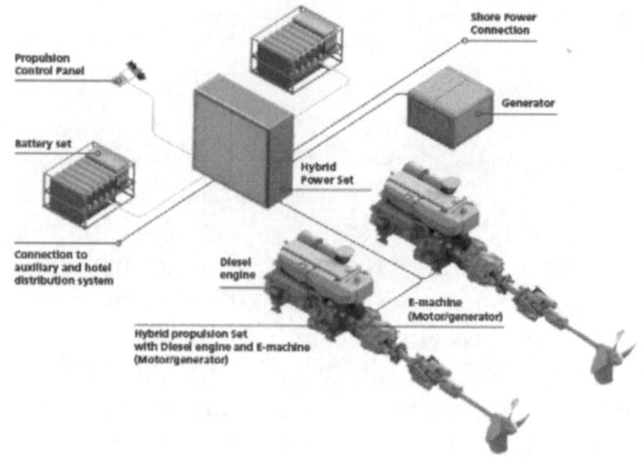

Figure 6 – Hybrid propulsion system

A shaft generator can also be used to form a hybrid propulsion system where the main diesel propulsion engine can be disconnected from the shaft and the vessel is powered using the shaft generator as a motor. The motor can be powered by the other main engine shaft generator, a separate diesel generator or batteries. Despite the various energy conversion losses in such a setup, these options allow much reduced engine running hours with better total efficiency when compared to main engines running at very low loads, such as during manoeuvring and slow trolling operation.

4.3 BATTERIES

Electrical energy storage using chemical batteries was an obvious technology for powering the future AIMS vessel when combined with a hybrid propulsion system and other energy systems such as wind or solar. Batteries can be used to store energy during good harvesting conditions from wind and solar for later use. Batteries can also enable more efficient power generation by creating a charging load that enables generators to run at higher more optimum operating loads for shorter periods instead of inefficient continuous running. This also reduces running hours on generators and therefore maintenance expenses.

The lithium-ion and lithium polymer batteries are key mature technologies that are enabling a clean and energy efficient future. With good energy density, cycle capability and a high round-trip-efficiency of about 90% they are clearly superior to other chemical batteries. The higher cost of lithium-ion batteries has been dropping quickly, driven by the increasing maturity and uptake of electric vehicles and on-site energy storage. From a price of over 1,000 USD per kWh in 2010, *Bloomberg* report that the cost of Li-ion batteries is reducing at 21.6% per year. A cost of under 200 USD per kWh is predicted by the early 2020's.

4.4 SOLAR POWER

Solar power through photovoltaic (PV) cells is another key technology that is enabling a clean and energy efficient future. Lightweight and modular, they are an obvious choice for the future AIMS vessel concept. Although the energy harvesting density is still very low at about 100-200 Watts per square meter, it is envisaged that solar panels can be fitted to the top of the deck house and could even be retractable using lightweight folding structures or rollable PV cells, although these have lower energy density compared to fixed panels.

Figure 7 – Rollable PV solar panels (Renovagen)

A 200 square meter installation would be enough to produce 20-40 kWe of power, which should be enough to cover the normal ship service load or low speed science operations. Excess energy will be stored in batteries for use at night, significantly reducing inefficient generator runtime, fuel consumption and noise. An added benefit of solar energy harvesting is the heat shading it provides, which reduces the heat load on the vessel and lowers the air conditioning load.

The previous high cost of photovoltaic cells has been dropping quickly, driven by the increasing maturity and global uptake of solar power both on-site and mass energy harvesting and distribution. From a price of about 1100 USD per kW in 2010, *Bloomberg* report that the cost of photovoltaic cells is reducing at 24.3% per year. A cost of under 100 USD per kW is predicted by 2020.

4.5 WIND PROPULSION

A number of sail systems designed for auxiliary propulsion were evaluated during the concept study. Such systems reduce loads on conventional propeller driven propulsion systems and thereby save energy. All sail systems have some degree of technical, functional or operational compromises when being applied to an oceanographic vessel. For example, a traditional soft sail Bermuda rig requires a large mast with bracing wires which limit the reach of shipboard handling cranes. The sail systems that provide the least compromises are kite sails and Flettner rotors.

Kite sails, such as those developed by SkySails™, fly dynamically ahead of the vessel providing a towing force. The sail operates at a height of 100-300m above the sea and can operate effectively in winds between 5-20 m/s. Installation requires a telescopic launch and recovery mast and winch system at the vessels bow.

Figure 8 – SkySails wind propulsion system (Skysails)

SkySails claims that a 50m^2 kite is able to supplement up to 200 kW of main engine propulsion power in good wind conditions. However, studies have shown that actual average power achieved over outgoing and returning voyages is much lower, varying between 10 and 25% in real world conditions, since kites cannot operate effectively in head winds.

Other potential benefits of kite sails are the capability to aerially deploy lightweight cameras and other sensors with the kite for observation or other scientific purposes. The kite can also prominently display logos and branding for public outreach promoting green technologies and the work of AIMS.

Flettner rotors are vertical rotating cylindrical sails that create a lifting force normal to the wind direction. Simple in concept, they are one of the least obtrusive sail technologies due to their compact footprint and rigid construction. A Flettner rotor of about 20m^2 in cross section can provide up to 200 kW propulsion power in good wind conditions. However, a significant ratio of the power is required to drive the motors that spin the sails which varies with the wind strength and direction.

Figure 9 – Flettner rotor sail system (Enecon)

Although they provide powerful lifting forces for their size, the Flettner rotors main limitation is that they can only operate effectively when the wind is from a beam direction. In head winds the rotors have a negative effect since they cannot be reefed (sail in stowed condition) and add air resistance to the vessel. Studies have shown that the actual average power that can be achieved is usually only about 12% of the maximum in real world conditions. However, compared to kite sails, Flettner rotors will provide more consistent performance.

4.6 VARIABLE FREQUENCY DRIVES

Variable frequency drives (VFDs) are electrical devices for controlling the supply of power to electric motors. They enable precise control over motor torque and speed and greatly reduce the starting current. A VFD can save considerable energy by enabling pumps and fans to run at a speed that meets the demand of the system instead of a fixed speed.

For example, engine room ventilation fans can be speed controlled by sensors to meet the required temperature and pressure conditions in the engine room. The air flow required will vary considerably depending on ambient temperature and engine running load. A fan without VFD needs to run at full load all the time to meet a design condition that may only happen occasionally, if at all.

Another benefit of VFDs is in managing starting loads. Typically, a motor draws about six times the full load current when starting up DOL (Direct On-Line) without a starting device. This is not normally a problem for smaller motors, however, larger motors driving big pumps or thrusters need to have a starting device to prevent overloading of the vessels electrical generator system.

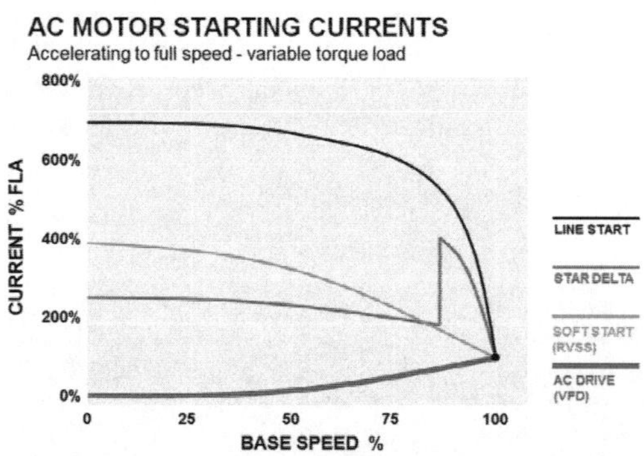

Figure 10 – AC motor starting devices

A VFD allows starting of larger consumers such as big pumps and compressors without oversizing of the generators or needing to bring another generator on-line to cover the starting current. Therefore the generators and engines can be sized for maximum efficiency rather than needing to consider starting loads.

5 BASELINE DESIGN

The Baseline Design is a like-for-like replacement or addition to the current AIMS fleet. As with the existing AIMS vessels, RV Cape Ferguson and RV Solander, the base design is a steel monohull similar to fishing trawler.

The requirements from AIMS vessel user groups has strongly indicated a desire for a larger vessel. The length of the vessel has been kept to under 35 meters in order to limit the crewing numbers and qualifications in accordance with AMSA requirements. Although the shallow water requirement is not particularly important for future requirements, the draught has also been kept to under 2.8 meters.

The Base Design has a conventional twin diesel mechanical propulsion system with controllable pitch propellers similar to the existing vessels. To enhance manoeuvrability, a bow tunnel thruster is included.

Design Length	34.9m
Moulded beam	9.0m
Hull depth	4.4m
Design draught	2.8m
Displacement	450 tons
Deadweight	110 tons
Gross tonnage	480 GT

Table 3- Base design principle particulars

Compared to the RV Solander, the vessel beam has been increased slightly from 8.6 to 9.0 meters to provide more useable space and stability. The upper deck is increased slightly to accommodate two more crew rooms and the wheelhouse moved up half a deck.

The resulting increase in useable volume has allowed the following improvements when compared to the existing vessels:

- All cabins have own toilet and shower facilities
- The lounge and mess have separate seating
- The internal science area is increased by 30% (from 50 to 65m^2)
- Provisioning from the either end of the vessel

Other internal improvements include:
- Stairs instead of ladders
- Circular access to lower accommodation
- Raised bridge for increased visibility

Exterior improvements include:
- 5 meter rescue boat with own davit
- Larger crane – (from 10 to 30 ton meters)
- Safer tender boarding from midship
- Large hatch for engine removal
- Dropkeel for acoustic instruments (sensors up to 300mm diameter)

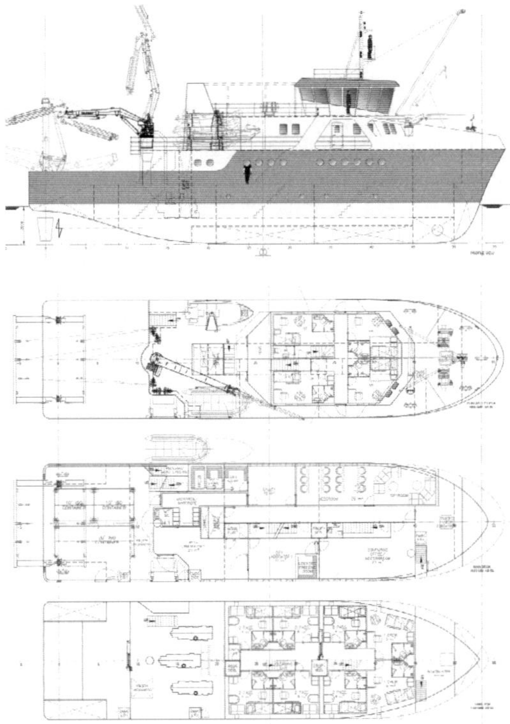

Figure 11 – Baseline Design General Arrangement

6 GREEN DESIGN

The Green Design option has the same platform and arrangement as the Base Design with very similar principle particulars (see previous table), however it includes addition equipment to enable more efficient and environmentally friendly operation.

6.1 OPERATING COSTS

The Green Design has similar crewing requirements compared to the Base Design and RV Solander and therefore no significant changes to crewing costs.

Fuel costs and maintenance costs can be expected to be considerably lower with the energy saving systems installed in the Green Design.

6.2 POWER GENERATION SAVINGS

Shaft generators will enable the vessel to generate electricity from the main engines while cruising. This has three benefits; enabling the main engines to run at a higher and more efficient load, using main engines to generate electricity is more efficient than the smaller genset engines and reducing the running hours on the generator engines.

Batteries can be charged from the shaft generators while cruising or by the gensets when at anchor. This ensures that the gensets can be always loaded highly for more efficient running and gensets can cycle off periodically to reduce running hours and maintenance costs. A battery system of 200 kWh should allow the vessel to be powered over night without needing to run the gensets.

Solar photovoltaic panels will significantly reduce the power that is needed to be generated using diesel fuel. With the monohull Base Design configuration, about 10 kWe of fixed panels is possible and an additional 10 kWe of rollable panels deployed with the A-frame for a total of 20 kWe. On a typical day, it should be possible to harvest between 80 and 100 kWh. This is equivalent to a fuel saving of 10,000 to 12,000 litres per year. This should payback in about four years at an installation cost of 2 kAUD per kWe.

6.3 POWER CONSUMPTION SAVINGS

An additional benefit of solar panels will be to insulate the vessel from solar radiation and reduce the solar heat gain that would need to be dissipated by the air conditioning system. This is expected to reduce the power required by the air conditioning system significantly. Rollable solar panels deployed by the A-frame will shade the open deck areas reducing sun exposure and improving working conditions.

Inverter control on heavy consumers such as thruster motors, air conditioning compressors and large pumps will allow lower starting currents and capacity control to better meet load demands. The main benefit, besides lower power usage, is to enable smaller size generators to be fitted which will operate at closer to their limit at a more efficient specific fuel consumption. For example, the generator sets on the existing RV Solander are rated at about 135 kWe, compared to the maximum continuous running load of about 95 kWe and a typical daytime load of 55 kWe and night load of 15 kWe.

7 X-FACTOR DESIGN

A number of different hull platforms other than monohulls were investigated, notably catamarans,

SWATH and trimarans, since they offer increased deck space desirable for science operations. Low motions and accelerations are crucial for the science work and standard catamaran hulls were eliminated due to motions and accelerations in the expected wave spectrum of tidal littoral waters.

The SWATH platform has very good motion characteristics for a research vessel however, the required deadweight of about 100 tons would be difficult to meet. The light and loaded draught variation would be nearly two metres with the deepest at over 4 meters. Another limitation of the SWATH platform for a research vessel is the limited useful volume in the hulls. For this size vessel, all accommodation and science spaces would need to be located above the main deck resulting in a significantly larger superstructure.

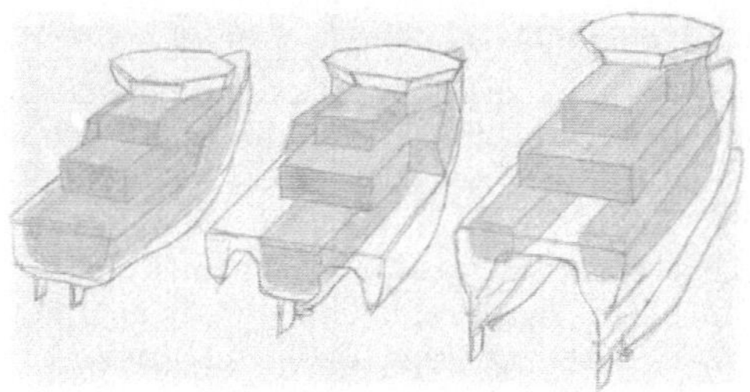

Figure 12 – Hull platform useable volume (left to right: monohull, trimaran and SWATH)

A trimaran hull form and aluminium construction has been chosen for the X-Factor Design to provide both a larger and lighter platform for increased capability within the 35 metre length constraint. The trimaran platform provides the large deck area and stability of a catamaran with significantly better options to control motions in waves by tuning of the side hull shape and location. Furthermore, unlike catamarans, anti-roll tanks or stabiliser fins can be applied effectively to a trimaran platform if active stabilisation is desirable.

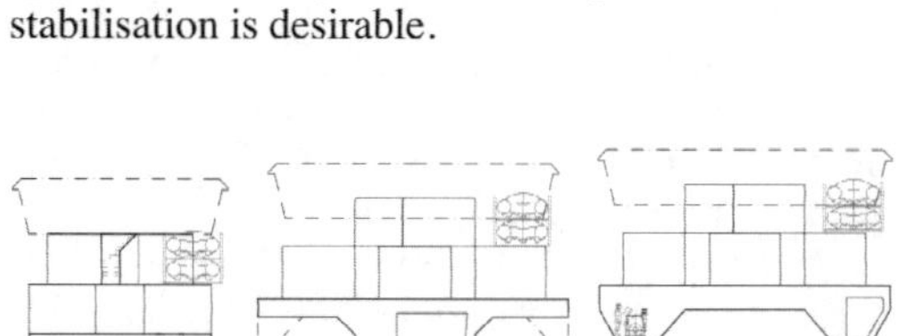

Figure 13 – Hull platform cross section (left to right: monohull, trimaran and SWATH)

It is also recognised that there is likely be a requirement for the vessel to be built in Australia and there are a number of Australian builders that have experience with trimaran design and construction.

Design Length	34.9m
Moulded beam	15.6m
Hull depth	6.4m
Design draught	3.0m
Displacement	325 tons
Deadweight	85 tons
Gross tonnage	605 GT

Table 4- X-Factor design principle particulars

7.1 PLATFORM IMPROVEMENTS

The trimaran hull form provides for a more stable science platform in the light to moderate wave conditions that the vessel will predominately operate in. The tunnel height of over 2.5 meters will allow transit in Sea State 4 without significant tunnel slamming. The finer bow form will reduce bow slamming when penetrating heavier seas. Further seakeeping analysis however, is required to ensure that accelerations and motions due to coupled roll and pitching motions in heavier seas are acceptable.

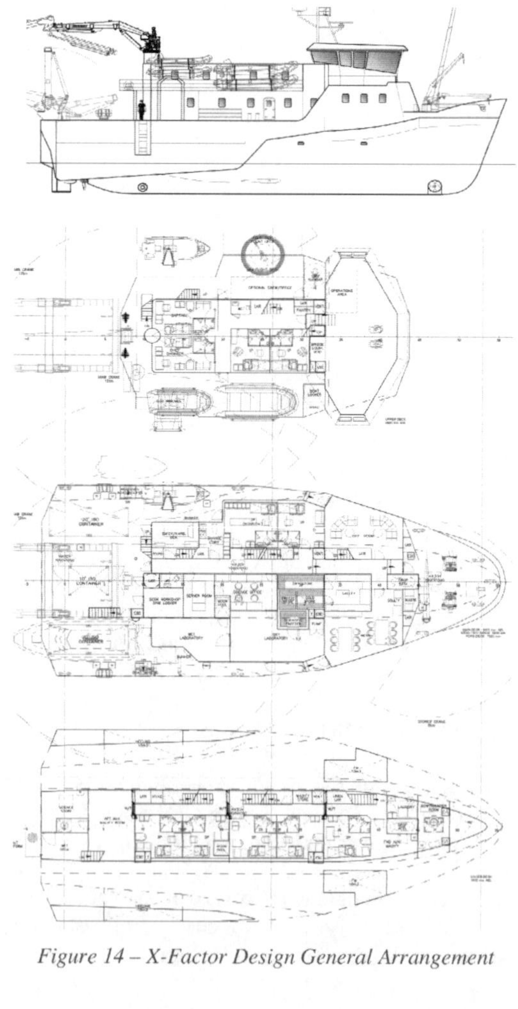

Figure 14 – X-Factor Design General Arrangement

Compared to the existing RV Solander and Base Design, the X-Factor Design provides increased capability as can be seen in the following table.

	RV Solander	Base Design	X-Factor Design
Internal science area	59m^2	68m^2	79m^2
Aft working deck	70m^2	70m^2	100m^2
Container bays	1 x 10 ft	2 x 20 ft	3 x 20 ft
Rescue boat (with davit)	None	1 x 5.0m	1 x 5.0m
RHIB tenders	2 x 4.6m, 2 x 5.8m	3 x 5.8m	2 x 4.6m, 2 x 5.8m
Main crane	1t @ 9m	2.5t @ 13m	4.0t @ 15m
Moon pool	0.9 x 0.9m	1.0 x 1.0m	1.0 x 1.0m
Drop Keel (length)	None	1.0 metre	2.0 metre
UAV flight deck	None	None	6 x 6m
Cabins	2 single, 8 double	6 single, 6 double	6 single, 6 double
Sick bay	None	None	Main deck
Cruise speed	11 kn	11 kn	13 kn

Table 5- Comparison of vessel capabilities and facilities

The larger deck area also allows for the sick bay to located on the main deck with level access to working areas for improved trauma response.

7.2 OPERATING COSTS

The X-Factor Design has similar crewing requirements compared to the Base Design and RV Solander and therefore no significant changes to crewing costs are expected.

Fuel costs can be expected to be considerably lower with the same energy saving systems installed as the Green Design. With the larger trimaran platform the amount of solar panels can be doubled to about 40 kWe reducing fuel consumption further.

In addition, the lighter weight of the X-Factor aluminium structure and the reduced wave resistance of the trimaran hull form will also reduce fuel consumption significantly.

It is estimated that diesel fuel savings will in the order of 30 to 40% less than AIMS existing vessel RV Solander as calculated in the table below.

	RV Solander	Green Design	X-Factor Design
Electrical load (kWe)	35	29	28
Avg solar power (kWe)	0	4	8
Operating hrs per year (h)	6,000	6,000	6,000
Generator SFC (L/kWeh)	0.32	0.30	0.28
Generator fuel per year (L)	67,200	45,000	33,600
Avg propulsion (kW)	450	480	420
Avg kite sail assist (kW)	0	35	35
Propulsion SFC (L/kWh)	0.29	0.26	0.26
Steaming per year (nm)	20,000	20,000	20,000
Steaming per year (h)	2,400	2,400	2,000
Avg speed (kns)	8.3	8.3	10
Propulsion fuel per year (L)	313,000	278,000	200,200
Total fuel per year (L)	380,000	323,000	233,800
Fuel saving per year (%)	0	15%	39%

Table 6- Comparison of fuel consumption

Maintenance costs can also be expected to be considerably lower with the same energy saving systems installed as the Green Design. By utilising shaft generators powered by the main engines, the auxiliary engines do not need to run when the vessel is steaming. When the vessel is anchored or in port, battery storage allows gensets to run at a higher load for greater efficiency and for only half the time. A further benefit is not needing to run engines while at anchor overnight.

The X-Factor Design propulsion system, which has two propulsion motors driving a single shaft, can run efficiently on a single motor whereas the existing RV Solander and the Base Design must run both shafts for efficiency. This allows the X-Factor Design to run more efficient single main engine operations when steaming at lower speeds thus reducing engine running hours significantly. The battery system also has the capacity to allow a number of hours of low speed silent steaming without any engines running.

	RV Solander	Green Design	X-Factor Design
Aux. engine hours per year	6,000	3,600	1,800
Main engine hours per year	2,400	2,400	1,500

Table 7- Comparison of engine running hours

8 3D MODELLING & DESIGN REVIEW

Development of the hull form was undertaken in Rhinoceros™ in order to better understand the greater complexity of the trimaran hull form and its structural requirements. Main structure was also modelled simply so that a good estimate of the vessel lightweight and centre of gravity could be calculated to assist in the development of the hull form.

The basic 3D model was then brought into ShipSpace™, a virtual reality design review and collaboration tool, to check the usability and accessibility of the many compartments and voids. This was particularly useful in the foreship and aft shoulders regions where waterlines change significantly between decks. Access into the side amah hulls and joining tunnel structure was also checked.

Once the viability of the hull form was confirmed, the 3D model was developed further in 3D Studio Max™ into a full concept design model with representative exterior details. This concept model will be used by AIMS for various internal and external communication and promotion purposes where a realistic visual impression of the future vessel is needed.

Interior details have also been added to the vessel concept model to enable AIMS scientists and other stakeholders to review and comment of the vessel arrangement from a human factor engineering perspective. The ShipSpace™ system enables users to explore and investigate the entire vessel in virtual reality, which has proven to enable more accurate, thorough and much richer feedback from stakeholders and subject matter experts when compared to looking at drawings or pictures of the vessel. Users often comment that the convincing experience provided by the ShipSpace™ system is akin to being on board a real vessel.

9 FUTURE WORK

Investigations into hull platforms suitable for littoral research vessel has shown that the trimaran platform offers a larger deck area when compared to standard monohulls and without the deadweight and draught issues of a SWATH platform.

The next phase of the project will involve developing the concept to a tender design package. Further calculation on the structure using FEM will inform lightweight and hull form. CFD analysis of the hull form will be used to refine main hull lines and location of the side hulls.

Workshops using ShipSpace™ are proposed to enable stakeholders to provide feedback into the vessel arrangement and layout of the work spaces.

10 LIST OF TABLES & FIGURES

11 AUTHORS BIOGRAPHY

Ken Goh holds the current position of General Manager, at Knud E. Hansen Australia and specialises in mechanical engineering and advanced concepts. His project experience includes icebreakers, research and naval vessels. He is also active in the development of autonomous and wind powered vessels.

DAMEN SHIPYARDS – HARDWARE IN THE LOOP – ADVANCED SIMULATION TECHNOLOGY

Damen is utilising advanced simulation techniques to digitally optimise the ship development and design process. The technology is called Hardware in the Loop (HIL), and the company is using it as a tool to improve system integration and controls design before construction even begins.

"During commissioning and sea trials of complex vessels we have often experienced problems with interfacing and software functionality. The main reason for these issues is that the first time that all systems come together is on-board. Integrated functionality is most of the time not checked or to a very limited extent.

A HIL platform offers us a way to face such challenges at an earlier stage, thereby resolving issues quicker, in a better manner and in the comfort of our office.

PRECISION SIMULATION

HIL is a platform that enables testing of control systems. The machinery and physical systems connected to the controller that is being investigated are replaced by simulations.

"We can select a control system from a vessel and 'put it in the loop' – a simulation environment that allows us to fully investigate the behaviour and performance of that component before we apply it to a full-scale application. The HIL simulation shows us how the controller responds to real operational scenarios."

"By using simulation technology that can predict performance during defined conditions and operations, we are able to offer our clients extra value. This will allow them to differentiate themselves in their specific market by adding additional workability."

Looking at other production-based industries, such as automotive and aeronautical, the use of HIL has been seen to save substantial amounts of time and money. "We saw this and it worked well – a little bit like Kommer Damen's original approach to standardisation after seeing its virtues in the automotive industry."

Damen very quickly understood HIL's potential in its own operations: "Initially, our idea of applying HIL was for risk management purposes, but we realised that we could also apply this not just to avoid errors, but to improve performance," she notes.

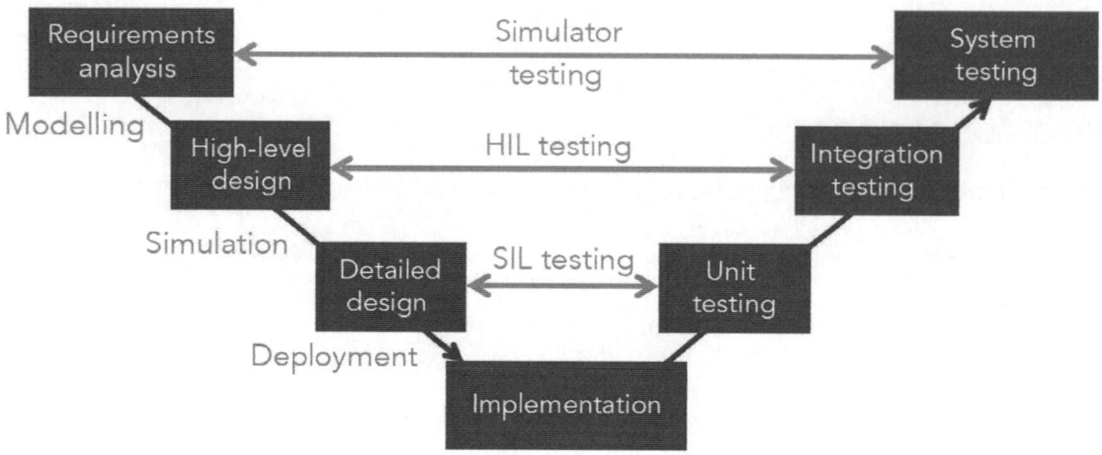

FIRST PROJECTS

And so began Damen's work with HIL. "One of the first applications was on Bibby Marine Services' Services Operations Vessel *Bibby WaveMaster 1* implemented in MARIN's simulator. In fact, this has been one of the key aspects of this project – to collaborate closely with suppliers and clients, as well as institutes like MARIN, TU Delft etc."

In parallel with this 'walk to work' application, Damen decided to invest in its own HIL set-up and a research project was launched to investigate the potential of a real-time simulation environment – making the company one of the few shipbuilding companies in the world to apply HIL.

"The pilot application of the HIL platform was on the Multi-role Aviation Training Vessel *MV Sycamore* for the Royal Australian Navy, from which we had a good response. Trials went well and everyone was very enthousiaste."

The simulation model runs on the processor hardware of dSPACE. The simulator generates and measures I/O signals via integrated I/O boards. The number of components, and their types, depend on the specific requirements of the project and are configured to fit each application. The implementation software where the interfacing takes place is the Real-Time Interface (RTI). This software is the link between the dSpace hardware and the modelling software of Matlab Simulink.
The first commercial application involved an Offshore Patrol Vessel: "This vessel has two engines per gearbox and a whole array of operational modes and changeovers in addition to two main controllers on the propulsion train from different suppliers. These were brought to Damen and hooked up to the simulator in order to perform

software logics checks and test the integrated functionalities." The initial findings from these testing sessions were positive: a large number of issues were identified and resolved, all of which would have significantly hindered the commissioning procedure and acceptance trials.

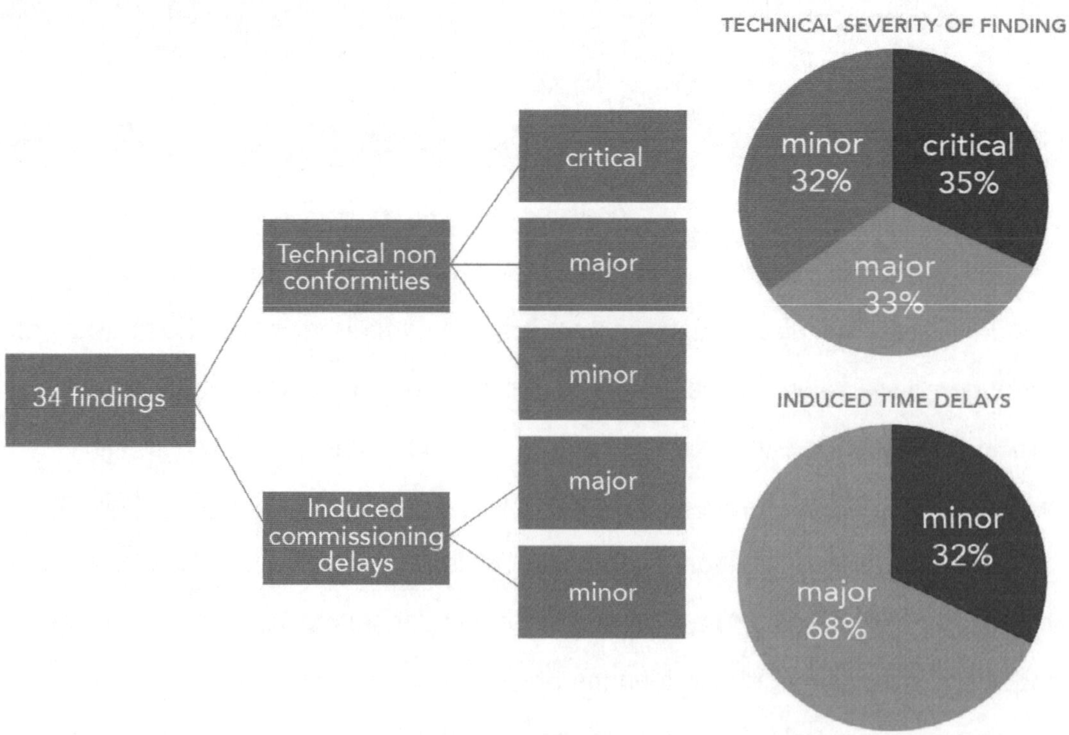

Charter strategies

Focusing clearly on system integration, these projects illustrate how broad HIL technology can be applied. Further examples include:

- Propulsion control systems
- Engine limits/ load control
- DP systems
- System integration testing
- Power management systems
- Hybrid systems

In terms of potential benefits to Damen's customers, HIL technology has three major advantages.

1. Increased fail safe capability of the ship's systems by testing failures and off design conditions and tuning the responses.
2. Enhance quality, save time and money during testing: specified weather conditions can be simulated like in the case of the W2W.
3. Human factor: training ad familiarisation if commissioning engineers and other system users.

PERFORMANCE CONTRACTING

"Performance optimisation is also a part of this; we can offer this to clients wanting a different type of vessel. BC Ferries, for example, were looking to do something more environmentally friendly, and they wanted to make sure it would also be cost-effective. Their question was 'can you guarantee the predictability of a complex system?' The HIL simulator was the way to answer this."

With an eye on future projects and future clients, Ms Xepapa highlights Damen's intention to move towards so-called '**performance contracting**'.

"We often contract on technical specifications, but in the future we could offer a more targeted contract based on the operator's requirements. This will be a business model of the future."

HARDWARE-IN-THE-LOOP SIMULATION

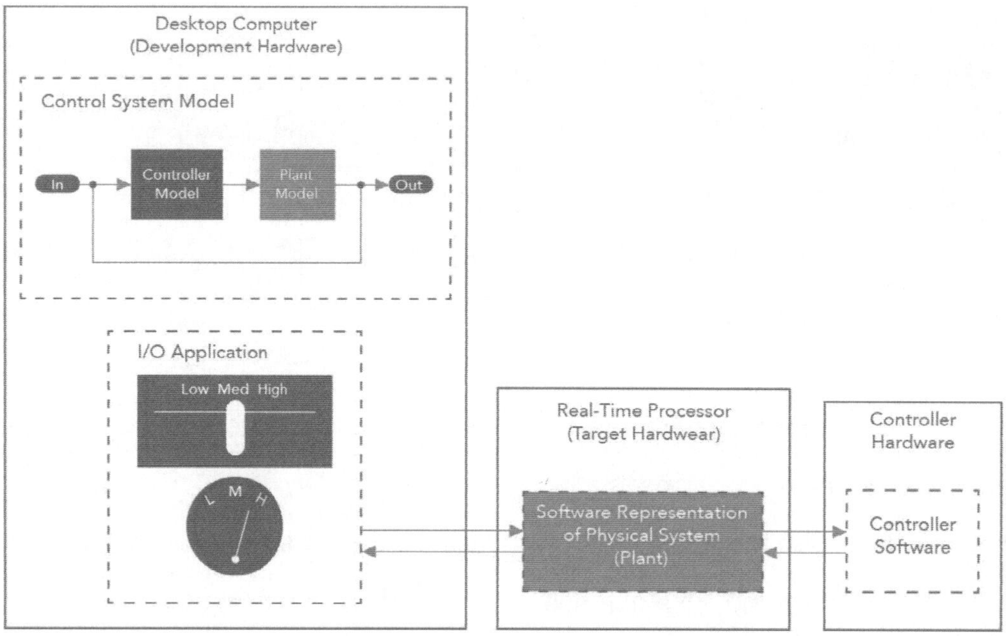

Multi-vessel Sea State Estimation Utilising Swarm Shepherding

Nathan K. Long[1,2], Daniel Sgarioto[1], Matthew Garratt[2], Karl Sammut[1,3], Hussein Abbass[2]
[1]Defence Science and Technology Group, Melbourne, Australia
[2]UNSW Canberra, Canberra, Australia
[3]Flinders University, Adelaide, Australia
nathan.k.long.91@gmail.com

ABSTRACT

Sea State Estimation (SSE) remains critical to maritime operations worldwide. Presently, data on sea states is primarily collected by moored wave buoys in a relatively limited number of locations. Recently, research into the use of Uninhabited Surface Vessels (USVs), acting as an on-site wave buoy, has emerged. By measuring the response of the USV to wave motions and implementing a wave buoy analogy model, directional wave properties can be estimated.

This study investigates the benefits of utilising an intelligent swarm of USVs (greater than two), with each member acting as a wave buoy, to estimate an aggregated sea state in real-time. Swarm intelligence can be defined as an interaction between relatively simple agents leading to the emergence of complex behaviours. It is hypothesised that a USV swarm can collaborate to enhance sea state characterisation. The challenge of swarm guidance is addressed by introducing a shepherd which guides the swarm of USVs with the intent of achieving the desired objective estimation. This method of control mimics the way a sheepdog is able to herd sheep to a target destination.

A heuristic is presented which describes how a series of influential force vectors can generate swarm behaviours with the aid of a shepherd, and be used to distribute a network for the purpose of estimating the properties of a complex sea environment.

The ability to estimate wave properties, with a higher degree of confidence, in-situ in any geographic region would allow maritime vessels to enhance operability, improve safety, and extend service life.

1 INTRODUCTION

The effectiveness, efficiency, and safety of global maritime operations are reliant on having accurate knowledge of the prevailing sea states which affect marine platforms. Wave conditions have a direct impact on a vessel's fuel efficiency [1], crew safety, security and comfort [2], and cause damage to marine structures due to wave-induced loads [3]. Ship-to-ship, ship-to-coast, and ship-to-infrastructure interactions, in particular, require certain levels of sustained quiescence to avoid potentially catastrophic impacts. Additionally, with advances in automated control for maritime vehicles, it becomes vital for autonomous navigation systems to take into account ocean states to restrict the pitching and rolling motion of vessels, an acquired skill developed by experienced human pilots, for effective operation. Regions containing complex wave spectrum pose a particularly difficult problem for maritime navigation, especially those found along the coast or around ocean-based infrastructure. Examples of industries which could be affected by unknown rough sea states include cargo transportation, fishing, defence, oil and gas exploration and drilling, and offshore wind farming.

While numerous SSE techniques exist, such as the use of wave buoys [4], ship-based wave radar [5], and satellite altimetry [6], each encounter significant challenges as they rely on ad-hoc non-coordinated measurement or sensor distribution. Errors, uncertainties, and the lack of a clear approach to covering a range of local and regional sea spaces afflict current techniques. Therefore, a novel method for a dynamic, collaborative, and self-organising SSE mechanism would fundamentally augment the maritime industry's operational capabilities by permitting enhanced real-time, in-situ wave property data, especially for complex wave system identification.

To solve the problem of realising an in-situ, real-time comprehensive knowledge base of wave properties in complex environments, a network of SSE agents is proposed. The intricacies of sea states around coasts and ocean-based infrastructure elicits a cooperative approach which would permit an aggregated wave property distribution to be formed.

In order to distribute the network, a swarm (here defined as more than two agents) of self-propelling SSE devices offer a solution. Researchers have recently developed an SSE technique which utilises the ship as a wave buoy analogy; the motion of the ship itself is used to estimate the wave properties via a transfer function. Therefore, employing a swarm USVs, each utilising the wave buoy analogy, would enable the creation of a dynamic network of agile SSE devices.

However, a cooperative approach has its own costs. Cooperation could create severe demands on communication and complexities associated with multi-party negotiation. The latter could require high computational resources that are not expected to exist on a wave buoy. To scale the system, a level of centralisation of command is unavoidable, whereby airborne or surface parent vessels with sufficient computational powers could govern the low-cost distributed sensor network.

Therefore, the challenge of swarm guidance can be addressed by introducing a shepherd which can be used to effectively oversee and organise the network of sea state estimating agents. Shepherding guidance is a bio-inspired swarm control method derived from the behaviour exhibited by sheepdogs when herding sheep. Shepherding is suited to the problem due to its control technique being reliant on a simple library of primitive behaviours that collectively could generate complex guidance patterns. In its biological form, shepherding relies on its evolution based on collecting and driving a herd of agents (in nature, sheep) to an objective location, as well as introducing a simple distributed swarm control mechanism that can be implemented autonomously, in real-time, react to a dynamic environment, and permit effective agent self-organisation.

It is further proposed to use an Uninhabited Aerial Vehicle (UAV) as the shepherd to guide the swarm of USVs in order to obtain a combined SSE of a selected area. Benefits of utilising a UAV, alongside the USVs, include that situational awareness of the desired region can be maintained, communication with swarm members can be sustained, and it also enables the shepherd to fly over the top of obstacles throughout the SSE task. Figure 1 shows an example illustration of the UAV shepherd flying above obstacles in an ocean wave environment, while monitoring and guiding the USV swarm. Such a UAV does not necessarily need to be a low-flying platform as this is problematic in environments with harsh weather conditions. Moreover, the concept is extendible to a surface vessel acting as the sheepdog in place of the UAV.

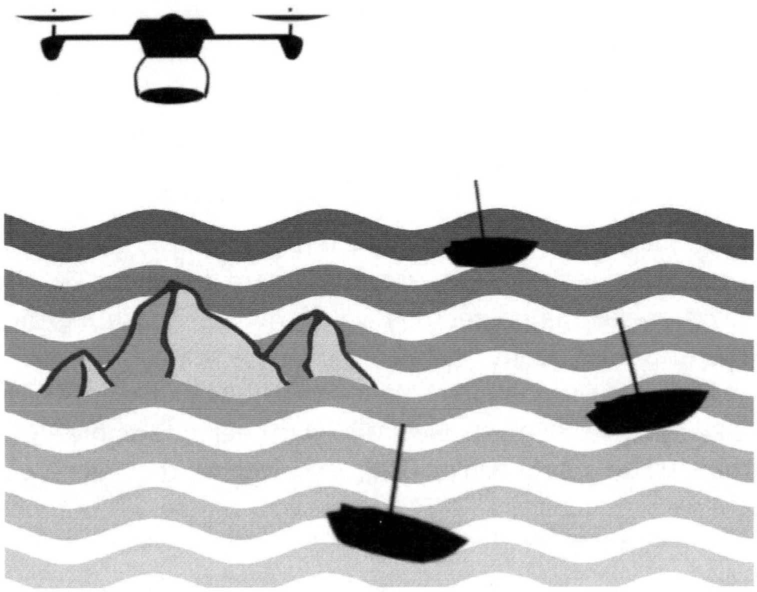

Figure 1 Illustration of a UAV shepherd guiding a swarm of USV agents on the ocean, while flying over an obstacle in the environment.

We hypothesise that a coordinated, distributed network of SSE devices could better capture complex wave system properties than any single- or multi-vessel non-coordinated SSE setup. This study gives a description of how the multi-vessel SSE concept of operation can be effectively carried out by employing a swarm shepherding heuristic.

The background literature contributing to this research problem definition is given in Section 2, followed by a description of the research gap and an outline of the SSE procedure in Section 3. Section 4 presents the design of the heuristic, and envisioned future research and the outcome if successfully implemented is explored in Section 5.

2 BACKGROUND RESEARCH

2.1 Sea State Estimation

SSE is a problem concerned with the estimation of wave properties such as wave height, frequency, and direction. Numerous methods for SSE are already being employed globally to improve maritime operations, aid in weather forecasting, for the study of oceanography, and to help understand coastal degradation. Popular techniques include the use of wave buoys, ship-based radar, satellite remote sensing and Global Navigation Satellite System (GNSS) geodesy, and an emerging method based on the ship as a wave buoy analogy.

The most common method of taking sea state measurements is via wave buoys; floating sensor-mounted platforms which are primarily located along continental margins [7]. Although wave buoys are useful for monitoring global sea conditions, they are few and far apart, and contain no means of locomotion, thus making them impractical for dynamic in-situ operations. Ship based radar systems solve this problem by allowing measurements to be taken during ocean transit, however, the radar requires frequent calibration, and often requires computationally heavy algorithms to interpret the radar data [8].

Another category of SSE methods employs the use of satellites to remotely detect sea surface properties, with two sub-branches being satellite altimetry (e.g. [6]) and GNSS geodesy (e.g. [4]). Satellite altimetry encompasses the procedures which involve measuring the time taken for a radar pulse, transmitted from a satellite, to reflect from the sea surface and travel back to the satellite or a separate receiver. The spatial resolution of satellite altimetry, however, is generally quite crude (on the order of 10-100km), as well as having coarse temporal resolutions [9]. GNSS geodesy uses the triangulation between a GNSS receiver at the ocean surface and global navigation satellites, however, requires a GNSS reference station for calibration, or an abundance of information with respect to the geoid and tides [10].

The use of the ship as a wave buoy analogy, whereby ship motions are passed into a transfer function able to estimate the causal wave properties, is a method for SSE which shows promise due to its adaptability to autonomous systems and flexibility in application. If wave conditions are assumed to be moderate, then the 6-degree-of-freedom motion of a ship, and resulting loads imparted on the hull can be described by an approximately linear relationship with respect to the causal incident waves [11]. This relationship can be elaborated to consider wave systems with multiple wave directions and properties. Using this information, a ship-specific transfer function can be formulated which takes into account ship characteristics such as form and loading condition.

While most of the work focussed on the ship as a wave buoy method has centred around wave analysis in the frequency domain, for real-time SSE to be made in-situ and on-demand, solving SSE in the time domain remains more viable as it only requires current vessel motion measurements (the frequency domain solution requires previous response history) [11]. Different approaches to estimating sea states in the time domain have been undertaken. Pascoal and Soares [12] developed a Kalman filtering-based algorithm for SSE, while Nielsen et al. [13] created a stepwise procedure for obtaining sea state information in the time-domain. This paper assumes that the SSE technique is refined enough to obtain sufficient sea state estimates and, rather, focusses on the USV swarm control and distribution framework. For more information on SSE methods, please refer to a survey written by Nielsen [11] on SSE techniques using the ship as a wave buoy analogy.

A further study by Nielsen et al. [14] is the only one to date to utilise multiple vessels for SSE. The authors mention the use of a network of SSE platforms to improve SSE performance; however, they focus on the use of multiple ships for the wave buoy analogy in their study. Nielsen et al. considered the SSE task to be performed by ships already at sea completing different operations, and as a result, cannot be dynamically employed for localised estimations. Simulations involving three different types of ship encountering a temporally fixed wave system were run (see Figure 2), where the SSE of each ship was weighted based on its dimensions and geometry, then combined. Overall, the weighted aggregate estimates performed better than the non-weighted or single vessel weighted estimates. The study, thus, supports the notion that multiple vessels can be used to improve upon in-situ SSE while at sea.

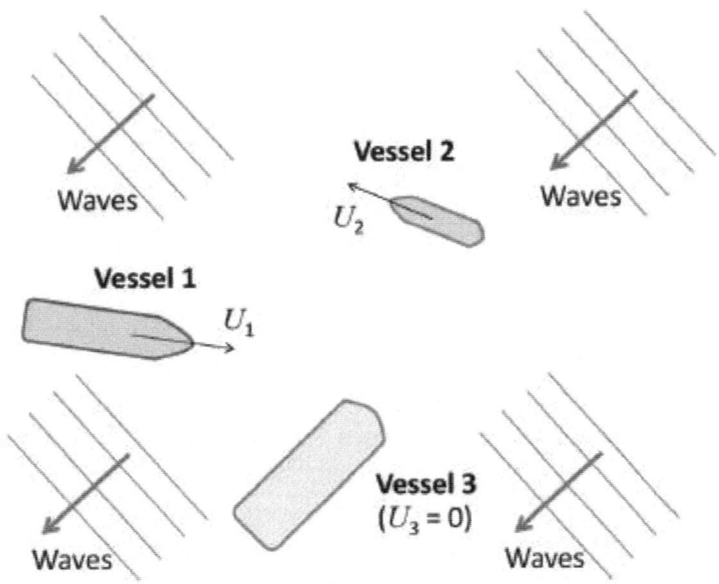

Figure 2 Multi-vessel SSE operational setup by Nielsen et al. [14]. Reprinted from Applied Ocean Research, Vol 83, U. D. Nielsen, A. H. Brodtkorb, and A. J. Sørensen, Sea state estimation using multiple ships simultaneously as sailing wave buoys, Page 71, Copyright (2019), with permission from Elsevier.

2.2 Swarm Shepherding

Swarm intelligence arises when a series of local interactions between a group of relatively simple individuals exhibit complex nonlinear behaviour at the global level. The intelligence of a swarm is normally defined as an emergent property; it manifests without any external control or regulation [15]. The benefits of a swarm intelligence system, such as in a swarm robotics system (in this case, the robots are USVs), include that each agent is simple to create and program, the system is highly robust as the failure of one agent does not necessarily lead to the failure of the system. The unit cost of each agent is relatively cheap to allow the system to be scalable, and permit a high degree of flexibility to the system.

Reynolds [16] introduced one of the first swarm control frameworks in 1987 for the purpose of controlling a flock of simulated birds. Reynolds developed an advanced particle system model which represented each bird agent as a boid; an object which has coordinate and orientation attributes. The boids' motion was then dictated by several behaviours: collision avoidance, velocity matching, and flock centering. Collision avoidance is needed so that flock members do not collide with each other while swarming, as well as to navigate around obstacles. Velocity matching was used to keep the flock of bird agents moving in unison, and flock centring stopped the birds from scattering apart. These behaviours were enacted using a series of acceleration vectors.

A swarm control technique, known as swarm shepherding, was later developed based on the sheep herding method learnt by sheepdogs. Strömbom et al. [17] created a heuristic which mimics the approach taken by a sheepdog when herding sheep to a target destination. Two main herding actions are defined: driving and collecting. Driving is done when the sheep herd is dense enough, where the sheepdog positions itself behind the sheep to guide them in the opposite direction. Collection is undertaken when a sheep strays too far from the centre of mass of the herd. Strömbom et al. implement Reynolds' boid model in their study, where each herd agent is defined as a boid, and manoeuvred by behaviours which are enacted via force vectors. Three behaviours are described for sheep herding: shepherd-to-sheep repulsion, sheep-to-sheep repulsion, and sheep attraction to the local centre of mass. The results of a series of simulations showed that the simulated shepherd and sheep motion was similar to that of the real-life herding task.

Shepherding, as a swarm control technique, has been expanded beyond the sheep herding task. Gee and Abbass [18] define shepherding as "externally influencing a swarm or a group of agents to guide them towards the Shepherd's goal". More abstract swarm control problems have, thus, emerged which take advantage of swarm shepherding. Chaimowicz and Kumar [19] ran a number of experiments involving a swarm of Uninhabited Ground Vehicles (UGVs) which explored an unknown urban terrain. In order to control the swarm, multiple UAVs were employed as 'aerial shepherds', such that both aerial and ground views (taken from vehicle-mounted cameras) could be combined to create a more comprehensive image of the environment. Two main behaviours were used: merging and splitting. Splitting involved a larger group of UGVs splitting into smaller sub-groups, and merging was the converse, where sub-groups merged into a larger group. Each group or sub-group was designated its own aerial shepherd which controlled the splitting and merging behaviours, which allowed for more robust navigation of the environment.

Clayton and Abbass [20], however, discuss how methodologies attempting to describe shepherding tasks (such as that of Strömbom et al. [17] and Chaimowicz and Kumar [19]) oversimplify the complex shepherd-swarm interactions, and are unable to be applied to general herding problems. Therefore, the application of shepherding to the SSE problem in a complex environment requires a new heuristic to be developed for that purpose.

3 HEURISTIC DESIGN

A boid representation of the swarm agents, as developed by Reynolds [16] and used by Strömbom et al. [17], is proposed for this study to define the USV swarm agents. Each USV will be represented as an object with three-dimensional coordinates and an orientation. The USV boids will react to four main behaviours: shepherd-to-swarm-agent repulsion, swarm-agent-to-swarm-agent repulsion, obstacle repulsion, and target location attraction. The relative location and status of each swarm member dictates what actions are taken at each time step. Each behaviour is enacted via a force vector which is activated if a swarm member falls within a radius of influence surrounding the shepherd, other swarm agents, obstacles, or the SSE target locations. Figure 3 shows an example of the shepherd and swarm agents interacting within an environment. The radii of influence (here, repulsion radii) are represented by the circular shadows surrounding the shepherd and swarm agents. This example gives a simplified approximation of the interactions between the shepherd and swarm agents. In the real case, the radii of influence will extend much further from each agent, and the agents will not remain stationary across time steps.

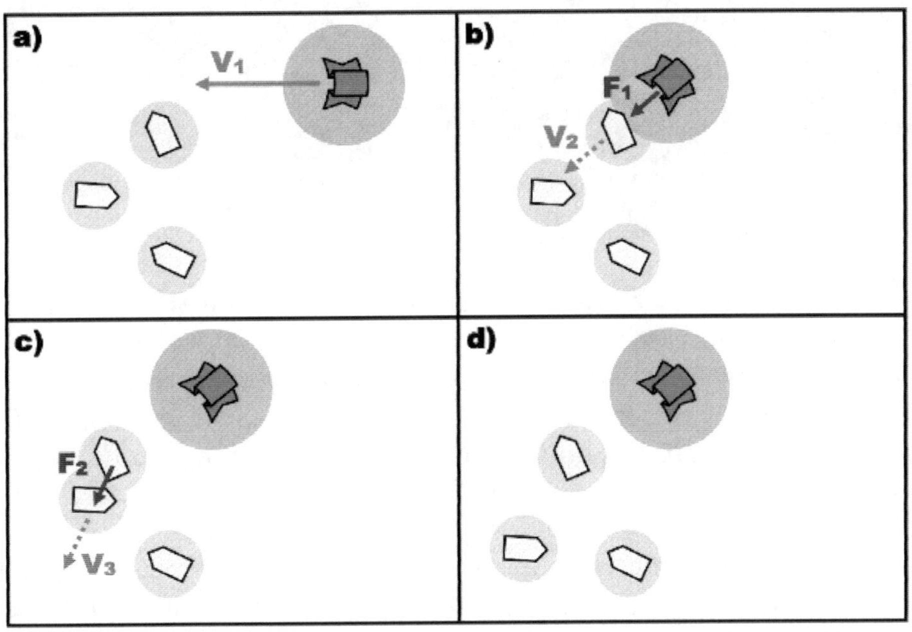

Figure 1 Repulsion radii between shepherd (pink) and swarm agents (yellow) are shown. **a)** Shepherd moves with initial velocity V_1. **b)** Swarm agent within repulsion radius of shepherd so force F_1 imparted on it, then agent moves away with velocity V_2. **c)** Swarm agent within swarm agent repulsion radius so force F_2 imparted on it, then it moves away with velocity V_3. **d)** Final position of agents.

Multiple target destinations replace the single target destination of the normal dog herding sheep exercise. Further, as opposed to an entire herd being guided to a common destination, the multiple targets in the environment require only one swarm member to reach them.

To begin the SSE task, an image of the environment is taken by a UAV, and locations for an initial measurement of wave properties are selected by a human operator. Figure 4a shows an example of an image as would be taken by the UAV, as well as the initial goal destinations to commence building a map of the wave system. The swarm of SSE agents are then deployed into the environment. In this case, the environment is the waters surrounding an island, where a ship may wish to dock during rough seas, thus, requires an SSE to ensure it is safe to proceed.

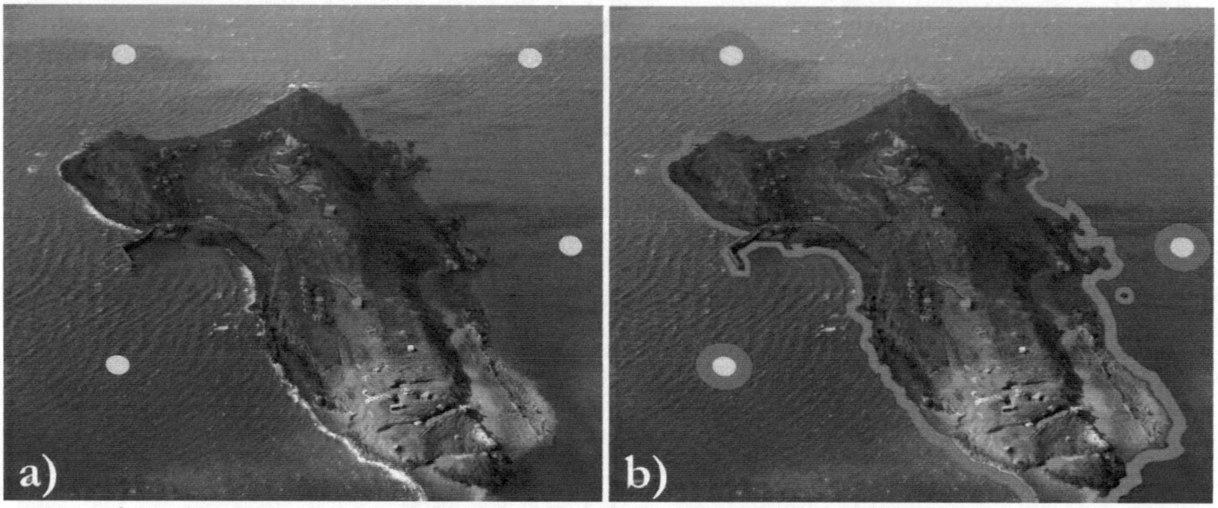

Figure 2 **a)** Initial SSE target locations (yellow dots) in an example environment. **b)** Target location attraction radii (orange shadows) and obstacle repulsion radii (pink shadow). The image has been modified from a picture taken by Gyula Péter, available on Wikipedia Commons [21].

In order to ensure that a swarm member reaches a target destination, the attractive force vector is introduced, radiating out from the goal locations (seen as the orange shadows surrounding the goal locations in 4b). Obstacles within the environment are avoided by repulsing any agent which falls within their own radii of influence (as seen in red in the example given in Figure 4b, where the coastline and rocks are obstacles present, assuming that the sea is sufficiently deep across the environment).

The USV swarm will move throughout the environment, being repulsed by other agents, attracted by the target locations, repulsed from the obstacles, and guided by the repulsive force of the UAV shepherd. A "random walk" term will be incorporated to stimulate interactions between the agents and the environment, and to help to reduce the chance of a swarm member becoming stuck.

Once any single agent reaches a target destination, *all agents stop* (after force reactions are complete) and each makes an SSE. The SSE of adjacent agents are compared, and if a discrepancy tolerance is reached (for wave properties: height, direction, or frequency), then a new target destination will be generated between the neighbouring swarm agents. The SSE task then continues, new target positions radiate out a new attractive force and, once a swarm agent is ensnared, a new estimate can be taken to validate or improve upon the previous estimates. This process continues until all target locations have been reached and smooth transitions between adjacent SSEs exist, or until a time limit is reached.

The simplicity of each swarm agent is evident in the fact that they do not have any information about their objective, only the shepherd 'knows' their objective The shepherd's purpose is to coordinate the swarm, improve the efficiency of the SSE task, and to enable cooperative swarm behaviour without the need for communication between swarm agents. The shepherd is able to sense each swarm agent's location and orientation and, thus, able to infer which forces are being activated at each time step. When a critical situation arises, the shepherd acts as a referee. An example of this is if two agents are at the same distance from a target location and within its radius of influence. In this case, the shepherd will move to disperse one of the agents away from the goal or push one into it. Further, if agents reach a local solution, then the shepherd will scatter agents across the environment to ensure a global solution is attained.

Variables in the SSE task include how much time is allocated for making an SSE, how much time is required to create a complete SSE map of a given environment, how many swarm agents are needed to attain the SSE, and which force and velocity magnitudes produce the best results.

4 CONCLUSION

A multi-vessel SSE heuristic utilising swarm shepherding has been described in the preceding section. The heuristic framework expands on the multi-vessel SSE described by Nielsen et al. [14] by introducing a coordinated swarm of self-propelling SSE devices which could be better implemented during dynamic operations for complex wave property measurement. Using the ship as a wave buoy analogy for SSE, a network of USVs would support the effort to create a controllable, aggregated sea state map for enhanced situational awareness. Addressing the control of the USV swarm, shepherding shows promise when it comes to a simple and autonomous solution. Chaimowicz and Kumar [19] give an analogy to the way in which shepherding could distribute the USV network around land masses and ocean-based structures. The heuristic describes a boid-based swarm agent system, governed by several behaviours, which can act as a basis for a distributed SSE under the guidance of the shepherd, and the procedures required to achieve the SSE task are defined.

An initial two-dimensional simulation of a shepherd guiding swarm agents to goal locations within an environment is under development, with results expected to be published in the near-future. Further experiments will then elaborate upon these, incorporating full three-dimensional wave system simulations and USV geometries. Should our hypothesis be proven true, then the safety, reliability, and effectiveness of maritime operations can be improved by obtaining in-situ, real-time knowledge of the prevailing sea state in order to evaluate how to best navigate the oncoming marine environment. Moreover, the viability of shepherding as a swarm control technique can be further substantiated or discredited.

REFERENCES

[1] N. Degiuli, I. Ćatipović, I. Martić, A. Werner, and V. Coric, "Increase of the ship fuel consumption due to added resistance in waves", *Journal of Sustainable Development of Energy, Water and Environment Systems*, vol. 5, pp. 1–14, Mar. 2017. DOI: 10.13044/j.sdewes.d5.0129.

[2] S. C. Stevens and M. G. Parsons, "Effects of motion at sea on crew performance: A survey", *Marine Technology and SNAME News*, vol. 39, no. 1, p. 29, Jan. 2002.

[3] X. Gu and T. Moan, "Long-term fatigue damage of ship structures under nonlinear wave loads", *Marine Technology and SNAME News*, vol. 39, no. 2, pp. 95–104, Apr. 2002.

[4] S. Tyrberg, O. Svensson, V. Kurupath, J. Engstrom, E. Stromstedt, and M. Leijon, "Wave buoy and translator motions—on-site measurements and simulations", *IEEE Journal of Oceanic Engineering*, vol. 36, no. 3, pp. 377–385, Jul. 2011. DOI: 10.1109/JOE.2011.2136970.

[5] W. Huang, X. Liu, and E. W. Gill, "Ocean wind and wave measurements using x-band marine radar: A comprehensive review", *Remote Sensing*, vol. 9, no. 12, 2017. DOI:10.3390/rs9121261.

[6] M. Martin-Neira, M. Caparrini, J. Font-Rossello, S. Lannelongue, and C. S. Vallmitjana, "The PARIS concept: An experimental demonstration of sea surface altimetry using GPS reflected signals", *IEEE Transactions on Geoscience and Remote Sensing,* vol. 39, no. 1, pp. 142–150, Jan. 2001. DOI: 10.1109/36.898676.

[7] V. J. Cardone, B. T. Callahan, H. Chen, A. T. Cox, M. A. Morrone, and V. R. Swail, "Global distribution and risk to shipping of very extreme sea states (vess)", *International Journal of Climatology*, vol. 35, no. 1, pp. 69–84, 2014. DOI: 10.1002/joc.3963.

[8] A. H. Brodtkorb, U. D. Nielsen, and A. J. Sørensen, "Sea state estimation using modelscale dp measurements", in *OCEANS 2015 - MTS/IEEE Washington*, Oct. 2015, pp. 1–7. DOI: 10.23919/OCEANS.2015.7404402.

[9] M. A. Morales Maqueda, N. T. Penna, S. D. P. Williams, P. R. Foden, I. Martin, and J. Pugh, "Water surface height determination with a GPS wave glider: A demonstration in Loch Ness, Scotland", *Journal of Atmospheric and Oceanic Technology*, vol. 33, no. 6, pp. 1159–1168, 2016. DOI: 10.1175/JTECH-D-15-0162.1.

[10] J. P. R. Marreiros, "Kinematic GNSS precise point positioning: Application to marine platforms", PhD thesis, University of Porto, 2012. [Online]. Available: https://core.ac.uk/download/pdf/143396236.pdf.

[11] U. Nielsen, "A concise account of techniques available for shipboard sea state estimation", *Ocean Engineering*, vol. 129, pp. 352–362, 2017. DOI: 10.1016/j.oceaneng.2016.11.035.

[12] R. Pascoal and C. G. Soares, "Kalman filtering of vessel motions for ocean wave directional spectrum estimation", *Ocean Engineering*, vol. 36, no. 6, pp. 477–488, 2009, DOI: 10.1016/j.oceaneng.2009.01.013.

[13] U. D. Nielsen, R. Galeazzi, and A. H. Brodtkorb, "Evaluation of shipboard wave estimation techniques through model-scale experiments", in *OCEANS 2016 - Shanghai*, 2016, pp. 1–8. DOI: 10.1109/oceansap.2016.7485701.

[14] U. Nielsen, A. Brodtkorb, and A. Sørensen, "Sea state estimation using multiple ships simultaneously as sailing wave buoys", *Applied Ocean Research*, vol. 83, pp. 65–76, Feb. 2019. DOI: 10.1016/j.apor.2018.12.004.

[15] J. Harvey, "The blessing and curse of emergence in swarm intelligence systems", in *Foundations of Trusted Autonomy*, H. A. Abbass, J. Scholz, and D. J. Reid, Eds. Cham: Springer International Publishing, 2018, pp. 117–124. DOI: 10.1007/978-3-319-64816-3_6.

[16] C. W. Reynolds, "Flocks, herds and schools: A distributed behavioural model", *SIGGRAPH Computer Graphics,* vol. 21, no. 4, pp. 25–34, Aug. 1987. DOI: 10.1145/37402.37406.

[17] D. Strömbom, R. P. Mann, A. M. Wilson, S. Hailes, A. J. Morton, D. J. T. Sumpter, and A. J. King, "Solving the shepherding problem: Heuristics for herding autonomous, interacting agents", *Journal of The Royal Society Interface*, vol. 11, no. 100, p. 20 140 719, 2014. DOI: 10.1098/rsif.2014.0719.

[18] A. Gee, and H. A. Abbass, "Transparent Machine Education of Neural Networks for Swarm Shepherding Using Curriculum Design", *International Joint Conference on Neural Networks (IJCNN),* Budapest, Hungary. 2019. [Online]. Available: http://arxiv.org/abs/1903.09297.

[19] L. Chaimowicz and V. Kumar, "Aerial shepherds: Coordination among UAVs and swarms of robots", in *Distributed Autonomous Robotic Systems 6*, R. Alami, R. Chatila, and H. Asama, Eds., Tokyo: Springer Japan, 2007, pp. 243–252.

[20] N. R. Clayton and H. A. Abbass, "Machine teaching in hierarchical genetic reinforcement learning: Curriculum design of reward functions for swarm shepherding", . *IEEE Congress on Evolutionary Computation,* Wellington, New Zealand. 2019. [Online]. Available: http://arxiv.org/abs/1901.00949.

[21] G. Péter, *Az Inchkeith-sziget a Forth-öbölben, Edinburgh közelében panoramio*. Wikipedia Commons, Jul. 2016. [Online]. Available: https://commons.wikimedia.org/wiki/File:Az_Inchkeith-sziget_a_Forth-%C3%B6b%C3%B6lben,_Edinburgh_k%C3%B6zel%C3%A9ben_(The_Inchkeith_Island_in_the_Firth_of_Forth,_near_Edinburgh)_-_panoramio.jpg.

Characterising the Southern Ocean and Ross Sea wave climate

Sally Garrett [1] Tom Durrant[2]

[1] Defence Technology Agency, New Zealand Defence Force
[2] Oceanum Limited

ABSTRACT

The waves of the Southern Ocean and the Ross Sea are largely unstudied. The New Zealand Defence Force (NZDF) routinely operates in these areas and is currently engaged in a shipbuilding programme which requires a detailed understanding of the wave climate for sea keeping analysis and ice belt design. Unlike other areas, the Southern Ocean and the Ross Sea has limited ship traffic and therefore limited wave observations from volunteer observing ships. Moreover, due to the difficult conditions and remote location limited scientific measurements of waves have been completed. In 2017, the NZDF deployed the first wave buoy in the open ocean south of 47 S anywhere in the world. In addition, 21 free floating buoys were also deployed between 42 S and 67 S. This array has provided an understanding of wave characteristics across the Southern Ocean and the Ross Sea.

The data from these platforms have been used to optimise the WaveWatch III wave forecast model. The optimised setup is then used to create a 24-year hind-cast wave atlas for the ice-free areas south of 31 S.

In this paper, this previously unpublished wave atlas for the Southern Ocean and Ross Sea will be presented. A limited comparison will also be made between the wave statistics from both wave buoys observations and the wave atlas in these regions with the bivariate frequency wave height-period occurrence tables recommended for the North Atlantic by the International Association of Class Societies (International Association of Classification Societies, 2001).

INTRODUCTION

The New Zealand Defence Force (NZDF) routinely operates in Southern Ocean and Ross Sea and is engaged in a shipbuilding programme which requires a detailed understanding of the wave climate for sea keeping analysis and ice belt design. This characterisation of the Southern Ocean and Ross Sea waves is required for all sea areas from 31 S to 78 S year round.

Knowledge of the wave climate of an area is a critical element for the design of vessels and offshore structures. Several studies have attempted to quantify the effect on design of uncertainties caused by a lack of knowledge in the wave climate (Bitner-Gregersen & Guedes Soares, 2007) and have found differences in long-term ship responses of up to 150 % of a nominal value, e.g. Guedes Soares & Travoa, (1991). This high uncertainty may lead to overdesign/under design of ships, with significant economic and risk-related consequences.

The need to improve the availability, quality and reliability of wave databases has been identified by several professional organisations (Bitner-Gregersen et al., 2009). In addition, the evolution of design processes has increased the detail required of not only the characteristic wave statistics and bivariate frequency wave height-period occurrence tables (scatter diagrams), but also of the full directional wave spectrum (Cardone et al., 2015).

Scatter diagrams based on the historical visual observations of sea state by merchant vessels participating in the Voluntary Observing Ships (VOS) form the basis of the Global Wave Statistics (GWS) atlas (BMT Group, 1986). Products derived from the GWS atlas include the International Association of Class Society recommended wave data (International Association of Classification Societies, 2001). The atlas divides the globe in 104 regions, as illustrated in Figure 1, each with a scatter diagram of historical observations. This data is still employed for ship motion and design studies. However, while this database has been used for many years its accuracy has been questioned by researchers, especially concerning wave period and steepness (Bitner-Gregersen & Guedes Soares, 2007).

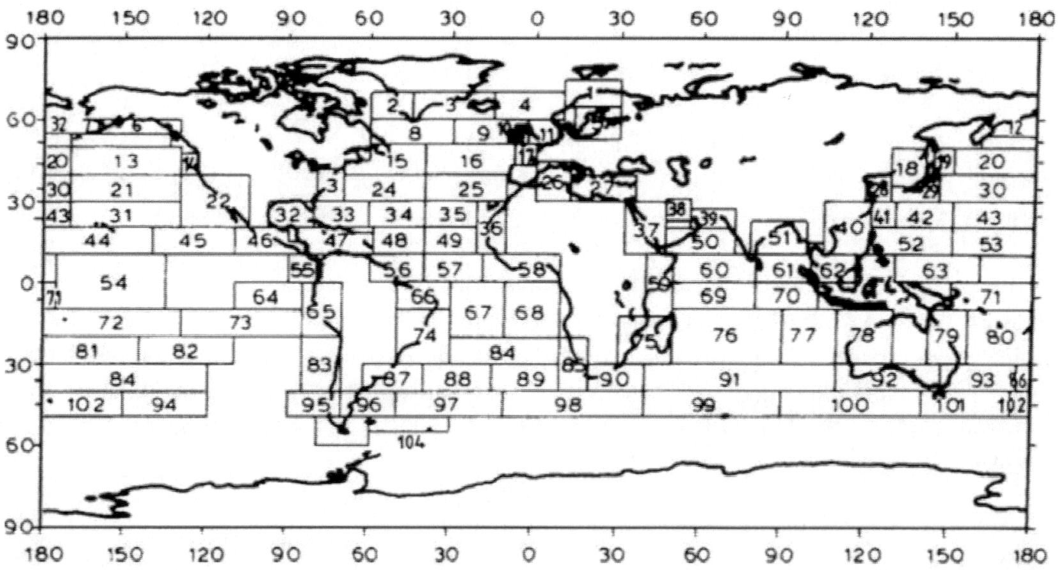

Figure 1 Map of the GWS sea areas. Source:(BMT Group, 1986)

Another shortcoming of the GWS atlas is that is does not have global coverage and does not define the wave climate for the Southern Ocean area. The southernmost latitude of the available GWS data is 50 S (except for the Drake Passage sea area -104); well short of the 77 S ships are able to navigate to in the summer. Professional practise currently recommends the use of a refined GWS scatter diagram for the entire North Atlantic to be used for designing ships operating globally including those operating in the Southern Ocean (International Association of Classification Societies, 2001).

To test if this practise adequately describe the conditions of the Southern Ocean, Cannon, Guzsvany, & Turner (2004) compared the ship motion modelled for an ANZAC-class frigate ship for both the Southern Indian Ocean (GWS sea areas 91, 92, 99, 100, 101) and North Atlantic (GWS sea areas 3, 4, 8, 9). Southern Indian Ocean data from 30 S – 50 S was used due to the absence of data in the Southern Ocean. Overall the global bending moment of a frigate operating in the Southern Indian Ocean was 10% greater than for the same vessel in the North Atlantic.

Slamming calculations were made for a frigate travelling at 7.5 knots in all sea states for head seas only. The number of slams for the same length of simulation was 1.5-3 times greater for the Southern Indian Ocean sea areas compared to the North Atlantic sea areas. The greatest number of slams was modelled for Southern Indian Ocean sea area 100. The pressure of these slams was also modelled, with sea area 100 having the greatest slamming pressures and the average Southern Indian Ocean slamming pressure being 1.2 times higher than when operating in the North Atlantic.

A limitation of this study that was identified was the lack of winter Southern Indian Ocean data due to a lack of shipping in the area, making the results a conservative estimate. However, even considering this limitation the larger global bending moment estimated for the Southern Hemisphere brought doubt to the assumption that North Atlantic sea areas expose ships to the worst sea conditions.

Moreover, a comparison of the GWS data with available satellite altimeter data showed that the Southern Indian Ocean sea areas had a greater peak significant wave height than any area described in the GWS database. While for North Atlantic sea areas the peak significant wave height was similar for GWS and satellite databases, highlighting the GWS data for the Southern Indian Ocean was underestimating the conditions that would be experienced at sea.

At the time of the study wave period data was unable to be collected by satellite and therefore no comparison between satellite wave period and the GWS datasets was made. The inability to accurately sense wave period from space continues today. Synthetic aperture radar (SAR) is the only direct source of wave period estimates available from space. High errors have been identified in this data due to the SAR only capturing the long-wave part of ocean wave fields. Violante-Carvalho (2005) noted that the high-frequency cut-off is sea state dependent but, in general, waves shorter than 10 seconds propagating parallel to the satellite track are not mapped.

The New Zealand Defence Force operates in the Southern Ocean annually. Ship's staff have reported high sea sates on all voyages. This study aims to measure and then model the sea conditions in the Southern Ocean and compare these to the International Association of Classification Societies (2001) scatter diagram for the North Atlantic. This paper provides the initial findings of this new wave characterisation by measurement and modelling, with papers presented by Marsland and Ballard providing details of the implications for ship design.

WAVE CLIMATE MEASUREMENT AND MODELLING

The Southern Ocean is the southernmost part of the global ocean and represents around 22% of the sea surface area. The combination of persistent westerly winds, and the largely unbroken expanse of sea, produces potentially enormous fetches, resulting in the Southern Ocean experiencing higher wave heights for longer periods than any other body of water (Young, 1999). Due to the harsh ocean environment and remote location it is also the least observed of any ocean body.

The NZDF ship building program requires characterisation of the wave conditions year-round in the Southern Ocean and Ross Sea to 78 S. Initial studies of waves in the area to support this aim focussed on three main data sources; ship observations from the Volunteering Observing Ships (VOS) program (Kent et al., 2011), ship motion measurements from Royal New Zealand Navy (RNZN) vessels to enable estimation wave characteristics using the techniques of Nielsen & Stredulinsky (2012) and satellite observations using the Globwav database (Ash et al., 2010). Each of these data sources were unable to provide year-round characteristics for the required spatial extent.

Ship voyages in the Southern Ocean and Ross Sea are currently limited both temporally and spatially (Summerhayes, Dickson, Meredith, Dexter, & Alverson, 2007). Unlike similar latitudes in the Northern Hemisphere ship-based observations both visual and using ship motion sensors in the Southern Ocean currently do not provide year-round wide area coverage of wave conditions.

An example of these spatial and temporal limitations is the cruises completed by the fleet of New Zealand registered ships that operate in the Southern Ocean and Ross Sea routinely. The fleet includes the RNZN Offshore Patrol Vessels, the National Institute of Wave and Atmospheric research (NIWA) research vessel RV TANAGAROA and the fishing vessels conducting toothfish fishing in the Ross Sea. The vessels from each of these groups operate in the Southern Ocean and Ross Sea between November and March, returning to similar areas on each voyage, all routinely make observations. However, no observations are made by the New Zealand Fleet in the winter months and locations where these vessels do not routinely operate.

Tedeschi, Rizzo, & Carrera (2001) study of the route between New Zealand and Terra Nova Bay in the Ross Sea highlights the low number of observations in the winter months. Only 88 observations south of 60 S exist between 160 E and 180 E for wind speed or wave height or wave period summed across the period of April to September for all years from 1960 -2000. For the same 40-year period and longitudes, 5778 observations were made between 50 S and 60 S.

These very low numbers of observations have been associated with high sampling biases (between ship based measurements and models) in the Southern Ocean, where wave height may be underestimated by the VOS fleet by as much as 1–1.5 m because of poor sampling, primarily associated with a fair weather bias of ship routing and observation (Gulev, Grigorieva, Sterl, & Woolf, 2003) . Ship based observations were therefore deemed unsuitable as a sole source of data for developing a ship building wave climatology.

Satellite observations in open water provide a year-round unbiased method of characterising wave height. However altimeters, the most common sensor used, cannot estimate wave height in the presence of sea ice or other non-Gaussian surfaces within the sensor footprint (Ardhuin et al., 2019). For SAR collected images ocean waves are clearly visible in the imagery of ice-covered waters, however their quantitative analysis is presently limited to the estimation of a dominant wavelength and direction (Liu, Holt, & Vachon, 1991; Schulz-Stellenfleth & Lehner, 2002) for wave periods less than 10 seconds (Violante-Carvalho 2005). As this study required wave characteristics within areas with sea-ice coverage, and requires detailed understanding wave period, these limitations meant the use of satellite data was limited to open-water areas for the validation of the wave model outputs.

To enable a year-round understanding of wave conditions in all locations to 78 S a three-part approach was used: in-situ measurement of waves were completed using wave buoys; the data from these buoys along with satellite altimeter measuements was then used to validate a Southern Ocean and Ross Sea optimised setup of the WaveWatch III wave model; this model was then used to create a 24-year hindcast for the area.

The first long-term wave buoy deployment was Southern Ocean Flux Station (SOFS), established as part of the Southern Ocean Time Series project (Trull et al., 2010). Five deployments have been performed since March 2010, providing measurements of wave characteristics and spectra spanning a period of over 700 days. The approximate coordinates of the mooring location were 47 S and 142 E. SFOS was the southernmost directional measurement of waves in the South Pacific sector of the Southern Ocean.

To enable measurements of waves throughout the South Pacific sector of the Southern Ocean an array of free floating and moored wave buoys was deployed from 53 S to 67 S. A combination of free floating and moored wave buoys were used due to the cost and difficulties associated with the deployment of moored wave buoys.

Wave buoys provide point measurements of waves for a short time period. The second part of this study was to hind-cast the wave conditions of the Southern Ocean for a 24 year period. The model outputs were then validated against both the collected wave buoy data and intercalibrated altimeter data.

Observation - Wave Buoy Array

Moored and free-floating wave buoys were used to measure waves from 42 S to 67 S between February 2017 and present. A time series of the latitude of the deployed buoys is shown in Figure 2.

The wave buoy array consisted of the following buoys:

- Two TRIAXS Directional Wave Buoys, the first of which was deployed 11 nm miles south of Campbell Island (52°45.71'S, 169°02.54'E) in February 2017. This buoy broke free in July 2017and was replaced by the present buoy in March 2018.

- Nineteen Miniature Wave Buoys, developed by Scripps Institution of Oceanography (SIO) (Coastal Observing Reserach and Development Center, 2016), deployed from

Royal New Zealand Navy ships and the RV TANGAROA across the Southern Ocean and in the Ross Sea.

- Four Spoondrift buoys deployed in the Southern Ocean from RV TANGAROA during the 2018 Ross Sea cruise.

The moored TRIAXYS buoys were deployed within New Zealand territorial seas, with the location decided upon due to the relatively shallow water depth (~140m) in an area with limited other sea bed features that would cause shoaling of waves.

The latitude of the free-floating buoys was able to be planned with the deploying ships, however the longitude of deployment was decided by the ship.

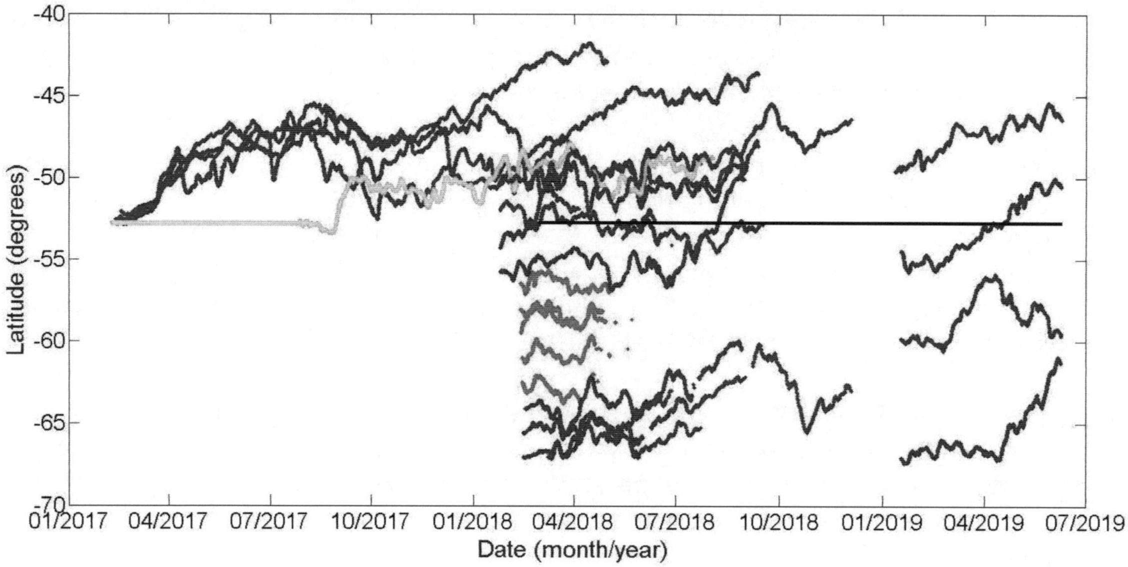

Figure 2 Time series of the latitudes of wave measurements made by SCRIPPS (blue), Spoondrift (red), TRIAXYS 2017 (green) and TRIAXYS 2018 (black) wave buoy deployments

Wave measurements for all buoys were completed using a standard ocean wave measuring regime of sampling at 2 Hz for a sampling burst of 20 minutes. For the TRIAYXS and Spoondrift buoys these sampling bursts occur 3-hourly with hourly sampling for all the SIO buoys. Onboard processing is then completed and wave characteristics including significant wave height, wave peak period, wave peak direction as well as directional spectra are telemetered to shore by satellite.

The TRIAXYS and Spoondrift buoys measure waves using accelerometers (Raghukumar et al., 2017) while the SIO buoys measure waves using GPS (Coastal Observing Reserach and Development Center, 2016).

The data collected from the TIAXYS and Spoondrift buoys are available from https://www.metocean.co.nz/southern-ocean .

Modelling - Southern Ocean Wave Atlas

In order to enable a greater spatial and temporal characterisation of waves a 24 year hind-cast was completed for all ice-free areas south of 31 S. This hind-cast was modelled using the WaveWatch III version 5.16 (Tolman, 1991). The details of the setup of the model is detailed in Southern Ocean Wave Atlas report (MetOcean Solutions Limited, 2019) and summarised in Table 1.

Table 1 Setup parameters for WaveWatch III Southern Ocean hind-cast.

Parameter	Hind-cast setup
Resolution	0.25 deg x 0.2 deg
Extent	31 S – 77.5 S
Period	January 1993 – December 2017
Wind field	ERA5 reanalysis from the European Centre for Medium-Range Weather Forecasts
Current input	GLORYS reanalysis from Mercato Ocean European Copernicus Marine Environment Monitoring Service

The validation of the wave hind-cast was based on intercalibrated altimeter data from 1993 to 2016 from the Globwav multi-altimeter dataset (Ash et al., 2010). The average bias in wave height was less than 0.05 m with a slight overestimation of the smallest waves and a slight underestimation of the largest waves. The root-mean-square deviation was also relatively low with overall values of ~ 0.4 m. Regions covered by ice during part the year had the largest errors, including the coastal areas near Antarctica between 60 E and 120 E and between 180 E and 240 E.

Seasonal bias was calculated for the 1993-2016 period. These biases were generally positive, especially south of 50 S. The largest positive bias was found in the summer season from December to February, with values ~0.2 m throughout the domain and up to 0.4 m in some areas. During the March to May there was a smaller positive bias around 0.15 m on average, with less areas of higher bias compared to summer months. Winter (June to August) had the smallest bias on average, with few regions showing positive bias around 0.05 m, but with surprisingly high negative bias in the ice marginal zone. Spring biases were like those in the March to May period.

Full details of the validation are provided in the Southern Ocean Wave Atlas report (MetOcean Solutions Limited, 2019) and, as this hindcast was only recently completed (June 2019). further analysis will be presented in a series of forthcoming papers.

The data from the 24-year hindcast is available from https://metoceanview.com/ .

The GWS atlas divides the area covered by the atlas into 104 sea areas. These sea areas as shown in Figure 1 are rectangular in shape and in some cases cover locations where bivariate frequency wave height-period occurrence may have different distributions. For example, area 104 covers both the western and eastern approaches to Cape Horn.

The intent for the newly developed wave atlas was to divide the hind-cast modelled area in to a series of regions with a similar distribution of wave heights and periods. This was completed by calculating the mean significant wave height, the standard deviation of the significant wave height, the mean wave period and the standard deviation of the wave period for each of the 300 months in the hind-cast and then clustering using a K-means algorithm to find areas of similar wave characteristics (MetOcean Solutions Limited, 2019).

The resulting clustering is shown in Figure 3. To increase the usability of the data the grid was divided in to 5 degree x 5 degree regions and the cluster for each region applied from the K-means analysis as shown in Figure 4.

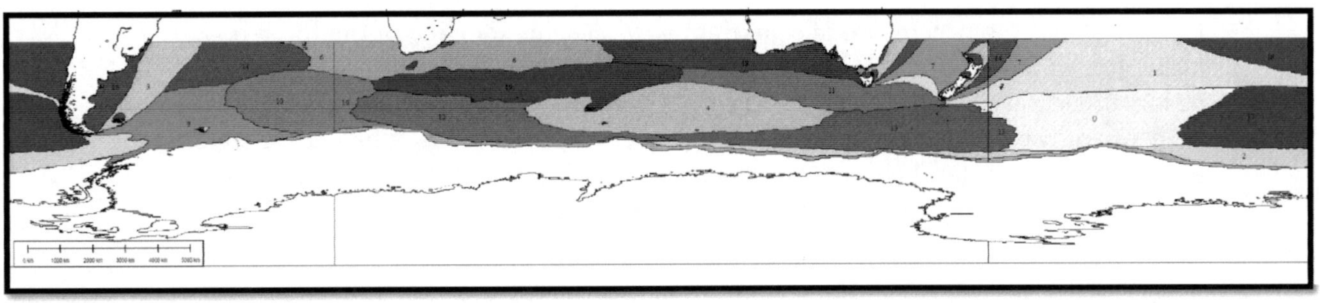

Figure 3 Classification of Southern Ocean zones obtained from K-means clustering

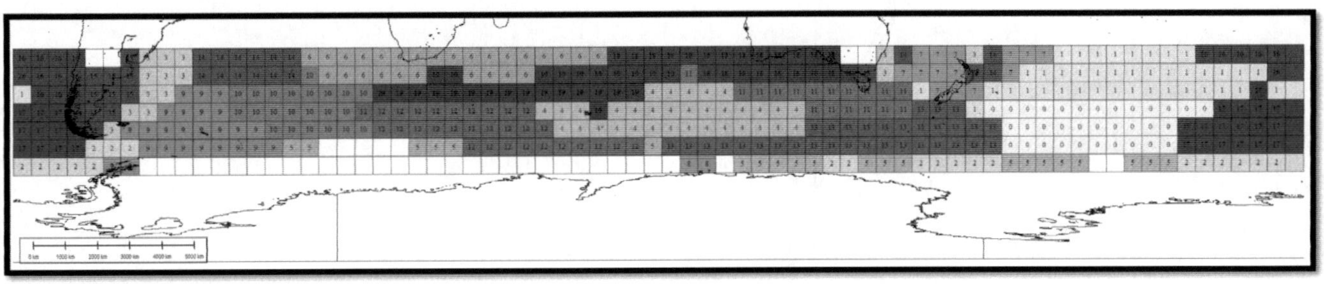

Figure 4 Simplified classification of Southern Ocean zones.

SOUTHERN OCEAN AND ROSS SEA WAVE CLIMATE

Wave Measurements

In total 5622 buoy days (at June 1st 2019) of wave data has been collected at latitudes between 42 S and 67 S in the South Pacific sector of the Southern Ocean. The only buoy to have collected data for a 12-month period south of 50 S (the area outside the GWS sea areas) is the TRIAXYS buoy deployed in March 2018. This dataset will be used for the comparison against the IACS North Atlantic data.

In the Ross Sea only free floating buoys were deployed. The longest wave buoy record south of 60 S is currently 142 days in duration; after this time the buoy drifted into the Southern Ocean. 30-day records from each of four wave buoys were used in addition to the 142 day time series to develop an understanding of waves in the area.

Figure 5 shows the occurrence of significant wave heights for each of the three regions. For measured wave data in both the Southern Ocean and the Ross Sea the largest number of observations was between 4-5m. For the IACS North Atlantic data this peak is for waves with a significant wave height between 3–4m. However, the occurrence of higher sea states in the IACS table was greater than the Ross Sea observations for all wave heights greater than 5m and greater than the Campbell Island buoy for all wave heights greater than 7m.

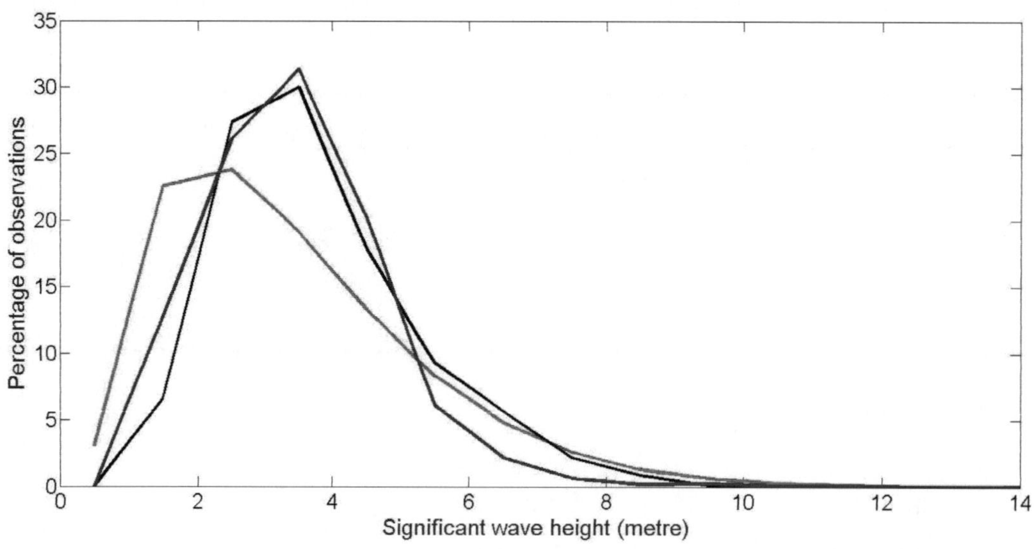

Figure 5 Distribution of significant wave heights for International Association of Class Society North Atlantic sea area (red), Campbell Island TRIAXYS wave buoy (black) and Ross Sea SCRIPPS wave buoys (blue).

Figure 6 shows the occurrence of peak wave period for each of the three regions. Wave period was greater for the observational datasets, with the Ross Sea data having the greatest peak wave period. The Ross Sea dataset was collected with free floating GPS buoys, however the ability of these buoys to measure long period waves has not been extensively researched and it is possible the variations between the two observed datasets is associated with sensor limitations. The Campbell Island wave buoy has a peak at 8.5 seconds, further analysis is required to determine the reason for this peak, however the majority of the waves at the period come from a North West direction where the fetch is affected by Campbell Island.

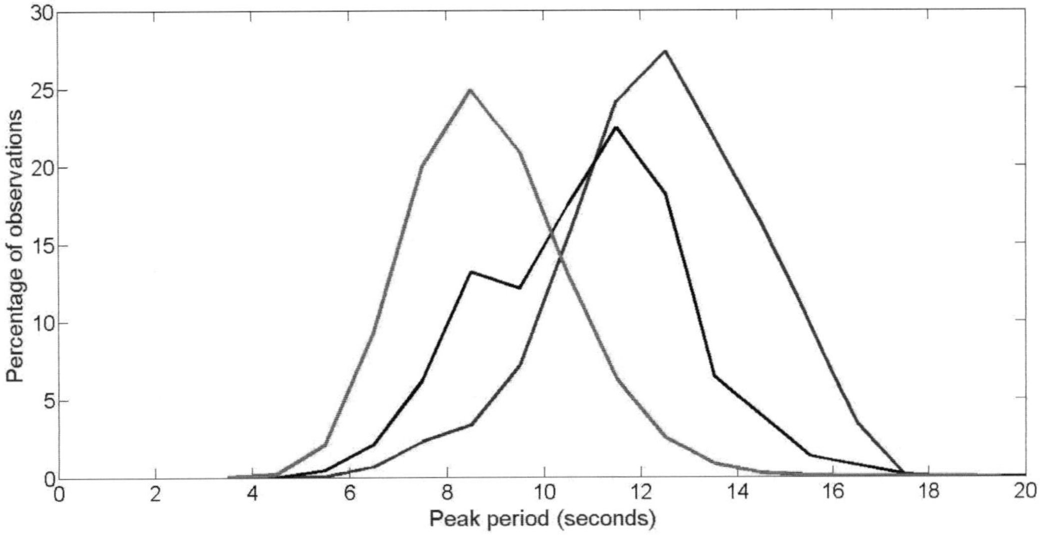

Figure 6 Distribution of peak period for International Association of Class Society North Atlantic sea area (red), Campbell Island TRIAXYS wave buoy (black) and Ross Sea SCRIPPS wave buoys (blue).

The IACS table's largest wave heights were between 16-17m with periods between 10-16 seconds. For the Campbell Island observation, they were 15-16m at 16-17 seconds and for the Ross Sea the largest wave height was between 12-13m. However, as the wave buoys are a point measurement of waves and the length of each observational time series is short, it is difficult to compare these maximum values to the long-term time series of the IACS data. For this reason, the hind-cast was completed.

Wave Atlas

Wave height distributions for the 20 zones of the Southern Ocean Wave Atlas (SOWA) compared to the IACS North Atlantic dataset are shown in Figure 7. The significant wave height distributions for the SOWA zones vary in terms of peak significant wave height and also in distribution of wave height greater than 6m. 12 of the SOWA zones have a peak significant wave height greater than the peak of between 3-4m of the IACS. Moreover, 8 zones have a higher proportion of waves greater than 6m than the IACS, and 2 zones have a higher proportion of waves greater than 8m.

At high sea states the largest wave heights for the 20 zones is between 15-16m, while for the IACS data it is between 18-19m. However, only 0.00198% in the IACS data set are greater than 16m. Additional understanding is required of the distribution of extreme sea states in the Southern Ocean, in particular whether the SOWA zones should include calculated wave height extremes or remain only distributions of significant wave height.

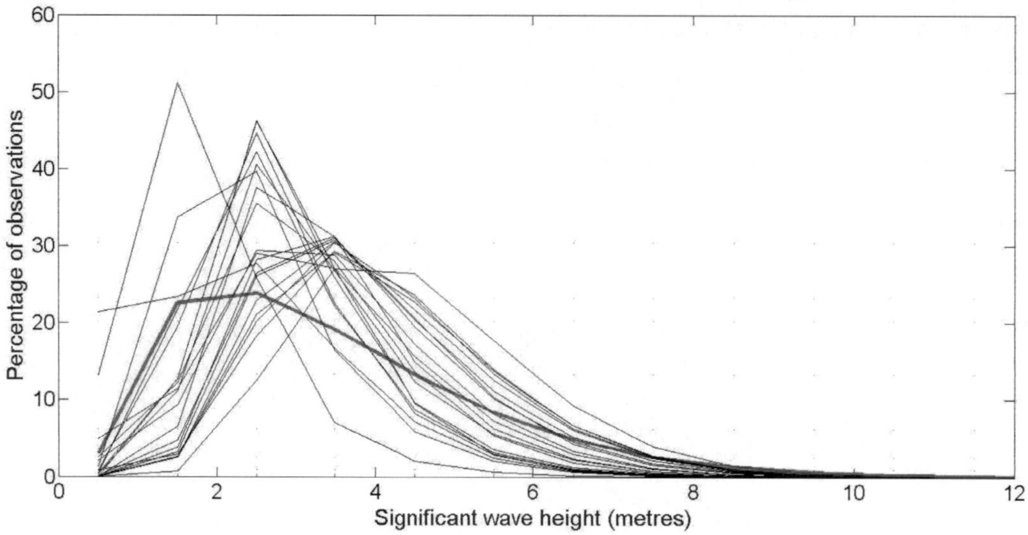

Figure 7 Distribution of significant wave heights for International Association of Class Society North Atlantic sea area (red), wave hind-cast zones 0-19 (black)

Wave peak period distributions for the 20 zones of the SOWA compared to the IACS North Atlantic dataset are shown in Figure 8. For 18 of the 20 SOWA sea areas the peak of the peak period distribution is greater than the IACS with all SOWA sea areas having a greater percentage of wave periods than the IACS for periods greater than 12 seconds. This is consistent with the observational data presented above and also with the work by (Bitner-Gregersen & Guedes Soares, 2007)

A bimodal distribution in wave peak period is shown for SOWA sea area 9 located to the east of the Drake Passage and the Antarctic Peninsula. This sea area also has a large percentage of lower significant wave height in its distribution, indicating the effects of land on the seas in the area.

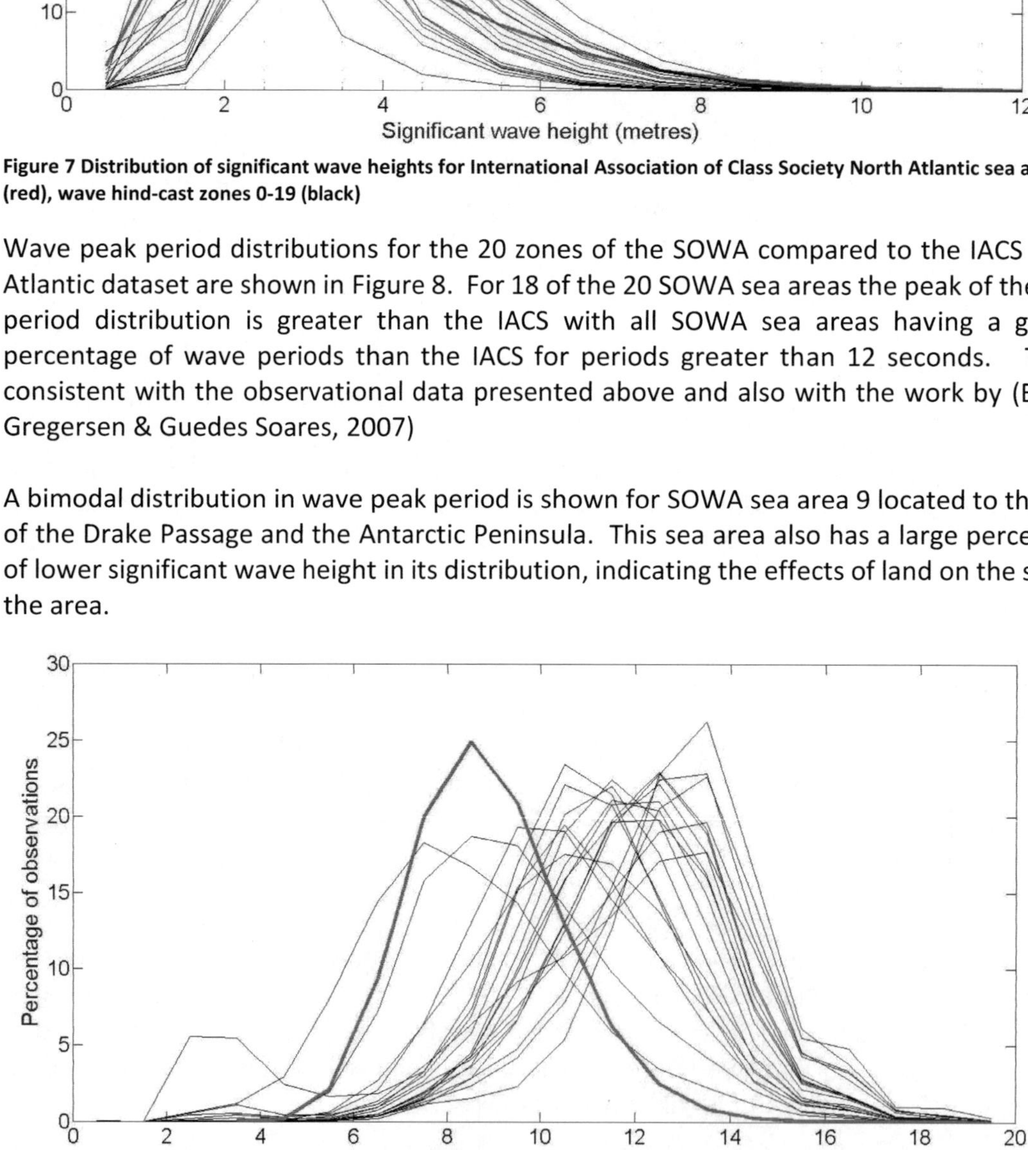

Figure 8 Distribution of peak wave period for International Association of Class Society North Atlantic sea area (red), wave hind-cast zones 0-19 (black)

CONCLUSIONS

In this study a series of short-term wave buoy deployments have been used to characterise the waves across the Southern Ocean south of New Zealand and the northern Ross Sea, to enable comparison to the IACS wave data for the North Atlantic. The aim of this comparison was to gain an understanding of whether the IACS recommendation 34, to use the North Atlantic data for worst case conditions for design, is appropriate for Southern Ocean ships. The observational datasets indicate that the average wave conditions experienced in this area represent greater significant wave heights and longer periods than those specified by IACS. However, higher sea states occur at greater frequency in the IACS tables.

The collected wave data has several limitations including the short length of each deployment and the limited spatial coverage. The TRIAYXS wave buoy data from Campbell Island included 12 months of data, while the Ross Sea collection period was 262 days. Spatially, all buoy deployments occurred in the Southern Ocean Pacific Ocean sector.

To address these limitations a 24 year hind-cast for all ice-free sea areas south of 30 S globally was developed. This data was then divided into sea areas based on similar significant wave height and peak period. As with the observational datasets, the modelled datasets indicate that the average wave conditions experienced in the Southern Ocean have a higher significant wave height and longer period than those specified by IACS. However, they also indicate once again that higher sea states occur at greater frequency in the IACS tables. This is an area for further research.

The findings of this study conclude that bivariate frequency wave height-period occurrence tables for the Southern Ocean and Ross Sea differ from the recommend IACS tables. These differences of a greater average significant wave height and period are the same as that noted from the Southern Indian Ocean by Cannon (2004), who linked the greater average significant wave height and wave period caused a 10% greater global bending moment in an ANZAC frigate.

This programme of work continues with the characterisation of wave spectra using the observations and hind-cast in this paper to compare to industry standard spectra and, if appropriate, the development of specialised spectra for the Southern Ocean.

REFERENCES

Ardhuin, F., Stopa, J. E., Chapron, B., Collard, F., Husson, R., Jensen, R. E., ... Young, I. (2019). Observing Sea States. *Frontiers in Marine Science*, *6*, 124. https://doi.org/10.3389/fmars.2019.00124

Ash, E., Busswell, G., (2010). *Amendment History*. Globwave . Retrieved from http://www.globwave.org/content/download/16022/104639/version/1/file/GlobWave_D.7_PUG3_v1.0.pdf

Bitner-Gregersen, E. M., Ellermann, K., Ewans, K. C., Falzarano, J. M., Johnson, M. C., Nielsen, U. D., ... Waseda, T. (2009). *17 th INTERNATIONAL SHIP AND OFFSHORE STRUCTURES CONGRESS 16-21*. Retrieved from http://www.issc2018.org/images/issc2009/ENVIRONMENT.pdf

Bitner-Gregersen, E. M., & Guedes Soares, C. (2007). Uncertainty of average wave steepness prediction from global wave databases. *Advancements in Marine Structures - Proceedings of MARSTRUCT 2007, The 1st International Conference on Marine Structures*, (January 2007), 3–10.

BMT Group. (1986). *Global Wave Statistics*. London .

Cannon, S., Guzsvany, G., & Turner, T. (2004). A Study of the Effect of Wave Loading on a Frigate Operating in the Southern Ocean Versus the North Atlantic. In *Proceedings of Pacific 2004 International Maritime Conference* (p. 543). Pacific 2004 International Maritime Conference Managers.

Cardone, V. J., Callahan, B. T., Chen, H., Cox, A. T., Morrone, M. A., & Swail, V. R. (2015). Global distribution and risk to shipping of very extreme sea states (VESS). *International Journal of Climatology*, *35*(1), 69–84. https://doi.org/10.1002/joc.3963

Coastal Observing Reserach and Development Center. (2016). *Miniature Wave Buoy Operating Manual*.

Guedes Soares, C., & Travoa, C. (1991). Influence of wave climate modelling on the long-term prediction of wave induced responses of ship structures. (p. 338). Dynamics of marine vehicles and structures in waves : proceedings of an International Union of Theoretical and Applied Mechanics Memorial Symposium on the Dynamics of Marine Vehicles and Structures in Waves, held at Brunel University, Uxbridge, U.K., 24-2: Elsevier. Retrieved from https://trid.trb.org/view/439995

Gulev, S. K., Grigorieva, V., Sterl, A., & Woolf, D. (2003). Assessment of the reliability of wave observations from voluntary observing ships: Insights from the validation of a global wind wave climatology based on voluntary observing ship data. *Journal of Geophysical Research*, *108*(C7), 3236. https://doi.org/10.1029/2002JC001437

International Association of Classification Societies. (2001). Standard Wave Data (North Atlantic Scatter Diagramm), *34*(34), 1–4.

Kent, E. C., Ball, G., Berry, D. I., Fletcher, J., Hall, A., North, S., & Woodruff, S. D. (2011). THE VOLUNTARY OBSERVING SHIP (VOS) SCHEME. In J. Hall., D. E. Harrison, & D. Stammer (Eds.), *Proceedings of OceanObs'09: Sustained Ocean Observations and Information for Society*.

Liu, A. K., Holt, B., & Vachon, P. W. (1991). Wave propagation in the marginal ice zone: Model predictions and comparisons with buoy and synthetic aperture radar data. *Journal of Geophysical Research*, *96*(C3), 4605. https://doi.org/10.1029/90JC02267

MetOcean Solutions Limited. (2019). *Southern Ocean Wave Atlas*. Raglan, New Zelaland.

Nielsen, U. D., & Stredulinsky, D. C. (2012). Sea state estimation from an advancing ship – A comparative study using sea trial data. *Applied Ocean Research*, *34*, 33–44. https://doi.org/10.1016/J.APOR.2011.11.001

Raghukumar, K., Chang, G., Spada, F., Jones, C., Gans, W., & Janssen, T. (2017). Wave-Measuring Performance Characteristics of Spoondrift Spotter. *Ocean Waves Workshop*. Retrieved from https://scholarworks.uno.edu/oceanwaves/2017/posters/2

Schulz-Stellenfleth, J., & Lehner, S. (2002). Spaceborne synthetic aperture radar observations of ocean waves traveling into sea ice. https://doi.org/10.1029/2001JC000837

Summerhayes, C., Dickson, B., Meredith, M., Dexter, P., & Alverson, K. (2007). Observing the polar oceans during the International Polar Year and beyond. *World Meteorological Organisation Bulletin* , *54*(4). Retrieved from https://ane4bf-datap1.s3-eu-west-1.amazonaws.com/wmocms/s3fs-public/article_bulletin/related_docs/oceans.pdf?6DGXFFWhtEF5kVJ0lV8ZkeU0if6i.8P8

Tedeschi, R., Rizzo, C., & Carrera, G. (2001). *Weather conditions along the route New Zealand - Terra Nova Bay*. Genova.

Tolman, H. L. (1991). A Third-Generation Model for Wind Waves on Slowly Varying, Unsteady, and Inhomogeneous Depths and Currents. *Journal of Physical Oceanography*, *21*(6), 782–797. https://doi.org/10.1175/1520-0485(1991)021<0782:ATGMFW>2.0.CO;2

Trull, T. W., Schulz, E., Bray, S. G., Pender, L., McLaughlan, D., Tilbrook, B., ... Lynch, T. (2010). The Australian Integrated Marine Observing System Southern Ocean Time Series facility. In *OCEANS'10 IEEE SYDNEY* (pp. 1–7). IEEE. https://doi.org/10.1109/OCEANSSYD.2010.5603514

Violante-Carvalho, N. (2005). On the Retrieval of Significant Wave Heights from Spaceborne Synthetic Aperture Radar (ERS-SAR) using the Max-Planck Institut (MPI) Algorithm, *77*(4), 745–755. Retrieved from www.scielo.br/aabc

Young, I. R. (1999). Seasonal variability of the global ocean wind and wave climate. *International Journal of Climatology*, *19*(9), 931–950. https://doi.org/10.1002/(SICI)1097-0088(199907)19:9<931::AID-JOC412>3.0.CO;2-O

Propulsive Efficiency of Submarines and Surface Warships over their Operational Speed Ranges

Gregory J. Seil, Advanced VTOL Technologies, gregory.seil@dst.defence.gov.au
Chris Gargan-Shingles, Defence Science & Technology Group, christopher.shingles@dst.defence.gov.au
Dev Ranmuthugala, Defence Science & Technology Group, dev.ranmuthugala2@dst.defence.gov.au
Martin Renilson, Pacific ESI, martin@renilson-marine.com

ABSTRACT

Surface warships and submarines are required to operate over a range of speeds, unlike merchant ships that are usually optimised for one particular cruising speed. For surface warships, the advance ratio is a function of speed due to the change in the resistance coefficient at different vessel speeds caused by wavemaking resistance and sea conditions. Surface warships are required to operate over a range of speeds, e.g. low speed cruising and high speed sprinting. Thus, their propulsors need to be efficient over a wider range of advance ratios and flow conditions, and the design process must take into account the efficiency at low and high speeds.

Deeply submerged submarines maintain a relatively constant advance ratio, and hence constant propulsion efficiency, over their operational speed range. Thus, the non-dimensionalised flow field is essentially the same and independent of speed, enabling the propulsor to be optimised for this condition.

This paper uses generic data for submarine and surface warships to show the change in propeller advance ratios with speed. It is seen that changes in a vessel's propulsive efficiency with speed are related to the resistance coefficient, which is influenced by the wave making resistance, propulsor-hull interaction and the propulsor operating point. The paper provides trends on these effects for submarines and surface warships over typical operating conditions. Unlike surface warships, the change in advance ratio for deeply submerged submarines is small over the operating speed range. Thus, the selection and optimisation of the propulsors require different strategies for surface warships and submarines.

INTRODUCTION

Surface warships and submarines are required to operate over a range of speeds, unlike merchant ships that are usually optimised for one particular cruising speed. For surface warships, the propeller advance ratio (J) is a function of speed due to the change in the resistance coefficient (C_T) at different speeds caused by wavemaking resistance and sea conditions. Surface warships are required to operate over a range of speeds, for example low speed cruising and high speed sprinting. Thus, their propulsors need to be efficient over a wider range of advance ratios (and thus speeds) and flow conditions, and the design process must take into account the efficiency at low and high speeds.

Deeply submerged submarines, where the influence of the free surface is negligible, maintain a relatively constant advance ratio, and hence constant propulsion efficiency, over their operational speed range. Thus, the non-dimensionalised flow field is essentially constant and independent of speed, enabling the propulsor to be optimised for this condition.

The propulsive performance of a generic frigate and a generic submarine is compared in this paper to examine differences in resistance components, propulsor-hull interaction and the variation in propulsor operating point and performance over the speed range of the vessel. Both vessels utilise conventional propellers as their propulsion device for the purposes of the comparison.

The generic frigate is based on DTMB 5415 hull form [1]. Experimental studies of a geosim of the DTMB 5415 hull form (i.e. INSEAN 2340, shown in Figure 1) were performed by Oliveri et al. [1]. Similar studies of a smaller scale geosim of the DTMB 5415 were performed by the Iowa Institute of Hydraulics Research (IIHR) at the University of Iowa. Data for this geosim (IIHR 5512) is available from that University's website [2] and has been used as the basis of the analysis presented in this report. Note: the IIHR 5512 is without appendages such as bilge keels, shafting, rudders and brackets.

The generic submarine utilised in this study is the Defence Science and Technology (DST) Group BB2 geometry, which represents a conventional diesel-electric (SSK) submarine. A model of this submarine, which has been built and tested by MARIN, is shown in Figure 2 taken from Overpelt et al. [3]. The BB2 is modelled with all its appendages.

The propulsive estimates are based on data for the Wageningen B-Series propeller given in [4]. Using the required thrust and J for each vessel, five blade propeller variants were selected using the curves provided in [4]. Full-scale vessel and propeller data for both vessels are presented in Table 1. The propeller curves used in the analysis presented in this paper have been scaled to full-scale conditions.

RESISTANCE OVER SPEED RANGE

A vessel operating on the surface has three primary components of resistance:

1) Viscous resistance: these include skin friction and pressure drag arising from the development of the boundary layer around the hull. Hull roughness increases the frictional resistance.
2) Wavemaking resistance: this is associated with pressure changes on the hull due to interactions with the free surface and flow separation (such as off the transom). This results in wavemaking, sinkage and trim changes.
3) Air resistance – this is the basic aerodynamic drag of the hull and superstructure above the water.

For a submarine operating deeply submerged, there is no wavemaking (due to no interaction with a free surface) or air resistance.

The resistance coefficient (C_T) can be calculated as,

$$C_T = \frac{R_T}{\frac{1}{2}\rho S_W V^2}$$

(1)

where R_T is the resistance of the vessel, ρ is the density of the fluid, V is the speed of the vessel and S_W is its wetted surface area.

For the surface ship, neglecting air resistance, the resistance curve of the IIHR 5512 scaled to full-scale is calculated from the model-scale data of [2] using,

$$C_{TS} = C_{FS} + C_R + \Delta C_F$$

(2)

where C_{TS} is the total resistance coefficient at full-scale, C_{FS} is the frictional resistance coefficient at full-scale, C_R is the residuary resistance coefficient and ΔC_F is the roughness allowance accounting for the fact that the hull is not hydraulically smooth.

The residuary resistance coefficient is generally constant for the model and full scale. The frictional resistance coefficient is calculated using the ITTC 1957 correlation line [5],

$$C_{FS} = \frac{0.075}{(\log_{10} Re - 2.0)^2}\ , \quad Re = \frac{\rho V L_{WL}}{\mu}$$

(3)

where Re is the Reynolds number, L_{WL} is the length of the vessel and μ is the dynamic viscosity of the fluid.

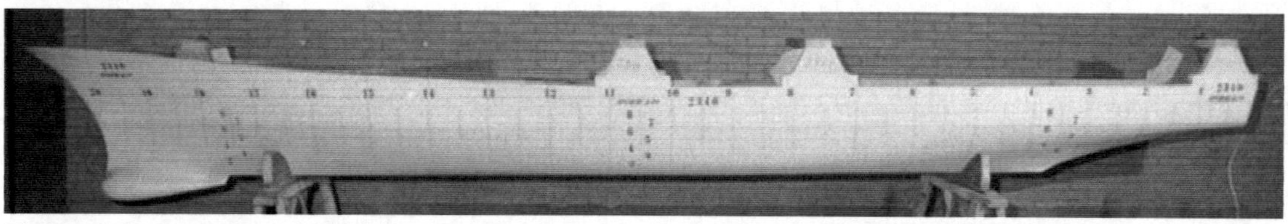

Figure 1: Geometry of INSEAN 2340, photo taken from [2]. Both the INSEAN 2340 and the IIHR 5512 are geosims of DTMB 5415. Scale ratio (λ) of INSEAN 2340 is 24.824 and for the IIHR 5512 is 46.6.

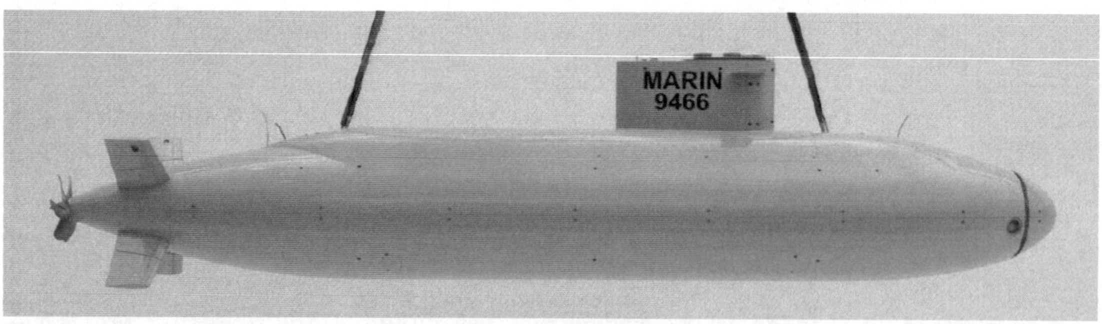

Figure 2: Free running model of BB2. Photo taken from [3].

As there is no wavemaking resistance for a deeply submerged submarine, the wavemaking resistance coefficient, $C_W = 0$. Thus the resistance of the deeply submerged submarine BB2 can be calculated as,

$$C_{TS} = (1 + k)C_{FS} + C_R + \Delta C_F \tag{4}$$

where $(1 + k)$ is a whole-of-boat form factor which includes the effect of both friction and pressure (form) drag. For the BB2 geometry, a form factor of 1.2 is used based on CFD results and C_{FS} is given by equation (3). The residuary resistance coefficient (C_R), which includes other effects such as induced drag, is relatively small and can be neglected in this case.

The full-scale resistance coefficient curves for the IIHR 5512 in calm water and the BB2 deeply submerged are shown in Figure 3 and Figure 4 respectively. For this study, $\Delta C_F = 0.0004$, in accordance with the information provided in Renilson [6].

	Symbol	Frigate	Submarine
Waterline length (or Length)	L_{WL} or L	142 m	70.2 m
Beam	B	18.9 m	9.6 m
Draft	T	6.16 m	NA
Volume	∇	8425.4 m^3	4349.7 m^3
Wetted Area	S_W	2949.5 m^2	2143.5 m^2
Propeller Series	-	Wageningen B-Series	Wageningen B-Series
No. of Propellers	N_p	2	1
No. of Blades per Propeller	Z	5	5
Pitch-to-Diameter Ratio	P/D	1.2	0.8
Expanded Area Ratio	EAR	0.7	0.6
Propeller Diameter	D_S	4.5 m	5.0 m

Table 1: Vessel and propeller data for the IIHR 5512 frigate and the BB2 submarine at full-scale

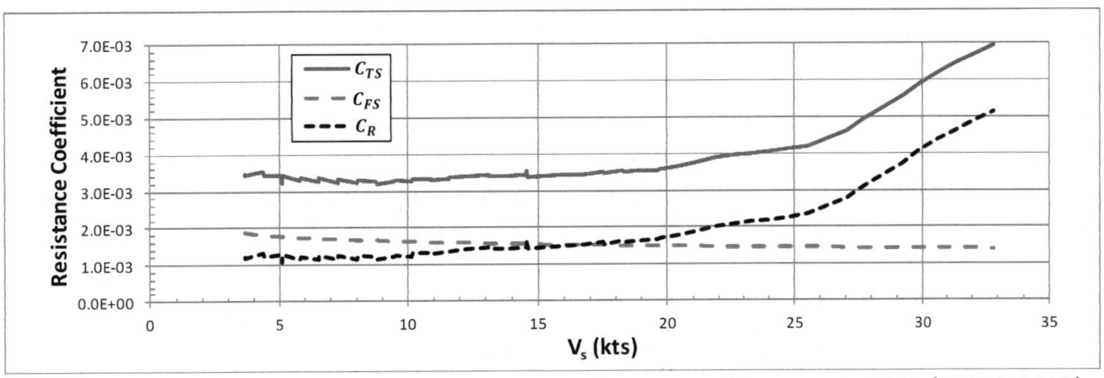

Figure 3: Calm water resistance coefficient curve for generic frigate IIHR 5512 (DTMB 5415) at full-scale

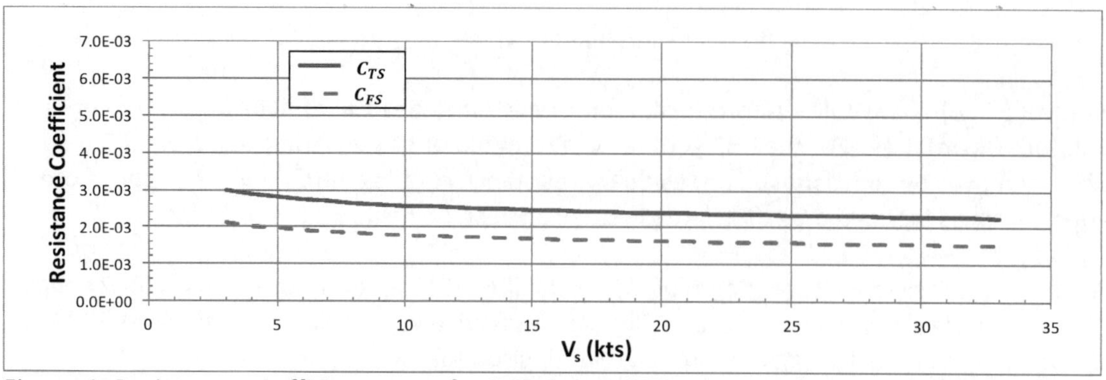

Figure 4: Resistance coefficient curve for DST Group BB2 generic submarine at full-scale

The resistance coefficient curves of both vessels show opposite trends with increasing speed. The general trend of increased C_{TS} with increased speed for the frigate, particularly for speeds greater than 20 kts, is driven by the increase in C_R. This is mainly due to the wave making resistance coefficient that increases at higher vessel speeds. The C_{TS} for the submarine on the other hand decreases slightly with increased speed due to Reynolds number dependency.

PROPELLER CURVES

The operating point of the propulsor can be characterised by its advance ratio (J),

$$J = \frac{(1-w)V_S}{nD_S} \tag{5}$$

where n is the propeller or rotor speed (in revolutions per second), D_S is the diameter of the propeller or rotor and w is the effective wake fraction. At each value of J, the performance of the propulsor is given by its thrust coefficient (K_T), torque coefficient (K_Q) and open water efficiency (η_o) defined by,

$$K_T = \frac{T}{n^2 D_S^4}, \quad K_Q = \frac{Q}{n^2 D_S^5}, \quad \eta_o = \frac{J}{2\pi}\frac{K_T}{K_Q} \tag{6}$$

where T is the thrust produced by the propeller and Q is the rotor torque. The propeller curves for the propellers used in this study and based on the Wageningen B-Series [4] are presented in Figure 5. The pitch-to-diameter ratio was chosen to provide a design advance ratio close to the peak propulsor efficiency.

PROPULSIVE PERFORMANCE

The thrust of the propulsor (T) is related to the total resistance of the vessel (R_T) by the thrust deduction fraction (t),

$$t = 1 - \frac{R_T}{T} \tag{7}$$

The propulsive efficiency (η_D) is given by,

$$\eta_D = \eta_H \eta_R \eta_o \ , \qquad \eta_H = \frac{1-t}{1-w_s}, \qquad \eta_R = \frac{K_{QT}}{K_Q} \tag{8}$$

where η_R is the relative rotative efficiency, η_H is the hull efficiency and η_O is the open water efficiency of the propeller. The relative rotative efficiency is determined in this case by the ratio of the actual torque coefficient (K_Q) to the torque coefficient determined from the propulsor curve (K_{QT}) when using the thrust identity [7]. It can be shown that the thrust coefficient can be related to the operating point of the propulsor by [7],

$$\frac{K_T}{J^2} = \frac{1}{N_p} \frac{S_W}{2D_S^2} \frac{C_{TS}}{(1-t)(1-w_s)^2} \tag{9}$$

where N_p is the number of propulsors. Calculation of the right-hand side of (9) allows the value of K_T/J^2 to be determined. Using this value in conjunction with the propulsor curves, the advance ratio (J) and the propulsor open water efficiency (η_o) can be determined. For more information on these calculations see Renilson [6] and ITTC [7].

An analysis of the self-propulsion performance of the deeply submerged submarine and the frigate was performed for their respective speed ranges. A propeller diameter of 4.5 m was chosen for the frigate and a diameter of 5.0 m for the submarine. Note that the frigate has two propellers while the submarine has a single propeller. It is assumed that that effective wake fraction (w), thrust deduction fraction (t) and relative rotative efficiency (η_R) remain constant over the speed range.

Note that no account has been made of the effect of cavitation inception on the propeller efficiency when the frigate is operating at high speed. The variations of the advance ratio (J) of the propeller operating point for the full scale IIHR 5512 frigate and the deeply submerged BB2 submarine over their respective speed ranges are marked up on the propeller open water diagrams in Figure 5. The propulsive efficiency of both vessels is compared in Figure 6. The variation of propeller speed over the speed range of each vessel is compared in Figure 7.

For both vessels, there is relatively little change in the propulsive efficiency (η_D) over the lower speed range, however, it drops offs significantly for the frigate as the speed increases. In both cases, the change in propulsive efficiency is driven by changes in the propulsor open water efficiency (η_o), since the hull efficiency (η_H) and relative rotative efficiency (η_R) remain constant. The trend in propulsive efficiency for each vessel follows the trend in their respective resistance curves.

The propulsor on the frigate becomes more heavily loaded at speeds greater than 20 kts resulting in a significant reduction in the advance coefficient (J), an increase in thrust coefficient (K_T) and a drop in absolute propulsor open water efficiency (η_o) of around 5%. Note that there is a very small variation in advance coefficient and propulsor efficiency between 3.63 kts (the first data point) and 20.40 kts due to the relatively constant resistance coefficient between these speeds. It is beyond this speed that the wave making resistance effects the efficiency as the variation in J is of sufficient magnitude to adversely affect η_o even at the 'flatter' section of the curve, i.e. close to its peak.

For the deeply submerged submarine, the propulsive efficiency increases with increased speed as the propulsor becomes less loaded (i.e. increased J) and η_o increases. Above 10 knots, the propulsive efficiency (η_D) is almost constant as the change in J is extremely small and the propulsor efficiency (η_o) is close to the peak value on the efficiency curve, i.e. in the flatter section of the curve.

The frigate shows a linear variation in propeller speed with vessel speed (i.e. Turns Per Knot, TPK) up to 20 knots, however above this speed the variation becomes non-linear (rising above the linear trend) due to the increasing wavemaking resistance coefficient. The deeply submerged submarine shows a linear variation over its speed range due to its much smaller variation in the resistance coefficient.

Note that the changes in the advance ratio and thrust coefficient over the speed range are much greater for the frigate than the deeply submerged submarine. Hence the operating range of advanced ratios (J) of the propeller is smaller for the deeply submerged submarine than the frigate. In other words, the propulsor on the frigate must operate over a greater range of off-design conditions, whereas the propulsor on the submarine operates over a narrow band of operating conditions. From Figure 6 it is also seen that the change in propulsive efficiency for the frigate over its speed range is much greater than the change for the deeply submerged submarine.

DISCUSSION

The propulsive efficiency of the frigate dropped above a vessel speed of 20 kts due to the higher propulsor loading resulting from the greater wavemaking resistance experienced by the vessel. This reduction would be even greater if the added resistance due to adverse sea conditions, propulsor cavitation, or unsteady propulsor inflow effects due to the wake generated at the aft end of the frigate hull were included. On the other hand, the propulsive efficiency of the submarine was, for all intents and purposes, insensitive to the operating condition due to the small changes in propeller efficiency over the speed range.

If the efficiency of the propulsor is more sensitive to loading (i.e. if the vertical lines in Figure 5a and Figure 5b, denoting the operating points, moved proportionally further apart), then a greater change in propulsor efficiency (η_o) would be evident, resulting in a greater change in propulsive efficiency at off-design conditions. However, due to the greater dependency of the frigate resistance coefficient on the vessel speed (i.e. due to the wavemaking resistance), this would be particularly pronounced for the frigate and much less so for the submarine.

If the thrust and torque coefficient curves were 'flatter' (for example due to slightly different assumptions in design parameters), then a change in resistance coefficient (C_{TS}) over the speed range will result in greater spread in the advance ratio (J) about its design point, resulting in a larger change in the propulsive efficiencies. Again, the spread in the advance coefficient would be more pronounced for the frigate than for the submarine.

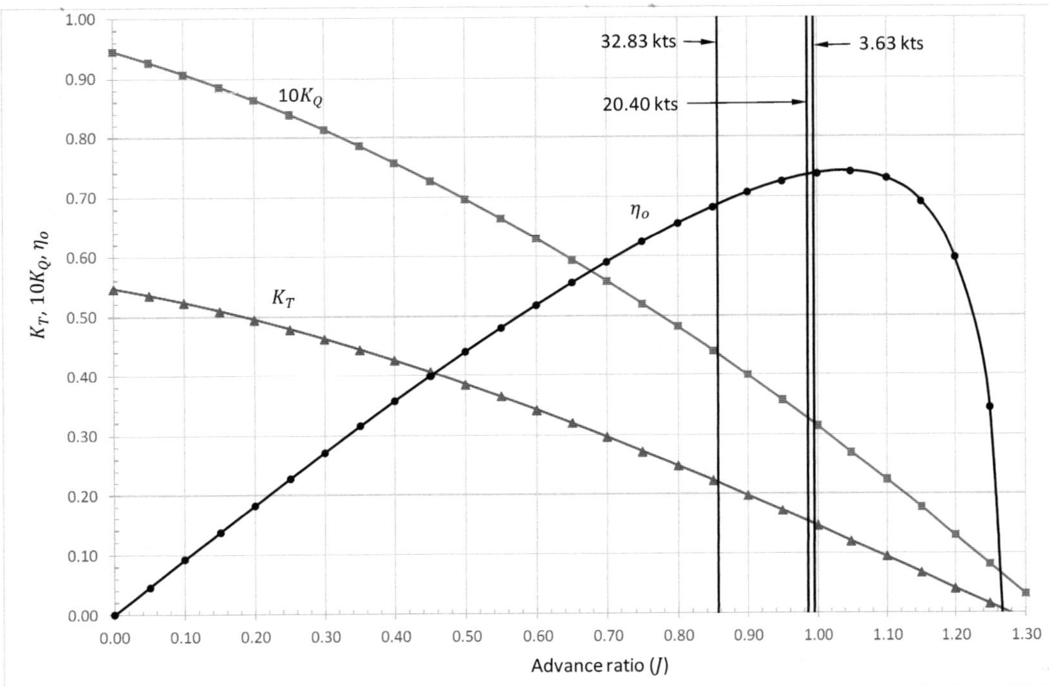

a) Performance curves for the full scale IIHR 5512 frigate
$t = 0.1, w = 0.03, \eta_R = 0.98$

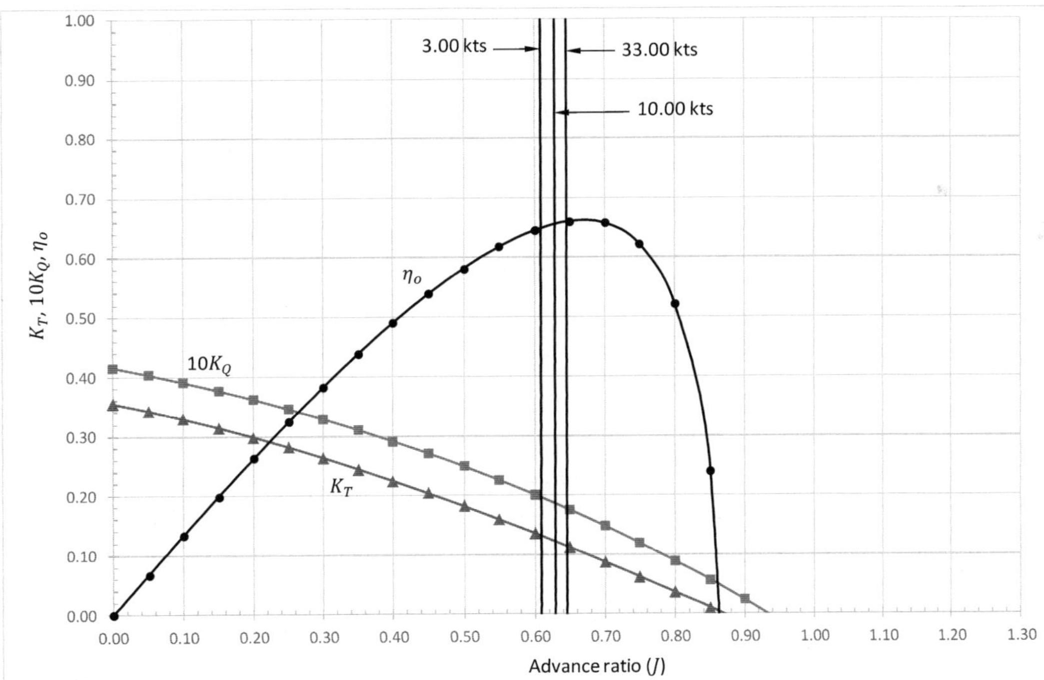

b) Performance curves for the full scale deeply submerged BB2 submarine
$k = 0.2, t = 0.22, w = 0.32, \eta_R = 1.01$

Figure 5: Propulsive performance curves showing changes in operating point across the speed range.

In the analysis a constant value of roughness allowance was assumed. In reality this may not be the case and the roughness allowance will change as a function of speed. For the submarine, as the speed of the vessel increases, the boundary layer becomes thinner relative to the size of the vessel. This increases the resistive effect of surface coatings, protuberances, and openings on the hull, and therefore the contribution of these factors to the overall resistance coefficient will be larger at greater speeds. Hence, once these additional contributions are added, it may be expected that the resistance coefficient will increase for both the frigate and the submarine. However, due to the shape of these curves (see Figure 3 and Figure 4), the effect will be more pronounced for the frigate, thus increasing the spread of J as the vessel speed increases, especially at the higher end of the speed range.

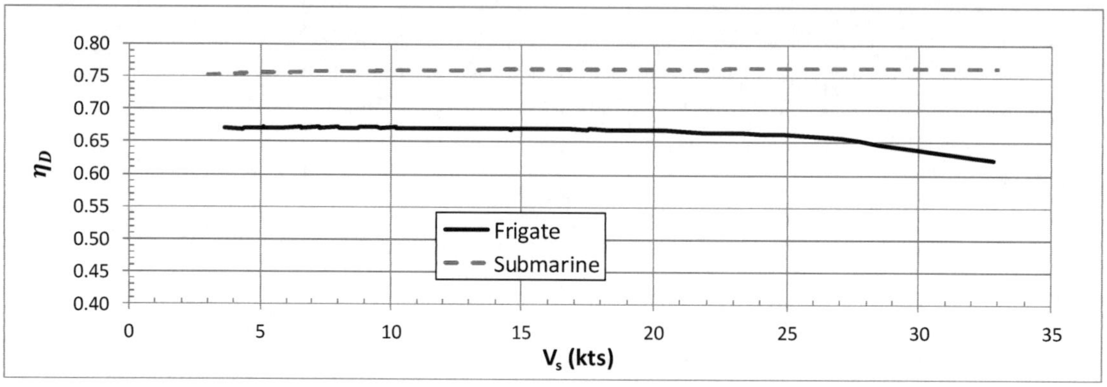

Figure 6: Propulsive efficiency over the speed range for the IIHR 5512 frigate and the deeply submerged BB2 submarine.

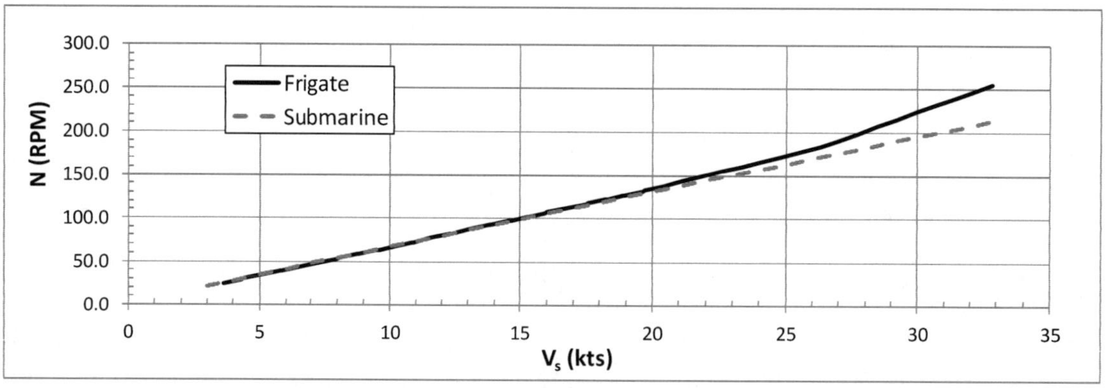

Figure 7: Variation in propeller speed over the speed range for the IIHR 5512 frigate and the deeply submerged BB2 submarine.

The above analysis for the frigate has not considered waves (sea state), dynamic sinkage and trim effects which all affect the vessel's resistance, thrust deduction fraction (t) and wake fraction (w), which in turn affects the hull efficiency (η_H) and the propulsive efficiency. This is obviously not applicable to a deeply submerged submarine.

CONCLUSIONS

From the analysis presented above, the propulsive efficiency of a deeply submerged submarine was shown to vary little over its speed range. The frigate showed a much greater variation, with an absolute variation in the propulsive efficiency of just under 5%, which can be significant.

The propulsor designed for a frigate must be able to operate successfully over a wider range of operating conditions, and hence off-design conditions, than a submarine. In other words, the frigate propulsor must represent a design compromise between low-speed cruising and high-speed sprint. However, the relatively narrow band of operating conditions of the deeply submerged submarine, allows the submarine propulsor to be optimised around a single operating point. There are other factors that will also adversely affect the efficiency of a surface vessel, such as the surface sea state conditions, dynamic sinkage and trim changes that do not apply to a deeply submerged submarine.

Thus, the selection and optimisation of the propulsors for surface warships and deeply submerged submarine require different approaches and strategies due to:

- Opposite trends in resistance coefficient as functions of speed: the resistance coefficient of a surface warship increases rapidly at high speed due to the rapid increase in the wavemaking resistance coefficient as a function of speed, whereas for a deeply submerged submarine the resistance coefficient remains relatively constant.
- The submarine propulsor operates over a significantly narrower range of advance ratios than for a surface warship.

REFERENCES

1) Oliveri, A., Pistani, F., Avanzi, A., Stern, F. and Penna, R. (2001), "Towing Tank Experiments of Resistance, Sinkage and Trim, Boundary Layer, Wake, and Free Surface Flow around a Naval Combatant Insean 2340 Model", IIHR Technical Report, No.421, September 2001.

2) https://www.iihr.uiowa.edu/shiphydro/efd-data/5512-steady/ . Accessed 29/04/2019

3) Overpelt, B., Nienhuis, B., Anderson, A. (2015), "Free Running Manoeuvring Model Tests on a Modern Generic SSK Class Submarine (BB2)", *Proc. Pacific 2015 International Maritime Conference*, Sydney, 6-8 October 2015.

4) Bernitsas, M. M., Ray, D. and Kinley (1981), "K_T, K_Q and Efficiency Curves for the Wageningen B-Series Propellers.", Report No. 237, Department of Naval Architecture and Marine Engineering College of Engineering, The University of Michigan Ann Arbor, Michigan.

5) ITTC 7.5-02-02-01 Rev. 04 – Resistance Test.

6) Renilson, M. (2018), *Submarine Hydrodynamics*, 2nd edition, Springer.

7) ITTC 7.5-02-03-01.4 Rev. 04 – 1978 ITTC Performance Prediction Method.

Naval Hybrid Power Take-Off and Power Take-In Solutions - Enabling Technologies and Intricacies of Power Systems Operations

Author: Dr Makhlouf Benatmane, BSc (Hons) PhD CEng FIMarEST FIET.
Email: Makhlouf.Benatmane@ge.com

Author biography:
Makhlouf Benatmane has extensive experience in electrical systems engineering, in industrial and marine applications. He has been a University lecturer in Power System Design, Electrical Power Station Design and Power Electronics. He holds a PhD in Electrical Engineering from The University of Nottingham and a BSc(Hons). He is a Chartered Engineer, Fellow of The Institute of Engineering and Technology, Fellow of The Institute of Marine Engineering, Science & Technology. He is currently the Marine Naval Solution Leader Marine for GE Power Conversion.

Affiliation: GE Energy Power Conversion UK Ltd, Rugby, UK. © GE

SYNOPSIS

Naval vessels have increasingly diverse operational profiles and governments face tightening of budgets and new environmental legislation. This leads to challenges for new vessel programmes to improve efficiency - both operational costs and energy consumption, deploying technologies that meet regulations whilst still enabling the naval mission to be carried out.

A variety of power and propulsion architectures have been used over the years. Each has its own benefits, but selection should consider what is most suited to the operational role and mission effects. Amongst propulsion systems available are several hybrid-electric applications, either in service or currently in build, such as the UK Type 26 City Class and the future Australian Hunter Class ASW Frigates.

Hybrid propulsion is not new; the latest hybrid solutions combine main diesels or a gas turbine engine with an active-front-end converter in combination with an induction motor, that can operate as both a motor (PTI) and a generator (PTO), to drive a controllable or fixed pitch propeller, either through a gearbox or direct. PTO/PTI can improve fuel efficiency by harnessing excess mechanical energy from a vessel's propulsion shaft.

The focus of this paper is to:

- Provide a hybrid propulsion concept overview, key electrical, mechanical qualities and issues leading to the shift towards such systems
- Describe different configurations and performances of hybrid propulsion from demonstrated and operational systems in the commercial and naval world
- Describe technologies used and systems modes of operation
- Cover lessons learnt in technologies and controls used of such systems
- Examine future architectures including energy storage and explore the benefits and flexibility these can bring to the hybrid propulsion sphere.

1. INTRODUCTION

The industrial revolution was driven by one of the greatest inventions of the 20th century: the combustion engine. For the majority of applications, it has been the power and propulsion behind almost the entire marine ecosystem. However, technology advances, coupled with a more challenging economic environment and the need to reduce carbon emissions, are now changing the way we can operate as an industry. While there is still a place for combustion engines, the marine ecosystem is more widely embracing electrification solutions, in a bid to become cleaner, smarter and more efficient.

Some key objectives for a design of a complex navy ship are to have versatility - the ability to have enough installed power *and* meet the trade-off between efficiency, adaptability of diverse operational profiles. Ship loads are increasing to the point where mission power requirements are close to that of the propulsion system. Therefore, a departure from the traditional mechanically driven propulsion with prime movers via a gearbox must be considered, such as integrated electric power and propulsion (IEP) or a hybrid electric system that may be better aligned to address such needs. Indeed, hybrid propulsion is not new to the navy world, it is being increasingly employed for frigates operating a low proportion of time at full speed and spending significant periods at low power, e.g. below 40-50% of their top speed.

2. SELECTION PRINCIPLES OF POWER AND PROPULSION SYSTEMS

The selection and design of a mechanical or an IEP[9] propulsion system follow a similar process and can be complex. The key points to consider are:

- Initial Concept of Operations (CONOPS)
- Propulsion power requirements, per shaft, per motor including torque/speed curves at rated and degraded parameters.
- Maximum and cruising speed powering requirements – link requirements to mission profile, systems and total ship mission effectiveness.
- Type and size of prime movers, e.g. gas turbine (GTG) or diesel generators (DG), allowing most economical fit to meet redundancy and powering requirements (total generation, overloads, ship services including end of life growth estimates).
- Environmental requirements especially relating to operation in higher temperatures and graceful degradation permitted. Both for cooling water (fresh/sea/chilled) and ambient air temperatures.

- Quality of power supply requirements.
- Power requirements and system fault level at high, medium and low voltage to determine most appropriate system voltage and equipment fault level capabilities.
- Redundancy groups, distributed system architectures, how the system will be split for resilience, i.e. graceful degradation after failure/damage.
- Simplified post shock event operational requirements.

The use of mechanically driven propulsion systems has its advantages where other loads of the ship are relatively low compared to the propulsion loads (less than 10%), but its main drawbacks are:

- Its power is dedicated 100% to use for propulsion
- Propulsion diesel or gas turbines must run at high speed for best efficiency, so a high ratio reduction gear is essential to obtain any economical propeller speed. Efficiencies of such engines are generally low at slow speed/power.
- Less flexibility to scale for future power and energy needs which are not defined to influence the 30-year ship acquisition plan.
- Complex shaft-line with large gearboxes

2.1 Integrated Electric Power and Propulsion

Many IEP navy platforms are now in service, with the Royal Navy, the US Navy and the French Navy, leading the way.

A typical IEP system will comprise prime movers (gas turbine or diesel engines) driving generators to produce electric power.

Most generating systems employ more than one generator. Such generators may be of equal or different ratings and speeds. However, they all must work in parallel and share load proportionally. Droop characteristics of voltage and frequency are employed with the voltage regulator reference drooping pro-rata with reactive current and the governor speed reference drooping with applied shaft load. This facilitates load sharing of the parallel connected generators particularly when they are of unequal ratings or mixed engine types. Electrical design of generators must be co-ordinated with the switchgear to ensure fault levels and protection are achieved.

This electric power is then fed to switchboards which distribute the power to various ship consumers. IEP interfaces with the vessel's control and automation system.

The advantages of a central power plant configuration of an IEP are well documented for vessels ranging from auxiliary and support vessels through to surface combatants and even aircraft carriers e.g.:

Increased vessel safety: Redundancy in the power generation plant and propulsion drive trains enable the system to continue to supply power to the propellers even in the event of multiple failures. Electrical drive trains designed to be quiet and shock capable.

Higher flexibility: In addition to facilitating flexible machinery layouts, for example to contain fire and flood damage; the system allows the power plant to be electrically configured to suit the vessel's different operating profiles. Worth also noting that higher energy power demands for future sensors, radars and advanced weapon systems continue to be primarily electrical and such advanced features will only increase over time and may be called to operate simultaneously with propulsion systems.
Lower operating, through life costs: Achieved from savings in fuel and maintenance costs by running the minimum number of generator-sets at their optimum loads.
Electric drive train provides for a simplified shaft line and eliminates the need for costly complex gear boxes.
Enhanced availability, reliability and maintainability: Achieved by operating the inherently robust plant most efficiently and by permitting preventative maintenance work to be carried out during voyages. IEP systems can be reconfigured for efficiency and battle resilience, recovery and availability.

As in almost all industries, there is a push to achieve more with less for marine organisations. Some vessel operators are realising the value that electrical solutions bring in terms of increasing asset reliability and vessel longevity.

2.2 Hybrid Electric Power and Propulsion

Hybrid propulsion systems, introduced almost 40 years ago, take advantage of the best of both worlds of propulsion, conventional mechanical and full electric, to make a system better suited for some specific type of vessels, especially where the objective is to focus on efficient and where necessary quiet operation.

There are a wide range of hybrid configurations [5,7]; such as a high-speed motor to a gear box Figure 1a, or motors directly on the propeller shaft line, Figure 1b.

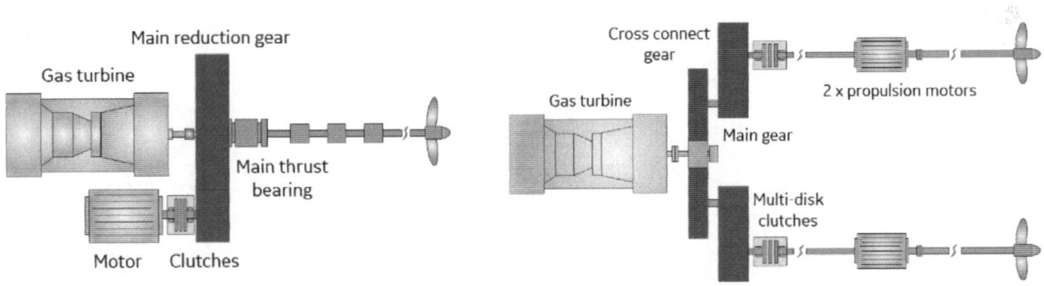

Figure 1: CODLAG Hybrid Propulsion Configuration – High and Slow Speed Electric Motor

The main benefits of hybrid electric propulsion are:

- Fuel savings by using the smaller electric motor for speeds up to 15-20 knots thereby reducing the need for large main engine operation.
- Reduced maintenance on the main engines due to reduced run time by using the hybrid electric motors (HEM).
- Main engines and hybrid electric motors can be operated independent of main gear.
- Enhanced fuel economy in all modes of operation, especially for vessels with fluctuating load demands and a mixture of transit and slow speed operation.

- Lower noise signature for Anti-Submarine Warfare (ASW) operations.
- Additional installed power for future power availability, accomplished by powering the ship using the main mechanical engine and using the hybrid electric machine system to slow down the ship enabling the hybrid motor to act as a generator power take off (PTO) that can feed pulsed power applications. This power take-off capability is not used on all ships.

3. HYBRID PROPULSION SYSTEMS CONCEPT CONFIGURATIONS

The focus of this paper is on a **CO**mbined **D**iesel e**L**ectric **A**nd **G**as (CODLAG) configuration, as represented in the single line diagram (SLD) at Figure 2. This hybrid system with one gas turbine via a reduction gearbox with two shafts, each with a hybrid electrical motor (HEM), drive a controllable pitch propeller (CPP) and could be a simplistic representation of an Anti-Submarine Warfare Frigate (ASW) or the Global Surface Combatant currently on order by the Royal Navies of UK, Australia and Canada.

The paper does not give any descriptions of the systems' operations for such frigates or details of equipment but covers concepts or possible concepts of operations and beneficial additions of typical CODLAG systems.

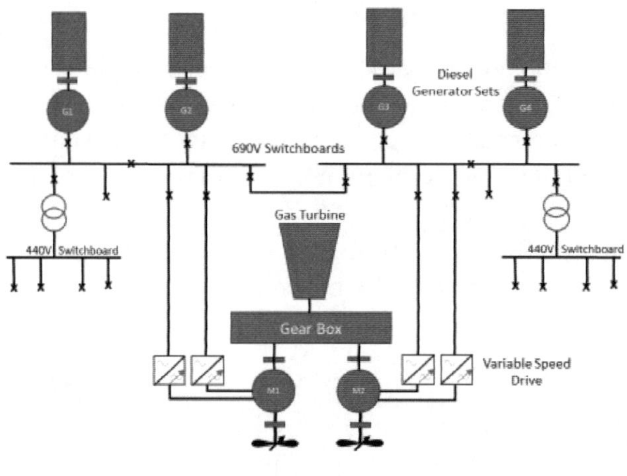

Figure 2: CODLAG Hybrid Propulsion

The propulsion and the electrical power systems have multiple operating configurations. The chosen operating mode primarily depends on a combination of the desired ship speed, ship mission and its effectiveness and the total electrical load demand on the system. The actual design of the system and the ratings of the prime movers requires careful consideration, such as total required propulsion power, desired 'efficient and quiet' electric propulsion load, maximum electrical load (including sufficient through life design margin), and often availability of products to meet most requirements at affordable cost.

The HEM is of induction machine design, with multiple phases and each group of three phases has its own dedicated pulse width modulated power variable frequency drive (VFD) to vary its frequency and therefore speed. The compact induction motor has been selected as it is inherently simple, has fewer components, has robust, reliable, well-proven technology and is successfully used on many navy platforms (destroyers and aircraft carriers).

The VFD is able to provide accurate and smooth control of both the voltage and the frequency of the waveform across a wide range of motor speeds. The use of an active-front-end (AFE) converter i.e. bidirectional, results in a very high-quality generated waveform suitable for quiet mode operation of the HEM and direct connection to the main distribution network, without the need for isolation transformers.

Both HEM and VFD as a drive train and within the full-scale electrical system have been through extensive proving testing to demonstrate the ultra-quiet function at various speeds and loads and have been barge shock tested. On some arrangements the electrical motor has the added flexibility to operate as both a motor (Power Take In – PTI) and a generator (Power Take Off – PTO) with the excitation from the VFD, providing additional electric power for the ship, and it is connected to the main distribution network via a bi-directional power electronic converter. This, in addition to diesel generators sets (DGs), can supply power for either electric propulsion or the ship's service loads. Note that the PTO mode is not selected to ASW platforms currently in build.

3.1 Full and Top Speed:

The arrangements below give possible hybrid systems configurations and operating modes:

Figure 3a. GT fully driving both propellers; HEM off; DGs providing ship's service electric loads

Figure 3b. GT can be boosted by HEMs driving both propellers for maximum speed; DGs providing ship's service electric loads.

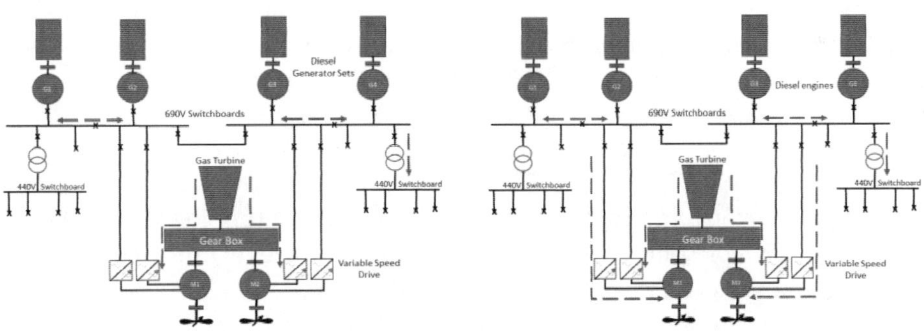

Figure 3: (a & b) CODLAG Propulsion by GT and topped up by HEM

3.2 Quiet Mode:

Figure 4 HEM driving propellers in PTI (mode); DGs providing propulsion and ship's service electric loads. GT / gear box not running

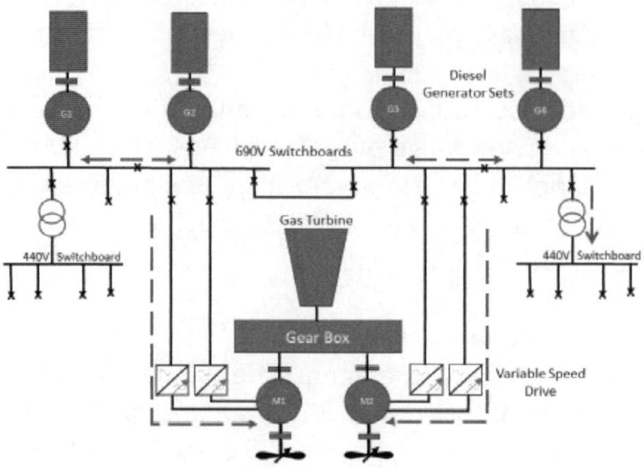

Figure 4: CODLAG Propulsion by HEM

Naval vessel design can be driven by ASW requirements, with a need for low noise and vibration (N & V) signature and very quiet operations at slow to medium speeds: "Detect the enemy before they detect you". Electric propulsion systems can meet highly demanding acoustic signature specifications, in part because gearboxes and their auxiliaries are not required. This allows naval vessels to operate with lower risk of detection.

Advances in electrical machines and their associated converter drives in electromagnetic design, waveform smoothing, anti-vibration technology along with shock proof motors achieve a much quieter operation than can easily be obtained by traditional mechanical systems.

The ability of the system to operate without a gearbox improves its maintenance regime, the efficiency of the drive train and the noise signature (highly relevant to ASW frigates).

3.3 Electrical System Complemented by Energy Storage

As stated above, arrangements below do not form current topologies of ASW electric hybrid systems but could easily be envisaged in the future. However, such configurations need to consider type and size of prime movers, ship services' power requirements, system architectures, how the system will be split for resilience and space ...etc.

Energy Storage System (ESS) [11, 14] can provide multiple additional benefits [12] such as:

- Running lower numbers of generators online at higher load due to reducing limits for spinning reserve. ESS acts as online backup or additional power reserve instead of generator power reserve, providing available power in case of a generator's failure for a predetermined duration.
- Improves system stability to accommodate response time of engines to load changes.
- Enables reducing load ramp limits at thrusters and quickly available thrust force. This enables fast responsiveness to environmental forces, enhances the manoeuvring capabilities of the vessel, thus increasing safety against collision and grounding.

- ESS can reduce the power system vulnerability to faults in externally supported systems, provide power to essential auxiliary systems, offer additional benefits if self-supported and not dependent on external systems, especially during possible power system blackout.
- Reduces operating costs due to enhanced fuel consumption and lowers maintenance costs when engine(s) cyclically loaded.
- Reduces the risks of blackout by operating ESS as 'Uninterruptable' Power Supply (UPS), and in case of blackout enables quick power system recovery, faster than when emergency generators are used.
- Depending on the operating load profile, ESS can reduce the number of generator start/stops and make power available during start-up of stand-by generator(s).

In addition to above advantages, ESS could power HEM providing an even quieter mode of operation in naval applications without any GT or DGs running.

In light of new, up-coming regulation, emission free modes, taking power from ESS, cannot be ignored when the vessel operates in Emission Control Areas (ECAs), close to harbour, or entering/leaving harbour or during short stays in harbour where shore supply is not available.

The ESS duration of support could typically vary from just a few minutes in vessel manoeuvring, up to 20-30 minutes in transit, and could be extended up to one or two hours if the battery can be recharged at shore.

With the flexibility of electrical systems, ESS could be connected to a main switchboard through a power electronic rectifier or to the propulsion VFD DC link or through another optional DC/DC rectifier.

Configurations with ESS are shown at Figure 5.

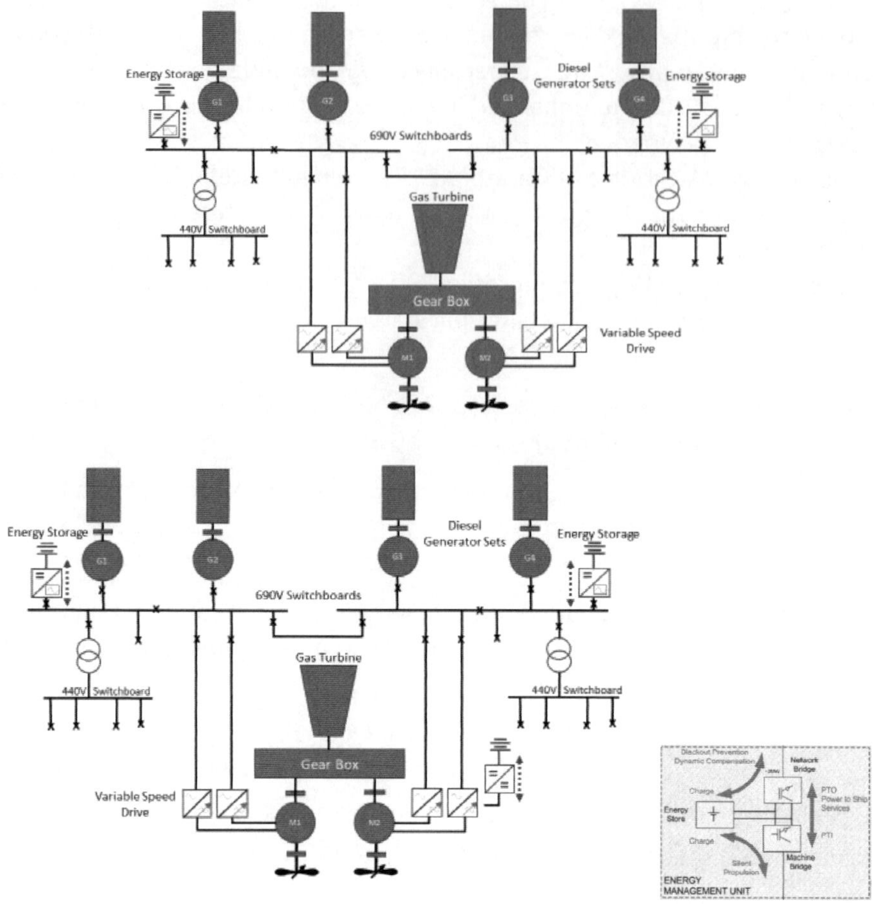

Figure 5: Electrical system complemented by energy storage

3.4 Power Take Off Generator (PTO) – Figure 6:

GT driving both propellers; HEM on as PTO; reduced number of running DGs providing ship's service electric loads. This mode normally used in transit sailing harnesses the excess mechanical energy from the vessel's propulsion shaft. The GT not only powers ship propulsion but could supply electric power for the ship's domestic consumers. This mode will lead to high loading of the GT with low specific fuel oil consumption, minimal emissions and impact on the maintenance period of the DGs as they can be switched off when not needed.

Additionally, with a suitable bi-directional converter sized converter for the hybrid PTI/PTO , an option exists to make use of this as a frequency changer for the shore supply to the ship of different grid frequency The hybrid converter in this case can be used as a static frequency converter (SFC) on board allowing the seamless import of power from shore, or export from ship; helpful functionality can be in a disaster relief situation.

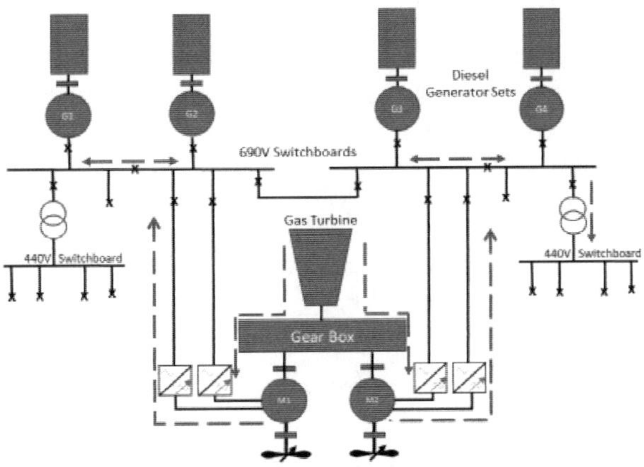

Figure 6: Power Take Off Generator (PTO)

4. OVERVIEW OF ADVANTAGES AND THEIR APPLICATION

There are several hybrid applications that have either recently been brought into service or are currently in build and these illustrate the range of potential benefits that a hybrid system can bring. No single application makes use of all those features, but each selects the range of benefits that is most suited to their operational role. A summary of benefits and the vessel types are illustrated in Table 1**Error! Reference source not found.**:

	Direct Drive Frigate Anti-Submarine Warfare (ASW)	**Direct Drive Commercial**	**Geared Auxiliary Vessel**	**Geared Light Frigate**	**Geared Assault Ship LHD/LHA**
Reduced Engine Count		✓	✓		
No Gearbox	(✓)	✓			
Efficiency	✓	✓	✓	✓	✓
Reduced N&V	✓			(✓)	
Electrical Energy	(✓)	✓	✓	✓	✓
Survivability	✓			✓	✓
Shore Supply / Disaster Relief				✓	

Table 1: Example Recent Hybrid Systems

4.1 Shock-Rated Electric Limp Home Mode

Frontline Naval ships typically demand shock-rated equipment for their propulsion systems; however, adding shock-rating to an entire shaft line is an expensive option. This would be especially true for a hybrid system if both the electrical and mechanical propulsion systems required shock hardening.

An alternative option is to only shock harden the HEM drive train, along with another generator(s). This greatly reduces the cost and burden of a complete shock hardened overall propulsion system, whilst providing an electric 'limp-home' mode which can still be used following a significant shock event.

5. FUNCTIONALITY CHALLENGES AND ADVANTAGES

5.1 Propulsion Control System

The complexity of different modes of operation of a CODLAG system requires a propulsion controller able to:

- Perform several dedicated functions to safely manage the various propulsive modes of the vessel.
- Consider the different propulsion modes and their changeover, and in different quadrants in the case of PTI/PTO, as illustrated in this paper.
- Represent the dynamic behaviour of the machinery performance and manoeuvrability of the vessel.
- Select the proper matching between prime movers and propellers loads, take advantage of the simultaneous use of GT and HEM with the ship navigation and manoeuvring conditions in mind.

5.2 Shaft Load Dynamics

Significant torsional load changes[3] on a propulsion shaft line are typically gradual with the changes taking place over a number of seconds. However, torsional load on a *generator* shaft line can change very suddenly with a large load application or rejection on the electrical distribution system. For this reason, engines that are designed to operate as the prime mover of a generator have demanding performance standards, applied by the class societies, that are not normally applied to propulsion engines and it may be expensive to do so.

5.3 System Component Selection

A key difference between sizing a converter for duty as a PTI motor drive and sizing it for use as a PTO generator [1,2] is that a PTI drive has its input voltage sustained by the distribution network, but a PTO generator must sustain the voltage downstream of its network filter. This means that the power electronics must counteract the effect of volt drop in that filter. A key factor that gives rise to volt drop is the amount of reactive current that must be supplied by the converter i.e. the power factor of the system. An AFE drive has the luxury of always being able to draw its power at near unity power factor but a PTO converter must supply its power at the power factor dictated by the distribution network and the connected loads.

5.3.1 Protection Implications of PTO

There are a number of system related issues that arise when integrating PTO with a vessel. A common certification class requirement is that for the PTO generators to be classed as primary generators and included in the ship's load list they have to provide full discrimination for protection. If they were not primary generators then their supply of power to the system could not be counted when looking at electrical loading for given operational scenarios, which would result in extra DGs being required. In order, for a PTO generator to provide discrimination it needs to be able to provide sufficient short circuit current in the event of a fault.

Some of these issues arise out the fact that PTO is tying one system, the propulsion system, to another, the electrical distribution system. Others relate to the fact that a new technology is being used as a source of energy (i.e. based on solid state silicon) that has different underlying behaviours from the conventional technology (based on rotating copper and iron). Comparatively a power electronic converter has very limited overload capability since its thermal mass is much less, and beyond the limit the controller will normally trip the converter to protect the devices. [(1,2))].

5.3.2 Avoiding Hybrid Converter Trips

VFDs are important assets and protect themselves from a faulty distribution network often by tripping and shutdown requiring an operator to reset the controller and bring it back online, e.g. a significant fault that causes a potential fault current greater than maximum overload capability. Such an event can be lengthy, and, depending on the operational scenario, might have a major impact. This philosophy cannot be followed when the VFD is acting as the power source (i.e. the PTO generator), a process of current management with the controller software code needs to be applied to maintain the converter connected.

5.3.3 Current Management for Selectivity

A conventional generator has the natural property of delivering a significantly higher current than its maximum load current when a short circuit is applied. Sensing-circuit and protection relays deduce that there is a downstream fault and open associated circuit breakers to isolate this fault. A converter, however, does not have this natural property. It is normally sized to deliver its load current with only a small overload capability. Where the protection settings can be integrated with the electrical power system configuration this can be managed with a simple current clipping feature in the converter [(1,2)]. However, where a simple, conventional protection scheme is used the converter must be designed to deliver the correct fault current in the event of a fault. This is illustrated in Figure 7. This shows current and voltage traces during a fault event.

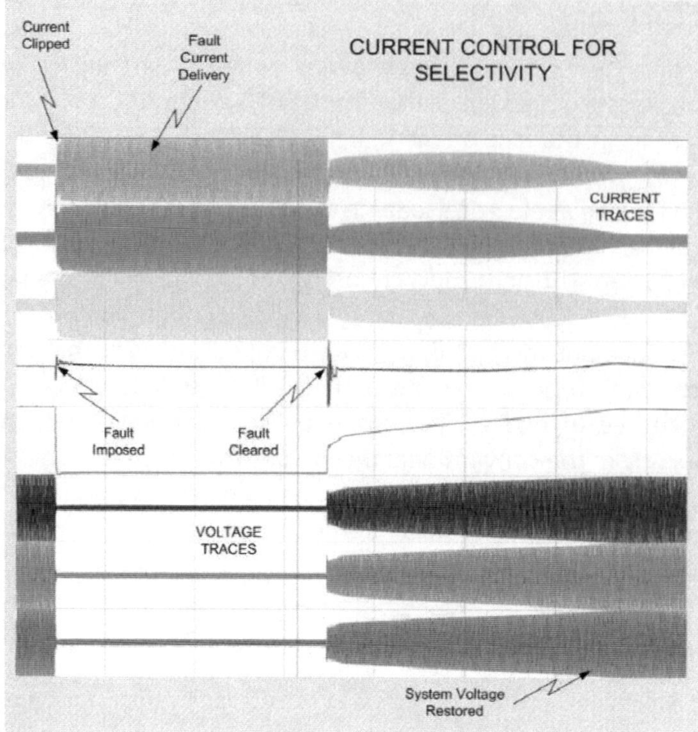

Figure 7: Current Control for Selectivity

5.4 Harmonics

A converter imposes time harmonics on the network to which it is connected. On a three-phase system where a converter has a Diode Front End (DFE) network bridge then those harmonics occur at $F_O(6N\pm1)$, where F_O is the fundamental frequency and N is an integer. For a modern PTO system with an Active Front End drive (AFE) the network bridge uses transistor switching and the harmonics are dominated by the Pulse Width Modulation (PWM) switching frequency and its multiples, of 10 to 50 times F_O or more. However, the AFE will incorporate a network filter which will reduce the PWM element to a small percentage of the waveform. This is typically a few percent and is therefore normally within the class society voltage Total Harmonic Distortion (THD) limits.

The electrical system load, however, may include DFE converters and these are typically unfiltered. Where a system is generator fed, the sub-transient reactance of the generator will tend to reduce the impact of these harmonic currents on the system voltage. This is not the case for a PTO system and the PTO filter is unlikely to be targeting the DFE harmonics. This means that for a PTO fed system the total amount of distorting load on the system must be carefully managed to ensure the system stays within limits in the worst-case configuration of the power system.

5.5 Black-start Recovery

Another challenge is using the PTO generators for black-start recovery operation, whereby the entire ship has lost power and the hybrid system is used to restore power. This could be the case operationally if the DGs are unavailable. If an induction machine is used for the hybrid it causes a challenge when it comes to black-start recovery, as it requires excitation voltage to be applied for magnetising before it can produce power. Normally when the hybrids become PTO generators this voltage can be simply provided by the converter which is connected to a live distribution network (fed by the DGs). In a blackout situation this is not possible and so a restart module is needed to connect to the hybrid converter. ESS is the ultimate solution.

5.6 Load Sharing

In a traditional ship's supply system, the frequency is controlled by a relatively low bandwidth engine governor and the voltage by a generator automatic voltage regulator (AVR) acting through the (relatively slow response) of the generator field windings. For a PTO system, the control bandwidth and precision are vastly superior to the traditional generator. Exact, high response control of frequency, voltage, active and reactive power are possible. In this context it is possible to shape the response of the PTO system depending on whether it is sharing with a generator and what the overall state of the system is.

When this capability is combined with a converter energy store that is independent of the dynamic limitations of the mechanical system then further system capability benefits can be released.

5.7 Propulsion Converter as Energy Management Unit

The listed benefits of energy storage come at a price in terms of volume and vessel infrastructure and its energy is not stored in convenient 60Hz units that can match a normal AC distribution bus. Therefore, all energy stores need some form of power electronics to interface to the DC or AC distribution bus. Such power electronics add cost, take up volume, need auxiliary services and are typically connected to the distribution bus via a dedicated protective device, such as a circuit breaker. All these factors need to be considered.

6. CONCLUSION

Recently completed projects with modern hybrid propulsion have demonstrated advantages as well as some challenges encountered during system design, testing and operation. Electric hybrid propulsion systems need to be tailored at the design stage to meet the operator's requirements and what is needed from the ship.

Whilst the new advanced hybrid systems are very flexible, the main benefits come when the complete system combining the needs of propulsion, ship's service power supply and equipment type are fully optimised and integrated to the platform. Integration of the hybrid system as a whole is essential.

The hybrid concept continues to evolve to give greater benefits and flexibility, and the use of its major components, particularly the hybrid propulsion converter, is emerging as the heart of the energy management system of vessels of the future.

ACKNOWLEDGEMENTS
The author is grateful for the permission of GE Power Conversion to publish this paper.

REFERENCES

[1.] O. J. Simmonds, Advanced hybrid systems and new integration challenges. Bristol INEC 2016
[2.] Simmonds O.J.: "A History of Naval Electric Propulsion and Today's Choices for New Vessels", Proceedings of the International Workshops, Conferences and Expo for Military and Marine Applications, Pune, India, May 18-20 2013.
[3.] Dr Makhlouf Benatmane, Author: Benjamin Salter. Naval Hybrid Power Take-Off and Power Take-In – Lessons Learnt and Future Advances. Glasgow INEC 2018
[4.] G. Bellamy, B. Salter. Benefits of integrated shore based testing. Bristol INEC 2016
[5.] Castles G, Bendre A. Economic benefits of hybrid drive propulsion for naval ships. In: Proceedings of the 2009 IEEE electric ship technologies symposium. Baltimore, Maryland, USA; 2009. p. 515–20
[6.] DNV-GL Rules for Classification of Ships, Part 4 October 2015
[7.] D Greertsma, RR Negenborna, K Visserab, J J Hopmana. Design and control of hybrid power and propulsion systems for smart ships: a review of developments. Applied Energy 194 30-54. 2017
[8.] Martin Southall, Kamesh Ganti. Battery & ultra-capacitor based energy storage vessel integration, capabilities, considerations and challenges
[9.] Capt. Norbert H. Doerry1 and Howard Fireman. DESIGNING ALL ELECTRIC SHIPS. Proceedings of the Ninth International Marine Design Conference, 16-19 May, 2006, Ann Arbor, MI
[10.] Sulligoi G, Castellan S, Aizza M, Bosisch D, Piva L, Lipardi G. Active front-end for shaft power generation and voltage control in FREMM frigates integrated power system: Modelling and validation. In: Proceedings of the 21st international symposium on power electronics, electrical drives, automation and motion. Sorrento, Italy; 2012. p. 452-7
[11.] Dr Damir Radan, Martin Southall, Dr Makhlouf Benatmane, Martin Butcher. Integration, optimisation and benefits of energy storage for marine applications. Bristol INEC 2016
[12.] DNV-GL Rules for Classification of Ships, Part 6 Additional class notations, Chapter 2 Propulsion, power generation and auxiliary system, October 2015
[13.] DNV-GL Guideline for large maritime battery systems, Joint project between ZEM, Greenland Energy and DNV-GL, Supported by Transnova, 2013
[14.] Lloyd's Register, Large battery installations – a Lloyd's Register Guidance Note, 2015
[15.] Smith, T., D. Argyros, C. Raucci, N. Sabio, Global Marine Fuel Trends 2030, Lloyd Register& UCL Energy Institute, 2014
[16.] Radan D., Asgeir J. Sørensen, Alf Kåre Ådnanes, Tor A. Johansen (2008), Reducing Power Load Fluctuations on Ships Using Power Redistribution Control, Marine Technology Jounal-SNAME
[17.] MAN B&W L27/38 Project Guide – Marine, Four-stroke GenSet, 2014

Australia's part in the age of high-efficiency vessels

W Peter Symington, Hung Nguyen and Jonathan R Binns
Australian Maritime College, University of Tasmania
eMail: william.symington@utas.edu.au

ABSTRACT: *The International Maritime Organization (IMO) MARPOL annex VI with its recent amendments requires that emission reduction must be considered. Regional requirements such as Emission Control Areas and the Norwegian ban on shipping emissions in its World Heritage listed fjords by 2026 indicate the same. Electric propulsion with battery storage is increasingly being used to reduce fuel costs and emissions, by enabling increased use of shore electrical power and solar photo-voltaic (PV) modules.*

European ferry operators have demonstrated operational savings with marine electric propulsion by completely replacing diesel fuel with mains electricity. For example, the Ar Vag Tredan ferry in France uses shore electrical power stored in super-capacitors and the Norwegian car ferry Ampere uses shore electrical power stored in lithium-ion batteries.

It is no coincidence that Ar Vag Tredan and Ampere have aluminium hulls. Ampere is half the mass of a similar sized steel ferry, improving the viability of its electrical propulsion system.[1] *This has important implications for Australia, because Australian aluminium shipbuilders have a significant share of the world market in contracted aluminium vessels. Australia is therefore in a unique position globally with regard to the development of high-efficiency vessels. This paper investigates marine electric propulsion development in Australia and suggests how and why we might do better.*

1. INTRODUCTION

The 62nd session of the International Maritime Organization (IMO) Marine Environment Protection Committee (MEPC) introduced mandatory measures, including the Energy Efficiency Design Index (EEDI) for new ships and the Ship Energy Efficiency Management Plan (SEEMP) for all ships, in order to accelerate efficiency improvements and reduce emissions.[2] Furthermore, in order to continue to combat climate change and to contribute to the UN Sustainable Development Goals, IMO has continued its development of MARPOL annex VI, the latest amendment being the introduction of the 0.5 % global sulphur cap which will enter into force in 2020. Electric and hybrid propulsion is increasingly being used to meet these new requirements.

Marine Electric Propulsion (MEP) is not a new concept. It has been used since the early 20th century, mainly in diesel-electric propulsion systems in submarines, warships, cruise ships and ice-breakers. In addition to meeting new efficiency and emission requirements electric propulsion facilitates:[3]

- more flexible machinery arrangements;
- low noise operation;
- improved vessel efficiency when partially loaded or with large auxiliary loads;
- increased manoeuvrability, for example via pod propulsors and bow thrusters;
- increased redundancy; and
- use of solar energy and shore mains electricity for propulsion.

Electrical systems are playing an increasingly important role in vessel efficiency in the following areas:

- energy sources, such as shore electrical supply and alternators;

- rechargeable energy storage systems (ESS), such as batteries and supercapacitors;
- electric motors used in propulsion drives; and
- power management systems for vessel propulsion and electrical auxiliary power.

2. EUROPEANS HAVE TAKEN THE LEAD

European ferry operators were first to demonstrate operational savings with marine electric propulsion, by completely replacing diesel fuel with mains electricity. The Ar Vag Tredan ferry in France was first to use shore electrical power stored in super-capacitors. The Norwegian car ferry Ampere uses shore electrical power stored in lithium-ion batteries and commenced service in March 2015. By changing from diesel to all-electric propulsion, Ampere's operator Norled reduced their fuel cost by 60 percent.[1]

It is no coincidence that Ar Vag Tredan and Ampere are made of aluminium. Ampere is half the mass of a similar sized steel ferry and has double the expected hull lifetime. This has important implications for Australia, because Australian aluminium shipbuilders have a strong market share as evidenced by the success of Austral, which is contracted to manufacture 15% of the US Navy's fleet.[4] Australia's success in aluminium shipbuilding provides a strong engineering and industrial base that could be leveraged into the development and manufacture of high-efficiency vessels.

3. THE FJORDS TRANSITION TO All-ELECTRIC AND URBAN

The all-electric ferry Future of the Fjords was launched in May 2018, two years after its hybrid diesel-electric sister ship, Vision of the Fjords (Figure 1). Both vessels ferry tourists between Gudvangen and Flam in the scenic Sognefjord, Norway. Aside from reduced emissions, the quieter running electric vessels avoid disturbing the tranquility of the fjords.

Figure 1: Vision of the Fjords in Sognefjord, Norway. Photo by P Symington 2018.

Norway is banning all emissions in world heritage listed fjords by 2026. An emerging problem with meeting the deadline is the limited mains power infrastructure between Norway's hydro-electric power stations and the fjords. One solution to this is a floating PowerDock equipped with batteries

(Figure 2). The dock's batteries are slowly charged from mains power and are then used to fast charge the ferries when they dock. A similar concept could be used with solar charging.

Figure 2: Floating PowerDock (foreground) in Sognefjord, Norway. Photo by P Symington 2018.

The Future of the Fjords' Norwegian operator, The Fjords, recently announced construction of a new all-electric vessel, "The Legacy of the Fjords", for operation in Norway's capital city Oslo by May 2020.[5] The company describes this as a natural progression of their business operations. It demonstrates the viability of all-electric vessels in two niche markets:

1. sightseeing routes in sheltered waters and
2. ferry routes in metropolitan areas.

The Norwegian industry is benefitting from its lead in electric propulsion. For example, there have recently been substantial investments in high-tech manufacturing capability in Norway by leading electric propulsion companies such as Siemens[6] and Corvus Energy[7].

With its established lead in aluminium vessel manufacturing, the Australian shipbuilding industry could generate similar national benefits, by actively leveraging into the early stages of high-efficiency vessel development. Conversely, without an active approach by industry and government, European shipbuilders could displace Australian aluminium shipbuilders in the world market.

4. A WORLD LEADER IN ALUMINIUM SHIPBUILDING

Australia is a world leader in high-speed aluminium shipbuilding. Australian aluminium shipbuilders built more than 70% of the world's contracted high-speed vessels in 2008.[8] In second place was Italy with less than 10% and third was China. In 2019, Australian aluminium shipbuilders still maintain a solid 30% per cent share of the international market.[9] Australian companies contributing to this remarkable achievement are located in the states of Tasmania, Western Australia and Queensland and include:

- Incat in Tasmania,
- Richardson Devine Marine in Tasmania,
- Austal in Western Australia,
- Strategic Marine in Western Australia,
- Echo Yachts in Western Australia,
- Aluminium Marine in Queensland, and
- Aus Ships Group in Queensland.

The technology leadership underpinning this market leadership was clearly demonstrated by Austal winning a US Navy contract to build a high-speed Littoral Combat Ship (LCS) valued at US$432.1 million, with options for nine additional LCS ships over five years. This is in addition to US

Department of Defence contracts to build Joint High Speed Vessels (JHSV) as part of a 10-vessel program valued at up to AUD$2 billion.[10]

Key features of high-speed vessels, namely sleek, lightweight multi-hulls, are also applicable to high-efficiency vessels. The Australian shipbuilding industry is therefore well-positioned to supply the emerging world market in high-efficiency vessels.

5. SOME EXPERIENCE IN RENEWABLE ENERGY PLOPULSION

Australia was a leader in fuel, solar and wind hybrid propulsion vessels at the end of the 20th century. For example, sailing vessel technology honed on recreational and racing sailing vessels, such as Australia II, was adapted to commercial sailing catamarans designed by Incat Crowther (Figure 3) and Grahame Parker. These 25 to 32 m sailing catamarans still provide tourist trips to the Great Barrier Reef from Cairns and Port Douglas. They are especially significant as they have been operating successfully in competition with diesel-powered vessels. They are hybrid vessels, with diesel engines for travelling in adverse winds or calm conditions.

Figure 3: Commercial sailing catamaran Wavedancer designed by Incat Crowther (1986). Photo courtesy of Quicksilver Cruises, http://quicksilver-cruises.com.

Solar Sailor Holdings Ltd was established in the late 1990s to commercialise a patented solar wing. With government assistance it built the Solar Sailor ferry for the Sydney 2000 Olympics. In 2011, Solar Sailor Holdings Ltd built the Solar Albatross for the Hong Kong Jockey Club.[10] Solar Sailor was renamed to Ocius in 2014 and pivoted its development focus to autonomous maritime surveillance vessels.

The all-electric ferries launched in Europe since 2015 show that it is now leading the world in marine electric propulsion development for manned vessels. Nevertheless, several small, electric and hybrid vessels have been designed and built in Australia in the past five years:

- "Adels Grove" solar boat operating in the Lawn Hill Gorge National Park in north-west Queensland.

- "Ellie J" solar/electric ferry built in Queensland by Scruffie Marine in 2016 and operated on the Swan River in Perth by the Little Ferry Company.[11] This vessel follows the European examples of "Ar Vag Tredan" and "Ampere" in demonstrating the technical and commercial viability of marine electric propulsion for short-run ferry routes.
- "Kato", a 20m solar/electric, carbon-fibre sailing catamaran built in Queensland by Noosa Marine in 2017.[12] This vessel is capable of electric motoring at speeds of 10 knots and sailing speeds in excess of 25 knots, while regenerating up to 10 kW under sail power.

These examples show that Australia has some capability in adapting marine electric propulsion systems for small vessels. However, rapid development of skills, technology and systems will be required to integrate this with Australia's extensive capability in aluminium shipbuilding at the speed and scale required.

6. TOWARDS ENERGY EFFICIENT SHIPPING

A decade ago, two major steel shipbuilders developed concept designs for 'super' ships propelled by a hybrid system of wind, gas fuel and solar energies. STX Europe's EOSEAS concept comprised a 5-mast, 305m, multi-hull cruise ship utilising LPG, wind and solar power.[13]

In 2009 NYK of Japan published the Super Eco Ship 2030 concept design utilising a traditional monohull, gas fuel cells, wind and solar power. NYK suggested that the Super Eco Ship 2030 may be able to be built by 2030 and in 2019 published its new Super Eco Ship 2050 concept.[14]

The premise that rapid efficiency improvements can be easily achieved by adapting new technology to super ships may be optimistic because:

1. The complex structural design of existing super ships is not readily adapted to the very different stresses of wind propulsion.
2. The timeframe for building a super ship is approximately 3 years, so the design-build-operate-feedback-design cycle for super ships does not facilitate rapid evolutionary change.
3. The premise relies on international shipping operators and super ship manufacturers taking the significant risks and costs involved in making significant changes to super ship design.
4. After 2008 the pace of implementing super ship efficiency improvements was hampered by an increase in spare shipping capacity and a shipbuilding downturn due to the Global Financial Crisis and the subsequent protracted recovery.[10]

Progress in low-fuel and low-emissions propulsion will proceed more rapidly via an evolutionary path, commencing with smaller vessels operating within enforceable national boundaries, for which the transition steps are smaller, faster and lower risk. The evolutionary paths used in the development of fossil-fuelled vessels between 1788 and 1950 and in the development of high-speed aluminium vessels between 1983 and 2010,[8] show that increases in scale evolve with increasing confidence in technology, design and performance. With a good strategy, good engineering and sufficient funding, the evolutionary process can be accelerated (for example, Spacex's rapid rocket development and rise in the US space launch industry) but it cannot easily be skipped.

This supposition has important implications for Australia. It suggests that the future market for electric and hybrid ships may not belong to existing steel super ship manufacturers. Australia's established high-efficiency vessel design and build capability might be a more significant capability in gaining market share of the world's future shipbuilding industry.

7. USE OF SOLAR ENERGY FOR MARINE PROPULSION

In Australia we have many tourist areas where electric propulsion could provide similar benefits to those in Norwegian fjords, but the mains electricity infrastructure in remote areas is as restrictive as it is in Norway. Australia does however have an advantage in many of its remote areas due to an abundance of solar power.

Figure 4: Zone of peak, year-round solar energy, between 30⁰ N and 30⁰ S.

The map in figure 4 shows regions of the world lying within +/- 30⁰ latitude of the Equator. This is where solar energy is most useful over the full year, a critical factor in the adaptation of solar energy for use on vessels.[14] Northern parts of Australia are clearly within the zone, as are countries in Central America, South America, Africa, Southern Asia and South-East Asia.

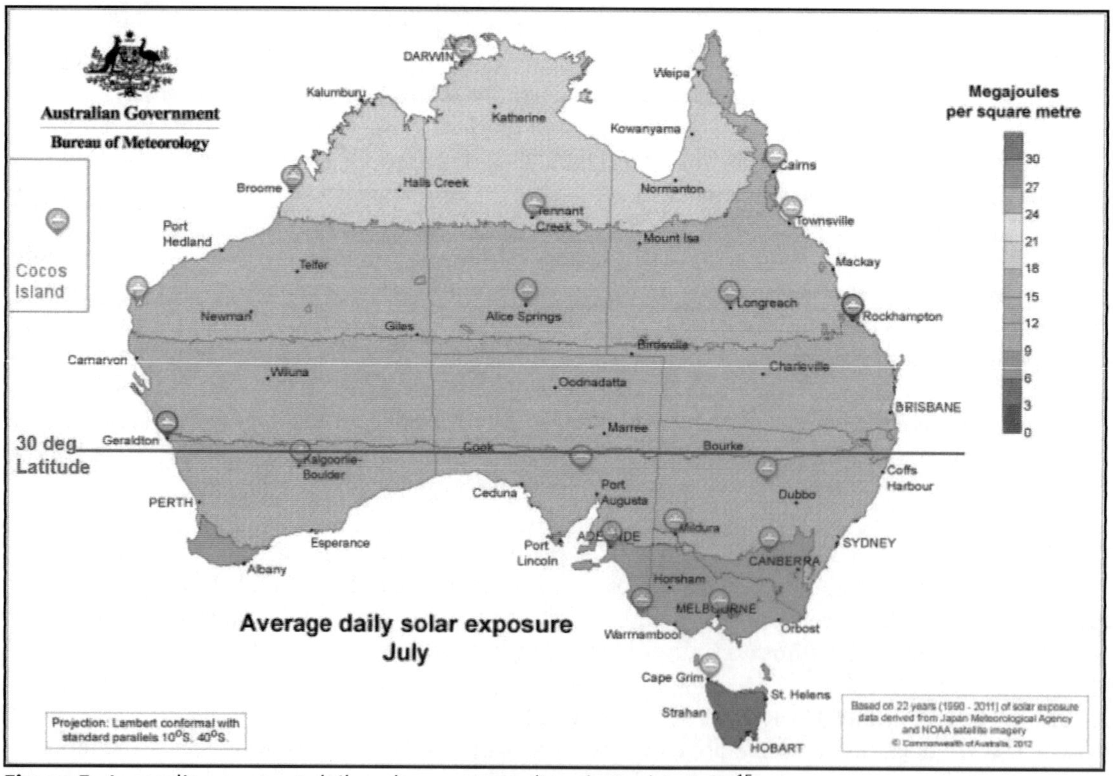

Figure 5: Australian average daily solar exposure (insolation) in July.[15]

The 30⁰ parallels approximate the dividing line between useful year-round energy and strong seasonal variation. As Figure 5 shows, in Australia the 30⁰ south parallel runs just south of Geraldton in Western Australia and just north of Coffs Harbour in New South Wales. It equates to insolation of approximately 12 megajoules per square metre per day in winter.

Solar photovoltaic (PV) panels have been used to provide auxiliary power in recreational sailing vessels for several decades. The average price for US residential solar PV systems is graphed in Figure 6. Residential PV system costs are a relevant analogy for PV system prices in the marine vessel context due to their similar scale. The potential for solar energy to contribute to vessel propulsion was demonstrated by the first global circumnavigation of a solar-powered vessel, the Turanor PlanetSolar, in May 2012.

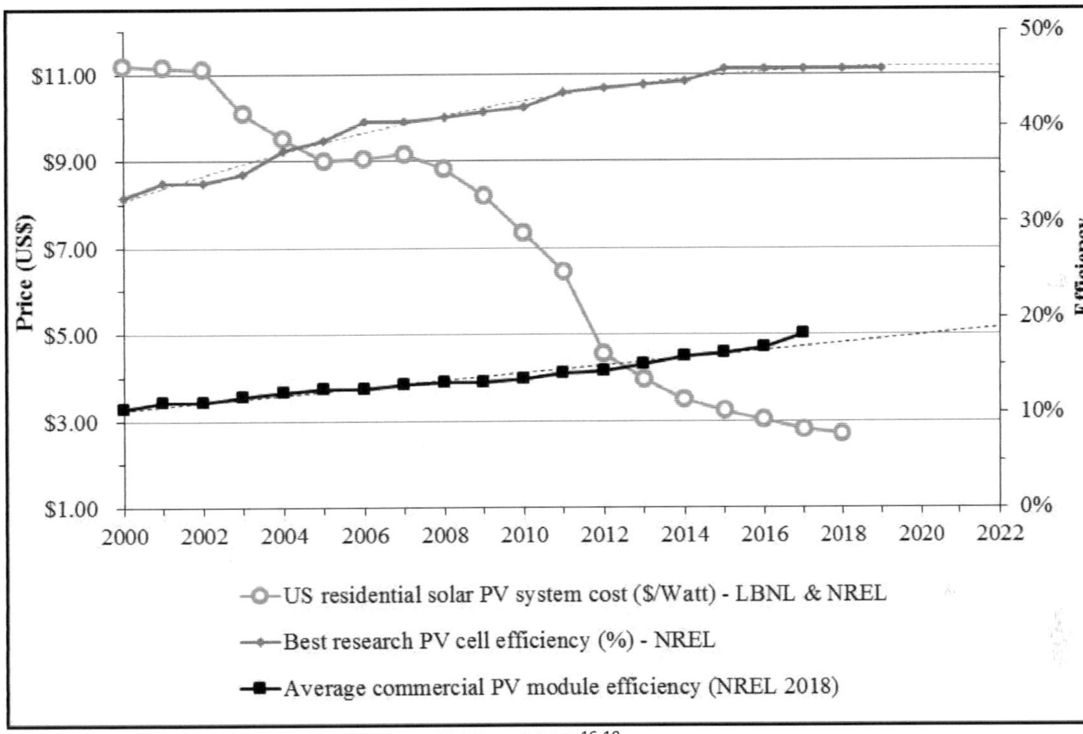

Figure 6: Solar PV efficiency and price, 2000 to 2018.[16-18]

Ongoing improvements in solar PV efficiency and reducing costs mean that solar power is an increasingly viable energy source for vessels.[19]

Figure 7 shows the typical output of a small solar array in Maryborough, Queensland across two days. The first day was sunny and the second day was partly cloudy. The figure includes the total daily insolation energy measured by two nearby Bureau of Meteorology weather stations and the PV energy stored in batteries by the solar system.

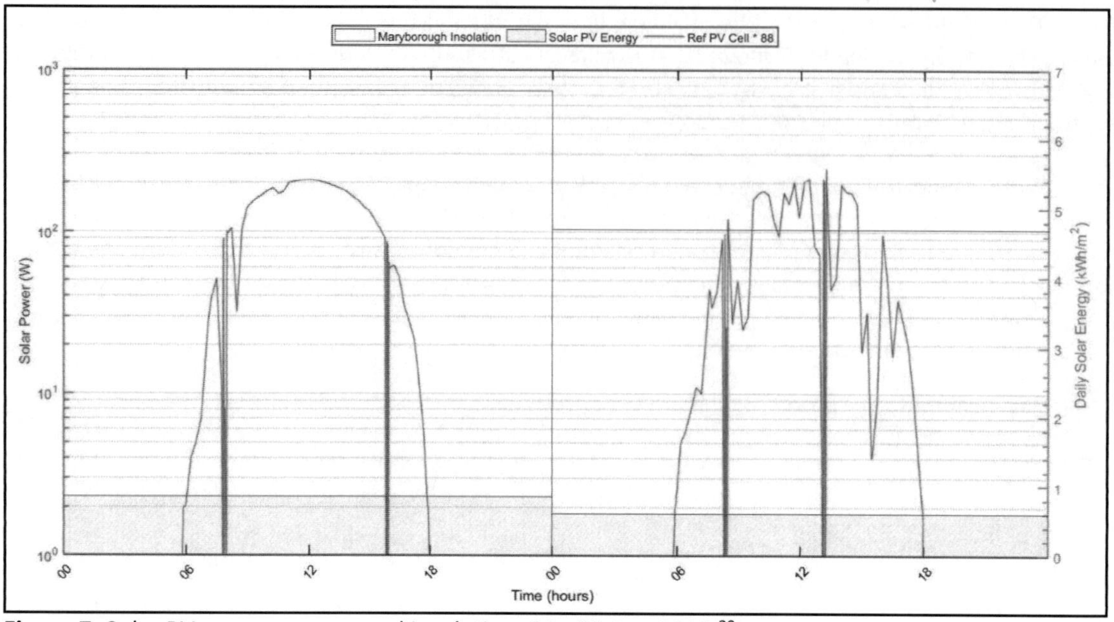

Figure 7: Solar PV power, energy and insolation, 24 - 25 Sep 2016.[20]

Analysis of this system has led to findings on the relative significance of solar PV system losses shown in Figure 8. These findings will be used to develop guidelines for solar system design for vessels operating in sub-tropical zones.

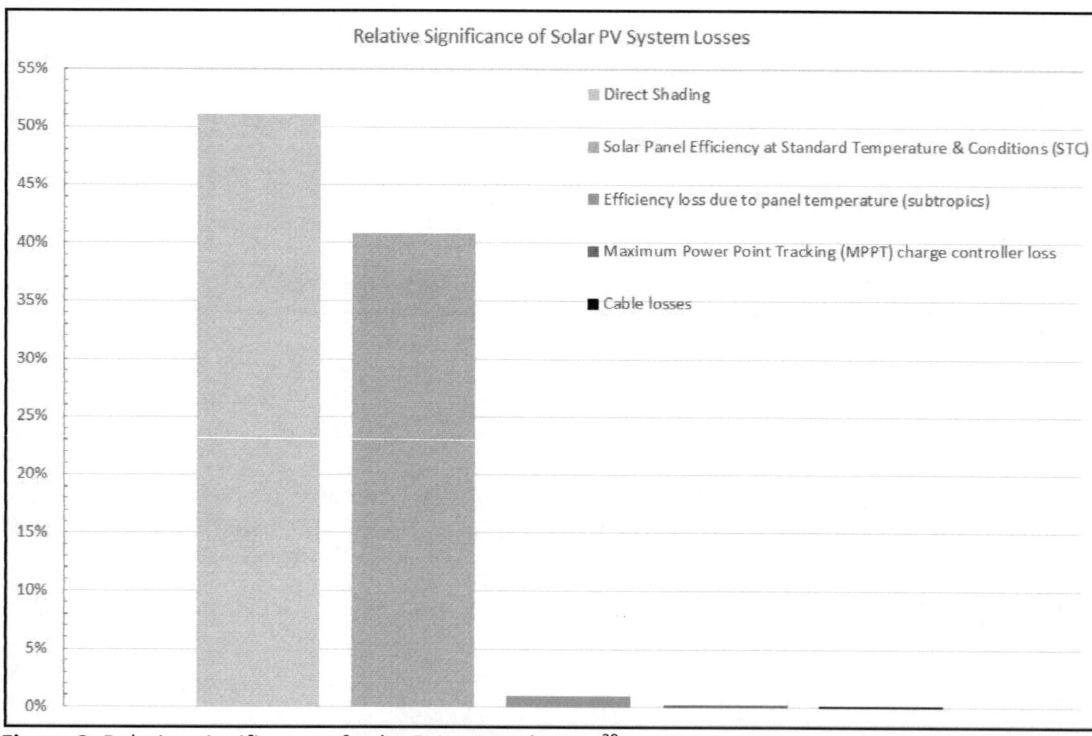

Figure 8: Relative significance of solar PV system losses.[20]

An interesting phenomenon identified in BOM data is that of irradiance amplification due to partial clouding.

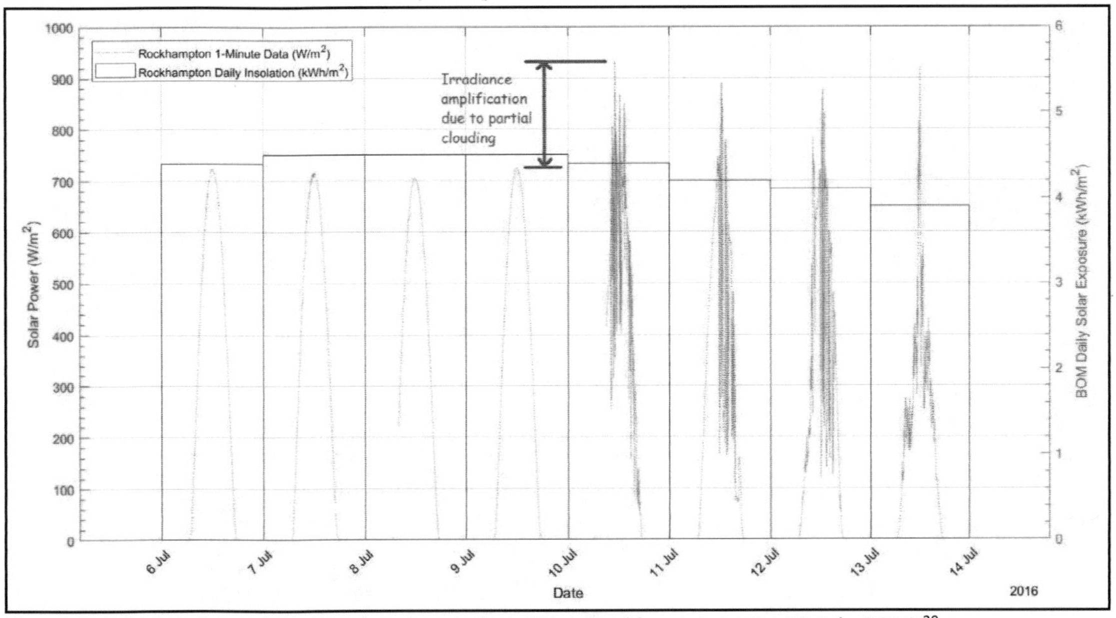

Figure 9: BOM daily and one-minute insolation data, Rockhampton, 6 - 13 July 2016.[20]

Figure 9 shows that on 10 July 2016 in Rockhampton, the amplification effect of reflection from clouds during partial clouding amplified one-minute insolation power levels by up to 28%. Diffuse irradiation is also an important contributor to total solar irradiance at European latitudes during winter.[21]

Diffuse irradiation and the irradiance amplification effect of partial clouding has been described[22] and modelled,[23] however its consideration in the design of marine vessel solar systems has been limited to date.

8. FERRY ROUTES IN METROPOLITAN AREAS

Metropolitan areas are an ideal application for electric propulsion, due to the close proximity of mains electricity infrastructure and the need to reduce SOx, NOx and particulate emissions in cities. The 11-passenger, all-electric ferry, "Ellie J" (Figure 10) has been operating for three years on the Swan River in Perth, Western Australia.

"Ellie J" operates from mains electrical energy stored in 21 kWh of lithium-ion batteries. The vessel also has 1.6 kW of solar panels on its roof that provide up to 35% of its energy requirements. The vessel is propelled by two 4 kW electric outboard motors.[11]

Figure 11 shows "Ellie J"'s powering curve predicted using several methods and as measured during sea trials, where it achieved 7.3 knots at its full 8kW power. Most impressively, the vessel can cruise continuously on a sunny day at up to 4 knots on solar power alone. This shows that electric propulsion systems are predictable and can be readily measured, even in the marine environment.

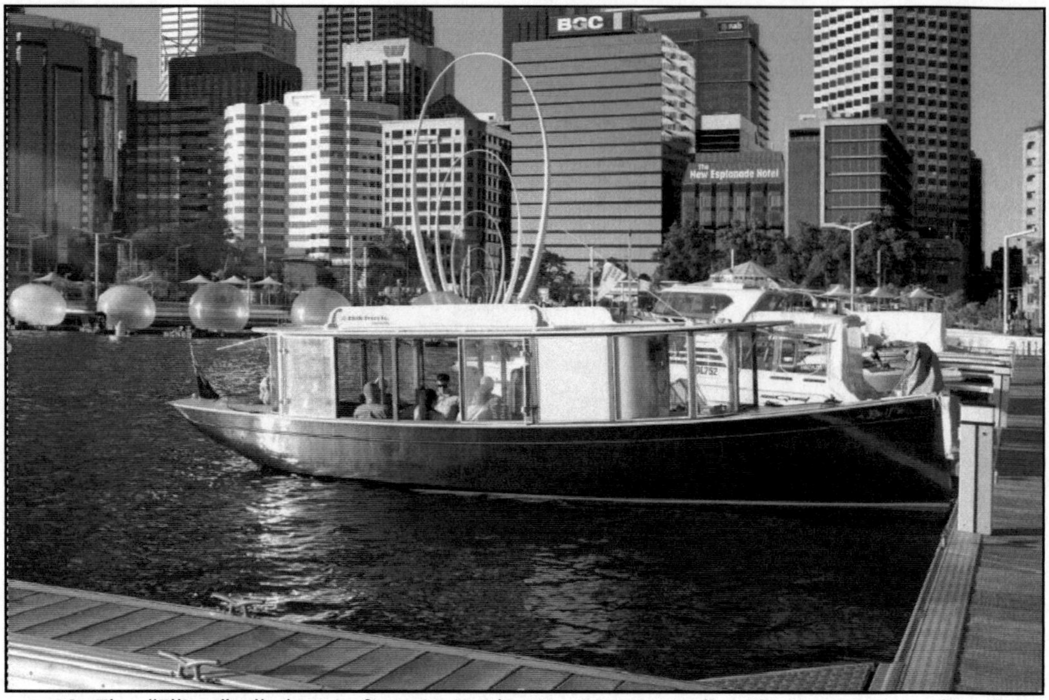

Figure 10: The "Ellie J" all-electric ferry in Perth, Western Australia. Photo courtesy of Little Ferry Company, https://www.littleferryco.com.au

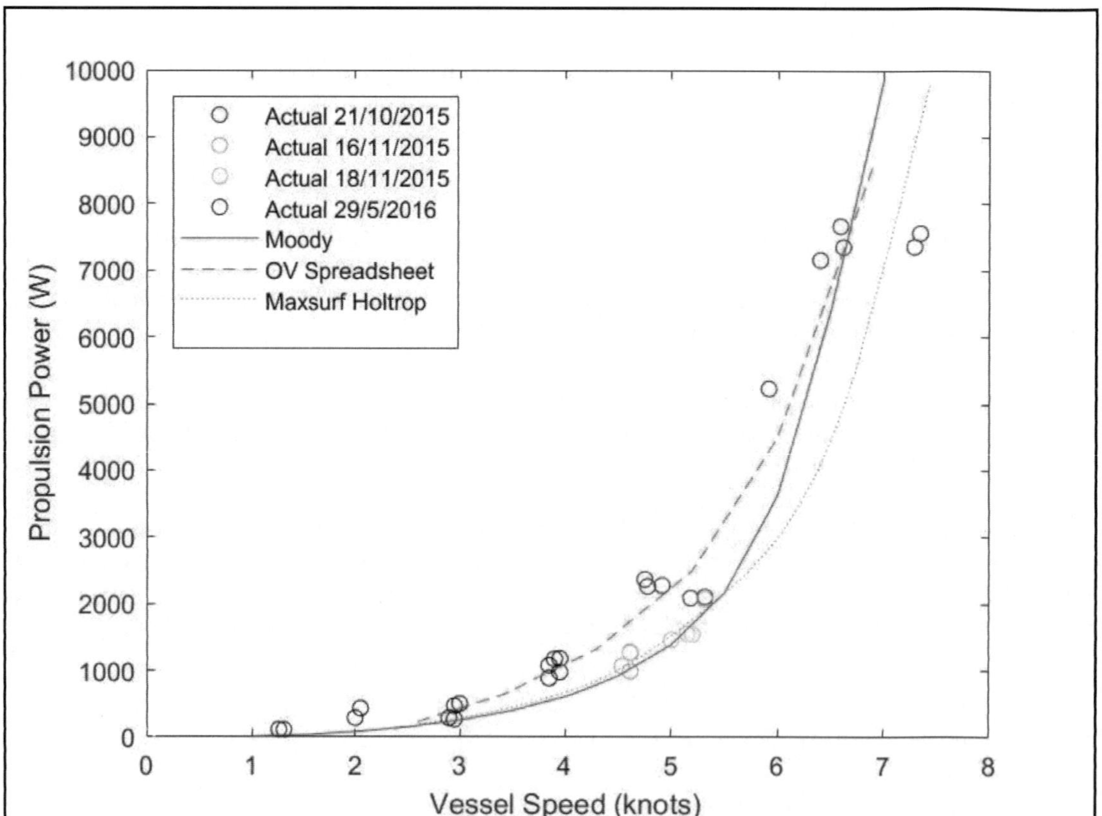

Figure 11: Modelled and actual powering performance of all-electric ferry "Ellie J".[20]

A top speed of 7.3 knots is ample for ferry routes on the Swan River near Perth's CBD, where the speed limit is 8 knots in most areas and drops to 5 knots near bridges and quays. A key feature of "Ellie J", like that of "Future of the Fjords" in Norway, is its lack of engine noise. Passengers can converse at normal levels, needing only to compete with the sound of water lapping against the hull. Unlike diesel ferries that envelop passengers and bystanders in clouds of diesel exhaust as they accelerate from jetties, the "Ellie J" slips quietly away with no emissions.

Perth's Little Ferry Company has demonstrated that electric propulsion is viable for short-run ferry routes and the company launched an identical sister ferry, the "Jessica Leigh" in 2017. Importantly, these ferries are showing that the public is keen to support quiet, low-emissions transportation. For example, Little Ferry Company's ferries were rated by TripAdvisor in April 2017 as #3 out of 178 things to do in Perth and in May 2017 as #1 out of 25 Boat Tours & Water Sports in Perth.[24]

9. A PLAN FOR MARKET DOMINANCE

Australia's world leadership in aluminium high-speed vessel manufacturing will not, by itself, lead to a significant market share in high-efficiency vessels. Additional factors for success, include a national investment strategy and supportive marine industry policy framework for fully applying these capabilities to national benefit and ultimately, global benefit.

Harnessing Australia's world-leading marine design and manufacturing capabilities to improve market share could involve replacement of fossil-fuelled engines in metropolitan ferries and remote tourism vessels, by a systematic process of retrofits and new-builds. With development of technologies and increasing scale, this process could mimic the growth and dominance of Australian aluminium shipbuilding. The path to success could look something like this:

1. Leading-edge high-efficiency vessel technologies and shipbuilding capability mobilised to reduce vessel fossil fuel consumption and reduce emissions of remote tourism vessels and metropolitan ferries.
2. Apply this emerging capability in mid-sized vessels to reduce the diesel dependence of Pacific island nations.
3. As technologies and scale evolve, apply honed capabilities to military applications and inter-regional freight transport, particularly in niches where silent and efficient marine electric propulsion is the most effective.

With rapid and well-coordinated action by industry, researchers and policy makers, the rate of evolution for high-efficiency vessel production in Australia could be (and will need to be) at least as fast as it was for aluminium high-speed vessels.

An important difference between high-efficiency shipbuilding and high-speed shipbuilding will be in market scale. IDTechEX predicts that that the world market for (non-military) hybrid and pure electric boats and ships will rise rapidly to around US$ 20 Billion (AU$27 billion) in 2027.[25] Figure 12 shows the potential opportunity if Australia was to target 50% of the world market in electric and hybrid vessels by 2027 and compares this to a "business as usual" compound growth rate of 2.2% projected by the Australian Trade and Industry Commission for Australia's shipbuilding and maintenance, repair and overhaul (MRO) sector.[26]

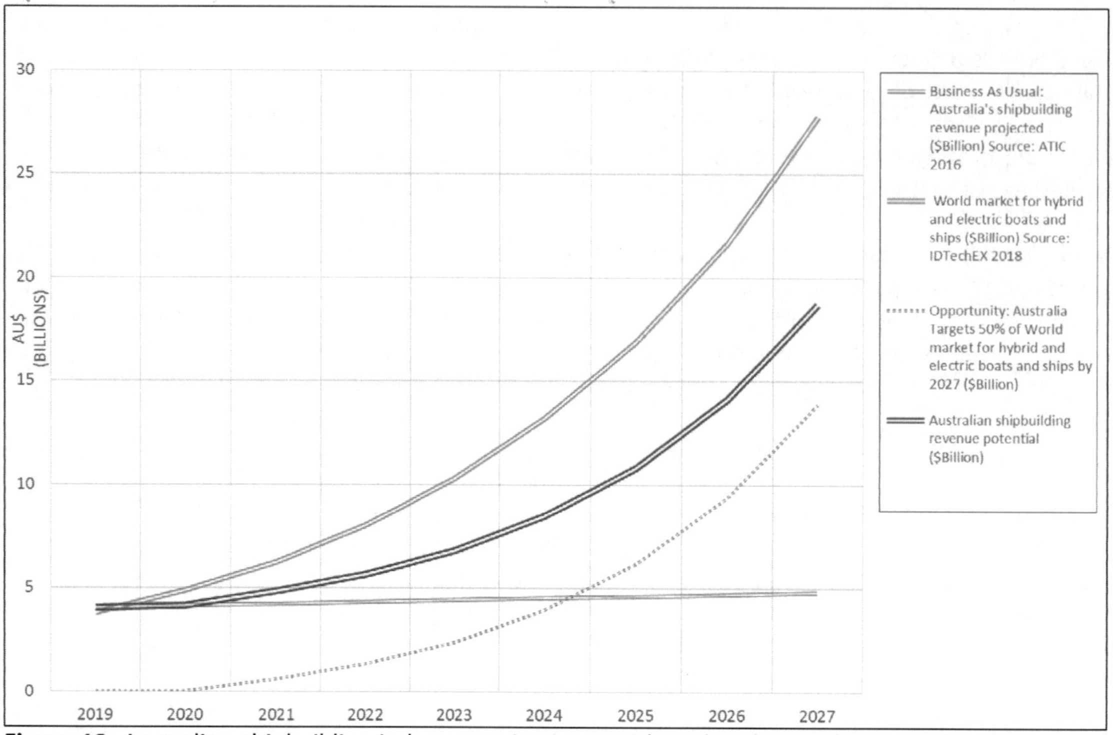

Figure 12: Australian shipbuilding industry projections, with and without a focus on the electric and hybrid vessel market.[25, 26]

Figure 12 indicates that the 2019 world market in electric and hybrid vessels is close to Australia's 2019 shipbuilding revenue, of around AU$4 billion per annum. However, the world market in electric and hybrid vessels has a far higher growth rate. For example, 50 electric and hybrid vessels were under construction or in operation worldwide in 2015 and DNV-GL projects that there will be 200 in 2020.[27]

The Australian aluminium shipbuilding industry demonstrated its capability to evolve and build significant market share of the world market in aluminium high-speed vessels. A 50% share of the world market in electric and hybrid vessels will require rapidly leveraging this capability, while accelerating systems development. The result could be an Australian shipbuilding and MRO sector with annual revenue of AU$17 billion by 2027, compared with projected "business as usual" revenue of less than AU$5 billion.

10. CONCLUSIONS

The dawning of the age of high-efficiency vessels is accelerating as technology development facilitates emissions reduction and operational savings. There is a concurrence of factors that place Australia in an important position globally with regard to the development and use of high-efficiency vessels. Those factors include world-leading aluminium shipbuilding capability, strong systems engineering capability, suitable locations and ideal applications for development of relevant technologies.

The Australian aluminium shipbuilding industry's share of the world market for high-speed vessels has declined over the past decade, but its current 30% share provides a solid foundation for pivoting into the high-growth market for high-efficiency vessels. Development and adaption of electric propulsion in Australia has been slow and is trailing Europe. Accelerated development and implementation of electric and hybrid propulsion technologies in Australia will have wider benefits to industries outside aluminium shipbuilding, such as defence, the tourism industry, crew training and establishment of support industries for operating future vessels.

Increased recognition of the potential opportunities by Australian governments is needed. Development of an active and supportive policy framework could ensure that national benefit is derived from the window of opportunity presenting itself to the Australian shipbuilding and MRO sector. With coordinated action, Australia has the capability to take the lead in the emerging high-efficiency vessel market. In addition to economic benefits, this could contribute to UN Sustainable development goals and the Australian 2030 climate change target.

Recognition of these factors and coordinated action by Australian policy-makers, researchers and key industry players will be required to build a future industry of global significance.

REFERENCES

1. Siemens. First electrical car ferry in the world in operation in Norway now. Siemens AG, 2015.
2. International Maritime Organization. Marine Environment Protection Committee (MEPC) - 62nd session: 11 to 15 July 2011. IMO MEPC, 2011.
3. Molland A. *The Maritime Engineering Reference Book: A guide to ship design, construction and operation*. UK: Elsevier Ltd, 2008.
4. Austal Ltd. ASX Announcement: US Navy Awards Austal Multi-Vessel LCS Contract. 2010.
5. Marex. The Fjords Announces Construction of New Zero-Emissions Vessel. *The Maritime Executive*2018.
6. Siemens. Siemens opens advanced robotized and digitized battery module factory in Norway. Siemens AG, 2019.
7. Corvus. Corvus Energy strengthens its product portfolio for maritime batteries with the acquisition of Grenland Energy. Corvus Energy, 2019.
8. Armstrong A. Engineering Excellence on the Move - the Development of High-Speed Ships. *Institution of Engineers Australia Eminent Speaker Series*. Engineers Australia Webcasts: Engineers Australia, 2009.
9. Austrade. Transport Industry Capability Reports. Australian Trade and Investment Commission, 2019.
10. Symington WP, Binns JR. The dawning of the age of high-efficiency vessels. *Australian Journal of Mechanical Engineering*. 2015; 13: 154-62.
11. Symington WP, Desjardins C. The Little Ferry Setting a Big Example in Urban River Transport. *IEEE Transportation Electrification Community newsletter*. US: IEEE, 2017.
12. Lynch P. Kato G-Force 2000 Speciale. *Australian Multihull World*: P&G Lynch & Assoc Pty Ltd, 2017.
13. Tincelin T, Pelerin E. EOSEAS Green Cruise Ship Concept. *7th Annual Green Ship Technology conference*. Copenhagen: Stirling Design International, 2010.
14. NYK. Super Eco Ship 2050. 2019.
15. BOM. Climate Data Online: Stations 040443, 040126 & 039083 Rainfall, Temperature, Solar exposure. Australian Bureau of Meteorology, 2017.
16. Barbose G, Darghouth N. Tracking the Sun: Installed Price Trends for Distributed Photovoltaic Systems in the United States - 2018 Edition. Lawrence Berkeley National Laboratory, 2018.
17. NREL. Best Research-Cell Efficiencies. NREL, 2019.
18. Fu R, Feldman D, Margolis R. U.S. Solar Photovoltaic System Cost Benchmark: Q1 2018. NREL, 2018.

19. Symington W, Belle A, Nguyen HD, Binns JR. Emerging technologies in marine electric propulsion. *Proceedings of the Institution of Mechanical Engineers, Part M: Journal of Engineering for the Maritime Environment*. 2014.
20. Symington WP. Modelling and solar PV contribution to marine electric propulsion. *Australian Maritime College*: UTAS, 2019.
21. Deutsche Gesellschaft Für Sonnenenergie. *Planning and Installing Photovoltaic Systems : A Guide for Installers, Architects and Engineers*. UK: Earthscan, 2012.
22. Liu BYH, Jordan RC. The interrelationship and characteristic distribution of direct, diffuse and total solar radiation. *Solar Energy*. 1960; 4: 1-19.
23. Tapakis R, Michaelides S, Charalambides AG. Computations of diffuse fraction of global irradiance: Part 1 – Analytical modelling. *Solar Energy*. 2016; 139: 711-22.
24. TripAdvisor - Little Ferry Co.: TripAdvisor LLC, 2017, p. Little Ferry Co. rated #3 out of 178 things to do in Perth.
25. Gonzales F. Electric Boats and Ships 2017-2027 Technologies, Markets, Forecasts, Key Players and Opportunities. IDTechEx, 2017.
26. Austrade. Commercial Shipbuilding and Services. Australian Trade and Investment Commission, 2016.
27. Henningsgård S. Someone has to go first. DNV-GL, 2019.

TITOMIC

Titomic Limited

Paper Submission for PACIFIC 2019

3/270 Ferntree Gully Rd,
Notting Hill VIC 3168

titomic.com

+613 9558 8822

info@titomic.com

ABN 77 602 793 644

About Titomic

Titomic (ASX: TTT) is an Australian public company specialising in Digital Manufacturing Solutions for industrial scale metal additive manufacturing using its patented Titomic Kinetic Fusion® (TKF) technology. TKF technology provides unique capabilities for producing commercially viable additively manufactured metal products, competing directly with traditional manufacturing methods. Titomic provides OEM production and R&D services from its TKF Smart Production Bureaus to the global Aerospace, Defence, Shipbuilding, Oil & Gas, Mining and Automotive industries. Titomic also provides an extensive range of metal powders for 3D Printing, especially titanium and super alloys, and provides sales and support services for their TKF production systems.

Technology – Titomic Kinetic Fusion®

Co-developed with the CSIRO, Titomic has exclusively licenced the patented technology Titomic Kinetic Fusion®. Titomic Kinetic Fusion® deposits solid metal powders 5-75 micrometres in diameter, accelerating them in a supersonic gas jet. During impact, particles undergo plastic deformation and fuse to each other. Multiple metals, including titanium, ceramics and metal composite materials can be deposited using Titomic Kinetic Fusion®.

Titomic Kinetic Fusion® is the only additive manufacturing method capable of fusing dissimilar metals into one seamless part, as well as engineered high-performance coatings to enhance corrosion and erosion resistance and mechanical properties. The Titomic process can also create hybrid materials such as composite ceramic metal matrices, anti-corrosion and ballistics coatings. This capability is especially suited for marine applications, allowing weight savings and high corrosion resistance.

Unlike many other additive manufacturing methods, Titomic Kinetic Fusion® does not melt metals. This allows Titomic to create large, industrial scale, seamless metal structures larger and faster, rivalling all other manufacturing methodologies.

Case Study

Titomic Kinetic Fusion® is used to improve the lifespan and manufacturing process of tooling used in the production of defence componentry.

	Traditional Manufacturing	Titomic Kinetic Fusion™
Challenge | Titomic solution	Tooling manufacture may experience lead times longer than 12 months. Parts may last for less than 2 years before replacement or is required.	TKF allows tooling to be made significantly faster, reducing lead time by up to 50%. TKF tooling allows for tool design optimisation.
Process	Machined invar, welded to size	Sprayed as a single, near net piece
Comparison	• Weld porosity • Weld cracking • Hard to repair	• No welding • Low porosity • Repairable
Manufacturing Factors	• Labour intensive • May take months to manufacture • Significant capacity constraints	• Largely unattended production • Weeks to manufacture • Minimal capacity constraints

Effects of multiple simulated weld repair cycles on the mechanical properties of a high strength, quenched and tempered steel

P. Dissanayaka[1,2] (spd957@uowmail.edu.au), K. Carpenter[1,2], H. Li[1], S. van Duin[1,2],
Z. Sterjovski[3], J. Donato[3], D. Miller[4], M. Johansson[4]
[1]Research Training Centre for Naval Design and Manufacturing (RTCNDM),
[2]University of Wollongong,
[3]Defence Science and Technology Group,
[4] ASC Pty. Ltd.

ABSTRACT

Corrosion damage to sea-going vessels can be repaired by removing the damaged metal and rebuilding the area with weld deposits. Over the life of a vessel, repair welding at the same location is often unavoidable and preferable to plate replacement, as it is more cost effective and timely procedure.

In this study, the influence of the number of simulated weld repair thermal cycles (i.e. up to three repeated weld thermal cycles) on the microstructure and mechanical properties of a quenched and tempered (Q&T) steel was assessed. This assessment was undertaken through thermo-mechanical heat affected zone (HAZ) simulations, so as to isolate and assess the toughness of the different sub-zones in the HAZ over multiple thermal cycles. On all of the simulated samples, Charpy impact toughness and hardness tests were performed, followed by microstructural and fractographic analysis. The results attained were benchmarked against data from the original coarse grain (CG) and intercritically reheated coarse grain (CG-IC) HAZ thermal cycles. Although subtle differences in microstructure were observed, there was no significant degradation in toughness or hardness.

1. INTRODUCTION

Corrosion damage is one of the main factors that can affect structural integrity of sea-going vessels, which requires repairs to rectify [1]. It is inevitable that a repaired region may suffer corrosion damage again and it is preferable to carry out a second or third weld repair instead of hull section replacement due to economic viabilities and time restrictions. Repair welds involve multi-pass welds and in re-occurring repair welds, there is a possibility of the base plate experiencing overlapping thermal cycles which is not possible in multi-pass welds, such as CG-CG-IC.

Thermal cycles in welding can create adverse effects on the microstructure and mechanical properties of materials. In particular, the combination of high strength and toughness of Q&T steel can be degraded in the HAZ because of weld thermal cycles [2-4]. The intercritically reheated coarse grain (CG-IC or commonly referred as IRCG or ICCG) region typically has the lowest toughness, followed by coarse grain (CG) region. An intercritically reheated CG HAZ occurs in multi-run welds, when the first cycle is heated to the CG temperature region (usually ≥ 1200°C) and the second to the intercritical temperature, between Ac_1 and Ac_3 temperatures.

The low toughness of the CG region is mainly due to the transformation of as-quenched martensite from large prior austenite grains [5]. The toughness of the CG-IC region is controlled by both the large grain size and martensite-austenite constituents (MA) [3]. It is well understood that the formation of coarse MA are linked to low toughness in welded steels since they act as fracture initiation sites due to their inherent brittleness [6-8]. In addition, other researchers found that fracture can initiate near the MA/matrix interface (i.e. particle de-bonding) due to high stress concentration [8-13].

Moeinifar, et al. [14] reported that the size of MA strongly influences the impact toughness of weldments in X80 line-pipe steel. Moeinifar, et al. [9], also found that a CGHAZ reheated to a mid-inter-critical temperature of 800°C had the lowest impact toughness compared to subcritical reheating. Zhu et. al. [15] report that high volume fractions of MA particles formed along the prior austenite grain boundaries were unfavorable for toughness in the ICCGHAZ of X70 line-pipe steel weldments. Lan, et al. [6] reported that MA particles greater than 3 μm in diameter reduced the fracture initiation energy, thus leading to reduced toughness in the CGHAZ of low-carbon bainitic steels. Chen also found that the size of MA constituents was the controlling factor in the promotion of cleavage cracking in CGHAZ simulated specimens for high strength HT80 steels [7].

Particularly low toughness has been observed when the intercritical reheat temperature is close to the A_{c1} temperature, due to the formation of necklace-like MA constituents along the prior austenite grain boundaries[16]. It was observed that toughness can improve with increasing intercritical reheating temperature [5, 17], due to the increased degree of austenite transformation allows refined grains martensite and/or bainite to transform on cooling, which is beneficial for toughness.

The effect of multiple repeated thermal cycles on high strength Q & T steels has not been comprehensively studied, but some recent work has been conducted on HSLA steels. In a study with HSLA100 steel, Duch and Dupont [18] concluded that multiple applications of CG thermal cycles did not affect toughness, but toughness decreased with multiple IC cycles. Vega et al [19] assessed the effects of multiple repairs in girth welds for API X-52 pipeline steel, using shielded metal arc welding (SMAW). The mechanical properties and microstructures for one, two, three and four SMAW repairs in the same area were evaluated. Their results found that after multiple repair welds the microstructural constituents of the HAZ had not significantly changed, but grain growth in the HAZ at the mid-thickness did result in a deterioration of toughness properties.

Previous research identified the CGHAZ and CG-ICHAZ with low intercritical temperature had the lowest toughness, so this work focused on the effects of multiple thermal cycles on these above HAZ regions [5]. The microstructural evolution, hardness and toughness of multiple reoccurring weld thermal cycles of a 690 MPa, high strength Q&T steel was investigated.

2. MATERIALS AND METHODS

The material used in this study was a micro-alloyed, quenched and tempered steel (BIS812EMA). The chemical composition and typical mechanical properties of the steel are shown in Table 1 and Table 2, respectively.

Table 1: Typical chemical composition of BIS812EMA steel [20]

C	Si	Mn	P	S	Cr	Ni	Mo	V	Ti
0.13	0.29	0.87	0.013	0.002	0.45	1.2	0.35	0.015	0.009
Cu	**Al**	**Nb**	**B**	**N**	**Ca**	**O**	**Fe**	**CE_N**	**P_{cm}**
0.19	0.077	0.01	0.0015	-	<0.0003	0.009	Bal.	0.409	0.22

Table 2: Typical mechanical properties of BIS812EMA steel

0.2% Proof stress	UTS/MPa	Minimum yield strength/MPa	Elongation/ %	Charpy impact energy/J (at -20°C)
754[21]	845[21]	690[21]	18[21]	202

HAZ simulations were conducted to replicate the effect of thermal cycles in real welds in order to isolate and test the mechanical properties of different HAZ microstructures. Thermal cycles were applied in Gleeble 3500 thermo-mechanical simulator.

Simulated HAZ samples were 10 x 10 x 80 mm in size (to allow full-sized Charpy samples) and were cut from the long-transverse (LT) direction from 30 mm thick plate and 2mm below from the surface of the plate, avoiding closer to edges and the mid-section of steel plate. A thin K-type thermocouple was spot welded at the middle, on one facet to control and monitor the temperature. Samples were heated with resistance heating in a vacuum. The jaw gap was maintained at 15 mm for all test samples to produce a uniform HAZ at the mid-section. Pressurised air cooling was used to maintain the required cooling rate.

Dilatometer experiments were conducted to determine the lower and upper transformation temperatures upon heating (A_{c1} and A_{c3}), which were 740°C and 860°C respectively, at a heating rate of 150°C/s (maximum stable rate). Actual weld trials were conducted with MMAW process on 30 mm thick plate with 1 kJ/mm heat input per-pass. Welding procedures were pre-qualified (guidelines on heat input range, V, I and speed ranges per welding rod diameter, preheat and interpass specifications) in order to minimize hydrogen assisted cold cracking. Temperature profiles obtained from instrumented weld trials (with imbedded thermocouples to record the temperature history) were used to define the experimental variables for simulations; which are shown in Table 3. All samples were pre-heated to 150 °C before applying the thermal cycle, to replicate the pre-heat in a real weld. Cooling curves were simulated with a Rykalin-3D mathematical model, which is used to simulate the welding of thick plates, based on time to cool down from 800°C to 500°C ($t_{8/5}$). When the temperature of the cooling process was less than 160°C, the temperature control was switched off and specimens were either left cool to room temperature or the next cycle was started. This cut-off temperature was chosen to limit test time and any microstructural changes below this temperature are negligible. At least three samples for each thermal cycle were prepared for subsequent Charpy impact testing.

Table 3: Characteristics of thermal cycles used to simulate different HAZs

Simulating HAZ region	Maximum temperature of a cycle/ °C	Heating rate/(°C/s)	$t_{8/5}$ /s	Hold time at peak temp/s
CG	1350	200	4.3	0.5
IC	750	130	6.7	0.5

Combinations of multiple thermal cycles were applied to various sub-zones of the HAZ, as shown in Table 4. Thermal cycles were repeated up to three times to simulate multiple repair welds.

Table 4: List of combination of thermal cycles tested for the study

Notation	Maximum temperature of a thermal cycle /°C					
	1st	2nd	3rd	4th	5th	6th
CG	1350					
CGx2	1350	1350				
CGx3	1350	1350	1350			
CG-IC	1350	750				
CGx2-IC	1350	1350	750			
CGx3-IC	1350	1350	1350	750		
CG-ICx2	1350	750	750			
CG-ICx3	1350	750	750	750		
(CG-IC)x2	1350	750	1350	750		
(CG-IC)x3	1350	750	1350	750	1350	750

In earlier research, it was found that when IC temperature in ICCG region which closer to Ac_1 temperature showed the lowest toughness but at higher IC temperature, the toughness was considerably improved [5, 17]. Therefore, in this research, multiple thermal cycles of CG

(1350°C) and lower IC temperature (750°C) were considered. The notation of thermal cycles is given as the order it was applied to the sample. As an example CG-IC refers to CG thermal cycle followed by IC cycle, CG-ICx2 refers to CG cycle followed by two IC cycles and (CG-IC)x2 refers to the CG-IC cycle was applied twice.

3. RESULTS

3.1 Microstructure

Microstructure evolution was studied for multiple simulated thermal cycles. The microstructures for the CG thermal cycles, repeated up to three times, are shown in Figure 1. The microstructures consisted of lath martensite for all CG cycles and there was no noticeable change in the microstructure with repeated cycles.

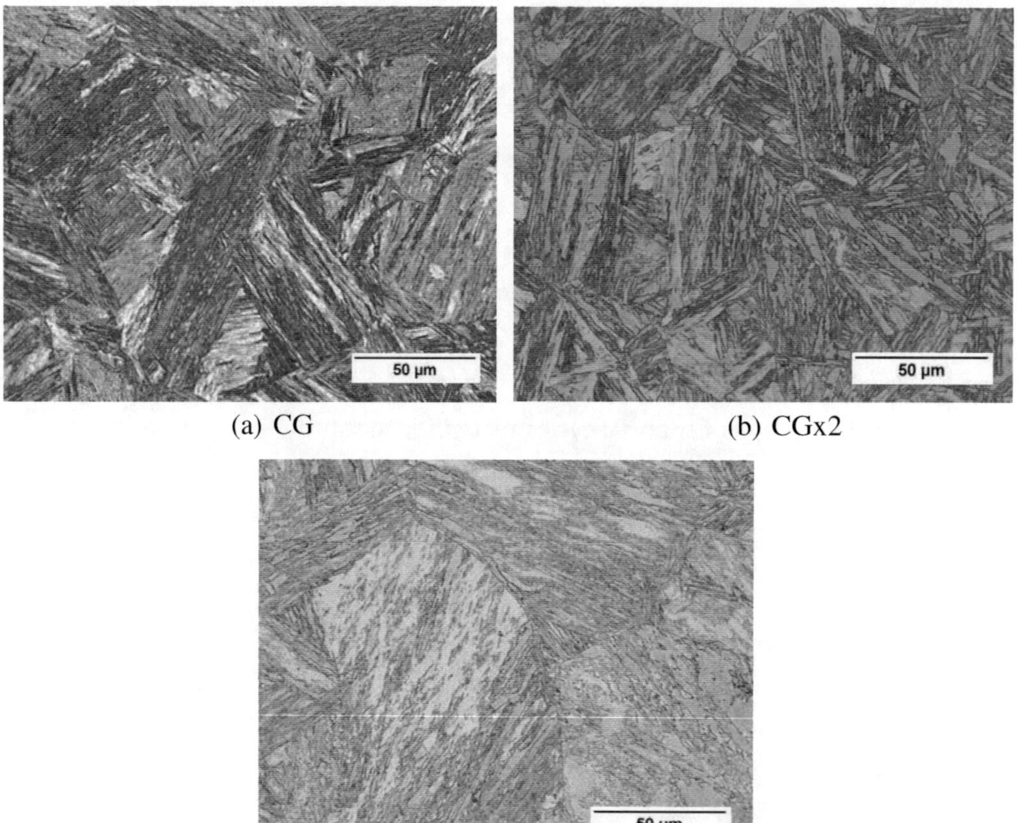

(a) CG (b) CGx2

(c) CGx3

Figure 1: Microstructure with different consecutive CG thermal cycles. (a) single CG cycle, (b) double CG cycles and (c) three CG cycles. 2% nital etch.

Microstructures of multiple CG-IC thermal cycles are shown in Figure 2. Regardless of whether the initial CG cycle was repeated, the subsequent IC cycle was repeated or the 'CG-IC' cycle was repeated, a similar microstructure consisting of coarse grained lath martensite with fine MA particles at prior austenite grain (PAG) boundaries was obtained.

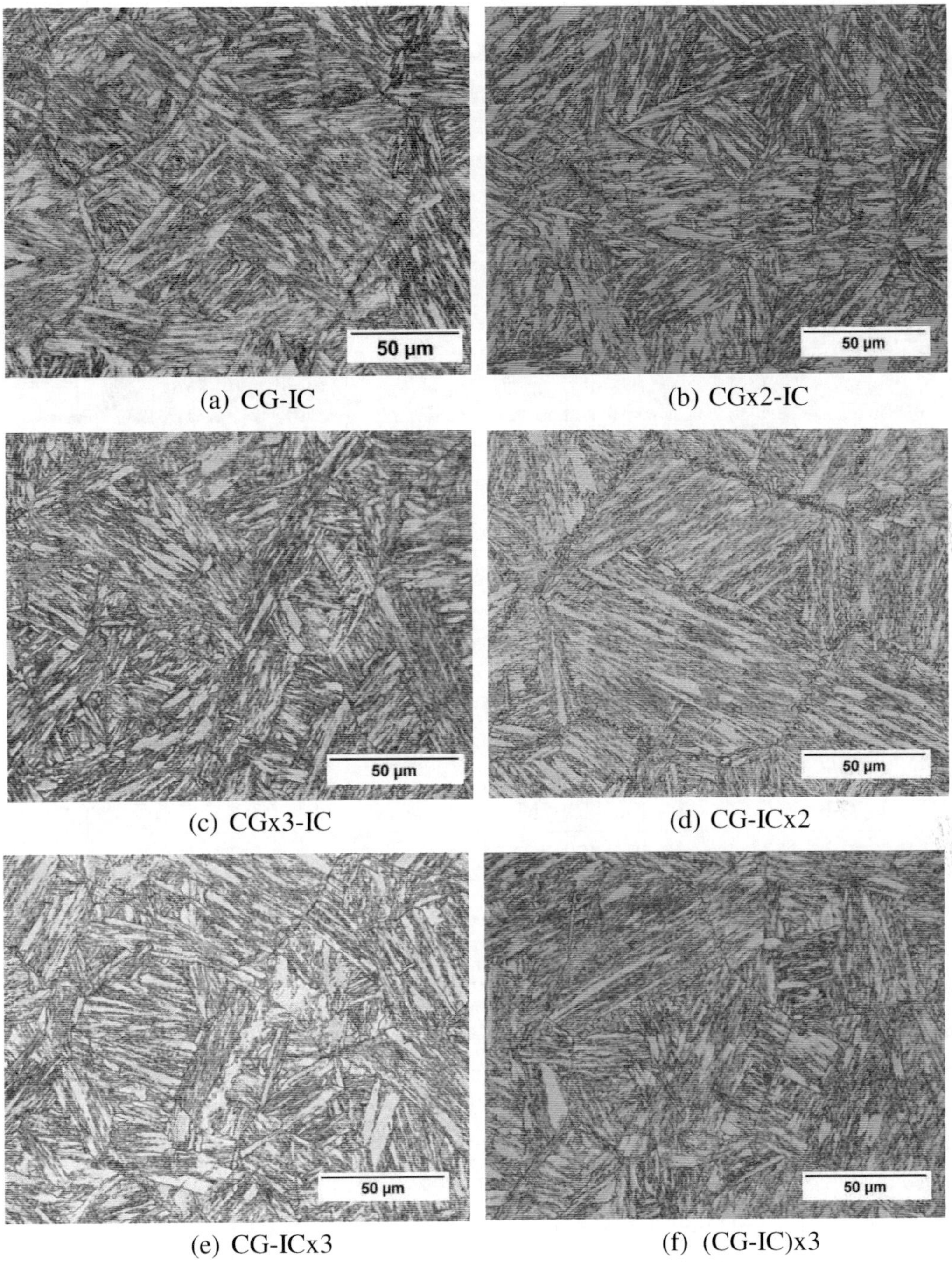

Figure 2: Microstructures of samples with different variations of multiple CG and IC thermal cycles. 2% nital etch.

In order to reveal the MA constituents more clearly, LePera etchant was used for the interctirically reheated specimens. Figure 3 shows the CG-IC and CG-IC-IC specimens, where both show a network of MA constituents along PAGB's, along with small slivers at inter-lath locations. Image analysis (with ImageJ) was used to determine the volume fraction of MA constituents and the results are shown in Table 5.

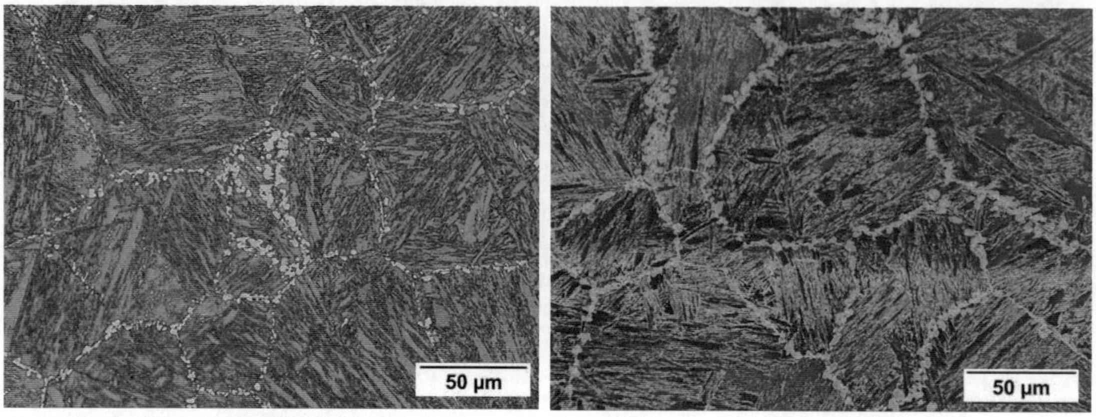

(a) CGIC (b) CGICIC

Figure 3: Microstructure of CG region with single and double IC cycles showing MA on grain boundaries. (a) CG-IC and (b) CG-ICx2. (LePera etch).

Table 5: Measured area percentage of MA constituent with different thermal cycles

Notation	Thermal cycle	% of MA
CG	1350°C	0
CG-IC	1350-750°C	4.4
CG-ICx2	1350-750-750°C	5.4
CG-ICx3	1350-750-750-750°C	5.3
CGx2-IC	1350-1350-750°C	4.8
CGx3-IC	1350-1350-1350-750°C	4.6
(CG-IC)x2	1350-750-1350-750°C	3.9
CGx2-ICx2	1350-1350-750-750°C	4.3

3.2 Hardness

The hardness measurement with single and multiple thermal cycles indicated that there were no considerable changes of hardness with application of multiple thermal cycles of CG or multiple CG cycles followed by IC cycle(s), as shown in Figure 4. Application of IC thermal cycle on top of a CG cycle decreased the hardness on average from 380HV5 to 325HV5 (14.5%).

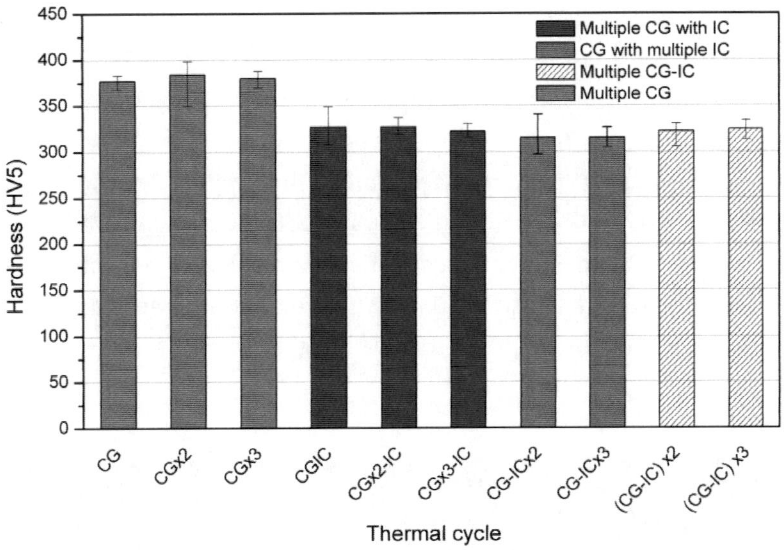

Figure 4: Variation of hardness of simulated HAZ samples with different thermal cycles.

3.3 Toughness

Charpy impact toughness with various multiple thermal cycles is shown in Figure 5. The toughness of multiple CG cycles was comparable. For the CG-IC cycles, regardless of which cycle was repeated (only the first cycle, only the second cycle or both cycles together) toughness remained similar. In all cases, lower toughness was recorded when the final cycle was an IC cycle, compared to CG cycles.

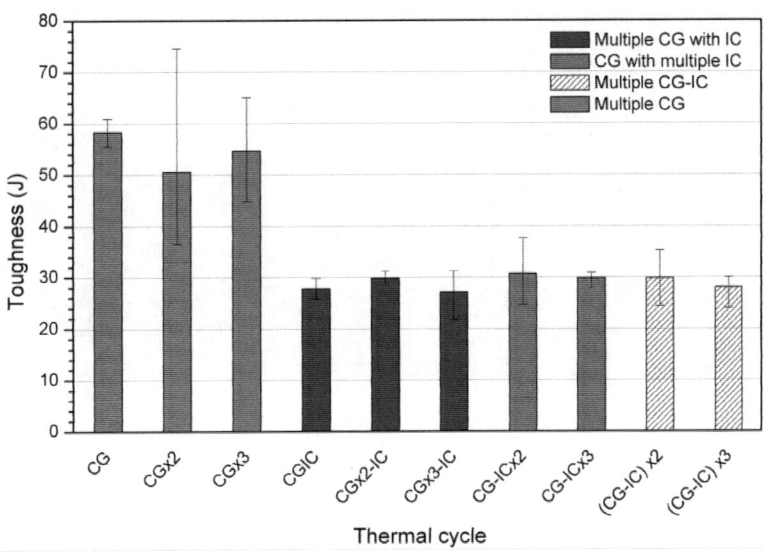

Figure 5: Variation of Charpy toughness of simulated HAZ samples with different thermal cycles. Tested at -20°C.

3.4 Fractography

Secondary electron images were taken from fractured simulated HAZ samples to assess the fracture surfaces. Fracture appearance of samples with single CG cycle and three consecutive CG cycles are shown in Figure 6 and Figure 7 respectively. Both showed similar cleavage fracture facets, typical of brittle fracture, reflecting the similar Charpy impact toughness results. The triple weld thermal cycles, CG-CG-IC and CG-IC-IC also showed cleavage fracture facets, as shown in Figure 8 and Figure 9, respectively. The cleavage fracture facets from the CGx3-IC sample are shown in the Figure 10. The images were taken from a region showing significant secondary cracking, highlighting the low toughness of coarse grained, intercritically reheated specimens.

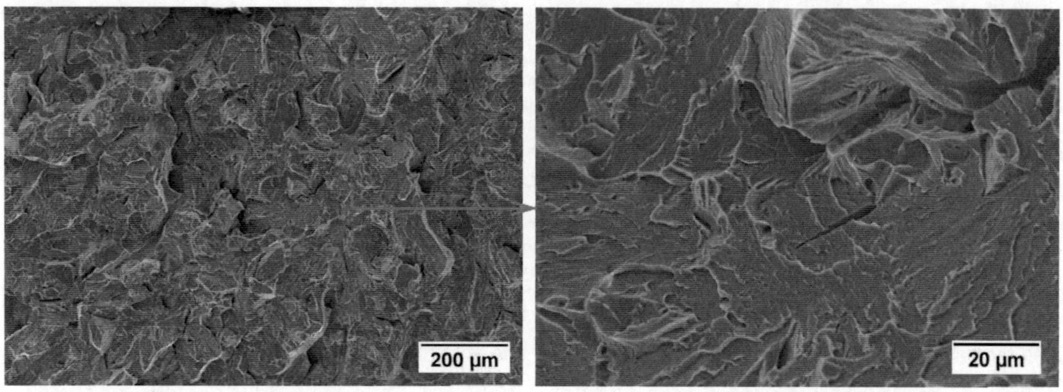

Figure 6: SEM fractographs of simulated HAZ of single CG cycles (CG), (1350°C)

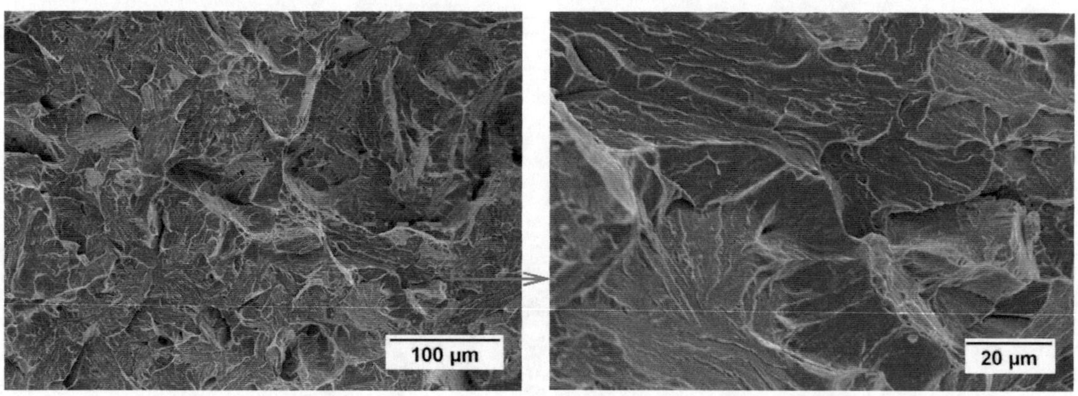

Figure 7: SEM fractographs of simulated HAZ with three CG cycles CGx3 (1350°C-1350°C-1350°C).

Figure 8: SEM fractographs of simulated HAZ with three thermal cycles CGx2-IC (1350°C - 1350°C - 750°C).

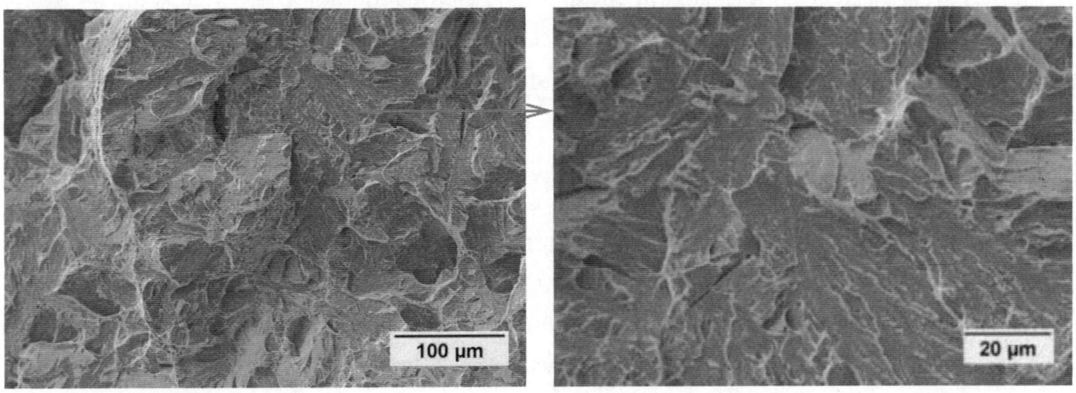

Figure 9: SEM fractographs of simulated HAZ with three thermal cycles CG-ICx2 (1350°C - 750°C - 750°C).

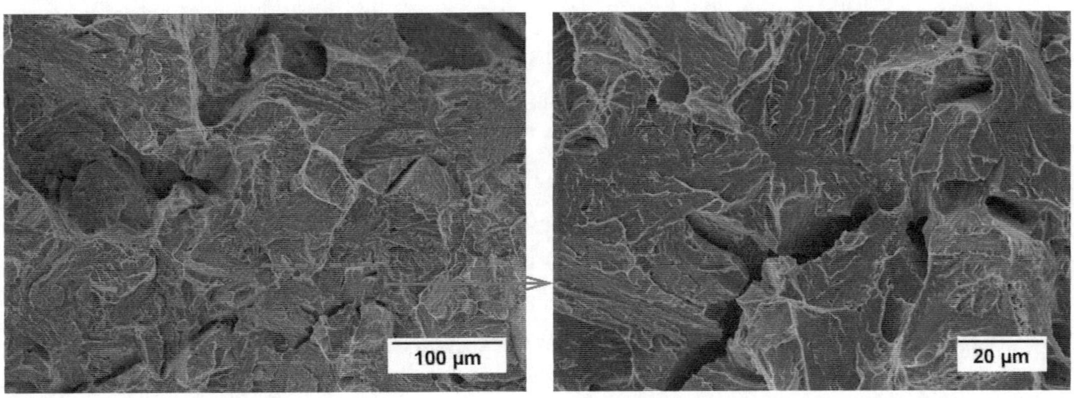

Figure 10: SEM fractographs of simulated HAZ with four thermal cycles CGx3-IC (1350°C - 1350°C - 1350°C - 750°C).

4. DISCUSSION

The effects of multiple thermal cycles to the microstructure of BIS812EMA steel was observed with optical microscope and were shown in Figure 1 andFigure 2. All samples with CG thermal cycles showed lath martensite structure. Samples with an initial CG cycle and finished with an IC cycle showed lath martensite in main matrix and secondary MA constituent along grain boundaries. There was no noticeable effect on the microstructures due to multiple cycles. For multiple CG thermal cycles (1350°C), the previous microstructures were completely replaced with a new, similar coarse martensitic microstructure. Grain growth was negligible with multiple consecutive CG cycles (Average PAG size was 64 μm for CG and 69 μm CG-CG) because of the hold times at the peak temperature were significantly low (0.5 s). For repeated IC cycles, while the microstructure was not replaced, the very low intercritical temperature, limited the formation of MA constituent to the prior austenite grain boundaries and the short time at peak temperature minimised any significant increase in volume fraction of MA at the boundaries. Examining Table 5, the volume fraction of MA varied by less than 1% on average, between all the different cycles, confirming that there was insufficient time during repeated cycles for extra MA constituents to form.

The hardness of the CG cycles was similar for each cycle, which is unsurprising as a similar microstructure was produced for single and multiple cycles. The hardness of samples finished with an IC cycle was lower in average by 14% compared to the CG samples (see Figure 4), even though the hardness of MA constituents is expected to be high. Research by Kim, et al. [22] found that the carbon concentration in MA was significantly higher than the rest of the matrix, hence the higher hardness of MA. Carbon has higher solubility in austenite, so diffuses into the newly formed austenite. Due to the rapid welding cycle, the carbon is trapped and a higher carbon martensite, austenite complex forms [16]. The higher hardness of the MA constituent did not result in higher hardness of the microstructure. This can be explained by 1) the increased carbon concentration in the MA constituent would slightly decrease the carbon in the adjacent regions and 2) more importantly, the IC cycle would have tempered the existing matrix martensite and this would significantly decrease the hardness.

The application of multiple CGHAZ cycles, CGx3, did not cause any significant decrease in toughness. This was due to the transformation of similar microstructures after each cycle, as previously explained. This finding matches that found by Duch and Dupont [18], where toughness did not degraded with multiple simulated CGHAZ cycles (three cycles) for a HSLA100 steel. Toughness was significantly reduced, with an average reduction of 50%, by the introduction of IC thermal cycle(s) on top of CG cycle(s) (see Figure 5). This toughness reduction was attributed to the formation of MA phase along grain boundaries, which has previously been reported to lower toughness [22, 23]. Toughness of the HAZ is generally linked to hardness, where higher hardness typically corresponds to lower toughness. This was not the case for intercritically reheated specimens, where toughness decreased in spite of the decrease in hardness, due to the overriding influence of MA constituent. Thus, it is important to assess both hardness and the microstructure for multi-pass welded structures. Finally, the loss of toughness for intercritically reheated specimens did not have any correlation with the number of previous CG cycles or number of IC cycles. This is understandable, as for the given welding conditions, similar final microstructures were obtained and the microstructure controls the toughness.

Fractography showed cleavage fracture facets for both multiple CG HAZ cycles and multiple CG-IC HAZ cycles, typical for low toughness conditions. The fracture facet sizes appear similar for both the multiple CG cycle samples and CG-IC samples. Research carried out on martensitic steel by Luo, et al. [24] reported that fracture facet sizes were similar to the size of martensite blocks. As the matrix microstructure remained similar for all multi-cycles, it is reasonable to assume that the martensite block sizes were not altered with multiple thermal cycles. Due to the similar coarse grain size of all specimens, cleavage crack propagation behaviour would be expected to be similar. As discussed in the introduction, MA constituents forming at PAG boundaries decrease toughness by acting as cleavage initiation sites. Therefore, the addition of an intercritical cycle to a CG HAZ, lowers toughness by decreasing cleavage initiation resistance.

The simulated HAZ program was used to evaluate multiple weld repair via the weld metal build up approach, where defective metal due to corrosion damage is removed and replaced by weld metal. In such a scenario, the weld layers can be removed by grinding and the plate can be repaired again. This technique is not affected by high constraint conditions as per weld repairs done to remove defects in a weld, such as for repair of girth welds [25] in pipes or butt welds [26, 27]. Increased residual stresses can increase the risk of failure for such cases.

5. CONCLUSIONS

The microstructural and mechanical properties of a 690 MPa yield strength quenched and tempered steel (BIS812EMA) were investigated by Gleeble HAZ simulations. The following conclusions were made from the study:

1. There were no observable microstructural changes with multiple CG thermal cycles or multiple combinations of CG and IC thermal cycles for the given simulated welding conditions.
2. Hardness was not affect by multiple CG HAZ cycles or multiple GC-IC HAZ cycles. However, hardness was reduced when the final cycle was an IC thermal cycle, compared to CG cycles, due to tempering of the matrix martensite.
3. Toughness was always lower when the final thermal cycle was an IC cycle, compared to CG cycle(s), due to the detrimental effects of MA constituents formed at prior austenite grain boundaries on initiating cleavage cracking.

4. Overall, toughness was not further degraded by multiple CG thermal cycles or for multiple combinations of CG and IC thermal cycles for the given simulated welding conditions.

ACKNOWLEDGEMENTS

The Authors acknowledge the support of the Research Training Centre for Naval Design and Manufacturing (RTCNDM) in this investigation. The RTCNDM is a University-Industry partnership established under the Australian Research Council Industry Transformation grant scheme (ARC IC140100003). The Authors would like to express their gratitude to DST Group and ASC Pty Ltd for use of their facilities.

REFERENCES

[1] Z. Sterjovski, "Pad-weld repairs of in-service high-strength steel plate used in seawater environments," *Welding in the World,* vol. 54, pp. R173-R181, 2010.

[2] S. Lee, B. C. Kim, and D. Kwon, "Fracture toughness analysis of heat-affected zones in high-strength low-alloy steel welds," *Metallurgical Transactions A,* vol. 24, pp. 1133-1141, 1993.

[3] J. Hu, L.-X. Du, J.-J. Wang, H. Xie, C.-R. Gao, and R. D. K. Misra, "High toughness in the intercritically reheated coarse-grained (ICRCG) heat-affected zone (HAZ) of low carbon microalloyed steel," *Materials Science and Engineering: A,* vol. 590, pp. 323-328, 1/10/2014.

[4] J. R. Loureiro and A. A. Fernandes, *Toughness of CG HAZs of Welds in Q-and-T Steels*: American Welding Society, 1994.

[5] P. Dissanayaka, K. Carpenter, H. Li, S. van Duin, Z. Sterjovski, J. Donato*, et al.*, "Effect of prior austenite grain size and martensite-austenite constituent on toughness of a high strength quenched and tempered steel subjected to single and double HAZ simulations," unpublished.

[6] L. Lan, C. Qiu, D. Zhao, X. Gao, and L. Du, "Microstructural characteristics and toughness of the simulated coarse grained heat affected zone of high strength low carbon bainitic steel," *Materials Science and Engineering A,* vol. 529, pp. 192-200, 2011.

[7] J. H. Chen, Y. Kikuta, T. Araki, M. Yoneda, and Y. Matsuda, "Micro-fracture behaviour induced by M-A constituent (Island Martensite) in simulated welding heat affected zone of HT80 high strength low alloyed steel," *Acta Metallurgica,* vol. 32, pp. 1779-1788, 1984/10/01 1984.

[8] Y. Li and T. N. Baker, "Effect of morphology of martensite–austenite phase on fracture of weld heat affected zone in vanadium and niobium microalloyed steels," *Materials Science and Technology,* vol. 26, pp. 1029-1040, 2010/09/01 2010.

[9] S. Moeinifar, A. H. Kokabi, and H. R. Madaah Hosseini, "Influence of peak temperature during simulation and real thermal cycles on microstructure and fracture properties of the reheated zones," *Materials and Design,* vol. 31, pp. 2948-2955, 2010.

[10] C. L. Davis and J. E. King, "Effect of cooling rate on intercritically reheated microstructure and toughness in high strength low alloy steel," *Materials Science & Technology,* vol. 9, pp. 8-15, 1993.

[11] C. L. Davis and J. E. King, "Cleavage initiation in the intercritically reheated coarse-grained heat-affected zone: Part I. Fractographic evidence," *Metallurgical and Materials Transactions A,* vol. 25, pp. 563-573, 1994.

[12] L. Lan, C. Qiu, H. Song, and D. Zhao, "Correlation of martensite-austenite constituent and cleavage crack initiation in welding heat affected zone of low carbon bainitic steel," *Materials Letters,* vol. 125, pp. 86-88, 06 / 15 / 2014.

[13] P. Mohseni, J. K. Solberg, M. Karlsen, O. M. Akselsen, and E. Østby, "Investigation of mechanism of cleavage fracture initiation in intercritically coarse grained heat affected zone of HSLA steel," *Materials Science and Technology,* vol. 28, pp. 1261-1268, 2012.

[14] S. Moeinifar, A. H. Kokabi, and H. R. M. Hosseini, "Effect of tandem submerged arc welding process and parameters of Gleeble simulator thermal cycles on properties of the intercritically reheated heat affected zone," *Materials & design,* vol. 32, pp. 869-876, 2011.

[15] Z. X. Zhu, L. Kuzmikova, H. J. Li, and F. Barbaro, "Effect of inter-critically reheating temperature on microstructure and properties of simulated inter-critically reheated coarse grained heat affected zone in X70 steel," *Materials science and engineering a-structural materials properties microstructure and processing,* vol. 605, pp. 8-13, 2014.

[16] H. K. D. H. Bhadeshia, "About calculating the characteristics of the martensite–austenite constituent," presented at the Proc. Int. seminar on Welding of high strength pipeline steels, Araxa, Brazil, 2013.

[17] X. Li, Y. Fan, X. Ma, S. V. Subramanian, and C. Shang, "Influence of Martensite-Austenite constituents formed at different intercritical temperatures on toughness," *Materials and Design,* vol. 67, pp. 457-463, 2015.

[18] E. Duch and J. N. Dupont, "Effect of multiple weld thermal cycles on hsla-100 steel," *Welding Journal,* vol. 98, pp. 88-S-98-S, 2019.

[19] O. E. Vega, J. M. Hallen, A. Villagomez, and A. Contreras, "Effect of multiple repairs in girth welds of pipelines on the mechanical properties," *Materials Characterization,* vol. 59, pp. 1498-1507, 1/1/2008 2008.

[20] B. Dixon and K. Hakansson, "Effects of welding parameters on weld zone toughness and hardness in 690 MPa steel," *Welding Journal (Miami, Fla),* vol. 74, pp. 122-132, 04 / 01 / 1995.

[21] L. Chen, "Characterisation of transverse cold cracking in weld metal of a high strength quenched and tempered steel," Doctor of Philosophy, Department of materials engineering, University of Wollongong, Australia, 2000.

[22] I. Kim, H. Nam, M. Lee, D. Nam, Y. Park, and N. Kang, "Effect of Martensite–Austenite Constituent on Low-Temperature Toughness in YS 500 MPa Grade Steel Welds," *Metals,* vol. 8, p. 638, 2018.

[23] H. Ikawa, H. Oshige, and T. Tanoue, "Effect of Martensite-Austenite constetuent on HAZ toughness of a high strength steel," *Transactions of the Japan Welding Society,* vol. 11, pp. 87-96, 1980.

[24] Z. J. Luo, J. C. Shen, H. Su, Y. H. Ding, C. F. Yang, and X. Zhu, "Effect of substructure on toughness of lath martensite/bainite mixed structure in low-carbon steels," *Journal of Iron and Steel Research International,* vol. 17, pp. 40-48, 2010.

[25] P. J. Bouchard, D. George, J. R. Santisteban, G. Bruno, M. Dutta, L. Edwards, *et al.*, "Measurement of the residual stresses in a stainless steel pipe girth weld containing long and short repairs," *International Journal of Pressure Vessels and Piping,* vol. 82, pp. 299-310, 2005.

[26] P. Dong, J. K. Hong, and P. J. Bouchard, "Analysis of residual stresses at weld repairs," *International Journal of Pressure Vessels and Piping,* vol. 82, pp. 258-269, 1/1/2005 2005.

[27] M. Zeinoddini, S. Arnavaz, A. P. Zandi, and Y. A. Vaghasloo, "Repair welding influence on offshore pipelines residual stress fields: An experimental study," *Journal of Constructional Steel Research,* vol. 86, pp. 31-41, 7// 2013.

Pacific International Maritime Conference 2019

Functionally Seaworthy: How functional alignment of integrated logistic support can deliver improved seaworthiness

Tim M Lee, AMOG Consulting, tim.lee@amog.consulting

Daniel Johnstone, AMOG Consulting, daniel.johnstone@amog.consulting

Cherie Holland, AMOG Consulting, cherie.holland@amog.consulting

Dan Washington, AMOG Consulting, dan.washington@amog.consulting

RADM Peter Marshall RANR, peter.marshall@amog.consulting

ABSTRACT

Integrated Logistic Support (ILS) packages for maritime mission systems have traditionally been based on a physical system centric view of the maritime system. The analyses upon which these ILS packages are built, such as Reliability Availability Maintainability (RAM) and Level-of-Repair Analysis (LORA), are generally developed by considering the equipment items that make up the system and their physical architecture. This approach tends to drive the sustainment of a maritime system towards achieving the desired availability of the individual equipment items that make up the system.

The maritime mission system, however, delivers its operational effect through the functions it performs. These functions are often implemented by multiple systems, which can lead to a purely system-centric based ILS package being poorly adapted to support the efficient delivery of the operational effect. Furthermore, taking a functional view of sustainment, allows the better linkage between the operational effect of the maritime mission system, and its ongoing support.

This paper aims to demonstrate, by means of example, how applying functional thinking, and developing the ILS package on the basis of a functionally-aligned RAM and LORA analysis of the maritime mission system, can realise efficiency gains in the allocation of sustainment resources. It will also discuss how this approach can result in an increased likelihood of efficiently achieving the operational effect, whilst at the same time satisfying the seaworthiness goals of the Royal Australian Navy (RAN).

INTRODUCTION

The Royal Australian Navy (RAN) has the mission "to fight and to win in the maritime environment." [1] The tenets to achieve this mission are described in the capstone document Australian Maritime Doctrine [1], which describes the underlying strategic and operational concepts required to achieve the necessary sea power to successfully achieve this mission. Execution of these strategic and operational concepts is enabled by an effective support system, within which logistics, both afloat and ashore, is a key element. It is aspects of the initial development of the logistics support, that is to say, the Reliability Availability Maintainability (RAM) program, along with the System Safety Program, that are the focus of this paper.

NOTE ON TERMINOLOGY

This paper employs systems engineering concepts in the description of a warship, where terminology is in accordance with that used in the Systems Engineering Body of Knowledge (SEBoK) [2] and the INCOSE Systems Engineering Handbook [3].

A **system** is a *"combination of interacting elements organized to achieve one or more stated purposes". [2]*

The warship mission system is a **system of systems**, made up of systems such as the propulsion system, auxiliary systems, combat system, weapons systems etc. Each of these systems may be further broken down to subsystems.

A **function** *"is a characteristic task, action or activity that must be performed to achieve a desired outcome."* [3]

Safety is *"freedom from conditions that can cause death, injury, occupational illness, damage to or loss of equipment or property, or damage to the environment"*, whilst **system safety** is *"The application of engineering and management principles, criteria, and techniques to achieve acceptable risk within the constraints of operational effectiveness and suitability, time, and cost throughout all phases of the system life-cycle."* [4]

Reliability is defined as *"the probability that an item will perform its intended function for a specified interval under stated conditions."* [5], and **availability** is *"a measure of the degree to which an item is in an operable and committable state at the start of the mission, when the mission is called for at an unknown (random) time."* [5].

Maintainability is defined as:

> *"The relative ease and economy of time and resources with which an item can be retained in, or restored to, a specified condition when maintenance is performed by personnel having specified skill levels, using prescribed procedures and resources, at each prescribed level of maintenance and repair. Also, the probability that an item can be retained in, or restored to, a specified condition when maintenance is performed by personnel having specified skill levels, using prescribed procedures and resources, at each prescribed level of maintenance and repair."* [5]

SEA POWER AND SEAWORTHINESS

Sea power has particular characteristics that differ from air power and land power, and these characteristics make sea power a valuable contribution towards the overall strategic capability of the Australian Defence Force. Within the context of the logistical support, some of these particular characteristics of sea power may be highlighted, notably:

- Readiness: warships can be ready to rapidly deploy and conduct operational tasks; and
- Resilience: a warship has the ability to remain operational in the face of damage and defects. [1],

The highlighted characteristics of sea power are highly reliant upon both the warship mission system, via the system design for reliability and availability (frequently by the incorporation of redundancy), and the support systems, including the maintenance and spares logistics.

The RAN's Australian Maritime Operations [6] further expands on this, by the introduction of the concept of seaworthiness. Seaworthiness is defined as "A judgment of mission system(s) to be supported in becoming and thence remaining seaworthy utilising the services of the support system". [6] The concepts of seaworthiness are that mission systems need to be designed and maintained to achieve their operating intent, with a risk-based management system applied to eliminate or reduce risks to the seaworthiness outcome.

Seaworthiness is implemented within the RAN under the Defence Seaworthiness Management System (DSwMS). This system guides the achievement of the seaworthiness outcome, which is to:

> *"ensure that the operation of a maritime mission system, in accordance with its Capability Manager's operating and support intent and enabled by its support system:*
>
> 1. *maximises the likelihood of achieving the specified operational effect for the defined tasking, where*
> 2. *efforts have been made to eliminate or minimise so far as is reasonably practicable (SFARP) hazards/risks to personnel, the public and the environment.* [7]

The key concept introduced by the DSwMS is the closer integration of hazard and risk management of a maritime mission system with targeting the achievement of the maritime mission system operational effect. This requires the confluence of two different streams of specialist systems engineering activities: System Safety (which is focused on the elimination, or minimisation SFARP, of risks to safety of personnel, the public and environment), and Logistic Support Analysis, which is in place to develop the Support System that is necessary to ensure a Mission System can achieve its operational effect.

Historically, the System Safety and Logistic Support engineering activities have been developed in parallel but separately, with the point of interface between the two programs at the subsystem and equipment level, using bottom-up analysis techniques. Whilst this can result in a practical and robust outcome, the independent commencement of the two strands of analysis means that when these are integrated to assess the seaworthiness outcome, there may well be misalignments in approach above the subsystem or equipment level. This could be rectified by the overlay of a top-down analysis approach, as is demonstrated by the example later in this paper.

Furthermore, the engineering analysis that supports both of these activities is generally performed on a physical architecture basis of the system; that is to say, the system is represented through the physical configuration items within subsystems, and the analysis is performed by assessing the reliability, availability, maintainability and safety of the systems on the basis of each of these subsystems and configuration items. Functional representations of the system are more often restricted to specific analysis of support of the design of particular system elements, such as the hardware architecture and software aspects of control, monitoring and combat systems.

The operational effect of a maritime mission system is frequently achieved by functions that extend across the boundaries of multiple physical configuration items or are part of multiple subsystems. For some functions, which reside within a single subsystem, and where that subsystem has a closely defined purpose, there may be a high degree of correlation between a physical and functional representation of the subsystem. An example of this would be in the case of propulsion, which is a clearly defined function (to propel the ship), and this function is provided (for a conventional ship) from a single system (the propulsion system) Accordingly the functional architecture and the physical architecture closely correspond (i.e. via the drive-line).

However, for other functions of a warship, these representations may be more divergent, with the functional model crossing the physical system boundaries. An example of such would be a warship with an integrated electric propulsion system, where the propulsion function crosses system boundaries to the electrical generation system.

Functions-based systems engineering (FBSE) employs functional models to develop a functional architecture.[3] For design purposes, the functional architecture is used to inform the physical architecture of the system. However, in the context of system safety engineering and logistic support analysis, where there will typically have been an element of design and definition of the physical architecture already in place, a functional architecture may be mapped to the physical architecture. Such mapping will reveal where functions cross physical system boundaries, and are dependent on these other systems to achieve the operational effect.

WHY APPLY FUNCTIONS-BASED TECHNIQUES?

As stated above, the aim of DSwMS is to ensure that maritime mission systems maximise the likelihood of achieving their operating intent, whilst eliminating or minimising SFARP risk. This implies that similar metrics should be used for assessing the operational performance (informed by the reliability, availability and maintainability of the system) and the safety of the systems.

Method of functional analysis of aircraft systems

Civil aircraft and systems are frequently developed in the context of the requirement to certify these systems to the airworthiness regulations of the Federal Aviation Administration (FAA) in the United States, and the European Union Aviation Safety Agency (EASA). There are multiple recommended design practices that have been created to guide the development of aircraft systems and demonstrate compliance with the airworthiness regulations, amongst which is Aerospace Recommended Practice ARP4754A *Guidelines for the Development of Civil Aircraft and Systems*. [8] This recommended practice is:

> *"directed toward systems that support aircraft-level functions and have failure modes with the potential to affect the safety of the aircraft. Typically, these systems involve significant interactions with other systems in a larger integrated environment. Frequently, significant elements of these systems are developed by separate individuals, groups or organizations. These systems require added design discipline and*

development structure to ensure that safety and operational requirements can be fully realized and substantiated."

The guidance, which explicitly links safety with operational guidance, forms an ideal basis from which to adapt for application to maritime mission systems.

The ARP4754A process focuses on the development phase of systems, from the functional definition to the architectural definition and into the physical design and implementation phases. It is based around an approach of analysing the top-level functions of an aircraft, via a Functional Hazard Analysis, and determining the consequences of top-level functional failures. Based upon the severity of the consequence of these functional failures, a Functional Development Assurance Level (FDAL) is assigned to each function. The FDAL represents the level of assurance that is required for the development, implementation and verification of the function, to ensure that there can be confidence that the function performs as intended, and that the potential for development errors in the function are minimised. As a top-level function is decomposed into subsidiary functions, and eventually mapped to the physical architecture and design of a system, the development assurance levels are allocated to these lower levels on a formal basis, to assign Item Development Assurance Levels (IDAL) to particular physical systems and system components. The intent is to use a formalised and structured process to design systems that can be used to achieve the required level of assurance without undue reliance on any single system.

Application to maritime systems

Whilst the approach of ARP4754A could, in theory, be applied to the development of maritime mission systems in its entirety, there are certain objections to this. These include that application of these processes wholesale to maritime projects would be a step change increase in the complexity of engineering assurance processes, with resultant impacts on cost. Secondly, in the Australian context, many maritime mission systems are not *de novo* developments, but rather evolutions of an existing system, whereby there is an existing physical design in place, and the design cannot be solely informed by the functional design. However, the ARP4754A may be adapted for application to the maritime environment, in support of achieving the seaworthiness objective.

APPROACH

As noted above, the ARP4754A commences with a Functional Hazard Analysis. The top-level functions of the mission system are analysed to determine the impact of their failure upon the mission system, and based upon the assessed criticality to safety, a FDAL is allocated against the function. Such an approach is focused towards safety and must be extended to address all aspects of seaworthiness.

To apply the approach to guide the development of safety and ILS deliverables during the design phases of a maritime mission system for a seaworthiness objective, the following approach is proposed:

1. Identify the principal functions of the maritime mission system under consideration. This functional identification should be based upon the desired operational effect goals from the Operating and Support Intent for the mission system.
2. Assess the consequences of a failure of these functions using the risk categories of safety (in accordance with the Defence Safety Manual [9] and ANP2200 Navy Safety Systems Manual [10]), environment (per ANP3201 Navy Environment Risk

Management and Compliance [11]) and Seaworthiness (in accordance with ANP3001 Navy Governance Application [12]). The ratings for assessment of consequence in each category are reproduced in

3. Table 1 to
4. Table 3.
5. Assign Development Assurance Targets (DAT) to each top-level function, based upon the assigned severities in each category. The DAT, equivalent to the FDAL of ARP4654A, would correspond to critical failure rates and reliability, availability and maintainability targets.
6. Develop a functional breakdown is developed through devolving the top-level function into supporting sub-functions. The functional breakdown should be developed to the extent that the contributory functions to the top-level function can be mapped to a grouping of single physical subsystems or configuration items. This will be an iterative process through the design development of the mission system.
7. Allocate the targets downwards through the functional breakdown, following the formal processes described in ARP4754A, whereby there is a requirement for independence and redundancy of supporting the functions in order to relax the target.
8. As the breakdown reaches the physical subsystem level, specific targets for the physical system design may be allocated following the process in Step 5, such that the conventionally applied methods for analysis of safety, reliability, availability and maintainability may be followed.

Table 1: Safety Consequence descriptors [10]

Rating	Consequence Description
Catastrophic	Multiple fatalities or 10 or more injuries/illnesses categorised as critical.
Critical	Single fatality and/or permanent total disability or 10 or more injuries or illnesses categorised as major.
Major	Serious injury or illness requiring immediate admission to hospital as an inpatient and/or permanent partial disability or 10 or more injuries/illnesses categorised as moderate.
Moderate	Injury or illness causing no permanent disability, which requires non-emergency medical attention by a registered health practitioner or 10 or more injuries or illnesses categorised as minor.
Minor	Minor injury or illness that is treatable in the workplace (first aid) or by a registered health practitioner, with no follow up treatment required.

Table 2: Environmental Consequence Descriptors [11]

Rating	Consequence Description
Catastrophic *Widespread serious permanent effect*	Incident is reportable to the regulator, serious permanent/persistent and irreversible damage is caused, significant public interest and media coverage and/or impacts not contained to Defence estate or area subject to the Defence activity which resulted in the incident.
Critical *Wider spread, moderate to long-term effect*	Incident is reportable to the regulator and notable damage is caused to an area or asset from which it will take more than 10 years to recover with long-term evidence of the incident resulting. OR Incident is reportable to the regulator and public concern raised.
Major *Localised, short-term to moderate effect*	Moderate but repairable damage that will take up to 10 years to recover. OR Incident is reportable to the regulator.
Moderate *Localised short-term effect*	Minor damage to the environment or heritage asset or area that is immediately contained on-site. It will take less than two years for the resource or asset to fully recover or it will only require minor repair. OR Disturbance to scarce or sensitive environmental or heritage resources.
Minor *No impact or no lasting effect*	Negligible damage that is contained on-site. AND The damage is fully recoverable with no permanent effects, taking less than six months to fully recover.

Table 3: Technical Seaworthiness Consequence Descriptors [12]

Rating	Consequence Description
Catastrophic ERM Cat 1	Loss of maritime system.
Critical ERM Cat 2	Loss of functional capabilities or ability to operate/control a maritime system such that continued safe operation is in jeopardy. Large reduction in safety margins that may result in loss of system.
Major ERM Cat 3	Reduction in functional capabilities or ability to operate or control the platform/materiel asset, or slight reduction in safety margins.
Moderate ERM Cat NA	Slight reduction in functional capabilities, including the ability to operate or control the platform/materiel asset, or a reduction in safety margins.
Minor ERN Cat NA	Loss of a system or function that does not affect the capability or safety of the maritime system.

The advantages of this approach are that:

1. The top-level functions are consistent between the desired operational effect, the system safety program and the ILS program. Therefore it is more straightforward to reconcile the two aspects of the seaworthiness objective (i.e. safety and operational performance).
2. In linking the platform's operational effect goals/functions to the structure used to analyse supportability, the effect of changes in sustainment strategies and activities on the platform's ability to deliver those functions can be readily shown.
3. Targets for the performance of each function are allocated from a top-down analysis so that there is consistency in their allocation to their contributory functions, based upon the potential for single points of failure to contribute to a reduction in safety or operational performance.
4. The process is agnostic of the system boundaries and allows the dependencies of a function upon other systems and subsystems to be incorporated into the analysis and subsequent development of requirements.
5. It enables consideration of the influence of other physical systems towards the elimination or minimisation of risks to safety SFARP, which is particularly important where the risk minimisation is reliant upon control systems or on independent layers of protection.
6. By using the top-level operational effect goals/functions of the platform to drive the analysis of the ILS and safety programs, progressive assurance of the seaworthiness outcome can be built from the earliest phases of an acquisition program.

EXAMPLE

Step 1: Identify Top-Level Functions

Consider a maritime mission system such as a notional minehunter ship. Three top-level functions of the ship would include:

- Locating mines;
- Classifying mines; and
- Prosecuting mines,

Such a ship may incorporate a number of physical systems that contribute towards these top-level functions, such as:

- A sonar system including a towed body which generates sonar signals for detecting mines;
- Remotely-operated Mine Detection Vehicle (MDV) for remote classification and prosecution of a detected mine;
- An auxiliary propulsion system for station keeping as well as reduced acoustic signature during mine hunting operations;
- A Clearance Diver Team, to manually prosecute mines;
- A degaussing system to reduce the magnetic signature of the ship; and

- An electro-optical system for remote identification of mines on the surface using video, thermal imaging and laser range-finding.

Each of these physical systems directly contribute towards the top-level functions. However, there are a number of other physical systems that support these systems, and thereby contribute to the operation effect of the top level functions, including, but not limited to:

- Electrical generation and distribution systems, which provide electric power to these systems;
- Davit and crane systems, that are used to deploy and retrieve the MDV;
- Winch systems, for deployment and retrieval of towed bodies;
- Support systems for the Clearance Diver Team, such as RHIBs (and their davit system), air compression systems, etc.;
- Internal communications systems, which pass target data between subsystems to facilitate data fusion; and
- Navigation systems, to determine position for the purpose of maintaining station and safe separation from the mine target.

Step 2: Risk Assessment

Taking the prosecution of the mine as the top-level function to analyse, it may be assessed for the consequences of a failure or loss of the function, resulting in an inability to prosecute a mine. The results of a notional risk assessment process are summarised in Table 4.

Table 4: Risk Assessment of Failure of Mine Prosecution Function

Category	Failure Condition	Consequence Rating
Safety	Failure of the overall ability to prosecute and dispose of mines may place this or other vessels in danger of exposure to a mine explosion, resulting multiple fatalities.	Catastrophic
Environment	Failure of the ability to prosecute a mine would result in the mine remaining in the marine environment, which would have negligible impact.	Minor
Technical Seaworthiness	Loss of the ability to prosecute a mine using the mine disposal system would result in the loss of a primary functional capability of the mission system.	Critical

Step 3: Allocate Development Assurance Targets

In the scheme, a DAT would be assigned to each top-level function. Whilst there is at present a gap in Navy (and indeed, wider Defence) practice at this time in respect of determining the appropriate targets to use, and how these correspond to the risk matrix consequence rating categories, for the purpose of this example the relationship given in Table 5 shall be used.

Table 5: Development Assurance Targets

Consequence Rating	Development Assurance Target
Catastrophic	1
Critical	2
Major	3
Moderate	4
Minor	5

Under this scheme, the DAT for the mine prosecution function is:

- Safety: DAT 1;
- Environment: DAT 5; and
- Technical Seaworthiness: DAT 2.

Step 4: Functional Breakdown

Taking the prosecution of the mine as the top-level function to investigate, we would draw on the functional modelling techniques to decompose the function into the supporting functions down to the level that these can be mapped to the physical configuration. An example of such a system decomposition, using a Fault Tree technique to map the relationship from the top-level function to the physical systems supporting the function, is shown in Figure 1.

Figure 1: Example of the Functional Decomposition from the Top-level Function

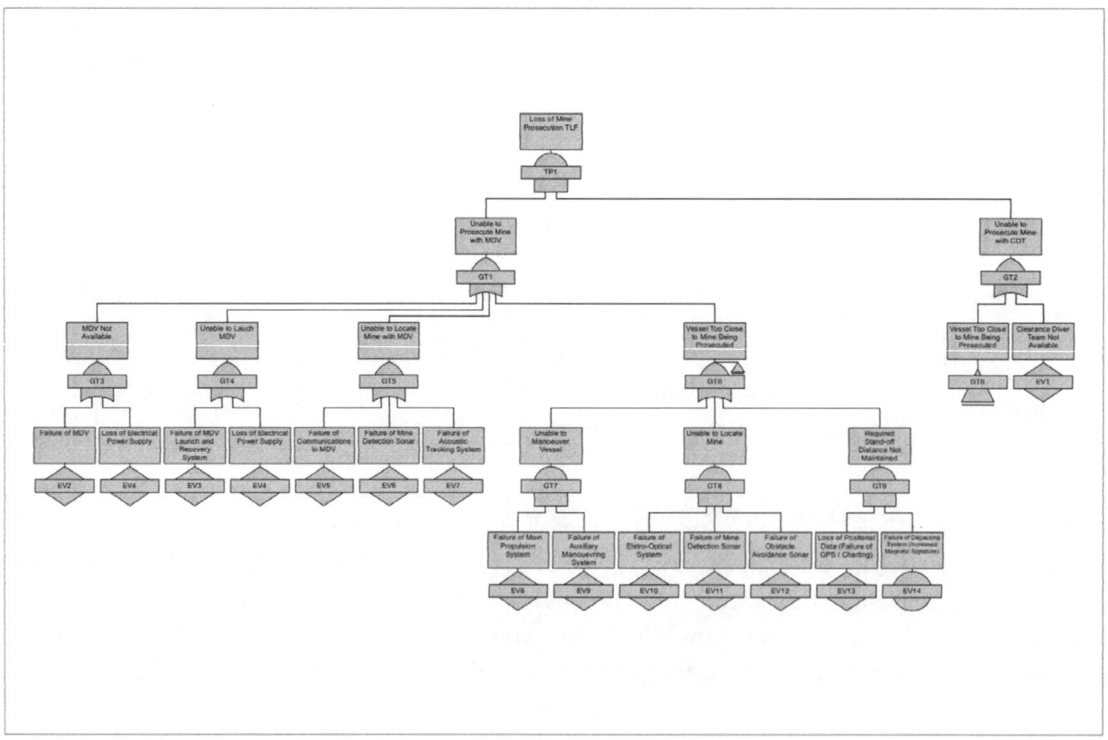

Step 5: Allocate DAT to Lower Level Functions

Following a process generally similar to that outlined in Section 5.2 of ARP4754A [8], DAT are then allocated to lower levels of the functional breakdown. Key aspects of the ARP5754A process for allocation of DAL, that need to be kept in mind for the allocation of DAT to these lower levels of the functional breakdown are:

- If the system architecture is not considered (i.e. the relationships between different system elements in contributing towards the top-level function), then all lower-level functions inherit the DAT of the top-level function.
- Where two or more lower-level (child) functions contribute to a higher-level (parent) function via an OR gate (i.e. failure of any of the child functions results in failure of the parent function) then the child functions inherit the DAT of the parent function.
- Where two or more child functions contribute to a parent function via an AND gate and the independence of these child functions can be demonstrated (i.e. the failure of the parent function requires failure of more than one of the independent child functions) then:
  - the DAT of the child functions may be reduced by one level from the DAT of the parent function; or
  - One of the child functions inherits the DAT of the parent function, and the DAT of the other child functions may be reduced two levels from that of the parent function.

An example of such an allocation of DAT using this scheme is shown in Figure 2.

Figure 2: Example of the DAT Allocation

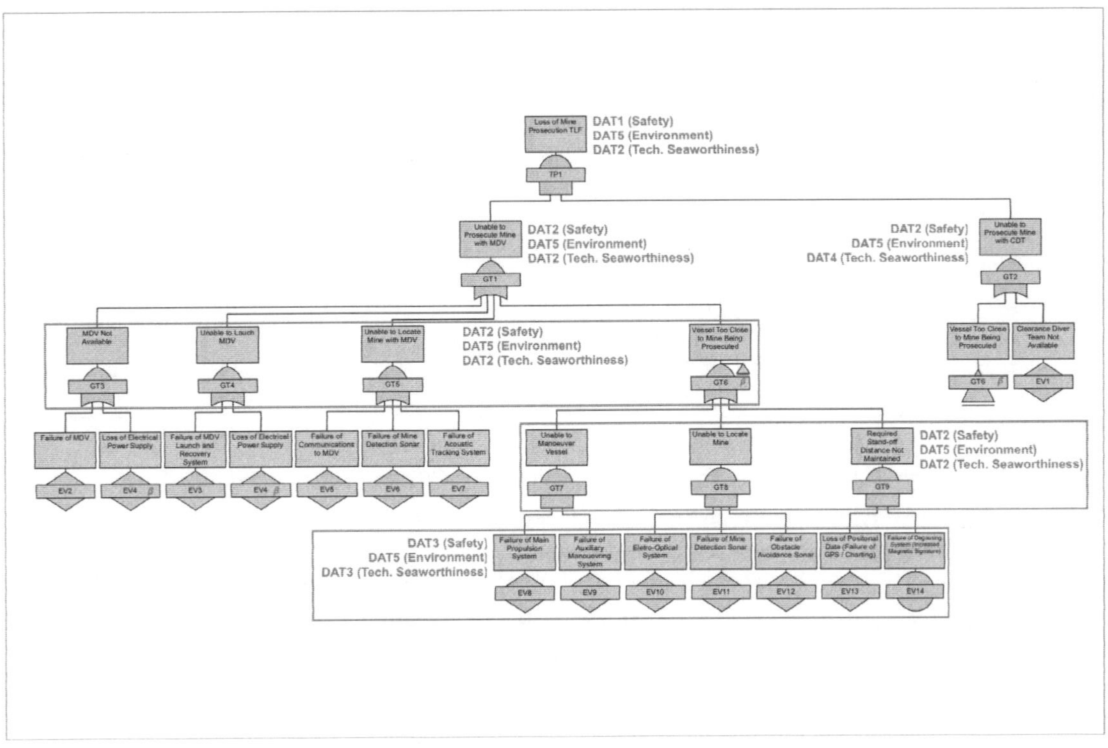

Step 6: Conduct Safety and RAM Analysis on these Functions

In order to achieve the benefit towards demonstrating a robust seaworthiness outcome, the bottom-up, physical system-oriented system safety and RAM analyses should be aggregated together on the basis of these functional breakdowns. To extend the example, for the mine detection function a functional block diagram of the contributory physical systems in the breakdown of Figure 1 may be developed; this is shown in Figure 3.

Figure 3: Functional Block Diagram

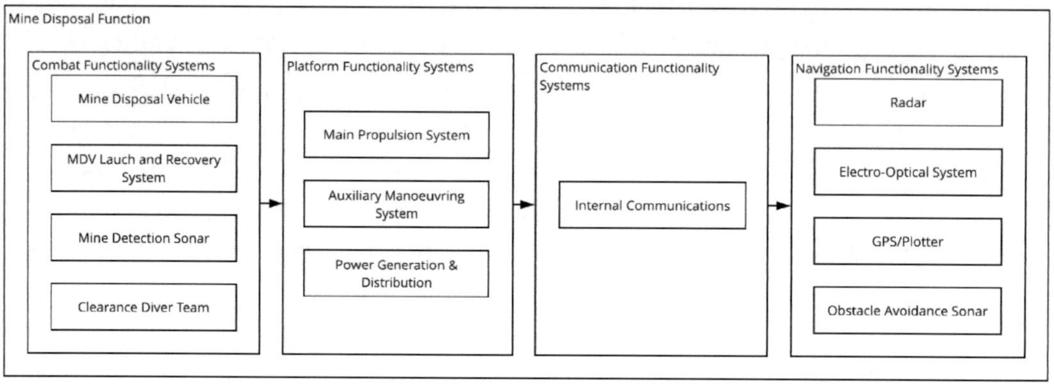

Then, using the outcomes of the bottom-up derived availability analyses of the physical subsystems that contribute to the overall function, an availability for the top-level function may be derived, shown via the availability block diagram in Figure 4.

Figure 4: Availability Block Diagram

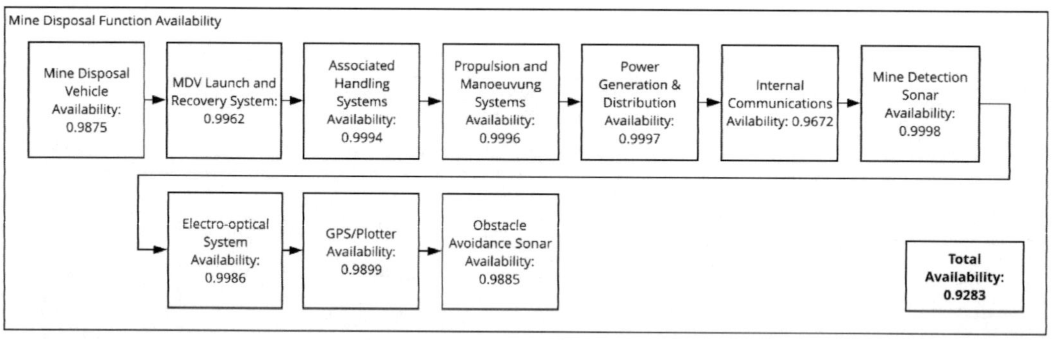

The reliability and maintainability of the function may be calculated likewise in a similar manner. The risk to safety or the environment may be calculated similarly, if quantitative risk analysis techniques are being used, however more commonly semi-quantitative arguments as to the level of risk associated with a function would be developed using the overall functional decomposition structure. In particular, the structured breakdown of the top-level function would indicate where risk mitigations may be best applied (i.e. to what contributory system) to reduce the overall level of risk SFARP.

The strength of the technique would be realised when conducting the subsequent logistic support analyses, such as the Level of Repair Analysis (LORA). With this overall structure in place, there is a framework to enable changes in the LORA to significantly impact on the overall mission availability and safety (by the availability, or otherwise, of physical risk mitigations that substitute, isolate or control by engineering, the safety risk), such that this can be optimised to best achieve the seaworthiness objective.

CONCLUSION

The approach proposed herein draws upon existing functional analysis and modelling techniques from systems engineering to more closely integrate aspects of the system safety and logistic analysis activities, in order to better address the objectives of the Defence Seaworthiness Management System. The proposed approach is, at present, conceptual, due to certain gaps in the existing RAN practices; in particular with respect to setting targets for reliability, availability and maintainability that are tied to the consequence levels associated with the loss or impairment of the top-level functions.

Whilst the proposed approach as described in this paper is focused on the development of a maritime-mission system, the outcomes of this approach would benefit the development and operation of the support system throughout the mission system's life cycle. Furthermore, it is adaptable to in-service mission systems, albeit without the potential benefits that may accrue from influencing the system design (i.e. by improving the reliability or availability of a contributory system through design modification). However, there are sufficient potential benefits to the improvement of the likelihood of achieving the desired operational effect, whilst better meeting the seaworthiness goals, that, in the opinion of the authors, that have value in further applied study of the application of these concepts to confirm that these benefits can be realised.

REFERENCES

[1] Sea Power Centre Australia, *Australian Maritime Doctrine: RAN Doctrine 1*, Second Edition. Canberra: Defence Publishing Service, 2010.

[2] BKCASE Editorial Board, *The Guide to the Systems Engineering Body of Knowledge (SEBoK)*, V 2.0. Hoboken, NJ: The Trustees of the Stevens Institute of Technology, 2019.

[3] D. D. Walden, G. J. Roedler, K. K. Forsberg, R. D. Hamelin, and T. M. Shortell, *Systems Engineering Handbook*, Fourth Edition. Hoboken, New Jersey: John Wiley & Sons, Inc., 2015.

[4] Department of Defense, 'System Safety', Department of Defense, MIL-STD-882E, May 2012.

[5] Defence RAM Working Group, *Defence Reliability, Availability and Maintainability Manual*, 2nd Edition. Canberra: Defence Publishing Service, 2015.

[6] Sea Power Centre Australia, *Australian Maritime Operations*, Second Edition. Canberra: Department of Defence, 2017.

[7] Defence Seaworthiness Authority, *Defence Seaworthiness Management System Manual*, Amendment 2. Canberra: Department of Defence, 2017.

[8] S-18 Aircraft and Sys Dev and Safety Assessment Committee, 'Guidelines for Development of Civil Aircraft and Systems', SAE International, ARP4754A.

[9] Department of Defence, 'Work Health and Safety Risk Management Policy'. Department of Defence, 2018.

[10] Royal Australian Navy, *ANP2200 Navy Safety Systems Manual*. Canberra: Department of Defence, 2019.

[11] Royal Australian Navy, *ANP3201 Navy Environment Risk Management and Compliance*, 1.0. Canberra: Department of Defence, 2019.

[12] Royal Australian Navy, *ANP3001 Navy Governance Application*. Canberra: Department of Defence, 2018.

Experimental Propeller Performance Characterisation behind a Towed Axisymmetric Body

Alexander Conway, DST Group, Australian Maritime College,
alexander.conway@dst.defence.gov.au
Chetan Kumar, DST Group, chetan.kumar@dst.defence.gov.au
Anthony Fowler, DST Group, RMIT University, anthony.fowler@dst.defence.gov.au
Rudaba Khan, DST Group, rudaba.khan@dst.defence.gov.au
Adam Rolls, Australian Maritime College, adam.rolls@utas.edu.au

ABSTRACT

The performance of a marine propeller is degraded by non-ideal operating conditions, including environmental disturbances and vehicle motion. A closed-loop controller can help to overcome these effects and maintain performance. As control law design and implementation requires a model of the system, the propeller must be characterised. To this effect, the performance of a propeller operating behind a towed axisymmetric body was captured via an experimental investigation at the Australian Maritime College (AMC) towing tank. All tests were performed on a generic underwater vehicle geometry with a generic 5-bladed propeller.

All experiments were undertaken with the model straight-ahead and deeply submerged. Computational Fluid Dynamics (CFD) and empirical methods were used to estimate the wake deduction factor to calculate the speed of advance.

The measured thrust, torque and efficiency curves were compared to open-water propeller curves undertaken on the same propeller geometry at the AMC Cavitation tunnel. Differences in experimental setup compared to the cavitation tunnel were accounted for using the ITTC guidelines regarding Reynolds independence. The carriage velocity and propeller rotational speeds were selected to ensure the same flow regime between both data sets.

The results indicate that the open-water propeller curves can provide reasonable estimates of propeller performance in straight-line, deeply submerged condition, provided the wake deduction factor is accurately estimated.

This data will be used to demonstrate real-time closed-loop thrust control techniques in future experimental trials. Further work will investigate the impact of model yaw and aft-control surface deflections on propeller performance.

INTRODUCTION

Marine vehicle propulsion systems often operate in far from ideal conditions that can result in significant thrust losses. These may result from the environmental state, vehicle motion, interactions between the vehicle and propeller and ventilation, which occurs when surface air or exhaust gases are drawn into the propeller blades due to a decrease in pressure [1]. In addition to potentially increasing the mechanical wear on the propulsion system, these thrust losses are detrimental to the manoeuvring performance of the vehicle.

The implementation of a closed-loop feedback control system to compensate for the losses in thrust performance can improve overall propeller performance in maritime applications [2]. By measuring relevant outputs from the propulsion system in real time, an appropriately designed control law can respond to disturbances acting on the propulsion system to maintain the desired propeller behaviour.

To facilitate control law design, a plant model describing the performance of the system to be controlled is necessary. Accurately predicting the thrust and torque generated by a marine propeller often involves the development of complex dynamic models, particularly if an accurate representation of the transient behaviour of the propeller is required [3]. For the purposes of control design, however, it is usually considered sufficient to use simpler empirical models, such as the conventional propeller curves based on non-dimensional thrust and torque coefficients [4].

Open-water experimental tests are commonly performed to characterise a propeller and generate its performance curves [5-7]. However, these curves no longer provide an accurate representation of the propeller's performance once it is placed behind a vehicle due to the resulting changes in flow conditions. For an underwater vehicle, a significant non-uniform wake is created by the hull, casing, sail and aft appendages [8], which leads to a reduction in the flow velocity into the propeller. For the purposes of control law design, it is therefore preferred to obtain a propeller model that captures the actual wake-affected behaviour of the propeller behind the vehicle, rather than in ideal open-water conditions. Thus, the open-water propeller characterisation approaches typically performed and described in the literature are not suitable for this purpose.

This paper describes an experimental study to characterise the performance of a marine propeller. In contrast to conventional open-water experiments, the behaviour of the propeller is assessed while located behind a towed axisymmetric body to generate performance data under realistic flow conditions. For this purpose, a generic geometry underwater body was fitted with the DSTG 115-1 fixed-pitch, 5-blade propeller [9]. Through an experimental program undertaken at the Australian Maritime College (AMC) towing tank, performance curves describing the propeller's non-dimensional thrust, torque and efficiency over the first quadrant of operation were generated. All experiments were undertaken with the model straight-ahead and deeply submerged.

The tests were performed using a newly designed and constructed test model. For verification and comparison purposes, a subset of the captured propeller performance data was compared to existing performance curves generated via open-water tests on the same propeller at the AMC cavitation tunnel [10]. To compensate for the effect of operating in the wake of the test model, Computational Fluid Dynamics (CFD) and empirical methods were used to estimate the wake deduction factor to calculate the actual speed of advance, allowing direct comparisons to be made between the data sets.

EXPERIMENTAL SETUP

The underwater vehicle model used in this study is a modified version of the Defence Science and Technology (DST) Group Joubert barehull model with aft control surfaces [11]. The axisymmetric model has a diameter of 0.3 m and a fixed length (L) of 1.3 m, resulting in a length to diameter (L/D) ratio of 4.33. It is axisymmetric for the first $0.07L$ from the nose tip, with the nose derived from a NACA-0018 forebody. The body is axisymmetric with a fared sail (NACA-0022) and X-form aft control surfaces. The model uses a 5-blade DST generic submarine propeller (shown in Figure 1) with a diameter of 0.15 m, located at $0.98L$. Details of the propeller are given in Table 1. For details on the X-form geometry, see Joubert [11] and Overpelt [12]. The model was mounted via a streamlined sail-shaped holding strut, as shown in the schematic drawing of the experiment model in Figure 2.

Figure 1 - DSTG 115-1 propeller with 0.15 m diameter.

Table 1 - Properties of DSTG 115-1 propeller used in the experiments [10].

Description	Value
Diameter	0.15 m
Number of blades	5
Expanded blade area ratio	0.60
Pitch ratio	1.20
Skew angle	44.5°

The tests were carried out at a submergence centreline depth of approximately 0.671 m, which is mid-depth in the towing tank. This ensures that the surface and bottom effects are minimised. The propeller diameter was sized to maximise thrust and torque measurements whilst minimising its free-surface impact. All measurements were carried out at approximately zero angles of yaw and pitch, with the accuracy for setting yaw and pitch being approximately 0.1°. The origin of the body coordinate system is located at the nose of the model.

The model was tested in the 100 m AMC towing tank [13] at freestream velocities from 2.25 m/s to 3.95 m/s, which correspond to propeller Reynolds numbers of 2.02×10^5 to 2.22×10^5 for the validation cases.

The propeller Reynolds number is given by:

$$Re_P = \frac{v_{res} c_{0.7D}}{v} \tag{1}$$

where

$$v_{res} = \sqrt{v_a^2 + (0.7\pi D n)^2} \tag{2}$$

and Re_P is the propeller Reynolds number, v_{res} is the velocity of the blade at $0.7D$, $c_{0.7D}$ is the blade chord length at $0.7D$, v is the kinematic viscosity, v_a is the speed of advance of the propeller, D is the propeller diameter and n is the revolutions per second of the propeller.

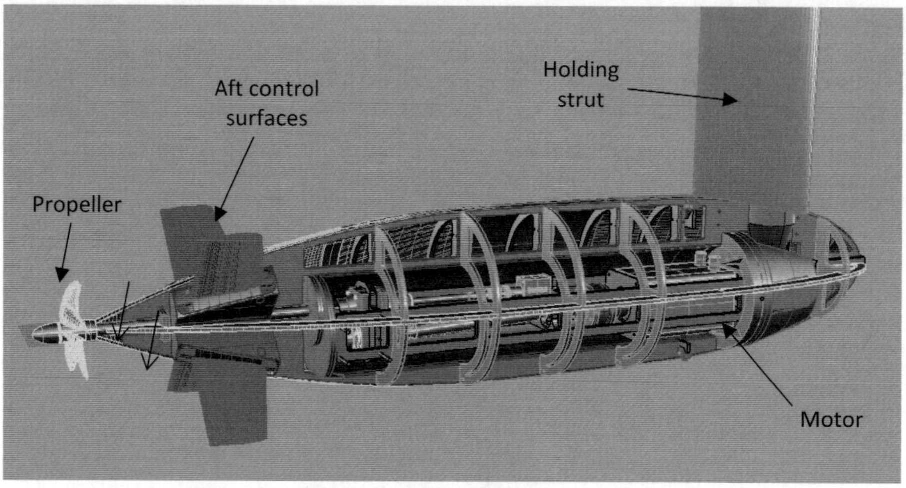

Figure 2 - Schematic drawing showing internal layout and external shape of the experiment model.

Thrust and Torque Measurement

The motor and data acquisition electronics for the thrust and torque measurements were housed inside a waterproof section, as shown in Figure 3. The propeller was driven by an Aerotech BM800 motor, and the propeller thrust and torque were measured using a MARIN 200 N seakeeping dynamometer attached to the propeller. The thrust and torque calibration were checked using a custom designed, in-situ calibration rig. The calibration data indicates a precision of 0.5 N for the thrust and 0.02 Nm for the torque. The total measurement uncertainty based on both systematic and random sources was estimated to be approximately 5% (confidence interval 95%).

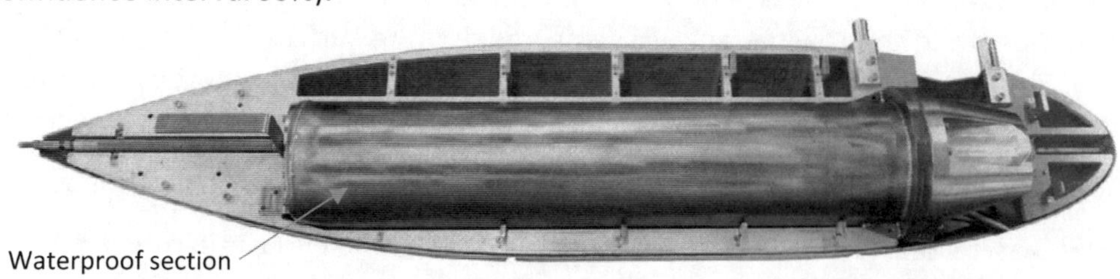

Figure 3 - Cross section of the towing tank model with waterproof section.

Hama Strip - Artificial Turbulence Stimulator (ATS)

Due to the size of the model and the carriage speeds required, laminar flow was predicted across the majority of the test cases. In order to allow the propeller to operate in conditions which represent those encountered on a full scale underwater vehicle, turbulence stimulation to generate a turbulent boundary layer was implemented.

The Hama strip consists of a span-wise array of triangular planform protuberances arranged in a repeating pattern. It was proposed by Hama [14] to provide a means of transitioning laminar to turbulent flow with low levels of parasitic drag in towing tank or wind tunnel model testing. The Hama strips were made from PVC electrical tape, with the protuberance edge defined using a pair of 5 mm pinking scissors. The Hama strip was attached to the model circumferentially and located at a streamwise location of x/L = 5%. The Hama strip was also attached to the holding strut and located at x/L = 10%, as shown in Figure 4. The aft-control surfaces were not tripped. The height of the Hama strips was chosen to be approximately 1 mm to ensure the flow remained turbulent across the test speed range. The trip height was sized based on previous studies in the towing tank, as discussed in Kumar [13].

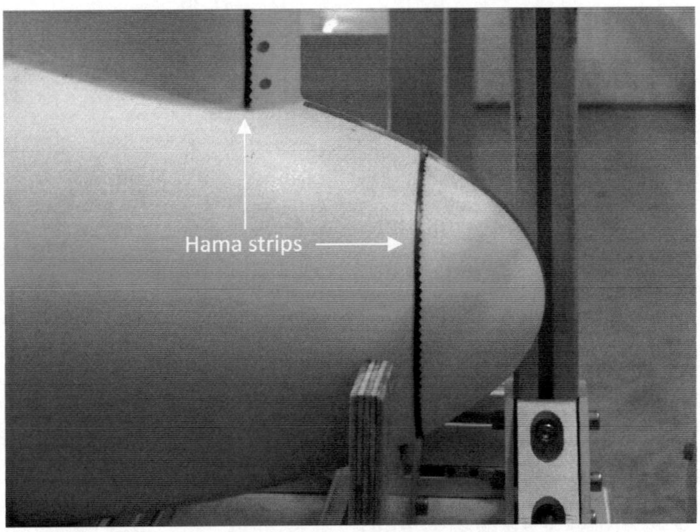

Figure 4 - Placement of Hama strip at the nose (x/L = 5%) and the sail (x/L = 10%).

VELOCITY OF ADVANCE CALCULATION

A critical factor in determining the thrust and torque coefficients is accurately predicting the speed of advance of the propeller. Therefore, both a numerical and an empirical approach were utilised to assess the velocity of the flow encountered by the propeller behind the axisymmetric body.

Using the empirical approach proposed by Renilson [8], adapted from Burcher and Rydill [15], the empirical Taylor wake fraction was predicted using Figure 5. As the prop/hull ratio for the model is 0.5 and it has a tail cone angle of 20°, the Taylor wake fraction is estimated as 0.31. The equation for Taylor wake fraction is,

$$\omega = \frac{v - v_a}{v} \tag{3}$$

where ω is the Taylor wake fraction and v is the forward speed of the vessel. Rearranging the equation and with a Taylor wake of 0.31 results in a speed of advance as,

$$v_a = 0.69v \tag{4}$$

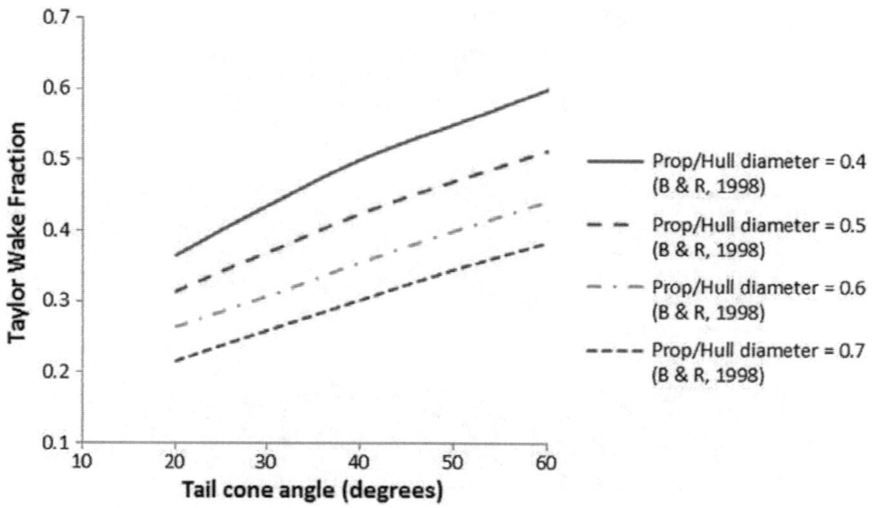

Figure 5 - Taylor wake fraction for underwater vehicles based on prop/hull diameter ratio and tail cone angle [8].

The speed of advance of the propeller was also predicted using CFD. The CFD software *Fluent* was used to perform the numerical simulations with a fully structured mesh generated in *Pointwise*. A grid independence study and validation was completed as discussed previously in Conway [16]. Figure 6 shows the axial velocity contour slice at the propeller plane. The free stream velocity for the simulation was 1.6 m/s and the average velocity was interpolated across the grid at $0.7D$, where D is the propeller diameter, as $0.7D$ is location of maximum chord length [17] and the maximum load on a propeller. The speed of advance was thus calculated as $0.68v$.

As both methods provided very similar predictions for the speed of advance, it was decided to use $v_a = 0.68v$, as the validated CFD simulations were assumed to have a higher accuracy than the empirical method.

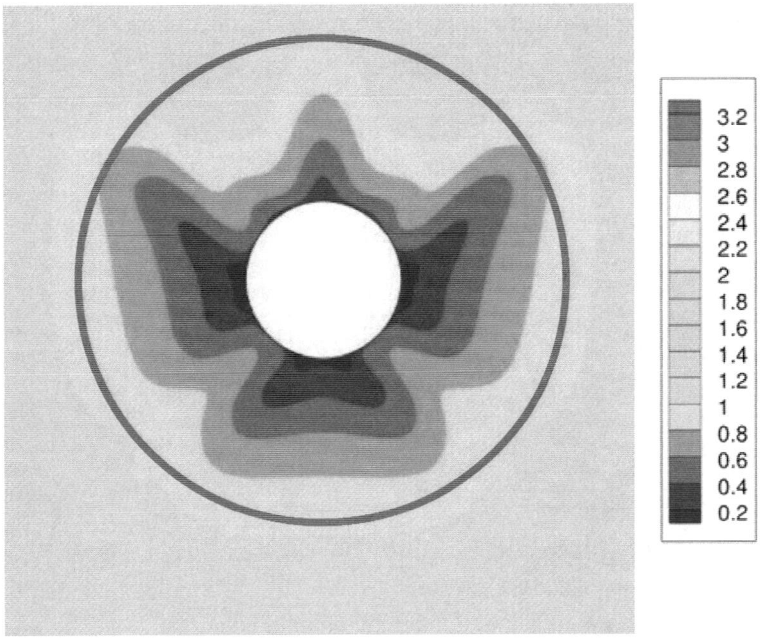

Figure 6 - Axial velocity gradient at the propeller plane for the BB2 submarine with a free stream velocity of 1.6 m/s. The red circle represents 0.7D for velocity prediction.

RESULTS/DISCUSSION

Experimental Data Validation

The initial test runs were performed for the non-dimensional advance coefficient (J) values in the range 0.7 to 1.3, with the resulting values for the thrust coefficient (K_T) and torque coefficient (K_Q) being shown in Figure 7. The J value was calculated using equation (7) below. The thrust and torque data is plotted in terms of K_T and K_Q coefficients calculated using equations (5) and (6). The efficiency (η) of the propeller was calculated using equation (8).

$$K_T = \frac{T}{\rho n^2 D^4} \tag{5}$$

$$K_Q = \frac{Q}{\rho n^2 D^5} \tag{6}$$

$$J = \frac{V_a}{nD} \tag{7}$$

$$\eta = \frac{K_T}{K_Q}\frac{J}{2\pi} \tag{8}$$

where T is the thrust, Q is the torque and ρ is the water density.

In order to validate the experimental results generated using the towing tank model, the data was compared to previous experimental data generated in the AMC cavitation tunnel using the same 5-bladed propeller geometry having a diameter of 250 mm in open-water conditions [10]. The propeller rotation rate and carriage speeds were carefully selected to maintain Reynolds independent flow when compared to the experiment conducted in the cavitation tunnel.

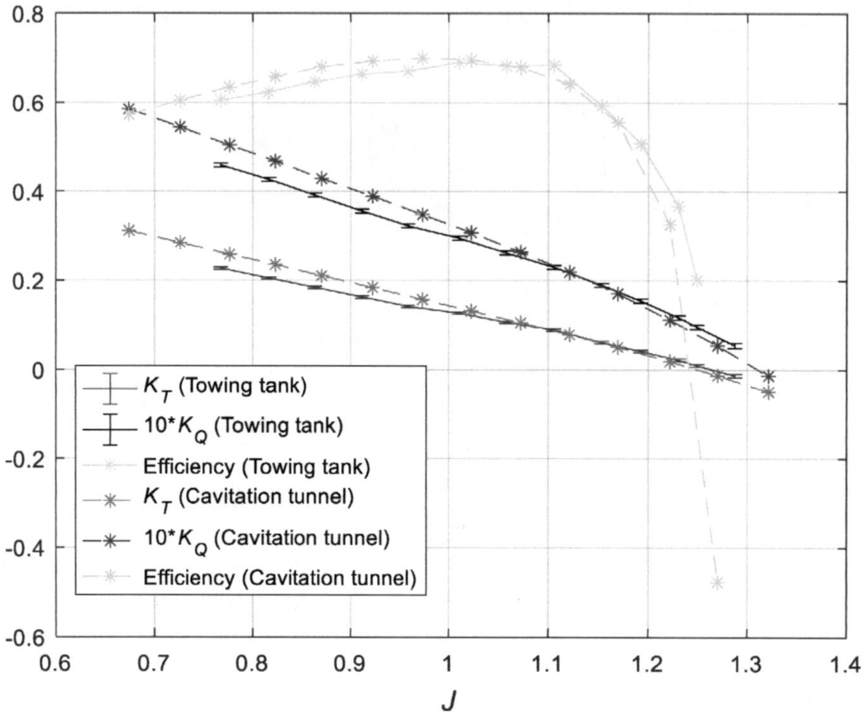

Figure 7 - Efficiency (η), thrust coefficient (K_T) and torque coefficient (K_Q) plotted against advance coefficient (J) values for the towing tank experiment and cavitation tunnel data [10].

As is seen in Figure 7, the correlation between the towing tank and cavitation tunnel data for thrust, torque and efficiency show similar trends, with a slight shift to the right for the towing tank data. Whilst all efforts were made to accurately predict the average inflow velocity of the propeller, as discussed earlier, it is hypothesised that this shift to the right is due to an over estimation in the speed of advance of the propeller by a small margin. However, due to the complex nature of the flow around the mounting strut and aft control surfaces, slight variations between the propeller operating in open water can be expected. Other variations in the data could be due to changes in Reynolds number. Although ITTC procedures suggest that Reynolds independent flow has been achieved, further investigation into the effects of varying Reynolds number would be required to confirm that this has not affected the results published above. A Reynolds dependency study will be conducted during the next phase of experiments.

The strong correlation between the two sets of data suggests that the experimental configuration and dynamometer used are appropriate to accurately capture the full quadrant thrust and torque values accurately.

First Quadrant Propeller Characterisation

For a fixed pitch propeller, it is possible to define four quadrants of operation based on an advance angle β as defined as [4],

$$\beta = \tan^{-1}\left(\frac{v_a}{0.7\pi nD}\right) \tag{9}$$

It is common practice to concentrate on first quadrant operation, where the propeller is spinning in the positive direction with forward advance velocity, as this represents normal operating conditions. However, designing a controller capable of handling all operating conditions ultimately requires characterising all four quadrants of propeller operation, including astern performance [2]. In this case, the use of J, K_T and K_Q when characterising propeller performance becomes mathematically problematic when approaching the quadrant boundaries.

A more flexible approach is to use the alternative non-dimensional thrust coefficient (C_T) and torque coefficients (C_Q) defined in [4] as,

$$C_T = \frac{T}{\left(\frac{\pi}{8}\right)\rho\left[V_a^2+(0.7\pi nD)^2\right]^2} \tag{10}$$

$$C_Q = \frac{Q}{\left(\frac{\pi}{8}\right)\rho\left[V_a^2+(0.7\pi nD)^2\right]^3} \tag{11}$$

These coefficients are well defined across all four operating quadrants (i.e. $0^\circ \leq \beta \leq 360^\circ$) and are therefore suitable for characterising propeller performance for control design and simulation purposes.

Having validated the obtained experimental data for J values in the range of 0.7 ($\beta = 17.7^\circ$) to 1.3 ($\beta = 30.6^\circ$), further propeller thrust and torque measurements were then obtained for the complete first quadrant from $\beta = 0^\circ$ (ahead bollard pull condition) to $\beta = 90^\circ$ (ahead motion with propeller stationary). The resulting propeller performance coefficients C_T and C_Q are plotted versus β in Figure 8. For comparison purposes, this figure also includes first-quadrant propeller data from Pivano et al. [2]. While the comparison data was obtained in open-water conditions and with a different propeller, it is nevertheless useful to compare the trends between the two data sets. The thrust and torque values produced by the propeller tested in this work are smaller at higher values of β, however it is seen that there are similar trends in the performance coefficients, therefore providing further validation for the experimental approach used.

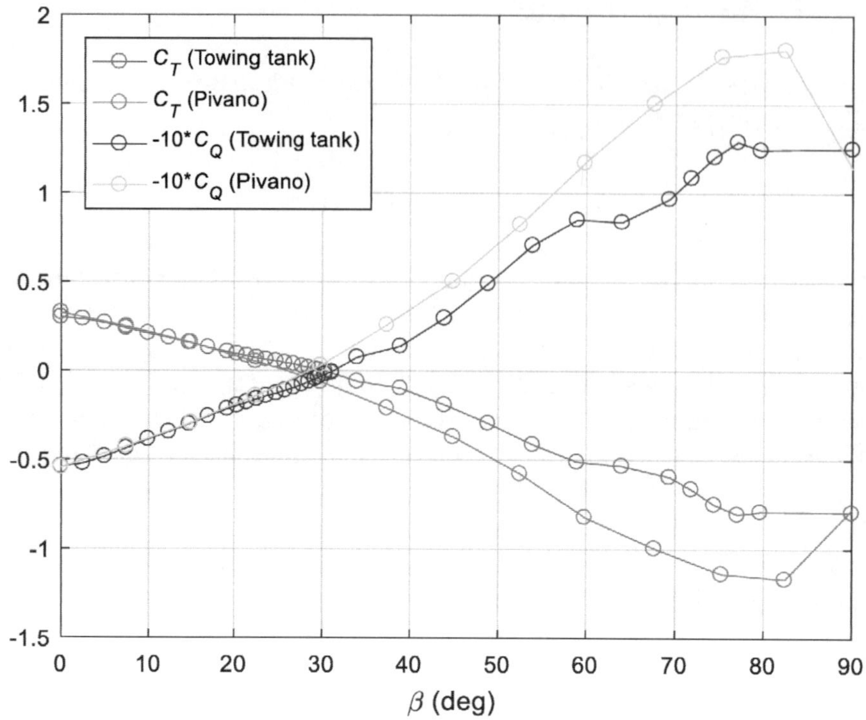

Figure 8 - Experimentally obtained non-dimensional thrust coefficient (C_T) and torque coefficient (C_Q) vs. advance angle (β) for first propeller quadrant. Results compared with data from Pivano [2].

CONCLUSION

The thrust and torque generated by a propeller operating behind an underwater vehicle have been investigated in this paper. Experiments were conducted in the AMC towing tank using a custom-made axisymmetric body with control surfaces. Using the estimated wake deduction factor calculated from CFD, the thrust and torque values recorded in the towing tank were validated against open-water cavitation tunnel data. The validation showed strong correlation between the two sets of data. This was followed by obtaining the propeller characteristics across the complete first quadrant from $\beta = 0^\circ$ to $\beta = 90^\circ$, providing an initial capability to develop control algorithms for standard operating conditions.

In contrast to the typical open-water experiments that are performed to assess propeller performance, this work allows a realistic propeller model to be generated that incorporates the non-ideal flow effects that arise from operating behind a body. Future work will involve characterising the performance of the propeller in the same manner across the remaining three quadrants involving astern, crashback and crashforward conditions. This will allow the generation of a complete empirical propeller model that will be used in the design of four-quadrant closed-loop thrust controllers.

ACKNOWLEDGEMENTS

The authors would like to thank Dr. Luca Pivano from DNV GL for his valuable suggestions and discussions. Many thanks to the AMC staff for their assistance during the trials and Liam Honeychurch for his excellent workmanship of the build. Thank you also to Alex Cameron and Dr Francis Valentinis from DST Group for their help during the trials.

REFERENCES

[1] E. Ruth, Ø. N. Smogeli, T. Perez, and A. J. Sørensen, "Antispin thrust allocation for marine vessels," *IEEE Transactions on Control Systems Technology*, vol. 17, no. 6, pp. 1257–1269, Nov. 2009.

[2] L. Pivano, T. A. Johansen, and Ø. N. Smogeli, "A Four-Quadrant Thrust Estimation Scheme for Marine Propellers: Theory and Experiments," *IEEE Transactions on Control Systems Technology*, vol. 17, no. 1, pp. 215-226, Jan. 2009.

[3] M. Blanke, K. P. Lindegaard, and T. I. Fossen, "Dynamic model for thrust generation of marine propellers," *5th IFAC Conference on Manoeuvring and Control of Marine Craft*, Aalborg, Denmark, 2000, pp. 363–368.

[4] J. Carlton, *Marine Propellers and Propulsion*, 3rd ed. UK: Butterworth-Heinemann, 2012.

[5] J. Dang, J. Brouwer, R. Bosman, and C. Pouw, "Quasi-Steady Two-Quadrant Open Water Tests for the Wageningen Propeller C-and D-Series," *Twenty-Ninth Symposium on Naval Hydrodynamics*, Gothenburg, Sweden, 2012, pp. 1–19.

[6] A. Vali, B. Saranjam, and R. Kamali, "Experimental and numerical study of a submarine and propeller behaviors in submergence and surface conditions," *Journal of Applied Fluid Mechanics*, vol. 11, no. 5, pp. 1297–1308, Sep. 2018.

[7] M. Tran, J. Binns, S. Chai, A. L. Forrest, and H. Nguyen, "A practical approach to the dynamic modelling of an underwater vehicle propeller in all four quadrants of operation," *Proc IMechE Part M: J Engineering for the Maritime Environment*, vol. 233, no. 1, pp. 333–344, Feb. 2019.

[8] M. Renilson, *Submarine Hydrodynamics*, 2nd ed. USA: Springer International Publishing, 2018.

[9] D. Norrison, W. Sidebottom, and B. Anderson, "Numerical Study of a Self-Propelled Conventional Submarine," *31st Symposium on Naval Hydrodynamics*, Monterey, CA, USA, 2016, pp. 1-20.

[10] D. B. Clarke, D. Butler, C. L. Ellis, R. T. Delaney, M. Popko, and P. A. Brandner, "Propeller Dynamometer Commissioning," DST Group, Australia, Technical Note TN-1777, 2018.

[11] P. N. Joubert, "Some Aspects of Submarine Design Part 2. Shape of a Submarine 2026," Defence Science and Technology Organisation, Australia, Technical Report TR-1920, 2006.

[12] B. Overpelt, B. Nienhuis, and B. Anderson, "Free Running Manoeuvring Model Tests On A Modern Generic SSK Class Submarine (BB2)," *Pacific International Maritime Conference*, Sydney, Australia, 2015, pp. 1-14.

[13] C. Kumar, Z. Q. Leong, D. Ranmuthugala, J. Binns, C. Carl, J. Davison, L. Hentschel, M. Courdier, and J. Duffy, "Boundary layer tripping study on a Generic Submarine Model in Water," *21st Australasian Fluid Mechanics Conference*, Adelaide, Australia, 2018, pp. 1-4.

[14] F. R. Hama, J. D. Long, and J. C. Hegarty, "On Transition from Laminar to Turbulent Flow," *Journal of Applied Physics*, vol. 28, no. 4, pp. 388-394, 1957.

[15] R. Burcher and L. J. Rydill, *Concepts in Submarine Design*, UK: Cambridge University Press, 1994.

[16] A. S. T. Conway, F. Valentinis, and G. Seil, "Characterisation of suction effects on a submarine body operating near the free surface," *21st Australasian Fluid Mechanics Conference*, Adelaide, Australia, 2018, pp. 1-4.

[17] ITTC, "Terminology and Nomenclature for Propeller Geometry," Denmark, Recommended Procedures and Guidelines 7.5-01-02-01, 2008.

Blockage Effect on the Drag of an Axisymmetric Underwater Body in Towing Tank Tests

Yieng Teen Huong, Australian Maritime College, yiengteen.huong@utas.edu.au
Zhi Quan Leong, Australian Maritime College, zhi.leong@utas.edu.au
Jonathan Duffy, Australian Maritime College, j.duffy@utas.edu.au
Alexander Conway, DST Group, Australian Maritime College, Alexander.Conway@dst.defence.gov.au
Dev Ranmuthugala, DST Group, dev.ranmuthugala2@dst.defence.gov.au
Martin Renilson, Australian Maritime College, martin@renilson-marine.com

ABSTRACT

Towing tanks are regularly used to measure the drag of underwater vehicles. Depending on the ratio of cross-sectional area between the vehicle model and the towing tank, blockage effects can be present that can significantly impact the accuracy of the measurement. To address blockage effects, blockage correctors have been developed over the years for surface ship experiments in towing tanks, and bluff bodies and aircraft in wind tunnels. However, the accuracy of these blockage corrector formulae is yet to be established for underwater vehicles tested in water channels. This paper investigates the blockage effects for an axisymmetric underwater vehicle model tested within a water channel with different blockage parameter ratios using Computational Fluid Dynamics (CFD).

Existing captive-model experimental datasets are used to validate the CFD model. The validated numerical model is then extended to investigate the relationship between the blockage parameter ratio, the velocity of the model, and its predicted drag. Results show that the blockage effect acting on the deeply submerged model is independent of Reynolds number for the speed range considered and the percentage increase in the drag coefficient is linearly proportional to the blockage parameter ratio. CFD predictions are then compared with ITTC and Wind Tunnel blockage correction formulae, and the applicability of modified Tamura's and the Wind Tunnel formula in correcting the blockage effect are demonstrated. It is intended to extend these findings to fully-appended underwater vehicle models tested in towing tanks.

NOMENCLATURE

Notation	Description	Units
A_M	Maximum cross-sectional area of axisymmetric model, $\pi D^2/4$	[m^2]
A_T	Cross-sectional area of water channel, b x d	[m^2]
b	Breadth of towing tank	[m]
C_D	Total Drag Coefficient, $D_T/0.5\rho V^2 S$	[-]
d	Water depth of towing tank	[m]
D	Diameter of the axisymmetric model	[m]
D_T	Total Drag force	[N]

Fr_{nd}^2	Depth Froude number, $V/\sqrt{gd}$	[-]
g	Gravity	[m s^{-2}]
H	Submergence depth from the model's centerline to the free surface	[m]
H^*	Non-dimensional submergence depth, H/D	[-]
L	Overall length of axisymmetric model	[m]
m	Blockage parameter ratio, A_M/A_T	[-]
M	Mach number	[-]
Re	Reynolds number, $\rho VL/\mu$	[-]
S	Wetted surface area	[m^2]
v_m	Model Volume	[m^3]
V	Fluid velocity	[m s^{-1}]
ρ	Fluid density	[kg m^{-3}]
μ	Kinematic viscosity	[m^2 s^{-1}]
x,y,z	Cartesian coordinates	[m]
y^+	Non-dimensional wall distance	[-]

INTRODUCTION

When a submerged body is tested in a fluid domain surrounded with rigid walls, it experiences blockage effects due to the reduction in the fluid cross sectional area around the body in comparison to the cross sectional area upstream and downstream of the body. The walls restrict the fluid displacement due to the presence of the body, which impacts the surrounding fluid velocity and the wake displacement of the tested body [1]. The blockage effect results in a difference between the measured data and that in open water. Therefore, quantifying the blockage effect or applying a blockage corrector for a submerged underwater vehicle is essential to ensure accurate measurement of its hydrodynamic characteristics.

Blockage effect is strongly correlated with the blockage parameter ratio, *m*, which is the ratio of the maximum frontal cross-sectional area of the model, A_M, and the cross-sectional area of tank test section, A_T. The work on blockage effect for surface ships tested in towing tanks [2-6] and bluff bodies in wind tunnel [7] has been extensively studied over the years. However, there is limited information in the open literature addressing the blockage effect acting on deeply submerged underwater vehicles tested in towing tanks as well as the formation of relevant blockage corrector formulae. The current ITTC guideline [6] provides the blockage corrector for surface ships tested in towing tanks. However, questions do arise on the applicability of the ITTC and wind tunnel blockage correctors in rectifying the blockage effect acting on the drag force of underwater vehicles obtained from water channels. Hence, the focus of this study is to investigate the applicability of the current surface ship ITTC and wind tunnel methods to the blockage correction for underwater vehicles.

To investigate the influence of blockage effect on the drag of deeply submerged axisymmetric underwater vehicles tested in water channels, RANS-based Computational Fluid Dynamic (CFD) simulations were used to quantify the drag under different domain blockage parameter ratios and Reynolds numbers. The objective was to develop a relevant blockage correction procedure as a function of the blockage parameter ratio, *m*. The results were also used to access the applicability of the surface ship based ITTC blockage correction standards to submerged bodies.

TEST PLAN

The geometry used in this study was the bare hull configuration of the generic Joubert model with a length, *L,* of 1.69 m and a diameter, *D,* of 0.23 m (see Figure 1). Further details of the geometry can be obtained from [8].

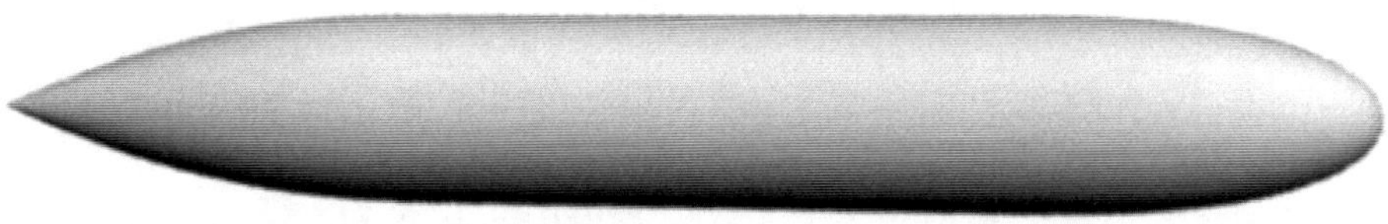

Figure 1: Axisymmetric bare hull Joubert geometry [8]

The model was located at the centre of the cross sectional area of the test section to ensure a symmetrical flow field along its body of revolution profile. A straight line analysis (zero drift and pitch angle) was conducted on the underwater vehicle at speeds ranging between 0.4 m/s to 1.2 m/s, giving a Reynolds number range of 0.6 million to 2 million.

Three tank breadth to depth (*b*:*d*) ratios (1:1, 2:1 and 3:1) were modelled to investigate the influence of tank confinement effect on the blockage effect acting on the model. Additionally, four blockage parameter ratios, *m*, were modelled for each tank breadth to depth ratio to study the blockage effect on the total drag coefficient for a fixed tank breadth to depth ratio. Table 1 lists the dimensions of the tank domain and the blockage parameter ratios investigated.

Table 1: Fluid domain size for different breadth to depth ratios and blockage parameter ratios

Tank's breadth (*b*) to depth (*d*) ratio	Tank Cross Section, (*b x d*) (m)	Blockage Parameter ratio, *m*
1:1	1 x 1	0.0415
	1.2 x 1.2	0.0289
	1.6 x 1.6	0.0162
	10 x 10	0.0004
2:1	2 x 1	0.0207
	2.4 x 1.2	0.0144
	3.2 x 1.6	0.0081
	20 x 10	0.0002
3:1	3 x 1	0.0138
	3.6 x 1.2	0.0096
	4.8 x 1.6	0.0054
	30 x 10	0.0001

SIMULATION SETUP

Simulations were conducted using commercial CFD code, ANSYS Fluent 19.3 to model the fluid flow and the fluid domain used for the investigation as shown in Figure 2. Due to the axisymmetric nature of the underwater vehicle, the domain was split about the *x-z* plane and only half of it was modelled to optimise the computational effort. Comparison between the half and full model simulations showed a difference in the computed drag coefficient of less than 0.1%. Thus, the former was deemed sufficient for this study. The inlet length from the underwater vehicle's centre of buoyancy (coinciding with the origin of the coordinate system) was two body lengths as recommended by the ITTC CFD verification and validation guideline [9]. The outlet dimension was determined based on the outlet sizing conducted to ensure the chosen outlet distance could resolve the submerged body's wake while avoiding the occurrence of any backflow. The distance to the outlet was sized to be four body lengths from the centre of buoyancy, as a further increase to seven body lengths only resulted in 0.4% change in the drag coefficient.

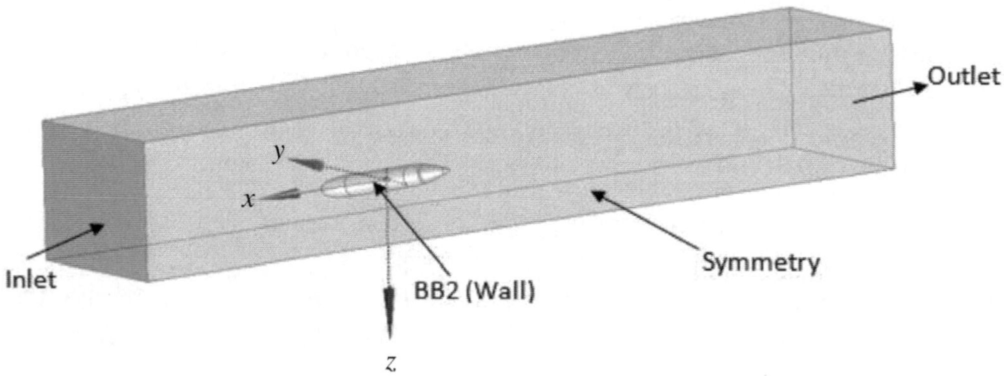

Figure 2: Computational fluid domain with coordinate system.

The computational mesh (Figure 3) was generated utilising ANSYS Fluent Meshing and adopted the Mosaic Meshing Technology [10]. A value of $y^+ = 30$ was used as the first layer prismatic height in combination with the Shear Stress Transport (SST) k-ω turbulence model. The SST k-ω model has been proven to model hydrodynamic coefficients of underwater vehicles in the translation motion with low drift and pitch angle accurately and was comparable with previous experimental results [11-15].

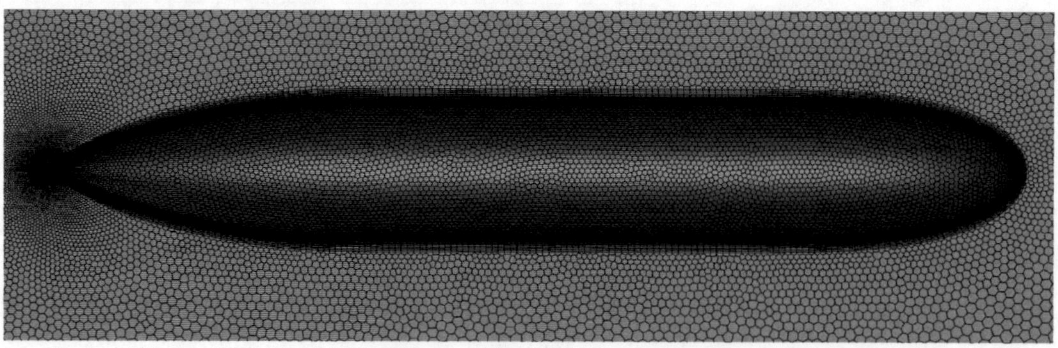

Figure 3: Mesh model of the axisymmetric bare hull Joubert geometry.

The ANSYS Fluent high-resolution scheme (PRESTO for pressure and Third Order MUSCL for momentum, turbulent kinetic energy and specific dissipation rate) was used throughout the simulations. The convergence criteria adapted for the study was that the root mean square value of the global residuals was to be less than 1e-04, with fluctuations in the predicted drag to be no more than three significant digits over 50 iterations.

VERIFICATION AND VALIDATION

The verification and validation phases of the CFD simulations were carried out against experimental data from the testing of the Joubert axisymmetric hull form in the Australian Maritime College (AMC) towing tank [16]. The computational domain size adapted for the validation study was constructed to replicate the AMC towing tank dimensions (breadth of 3.5 m and a water depth of 1.5 m) to ensure a like-to-like comparison between the numerical and the experimental results. The overall cell count used in the simulation was 1 million, which was deemed sufficient for this study as it provided deviations of less than 0.7% in the predicted drag when compared to the mesh refinement of 5.3 million (see Figure 4).

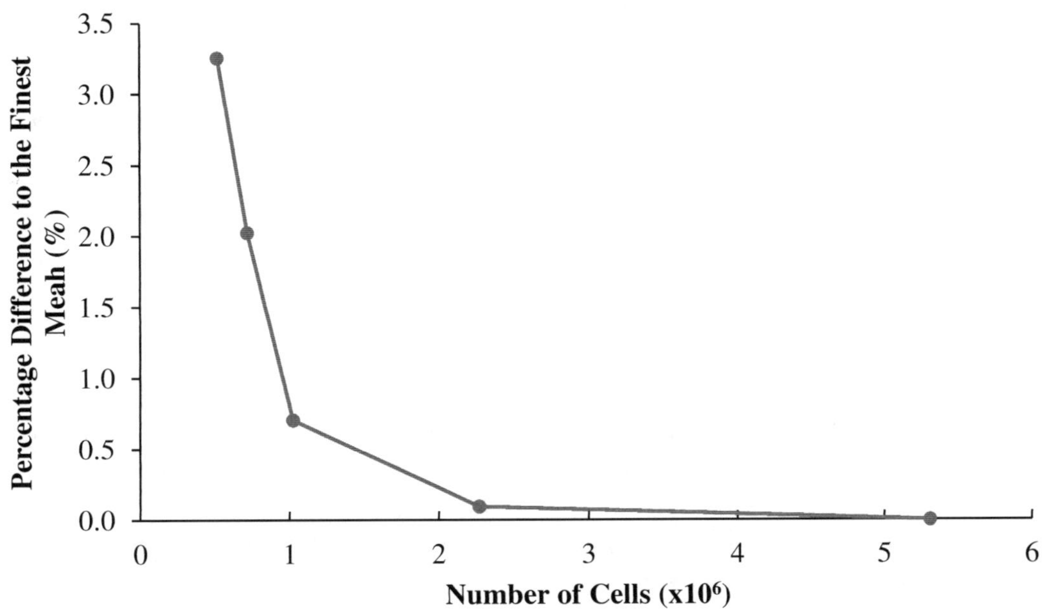

Figure 4: Grid independent study - percentage difference in the predicted drag compared to the finest 7.1 million cell simulation solution.

Validation of the CFD model was carried out using the experimental data obtained by Dawson [16] and is presented in Figure 5. The result shows a discrepancy in magnitude of around 10% between the numerical and experimental drag coefficient at the higher Reynolds Number. Although a higher discrepancy between the numerical and experimental results existed at the higher Reynolds Number, the results were deemed to be sufficiently accurate as most of the numerical results fall within the experimental uncertainty calculated by Dawson [16].

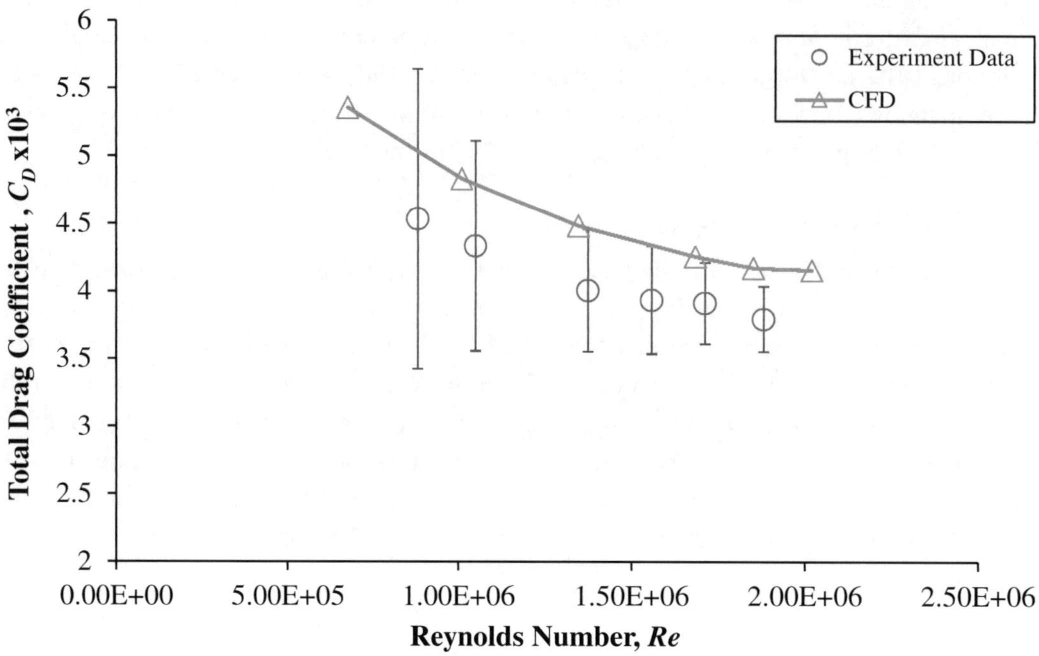

Figure 5: CFD predicted drag coefficients and experimental measurements [16] for the Joubert bare hull geometry in straight line tests, with error bars of maximum 23 % and minimum of 5%.

RESULTS AND DISCUSSION

Figure 6 shows the predicted drag coefficient of the underwater vehicle as a function of Reynolds number for the different blockage parameter ratios (*m*) and domain ratios (*b:d*). The trends of the predicted results show that an increase in the blockage parameter ratio will increase the drag coefficient acting on the submerged underwater vehicle. The results also show that the drag as a function of Reynolds Number is similar in shape for the different blockage domain ratios examined. The percentage increase in drag coefficient with respect to the largest domain as a function of Reynolds number for different block parameter ratios is given in Figure 7. The constant percentage difference along the Reynolds Number range tested indicating that the blockage effect is independent of Reynolds Number.

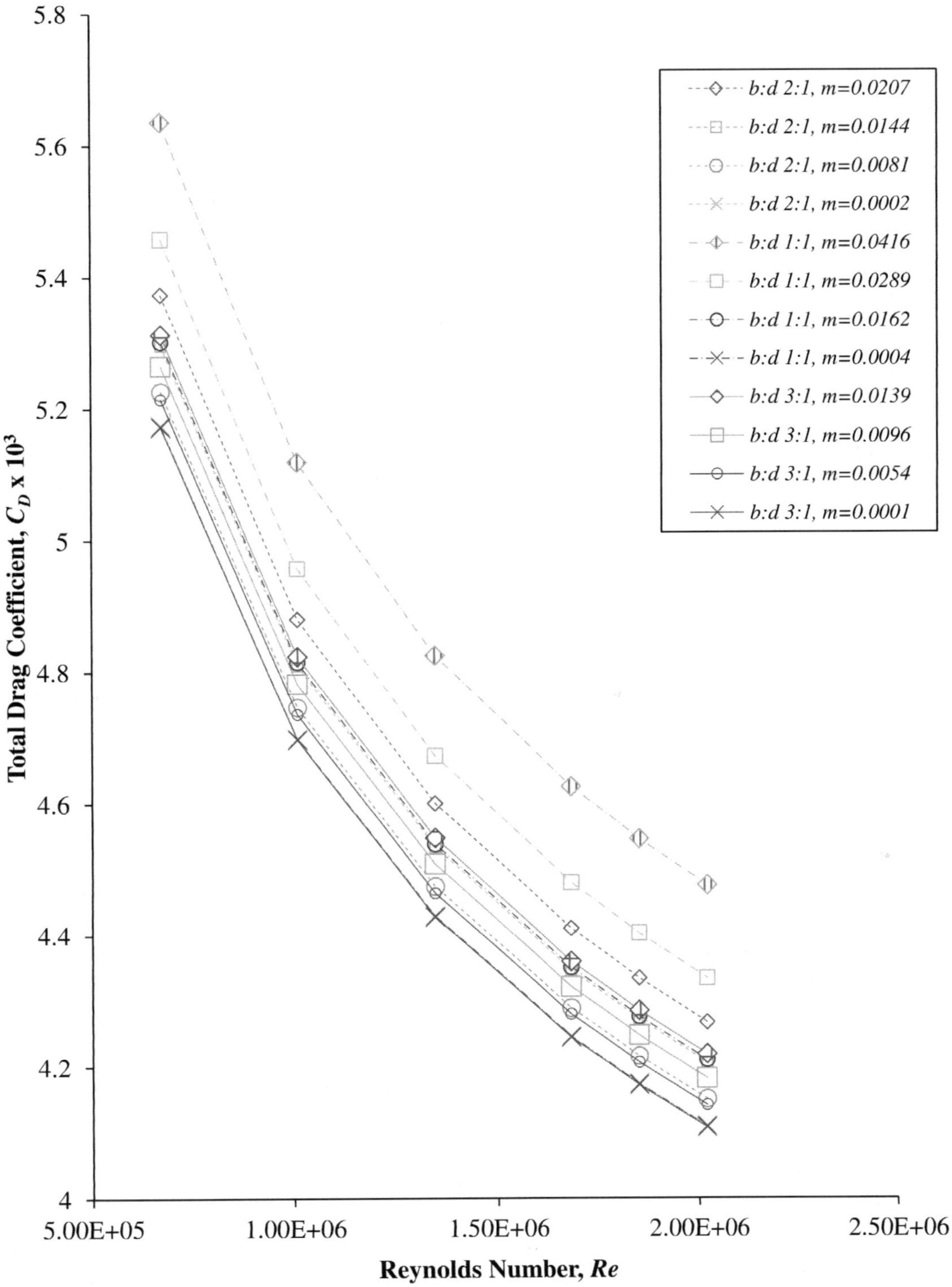

Figure 6: Total drag coefficient, C_D, as a function of Reynolds Number, *Re* for different domain ratios (*b:d*) and blockage parameter ratios, *m*.

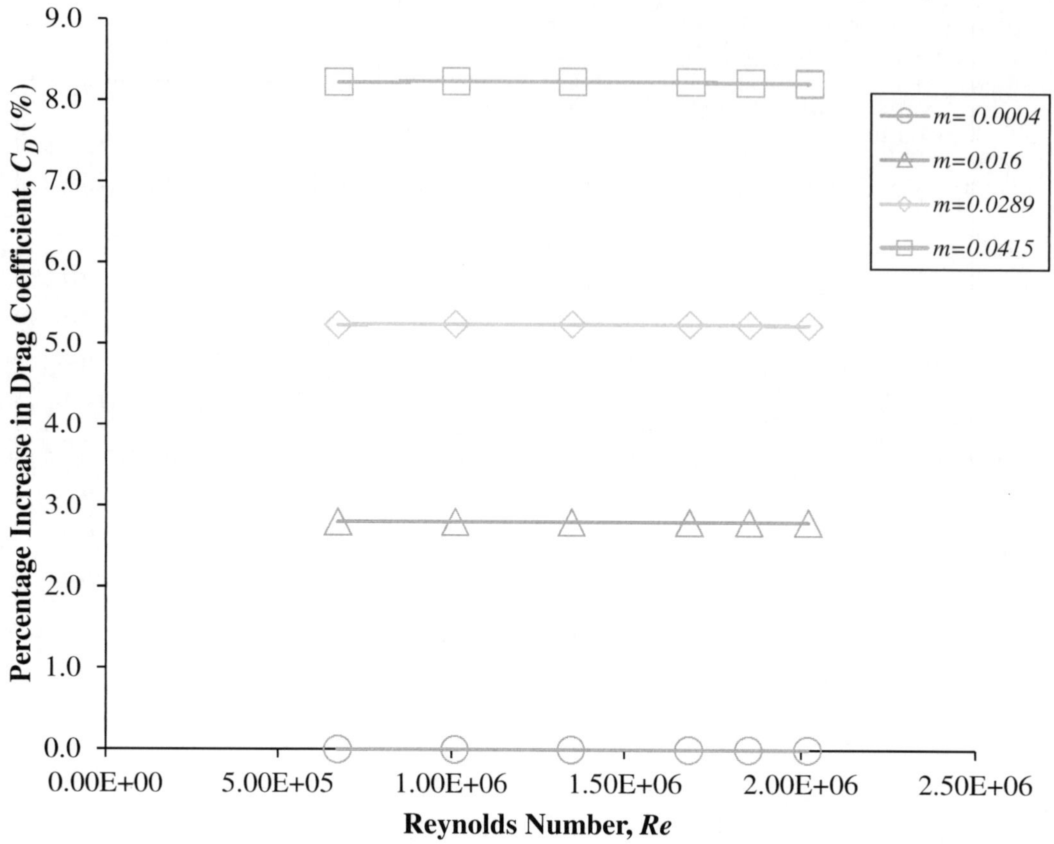

Figure 7: Percentage increase in drag coefficient, C_D, as a function of Reynolds Number, *Re*, for different blockage parameter ratios, *m* when compared to the largest blockage parameter ratio, *m*=0.0001

The independency between blockage parameter ratio and Reynolds Number enables the relationship between drag coefficient and the blockage parameter ratio to be quantified. Figure 8 shows the percentage increase in the drag coefficient for the different blockage domain ratios when compared to the largest domain (30 m breadth, 10 m depth, $m = 0.0001$). From Figure 8, it is seen that the percentage increase of the drag coefficient is linearly proportional to the blockage parameter ratio, *m*, therefore, an estimate of the blockage effect for a deeply submerged body can be obtained using a linear trend line. The increase in total drag coefficient is given by the following proposed empirical equation,

$$\text{\% increase in } C_D = 186.48\, m \tag{1}$$

The proposed blockage effect estimation for a deeply submerged body is shown to fit well with the numerical data, with a coefficient of determination, R^2 of 0.99.

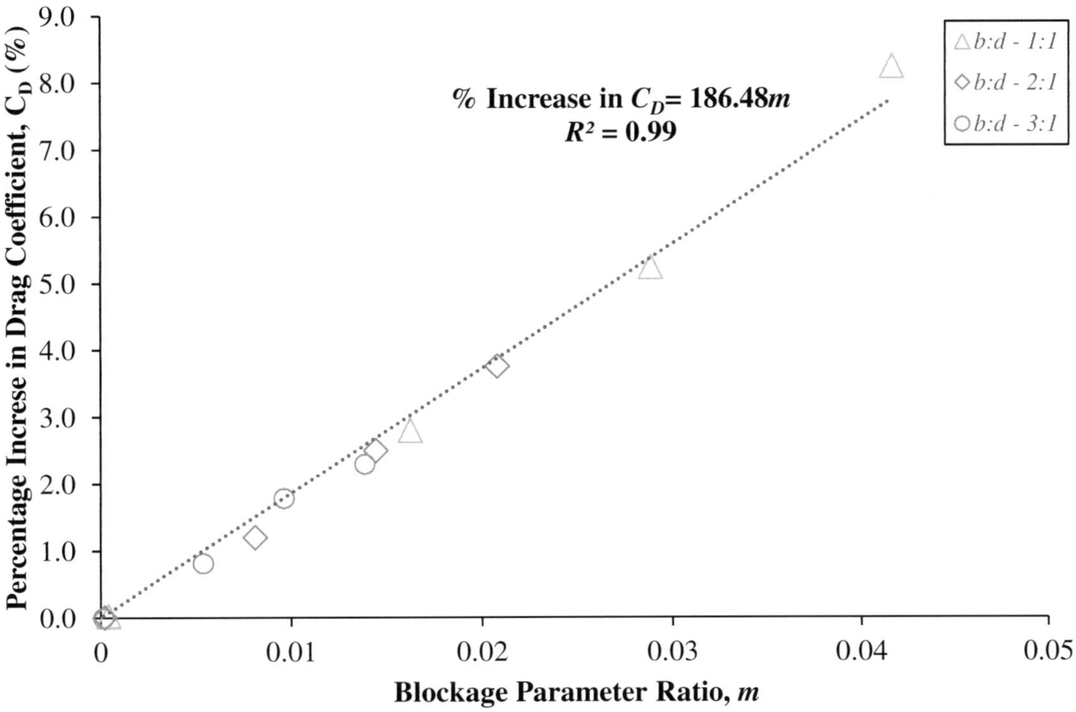

Figure 8: Empirical correction formula for the increase in drag coefficient, C_D, when compared to the largest domain at each blockage parameter ratio, *m*.

From Figure 8, it is seen that the domain sizing with the breadth to depth ratio of 1:1 displayed an overall higher increase in drag coefficient compared to other domain ratios. The equal distance between the model and the domain walls resulted in a stronger rigid wall interference with the underwater vehicle's flow field. This caused a higher increment in the fluid velocity thus, increasing the predicted drag coefficient in comparison to other domain ratios.

The current study did not include the modelling of the free surface, however, future work will include modelling the towing tank with the free surface at varying submergence depths to quantify the change in the drag coefficient as a function of blockage parameter ratio and submergence.

COMPARISON WITH BLOCKAGE CORRECTORS OF [6] AND [7]

The identification of an appropriate blockage corrector formula for a towing tank test domain is essential in order to accurately predict the drag coefficient of an underwater vehicle in an unrestricted domain. Most studies quantify the influence of blockage in term of velocity increase ratios [2-7]. The effective flow velocity, *V*, surrounding a submerged underwater vehicle in a water channel is assumed to be the same as the flow speed of the vehicle moving in an unrestricted domain with the velocity of *V+ΔV*. Using this assumption, the drag force acting on the vehicle moving in the channel is approximately equal to the drag force acting on the vehicle in an unrestricted domain at the velocity of *V+ΔV.* This increase in velocity can be obtained using available blockage correction formulae, which include the ITTC [6] and Wind Tunnel [7] blockage correctors.

Among the blockage correctors recommended in the ITTC guideline [6], Tamura's equation is the most applicable equation to correct the blockage effect acting on a deeply submerged underwater vehicle. Tamura's equation was developed based on two fundamental theories, the back flow theory and mean flow theory [3, 4] and is given as,

$$\frac{\Delta V}{V} = 0.67m\left(\frac{L}{b}\right)^{0.75}\frac{1}{(1 - Fr_{nd}^2)} \tag{2}$$

Mean flow theory describes the relationship between the blockage induced velocity increment and the depth Froude number, Fr_{nd}^2, for a surface ship [2, 3, 5]. However, the Froude effect does not apply for a deeply submerged underwater vehicle, and hence $Fr_{nd}^2 = 0$. This therefore, allows for the isolation of the mean flow based formula presented in the Tamura's equation, yielding the modified Tamura's corrector as,

$$\frac{\Delta V}{V} = 0.67m\left(\frac{L}{b}\right)^{0.75} \tag{3}$$

For the wind tunnel blockage correction, Thom's solid blockage corrector [7] was chosen to correct the velocity of the submerged underwater vehicle's flow field. The equation is given by,

$$\frac{\Delta V}{V} = \frac{k(v_m)}{A_T^{3/2}} \tag{4}$$

where k is 0.96. The performance of each blockage corrector method was compared between the three domain ratios. Figure 9 shows the comparison of the corrected drag coefficient (*Re* = 0.6 million) for different blockage parameter ratio, *m*, using the ITTC Tamura method [6] (equation (3)), Thom method [7] (equation (4)) and the proposed empirical equation (1) with the numerical predicted drag coefficient at an unrestrained domain, *m*=0.0001.

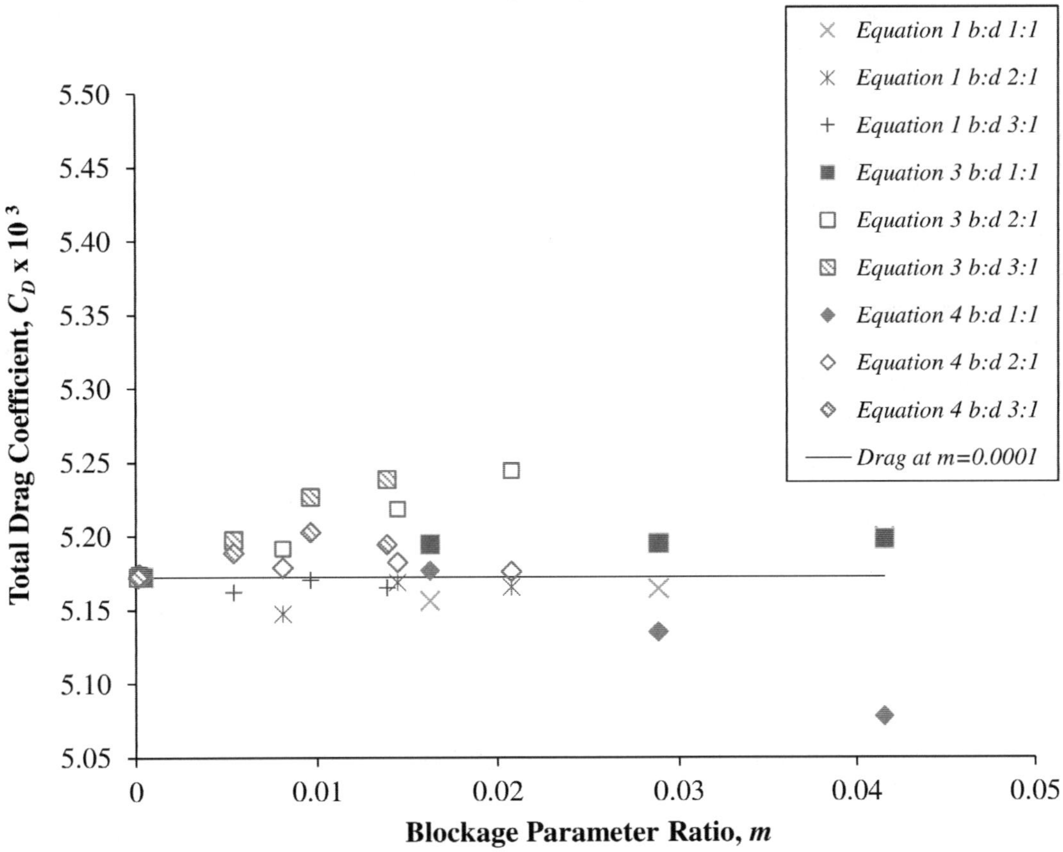

Figure 9: Total drag coefficient, C_D, predicted against blockage parameter ratio, *m*, at a Reynolds number, *Re*, of 0.6 million using different correctors.

Although the results show differences between the ITTC Tamura [6], the Thom [7] and the proposed empirical equation when compared to the numerical predicted drag coefficient at unrestricted domain, their discrepancies are within 2%. Thus, all three methods provide reasonable estimations of the blockage corrected drag coefficient based on the blockage parameter ratio for a deeply submerged underwater vehicle model tested in a towing tank for the specified speed range. It is seen that the proposed empirical formula provides the best prediction. However, further studies are required to include the effects of the free surface in predicting the blockage effect when testing underwater vehicles for drag coefficients in towing tanks.

CONCLUDING REMARKS

The paper presented a study conducted to quantify the blockage effect acting on a deeply submerged underwater vehicle tested in a water channel. The investigation utilised CFD simulations to extend the domain size such that the effects of the blockage parameter and the Reynolds Number on the drag coefficient of the underwater vehicle was obtained. Experimental data was utilised to validate the CFD simulation to predict the drag coefficient of the underwater vehicle model tested in the towing tank.

The results indicated that the blockage effect acting on a deeply submerged underwater vehicle model in the towing tank is independent of the Reynolds Number within the range investigated. As the increase in the drag coefficient is directly proportional to the blockage parameter ratio, a linear polynomial was fitted to obtain a proposed formula based correction thus allowing the prediction of the drag coefficient of the vehicle in an unrestricted domain. Comparison with the existing blockage correction methods [6, 7] showed good agreement.

However, as the free surface was not included in the simulations, further work is required to quantify the effect of submergence depth on the drag coefficient blockage corrector when an underwater body is being tested in a towing tank with free surface. Thus, future work includes investigating the influence of Froude number on the blockage effect acting on underwater vehicles tested in towing tanks. This will allow the establishment of a blockage corrector formula for underwater vehicles with variations in blockage parameter and submergence depths. The work will then be extended to fully appended underwater vehicles.

REFERENCES

1. El-Sherbiny, S.e.-S. (1972). *Effect of Wall Confinement on the Aerodynamics of Bluff Bodies.* Doctor of Philosophy, University of British Columbia,
2. Scott, J.,(1970), *On Blockage Correction and Extrapolation to Smooth Ship Resistance.* Transations of S.N.A.M.E, (7): p. 31.
3. Tamura, K.,(1972), *Study on the Blockage Correction.* Journal of the Society of Naval Architects of Japan. **131**: p. 17-28.
4. Kinya, T.,(1998), *Blockage Correctors and Their Appraisal.* The Bulletin of Nagasaki Institute of Applied Science. **39**(1): p. 1-18.
5. Hughes, G.,(1961), *Tank Boundary Effects on Model Resistance.* Transactions of I.N.A. **103**: p. 421-440.
6. ITTC. (2011). *Testing and Extrapolation Methods- Resistance* Resistance Test, (7.2-02-02-01). International Towing Tank Conference, UK.
7. Rae, W.H. and A. Pope,(1984), *Low-Speed Wind Tunnel Testing*. John Wiley.
8. Joubert, P., (2006), *Some Aspects of Submarine Design. Part 2. Shape of a Submarine 2026*. Defence Science and Technology Organisation Victoria, Australia.
9. ITTC. (2008). *Uncertainty Analysis in CFD Verification and Validation Methodology and Procedures*. (7.5-03-01-01). International Towing Tank Conference, UK.
10. ANSYS, (2018), *ANSYS Fluent Mosaic Technology Automatically Combines Disparate Meshes with Polyhedral Elements for Fast, Accurate Flow Resolution*.
11. Watt, G.D., C.R. Baker, and A.G. Gerber, (2006), *ANSYS CFX-10 RANS normal force predictions for the series model 4621 unappended axisymmetric submarine hull in translation*. Defence Reasearch and Development Canada: Ottawa.
12. Toxopeus, S.,(2008), *Viscous-flow calculations for bare hull DAPRA SUBOFF submarine at incidence* international Ship Building Progress. **55**: p. 227-251.
13. Phillps, A.B., S.R. Turnock, and M. Furlong,(2010), *Influence of Turbulence Closure Models on the Vortical Flow Field around a Submarine Body Undergoing Steady Drift.* Journal of Marine Science and Technology. **15**: p. 201-217.
14. Leong, Z.Q., et al.,(2012), *RANS-based CFD Prediction of the Hydrodynamic Coefficients of DARPA SUBOFF Geometry in Straight-Line and Rotating Arm Manoeuvres.* international Journal of Maritime Engineering. **154**: p. 41-52.

15. Huong, Y.T. (2018). *Numerical Analysis on the Effect of Submergence Depth on the Bare-hull Model Scaled BB2 at AMC Towing Tank.*, University of Tasmania, Launceston, Australia.
16. Dawson, E. (2014). *An Investigation into the Effects of Submergence Depth,Speed and Hull Length-to-Diameter Ratio on the Near-Surface Operation of Conventional Submarines.* Master of Philosophy, Australia Maritime College, Launceston.

The Influence of Sailplane Deflection on the Nominal Wake

David A. Pook[1,3], david.pook@dst.defence.gov.au
Gregory J. Seil[2,3], gregory.seil@dst.defence.gov.au
Chris Shingles [3], christopher.shingles@dst.defence.gov.au
Dev Ranmuthugala[3], dev.ranmuthugala2@dst.defence.gov.au

[1] Aerospace & Aviation, RMIT University
[2] Advanced VTOL Technologies Pty Ltd
[3] Maritime Division, Defence Science and Technology Group

ABSTRACT

This paper uses Computational Fluid Dynamics (CFD) simulations of the generic BB2 submarine geometry to investigate the effect of sailplane deflection on the nominal wake at the propeller plane. Studies of marine vehicle wakes typically consider the vessel heading straight-ahead with no deflection of the control surfaces. While this is the most-typical condition, vessels can operate for significant periods with control surfaces deflected. The wake of the control surfaces can flow downstream and alter the nominal wake seen by the propeller.

When the BB2 generic submarine geometry is simulated running straight-ahead, the wake of the sailplanes does not influence the nominal wake. However, deflection of the sailplanes can indirectly alter the nominal wake. The sailplane deflection modifies the pressure field in the vicinity of the sail, which alters the formation and downstream path of the sail-junction flow. The modification of the sail-junction flow then alters the nominal wake. Deflecting the sailplanes leading edge upwards, spreads the sail-junction flow away from the hull centreline, while an opposite deflection pulls the sail-junction flow towards the hull centreline.

INTRODUCTION

The nominal wake is defined as the flow at the propeller without including the influence of the propeller [1]. It is an important consideration when evaluating the hydrodynamics and propulsive performance of marine vessels. Adapting the propeller design for the incoming mean flow can increase the propulsive efficiency of the vessel [2, 3]. Spatial variation of the mean flow due to a non-axisymmetric hull and appendages causes unsteady loading on the propeller, creating undesired vibration. Typically, the nominal wake is evaluated for the straight-ahead condition. For surface ships, control surfaces (rudder) are often located aft of the nominal wake plane, hence they do not significantly influence the nominal wake.

The hulls of underwater vehicles are often (near) axisymmetric which minimises circumferential variation of the nominal wake. However, underwater vehicles typically feature relatively large control surfaces and appendages (e.g. sail, casing, and sensors) upstream of the propeller. For example, the geometry of the DST Group BB2 generic submarine geometry depicted in Figure 1 features a sail, casing, four aft-control-surfaces (ACS), and two sail mounted hydroplanes (sailplanes). Seil and Anderson [4] have previously examined the effect of sail geometry on the nominal wake. Due to the large size of underwater vehicle appendages and control surfaces, their potential influence on the nominal wake is of interest.

For the BB2 geometry, a portion of each ACS is located in the hull boundary layer. Therefore, their wakes will influence the nominal wake, and their deflection will alter the nominal wake. The portion of the sail located in the hull boundary layer will also alter the nominal wake. The sail-junction (horseshoe) vortex at the sail root will also influence the nominal wake. The sailplanes are located a significant distance outside of the hull boundary layer. Hence, the wake of the sailplanes will not typically enter the propeller, except possibly during extreme manoeuvres. Therefore, it might be assumed that sailplanes will not affect the nominal wake of the BB2 generic submarine geometry.

However, this paper will demonstrate that deflection of the sailplanes indirectly alters the nominal wake by influencing the sail-junction flow. Considering the size of the sail and its location, there is potential for significant changes in the nominal wake due to sailplane deflection. The possible influence of sailplane deflection on the nominal wake has not been previously reported in literature.

DST GROUP EVOLVED GENERIC SUBMARINE GEOMETRY (BB2)

The BB2 generic submarine geometry [5] is shown in Figure 1 with a body-fitted axes system (x, y, z) at amidships and a cylindrical coordinate system (r, θ) in the nominal wake plane. The boat length from bow to stern (L) is 70.2 m for the full-scale design. The BB2 geometry is constructed from an axisymmetric hull with a length to diameter (L/D) ratio of approximately 7.3. A casing is placed on the top side of the hull, modifying the axisymmetric profile of the hull. The sail is constructed from a NACA 0021 aerofoil section. The nominal wake plane is located at 98.1% of the boat length, measured from the bow $(x/L = -0.481)$. The BB2 propeller has a radius (R_p) of 2.5 m $(0.0356L)$.

The sailplanes are constructed from a NACA 0015 aerofoil. The sailplanes have a fixed region (headbox) adjacent to the sail, and an outer deflecting section (see Figure 1). The sailplanes rotate (deflect) about an axis located at approximately one-third of their root chord. A positive deflection moves the leading edge up and the trailing edge down.

COMPUTATIONAL FLUID DYNAMICS MODELLING

The flow was computed using the Reynolds Averaged Navier Stokes (RANS) equations with the ANSYS Fluent 14.5 pressure based, incompressible solver, using the Shear Stress Transport (SST) $k\omega$ turbulence model, where k is the turbulent kinetic energy and ω is the specific turbulence dissipation rate. All equations were differenced using second-order schemes. Pressure-velocity couple was carried out using the SIMPLE algorithm.

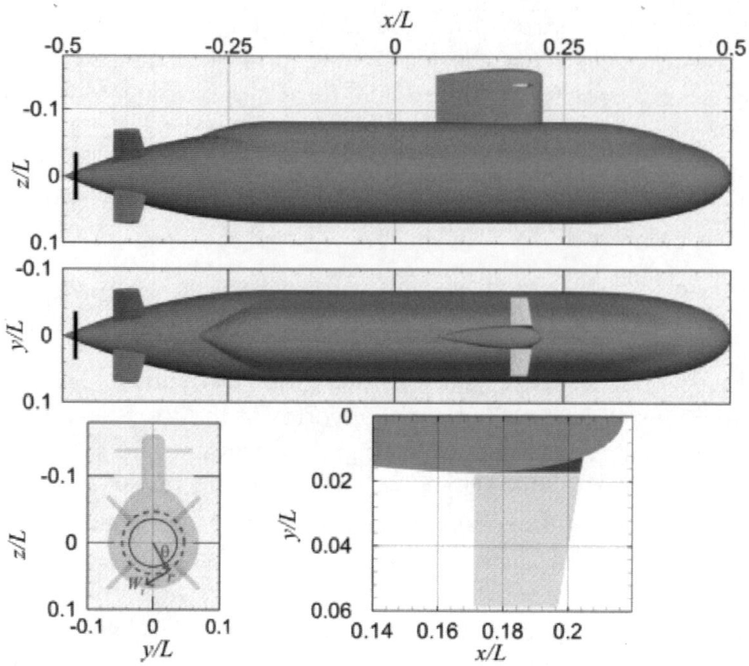

Figure 1 - DST Group BB2 generic submarine geometry, with the body-fixed coordinate system and the cylindrical coordinate system shown on the y-z plane. The outer edge of the BB2 propeller, at the nominal wake plane, is shown with the thick black vertical line. The moveable sailplane section (starboard-side) is shown in light blue. The fixed portion (headbox) of the sailplane is shown in dark blue.

The flow conditions were set to provide a Reynolds number based on length, L, of 5.2×10^6, as this Reynolds number is achievable in model-scale testing (e.g. [6]) and matches previous computational studies of the BB2 [7]. The BB2 geometry was modelled deeply submerged (no free-surface effects) and placed centrally in a rectangular domain having a length, width, height of $6.0L \times 3.0\,L \times 3.0\,L$, respectively. The distance to the domain boundary satisfies the recommended 1 to $2L$ guideline for CFD simulations of ship hull forms [8]. A uniform velocity inlet, with a velocity vector in the freestream direction, was specified on all domain boundaries. The turbulence intensity at the inlet was set as 1% and the turbulent viscosity ratio set as 10.

The grid was constructed in three distinct regions using Pointwise meshing software:

- Inner grid - containing the BB2 geometry and near flow, except the sailplanes;
- Outer grid - bounded by the outer domains and encompassing the inner grid; and
- Sailplane grid - contains on the sailplanes and near flow.

The Inner grid was a multi-block structured grid, shown in Figure 2. The cell count of the inner grid was approximately 40.2 million hexahedral cells. Seil et al. [9] performed a grid refinement study using this grid at the same Reynolds number with the same solver and settings. For straight-ahead conditions, increasing the cell count by a factor of four was found to change:

- the global resistance less than 0.06%;
- the nominal wake fraction less than 0.4%; and

- the standard deviation of the axial velocity circumferential variation (distortion coefficient) less than 3.1%.

The outer grid was also a structured grid of hexahedral cells, joined to the inner grid by a non-conformal interface boundary condition. The use of a non-conformal interface allowed a reduction in cell count in the outer grid.

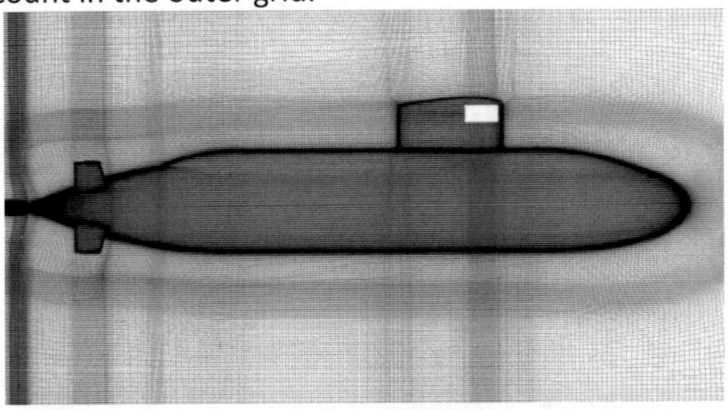

Figure 2 - Structured inner grid. The white rectangular region with a red border denotes the sailplane grid region.

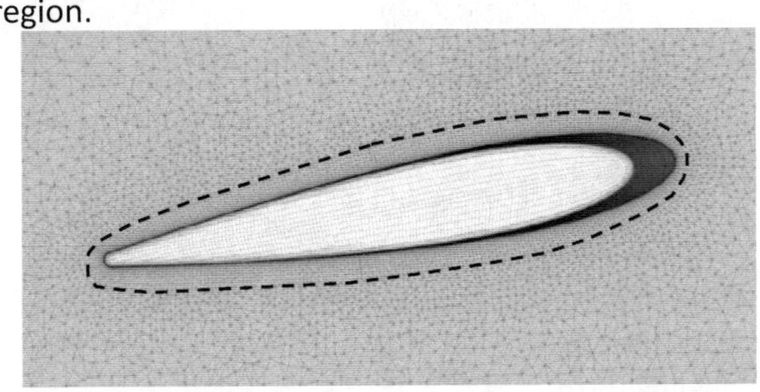

Figure 3 - Cut through a grid in the sailplane region. The dashed black line indicates the region where the Pointwise T-Rex tool recombines triangular cells to triangular prisms or hexahedra for boundary layer resolution.

The sailplane grid was unstructured and designed to simplify re-meshing of that region when the sailplane is deflected. The sailplanes were deflected to the desired angle, and the T-Rex tool within Pointwise used to grow layers of triangular prisms and hexahedra through an approximate height of the sailplane boundary layer (see Figure 3). Pyramid and tetrahedral cells then fill the sailplane grid region out to the interface boundary condition.

The wall-surfaces of the BB2 geometry were specified as non-slip walls. The first cell height was placed in the viscous sub-layer with a sub-layer scale distance (y^+) of approximately 1. The turbulence equations were integrated to the wall, i.e. no-wall function was used.

RESULTS

Overview

Figure 4 shows contours of the streamwise (x-direction) velocity component on planes normal to the hull, clipped at 95% of the freestream speed in order to represent the boundary layer thickness. The nominal wake at the propeller plane is distorted by the appendages (i.e. ACS and sail) which each create a circumferential variation (thickening/thinning) of the hull boundary layer. Each appendage creates a junction vortex at its base in the hull boundary layer. The junction vortex is created by skewing the vorticity of the approaching hull-boundary layer around the appendage leading edge, creating a streamwise oriented vortex. The hull boundary layer thickening and thinning shown in Figure 4 can be used to trace the path of a junction vortex to the nominal wake plane.

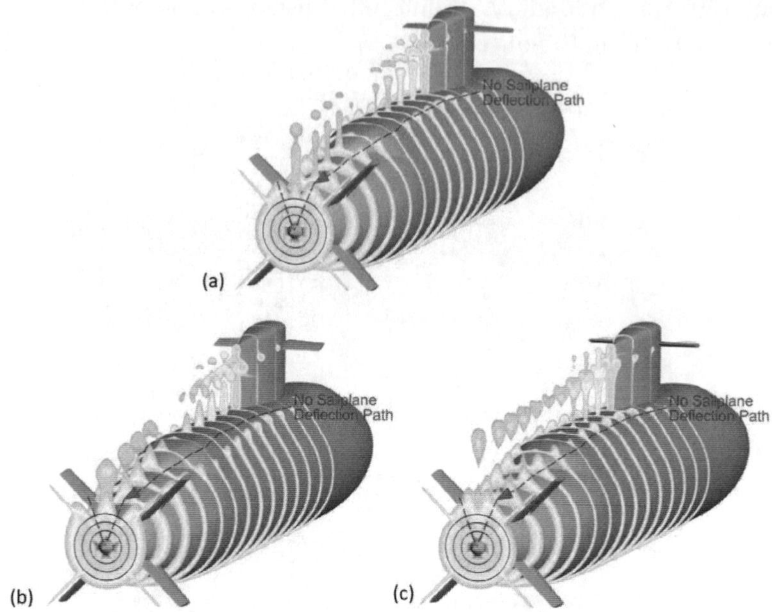

Figure 4 - Slices perpendicular to the centreline of the BB2 geometry showing contours of the streamwise velocity component. Contours cropped at 95% of the freestream speed to indicate boundary layer thickness. Sailplanes deflections: (a) undeflected; (b) leading edge up, $+10°$; (c) leading edge down, $-10°$.

Over the aft-hull section, the adverse pressure gradient compresses the streamwise oriented vorticity, possibly reducing the vorticity to a value lower than the strain. Hence Q-surfaces, typically used to identify vortices (regions where vorticity is greater than strain) cannot be used reliably to trace the sail-junction flow to the nominal wake plane.

From Figure 4, it is seen that the deflection of the sailplanes modifies the path taken by the sail-junction flow to the nominal wake plane. The red dashed line along the hull in Figure 4 indicates the path of the sail-junction flow with no sailplane deflection. Deflecting the sailplanes leading edge up (i.e. positive deflection), pushes the sail-junction flow away from the top of the hull. Conversely, deflecting the sailplane leading edge down moves the sail-junction flow towards the top of the hull.

Deflection of the sailplanes creates a `lift' force from the sailplanes, altering the pressure in the vicinity of the sail and sail-junction. Figure 5 shows contours of the pressure coefficient (C_p), defined as,

$$C_p = \frac{2(p-p_{ref})}{\rho U_\infty^2},$$

where p is the pressure, p_{ref} is a reference pressure upstream of the boat, and ρ is the fluid density. The pressure contours reveal that deflecting the leading edge of each sailplane upwards creates a higher pressure at a given distance from the sail, which pushes the sail-junction flow away from the sail as it passes downstream. Conversely, deflection the leading edge of each sailplane downwards creates a lower pressure closer to the sail, pulling the sail-junction flow inwards.

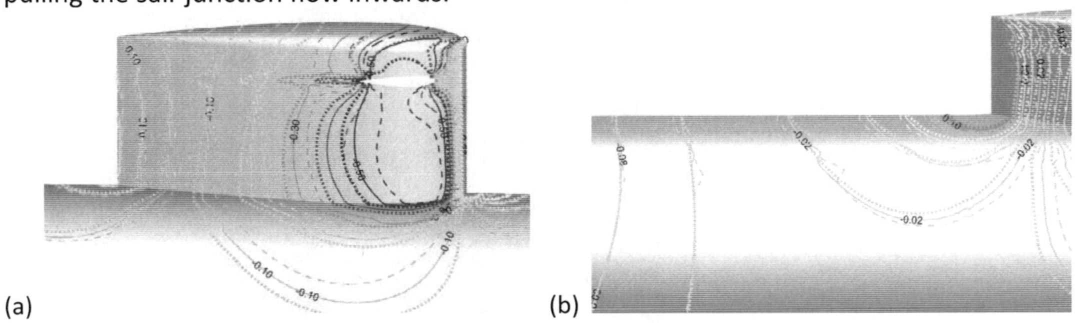

Figure 5 - Contours of the pressure coefficient for the sailplanes: undeflected, solid lines; $+10^{\circ}$ deflection, dotted lines; -10° deflection, dashed lines. Views: (a) sail region (sailplane fixed root position shown in white); (b) hull downstream of the sail.

Downstream of the sail, the pressure contours for the different sailplane deflections tend to coincide (Figure 5(b)), indicating the action of the sailplane deflection on the sail-junction flow is limited to the region near the sail.

Axial Velocity

Contours of the axial velocity component (U, in the x-direction) at the nominal wake plane are shown in Figure 6 for no sailplane deflection, -15° and $+15^{\circ}$ deflection. The tip vortices created by the sailplanes bypass the BB2 propeller (propeller diameter denoted by the outer concentric black ring in Figure 6) for all sailplane deflection angles examined. The wake of the sailplanes also bypass the BB2 propeller region with a significant margin for all cases, except the $+15^{\circ}$ sailplane deflection. For this deflection, the wake of the inner sailplane approaches the propeller outer diameter. Figure 6 also shows that sailplane deflection has little effect on the nominal wake in the underside region of the hull.

Contours within the propeller radius on the topside of the BB2 in Figure 6, are clearly different for the -15° and $+15^{\circ}$ sailplane deflections. For $+15^{\circ}$ deflection, there is a significantly stronger wake region on the hull centreline behind the sail. The changing wake directly behind the sail indicates the resistance of the sail is being modified by the sailplane deflection. Higher speed flow is also brought into smaller radial distances from the hull surface on either side of the stronger centreline wake. In contrast, for -15° deflection, the wake is strengthened (i.e. lower flow speed) in this region when compared to the case of no sailplane deflection.

The differences between positive and negative sailplane deflections are highlighted in the circumferential variation of the axial velocity shown in Figure 7, at 25% and 75% of the propeller radius. The circumferential variation of the negative sailplane deflection angles is similar to that with no deflection. Negative deflection increases the variation magnitude and shifts the circumferential peak location slightly. The $+10^{\circ}$ and $+15^{\circ}$ sailplane deflections, particularly at 75% of the propeller radius, produce a very different circumferential variation due to the relatively high speed flow brought in (around $\theta \approx -70^{\circ}$ and -110°) and the stronger wake behind the line of the sail (at $\theta = -90^{\circ}$).

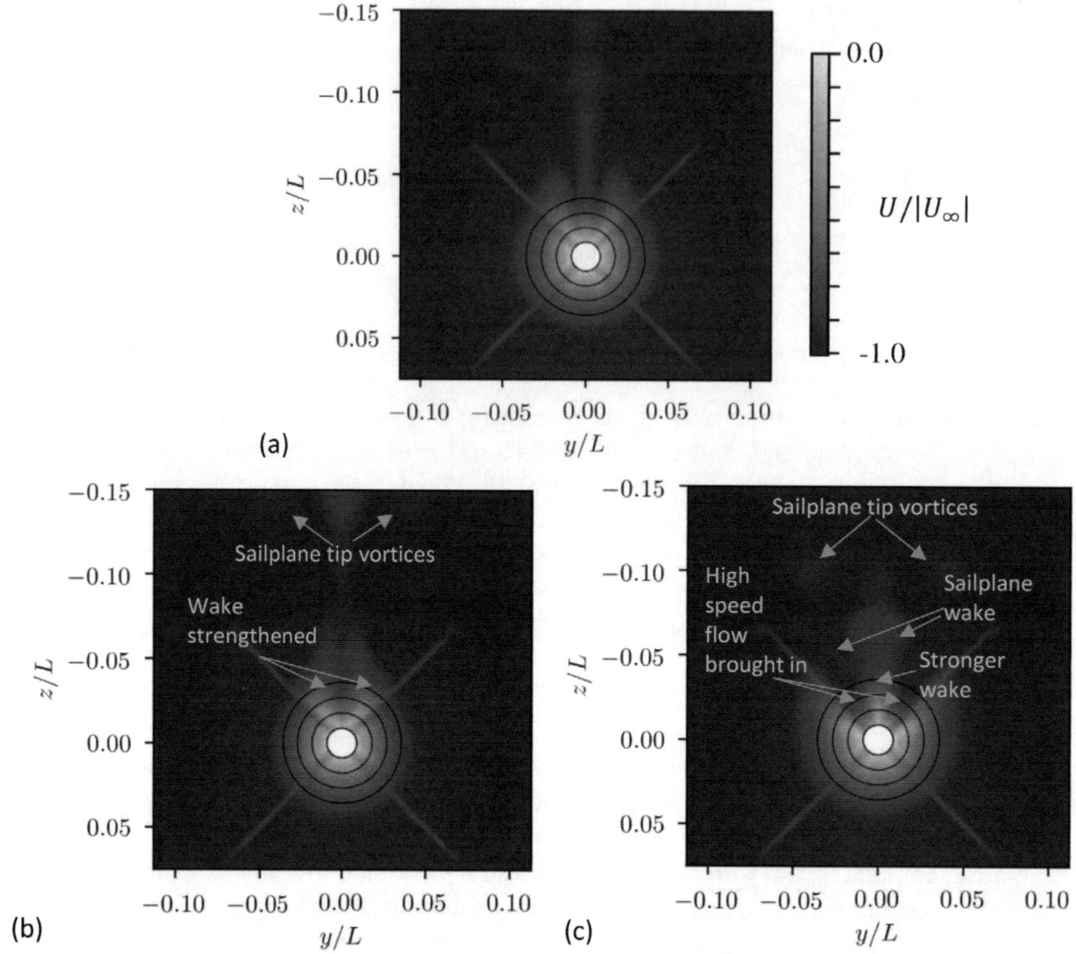

Figure 6 - Contours of the axial velocity (U) at the nominal wake plane for sailplane deflections: (a) undeflected; (b) $-15°$; (c) $+15°$. The black concentric rings are at a radius of $0.25R_p$, $0.5R_p$, $0.75R_p$ and $1.0R_p$.

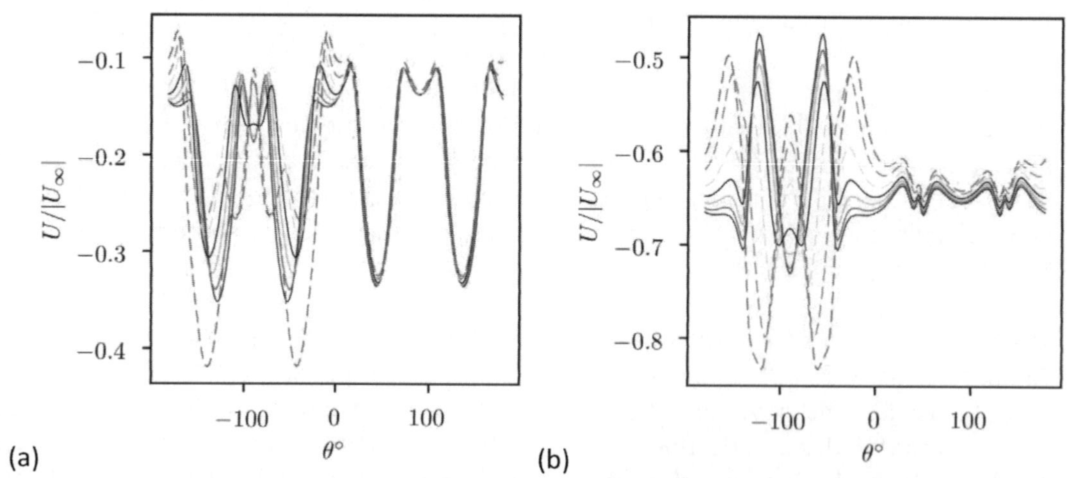

Figure 7 - Circumferential variation of the axial velocity at the propeller plane at a radius of: (a) $0.25R_p$; (b) $0.75R_p$. Sailplane deflection lines: undeflected, ——; −5°, ——; −10°, ——; −15°, ——; +5°, ——; +10°, - - -; +15°, - - -.. Sail located at $\theta = -90°$.

Radial Variation

The radial variation of the circumferential mean (μ) of the axial velocity at the nominal wake plane is shown in Figure 8(a). The mean axial velocity is altered slightly below 40% of the propeller radius, particularly for the $+15^\circ$ sailplane deflection. All sailplane deflections increase the net mass flow available to the propeller. The circumferential mean of the turbulent kinetic energy is shown in Figure 8(b). Deflection of the sailplane in either direction typically increases the turbulent kinetic energy over approximately 80% of the propeller radius.

The radial variation of the standard deviation in the circumferential direction (σ) of the axial velocity (distortion coefficient [10]) and the tangential velocity (W_t) are shown in Figure 8(c) and (d) respectively. The standard deviation gives a measure of the circumferential variation. Deflection of the sailplane leading edge by $+10^\circ$ and $+15^\circ$ significantly increases the axial velocity variation over most of the propeller radius, more than doubling it at some radii. The standard deviation of the tangential velocity component is also increased substantially.

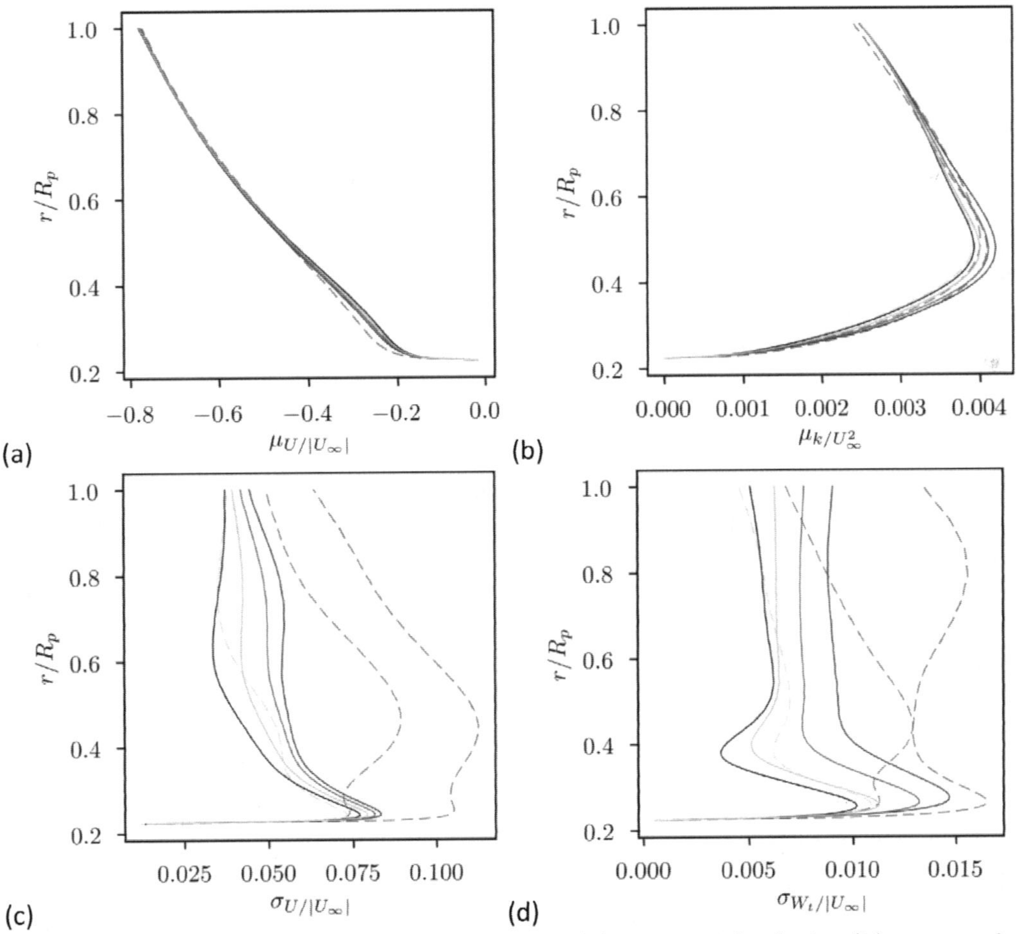

Figure 8 - Radial variation of the circumferential: (a) mean axial velocity; (b) mean turbulent kinetic energy; (c) standard deviation of the axial velocity; (d) standard deviation of the tangential velocity. Sailplane deflection lines: undeflected, ——; -5°, ——; -10°, ——; -15°, ——; $+5^\circ$, ——; $+10^\circ$, - - -; $+15^\circ$, - - -.

DISCUSSION

The results presented show that for the BB2 generic submarine geometry, deflection of the sailplanes can significantly alter the mean flow spatial variation of the nominal wake. Deflecting the leading edge upwards (positive sailplane deflection) had a greater effect on the nominal wake than downwards deflection. The nominal wake variation is due to the pressure field created by the sailplane deflection altering the sail-junction flow as it passes around the sail. As this is an indirect mechanism, the potential for sailplane deflection to alter the nominal wake will be design dependant. Reduction of the sail-junction vortex through improved sail design, filleting, etc. [11] might minimise the effect of sailplane deflection on the nominal wake. However, any geometry contouring in the sail region to minimise the sail-hull junction vortex will still be affected by sailplane deflection. In the absence of a significant sail-junction flow, sailplane deflection might still induce a lateral pressure gradient in the hull boundary layer that could create a local thickening/thinning that would flow downstream to the propeller. The sailplane geometry, and its relative distance to the sail-junction flow, will also alter the significance of any sailplane deflection.

The hull and casing shape in the vicinity of the sail, and downstream of the sail to the propeller, is a significant parameter that will affect the significance of sailplane deflection on the nominal wake. The BB2 geometry has a `flat' top casing in the vicinity of the sail. Thus, the sail-junction flow passes on a near horizontal surface and sailplane deflection shifts the junction flow inwards or outwards. However, hull shapes can be near axisymmetric or `boxy', i.e. with a near oval/square cross section. On an axisymmetric hull, deflection of the sailplane would be expected to shift the junction flow in/out in relation to the sailplane deflection. On a `boxy' hull, the junction flow could be pushed over (or brought back from) the edge of the casing, resulting in a strong non-linear relationship with sailplane deflection. The lack of a horizontal surface past the sail's maximum thickness might also alter the sailplane affect on the sail-hull junction vortex path. The arrangement of the ACS will also change the effect of the sailplane deflection if the sail-junction disturbance is moved near an ACS.

The CFD techniques used to model the BB2 geometry (e.g. SST $k\omega$ turbulence model) in this paper are known to have limitations accurately predicting junction flows (see [12] for a summary). However, the mechanism by which the sailplane can alter the nominal wake is captured by the modelling presented in this paper, and thus serves as a demonstration of the expected effects due to sailplane deflection. The results presented should not be interpreted as a prediction of absolute values. However, Chase [13] found predictions of the Suboff submarine geometry nominal wake, circumferential mean axial velocity, calculated using the using the SST $k\omega$ turbulence model (straight-ahead, Re$= 12 \times 10^6$) were within 3.85% of the experimental measurements made by Crook [14]. The nominal wake distortion due to the sail-junction wake was also consistent with measurements. Hence, the computations presented in this paper for the BB2 geometry are likely a reasonable prediction.

The simulations in this paper considered a steady flow with turbulent fluctuations (Reynolds stresses). The real junction flow to the propeller will possibly 'wander', particularly if it is pushed onto a highly curved region like the top edge of the BB2 casing. The simulations also do not model the transient effect on the BB2 motion due to sailplane deflection, nor the new steady-state condition the BB2 would achieve (i.e. relative angle of attack, vertical velocity component). These complications would alter the predicted effect of sailplane deflection on the nominal wake.

The results presented are for a model-scale Reynolds number. From model-scale to full-scale, the wake-fraction is known to change due to the relative thinning of the hull boundary layer. The sail-junction flow will still be present in the hull boundary layer at full-scale Reynolds numbers, although its characteristics (e.g. location, strength, streamwise decay) may change with Reynolds number. Thus, deflection of the sailplanes, which alters the sail-junction flow by changing the local pressure field, will still influence the sail-junction flow at full-scale Reynolds numbers. However, the sensitivity of the nominal wake to sailplane defection at full-scale Reynolds numbers requires further assessment.

CONCLUSION

Deflection of the BB2 generic submarine geometry sailplanes was shown to alter the path of the sail-junction vortex and modify the nominal wake. This effect was demonstrated using RANS-based CFD modelling. Deflecting the sailplane leading edge upwards pushed the sail-junction flow outwards (away from the top of the hull). A stronger wake on the hull centreline was also created. Deflecting the sailplane leading edge downwards pulled the sail-junction flow inwards, towards the top of the hull. For the BB2 generic submarine geometry and a given sailplane deflection angle, deflecting the leading edge of the sailplanes upwards (positive sailplane deflection) had a larger influence on the nominal wake.

REFERENCES

1. Carlton, J., *Marine Propellers and Propulsion*. 2nd ed. 2007: Butterworth-Heinemann.
2. Renilson, M., *Submarine Hydrodynamics*. 2nd ed. 2018: Springer.
3. ITTC, *The Specialist Committee on Scaling of Wake Field - Final Report and Recommendations to the 26th ITTC*. in *Proceedings of 26th ITTC*. 2011. Rio de Janeiro, Brazil.
4. Seil, G.J. and Anderson, B., *A Comparison of Submarine Fin Geometry on the Performance of a Generic Submarine*, in *Pacific 2012 International Maritime Conference*. 2012: Sydney, Australia.
5. Overpelt, B., Nienhuis, B., and Anderson, B., *Free Running Manoeuvring Model Tests on a Modern Generic SSK Class Submarine (BB2)*, in *Pacific International Maritime Conference*. 2015: Sydney, Australia.
6. Pook, D.A., Clarke, D.B., Jones, M., Quick, H., and Ranmuthugala, D., *RANS based CFD Prediction of Submarine Hydrodynamic Loads*, in *21st Australasian Fluid Mechanics Conference*. 2018: Adelaide, Australia.
7. Seil, G.J., Pook, D.A., Nguyen, M.C., and Leong, Z.Q., *The Influence of Appendages and their Stall on Submarine Hydrodynamic Loads*, in *PACIFIC 2017 International Maritime Conference*. 2017: Sydney, Australia.
8. ITTC, Recommended Procedures and Guidelines *Practical Guidelines for Ship CFD Applications*, **7.5-03-02-03 Rev. 01**, 2011
9. Seil, G.J., Nguyen, M.C., Pook, D.A., and Leong, Z.Q., *CFD Study of the Hydrodynamics of the Evolved DST Group Generic Submarine (BB2)*. 2019, Defence Science & Technology Group.

10. Seil, G.J., Nguyen, M.C., and Pook, D.A., *CFD Study of the KJ01 Submarine with Validation Against Wind Tunnel Measurements (FOUO, AUSTEO)*. 2019, Defence Science & Technology Group.
11. Toxopeus, S., Kuin, R., Kerkvliet, M., Hoeigmakers, H., and Nienhuis, B., *Improvement of Resistance and Wake Field of an Underwater Vehicle by Optimising the Fin-Body Junction Flow with CFD*, in *Proceedings of ASME 33rd International Conference on Ocean, Offshore and Arctic Engineering*. 2014: San Francisco, California.
12. Gand, F., Decl, S., Brunet, V., and Sagaut, P., *Flow dynamics past a simplified wing body junction.* Physics of Fluids, 2010. **22**: p. 115111.
13. Chase, N., *Simulations of the DARPA Suboff submarine including self-propulsion with the E1619 propeller*, 2012, Thesis, The University of Iowa
14. Crook, B., *Resistance for DARPA Suboff as Represented by Model 5470*, 1990, David Taylor Research Center, DRTC/SHD-1298-07

Simulating flood recovery manoeuvres using a free running submarine model

Paul Crossland and Philip Marchant, QinetiQ Ltd, pcrossland@qinetiq.com

ABSTRACT

The requirement to manage the safety of a submarine necessitates a number of key factors to be understood; one of which is to understand the manoeuvring performance of the submarine throughout the design process. A free running submarine model forms a fundamental component of this predictive capability so QinetiQ's Submarine Research Model (SRM) II was designed to support the UK MOD with their ongoing hydrodynamic assessment and modelling of the manoeuvring and control of the Royal Navy's current and future submarines. The SRMII design is based on a space-frame construction and uses COTS products where possible; the design is reconfigurable with critical components transferable between different hull shapes. Uniquely, the SRMII has been designed to have a large and capable ballast and trim system; a system that is capable of emulating the effects of compartment flooding incidents in submarines and moreover capable of representing the subsequent emergency recovery procedures including some aspects of blowing the main ballast tanks. This paper discusses how the ballast and trim system in the SRMII has been used to obtain model scale submarine manoeuvring data that were not obtainable from previous model tests. The experimental data are then used to validate the mathematical model used for predicting the response of a submarine to a compartment flood.

INTRODUCTION

Today's military submarine can typically only operate in a very restricted portion of the world's ocean depths; the typical collapse depths of submarines from World War II were around 280m [1], far less than the average depths of The World's Oceans. To provide assurance that a submarine design would be capable of operating safely within these rather tight boundaries means quantifying the manoeuvring characteristics early on in the design process; this is crucial in reducing the risks of producing designs that are unsuitable for the environment in which they are expected to operate.

The focus of manoeuvring and control studies is towards understanding the performance of a submarine operating in deep water, see [2] for example, and a number of approaches both numerical and experimental, reflecting the state-of-the-art at the time, can be applied throughout the design process. QinetiQ has an active role in assuring submarine safety using an approach, based on complementary numerical and experimental techniques that have been developed from its knowledge of hydrodynamics gained over a number of years, to provide a validated understanding of the manoeuvring and control performance of a submarine design, [3]. Irrespective of the operational requirements of a new platform, by and large, a submerged submarine will be required to manoeuvre safely in the vertical and horizontal planes, which can be translated, in generic terms, to be able to be able to accurately evaluate the performance of a particular hullform and appendage configuration at the design stage to:

- Determine measures of directional stability;
- Establish the size and power requirements of any control surfaces;
- Design suitable motion control systems;

- Determine that standard manoeuvres meet international maritime regulations or national design guidelines; and
- Demonstrate that the submarine can safely recover, within the boundaries of operational requirements, from the consequence of credible system level failure.

Therefore, a requirement for any nation that operates submarines, with due governance regarding safety, the ability to access validated tools and techniques to underpin timely advice, to accurately understand the behaviour of a submarine and to provide safe operating envelopes of current and future underwater platforms during normal operations and emergency conditions is key . This paper describes one aspect of mathematical model validation, free running model tests, particularly focussing on modelling flood events and the subsequent recovery of the submarine. The outcome of this research will provide the design guidance and validated evaluation toolsets for the benefit of:

- Future projects groups investigating submarine design at a concept level;
- Acquisition design groups carrying out more detailed project definition assessments; and
- In-service support to set safety and operational constraints for the submarine fleet.

HAZARD MITIGATION AND SAFE OPERATOR GUIDANCE

Through wider present-day awareness of safety of personnel, hazard mitigation modelling and the provision of Safe Operating Envelopes (SOEs) to submarine command is now considered as best practice by top-tier navies worldwide. Assuring the safety of submarines is paramount to the work that QinetiQ undertakes by considering the key elements that contribute to safety assurance; including design optimisation, provision of validated safety guidance and the development of robust emergency procedures for implementation by trained operators. This enduring support through the life of the submarine underpins the operational safety case that helps ensure that if an incident were to occur, the submarine would be able to recover safely and return to port.

The provision of this safety advice is based on two major incidents from which operators of submarines must be able to safely recover: a control surface jam at high speed and a flooding incident at low speed. The provision of safety guidance that mitigates the risk to the submarine following such incidents requires a detailed understanding of the behaviour of the submarine, a fully validated prediction capability, an understanding of the submarine systems that are crucial to safe recovery and the presentation of data in a way that is useful, unambiguous and easily understandable. Figure 1 provides an example of a Safe Operating Envelope (SOE) in the format of what is known as a Manoeuvring Limitation Diagram (MLD), in the UK, to provide guidance to Submarine Command, see [4]. This example provides boundaries of safe operation in terms of speed and depth; the slow speed boundaries (Flood Avoid Zone or FAZ) are present as a consequence of a flood and at higher speeds the restrictions are limited to mitigate against the risk of a plane jam (Jam lines). In order to generate such a robust set of curves it is crucial that the response of the submarine following such emergency scenarios is known; the only practicable means of doing that is to have a reliable mathematical model of a manoeuvring submarine.

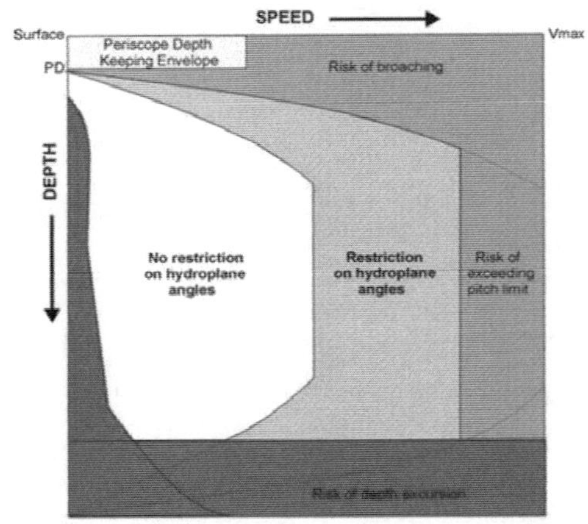

Figure 1: Example of Safe Operator Guidance [4]

The trajectory of a submarine at any instant in time is described in [5] in some detail but the assumptions and simplifications result in a set of 6 simultaneous non-linear equations for small disturbances and slowly varying motion, which are to be solved for the 6 unknown translational and rotational accelerations. Provided the forces and moments on the submarine are known, the accelerations can be found. The time integrations can be then undertaken to determine the translational velocity and rotational rates, and subsequently position and attitude. The key technical challenge is determining the hydrodynamic, and other non-hydrodynamic related, forces and moments on the submarine, at each time step. The essence of the mathematical model of submarine manoeuvring is the determination of the hydrodynamic forces and moments that are acting on the geometry. One mathematical approach assumes that the motion of the submarine is slowly varying [4], and that these quasi-steady state forces and moments on a manoeuvring submarine can be described by a series of empirical equations as described in [6] for example. The approach is then based on establishing a set of hydrodynamic coefficients that relate the state variables of the motion to the 3 forces and 3 moments acting on the submarine; current approaches include physical model tests of a constrained model, numerical methods or a combination of both, see [7] for an expanded explanation of the generic approach and see [8] for this process applied to a particular submarine design. Whether from physical model tests or Computational Fluid Dynamics (CFD), or a combination of both, once the hydrodynamic coefficient set has been obtained, the form of the mathematical model is known and can be used to develop a simulation capability for design studies including investigating the resultant SOE, in lieu of validation activities through free running model tests.

Once a mathematical model is available there are two features of the SOE that require validation through Free Running Model (FRM) tests, the FAZ and the jam lines.

It is more traditional to focus validation activities on the jam lines; a scenario where the submarine has high forward speed compared to its rate of change of depth. This can be seen in Figure 2a that shows how the calculated flow angle (or hydrodynamic angle of attack) varies with pitch angle for a number of hydroplane jams over range of speeds and jam angles; the maximum flow angle does not tend to exceed a non-dimensional pitch angle of ±1.0 in the example shown in Figure 2.

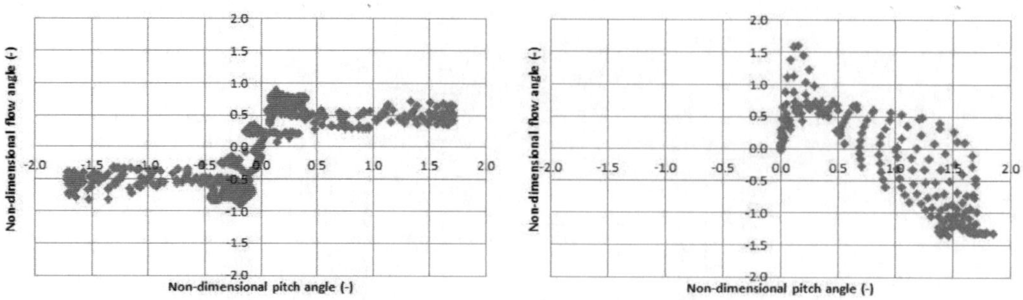

(a) Jam recoveries (b) Flood recoveries

Figure 2: Pitch angle vs *flow angle during jams and floods*[1]

When a submarine encounters a flooding incident, the flood water creates an out-of-trim condition which is then countered by increasing the speed of the submarine and if necessary blowing the Main Ballast Tanks (MBTs). These flood recovery manoeuvres can result in more significant flow angles on the submarine as a consequence of the low forward speed compared with the rate of change of depth and significant pitch on the submarine. It is likely that when the submarine is operating in an area that is adjacent to the FAZ boundary the flow angles experienced during a flood recovery would exceed those typically measured during the captive model tests.

Examples are shown in Figure 2b, that demonstrate how the calculated flow angles varies with pitch angle for a number of flood recovery manoeuvres for a range of different initial speeds, depths and flooding incidents. At the start of the flooding incident the submarine is heavy, sinking, and generates a pitch moment. This results in a rapid increase in the flow angle (in the positive direction) in excess of 1.5x the reference angle. As the submarine speeds up and gains hydrodynamic control over the flood mass, this reduces. However, as the MBTs are blown to create buoyancy, this results in the submarine ascending to the surface resulting in the flow angle becoming negative up to 1.5x the reference flow angle.

This implies that in order to fully validate the entire SOE boundaries, the mathematical model needs to be compared against measured responses that are typical of those around the jam lines and the FAZ. Furthermore, mathematical algorithms to account for the non-hydrodynamic aspects of the effects of flooding and blowing are also required in addition to the hydrodynamics model described earlier.

MODELLING FLOODING

Modelling floodwater entering a submerged submarine through a failed pressure hull penetration can be quantified with an adaptation of the Bernoulli equation for the velocity of fluid flow through a small orifice. The flow velocity through the hole is a function of the pressure inside the submarine, the hydrostatic pressure external to the submarine, the cross sectional area of the hole and a discharge coefficient.

If the total volume of flood water in the submarine compartment is significantly less than the total compartment volume, the pressure inside the submarine can be assumed to be

[1] Pitch and flow angles in Figure 2 have been non-dimensionalised using a typical maximum pitch angle that can be achieved during constrained model tests

at atmospheric pressure. In some cases (such as unsecured floods in small watertight compartments), it is more appropriate to take account of the rise in air pressure within the submarine.

The value of the discharge coefficient depends upon the type of flooding incident; for a direct penetration in the submarine pressure hull, a discharge coefficient of approximately 0.6, [9] to reflect the contraction of the flow through the flood hole. This can be considered appropriate for floods on a small diameter seawater system that is open to the sea, where the pressure hull fitting has failed. For systems that require larger pressure hull penetrations, it is normal to include some ability to secure the flood (through hull-mounted valves) should a failure occur. Therefore, for these types of securable systems, it is more likely that any failure is going to be within the pipework rather than the pressure hull penetration itself. In this case, a discharge coefficient of 0.6 may not be appropriate and may be somewhat less than 0.6 since there can be significant losses in the flow velocities due to the pipework system that the floodwater has to pass through.

Once the rate at which flood water enters the submarine is known, the impact of that flood water on the submarine response is required. The successful recovery of a submarine is sensitive to pitch angle, so the dynamic effect of the longitudinal centre of gravity of the flood mass may have a significant effect on the response of the submarine following a flood. Modelling this effect, taking account of the highly non-linear sloshing behaviour of the floodwater in the compartment would require complex CFD calculations to be coupled with a submarine manoeuvring simulation; this is not currently considered to be a practicable approach due to the significant computational cost associated with calculating the large number of manoeuvring trajectories required to define the FAZ. Therefore, it is considered appropriate to apply a quasi-steady state assumption for the flood water by taking the compartment geometry, and calculating the flood mass and its centroid for the instantaneous pitch angle.

MODELLING HIGH PRESSURE AIR BLOWING

Blowing high pressure air into MBTs can be divided into three parts: the flow of air from high pressure bottles, water flow out of from the ballast tank and the evolution of the pressure in the ballast tank, [9].

To model the flow of air from the bottle into the tank through a valve (that acts as a nozzle), [9] neglected pressure losses and heat transfer in the pipework that connected the bottle to the tank. A method was derived by [9] using a theory based on one dimensional steady flow of an ideal compressible gas. Since this method does not account for any pipework pressure losses, these algorithms are most suited to cases where the emergency bottle group is located adjacent to the blow nozzle in the MBT as any connecting pipework will be short. This would probably be the case where the HP air system has been designed to have a dedicated emergency blow bottle group, external to the pressure hull, which is independent from any normal HP air system, internal to the submarine, Figure 3. However, an alternative configuration is possible with the air from the emergency bottle group being supplemented with air from the main HP air system. To allow this arrangement to function, the pipework system is significantly more complex because the bottle groups supplying the main HP air system bottles (HPA) also have to supplement the emergency system, which has its own set of bottles, when required – see Figure 4. The other point to note is that the simplistic model is based on the assumption that the gas behaves as an ideal gas; [12], suggested that more sophisticated engineering software simulations methods coupled with the assumption that air behaves as a real gas would be better suited to model the complexities of an HP air blow that includes a main ring main. A number of commercial engineering simulation codes are available to model pneumatic systems; AMESim, [11], and Flowmaster are two examples. QinetiQ have developed AMEsim real gas models of representative HP air systems in submarines for blow modelling.

By using either the simplified blow model or the complex AMESIM model to determine the mass flow rate of air at the nozzle can be determined, and the pressure at the tank nozzle is then also known; it is the difference between the pressure in the tank and external water pressure that results in water being forced from the ballast tank through the flood grillages at the bottom of the MBT; [9] applied the Bernoulli equation at the ballast tank flood grillages to determine the volumetric flow of the water.

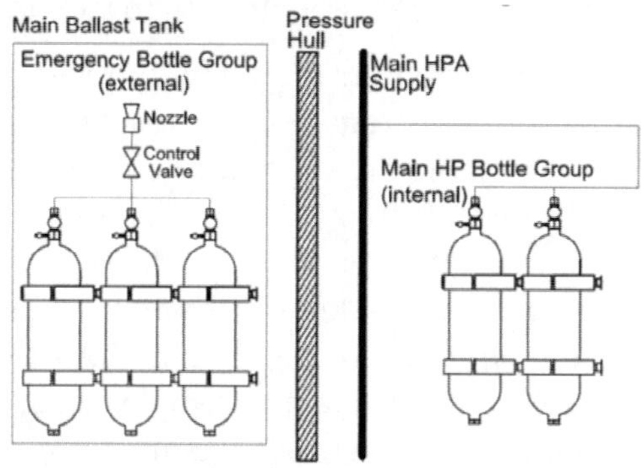

Figure 3: Arrangement for separate emergency HP air system

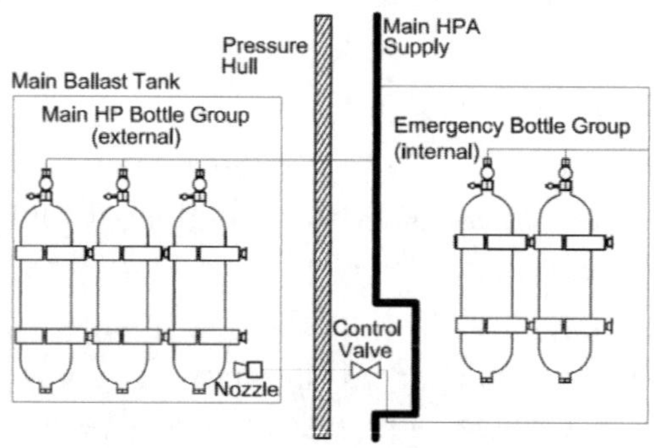

Figure 4: Arrangement for an integrated main and emergency HP air system

Once the flow of air to the ballast tank nozzle and the flow of water from the ballast tank are understood, the next step in the process is to understand how the air blown into the tank manifests itself as pressure. According to [9], the air is blown at high velocity, rapidly mixing with the water in the tank promoting good heat transfer from water to air (which will cause the air to expand); this process is considered isothermal and the ideal gas law can be used to determine the volume of the air in the tank. As the hydrostatic pressure decreases when the submarine drives towards the surface, the air volume within the MBT will expand. This results in an adiabatic process that decreases the air temperature in the MBT.

There is also an additional physical phenomenon to take into account during a flood recovery. During a recovery the submarine will adopt a positive pitch angle, which means that the forward MBTs are at a lower hydrostatic pressure than the aft ones. This phenomenon is considered to be significant when considering flood recoveries; whilst all MBTs may have equal masses of air blown into them, the air in the forward MBTs will have expanded to create a larger buoyant volume than the air in the aft ones. The impact of this is an additional pitch moment due to the differential expansion of air within the MBTs; this effect becomes more significant as the submarine approaches the surface.

FREE RUNNING SUBMARINE MODELS

A free-running model is a geosim of a full-scale submarine; it is used as a tool during the design and operation phases to gather data relating to the hydrodynamic manoeuvring and control performance of a submarine design. As part of the process for modelling the manoeuvring performance of a submarine, there are a number of reasons for undertaking FRM tests which include:

- Validation of the mathematical model;
- Manoeuvres suitable for System Identification, leading to improvements in the mathematical model predictions;
- To explore different control strategies;
- To investigate and validate the boundaries of the SOE
- Design of suitable motion control systems.

The UK's Submarine Research Model (SRM) capability has been employed in all the above areas, but is chiefly used for exploring the extremes of the manoeuvring envelope [4].
The SRM capability was first developed in the 1980's; it consists of an aluminium pressure hull that is 4.5m long with a diameter of 0.6m. Glass Reinforced Plastic (GRP) cladding is attached to this pressure hull to make it conform to the external shape of a range of submarine geometries with L/D ratios typical of SSNs or small SSKs, see [13]. The SRM is best suited to investigate hydroplane jams and subsequent recovery strategies, which are, by their nature, typically conducted at higher speeds. This first-generation design does not have active ballast control so is not particularly suited to slow speed operations, such as those required to investigate flood recovery scenarios.

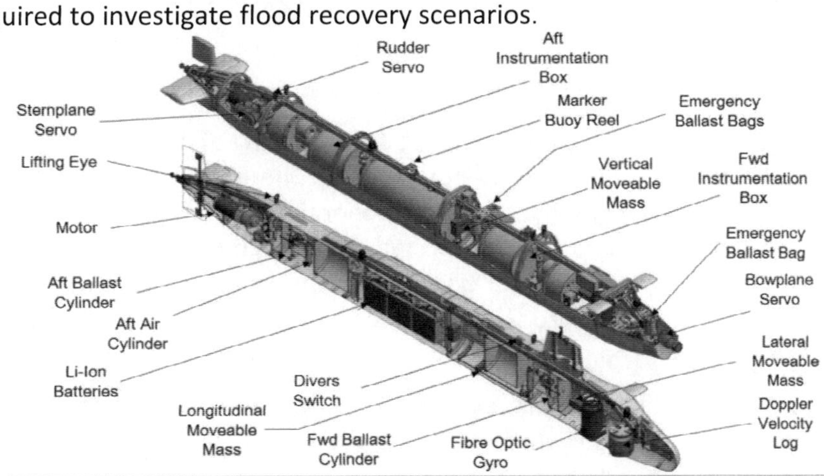

Figure 5: SRMII Free Running Model

In support of UK MOD studies to develop a replacement for the Vanguard Class SSBNs, this capability was upgraded, [13], to the Submarine Research Model II (SRMII) Figure 5. More detail about the design and operation of the SRMII can be found in [13] and [14], whilst the history of the development of FRM technology is discussed in [15]. The application of free-running models within the overall test and evaluation of a submarine's manoeuvring and control performance is detailed in [16], with further details as to why physical model experimentation of this type is still relevant explained in [17]. One of the most significant improvements in the SRMII design was the inclusion of an automated ballast and trim system that is described in the next section; this hugely capable system provides the means to investigate flood recoveries in submarines to inform on the validation of the entire SOE.

BALLAST AND TRIM SYSTEM OVERVIEW

The ballast system in the SRMII consists of two open ended cylinders, each containing an internal piston. The position of this piston is controlled by a stepper motor which changes the size of the 'dry' volume contained behind the piston. The capacity of each cylinder is just over 10 litres. The innovative approach to minimising the power requirement of the stepper motor was to keep the dry side (inside the cylinder) pressurised to offset the external hydrostatic pressure. The air is supplied from a small diver's air bottle inside the model. A regulator on the air supply line to the ballast cylinder ensures that the pressure behind the piston is maintained only slightly higher than the ambient hydrostatic pressure. The ballast cylinders are located at either end of the model and can be operated independently to provide a wide range of mass and moment changes; the components, assembly and locations within the model are shown in Figure 6.

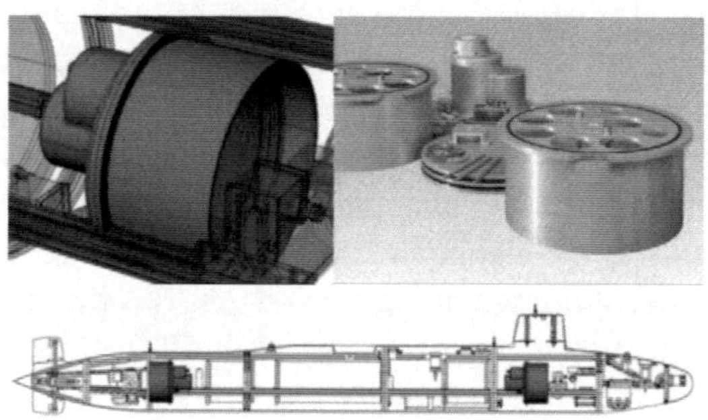

Figure 6: Ballast system components in SRMII

The design also incorporates a static trim system, consisting of three movable masses, each fixed to a lead screw, and driven by stepper motors. These align with the principal axes of the model such that the longitudinal mass can generate pitch moment, the transverse mass can generate roll moment, and the vertical mass changes the height of the centre of gravity (CoG) of the model. As a combined ballast and trim system the model can be either programmed with predetermined changes to the ballast and trim condition, or for the on-board software to order changes autonomously as required.

In the context of modelling flood recoveries, the ballast and trim system can replicate the change in mass and moment imparted on the submarine for a range of significant flood scenarios. This can be done in a number of ways, potentially by embedding the mathematical representation of the flood and blow algorithms in the on-board software as a closed loop system responding to the instantaneous depth and pitch of the model.

Alternatively the scenario can be treated as open loop, imparting a known change in mass and moment on the model. To provide validation evidence of the hydrodynamic models in a flow regime that is typical of flood recoveries, the most appropriate solution is to treat the scenario as open loop and use simulation to generate *a priori* time histories of changes in mass and moment, which are representative of a flood recovery. It was this approach that was adopted during the experiments described following section.

EXPERIMENT RESULTS

Flood and blow trajectories

The ballast and trim system was used to provide systematic changes in mass and moment, representative of different flood and blow scenarios, to achieve a wide range of incident flow angles (both positive and negative) for mathematical model validation. The flood scenarios replicated floods of increasing severity at a typical aft engine room location; where the size and complexity of seawater systems used are more likely to result in a significant flooding incident, should a breach to the watertight integrity occur. For the blow scenarios, the ballast and trim system was used to represent blowing of HP air into a forward MBT that made the model buoyant and created a bow up pitch.

For each flood scenario, the aft ballast cylinder was used to increase the mass of the model, whilst the forward cylinder was adjusted to ensure a representative moment. Once the flood had been represented, there was a short delay to replicate the time taken for the crew to react and execute their emergency recovery procedures. Following this time delay, the appropriate propulsor response was applied and the model was driven to the surface using a specifically designed autopilot that controlled pitch during the ascent.

All of the flood scenarios were initiated whilst the model was at low forward speed, and therefore the change in mass and moment created by the flood induced a significant pitch angle. Accelerating the model enables the hydroplanes to partially counter the flood mass and moment and reduce the pitch angle to an acceptable level. By varying the initial speed of the model and the recovery RPM, a range of peak pitch angles (the point at which the hydroplanes were able to reduce pitch) were obtained.

For each blow scenario, the forward ballast cylinder was used to make the model light, whilst the aft cylinder was adjusted to ensure a representative moment. Once the change in trim condition of the model had been achieved, the response of the model (in terms of demanded RPM and autopilot actions) was the same as that for the floods. Again, by varying the initial speed and the recovery RPM, a range of peak pitch angles could be obtained. Varying the initial conditions also enabled the model to undergo a range of hydrodynamic angles of attack (both positive and negative); comparisons of simulations with the experimental trajectories allows a greater level of assessment of the validity of the mathematical model for these type of manoeuvres.

Figure 7 shows typical flood and blow recovery sequences, taken from video stills, of the model during the run-up period (a), at the initiation of the flood or blow (b), and as the model drives to the surface (c). For the flood case, a large amount of air can be seen being vented from the aft ballast cylinder; for the initiation of the blow scenario, since air is added to the forward cylinder, it is only venting a small amount of air from the aft cylinder to represent the required pitch moment. The time sequence of the stills is the same for the flood and the blow so the figure shows clearly the model reaching the surface quicker for the blow case.

(a) run up to flood *(a) run up to blow*

(b) Flood initiated *(b) Blow initiated*

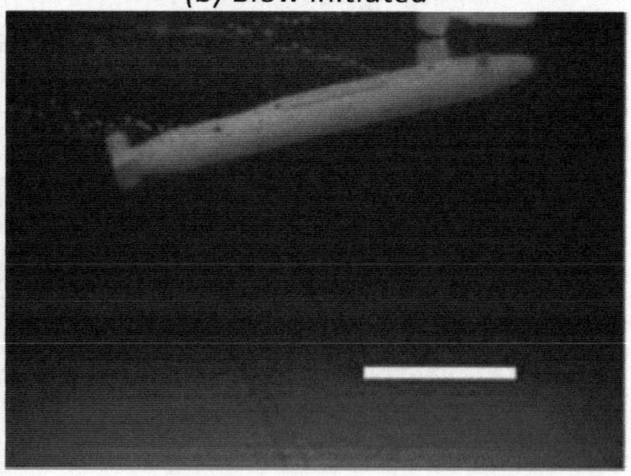

(c) Driving towards the surface (heavy) *(c) Driving towards the surface (light)*

Figure 7: Images of flooding and blowing

Figure 8 provides example measurements of flood and blow responses of the model (blue lines are the measured responses, orange are the demanded responses) compared with simulation (in this case red lines are the predicted responses and green are the demanded). In both cases, the predictions of the initial response in depth and pitch compare well with measurements, as do the ascent parts of the trajectory as the model drives to the surface.

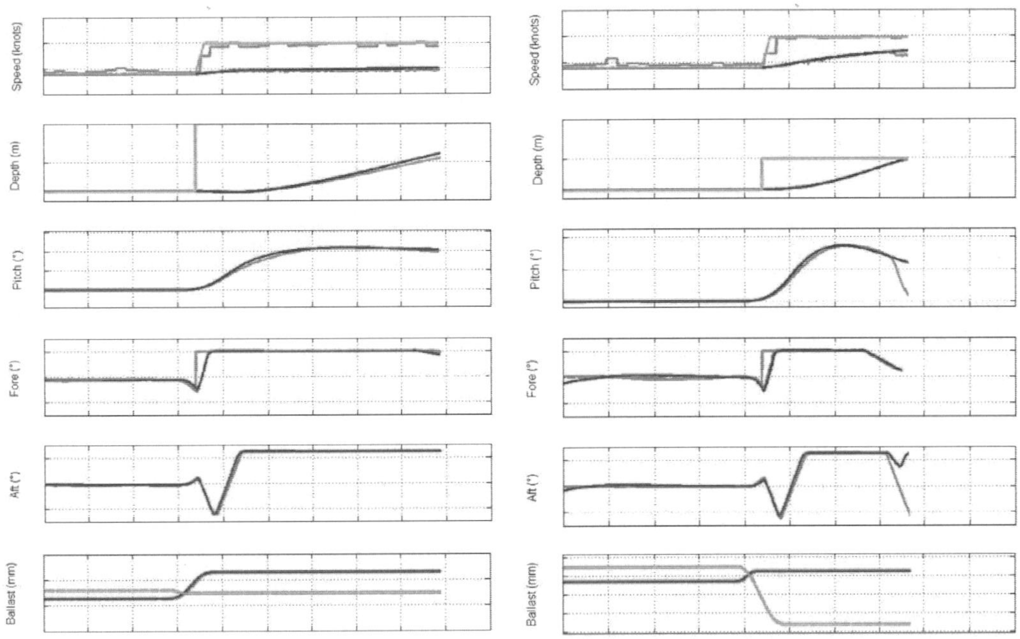

(a) Flood (b) Blow

Figure 8: Trajectories of flooding and blowing

Initial exploratory tests were conducted in QinetiQ's Ocean Basin at Haslar, where the water depth is around one model length. In order to allow the recovery trajectories to fully develop for the more extreme flood and blow cases, a deeper initial depth is required, so similar tests are planned for the deep-water reservoir facility.

Simulation correlation

To provide an overall view of the quality of the simulations when compared with the experiments, a correlation of the simulated peak pitch angles from the flood and blow trajectories with the measurements is shown in Figure 9. This is a similar concept to that used when assessing simulation of peak pitch angles during hydroplane jam recovery scenarios.

The results of the correlation show that predictions of maximum pitch angle for both (a) flood and (b) blow scenarios correlate very well with the measured pitch angle. However, also shown is a similar correlation plot for jam scenarios (c) which shows less scatter in the data compared with plots (a) and (b). This suggests either some limitations in the mathematical model, or an illustration of the increased level of uncertainty in conducting these types of experiments compared with hydroplane jam recovery manoeuvres.

To illustrate the extent to which the mathematical model has been validated, in the context of the FAZ, the combinations of the measured flow and pitch angles for the flood and blow scenarios generated during the experiment are shown in Figure 10. Also shown, for comparison, is the area indicated within Figure 2(b) as the flow and pitch angle combinations that are generated when calculating a FAZ boundary.

This shows that the extremes of flow angle that are predicted to occur were not sufficiently represented during the experiments and where within the range of the captive model tests. This is largely due to the inability to test the necessary combinations of floods and blows in the limited depth of the Ocean Basin. It is possible that it is these extreme flow angles where the accuracy of the mathematical model may be called into question.

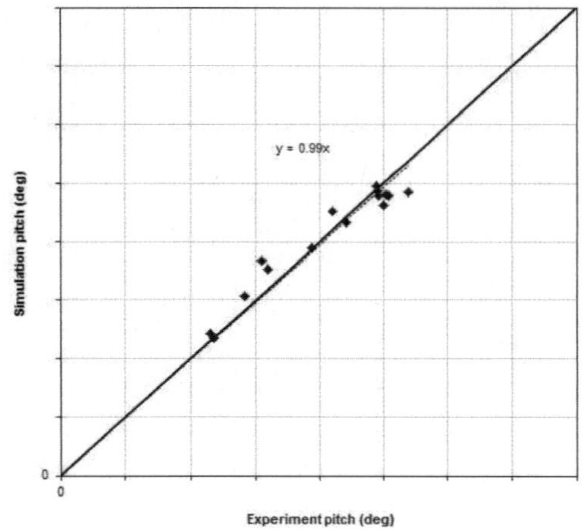

(a) Floods

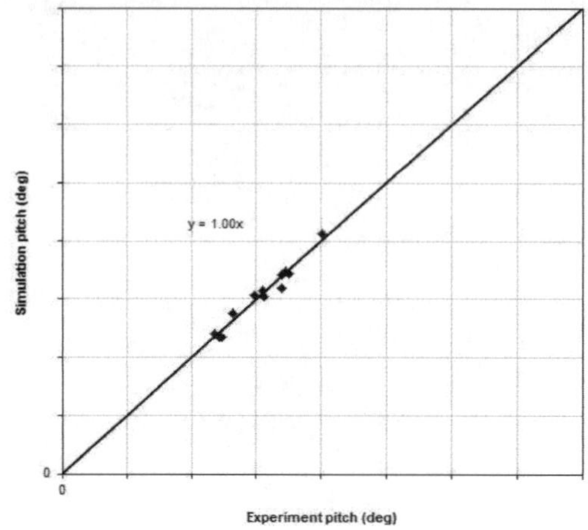

(b) Blows

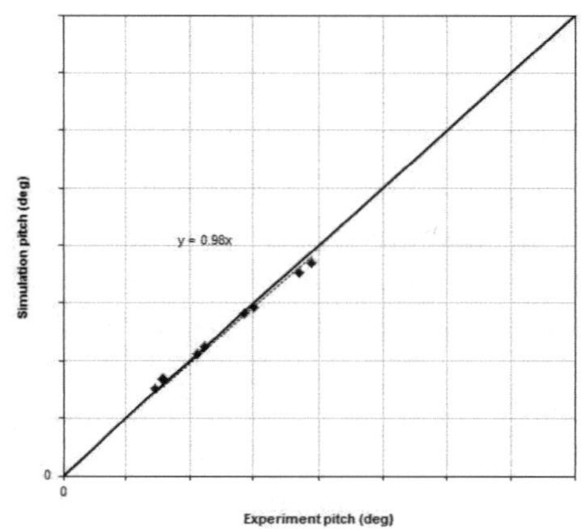

(c) Jams

Figure 9: Correlation of simulated peak pitch angles with measurements

Once the more extreme tests of combinations of flood and blows have been undertaken, then the full validation of the mathematical model can be performed. If the model requires improvements then this could be done by including higher order terms in the Gertler and Hagen mathematical expressions such as [18], who augmented the force and moment equations with up to fifth order terms derived from wind tunnel experiments on a captive model. However, the example data shown in [18] indicates that this approach does not take into account stall effects. In reality, high flow angles at low speeds are likely to result in stall and this may require a different approach. One alternative approach for capturing the non-linear effects due to high angles of incidence may be to introduce look-up tables that are derived from captive models tests. These look-up tables would be accessed at each time step of the simulation to provide a force and moment rather than through the coefficient based approach. One disadvantage of this approach is that the lookup table must capture the range of data required for simulation, as typically look-up tables cannot be used for extrapolation, whereas the coefficient based approach can extrapolate (with due caution) for scenarios outside of the fitted range of data.

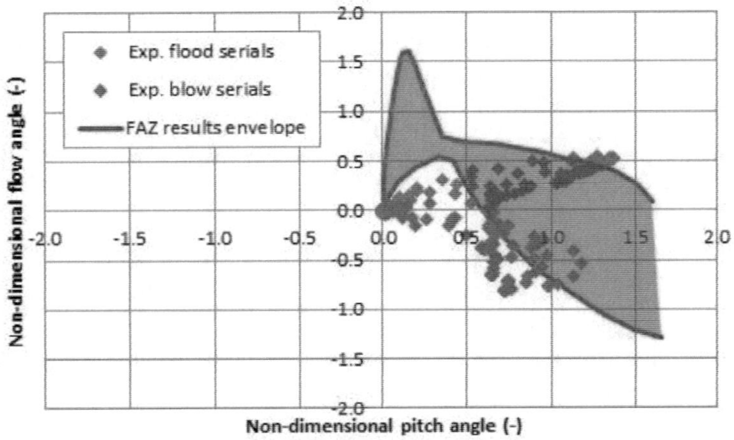

Figure 10: Hydrodynamic angle of attack vs pitch angle compared to FAZ

CONCLUDING DISCUSSIONS

This paper has discussed one specific application of a free running submarine model capability; to explore and validate responses that are typical of a submarine recovering from a flood or responding to a blow. The experiments successfully proved the methodology for conduct and analysis of validating trajectories for flood recoveries using QinetiQ's free running model capability.

The correlation that has been undertaken with the current data has shown that the predicted peak pitch angles compare well with those measured during the experiment. However, as a consequence of some of the limitations with this experiment, the flow angles achieved were not as extreme as those predicted to occur during flood scenarios around the FAZ boundary and moreover, were within the range over which a captive model would be tested (from which the hydrodynamic coefficients are derived). Therefore, to some extent the correlation was expected to be good.

Further tests will be planned using a deep-water test facility (reservoir) to provide the additional water space that would allow tests to be conducted at deeper initial depths. These conditions would generate more significant flow angles on the model thus providing validation data in these more extreme but still pertinent, areas.

These tests provide confidence that these validated tools are providing an understanding of the behaviour of a deeply submerged submarine that is subjected to a major flood as a consequence of a failure in a pressure hull penetration. This capability underpins the provision of safe operating envelopes of current and future underwater platforms during normal operations and emergency conditions which is a key requirement for any nation that operates submarines with due governance regarding safety.

REFRERENCES

[1] Gabler U., Submarine design, Bernard & Graefe Verlag, Koblenz, 1986, ISBN 3-7637-0124-9.

[2] Ray A, Seshadri V, Singh SN, Sen D, Manoeuvring studies of underwater vehicles – a review. International Journal of Maritime Engineering, 2008.

[3] Crossland P, Bayliss JA, Kimber NI, Submarine hull form design, safety and

operational guidance. Submarine Institute Australia Technology Conference 2013, Adelaide.

[4] Haynes D, Bayliss J, and Hardon P, Use of the Submarine Research Model to explore the manoeuvring envelope, RINA Warship Naval Submarines 7, June 2002, London,UK.

[5] Bishop RED, Parkinson AG, Directional stability and control of marine vehicles, Department of Mechanical Engineering, University College London, Nav. Arch. Rep. 3/70, 1970.

[6] Gertler M, Hagen GR, Standard equations of motion for submarine simulation, Naval Ship Research and Development Center Report No. 2510, June 1967, David Taylor Model Basin, Carderock, MD, USA.

[7] Renilson M, Submarine Hydrodynamics Second Edition, Springer International Publishing, ISBN 978-3-319-79056-5, 2018.

[8] Crossland P, Kimber NI, Thompson N, Understanding the Manoeuvring Performance of an X- plane Submarine in Deep Water and Near the Free Surface. Pacific 2012: International Maritime Conference, Sydney, January 2012.

[9] Franz DD, Melching CS, Full Equations Utilities (FEQUTL) Model for the Solution of the Full, Dynamic Equations of Motion for the Approximation of Hydraulic Characteristics of Open Channels and Control Structures During Unsteady Flow, U.S. Geological Survey Water Resources Investigations Report 97-4037, 1997.

[10] Font R, Garcia J, Ovalle D, Modelling and simulating ballast tank blowing and venting operations in manned submarines. 8Th IFAC conference on Control Applications in Marine Systems, Germany, September 2010.

[11] AMESIM online help documentation, Pneumatic Library, LMS Imagine S.A., Rev 10, Nov 2010.

[12] Marchant P, Thompson N, Cosserat D, Validation of submarine blow algorithms for flood recovery, Warship 2014: Naval Submarines and UUVs, June 2014, UK.

[13] Crossland P, Marchant P, Kimber NI, A Reconfigurable free running model submarine capability. Pacific 2015: International Maritime Conference, Sydney, October 2015.

[14] Crossland P, Nokes RC, Dunningham S, Marchant P and Kimber N, "SRMII – A reconfigurable free running model capability for submarines with large L/D ratio". RINA WARSHIP 2014, Bath,UK, June 2014.

[15] Marchant P and Kimber N, "The development of free manoeuvring submarine models for hydrodynamic research", UDT2015, Rotterdam The Netherlands, June 2015.

[16] Marchant P, Kimber N, "Assuring the safe operation of submarines with operator guidance", UDT2014, Liverpool, UK, June 2014.

[17] Marchant P, Kimber N, "Hydrodynamic Free Manoeuvring Models: Can't that all be done with simulation now?", UDT2016, Oslo, Norway, June 2016.

[18] Park JH, Kim N, Shin YK, "Experimental study on the hydrodynamic coefficients for high-incidence-angle maneuver of a submarine" International Journal of Naval Architecture and Ocean Engineering, Vol9, pp 100-133, 2017.

Analysis of Aft-Control Surface Interaction with an Axisymmetric Underwater Vehicle Hull

David Kirkby, Australian Maritime College & ENSE[3], David.Kirkby@dst.defence.gov.au
David A. Pook, RMIT, David.Pook@dst.defence.gov.au
Martin Renilson, Pacific ESI, martin@renilson-marine.com
Dev Ranmuthugala, DST, Dev.Ranmuthugala2@dst.defence.gov.au

ABSTRACT

Aft-Control-Surfaces (ACSs) provide underwater vehicles with the forces and moments required to maintain stability, and control manoeuvres. Thus, accurate prediction of ACS hydrodynamic loads is essential to adequately size the control surfaces and to predict the vehicle's stability and control characteristics. Undersized control surfaces may result in failure to meet manoeuvring requirements or provide insufficient directional and pitch stability. Over-sized control surfaces can adversely affect other aspects of the vehicle such as resistance and acoustic signature.

Sizing ACSs requires consideration of the hydrodynamic interaction with the vehicle's hull. The presence of the hull alters the normal force produced by an ACS when compared to the ACS geometry as an isolated wing exposed to the freestream. The ratio of these two conditions is referred to as the ACS efficiency (or tailplane efficiency). Lyons and Bisgood [1] provide a methodology to estimate hull effects on the ACS normal force curve slope. Dempsey [2] experimentally determined the effect of a tear-drop shaped hull on the ACS normal force curve slope for a range of ACS geometries. The Dempsey experiment showed a greater hull influence than predicted by the Lyons and Bisgood method. Bettle [3] and Mackay, Watt and Bohlmann [4] have suggested this might be due to Reynolds number effects.

In this paper, the possible Reynolds number effects were investigated by simulating the Dempsey [2] experiment using Computational Fluid Dynamics (CFD) that solved the Reynolds Averaged Navier-Stokes (RANS) equations. Standard RANS modelling techniques that assume all turbulent flow produced results in good agreement with the relatively low Reynolds number data of Dempsey, thereby discounting significant laminar boundary layer effects. Dempsey's experiment was simulated at higher Reynolds numbers and ACS efficiency was found to be dependent on the Reynolds Number. The choice of turbulence closure model in the RANS simulations was found to affect the prediction of ACS efficiency. The CFD results confirm the suitability of using Dempsey's data for predicting ACS efficiency at model scale Reynolds numbers for an underwater vehicle hull with relatively shallow stern angles.

INTRODUCTION

The control surfaces of an underwater vehicle, in conjunction with its movable ballast, and propulsor form the basis of the vehicle's control system. During the initial design phase, design of the control system is highly iterative and considers a large number of design alternatives [4]. Analysis of each iteration using model scale testing or numerical methods, such as Computational Fluid Dynamics (CFD), would be inefficient and prohibitively expensive. Consequently, it is advantageous that valid empirical predictions of the vehicle's control system are available to a designer during the preliminary design phase.

The positioning of the Aft Control Surfaces (ACSs) and their interaction with the hull results in reduced hydrodynamic loading when compared to the ACS geometry considered as an isolated wing exposed to the freestream. Integration of the ACS with the hull reduces the effective ACS normal force curve slope relative to an isolated ACS geometry. The ratio of the integrated ACS normal force curve slope and the ACS isolated lift curve slope is referred to as the ACS efficiency (or tailplane efficiency, K_{wb}),

$$K_{wb} = \frac{C_{Z,\alpha}}{C_{L,\alpha}} \quad (1)$$

where $C_{Z,\alpha}$ is the ACS normal force curve slope when integrated with the hull and $C_{L,\alpha}$ is the lift curve slope for the equivalent isolated ACS. Two widely used empirical methods for predicting ACS efficiency are Lyons and Bisgood [1] and Dempsey [2]. The results of Dempsey [2] show a much greater hull influence on the ACS lift curve slope and consequently a lower ACS efficiency, relative to the results of Lyons and Bisgood [1]. Bettle [3] and Mackay, Watt and Bohlmann [4] have suggested that the lower ACS efficiency predicted by the Dempsey [2] method is due to Reynolds number effects. Mackay, Watt and Bohlmann [4] proposed a critical Reynolds number (Re) threshold of approximately 10^6 (based on the ACS mean cord) above which efficiency tends toward the higher values of Lyons and Bisgood [1] and below which efficiency tends toward Dempsey's prediction. This proposed critical threshold, together with the lack of turbulence tripping on the ACS in Dempsey's experiment, suggests that laminar flow may be responsible for the reduced efficiency.

It is clear that there remains significant uncertainty in the empirical relationships used to predict ACS efficiency. This uncertainty limits their usefulness in the design of underwater vehicles. Appropriate sizing of control surfaces is critical for ensuring that the vehicle meets its manoeuvring requirements whilst minimising resistance and acoustic signature and it therefore remains an important area of research.

This study simulates the Dempsey [2] experiment using Reynolds Average Navier-Stokes (RANS) CFD to determine the suitability of Dempsey's empirical method for the prediction of underwater vehicle ACS efficiency. Investigations into the influence of turbulence modelling and Reynolds number on ACS efficiency are also presented.

THEORY

Dempsey [2] presented results in the body axis system with the origin at the body centre of buoyancy, as show in Figure 1. The normal forces were non-dimensionalised with,

$$Z' = \frac{2Z}{\rho L^2 U^2} \quad (2)$$

where Z is the normal force (in the body axis), L the hull length, ρ the fluid density, and U the freestream velocity.

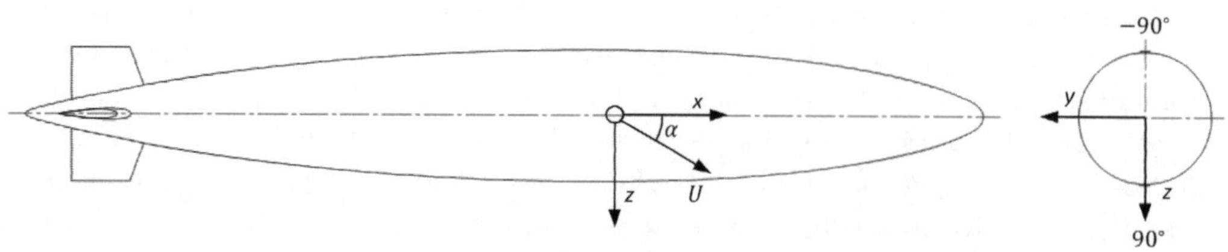

Figure 1: Appended Dempsey geometry and body fixed coordinate system with the origin at the centre of buoyancy.

Dempsey [2] provides a detailed explanation of the calculation methods used to determine ACS efficiency. The lift curve slope for an isolated ACS ($C_{L,\propto}$,) is calculated using the relationship presented by Whicker and Fehlner [5],

$$C_{L,\propto} = \frac{1.8\pi a}{1.8+\cos(\varphi)\sqrt{\frac{a^2}{\cos^2(\varphi)}+4}} \tag{3}$$

where a is the ACS aspect ratio, and φ is the ACS sweep angle at 25% chord.

To determine the ACS lift curve slope when integrated with the hull, Dempsey [2] measured the normal forces of the unappended hull (Z'_h) and the hull with ACSs (Z'_{ha}). The contribution of the ACS to a change of the normal force with respect to a change of vertical velocity ($\Delta Z'_w$) is,

$$\Delta Z'_w = \frac{\Delta Z'_{ha}}{\Delta w'} - \frac{\Delta Z'_h}{\Delta w'} \tag{4}$$

where w' is the vertical velocity (w) non-dimensionalised by U. For small pitch angles (α),

$$w' \cong -\alpha \tag{5}$$

where α is in radians. The normal force curve slope can be expressed as [6],

$$C_{Z,\alpha} = C_{L,\alpha} + C_D \tag{6}$$

where C_D is the vehicle's drag coefficient. For small pitch angles, C_D for the appended and unappended geometries is approximately equal and therefore cancel in the calculation of the ACS normal force curve slope in (4). Hence, for small α, the ACS lift and normal force curve slopes are approximately equivalent (as assumed by [2]). The ACS normal force curve slope when integrated with the hull is given by,

$$C_{Z,\alpha} = -\Delta Z'_W \frac{L^2}{2A} \tag{7}$$

where the reference area (A) has been changed to the planform area of the ACS (see Figure 2). The choice of pitch angles will affect the calculation of the ACS lift curve slope. Dempsey [2] does not explicitly state the pitch angles used. However, using provided data for pitch angles of 0° and 2° gives the Dempsey calculated ACS efficiency. ACS efficiency was then calculated as the ratio of $C_{Z,\alpha}$ to that of an equivalent isolated control surface, $C_{L,\alpha}$, as defined by (1).

NUMERICAL MODEL

Geometry

Dempsey's [2] towing tank experiments used the Model 4621 hull. The hull had a length (L) of 4.572 m and a maximum diameter (D) of 0.6228 m. The centre of buoyancy was located 2.0373 m from the bow. A sand strip was used at x/L = 0.396 to trip the hull boundary layer; however no turbulence trip was used on the ACSs. The Model 4621 hull differs somewhat from the design of a modern underwater vehicle due to its tear drop shape, with no parallel mid-body, and a relatively shallow tail half-angle of approximately 6.5°.

Dempsey [2] tested 20 different trapezoidal ACS configurations (a combination of five chords, and four spans) appended to the Model 4621 hull to parametrically assess the influence of ACS aspect ratio on the stability derivatives and ACS efficiency. Of the four different spans assessed by Dempsey, the present work focuses on the results of the 'B' span. The B span was the largest span that did not exceed the maximum hull diameter [6] and was therefore considered the most relevant to the design of an underwater vehicle. Two different cords lengths were assessed, the '3' and '11' cords, which represented the maximum and minimum cord lengths tested by Dempsey [2]. Details of the two configurations are defined in Table 1 and a plan view of the Model 4621 afterbody, and a generic trapezoidal ACS is shown in Figure 2.

The various geometry configurations were modelled in CATIA V5. Three simplifications were made to simplify generation of the CFD mesh and improve mesh quality: the two struts mounting the hull to the towing carriage were not included (geometry of struts was not specified by Dempsey); the tail of the Model 4621 hull was shortened (not tapered to a point) with a 5 mm radius applied to remove the need for an axisymmetric pole in the CFD grid; and the trailing edges of the ACS were truncated by 5 mm, increasing the trailing edge thickness to approximately 2mm (the physical trailing edge thickness was not specified by Dempsey).

Table 1: ACS configurations from Dempsey [2] simulated in this study.

Configuration (Dempsey 1977)	Span b (m)	Tip Cord C_{TIP} (m)	Reference Area A (m^2)	Aspect Ratio $(\frac{2b^2}{A})$	ACS Mean Body Diameter D_1 (m)
3B	0.3048	0.0762	0.04013	4.64	0.1551
11B	0.3048	0.2794	0.10206	1.82	0.2097

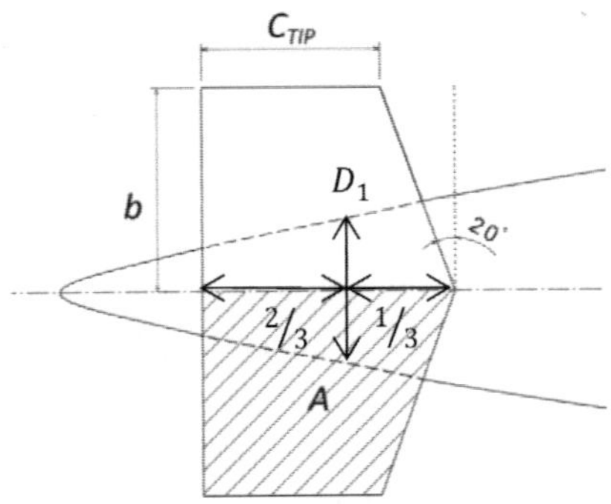

Figure 2: Plan view of Model 4621 afterbody showing a generic trapezoidal ACS with reference dimensions.

Fluid Domain

The fluid domain was composed of two sub domains, referred to as the inner and outer domains, joined by a non-conformal interface. This approach was used to reduce the mesh cell count and consequently the required computational resources. The inner domain extended $3D$ both forward and aft of the body and $3D$ away from the maximum diameter. The outer domain extended $3L$ forward of the body, $8L$ aft of the body, and $2L$ from the hull centreline. A mapped non-conformal interface boundary condition between the two domains was specified in the solver.

The boundary conditions at the extent of the outer domain are shown in Figure 3. Since the Dempsey experiment did not consider variation in yaw angle, the computational domain was symmetrical about the x-z plane. In order to reduce total mesh size a half model was simulated with a symmetry boundary condition imposed.

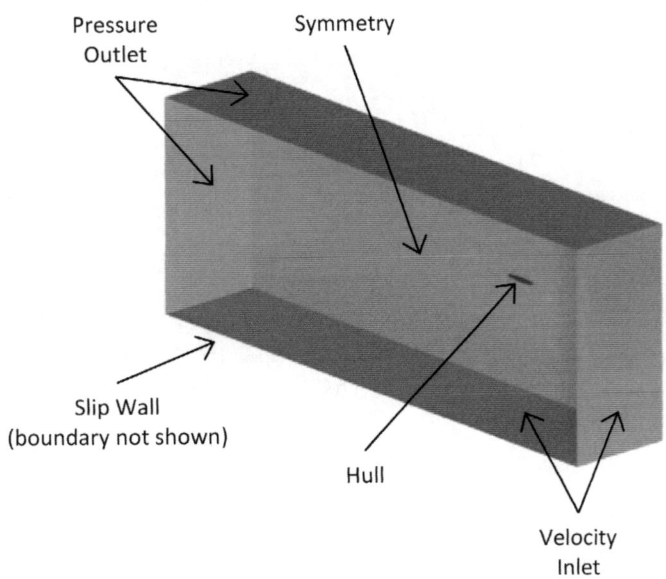

Figure 3: Fluid domain and boundary conditions

A velocity inlet was applied at the forward and lower boundaries of the domain. Specification of the u and w velocity components allowed for the hull to be considered at various pitch angles (0-16°) without the need to re-mesh the geometry. A velocity magnitude corresponding to a tow speed of 6 knots was imposed to model the conditions of Dempsey's [2] experiments. The tow speed provides a hull Reynolds Number of 12.5×10^6 based on L, and an ACS mean cord Reynolds Number of 0.32 to 0.86×10^6 for the 3B and 11B configurations respectively.

The downstream and upper boundaries were set as pressure outlets. The hull and ACS walls were treated as non-slip walls. The far-field wall opposite the symmetry plane was modelled as a slip wall to simulate free-stream conditions. In Dempsey's [2] experiments, the model was towed at a depth of 3.05 m which was considered sufficiently deep to ignore free surface effects.

Mesh

A fully structured grid was utilised to discretise the fluid domain around both the unappended and appended Model 4621 hull. This approach provided two main advantages. Firstly a structured mesh reduces solution time, due to the reduced cell count (relative to a tetrahedral mesh) and through improved multigrid acceleration. In addition, structured grids have demonstrated reduced discretisation error on well aligned flows thereby increasing solution accuracy [7].

The structured grid was developed using Pointwise (18.2 R1) meshing software. Figure 4 shows the surface mesh for the appended hull. An O-H grid topology was used for both appended and un-appended geometries. O-grid topology enveloped the body to a distance of $1D$ from the body in order to adequately resolve near wall behaviour. H-grid topology was then extended to the boundaries of the inner domain. The grid for the outer domain was consistent for all simulations, while the grid for the inner domain was varied to account for the different geometries. A non-dimensional wall distance (y^+) of order 1 was maintained for all RANS simulations.

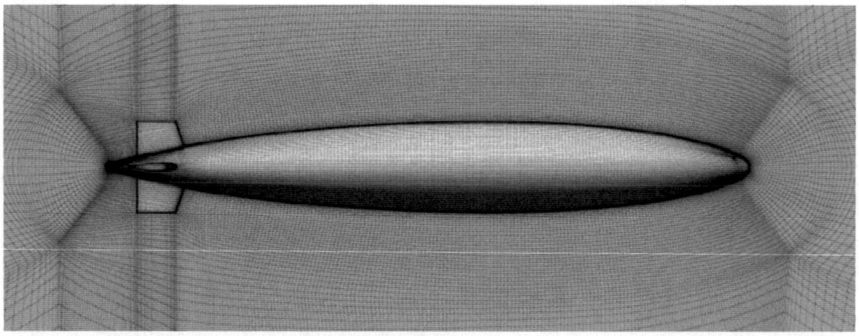

Figure 4: Surface mesh for the 11B appended Model 4621 hull and the symmetry plane mesh (every second gridline shown)

Solver

The incompressible RANS equations were solved using the ANSYS Fluent (18.1) finite volume CFD code. Both the k-ω SST and Baseline Reynolds Stress Model (BSL-RSM) turbulence closure models (TCM) were used to account for the Reynolds Stresses which arise from Reynolds averaging. Simulations using the BSL-RSM model were initialized from the corresponding k-ω SST solution to improve solution convergence and reduce computational requirements. First order discretization was used to establish the solution for the first 500 iterations. Second order discretization was then applied for all flow and turbulence quantities until convergence of the X and Z forces was achieved.

GRID CONVERGENCE STUDY

A grid convergence study was conducted for the 11B configuration. As the unappended hull and the 3B configuration used similar mesh topology and cell sizes, the grid convergence results for the 11B configuration were assumed to apply.

Three grids of increasing resolution were considered. The relative edge length and cell count for the three grids is shown Table 2. The grid refinement study considered the body at 2° pitch as this condition was critical for the determination of the ACS lift curve slope. The grid convergence study considered only the k-ω SST TCM and grid independence was assumed to apply to simulations using the BSL-RSM TCM.

Table 2: Relative edge length and cell count for the 11B appended Model 4621 hull grid convergence study.

Grid No.	Relative Edge Length	Cell Count ($\times 10^6$)
1	1.000	20.6
2	0.794	41.7
3	0.630	82.8

The percentage variation in Z' relative to the finest grid as a function of relative edge length is shown in Figure 5. The percentage variation decreased with increasing mesh resolution.

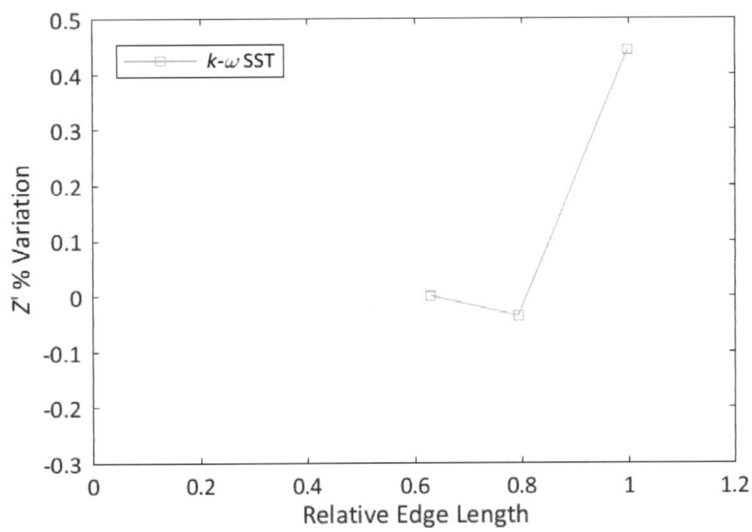

Figure 5: Percentage variation in normal force coefficient relative to finest grid

The percentage change in Z' from the coarsest to finest grid was 0.44%. Given that the error due to discretization was small relative to the large variations in ACS efficiency predicted by the Lyons and Bisgood [1] and Dempsey [2] methodologies it was decided to use the coarsest grid for subsequent simulations. Use of the coarsest grid enabled a larger parameter space to be considered with the available computational resources and time.

RESULTS

Normal Force

The CFD results for the k-ω SST and BSL-RSM TCMs were compared to Dempsey's [2] experimental results at various pitch angles to assess their accuracy. An assessment of TCM validity was made for both the unappended hull and 11B geometries.

The results of the validation study are shown in Figure 6 and Figure 7 for the unappended hull and 11B geometries respectively. Both TCMs captured the general evolution of hull normal force with angle of attack, however differences in absolute loading were seen.

For the unappended hull simulations both TCMs under predicted the normal force relative to Dempsey 's [2] experimental measurements for all pitch angles. At 2° angle of attack the k-ω SST model under predicted hull normal force by 15.9% compared to a 2.7% under prediction for the BSL-RSM TCM. This general trend was observed throughout the range of pitch angles assessed.

The calculation of ACS efficiency is dependent on the value of $C_{z,\propto}$ for both the appended and unappended configurations (see equation (4)). Consequently the large under prediction of $C_{z,\propto}$ for the unappended hull by the k-ω SST TCM results in significant over prediction of ACS efficiency (57%) relative to both the experimental data (47%) and the BSL-RSM TCM (53%). This result is in agreement with the conclusion of Phillips, Turnock and Furlong [8] which found that the BSL-RSM model is the most appropriate TCM for prediction of axisymmetric body loads. It also highlights the importance of TCM choice even at low angles of attack.

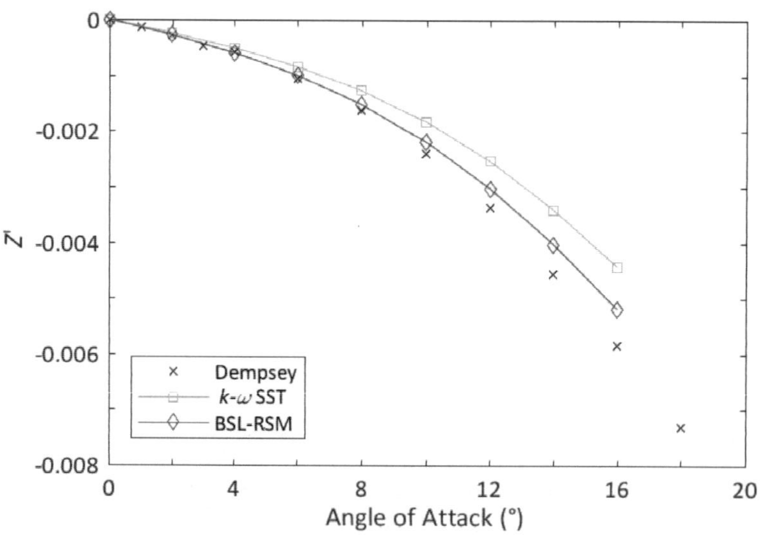

Figure 6: Variation of normal force with pitch angle for unappended Model 4621 hull.

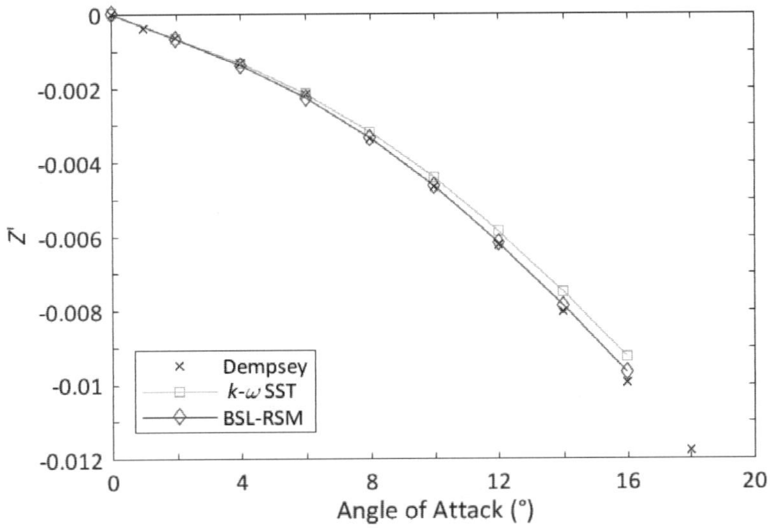

Figure 7: Variation of normal force with pitch angle for the 11B appended Model 4621 hull.

For the 11B configuration, both TCMs provided accurate predictions of the normal force up to pitch angles of approximately 6° compared to Dempsey [2] (see Figure 7). For the 11B configuration, the relative contribution of the axisymmetric body to the total normal force is reduced. Therefore, errors predicting axisymmetric hull loads are less apparent. This suggests that both models are adequate for predicting the ACS and hull normal force at low angles of attack. At pitch angles greater than approximately 6° both TCMs under predict the normal force as the axisymmetric body makes a larger contribution to total loading. At these higher pitch angles the BSL-RSM model demonstrated increased accuracy due to its improved prediction of the flow field around the axisymmetric body [8] as shown previously. This again highlights the importance of the TCM selection; however the results for high angles of attack are not relevant to the calculation of $C_{z,\propto}$ and ACS efficiency as the normal force curve slope was calculated between 0° and 2°.

Based on the results of the normal force study it was concluded that the baseline BSL-RSM turbulence model offered better prediction of the hull normal forces for both the appended and un-appended configurations. As a result it was expected that the BSL-RSM TCM would provide a better prediction of ACS efficiency. In addition, the validation study highlights the importance of turbulence model selection for accurate prediction of ACS loads and vehicle stability coefficients.

Displacement Thickness

At low angles of attack, the differing normal force predictions by the two TCMs for the unappended hull, and consequently $C_{z,\propto}$, was attributed to different boundary layer profiles and displacement thickness. Examination of surface wall-friction for both TCMs at 2° pitch did not show any flow separation from the hull; thereby discounting separation effects as the cause of reduced normal force.

The cumulative summation, from the bow to the stern, of the pressure force in the z-direction acting on the unappended Model 4621 hull (Figure 8) indicated that both TCMs gave similar force predictions over the forward section of the hull. However, the k-ω SST TCM predicted a greater positive normal pressure force over the aft section of the hull, leading to an overall smaller normal force for the unappended hull. This indicates differences between the two TCMs prediction of boundary layer growth in the adverse pressure gradient at the stern.

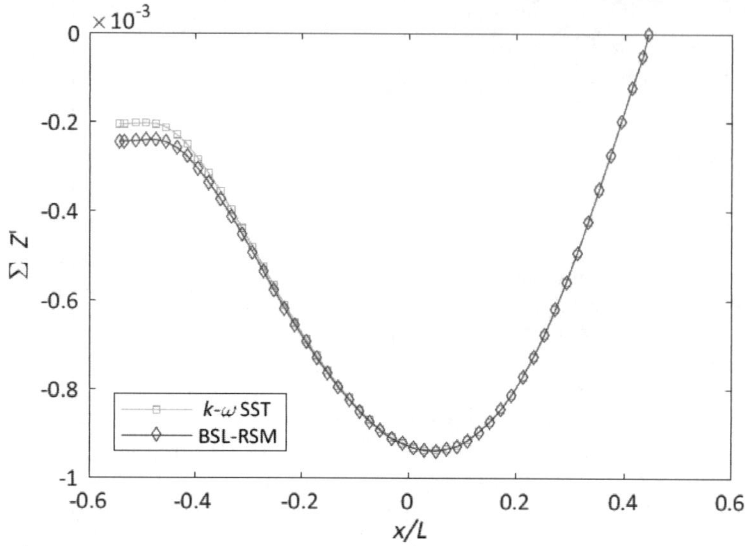

Figure 8: Cumulative summation (bow to stern) of the z-direction pressure force acting on the unappended Model 4621 hull.

The different prediction of the adverse pressure gradient was assumed to result in a different displacement thickness, and hence a different effective unappended hull shape. The displacement thickness, δ_1, was calculated as,

$$\delta_1 = \int_{s=0}^{s_e} \left(1 - \frac{u_p(s)}{u_i}\right) ds \quad (8)$$

where u_p is the velocity at a distance, s, normal to the wall and projected in the direction of u_i and u_i is the velocity of the inviscid flow past the unappended hull, which was determined from a separate inviscid simulation. The integral was computed from the wall at s = 0 to the edge of the boundary layer, s_e, which was defined as the distance where u_p was equal to 99% of u_i.

The azimuthal displacement thickness was calculated for both TCMs at x/L = -0.454 and is shown in Figure 9. The k-ω SST TCM predicted a thinner δ_1 over the high pressure side of the hull until the upper octant where it predicted a thicker δ_1. The net effect of this difference for the k-ω SST TCM was to create an effective hull shape with increased reflexive camber at the stern, relative to that from the BSL-RSM TCM. This effective hull shape resulted in reduced normal force and consequently increased $C_{z,\propto}$, compared to both Dempsey [2] and the BSL-RSM TCM.

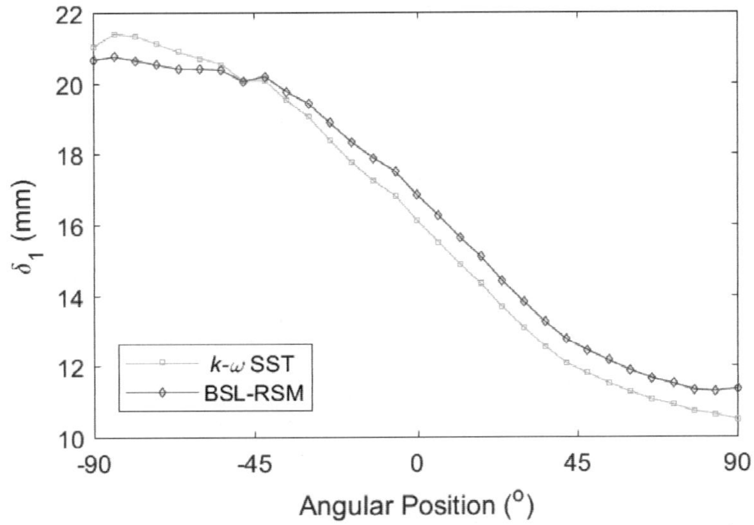

Figure 9: Azimuthal variation of the displacement thickness at x/L = -0.454 of the unappended Model 4621 hull for k-ω SST & BSL-RSM TCM.

Lift Curve Slope and Efficiency

The lift curve slope and ACS efficiency for both TCMs are shown in Table 3 & Table 4 respectively. BSL-RSM TCM accurately predicted the value of $C_{Z,\alpha}$ for the bare hull to within 3% of Dempsey's result. The BSL-RSM TCM also provided an accurate prediction of lift curve slope for the 3B and 11B geometries with the relative difference to Dempsey's experimental results less than 5% in both instances.

Table 3: Lift curve slope (rad^{-1}) for the unappended and appended Model 4621 hullform.

Configuration	Dempsey [2]	k-ω SST	BSL-RSM
Unappended	-0.008021	-0.006743	-0.007802
3B	-0.01518	-0.01496	-0.01569
11B	-0.01862	-0.01937	-0.01955

Table 4: ACS efficiency for the appended Model 4621 hullform.

Configuration	Dempsey [2]	k-ω SST	BSL-RSM
3B	50%	57%	55%
11B	47%	57%	53%

$C_{Z,\alpha}$ for the 3B and 11B configurations are plotted as functions of the ACS aspect ratio in Figure 10. Results for the BSL-RSM TCM are plotted together with $C_{L,\alpha}$ for an isolated ACS using the Whicker and Fehlner relationship in equation (3), $C_{Z,\alpha}$ assuming an ACS efficiency as predicted by Lyons and Bisgood [1], and Dempsey's [2] B span data points and proposed fit. The BSL-RSM results are in general agreement with Dempsey's fit, with the CFD predicting a slightly greater ACS efficiency relative to the experimental data.

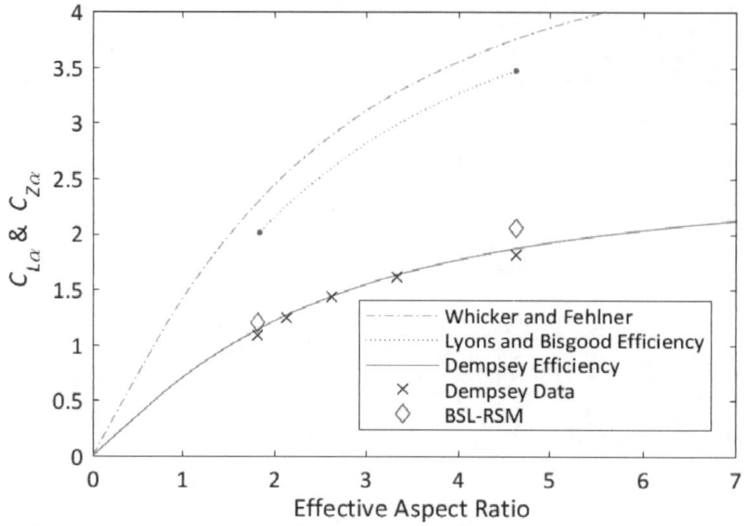

Figure 10: $C_{Z,\alpha}$ for 3B & 11B appended Model 4621 hull.

The BSL-RSM results, whilst not in absolute agreement with the proposed fit of Dempsey [2] are more closely aligned with Dempsey [2] as opposed to the equivalent prediction of Lyons and Bisgood [1]. For the 3B and 11B configurations , the Lyons and Bisgood [1] method predicts an ACS efficiency of 87% to 92% respectively as shown in Figure 10. The results of the CFD simulations provide strong evidence for the validity of the Dempsey [2] experiment and that the Dempsey [2] efficiency prediction is suitable for use at model scale.

Furthermore, solutions to the RANS equations assume a turbulent flow field. This assumption, together with the CFD results being in strong agreement with the experimental results for ACS efficiency suggests that any underestimation of tailplane efficiency is not attributable to laminar flow.

Reynolds Number Dependence

To investigate Reynolds number dependency, the flow over the 3B and 11B geometries was simulated at increased inlet velocities to simulate higher Reynolds numbers. The Reynolds numbers simulated encompassed the range that could be expected between model scale testing and a full scale submarine.

The results of this analysis are shown in Figure 12. In addition to the CFD model data, equivalent simulations for a geometry based on the DST generic BB2 submarine geometry [9] have been included. The BB2 hull has an L/D of 7.31 and an ACS aspect ratio of 4. The BB2 geometry was modified to match the configuration of Dempsey [2], i.e. the ACS layout was changed from an X-plane configuration to a cruciform arrangement, and the vertical rudders, sail, and casing were removed. The BB2 ACSs were designed within the hull bounding box and so rotation of the ACS by 45° to the cruciform configuration results in a larger span relative Dempsey's B span. This modified geometry is herein referred to as "mod-BB2" and is shown in Figure 11. Results for the mod-BB2 geometry were computed using a similar grid and solution methodology to the Dempsey simulations presented.

Figure 11: Plan and elevation views of the mod-BB2 geometry.

It is clear from Figure 12 that ACS efficiency is strongly influenced by Reynolds number. The ACS efficiency of both the Dempsey [2] and mod-BB2 geometries increased with Reynolds number. The mod-BB2 ACS efficiency increased more rapidly and attained a higher efficiency at full scale Reynolds numbers. The more rapid increase in ACS efficiency observed for the mod-BB2 might be due to the steeper stern half angle (16° for mod-BB2) relative to that of the Dempsey [2] (6.5°) geometry. The more rapid increase in efficiency might also be due to the increased span of the mod-BB2 geometry relative to Dempsey's B span.

ACS efficiency increases continuously with Reynolds number and critical/sub-critical behaviour was not observed. ACS efficiency appears to be independent of aspect ratio at higher Reynolds as both the 3B and 11B geometries converged to similar efficiencies at full scale Reynolds numbers, although this requires further simulations to verify.

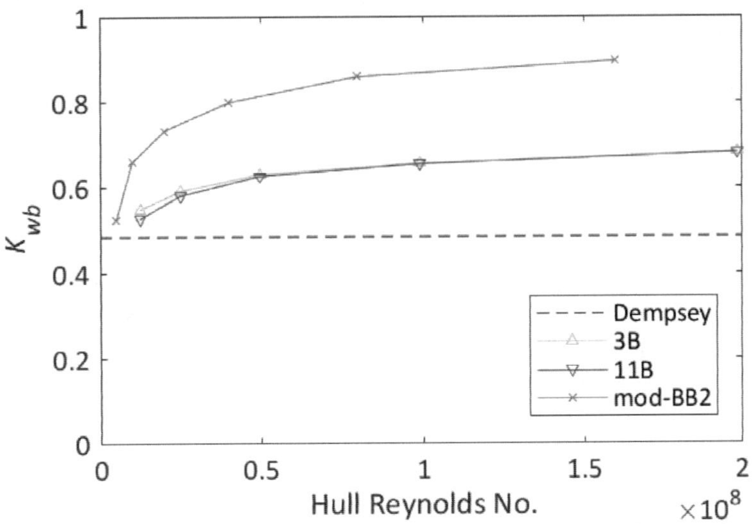

Figure 12: ACS efficiency variation with Reynolds Number for the appended Model 4621 hull and mod-BB2.

Mackay, Watt and Bohlmann [4] suggested that the increased efficiency as a function of Reynolds number is due to the effect of Reynolds number on the boundary layer thickness. The boundary layer thickness decreases with Reynolds numbers. At full scale the boundary layer at the ACS would be relatively thinner than at model scale, with an equivalent decrease in the momentum thickness. The reduced momentum thickness would result in a smaller proportion of the ACS span being influenced by the reduced velocity, low energy boundary layer and is therefore able to produce greater normal force. To confirm this, further investigation of the change in momentum thickness with Reynolds number is required.

CONCLUDING REMARKS

RANS CFD was used to numerically model the experiments of Dempsey [2]. Good agreement between the CFD model and experimental data was observed for the two different ACS geometries studied.

Both the k-ω SST and BSL-RSM TCMs were suitable for prediction of absolute normal loads for appended axisymmetric bodies. However the k-ω SST was found to be unsuitable for the prediction of loads for the unappended axisymmetric body at low (2°) pitch angles and is therefore unsuitable for prediction of ACS efficiency. This limitation was attributed to the k-ω SST TCM's different prediction of the adverse pressure gradient in the aft portion of the hull and consequently a different displacement thickness and equivalent hull shape.

The use of RANS CFD, which assumes fully turbulent flow, indicates that the reduced ACS efficiency of Dempsey's [2] experiments is not due to laminar boundary layer effects. Clear critical/subcritical ACS efficiency was not seen. However significant Reynolds number dependence was observed between model scale and full scale. It was suggested that this dependency was the result of the smaller boundary layer thickness at higher Reynolds numbers and therefore a greater proportion of the ACS span being subjected to the higher freestream velocity resulting in a greater normal force.

The CFD results confirms the suitability of using Dempsey's [2] data for predicting ACS efficiency at model scale Reynolds numbers and for geometries with relatively shallow stern angles. It highlights some of the limitations of empirical methods for calculating hydrodynamic coefficients and reinforces the importance of Reynolds number and geometry equivalence when applying empirical methodologies.

REFERENCES

[1] D. J. Lyons and P. L. Bisgood, "An Analysis of the Lift Slope of Aerofoils of Small Aspect Ratio, Including Fins, with Design Charts for Aerofoils and Control Surfaces," 1950.

[2] E. M. Dempsey, "Static Stability Characteristics of a Systematic Series of Stern Control Surfaces on a Body of Revolution," Bethesda, 1977.

[3] M. C. Bettle, "Validating Design Methods for Sizing Submarine Tailfins," in *Warship 2014: Naval Submarines & UUV's*, 2014.

[4] M. Mackay, G. D. Watt, and H.-J. Bohlmann, "Modeling Submarine Tailplane Efficiency," in *RTO SCI Symposium*, 2002.

[5] L. Whicker and L. Fehlner, "Free-Stream Characteristics of a Family of Low-Aspect Ratio, All-Movable Control Surfaces for Applications to Ship Design," Washington D.C., 1958.

[6] M. Renilson, *Submarine Hydrodynamics*, 2nd ed. Springer International Publishing, 2018.

[7] C. G. Armstrong, H. J. Fogg, C. M. Tierney, and T. T. Robinson, "Common themes in multi-block structured quad/hex mesh generation," in *24th International Meshing*

Roundtable (IMR24), 2015.

[8] A. B. Phillips, S. R. Turnock, and M. Furlong, "Influence of turbulence closure models on the vortical flow field around a submarine body undergoing steady drift," *J. Mar. Sci. Technol.*, vol. 15, no. 3, pp. 201–217, 2010.

[9] B. Overpelt, B. Nienhuis, and B. Anderson, "Free Running Manoeuvring Model Tests On A Modern Generic SSK Class Submarine (BB2)," in *Pacific International Maritime Conference*, 2015.

Identification of a Submarine Fin-tip Vortex Using POD Analysis

Y. Xue[1, 2], C. Kumar[1], S.K. Lee[1], M. Giacobello[1], and P. Manovski[1]

[1] Defence Science & Technology Group, Victoria 3207, Australia

[2] Australian Maritime College, University of Tasmania, Tasmania 7248, Australia

Yunpeng.xue.2016@gmail.com

Abstract

The meandering nature of the fin-tip vortex generated by a manoeuvring submarine can lead to a smearing of the vortex in the ensemble-averaged flow field. From stereoscopic particle image velocimetry (SPIV) measurements, it is possible to remove the smearing by shifting each instantaneous velocity field so as to produce a common centre for the vortex. However, it is difficult to identify the instantaneous vortex centre due to the presence of measurement noise and strong small-scale turbulent fluctuations. In this paper, a snapshot Proper Orthogonal Decomposition (POD) technique is used to capture the dominant large-scale coherent structures (from inspection of eigenvalue or energy distributions) to improve vortex identification and subsequent correction for meandering. The present findings suggest that, for the flow fields reconstructed from POD and subsequently corrected for meandering, the vortex flow exhibits steeper velocity gradients and is less noisy than the original results (inspection shows the reduction in turbulent fluctuation is as much as 16%).

Introduction

The terms vortex core meandering, vortex core wandering or vortex core precession are used to describe the instability of a swirling flow which is a time-dependent flow pattern, where the core of the system oscillates around the rotational axis. This phenomenon of the vortex instability exists in both open flow, such as wing-tip vortex wandering [1], and confined flow, such as vortex core precession in vortex tube [2] and vortex flow solar reactor [3]. The behaviour of vortex meandering/wandering or precession is of general interest because it can affect the transport of momentum and energy of the working fluid. For this paper, the interest is particularly on the meandering of an open-flow vortex trailing from the bridge fin-tip of a generic conventional submarine [4, 5] using different vortex identification criteria. The bridge fin-tip vortex is a major flow feature produced on the upper hull of the submarine during a pitch or yaw manoeuvre, and it can impact on the acoustic signature. Clarification of the meandering effect is important in the validation of the high-fidelity computational models.

The submarine geometry under consideration is known as the 'BB2' [5], which consists of an axisymmetric body with a casing, a bridge fin and control surfaces. Recently, a model scale of this geometry was built to allow stereoscopic particle image velocimetry (SPIV) testing of the model at 10° yaw in the Low-Speed Wind Tunnel at Defence Science and Technology (DST) Group [5]. The model has a length (L) of 2 m and a length-to-diameter ratio of 7.3. On the casing, the shape of the bridge fin is that of NACA-0022 with a height of 8%L and a chord length of 15.7%L. A nominal free stream velocity of 29 m/s was set resulting in a model-length Reynolds number of 4×10^6 (chord-length Reynolds number of 0.63×10^6). SPIV measurements have been conducted to obtain the 3-D velocity profiles at three selected measurement planes along with the model (51.1%L, 65.0%L and 81.5%L); see Figure 1. For further details on the submarine geometry, wind tunnel configuration, SPIV setup and parameters, and data processing, see [5].

The instantaneous SPIV measurements allow the possibility of shifting the image sequence to a common vortex core centre and effectively eliminating the meandering effect from the ensemble average and the flow statistics, which are important to characterise the flow field and hence to identify the impacts of the fin-tip vortex meandering.

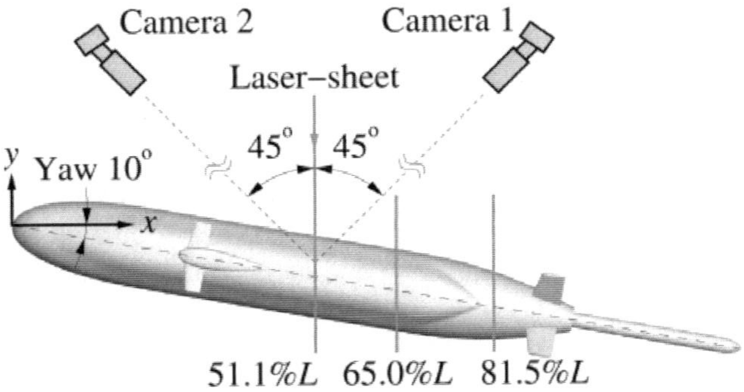

Figure 1. The BB2 generic conventional submarine model at 10° yaw and the SPIV measurement planes.

Proper Orthogonal Decomposition and Reconstruction

The objective is achieved by using a snapshot Proper Orthogonal Decomposition (POD) technique to capture the dominant large-scale coherent structure to improve vortex identification which then allows correction for meandering. The left image of Figure 2 is an instantaneous velocity field of the bridge fin-tip vortex at 65%*L,* which is the only measurement plane analysed in the current work. The noise or incoherent small-scale turbulent structures presented in the figure highlights the difficulty in the identification of the dominating turbulent structure and the vortex core. POD has been found to be effective in decomposing complex turbulent flow into a set of modes that represents the dominating flow structures [1, 6, 7].

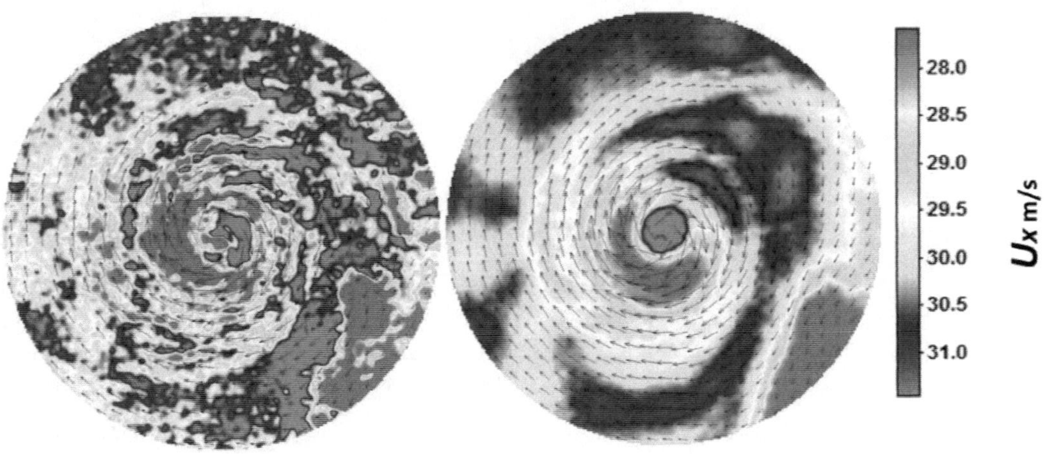

Figure 2. Left: an instantaneous velocity field of the bridge fin-tip vortex at 65% of the model length. Right: reconstruction of the same instantaneous velocity field based on the first three modes from the POD analysis. The contour levels represent the out-of-plane streamwise velocity (U_X).

The elementary idea of POD is to define a set of orthogonal functions (eigenmodes) with coefficients representing the flow field based on an energy-weighted calculation (eigenvalue or energy level). It is subsequently possible to identify the dominating large-scale coherent structures that contribute the most energy to the flow. Reconstruction of the velocity vector field using only the most energetic modes enables the dominant structures of the flow to be more clearly captured and improves the identification of the vortex core. Mode information of the snapshot POD and its mathematical process can be found in previous publications [1, 6, 7].

Instead of the POD analysis being applied to the whole velocity field of each measurement plane as presented in [5], the bridge fin-tip vortex (which includes the vortex and some wake component) and the vortex core are selected for accurate POD analysis. The centre of the bridge fin-tip vortex has been identified by the minimum the local out-of-plane vorticity (ω_x) with the normalized centre coordinates given in [5]. The fin-tip vortex and the vortex core are defined by the swirl velocity distribution as presented in Figure 3. A 30% of the maximum swirl velocity is selected to separate the fin-tip vortex ($D_{30\%Uy,\ max}$) from the whole velocity field and the forced vortex part within the central region is defined as the vortex core ($D_{vortex\ core}$). The diameter of the vortex core is about 0.01*L*, which is in a good agreement with the turbulent intensity defined vortex core in [5].

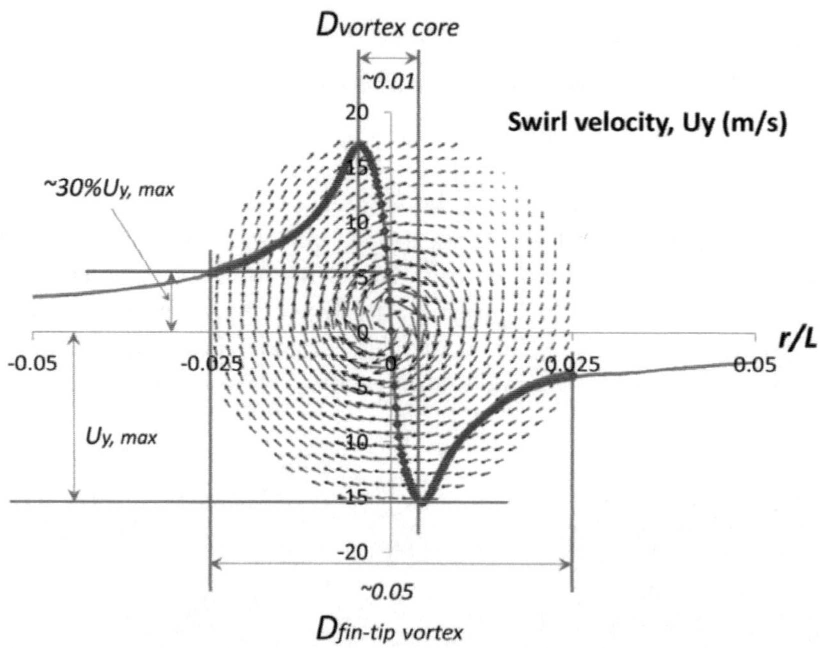

Figure 3. Swirl velocity distribution across the bridge fin-tip vortex and the selected views for POD analysis.

To perform the POD analysis efficiently, 1000 snapshots out of the available 3000 instantaneous velocity fields were used. Figure 4 presents the contributions from the POD modes to the total turbulent kinetic energy (energy level) in three configurations, i.e., 1000 snapshots of the bridge fin-tip vortex, 3000 snapshots of the bridge fin-tip vortex and 1000 snapshots of the vortex core. It can be seen from the figure that the lower modes contain most of the energy, while the contributions from the higher modes (≥5) are negligible for both the fin-tip vortex and the vortex core.

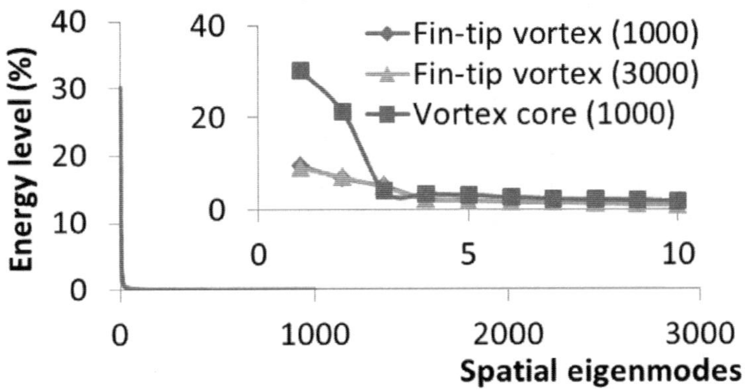

Figure 4. Contributions from the POD eigenmodes to the total energy for different cases with a particular focus on the first ten modes (1000 snapshots of the bridge fin-tip vortex, 3000 snapshots of the bridge fin-tip vortex and 1000 snapshots of the vortex core).

There is a maximum of 3% difference of the eigenvalues between the 1000 and the 3000 snapshots cases. This confirms the accuracy and acceptance of the POD analysis using 1000 snapshots of the vector fields instead of the 3000 vector fields. All the other POD analysis in this current work is based on the 1000 snapshots. Furthermore, the eigenvalues of the first three modes of the fin-tip vortex, i.e., 9.5%, 6.8%, and 5.1%, respectively, indicate the existence of dominating large-scale coherent turbulent structures compared to the remaining eigenmodes. The greater eigenvalues of the vortex core (30.3% and 21.1%) show more significant dominating large-scale coherent turbulent structures with the first two modes containing about 51.4% of the total turbulent kinetic energy (TKE). The difference of the eigenvalue distribution is due to the different regions selected for the POD analysis. For the selected bridge fin-tip vortex region, some small scale turbulent structures in the fin wake are included, while it is pure vortex flow in the selected vortex core region.

An instantaneous velocity field of the vortex core, the first three POD modes of the selected velocity vector and the reconstruction of the flow field based on the first three modes are presented in Figure 5. The two most energetic modes mainly consist of two counter-rotating vortex pair and owing to the symmetry along the axis of the mean flow; mode 2 is obtained from mode 1 by a 90 degree rotation. This is in good agreement with the POD analysis of a vortex flow in a cylinder duct [8] and the POD analysis of a wing-tip vortex in an open field [9]. Based on the most energetic modes, the velocity field can be reconstructed as shown in the figure which shows a clearer view of the velocity field. Figure 2 (right) shows the reconstruction of the same instantaneous velocity field (left) based on the first three modes from the POD analysis. The same improvement of the selected view is also seen from the top two images in Figure 5. Comparing with the original image, the small-scale fluctuations have been removed providing a much clearer view of the vector field, which will enable accurate identification of the dominating flow structure and location of the vortex centre.

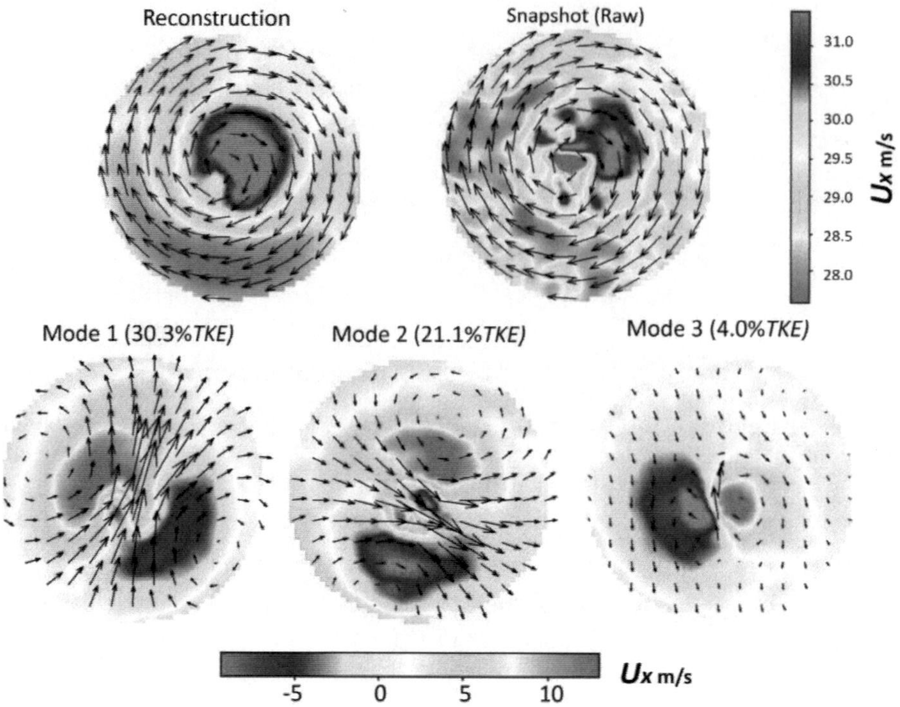

Figure 5. Top: an instantaneous velocity field of the vortex core and the reconstruction of the flow field based on the first three modes. Bottom: first three POD modes with their eigenvalues/energy level.

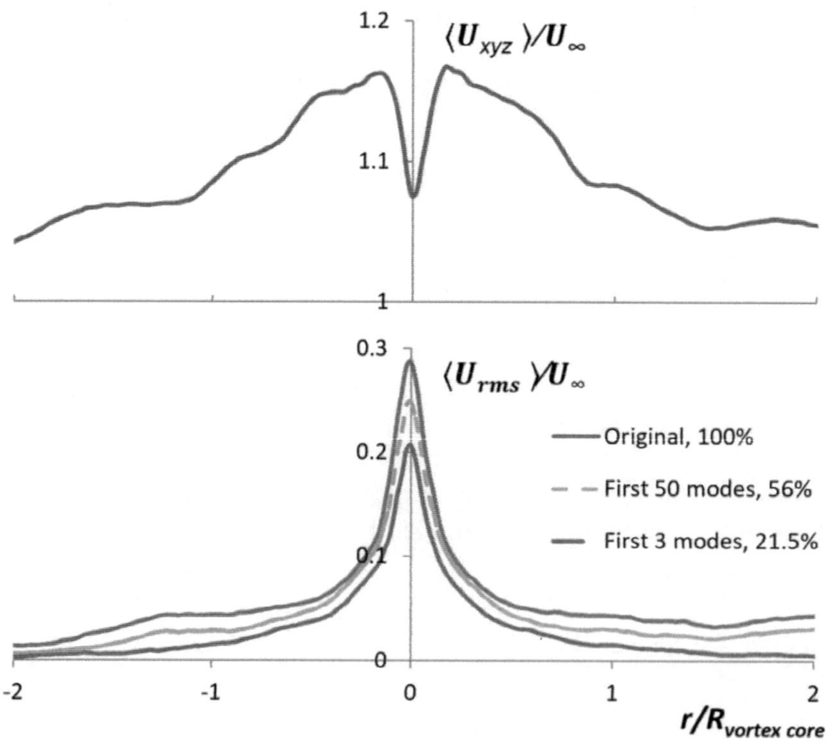

Figure 6. The normalized resultant velocity and the turbulence intensity of the original raw data, the POD approximation (first 50 modes, 56% of total turbulent energy) and the POD approximation (first 3 modes, 21.5% of total turbulent energy at 65%*L*.

Before further analysis of the bridge fin-tip vortex, assessment of the POD results is conducted. Figure 6 presents the normalized resultant velocity and fluctuating components (fin-tip vortex) averaged using the original velocity field, the approximated velocity field based on the first 50 modes of the POD analysis and the approximation based on the first three modes. It is clearly seen that the reconstructed average velocity field has no change from the original value and is independent of the modes used in the approximation. While, the significant impact of the POD approximation on smoothing the velocity field as shown in Figure 2 and Figure 5 is achieved by removing the small scale turbulent structure. Of course, the original velocity field has 100% of the turbulent energy, and the approximations based on the first 50 and 3 modes contain 56% and 21.5% of the total turbulent energy, respectively. With a decrease of the modes used in the reconstruction, the turbulence intensity drops significantly. The more modes used, the more turbulent energy will be kept and the more accurate approximation will be obtained, but the identification of the dominating flow structure will be more difficult. The characteristics of POD is that it does not change the average flow field but only impacts the turbulent components and therefore makes it a good pre-processing technique for further analysis.

Vortex core meandering correction

As mentioned above, the minimum local out-of-plane vorticity (obtained using a second order central finite difference scheme) is used to locate the vortex centre in the current work and in [5]. The 1000 snapshots of the vortex core defined in Figure 3 are used to analyse the meandering of the fin-tip vortex. A probability density function (PDF) of the instantaneous vortex centre of the 1000 selected raw and POD-corrected velocity fields are plotted in Figure 9.

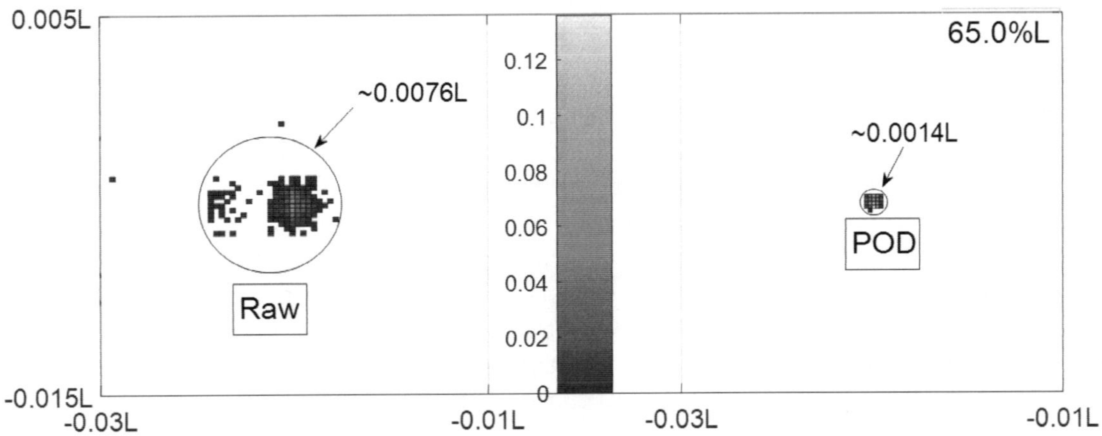

Figure 7. Probability density function (PDF) of the instantaneous vortex centre of the raw and POD-corrected velocity vectors identified by the minimum vorticity at three axial locations along the model.

The relative location of the instantaneous vortex centre can be seen from the figure and its probability density is indicated by the colour bar. The actual location of the vortex centre from the ensemble average velocity field is presented in [5]. A significant decrease of the vortex meandering area can be seen after each image sequence is shifted to a common vortex centre using POD (diameter of the estimated meandering area is indicated in the figure). The concentrated vortex centres found in the POD approximated velocity fields indicate the area that meandering occurs. The vortex centre that locates outside of the main cluster of the raw data should be considered as false results caused by the noise in the instantaneous velocity vectors. The secondary meandering centre of the vortex core that is indicated by the smaller cluster of the concentrated vortex centre away from the main region should also be considered as a result of the noise or small scale turbulent structure.

With the identified centre of the vortex core, the instantaneous velocity vectors are shifted to be coincident with the vortex centre in each vector field. In such a way, the effects of the vortex meandering on the flow field are removed and the meandering corrected velocity vectors are compared with the original results. The ensemble averaged resultant velocity (U_{xyz}) and fluctuation components (U_{rms}) for four cases are normalized using the free stream velocity and presented in Figure 8.

The 'Raw' and 'POD (3 modes)' are the raw data and the POD approximation using the first three modes. The 'Raw corrected' refers to the raw data corrected based on the vortex centre identified by the raw velocity field. The 'POD & Meandering corrected (PM)' refers to the meandering corrected using POD approximation. The 'Raw & Meandering corrected using POD (RM)' is the raw data processed in the following steps:

1. Use the vorticity to identify the vortex centre of the POD reconstructed velocity fields (first three modes);
2. Shift the raw velocity vector fields using these centre coordinates to remove the meandering effect;
3. Obtain the normalized ensemble-averaged velocity and turbulence intensity of the meandering corrected raw data.

The three meandering corrected resultant velocity results (Raw corrected, PM and RM) show an increased maximum velocity (~1%) at about $|r/L|=1.556\times10^{-3}$ and decreased velocity at the centre of the vortex (~3.7%). This sharpening and narrowing effect of the meandering corrected resultant velocities comparing to the raw and POD corrected cases is a result of the elimination of the meandering effect.

Comparing to the raw data, the normalized turbulence intensity ($\langle U_{rms}\rangle/U_\infty$) of the POD approximation shows an even drop across the vortex core, which is caused by removing the small-scale incoherent turbulent structure and measurement noise. Further removal of the meandering effect from the POD approximation (RM) results in a significant weakening of the fluctuation. The decrease of the flow fluctuation from the raw to the POD approximated and then to the POD and meandering corrected data implies the turbulence of the original vortex core mainly consists of two components, i.e., the incoherent turbulent and the fluctuation induced by the meandering. The maximum fluctuation in the PM case is about 0.1 and only consists of the fluctuation from the first three dominating modes. All the small scale turbulence and meandering induced turbulence are removed. The 'Raw and Meandering corrected using POD' case also has a significant drop in the central region (16.1%) and this decrease of the fluctuation is getting smaller in the outer region of the vortex core. The peak fluctuating core region is defined as $\langle U_{rms}\rangle/U_\infty \leq 0.1$. A significant decrease of this region is observed after the correction of meandering, i.e., from 0.44 $_{Dvortex\ core}$ to 0.26 D $_{Dvortex\ core}$. While, the 'Raw corrected' method shows a higher fluctuation in most part of the vortex core and this increased fluctuation is observed in all configurations. A bimodal peak in the central region is also observed here, which is likely a result of the false centre due to small scale turbulence, measurement noise and 'holes'.

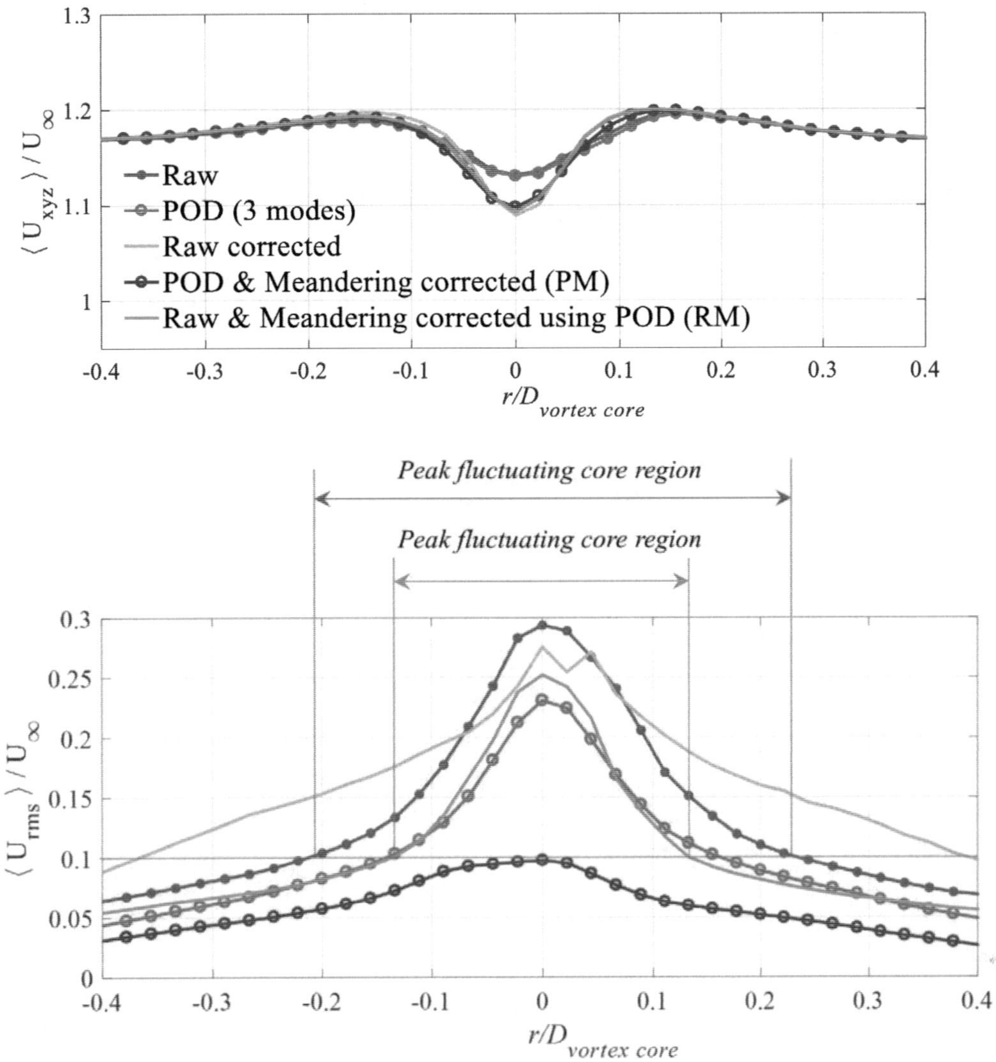

Figure 8. The normalized resultant velocity and fluctuation of Raw, POD approximated, raw corrected, POD & Meandering corrected and Raw & Meandering corrected using POD results at 65.0%L.

Several other local vortex identification criteria are also used in the current work, including the out-of-plane velocity (U_x), the Q criterion, Lambda-2 criterion (λ_2) and Gamma-1 criterion (Γ_1). The detail definitions of these criteria can be found in [11-13]. The normalized resultant velocity and fluctuation of raw data and meandering corrected results using different vortex identification criteria are presented in Figure 9. The vortex meandering is corrected with the vortex centre identified from POD processed data. It can be concluded from the plot of resultant velocity (U_{xyz}) that all criteria give similar results.

The bottom plot of Figure 9 is the meandering corrected fluctuation using different criteria with the vortex centre identified by the POD reconstructed velocity vector fields. It highlights a good agreement of the fluctuations corrected using different criteria, while the slight difference of the Γ-1 criterion is due to the different identified vortex centre. This means, for noisy or highly turbulent data, a POD reconstruction is necessary for the identification of the vortex centre and all gradient-based vortex identification criteria tested here can result in accurate identification of the vortex centre based on the POD reconstructed velocity fields.

The decrease of the fluctuating component after correcting the meandering confirms the contribution of the vortex meandering to the total fluctuation.

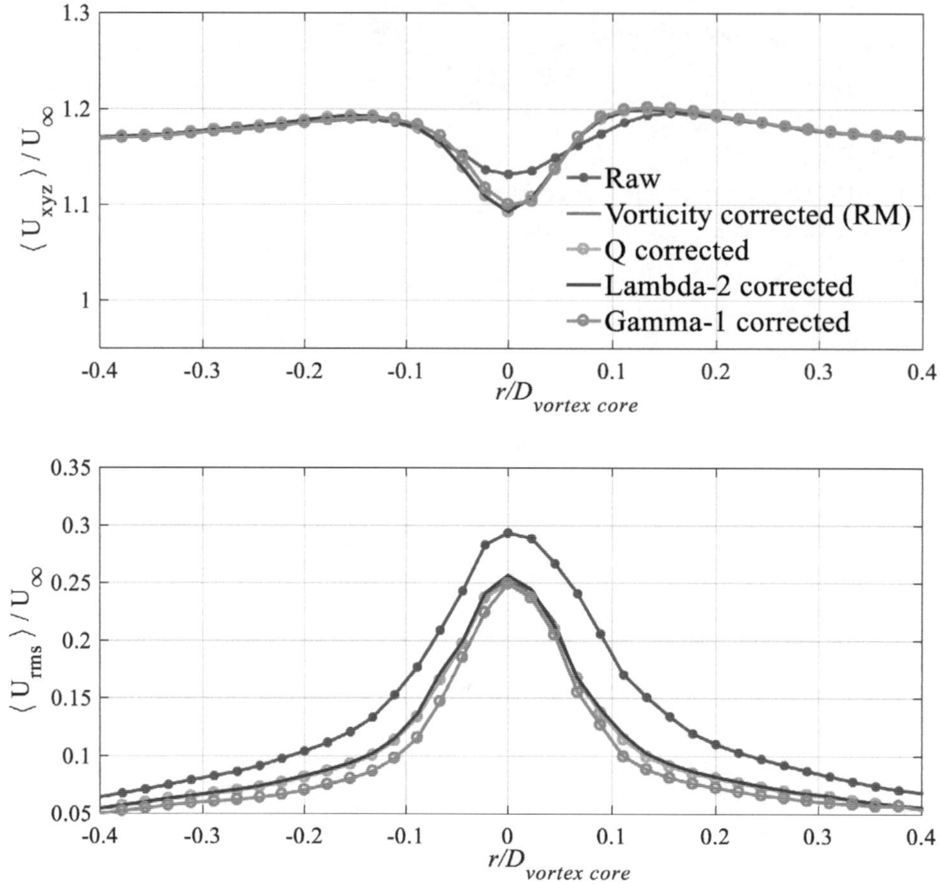

Figure 9. The normalized resultant velocity and fluctuation of raw data and meandering corrected results using different vortex identification criteria at 65.0%L.

Conclusion and recommendation

This paper presented a study of a submarine bridge fin-tip vortex using recently acquired SPIV measurements [5] from the DST Low-Speed Wind Tunnel. The vortex flow is analysed using a snapshot POD technique to establish the dominant large-scale turbulent structure. Several local vortex identification criteria were used to identify the vortex centre in the raw vortex field and POD reconstructed velocity field.

It is found the POD reconstruction of the velocity field can significantly improve the accuracy of the vortex centre identification, especially with the PIV measurement result that contains measurement noise, small scale turbulence, and missing data in the core region. For the POD reconstructed velocity vector fields, all criteria can generate a cleaner view to identify the vortex centre and the flow characteristics after meandering corrections have a good agreement. It is found that the vortex meandering has little impact on the ensemble-averaged resultant velocity with a maximum difference of 3.7%, but a significant influence on the fluctuating component with a reduction in the central region up to 16.1%. The meandering correction also leads to a significant decrease of the peak fluctuating core region.

Overall, this work serves as an initial step towards understanding the meandering nature of the fin-tip vortex. Another possibility is to extend the work to include time-resolved measurements to identify the frequency of vortex meandering. To obtain good measurement results in the core region of a gaseous vortex flow, neutrally buoyant tracer, such as helium-filled soap bubble, is recommended in the optical measurement.

Acknowledgement

The authors acknowledge the support and funding of DST's Maritime Division as well as the support of the Research Training Centre of Naval Design and Manufacturing (RTCNDM). The RTCNDM is a University-Industry partnership established under the Australian Research Council (ARC) Industry Transformation grant scheme (ARC IC140100003).

Reference

[1] A. Chatterjee, "An introduction to the proper orthogonal decomposition," Current Science, Article vol. 78, no. 7, pp. 808-817, 2000.

[2] X. Guo and B. Zhang, "Computational investigation of precessing vortex breakdown and energy separation in a Ranque–Hilsch vortex tube," International Journal of Refrigeration, Article vol. 85, pp. 42-57, 2018.

[3] A. Chinnici, Y. Xue, T. C. W. Lau, M. Arjomandi, and G. J. Nathan, "Experimental and numerical investigation of the flow characteristics within a Solar Expanding-Vortex Particle Receiver-Reactor," Solar Energy, vol. 141, pp. 25-37, 1/1/ 2017.

[4] C. Fureby et al., "Experimental and numerical study of a generic conventional submarine at 10° yaw," Ocean Engineering, vol. 116, no. Supplement C, pp. 1-20, 2016/04/01/ 2016.

[5] S. K. Lee, P. Manovski, and C. Kumar, "Wake of a DST Submarine Model captured by Stereoscopic Particle Image Velocimetry," in 21th Australasian Fluid Mechanics Conference, Adelaide, Australia, 2018.

[6] G. Berkooz, P. Holmes, and J. L. Lumley, "The proper orthogonal decomposition in the analysis of turbulent flows," in Annual Review of Fluid Mechanics vol. 25, ed, 1993, pp. 539-575.

[7] L. Sirovich, "Turbulence and the dynamics of coherent structures. I. Coherent structures," Quarterly of applied mathematics, vol. 45, no. 3, pp. 561-571, 1987.

[8] L. Graftieaux, M. Michard, and N. Grosjean, "Combining PIV, POD and vortex identification algorithms for the study of unsteady turbulent swirling flows," Measurement Science and technology, vol. 12, no. 9, p. 1422, 2001.

[9] A. Dahmouni, M. Oueslati, and S. B. Nasrallah, "Experimental Investigation of Tip Vortex Meandering in the Near Wake of a Horizontal-Axis Wind Turbine," Journal of Applied Fluid Mechanics, vol. 10, no. 6, pp. 1679-1688, 2017.

[10] A. M. Edstrand, T. B. Davis, P. J. Schmid, K. Taira, and L. N. Cattafesta, "On the mechanism of trailing vortex wandering," Journal of Fluid Mechanics, vol. 801, p. R1, 2016, Art. no. R1.

[11] J. Jeong, and F. Hussain, "On the identification of a vortex", Journal of fluid mechanics, 285: p. 69-94,1995.

[12] V. Kolář, "Vortex identification: New requirements and limitations," International journal of heat and fluid flow, 28(4): p. 638-652, 2007.

[13] Y. Zhang, et al., "A review of methods for vortex identification in hydroturbines," Renewable and Sustainable Energy Reviews, 2017. 81(1): p. 1269-1285.

Advanced Smart Escape Support System for Passenger Ship : Enhanced Intelligence with People Detection Technologies

James Choi, Ph.D., CORNERS Inc., jameschoi@corners.co.kr
Tony Kim, CORNERS Inc., tonykim@corners.co.kr
MinJee Son, CORNERS Inc., minjee@corners.co.kr
Chan-Su Yang, Ph.D., Korea Institute of Ocean Science & Tech., yangcs@kiost.ac

INTRODUCTION

SEGA (Smart Emergency Guiding Agent)-People is a kind of smart escape support system for passenger ships. The R&D project is sponsored by South Korea Ministry of Ocean and Fisheries starting from 2016 with 4 years of roadmap. SEGA-People comprises multiple active dynamic signage devices which communicate with real-time escape routing server system via LPWA (Low Power Wide Area) LoRa proprietary wireless network.

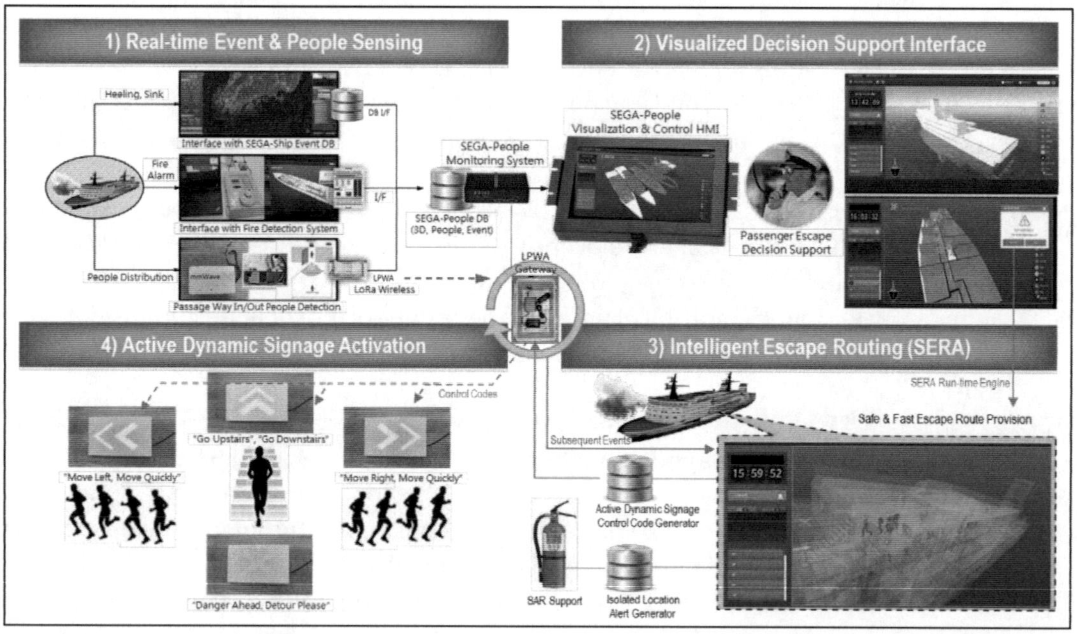

Fig. 1 SEGA-People Overview – Key Components and Closed-loop Control Process (SENSE-DECIDE-ACT)

Fig. 1 shows how SEGA-People works when escape or evacuation situation happens in a passenger ship such as abnormal heeling, fire breakout, and water leakage. SEGA-People can be regarded as a kind of closed loop control system which comprises continuous feedback cycle of "Real-time Sensing", "Precise Decision Making", and "Timely Actions" to minimize emergency response delay time and increase the escape routing efficiency.

Real-time Sensing

There can be options for gathering crisis event information from sensors – legacy sensor systems integration and new types of sensor network deployment. SEGA-People enables easy-to-install type of new IoT (Internet-of-Things) sensor network deployment by using LPWA LoRa wireless in-ship networking technology. Besides the hazardous location within a ship, SEGA-People can collect and utilize the people presence and congestion situation across the residential areas.

Precise Decision Making

SEGA-People automatically collects and analyzes sensor data streams to figure out eminent abnormality and people distribution snapshots across a ship. When SEGA-People detects an event, relevant data and information are visualized for captain's effective decision making whether escape of evacuation should be initiated or not.

Timely Actions

SEGA-People plays a critical role when escape or evacuation procedure initiated by captain. As soon as captain presses "Execute Escape!" button, SERA (Safe Escape Routing Algorithm) runtime engine generates a safe and fast escape route from every passenger location in a ship to designated emergency assembly stations just within a few seconds. Passengers can easily find their way by simply following the "Active Dynamic Signage – Voice & Arrow" which can be a promising alternative for LLL(Low-Location Lighting).

SEGA-PEOPLE TECHNICAL OPERATION PROCEDURE WITH ADVANCED FEATURES

SEGA-People can be implemented by integrating multiple hardware, software and networking resources. LoRa wireless network is used for collecting relevant events and people presence dynamics data. New type of active dynamic wireless signage device is developed. 3D data model of a ship is used to visualize events and to interact with crews. Run-time mathematical network problem solving algorithm engine generates routes within seconds. Visual escape simulation S/W is helpful for various crisis scenarios based training and escape drills.

1) Real-time Event Monitoring and People Presence/Congestion Detection

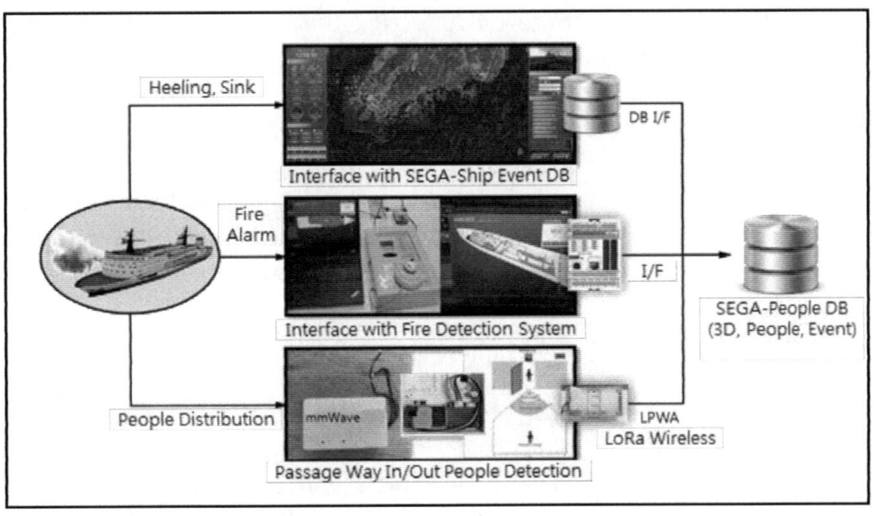

Fig. 2 SEGA-People Event Monitoring and People Detection Modules

In Fig. 2, SEGA-People receives abnormal ship status change events from SEGA-Ship, which is the name of neighborhood system to support captain's navigation decisions by integrating legacy ship navigation and management systems. One of SEGA-Ship's role is to identify ship stability changes caused by certain types of risk events and to deliver those events for SEGA-People. Fig. 3 shows how SEGA-People visualizes heeling status change.

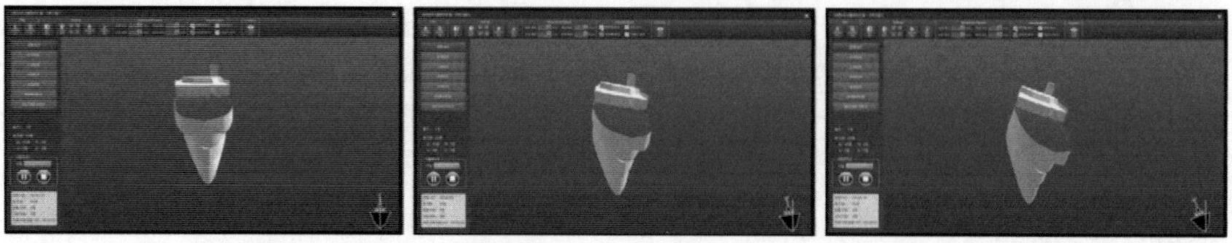

Fig. 3 Visualization of 3D Heeling Status Change Data with SEGA-Ship Interface

For the case of fire alarm system integration, SEGA-People maintains location information of every fire detectors installed within a ship. Each fire detector location is linked with more than one of nearby active dynamic signage devices. The interface with legacy fire alarm panel can be implemented via either TCP/IP software only communication API or general purpose interface device such as PLC (Programmable Logic Controller). Fig. 4 shows how fire alarm panel interface works.

Fig. 4 Test of Legacy Fire Alarm System Interface via PLC Device

The knowledge of people presence and congestion status is critical information in order to enhance the intelligence of SEGA-People. SERA (Safe Escape Routing Algorithm) can derive more effective escape route by reflecting the real-time people congestion data. SERA can identify isolated locations which should be notified by SEGA-People control center application. Now SEGA-People can notify not only the isolated locations but people presence data within those locations to increase SAR (Search and Rescue) effectiveness.

Multiple types of people detecting solutions and technologies have been tested. Fig. 5 shows the real-time CCTV video analysis type of solution to detect human shaped objects. As CNN (Convolutional Neural Network), a class of deep neural networks, is being used for variety of computer vision deep learning applications and CCTVs are everywhere, this solution is quite promising for detecting people in real-time. But, this solution requires high computing power for the purpose of real-time and simultaneous monitoring. Furthermore, the camera should be installed at least 2.5m above the bottom to perform properly. The other critical problem of generic CCTV is visibility issue, which means it is useless unless there are enough lighting – at night with no light or fire smoke filled areas. Due to the overlapping between people to people, the counting accuracy varied from 30%~79%.

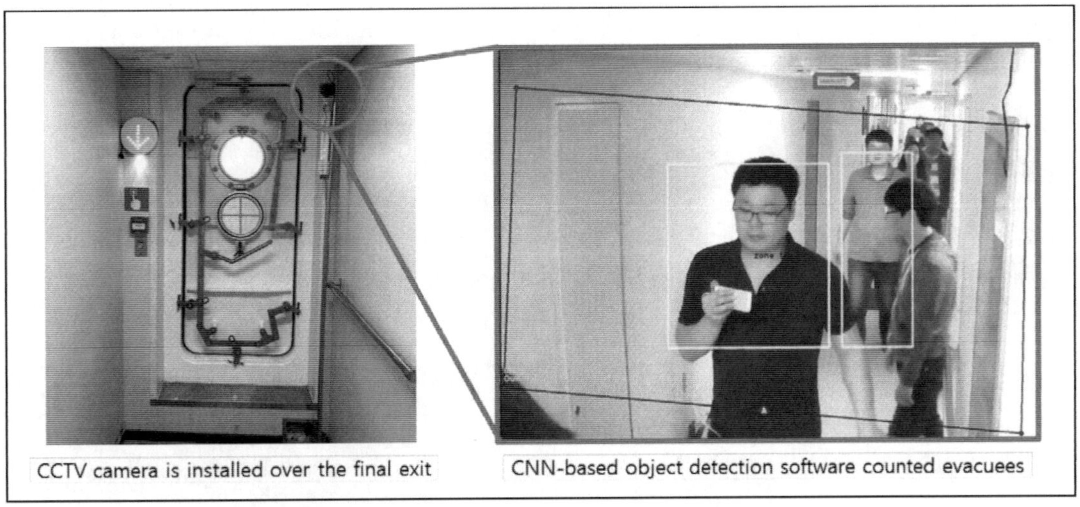

Fig. 5 People Detection Using Real-time CCTV Video Analysis Solution

CW (Continuous Wave) radar sensor was the second technology we tested. CW radar can detect the velocity and moving direction of an object. The price of CW radar chipset is relatively low and it is quite robust. Small size of the chipset is another advantage for developing a commercialized product. Combination of two CW radar sensors make it possible to determine In/Out direction when a person passes through a door. Fig. 6 shows the installed location of CW radar sensors and In/Out counting test. This type of solution can be effectively used when the door is narrow enough for a single person to enter. Otherwise the accuracy tends to be lower as the sensor cannot detect overlapped people passing through a wide open door.

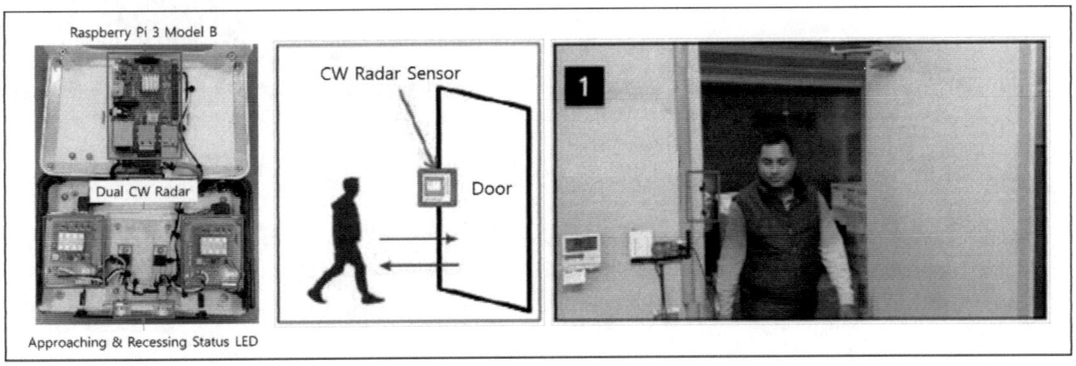

Fig. 6 Door-side In/Out People Detection Using CW Radar Sensors

CW radar sensor approach can be used to indirectly estimate people distribution across the ship. But, for the case of SAR activity to find people within isolated areas, there should be direct detecting solution – such as machine vision directly looking at people inside a room or area. To serve this purpose, we tested mmWave fmcw (frequency modulated continuous wave) radar sensor technology. This solution can be used for both door passing through people detection and people presence/congestion detection purposes. Fig. 7 shows test environment for the case of door looking type In/Out detection experiments. While the door side CW radar type is not suitable for a wide open gate, this 60GHz mmWave radar can detect multiple point cloud simultaneously – Max 32 point clouds with single radar.

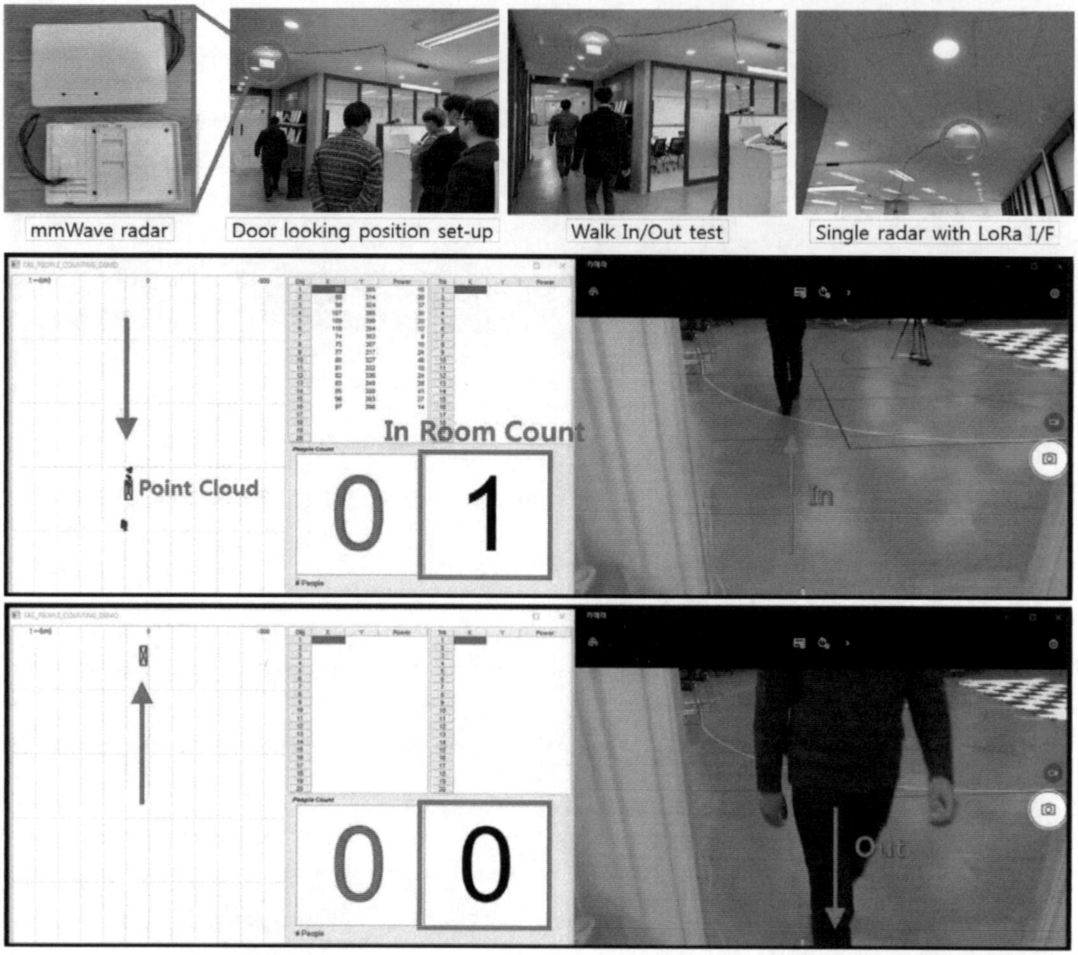

Fig. 7 Door looking In/Out People Point Cloud Detection Using Single mmWave Radar Sensor

Fig. 8 shows the second use case of mmWave radar sensor which detects people passing through a virtually bounded passage way area. Within the FoV (Field of View) of mmWave radar, we can pre-define areas of interest – just the point clouds within the area are detected and the signals reflected from outside of the area are discarded. Utilizing this type of detecting makes it possible to identify people congestion status in passage ways as well as people presence information. This congestion status data can be used by SERA real-time routing engine to determine the necessity of re-routing execution. For the case of a long passage way area, we can use multiple mmWave sensors to detect sequential area to area movement – each mmWave radar covers certain length of the passage way sequentially. This can also be effective for the case of cross roads type of locations – 'how many people came from where and went to which road?' The answer for this question results in people distribution map divided by the cross-road passage ways. This methods could be the cost effective way to derive maximum utility with minimum radar sensors.

Fig. 8 Passage Way Point Cloud Detection Using Single mmWave Radar Sensor

Fig. 9 shows the third use case of mmWave radar sensor which focuses on people presence knowledge for a bounded space areas such as room, hall, kitchen, etc. In contrary to the CCTV, mmWave radar has no issues concerning the privacy. mmWave radar can be installed inside of every room. Big rooms or area can be covered with multiple radars. Single mmWave radar sensor can cover 10m by 10m size of area to detect people presence status based on point clouds. (Max 32 point clouds – which depends on the design and performance of antennas)

Fig. 9 Room View Point Cloud Detection Using Single mmWave Radar Sensor

With these type of radar sensors, it is possible to acquire meaningful level of people presence and congestion data across the ship – door side type, door looking type, passage way looking type, and room view type can be used selectively considering the space characteristics. As current price of a single mmWave radar module is around 100 USD, and it seems to go down as the demand increases recently, it is quite promising cost effective technology with robust performance and low risk of adoption.

2) 3D-based Visualization UI for Captain and Crews' Decision Support

SEGA-People provides 3D ship model based visual user interface software which is able to display relevant alert and escape initiation command input interfaces to captain and crews. Fig. 10 shows the event monitoring and alarm visualization software, which supports quick and correct passenger escape decision by captain. 3D ship model can be used to visualize event alarms and escape execution status. It also shows current status of active dynamic signage devices. During the escape stage, it can display current directions each active dynamic signage guides and users can manually change status of specific locations from safe to danger or vice versa. This user interface software can be used for escape training mode or escape simulation mode also.

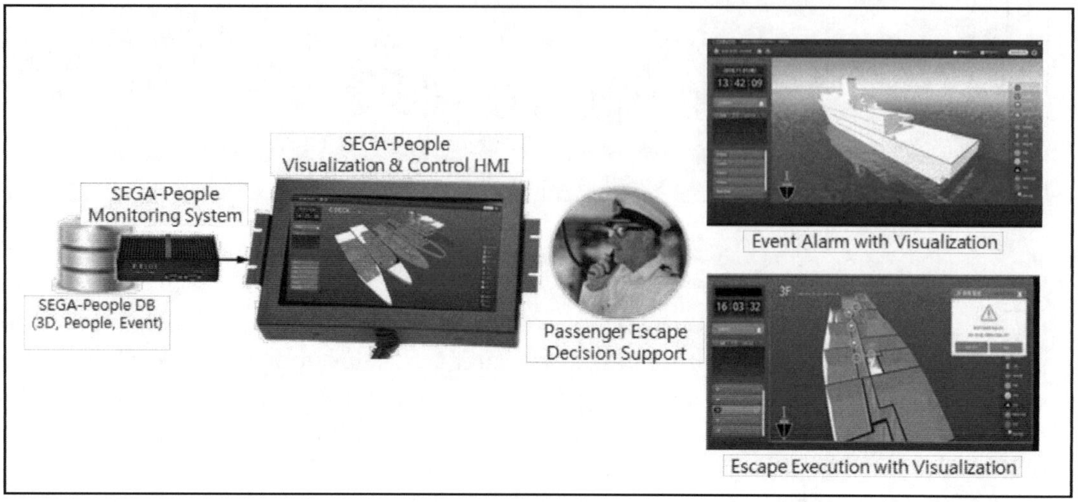

Fig. 10 SEGA-People 3D Visualization HMI Modules

In Fig. 11, 3D user interface software visualizes event data received from SEGA-Ship system. Initially, water leakage event happened but heeling status is still good. After a while, heeling status changed seriously and escape execution pop-up appeared. Captain and crews can select "Escape Now" or "Cancel".

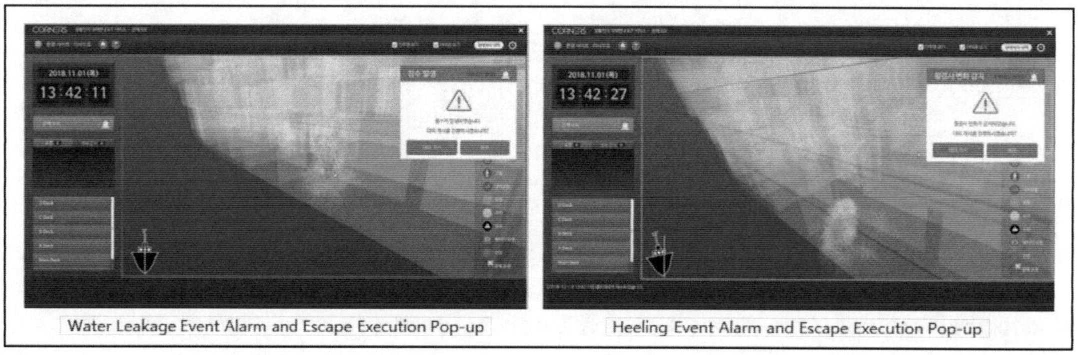

Fig. 11 Water Leakage and Heeling Event Visualization with Alert Pop-ups

Fig. 12 shows 3D user interface software visualizes fire alarm event data received from legacy fire control panel interface. Every fire detector location is managed by SEGA-People for visualization. Fire alarm event on A-Deck is visualized with semi-transparent mode and alert pop-up appeared. As soon as "Escape Now" button is selected, active dynamic signage device icons show current operation status. Red icon means the active dynamic signage shows red "X" to prevent people from approaching to the fire area. Green icons mean the active dynamic signage devices are normally operating to guide people. By changing the display mode into semi-transparent mode, current escape route can be seen clearly by green arrows. This route is the outcome from run-time SERA (Safe Escape Routing Algorithm) engine which calculates a safe and fast escape route from every node (the locations of active dynamic signage devices) to the mustering stations available.

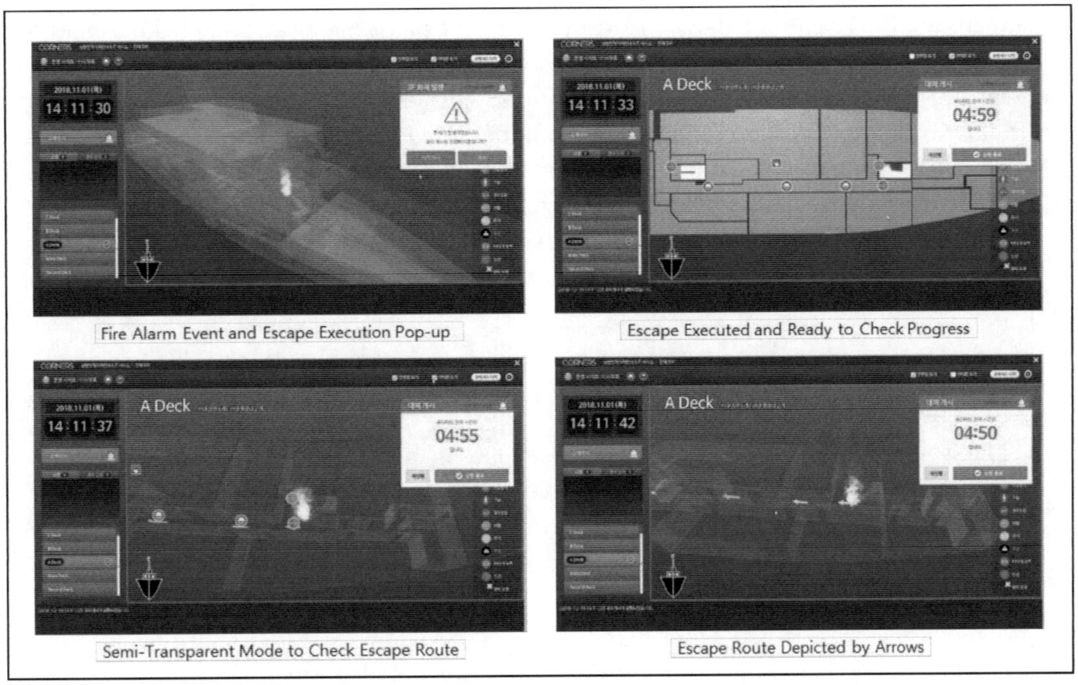

Fig. 12 Fire Alarm Event Visualization with Alert Pop-ups

3) Intelligent Escape Routing with SERA Run-time Engine

Fig. 13 shows the system flow when captain selects "Escape Now" command button.

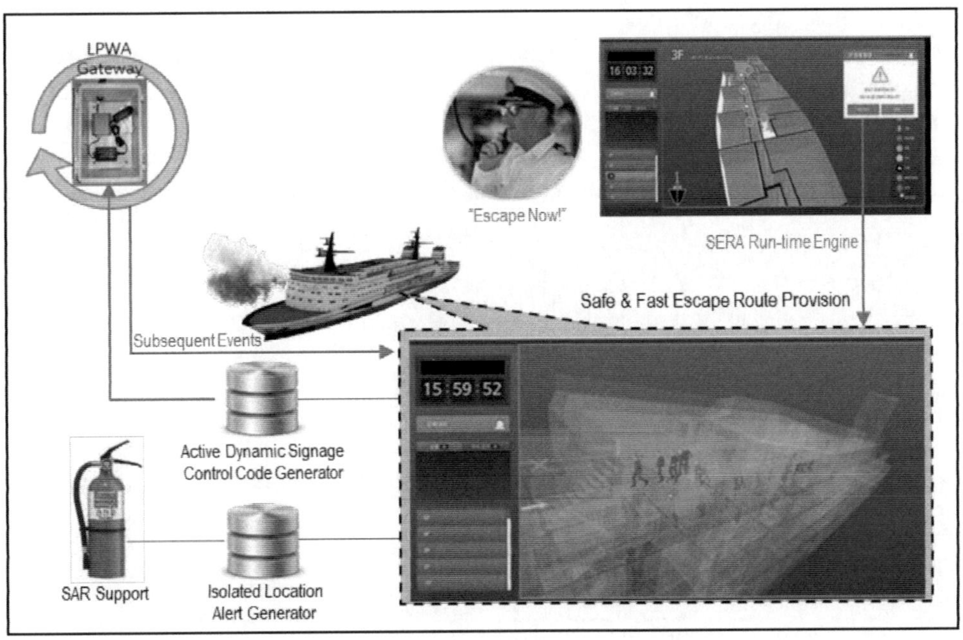

Fig. 13 Run-time Escape Routing Engine and

SERA (Safe Escape Routing Algorithm) can be described as Time-expanded, Min-Cost, Max-Flow type of heuristic network problem solving algorithm, which maximizes the number of escaping passengers while minimizing the total escape time in the escape route network. The input data include number of passengers at each node, location of mustering station (final exit), nodes and arcs attributes, and hazardous nodes identified currently. In a given escape routing problem, the travel time of each arc is defined as cost, and flow is defined as the number of moving passengers by using the arc for their escape. Fig. 14 shows an example of nodes and arcs based escape routing network problem. The left most node is a source node, two right most nodes are assembly stations, and the other nodes represents individual location of active dynamic signage devices.

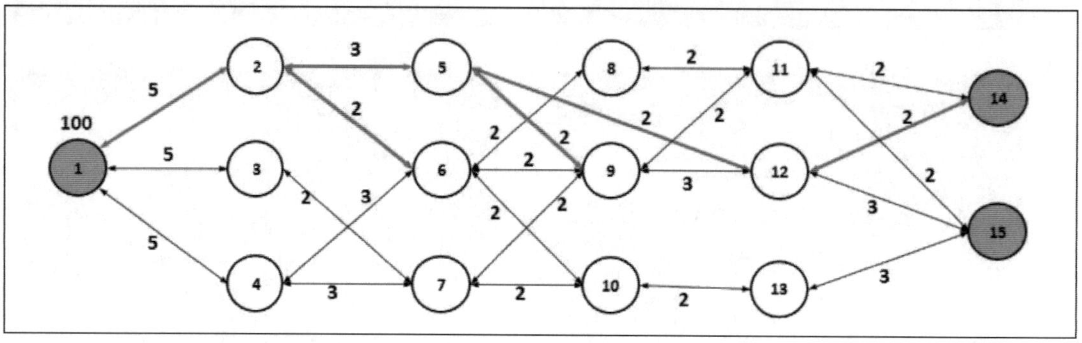

Fig. 14 Sample of nodes and arcs based escape routing network problem

By applying people presence and congestion detecting technologies, SERA can generate more realistic bottleneck minimizing escape route using time-expanded modelling technique. SERA can provide not only the isolated locations but also the people presence data for those locations. SERA is developed for being used as a run-time engine. Though there are high performance network problem solving engines such as CPLEX and Gurobi, their calculation time for finding optimal solution cannot be tolerated during the real disaster situation which requires calculation speed within a few seconds. Table 1 shows the calculation time comparison results which are derived by solving a problem size of 100 nodes, 232 arcs and 1,040 people.

Table 1 Comparison of SERA Performance with Other Solvers and Algorithm

Criteria	Solvers			
	SERA	ILOG CPLEX	Gurobi	Open Source (Edmonds-Karp)
Solving Time	Less then 1 sec	6.65 seconds	9.11 seconds	> 1.5 hour
Total Escape Time	193 seconds	193 seconds	193 seconds	193 seconds
N times Situation Change Case (Cumulative Delay Time Impact)	Max 1 sec x N	Min 6.65 sec x N	Min 9.11 sec x N	Min 1.5hr x N

(Computing Power: Intel i7-5500 CPU@2.4GHz, 8GB Memory, Windows 7 Ultimate K 64bit)

4) Wireless Control of Active Dynamic Signage

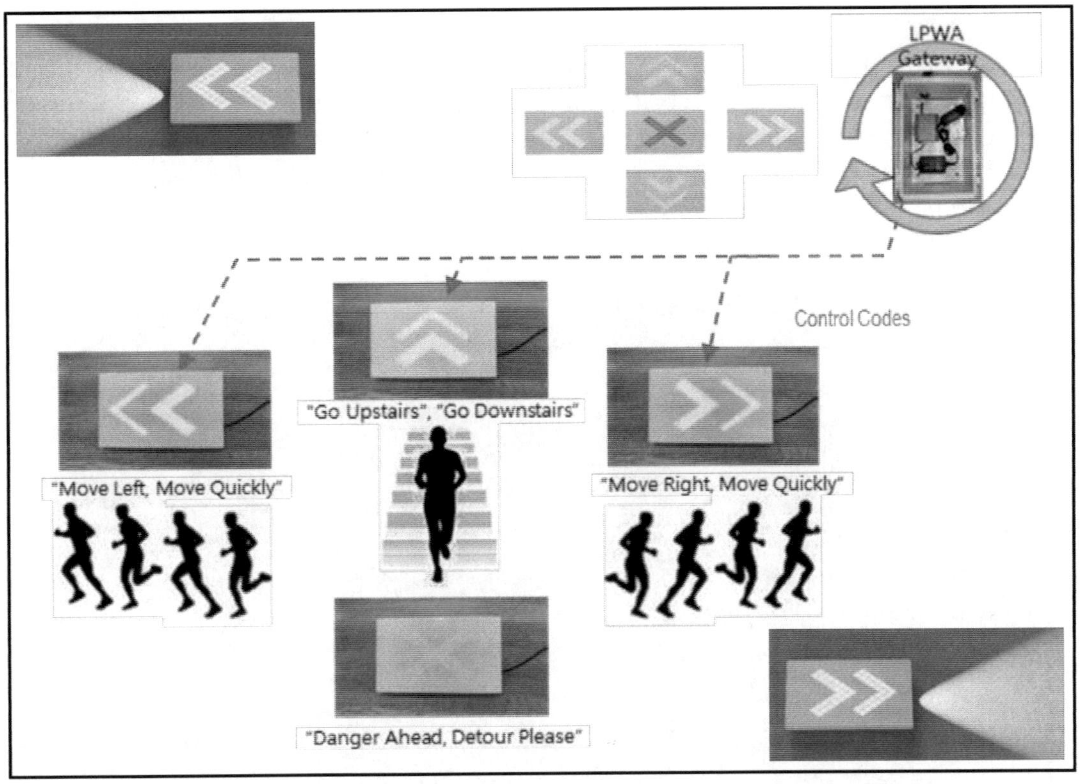

Fig. 15 Active Dynamic Signage Functions and Operation Scenarios

Fig. 15 shows how the active dynamic signage devices work – LED Arrow signs and corresponding voice instructions as well as beam flash lighting. Legacy exit signage and low location lighting is passive and static. They are always turned on and just waiting for passengers to watch them during escape or evacuation – passive. Meanwhile, they are always showing same symbols and same color of light, they cannot change the direction of sign and color of light – static. Imagining the critical role of crews in real escape or evacuation circumstances will be pictured by "as many crews as possible are dispatched to all over the passenger deck", "the crews are holding megaphone and radio communication device", "they communicate with captain and each other to clarify nearby hazard and find safe ways to mustering station", "guide passengers to follow certain lines and passages". The crews tell and passengers pay attention to them, the crews lead passengers and continuously find and change route to guarantee safe arrival to the mustering station – the crews are active and dynamic. Active dynamic signage mimics the crews' active and dynamic guidance activities. It has embedded loud speaker system to tell passenger to do what using pre-recorded voice announcement and guidance messages. It can show safe and fast escape direction (up, down, left, and right) with its front panel LED arrow display and edge-side beam light. ADS can also give warning message (do not enter) and visual sign (red "X" front led display) for passengers not to risk any nearby hazard. Normally ADS get electric power from AC power line and embedded rechargeable 2,600mA battery guarantees more than 1 hour operation under black-out situation.

"ISABU" SHIP ON BOARD ESCAPE DRILL

SEGA-People on board escape drill was conducted at "ISABU" – research vessel of KIOST (Korea Institute of Ocean Science & Technologies) on 17th Oct. 2017 to verify the value of SEGA-People under dynamic emergency escape scenarios. Total 20 participants were divided into four groups for B-Deck start scenario and five groups for Second-Deck start scenario. These two escape scenarios had been selected by running SEGA-People 3D simulation software showed in Fig. 16.

Fig. 16 SEGA-People 3D Passenger Ship Escape Simulation Software

Fig. 17 shows the location of starting points and mustering station. Total 5 decks had been used for this drills – Second-Deck, Main-Deck, A-Deck, B-Deck, and C-Deck. Mustering station is set on the C-Deck

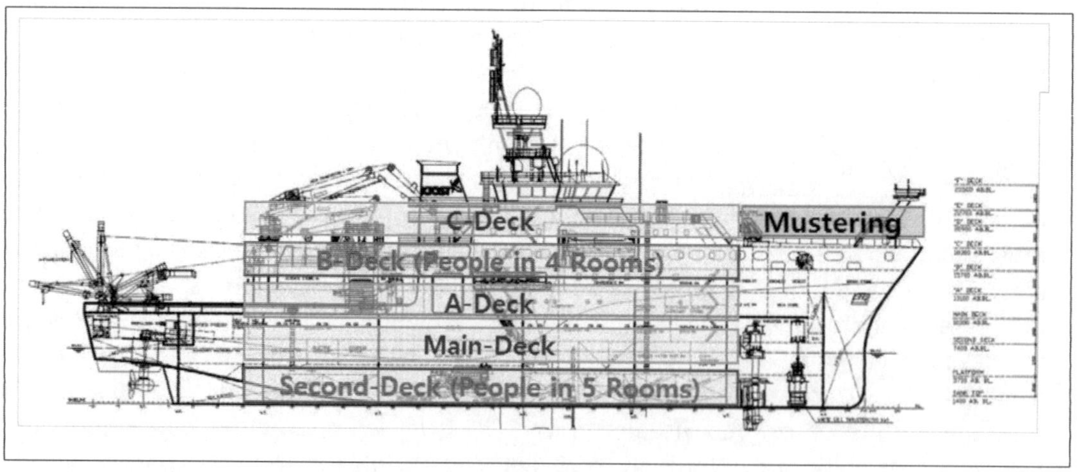

Fig. 17 "ISABU" Ship Decks – People starts B & Second-Deck to Reach a Mustering Station on C-Deck

Fig. 18 shows starting rooms on Second-Deck and B-Deck respectively.

Fig. 18 People Assigned in 5 Rooms for Second-Deck Scenario and 4 Rooms for B-Deck Scenario

Experiment settings are as follows.
- 4~5 participants distributed in 9 cabins
- Normal walking speed when escape from cabins without talking to each other
- Initial escape notification is given via ship area announcement system
- People should detour if they see danger area symbol ahead of their escape route

The experiment was conducted four times with two escape scenarios each of it has different danger area settings.

- Scenario 1: People starts escape from 5 rooms on Second-Deck.
  1-1. People escape without activating SEGA-People system
  1-2. People escape with SEGA-People system working
- Scenario 2: People starts escape from 4 rooms on B-Deck
  2-1. People escape without activating SEGA-People system
  2-2. People escape with SEGA-People system working

Measures for analyzing escape performance are

- Measure 1: Individual time consumed to reach at the mustering station
- Measure 2: Total time consumed till escape completion

Table 2 shows time stamps for the first and the last person escaped for each escape scenario. For the case of the first person escape time, SEGA-People reduced 25% and 31% for each scenario. In terms of total escape completion time, SEGA-People reduced 26% for scenario 1, and reduced 70% for scenario 2.

Table 2 SEGA-People Performance Analysis – Shortened Escape Completion Time

Scenario		People Start on	Description	First Escape Time	Gap (%)	Last Escape Time	Gap (%)
1	1-1	Second Deck	Self Escape / Danger Settings ON	1′ 53″	28″ (25%)	2′ 54″	46″ (26%)
	1-2		SEGA-People ON / Danger Settings ON	1′ 25″		2′ 8″	
2	2-1	B-Deck	Self Escape / Danger Settings ON	2′ 56″	57″ (31%)	11′ 6″	7′ 49″ (70%)
	2-2		SEGA-People ON / Danger Settings ON	1′ 59″		3′ 17″	

CONCLUSIONS AND NEXT STEPS

The mission of SEGA-People is to minimize casualties caused by improper or delayed escape execution. The results of "ISABU" ship experiment shows the total escape completion time could be shortened about 30% normally and up to 70% in extreme case. The bottleneck could be minimized with the use of people congestion detecting technologies such as mmWave radar sensors. The value of SEGA-People could be much greater during the early stage of escape when crews and passengers prone to get panic. Currently SEGA-People is in the middle of getting a Korean Register of Shipping (KR) certification as an alternative escape guidance system.

REFERENCES

1. IMO(2016), Guidelines for evacuation analysis for new and existing passenger ships, MSC.1/Circ.1533
2. IMO(2005), Functional requirements and performance standards for the assessment of evacuation guidance systems, MSC.1/Circ.1167
3. IMO(2005), Interim guidelines for the testing, approval and maintenance of evacuation guidance systems used as an alternative to low-location lighting systems, MSC.1/Circ.1168
4. SOLAS 2016 amend/ Chapter II-2 /Reg. 13.3.2, Means of escape in passenger ships
5. IMO SSE 4/INF.2 5th Jan. 2017, Information on the technological development of the "Escape Guiding System" for passenger ships.
6. Korea Maritime Safety Tribunal in Ministry of Oceans and Fisheries, Examples and lessons of major maritime accidents and incidents in (2015).
7. Eun-Gyeong Ryu, Chan-Su Yang, James Choi, Fundamental Research for Escape Guidance System Development of Passenger Ship, J. Navig. Port Res. Vol. 41, No. 2 : 39-46, April 2017
8. James Choi, Ji-Han Kim, Minjee Son, Chan-Su Yang, Smart Escape Support System for Passenger Ship: Active Dynamic Signage & Real-time Escape Routing, PACIFIC 2017, International Maritime Conference

Naval Assurance

Stefan Palm, Jason Steward, Christian v. Oldershausen

DNV GL Maritime

INTRODUCTION

Classification in its conventional understanding, is an important contributor to the assurance process of naval vessels. However, Classification needs to be adapted to navy's specific needs and the governance and regulatory environment within which they are operating. A navy shall serve the security interest of its respective nation and can therefore not be the subject of a prescribed regime as is the case for merchant vessels. This calls for assurance processes that, while covering Classification, is more diversified with a wider scope to fulfil specific Navy assurance requirements. Consequently, DNV GL's services for naval vessels are referred to as **DNV GL Naval Technical Assurance** indicating a more generic approach to the complete assurance process with respect to Materiel Safety, not limited to the conventional understanding of Classification. In contrast to the traditional Classification approach, the scope of assurance to be applied, including the particular standards to be referenced (as prescribed by the applicable government requirements), are defined together with the relevant naval authority for each project.

The key reason for acquiring a naval vessel is to achieve a specified operational effect. Basically, DNV GL's Naval Technical Assurance approach covers all platform related aspects, and thus is separate or complementary to any combat system assurance regime. Although the present paper is limited to Materiel Safety, DNV GL also takes a wider perspective in the services it offers for naval Ships through complementary Advisory Services. Examples of such services are analysis for assuring shock hardening, ensuring residual strength after specified damage events, etc.

DNV GL NAVAL TECHNICAL ASSURANCE

Due to its natural complexity, total assurance for a naval vessel covers various areas including amongst others, the platform and the weapon system interface, but also areas such as maintenance and logistics. The latter includes topics such as spare part management as well as munitions, crewing, training and victualling.

Ultimately naval total assurance is about ensuring the seaworthiness outcome and the required operational availability. Within the total assurance for naval vessels, technical assurance i.e. assuring the safety, reliability and availability performance of the platform are the key elements.

The objective of DNV GL Technical Assurance is to provide through-life technical assurance to navies that:

- builds on and incorporate state of the art assurance regimes and methods with proven track record for ensuring an adequate level of Materiel Safety;

- provides confidence in the process by external stakeholders;
- is cost efficient.

It is developed for and can be applied to naval vessels and vessels with similar modes of operation, for example vessels for coast guard duty, border control and police. DNV GL Naval Technical Assurance is not applicable to vessels for which the full SOLAS convention applies.

The naval technical assurance approach by DNV GL serves as a key element within a total assurance systematic for naval vessels by providing a solution for the technical assurance of the platform i.e. all systems except the combat or weapon systems respectively.

One of the main goals of the approach is to enable a systematic process that can be tailored to the individual needs and boundary conditions of the particular navy and project. This is realized by a modular set-up of the approach - expressed by two key documents - the standards plan and the certification matrix – both of which are endorsed and owned by the Naval Authority after a definition in close cooperation with and support by, DNV GL. The set-up for both the standards plan (SP) as well as the certification matrix (CM) follow the philosophy of the Naval Ship Code (NSC) and addresses similar assurance areas to be verified The SP is defined as the document, agreed by the Naval Administration, listing the standards, specifications, conventions etc. or other criteria that are being used to demonstrate that a vessel meets the Performance Requirements, Functional Objectives and Goals of the Technical Areas of Part 1 of the NSC. The CM is similar to Justification, Validation and Verification defined in the NSC.

The DNV GL Naval Technical Assurance process steps are summarised in the following **Error! Reference source not found.**

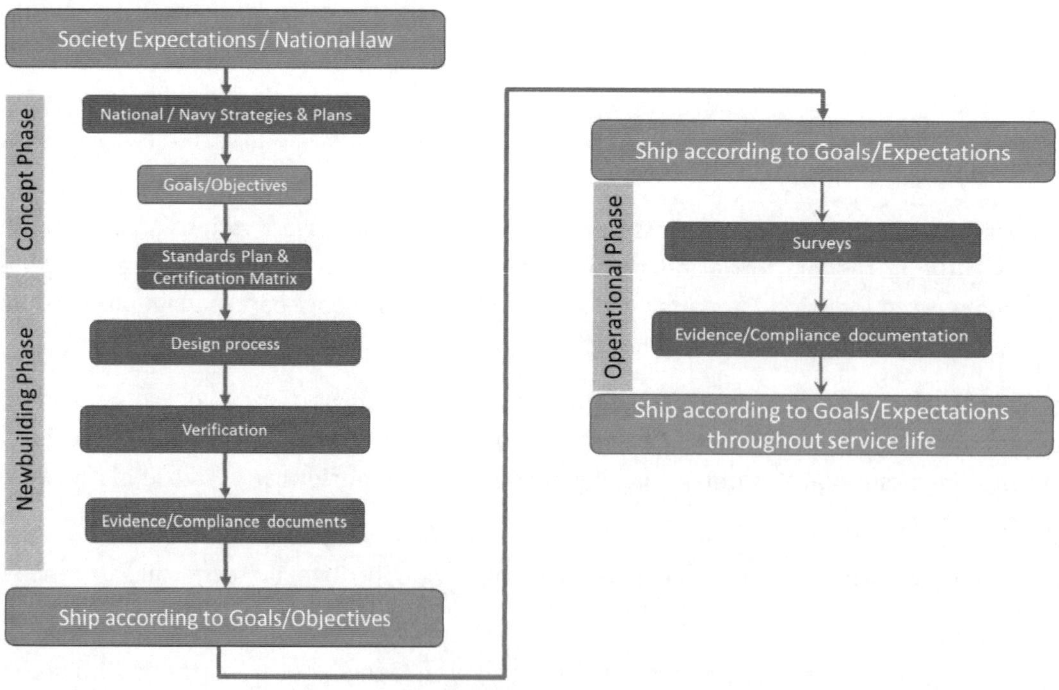

Figure 1: DNV GL Naval Technical Assurance process steps

The DNV GL Naval Technical Assurance is divided in three Phases: the Concept Phase, the Newbuilding Phase and the Operational Phase.

The objective of the Concept Phase is to develop a Specification of the vessel that ensures the specified Goals and Objectives are fulfilled. This Specification shall be detailed enough for the purpose of describing requirements for procurement and fabrication. DNV GL's involvement in this Phase is the subject of an agreement with the client and may cover Materiel Safety alone or in addition other attributes of the vessel.

There are three key deliverables from the Concept Phase:

- The Standards Plan;
- The Certification Matrix;
- Acceptance by the naval administration of the Standards Plan.

The Goals and Objectives for the vessel are normally outside of DNV GL's contribution. However, the definition of these have a large influence on how the DNV GL Naval Technical Assurance process is carried out. It is therefore imperative that full understanding is established between navy, yard and DNV GL on this matter.

It is advantageous to distinguish between Goals/Objectives concerning Materiel Safety in peace time and Goals/Objectives concerning the Operational Effect of the vessel in a wartime scenario as the applicable Standards are different and consequently a Standards Plan with regard to Operational Effect will have very little overlap with that for Materiel Safety.

DEVELOPMENT OF A STANDARDS PLAN

The Goals and Objectives are translated into the Standards Plan, whereby the ship and its equipment and systems are divided in groups that can be dealt with on an individual basis. DNV GL's default approach is to address this using the Safety Areas listed below, which follow the arrangement in ANEP-77 Naval Ship Cope:

- Hull structure
- Buoyance, stability and controllability
- Engineering systems
- Seamanship systems
- Fires safety
- Escape, evacuation and rescue
- Navigation
- Communication

- Dangerous goods

However, the way in which this division is done can vary between the particular regulatory systems and different nomenclature can be used: disciplines, safety areas, compliance obligations etc. In this process Goals and Objectives for each Safety Area are defined, followed by prescriptive technical solutions ensuring that the Goals and Objectives for each Safety Area are met, forming the Standards Plan.

The Standards Plan shall include the following:

- Specification of applicable standards;
- Links between the requirements in the standards to corresponding requirements in the navy's regulatory system;
- Justification of how the requirements in the Goals/Objectives and or regulatory requirements are fulfilled.

This Standards Plan shall then be agreed by the Naval Administration.

The Standards Plan may consist of any type of technical documentation specifying technical requirements supporting the Goals/Objectives:

- Navy proprietary standards;
- DNV GL Class Rules;
- IMO instruments;
- STANAG;
- ISO standards;
- IEC standards;
- Outcome of risk analysis;
- Other standards.

The most efficient option in this process is using existing risk controls, i.e. already available Standards, with the advantages that the solutions are based on well-known and proven technology for which ample operational experience is already available and that no additional resources are required for developing the solutions.

However, it is possible that a selected standard may not be suitable to its full extent, especially with respect to vessel operation under a naval regime. Tailoring is applied to adapt the existing standard to the required solution. Such a process includes developing technical solutions that can be demonstrated equivalent to the specified Standard or developing other actions to compensate for the increased risk, for example modifications of operational procedures, frequently referred to as mitigation. For such cases dedicated risk analyses may be applied.

THE CERTIFICATION MATRIX

The second key deliverable from the Concept Phase is the Certification Matrix. The purpose of the Certification Matrix is to control the execution of the activities providing evidence of compliance. The Certification Matrix specifies:

- how the verification with regard to the Standards Plan shall be carried out (verification of technical documents, test, sea trial etc.);
- by which organisation;
- the compliance documentation.

The Standards Plan and the Certification Matrix will be agreed by and thus eventually owned by the Naval Administration.

However, the drafting of the Standards Plan and the Certification Matrix can be carried out by parties other than the Naval Administration. For example, the responsibility for the work may be the remit of the procurement authority, the designer, the yard or third-party pre-project services. DNV GL can provide assistance in the process and has already done this for several naval newbuilding projects.

THE NEWBUILDING PHASE

The objective of technical assurance in the Newbuilding Phase is to generate the evidence based on which it can be concluded that the requirements defined in the Standards Plan are complied with and fulfilled, and to generate the specified compliance documentation. This phase is a straightforward verification process covering technical documentation, plan approval, analysis, testing or survey during the fabrication of the vessel.

THE OPERATIONAL PHASE

Similarly, the objective of the technical assurance in the Operational Phase is to generate the evidence based on which it can be concluded that the requirements defined in the Standards Plan are complied with throughout the service life of the vessel.

CONTRIBUTION TO THE TECHNICAL ASSURANCE PROCESS

DNV GL contributes in two ways in the Technical Assurance process:

- by providing Benchmarks in the form of technical standards (the Class Rules) and
- by providing verification and classification for the control of the assurance process.

Although naval vessels are different from merchant vessels in several aspects, there are also similarities. For example, many of the requirements for peacetime operation, global citizenship and occupational health and safety are similar for both merchant and naval vessels:

- the Float and Move out of Float, Move and Fight
- the vessels shall be a safe place to be
- The vessel shall be environmentally friendly

Differences between naval and merchant vessels required by combat operation are predominately naval survivability including susceptibility, vulnerability, and recoverability.

Considering separately the similarities and differences, the DNV GL systematics may provide a mature system of control delivering materiel assurance:

- For merchant ship scope, the DNV GL systematics of Class Rules and standards (ST), guidelines (CG), programmes (CP), statutory interpretations (SI), service specifications (SE) and recommended practices (RP) represent a large scope of standardised solutions for Materiel Assurance;
- For naval ship scope, the flexible approach using Navy proprietary standards, DNV GL Class Rules for naval ships, applicable IMO instruments, bespoke risk analysis and other standards, as described in this paper, may contribute Materiel Assurance.

DNV GL's assurance process has been developed and refined over many years and is the subject of continuing development with respect to quality, efficiency and, importantly, the creation of a user-friendly interface to yards, owners, flag administrations and navies.

The assurance process ensures efficiency and quality in the verification processes and also give the client access to DNV GL's technical competence and service experience in the form of direct technical support.

DNV GL NAVAL CLASSIFICATION

DNV GL Naval Classification is a process providing technical assurance services for the Newbuilding Phase and the Operational Phase and is applicable from the point when the Standard Plan has been specified and agreed upon by the Naval Administration.

DNV GL Naval Classification is applied when the Standards Plan consist mainly or in its entirety of DNV GL class rules and IMO Instruments.

The scope of DNV GL Naval Classification process steps are summarised in the following Figure 2:

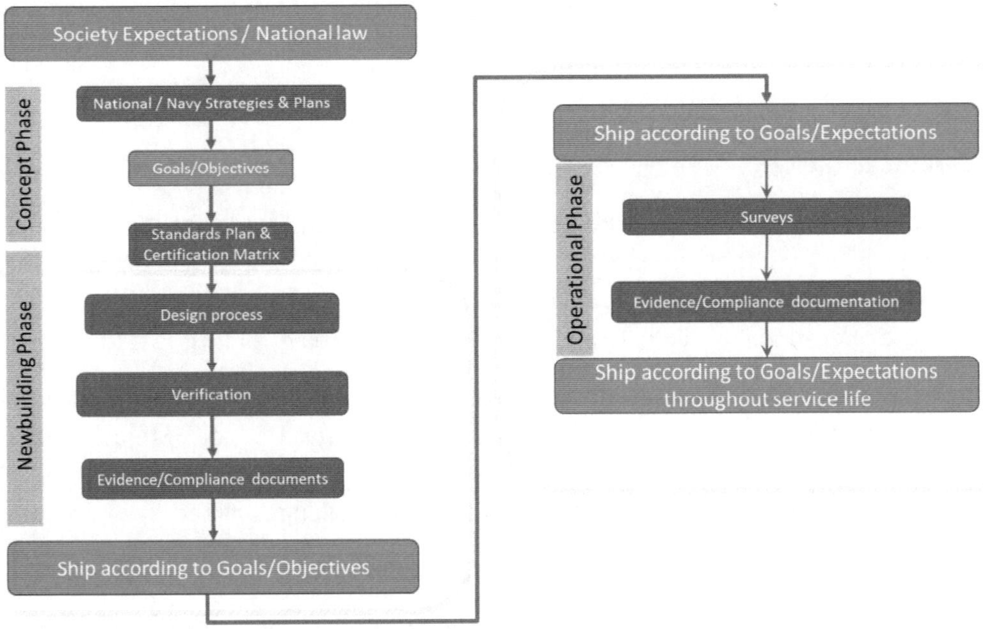

Figure 2: DNV GL Naval Technical Assurance – Scope of DNV GL Class

The objectives for DNV GL Naval Classification are the same as for the Newbuilding and Operational Phases as described above for DNV GL Naval Technical Assurance, ensuring that the vessel is fabricated and maintained in compliance with the specified standards.

DNV GL CLASSIFICATION RULES – THE BENCHMARK

The DNVGL Rules (RU-NAVAL, RU-SHIP, RU-HSLC) have been developed based on experience from more than 150 years of merchant shipping and more than 40 years of DNVGL cooperation with navies in different countries. They provide technical standards for the design and construction within all key areas relevant for the Materiel safety for practically any type of ship, merchant or naval. Having been developed and used for a long period they can be relied on as tried and tested solutions for providing adequate control of the Materiel safety.

These rules thus provide relevant Benchmarks for the design and equipment of naval vessels and for communicating the navy's objective with regard to safety and for the use in the Naval Technical Assurance process. A Benchmark is also a yard stick against which other technical solutions can be measured e.g. navy proprietary standards, or solutions based on risk analyses.

RESPONSIBILITIES

Although both DNV GL Naval Technical Assurance and DNV GL Naval Classification provide a systematic method ensuring the required safety and reliability of a naval platform, there are significant differences between both methods, which need to be considered when a decision is to be made on the preferable approach for the particular naval ship. Eventually it depends on the general procedures and preferences of the Naval Administration and may vary depending if it is new-building or an existing ship in operation.

FUTURE DEVELOPMENTS

Looking into the future it can be expected that despite of the understandable reluctance within defence organization regarding voluntary or involuntary information sharing, digitalization will significantly gain importance in terms of utilization of available information to increase the operational availability and decrease the related efforts time and cost wise.

The practical implementation and standard usage of digital twins along with condition-based maintenance procedures is seen as a future key element to get the goal. Standardization of data and interfaces as well as the achievement of sufficient data and information provision mainly from component manufacturers are seen as main challenges ahead.

CONCLUSIONS

DNV GL Naval Technical Assurance provides a number of advantages for the navy and their stakeholders.

DNV GL Naval Technical Assurance is a comprehensive and self-contained service. It covers the whole process from the inception of the development of the vessel once the operational intent has been specified as based on national interest. It is generic and can, as such, be applied to all aspects of the Materiel state of a naval vessel, being it relevant for Materiel safety or for attributes of the vessel ensuring that operational intents are fulfilled. DNV GL Naval Technical Assurance can support any navy regulatory system as well as provide a regulatory solution for navies who could find it useful.

For a navy this process provides an all-encompassing seamless process from inception to end of service life. The process ensures that everything with influence on the final performance of the vessel is addressed and that nothing is lost underway due to changes of responsibilities in different phases.

A government can obtain assurance that public resources are used as intended and that a project delivers to specified requirements. Similarly, the DNV GL Naval Technical Assurance process provides evidence that requirements for ensuring the health and safety aspects of a naval vessel as a workplace are adhered to. The same applies in case of the general public.

For all stakeholders in the Newbuilding process the DNV GL Naval Technical Assurance provides clear specifications and requirements for verification and certification providing direct input to their own work processes.

DNV GL Naval Technical Assurance has been proven to provide a sound solution for the verification of the compliance of the platform of a navy ship with the requirements i.e. fulfillment of the operational duties in a safe and reliable way. Therefore, the approach can be seen as a major element within the total assurance of a navy ship.

Due to its high flexibility in terms of the verification or reference basis i.e. the selected Rules and standards but also in terms of the scope the naval technical assurance approach appears to be suitable especially when applied for ships already in operation. When it comes to application for new-building projects it depends very much on the procedures, the strategy and the capabilities of the naval authority regarding which approach to adopt i.e. whether naval technical assurance or naval classification is the Materiel Assurance strategy to be realised.

The Technical Design Gap in Warship Environment Protection: Why 'Near Enough' is Often Better Than Perfect

John Polglaze CMarSci FIMarEST, PGM Environment,
john.polglaze@pgmenviro.com.au

ABSTRACT

Many navies aspire to acquire 'fully environmentally compliant' warships, but such statements indicate minimal understanding of what compliance entails, or realisation of the emasculation of ships' operational effectiveness inherent to such ambition. Additionally, inescapable technical realities void any possibility of achieving this aspiration in all but limited circumstances.

Compared with merchant ships, warships have very long lead times in design and build, and limited weight and space reserves for the incorporation of new design elements or equipment. International Maritime Organization (IMO) rules change regularly and are typically promulgated in advance of the technologies needed to comply with such rules, with the rule change acting as the catalyst for the development and production of requisite technologies. This paradigm works well for the timeframes associated with merchant ship construction but is totally unworkable for warship builds. In this context, even though a new or future rule may be known and well understood, no existing or reliable candidate technologies may be available to fulfil that (future) requirement at the time that a warship design or modification needs to be finalised. Even if such technologies exist, they may be immature and therefore represent risk, or otherwise be developed for and optimised for merchant ship application, and hence not be suitable for warships (eg. shock resistance, [inappropriate] capacity or operating parameters compared with merchant ship applications, signature implications, etc). These uncertainties introduce technical risk and invite unintended consequences.

Attempting to build a 'compliant' warship is a Sisyphean task of questionable purpose; this is better avoided by adopting smarter, focused solutions, based upon alternative compliance means using proven technologies and rational environmental risk evaluation.

BACKGROUND: A CONFUSION OF IDENTITY AND PURPOSE?

Protection of the environment is important, but warships are not built and maintained to demonstrate responsible or exemplary conformance with environmental protection rules – they are, or at least should be, built to fight and win at sea. In fact, the current Australian Chief of Navy, Vice Admiral Mike Noonan, advocated this line of thinking when he declared that the Royal Australian Navy (RAN) needed to 'think like a fighting navy'.

Australian strategic doctrine encompasses 10 fundamental principles of warfare, described as '*time proven and fundamentally important to achieving success*' [1, p. 6-2]. The keystone element of these is delineated as **selection and maintenance of the aim**. The same should apply to warship requirements and acquisition - through their governments, national populations invest very heavily in the development, maintenance and sustainment of naval capabilities, with these capabilities expressed through the ships which make up their navies. Both individually and collectively, navy ships – warships – exist, or at least should exist, to best be able to fight and win at sea – which manifests in the associated contexts of lethality and survivability, as being able to defeat an opponent and return home safely.

It is recognised that warships also typically undertake a host of other functions, particularly in peacetime, such as patrolling and surveillance, 'showing the flag' deployments in support of national diplomatic objectives, humanitarian assistance and disaster relief. Navies also exist to effect a standing deterrent, to make a potential adversary deeply contemplate the risks of taking action against the peer with the potent, well equipped navy. Consequently, even in most situations short of high intensity warfare, the effectiveness of a warship – symbolising national strength and resolve, or influencing strategic dialogue with friends and potential foes alike – is firmly embedded in having capable and credible maritime platforms, optimised for the warfighting role all hope will never eventuate. This has been summarised by others as the ability to 'deter, defeat, deny'; nowhere is this expressed as 'overawe with reduced diesel exhaust emission rates'!

Any national effort to acquire and operate warships must have the cardinal intent of providing an effective maritime warfighting platform. Ergo, everything to do with that acquisition should be focused upon that outcome – as an expression of **selection and maintenance of the aim**. There are many other design imperatives for a warship, but all are, or at least should be, subordinate to the primary aim. This includes matters of environmental compliance, especially when otherwise represented by any attempt at 'black letter' adherence to environment protection rule sets not intended to apply to warships, and the onerous, high risk or inappropriate technical 'solutions' which may be associated with such compliance.

Many navies aspire to acquire 'fully environmentally compliant' warships. Where procurement authorities may be tempted to deviate from or dilute this strict aim of providing an effective maritime warfighting platform, they would be well advised to properly consider the parallel risks to national defence. Rational consideration of functions that contribute to warfighting efficiency must not be displaced by false equivalence (ie. a 'fully compliant' warship is not equivalent to an efficient warfighting platform).

Even if such ambition had sufficient merit to pursue and a 'fully compliant warship' could be designed, which is a highly questionable proposition, to maintain such 'full compliance' over the time of build of a typical warship would be exceptionally difficult and expensive - even more so over the time of build of a class of sisterships [2]. Part of the challenge of maintaining warship compliance is due to the inevitable gaps which arise between the rules sets and the technologies intended to ensure compliance, with technical solutions focused upon merchant ship applications and merchant ship design and build chronologies. Given the unique mission requirements and characteristics of the warship design and build process, it is clear that 'perfect' technology for environment protection purposes should often be eschewed in favour of technology which is 'near to good enough'. This could also be labelled as the merit of the pragmatist over the purist.

This paper concentrates upon 'warships' as surface combatants, particularly destroyers and frigates, with the concepts and analyses presented being equally, if not acutely more applicable, to submarines. There is a lesser nexus with larger naval ships and auxiliaries such as afloat support ships, but the general concepts do still hold to varying degrees.

THE WARSHIP ENVIRONMENT PROTECTION TRADE SPACE: 'SLIM PICKINGS'

Like anything, warship designs exhibit finite boundaries – size limitations, space and weight restrictions, limited top weight margins, and cost ceilings – such that compromises need to be made in design selections and equipment fits and the resultant capability expectations. More so than most other ships, and particularly in the case of surface combatants and submarines, whatever is included in a warship design and fit means that something else needs to be left out – the unavoidable 'trade space' conundrum. Merchant ships can 'trade' space for revenue earning capacity per unit or offset increased operating costs by amending their cost recovery structures, neutralised by the fact that their competitors also need to do the same in an internationally regulated 'level playing field'. No such moderating factors apply for warships: nor should any be expected from opposition naval forces.

It is imperative, therefore, that navies adopt an agile, nuanced approach, applying lateral thinking to develop innovative, fit-for-purpose solutions focused upon managing <u>actual</u> environmental risks, rather than simply assuming those risks and their prescribed solutions (often not yet available) arising from the civil shipping sphere. This includes effective evaluation and astute adoption or rejection of the merchant ship-derived technologies which may supposedly be required to achieve environmental compliance, or at least be intended to do so.

THE ORIGIN, INTENT AND FOCUS OF SHIP ENVIRONMENT PROTECTION RULES AND THEIR INTRINSIC AND SYSTEMIC INCOMPATIBILITIES WITH WARSHIPS

The International Maritime Organization (IMO), by charter and precedent, focuses upon civil shipping: IMO environment protection rules, and their nationally legislated derivatives, are of, for and about merchant ships, their risks to the environment and their rational merchant-ship centric means of management. There are myriad reasons why many of these rules cannot be applied to warships, nor need to be.

The fact that these rules are made for and about ships in civil roles is the unavoidable root cause of the technical design gaps which arise in any endeavour to design and equip warships to comply with these merchant-ship derived rules for environment protection. Some of the reasons why these technical gaps arise include:

- The differentiation in timespans for the design and build of civil ships as opposed to those for warships, overlaid with the rate of introduction and change of rules.
- The availability and technical maturity of equipment intended to comply with these evolving rules.
- The minimal weight and space margins available in warships, compared with merchant ships, for the accommodation of equipment and fittings solely for the

purposes of environmental protection, including as may be necessary for new or upgraded equipment associated with amended rules.

- Marine pollution control equipment optimised for application to civil shipping, instead of warships.
- The 'level playing field' of regulatory requirements applying to civil shipping, in comparison with the competitive, critical requirements for national defence and the contest for maritime combat supremacy underlying warship design.

These civil ship / warship dichotomies also need to be considered within the paradox that warships present different sorts of risks to the environment, often of different scale and character, than do civil ships: these warship risk profiles are sometimes better, sometimes worse, and sometimes non-existent.

Warship v Merchant Ship Design and Build Periods

The IMO deliberates on environment protection issues via the Marine Environment Protection Committee (MEPC), which meets once or twice each calendar year; rules are consequently changed once or twice annually. New and amended rules span the range from the monumental to the mundane; from design rules changing the general arrangements of ships, to trivial intricacies such as record book layouts.

For each new or revised rule, the IMO also schedules a date for its entry into effect, usually around two to five years in the future. In this context, it is pertinent to consider timelines applicable to the design and build of merchant ships, characteristically short compared with warships. The period for merchant ships from design to keel laying to delivery can usually be measured in tens of months and rarely exceeds a few years. The normal window permitted by the IMO for entry into force of new regulations makes reasonable allowance for adoption by merchant ship designers, builders and operators, such that an infant new design / new build merchant ship can invariably be completed before the new rule enters force.

Merchant ship timeframes are incongruent with the gestation periods for warships, particularly combatants, with lead times and build programs often measured in decades. This has ramifications for achieving compliance during design and initial build, and a compounding problem if intended to maintain compliance throughout a build program and over life-of-type. During the warship design and build period many new environmental rules will emerge, with these nominally applicable to individual ships in a class.

Maritime environment protection rules display constant 'churn', becoming more expansive and rigorous over time, simultaneous with an increasing rate of promulgation and amendment. In its first 50 meetings, from 1974 to 2003, the MEPC issued 114 Resolutions, at around four per annum. Its next 23 meetings in the following 15 years generated a further 197 Resolutions, averaging around 13 per annum. This record indicates both the perpetual state of flux of the rule set and its accelerating rate of change.

In the context of this ever-changing rules framework, consider that the IMO deems the date of assignment of new or amended rules to any particular ship on the basis of one or more of a number of factors, with these variously deemed, *inter alia*, as:

- date of build / keel laying;
- date of delivery of the ship;
- date of installation of an individual item of equipment; or
- date of delivery for installation of an individual item of equipment; or
- date of defined 'conversion' for a ship in service.

While these milestones are easily recognisable and relatively straightforward for a merchant ship, warship 'compliance' within these sorts of parameters defies meaningful definition. For example, should an individual warship, or class, be 'compliant' according to regulations extant at the date of:

- Design finalisation? – noting that the design finalisation may be five years or considerably more before even the lead ship is laid down.
- Contract? – noting that the contract may be signed anywhere from five years to 25 years before the final ship is completed.
- Laying down the lead ship or each individual ship? – considering that the lead ship and final ship of the class may be 15 years or more apart.
- Commissioning (ie. 'delivery') of the lead ship or each individual ship? – considering that the lead ship may commission 10 or more years after the design was finalised, with final ship of the class 15 years or more years after that.
- Delivery for installation of an item of equipment for the lead ship or each individual ship? - noting that equipment sets for batches of ships are often purchased in blocks many years before being installed into an individual platform.
- Date of refit/upgrade (ie. 'conversion') for the first of the class upgraded, or for each individual ship? – even if such potentially irrelevant or inconsequential design or equipment changes can be accommodated.

In the situation of, for example, the Australian *Hunter* Class Frigate acquisition, based upon a design that matured around 2015 for a final ship to be delivered around 2040, it is reasonable to assume that IMO rule changes over this intervening period would number in the range of 250 or more. Not all new or amended rules would have direct relevance but all would need review and analysis. It is difficult to comprehend how a class of warships could maintain alignment with such a dynamic rule set during their build process. Furthermore, it is questionable whether (perceived) benefits would be worth the penalties, for what may be nothing more than incremental or inconsequential environmental gains, while introducing and compounding technical risks, design complexity and inflated costs.

It is axiomatic that even if a lead ship was, hypothetically and very improbably, fully compliant with applicable environment protection rules during build, the likelihood of retaining this status across sisterships is exceptionally unlikely, particularly with the more complex designs and lengthy build programs of surface combatants.

Any intention for strict compliance is further exacerbated throughout life-of-type as observation of IMO design and equipment rules can mandate replacement or enhancement of existing, or installation of entirely new, pollution control systems at different junctures in a ship's life. In accordance with many IMO rules, such requirements may be simply as a result of ship age, or be triggered as a consequence of other, potentially even unrelated, refit or upgrade work or modifications. For example, under strict observation of IMO rules, the mid-life upgrade of a frigate with new combat systems and improved diesel auxiliaries would precipitate the consequential need to also add complex exhaust treatment systems to comply with new emission rules. Even the hypothetical conversion of a frigate to a training ship, by the simple expedient of, for example, removing the main gun mount and replacing this with a deckhouse for a classroom may be argued as a change of role, and hence as a 'major conversion' for the purposes of some rules, which could result in the consequential need to make other modifications, potentially extensive, simply to comply with new or amended IMO environment protection rules which would not otherwise apply to that ship.

Some amelioration of inevitable obsolescence over ship build programs can be achieved by forecasting more stringent future requirements (albeit with the real prospect of introducing other risks, and assuming suitable technologies are available and proven). Indeed, some warship types (eg. afloat support ships, large amphibious units) have greater elasticity in terms of weight/space, power and ventilation margins for compliance modifications, than do warships of leaner, optimised design (eg. submarines, destroyers, frigates). Ship upgrades and technology insertions also provide some capacity to regain some level of compliance, but the inevitable and unavoidable trajectory will be of divergence between the nominal rules baseline and individual ship compliance as the build of class of sisterships progressed.

Ballast Water Treatment: An Example of a Merchant Ship-Centric Solution to a Common Risk

With good reason, the IMO has promulgated rules intended to minimise the likelihood that ship's ballast water will act as a transfer vector for marine pest species or human pathogens. The rules sets, and the technological solutions they spawned, are founded upon the way in which merchant ships typically take-up and discharge ballast water, which inversely reciprocates what they do with their cargo loads: rapid, block uptake in one port area; carriage to another port area; and rapid discharge in the new port area. Merchant ship ballast water capacities are typically measured in the thousands to tens of thousands of tonnes, with corresponding uptake and discharge rates in the thousands of cubic metres per hour. Obviously, commercial developers and providers of ship ballast water treatment systems have concentrated on the merchant ship customer, developing large capacity systems optimised for large volumes and rapid rates of ballast water movement.

Compare this with a typical surface combatant, which *may* (but not always) need to uptake or discharge a few tens or hundreds of tonnes (at most) of ballast water in any sequence, with the latitude to do this in a less time compressed manner than merchant ships needing to get alongside a commercial berth within a certain timeframe. The over-capacity merchant ship certified ballast water treatment system may be inappropriate for the warship, being oversized, consuming precious weight, space and power, requiring its own training and logistic support tail, and perhaps performing sub-optimally because it is rarely used as the design intended. This is not to suggest that small capacity ballast water treatment systems are not available, but just as there is merit in modern sailors and ships being able to multi-task, this is also a reasonable expectation for warship plant. Instead of adopting the merchant ship technical solution, warships can achieve the same objective outcome – sanitised water for use as ballast – by using freshwater generated by a ship's onboard reverse osmosis (RO) units.

Consider that in a typical merchant ship, ballast water use for very large cargo compensation systems is some orders of magnitude greater than the fresh water demand for their crews of around 20 to 30 personnel. This contrasts with a destroyer or frigate where large crews impose high demands for fresh water, while the ship presents minimal ballast water requirements. The warship ratio of ballast water to freshwater use is almost the inverse of the merchant ship equation, such that warship fresh water requirements and ballast water demand are commensurate in magnitude. It is intriguing to consider how the IMO's rules for ballast water treatment may have been articulated if the example they were considering was analogous to that in warships, where only a fraction of the water capable of already being treated in a ship was needed for ballast purposes. It must be considered that rather than mandating the development and installation of a totally novel treatment process, the IMO may have adopted the pragmatic approach of adapting existing systems and may have ruled that treated ballast water should be siphoned off from IMO-approved fresh water production systems.

The use of water produced by a warship's RO units, perhaps of augmented capacity to cope with periodic extra demand, may not strictly accord with any IMO-prescribed technical solution for ballast water management. It would, however, certainly result in the same objective outcome while lowering technical risks, reducing costs, simplifying designs and avoiding possible derogation of potential warfighting capability. The parallel use of RO units for ballast water production, rather than a single-use dedicated ballast water unit, also has the benefit of simplifying watchkeeping, maintenance, configuration management and logistic support tasks and costs, all while providing greater redundancy for a mission critical system.

This is a signal example of where declining to adopt a merchant ship tailored technical solution avoids collateral ship warfighting penalties, while still effectively and objectively addressing the subject risks to the environment. A 'near enough' albeit not 'perfect' technical solution which is more than fit-for-purpose.

WORKING WITHIN THE RULES: REALISING OUTCOMES AND NOT SIMPLY GETTING 'TICKS IN BOXES'

IMO conventions address ship-sourced pollution in a variety of ways. The IMO provides a relaxation for warships and does not anticipate that these rules can or should be strictly applied to warships. This is expressed in all of the IMO's environment-related conventions via a general exemption, tempered by an expectation of compliance as practicable:

> *The present Convention shall not apply to any warship, naval auxiliary or other ship owned or operated by a State and used, for the time being, only on government non-commercial service. However, each Party shall ensure by the adoption of appropriate measures not impairing the operations or operational capabilities of such ships owned or operated by it, that such ships act in a manner consistent, so far as is reasonable and practicable, with the present Convention.*

IMO conventions generally permit employment of exemptions and waivers when compliance would be 'unreasonable or impracticable'. The IMO also foresees avenues for 'equivalent' means of compliance, which may be thought of achieving the same objective outcome by mechanisms other than those imposed by the applicable regulation.

Two things are obvious from the IMO rules. The first is that the agency accepts that warships cannot and should not always comply strictly with their environmental rules. Secondly, is a dispensation for alternative means of achieving the intended objective, with leeway for either compliance with a rule's intent, or non-compliance if no practicable alternative exists.

The IMO's recognition of these realities can be further reinforced when viewed through a lens of objectivity. The fundamental rationale of any environment protection rule is not for ships to follow the nominated rule, *per se*, but for the environment to be protected from the subject risk – to achieve the same objective outcome. Realisation of this self-evident truth lends itself to focus upon achieving goal-based outcomes for warships, via alternative compliance measures if necessary, including the adoption or tailoring of appropriate and fit-for-purpose technologies, rather than simply cramming in those technologies which may be optimised and mandated for merchant ship compliance.

WHY 'NEAR ENOUGH' CAN BE AS GOOD, IF NOT BETTER, THAN PERFECT

It is germane to consider some examples to illustrate why the default application of merchant ship solutions to warships can introduce technical risk and/or result in the application of inappropriate and incompatible technologies. The absorption of these resultant risks and costs to warships may be hard to reconcile with the loss of combat effectiveness when considered in isolation, but is particularly difficult to accept when such penalties would be expected to be absorbed when the resultant environmental benefit is minimal, if not absent.

To illustrate and consolidate the hypothesis that 'near enough' environment protection technology is better for warships than ill-advisedly shoehorning in any supposedly 'perfect' merchant ship solution, there are considered to be six facets of potential incompatibility for candidate technological solutions. In terms of adoption by warships, prescribed or promoted solutions which may be perfectly suited to merchant ships may be categorised in one or more of the following groups, as technologies which are:

- inappropriate;
- inadequate;
- inopportune;
- inconsequential;
- inordinate; and/or
- inane.

These domains of unsuitability can be used to gauge the efficacy of candidate technologies for inclusion in warships. These themes are examined further below.

Inappropriate Technology: Sewage Treatment Plants and Why the 'Right' Stuff Won't Work Properly in the 'Wrong' Ship

IMO rules for the discharge of processed sewage stipulate a defined standard of treated effluent. The technology service providers have responded to this, in the main, by developing and marketing a range of sewage treatment plants (STPs) based upon membrane bioreactor (MBR) technologies. When properly installed and effectively operated, MBR STPs provide a proven and accepted means by which ships may demonstrate enduring compliance with IMO sewage treatment requirements.

As the name indicates, membrane bioreactors incorporate biological treatment processes. This treatment process relies upon a population of microorganisms to essentially 'eat' the raw sewage, generating a more stable and more environmentally benign 'waste' product from their own organic processes. These microbiological communities work best in steady state conditions, including quantity and quality of their 'food' supply – this translates to near constant input of sewage effluent from the ship in which they are installed. Commercial ships are intended to make money, and the only way to do this is to be operated very regularly, if not all the time. As such, these ships, be they offshore support vessels with 20 persons onboard (POB), or cruise liners with 5000 POB and rapid turnarounds in port, generally maintain the same number of people onboard, and hence a constant and steady upkeep of 'food' to the MBRs. This allows the MBRs to operate as intended, with minimal influent load fluctuations.

Warships, however, do not always operate, and a frigate, for example, can go from a situation of having 10 POB to 200 POB relatively instantaneously, as occurs on the morning of sailing. Naval in-service experience indicates that MBRs often have limited ability to accept such shock loadings, with the microbiological community which may well have died back significantly during the ship's period alongside unable to process the new load conditions. For certain, after a few days the system should have attained a new equilibrium at the higher load levels, but the irony of all this is that the STP will not have been operating at the intended performance level at the very time that it was most critical for it to do so – ie. when the frigate was transiting coastal waters immediately after sailing. This situation is exacerbated in larger warships, such as amphibious units, when 1000 or more extra personnel may be added in a period of 24 hours or so.

This is a clear example of an inappropriate technology for warships – something which may meet the performance specifications on paper and which is understood to work well in merchant ships, but which has limitations in a warship application. Arguably, it may be better to specify an STP for a warship that, hypothetically, is only partly as good as a perfectly functioning MBR unit, but that actually works more reliably when it really needs to - something that works near well enough all the time, rather than 'perfectly' only sometimes.

Inadequate Technology: Anti-Fouling Coatings

Anti-fouling coatings (AFCs) exist for a reason – to limit the amount of biofouling accumulated on a ship over the period between scheduled drydockings. An ineffective AFC is *bad* for the environment, as fouling on a ship increases drag with commensurate increase in fuel burn and emissions of greenhouse gases and other pollutants. Biofouling assemblages also present a risk of the transfer of potentially invasive marine species from one coastal area to another. An additional risk for warships is the increased platform self-noise induced by excessive fouling. These risks and costs can be controlled via the application of an effective AFC which is fit-for-purpose.

Most AFCs rely upon biocidal action, generated by toxic chemicals held within the paint structure, which leads to the release of these biocides into the marine environment. In previous times, some AFC biocides were overly toxic, poorly targeted and featured excessive release rates which resulted in collateral damage in the marine environment. International and national controls have now curbed these excesses, such that biocides now used in AFCs and available in developed nations like Australia are tightly regulated.

Some paint manufacturers now produce biocide-free AFCs, with these essentially working on the concept of a very smooth surface upon which biofouling organisms can find minimal purchase. On the right sort of ship with the appropriate operating profile these paints perform quite well. Essential to the correct functioning of these paints is for the ship to operate relatively regularly, at sustained, elevated speeds, to ensure that any fouling which may have attached is sloughed off. Failure to operate the ship in a matter conducive to effective paint functioning permits fouling to establish and adhere with some permanency. Once some fouling attaches and is not dislodged, not only does it become more securely fastened itself, but it acts as a substrate for the attachment of successive, denser, and more diverse fouling – exemplifying the axiom that 'fouling begets fouling'.

Clearly, specifying a non-biocidal fouling release coating for warships, which routinely spend many weeks or months alongside in extended maintenance or reduced activity periods may be considered as somewhat audacious. In such applications, fouling release coatings often perform poorly, because they are being used in a role for which they were not intended. Rather than derive supposed environmental benefit in abstaining from the use of an approved biocide in a regulated setting, the use of a biocide-free coating on a typical warship is more likely to realise a range of environmental disbenefits, relating to otherwise avoidable pollutant emissions, excess consumption of non-renewable resources, and accentuated marine biosecurity risks.

Thus, any navy seeking environmental cachet by specifying a non-biocidal AFC, not suited to their ships and their operating profiles, is actually likely to create more problems for the environment than it supposedly solves, as well as tactical ones for itself. This is an example of a technology which is inadequate for the fouling settlement pressures typically experienced by warships, while being perfectly suited for merchant ships with their more regular service profiles.

Inopportune Technology: 15 ppm Oily Water Treatment Systems
The current regulations controlling oil filtering equipment specify that water with an oil content of less than 15 parts per million (ppm) can be discharged to sea. This rule was first set by the IMO in 1977. Some argue that the standard is outdated and should be tightened, and the market now offers oily water management equipment purportedly capable of discharges of 5 ppm or less.

Although current regulations do indeed permit the installation of 15 ppm systems in ships, with no indication that the MEPC is considering any tightening of the standard, it is entirely open to speculation whether the same will be true in five years from now. The merchant ship designer and buyer could quite comfortably specify 15 ppm equipment when ordering a new build ship tomorrow, knowing that the ship will be delivered and in service in about 12 months from the date of ordering. The same does not hold true for the warship buyer.

In the realm of warship design, build and acquisition, five years from now, or even 10 years from now, is effectively 'tomorrow' – the design features and equipment configurations specified 'today' are pretty much what is going to be delivered and installed in five, 10 or 15 years from now. By extension, the warship buyer cannot specify 15 ppm oil processing equipment for the ship being ordered tomorrow with the same level of relaxation as his or her merchant ship counterpart.

In the context of warship design and acquisition, specifying equipment compliant with an extant although aged standard, when better performing alternatives are available, may be categorised as a reliance upon inopportune and borderline anachronistic technologies. Foregoing the established for the more capable, assuming no unacceptable technical risks, represents a sensible and responsible means by which to 'future proof' a warship against the advent of more stringent environmental compliance standards which may plausibly arise during that ship's lifetime.

Inconsequential Technology: Protected Fuel Tanks

The slab sided and flat bottomed profiles of merchants ships, coupled with single screw, single rudder fits and minimum crewing often meant the loss of fuel oil to the marine environment following collisions and groundings. The environmental consequences of any such fuel loss are further exacerbated by the persistent, polluting heavy fuel oils used by merchant ships. To address this legitimate risk to the environment, the IMO imposed rules on the size and location of fuel tanks in ships, with the assigned technical solutions centred upon minimum tank / hull separation distances or tank sizings intended to attenuate both the likelihood of loss of fuel oil in the first instance, and the volume of any fuel which may be lost as a result of such incident.

Contrast this with warships, which for other, mission-related purposes, have small, dispersed fuel tanks, use light refined fuels, and have hull forms which minimise the likelihood of hull breach in the event of collision and grounding. Combined with enhanced propulsion, steering and crewing arrangements, the result is that warships are less likely to get into this sort of trouble in the first instance, and less likely to lose oil in the event of such casualty. The IMO-prescribed rules are blind to this reality, however, as they are for good reason focused upon merchant ship risks and remedies – not warships.

In a 'space-poor' warship, any loss of space as would arise from tank/hull separations results in some reduction in weapon and/or sensor fit, range/endurance, and/or crew habitability, none of which can be readily compensated; mandated tank configurations may also compromise damaged stability condition. Application of the IMO rule via the oil outflow performance models, founded upon merchant ship risk factors, can derive similar penalties.

Imposition of the IMO rules for protected fuel tanks in merchant ships to surface combatants represents a clear case of the burden of an ill-considered and poorly targeted technological solution alien to warships for an inconsequential environmental risk [2]. This would be bad enough of itself, but the great irony is that the observation of these merchant ship rules, for an inconsequential risk, would actually more likely have the opposite of the intended effect, by introducing a greater risk of fuel loss from the surface combatant. Examination of the historical record indicates that the most frequent cause of the unintended release of fuel from warships is due to refueling and tank transfers [3]. Considering that compliance with the IMO rules would invariably result in fewer and/or smaller tanks, this would necessitate more frequent refuelling and tank transfers — with commensurate increase in the risk of fuel losses.

Thus, application of a rule not intended for and addressing an inconsequential risk for warships, would result in increased risk to the environment, in parallel with compromising a warship's operational effectiveness. This is an irredeemably perverse outcome, and one better avoided by recognising that warship design features are 'near enough' to what is intended as an objective outcome by the IMO, without the nonsensical penalty of imposing upon one's self the theoretically 'perfect' technical solution compliant with the letter of the IMO rule.

Inordinate Technology: NO_x Exhaust Reduction Systems

The IMO requires ships operating in comparatively small and specially declared nitrogen oxides (NO_x) Emission Control Areas (ECAs) to operate their diesels in NO_x Tier III mode. In this mode, ship diesel NO_x emission rates are reduced to about 20% to 25% of what is permissible in Tier II mode, which applies outside of the ECAs.

The most widely adopted Tier III compliance approach is urea-mediated selective catalytic reduction (SCR) systems. These Tier III systems occupy relatively significant volume, are expensive, of questionable shock rating, require diesels to operate within narrow performance envelopes [4, 5], and, most critically, usually have substantial, heavy components installed in the upper levels of the ship – meaning adverse effects upon stability and the surrender of irreplaceable surface combatant top weight margins. These are the very surface combatant design margins which are prized for the installation of weapons and sensors – ie. combat capability systems.

Noting that there are 80 000+ registered merchant ships in the global fleet, compared with about 600 or so surface combatants (frigates, destroyers, cruisers) in commission in the world's navies, it is easy to determine that merchant ships as a source of NO_x form the core of the IMO's concerns. Consider also, that the Tier II configured surface combatant running on diesel would emit around 20 kg to 50 kg per hour of NO_x, while the Tier III merchant ships sailing past her, ostensibly with more admirable 'environmental credentials', would each be emitting anything from 100 kg to 200 kg per hour due to their larger, low speed diesels [6]. On an airshed scale, the additional NO_x emissions from a non-Tier III warship would be inconsequential, and disproportionate to the 'public cost' from forfeiting irreplaceable combat capability.

This is a patent example of a compliance technology imposing inordinate design and operational penalties for a combatant warship with minimal environmental benefit. Adoption of SCR NO_x emission control systems inescapably adds cost and complexity to warship design, operation and sustainment, for nothing more than intangible, inconsequential and ephemeral reductions in NO_x emissions, against the permanent and substantial surrender of irreplaceable and precious top weight margins. These considerations should tilt the balance sheet towards non-adoption by warships, given that this technology provides minimal benefit but at the cost of combat capability – the actual reason for the substantial public investment in the ship.

Alternative technical means more consistent with warship requirements and operating parameters are available if desired to reduce NO_x, albeit more expensive. This is arguably preferable than 'making do' with incongruous and unwieldy merchant ship technologies entailing inordinate and disproportionate penalties for nugatory and illusory environmental benefit, but questionable whether it is worth the financial cost.

Inane Technology: Exotic Fuels and Other Energy Efficiency Initiatives

The international shipping community is seeking to reduce the rate of greenhouse gas emissions from ships per tonne of cargo per unit of distance carried. One of the main routes through which this is being achieved is by the derivation and implementation of incrementally more stringent energy efficiency standards for ships. These energy efficiency goals are being realised by a number of planned or proposed or implied means including, *inter alia*:

- Ship propulsive power limitations
- Restrictions on ship operating speeds
- Use of novel and exotic fuels, including LNG, methane and blended fuels
- Use of sails

It is patently clear that imposing a 15 knot speed limit on a frigate, or restricting a destroyer's available power to that required to barely make headway in a heavy seaway is untenable. In addition, the ship design complexities, not to mention the combat survivability hazards and afloat support penalties, inherent to any suggestion of using LNG, for example, should discount this fuel from any serious consideration for use in warships.

IMO energy efficiency aims, while sensible and apposite for merchant ships employed in the transfer of cargo from one geographic location to another, bear little relation to the operating requirements and typical taskings of warships, and not even to afloat support ships, the closest equivalent to merchant ships. Support ships are required to have the capability to loiter, 'sprint' as necessary, and undertake the full range of manoeuvre required of warships for operational and exercise purposes, and rarely sail directly between two points. By extension, any application of the full gamut of IMO ship energy efficiency requirements to a warship of any sort is inappropriate and nonsensical, at best. This is not to suggest that warships should not seek to optimise energy use, if for no other reasons as a means of enhancing endurance for any given amount of fuel, but the objective of adherence to IMO energy efficiency requirements is patently an absurd proposition.

Clearly, implementation of IMO promoted energy efficiency measures would result in the application of technologies to warships which would be inane at best, drastically limiting combat capability while significantly compromising ship lethality and survivability and operational utility. This is another example of where the rational adoption of sensible energy efficiency technologies in warships, while not necessarily implementing the IMO's prescriptions, would result in a 'fit-for-purpose' solution which is superior and eminently more preferable to the superficially 'perfect' technological solution.

SUMMARY

This paper has shown reasoning that contradicts extant navy policies which advocate 'full compliance' with IMO marine environment protection rules, without distinction. The real capability of a warship, whether expressed in anger during conflict or in peacetime deployments, is linked with its credibility as a capable friend or adversary in its designated warfighting role. All else is secondary to that, except to the extent that it may enable the primary function. While responsible management of environmental risks and compliance obligations is important, it is not, nor should it be, as critical a consideration as are those facets related to lethality, survivability, endurance, and signature. Accordingly, questions of ship design and equipment fit purely for the purposes of environmental compliance, especially if the risk to the environment from that platform is inconsequential, should never usurp the pursuit of optimising a ship's operational effectiveness and lethality.

There are many examples where myopic adoption by warships of technologies intended for merchant ships will incur penalties which are totally disproportionate to the actual risks, address risks which have little if any relevance to the warship, or which actually accentuate or introduce risks to the environment – the very thing which they are supposed to avoid. Furthermore, it is difficult to justify the adoption of technologies where not only are the actual risks to the environment modest, but where such adoption will incur significant penalties in ship combat capability – the whole reason why the warship exists.

Technologies developed for and focused upon merchant ships – and their typical design and build processes, ship characteristics and operating profiles – often have no sensible or reasonable application to warships. The unalloyed adoption of unsuitable technologies simply on the basis of 'this is what is used in merchant ships' does not withstand scrutiny on any level - legal, technical, operational or in terms of ameliorating actual environmental risks.

In many circumstances, the adoption by warships of alternative, parallel technological solutions represents a more responsible and balanced approach to environmental management than default adoption of the wrong technology. It is inarguable that achieving an outcome which is mostly right more often, is better than adopting an externally imposed, but inappropriate solution that may appear to be correct on paper, but which rarely, if ever, actually works properly. Adoption of the wrong technology for a specific application - use in a warship - not envisioned for that technology and which fails to address the intended purpose represents a folly. This is more the case when the actual risks to the environment, real or imagined, are not addressed and especially when any such risks are compounded and amplified. This is indefensible when such misguided endeavours also disproportionately detract from the combat capability of the subject ship.

The technological solution which works 'well enough' most of the time is eminently preferable to the wrong technology which rarely, if ever, works properly in the warship context. The warship designer should always seek the technology which works 'near enough' while simultaneously being compatible with warship design and operational imperatives, rather than simply adopting the superficially 'perfect' technology borrowed from others when such technology is not fit-for-purpose.

In the warship paradigm of limited space and weight availability and other design and operational constraints, 'near enough' is usually more than good enough when managing risks to the environment, even if short of what some may promote as a 'perfect' solution. Preference should be given in any warship 'trade space' to those modifications which will enhance operational effectiveness, rather than those focused solely upon limiting already minimal environmental risks, and especially not those done simply for a policy pursuit of 'black letter' compliance with a rule in the absence of any tangible risk to the environment. In essence 'near enough' in the warship design context can be expressed as 'fit-for-purpose', which is what all warship designs should strive for.

ACKNOWLEDGEMENTS
The author thanks Darren Burrowes and Michel Fontaine for their advice in the development of this paper. The views expressed herein are those of the author.

REFERENCES

[1] Defence Publishing Service (2012). Australian Defence Doctrine Publication - D: Foundations of Australian Military Doctrine, Edition 3. Australian Government Department of Defence, Canberra.

[2] PGM Environment (2015). Defence Seaworthiness Management System - Environment Protection: Analysis and User Guide. Report prepared by PGM Environment.

[3] URS: Oil Spill Contingency Plan (2002). Naval Waters - Fleet Base West 2002. Report prepared by URS Australia Pty Ltd, Perth.

[4] Fenske C. (2012). "Transforming Grey Ships to Green: Emission Technologies for Naval Propulsion Systems", Proceedings of the Pacific 2012 Conference, Sydney, Australia, 30 January – 3 February 2012.

[5] Laird, A., Ledingham L. & Goodship J. (2016). "A Future Green Navy - Managing Environmental Expectation on Naval Ships", Proceedings of INEC 2016, Bristol, United Kingdom, 16 – 28 April 2016.

[6] Polglaze, J.F. (2018). Instead of Simply Asking 'What?', Naval Engineers Need to Ask 'Why?': Environmental Compliance Challenges and Relevance in Warship Design. In Conference Proceedings of INEC 2 – 4 October 2018.

Systems Graphing Algorithm Method for Ship Evacuation Studies

Vincent Capizzi

SYSTRA: vcapizzi@Systra.com

Dr. R Marian

UNISA: romeo.marian@unisa.edu.au

Prof. P Campbell

UNISA: peter.campbell@unisa.edu.au

1. ABSTRACT

This paper presents an outline of the development of a novel means to conduct evacuation modelling for maritime platforms using multi-commodity network theory and graph search techniques.

SOLAS/MSC.1/Circ.1533 (2016)[1] presents guidelines to conduct simplified/quasi-advanced analysis methods for the evacuation analysis of ships. The techniques presented by SOLAS are prescriptive and are limited such that path options with alternatives are difficult to analyse and don't consider uncertainty associated with dynamic and fluid evacuation environments, visibility, fire, smoke, ship motion and hazard propagation.

A linear programming model using multi-commodity networking and graphing algorithm techniques has been implemented in Excel. This model specifies crew and passenger numbers onboard ship, allows for personnel specifics and simulates ships geometry and path traversal difficulty. The simulation checks viable paths and determines the shortest route to the nearest safe point/muster station based on path difficulty and environmental conditions. Additional iterations in time allows for the hazard to propagate. Monte Carlo is used to pinpoint uncertainty. Updates to confidence levels determine the final probable shortest path/time.

[1] Revised Guidelines On Evacuation Analysis For New And Existing Passenger Ships, IMO - MSC.1/Circ.1533 6 June 2016

Initial findings indicate this Excel technique is feasible. As a systems engineering tool, it can be used in trade-off studies for ship layout purposes.

2. INTRODUCTION

The research into the development of an intelligent tool for evacuation from Naval ship platforms is based on the premise that it is analogous to the puzzle presented by the Traveling Salesman Problem (TSP). A salesman must travel between several cities (nodes). The order in which he does this is not important as long as he visits each once during his trip and finishes where he started. Each city is connected to other close cities by links (designated as edges) by airplanes, or by road or railway. Each of those links between the cities has one or more weights (or the cost) attached. The cost describes how "difficult" it is to traverse this edge on the graph, perhaps by the length of the edge, or the time required to complete the traversal or its individual cost. The salesman wants to keep both the travel costs and the distance he travels as low as possible. This problem, used in operations research, focuses on optimization; to seek a better solution being cheaper, shorter, or faster. An algorithmic model is developed to solve the TSP and is based on nodes and edges exhibiting a network like structure typically referred to as a Graph algorithm. The Traveling Salesman Problem is one of the most intensively studied problems in computational mathematics and has been in existence since the 1800s by the Irish mathematician Sir William Rowan Hamilton and by the British mathematician Thomas Penyngton Kirkman[2].

When the Graph algorithm is applied to ship platforms, the following similar traits are identified:

a) The ship layout can be represented as a TSP with constraints identified and "difficulties" along links modelled;
b) The ship is represented by a routing network using nodes describing originating and transition points (compartments, rooms) and edges (links – corridors, passageways, stairs) connecting the nodes; and
c) Adopting an optimisation algorithm such as that provided by a technique known as the multi-commodity flow allows for some certain degradation with time and allows for optimising routes taken by the crew in an emergency. There is a notable difference with the TSP: the start and the end of the travel are different and set for each room.

This paper covers the above points, the constraints for both the platform and using the Multi-commodity flow algorithms, as well as their advantages.

Note that there is a plethora of evacuation models on the market[3]; the development of this application is seen as an alternate view to the standard approaches available.

---

[2] History of the TSP; http://www.math.uwaterloo.ca/tsp/history/index.html

[3] The current status and future issues in human evacuation from ships, D Lee, H Kim, JH Park, BJ Park, Safety Science 41 (2003) 861–876.

3. BACKGROUND

In response to major ship disasters, Nation States have moved towards internationalization of laws to be introduced that utilise local regulations, agreements or understandings among the leading maritime nations. A conference convened by the United Nations in Geneva in 1948 eventually led to the establishment of the International Maritime Organisation (IMO), a body responsible for the safety and security of shipping and the prevention of marine pollution by ships by adopting Safety Of Life At Sea (SOLAS) guidelines. As part of the IMO's "raison d'être", emergency evacuation simulation now plays an important part in the consideration of design of vehicles, ships, aircraft or buildings with interest coming from academic, industrial and scientific fields[4].

The implementation of SOLAS Regulation II-2/28-1.3, called for an appointed Sub-Committee on Fire Protection (FP) to develop guidelines for evacuation analysis for passenger ships and high-speed passenger craft. This promulgated a series of circulars, the most important being MSC/Circ.1033 (2002) & MSC.1/Circ.1238 (2007) which provide guidelines for the evacuation analysis for ship platforms. Currently MSC/1533 (2016)[5] replaces the previous two circulars and provides similar guidelines. To make correlation uniform, benchmark scenarios and relevant data are specified which are used to determine the evacuation performance of the ship. Most data and parameters stated in the guidelines are based on well-documented data coming from civil building experience. These guidelines provide performance criteria in terms of total calculated evacuation time compared to the allowable evacuation time.

4. EVACUATION EXAMPLE FROM MSC/1533

The MSC/1533 for ship evacuation provides:

1. the use of guidelines when conducting evacuation analyses, early in the design process, on new passenger ships to mitigate the risks of safe evacuation and put in place structural and procedural practices to deliver a safe ship; and
2. the encouragement to conduct evacuation analyses on existing passenger ships using these guidelines to uncover inherent evacuation difficulties in order to effect a perceived improvement to evacuation performance; to identify congestion points and/or critical areas and to provide recommendations as to where these points and critical areas are located onboard[6].

MCS/1533 provides an example of the use of the guidelines; the example refers to an early design analysis of arrangements of a hypothetical new cruise ship. The performance standard is assumed to be 60 mins, as for ro-ro passenger ships. It was noted that, at the time this

---

[4] DiNenno, P.E. et al. The SFPE Handbook of Fire Protection Engineering, 3rd Ed., 2002.

[5] Revised Guidelines On Evacuation Analysis For New And Existing Passenger Ships, IMO - MSC.1/Circ.1533 6 June 2016

[6] MSC 1533 states "it is the company's responsibility to ensure passenger and crew safety by means of operational measures, if the result of an analysis, conducted on an existing passenger ship shows that the maximum allowable evacuation duration has been exceeded, then the company should ensure that suitable operational measures (e.g. updates of the onboard emergency procedures, improved signage, emergency preparedness of the crew, etc.) are implemented"

example was developed, no such requirement is applicable for passenger ships other than ro-ro passenger ships.

The example ship exhibits seven decks (Deck 5 to 11); Deck 8 is the muster station. Corridors are connected to stairs that allow passages through to Deck 8. The lower decks are Decks 5,6 and 7. The upper decks are 9, 10 and 11. A diagrammatic view is shown in MSC/1533 and is presented in Figure 1 as a "hydraulic diagram".

The corridors connect to doors, stairs or openings; the goal to reach the Muster Station on Deck 8. Each corridor/door has properties relating to length and width in metres (Note that a door has no length, only width).

Corridors specify the number of crew or passengers that occupy the corridor space at a specific time as a distribution, comprising 449 passengers/crew in total.

The performance standard is stated below. It defines the important parameters and puts into perspective the expected conditions and evacuation response results.

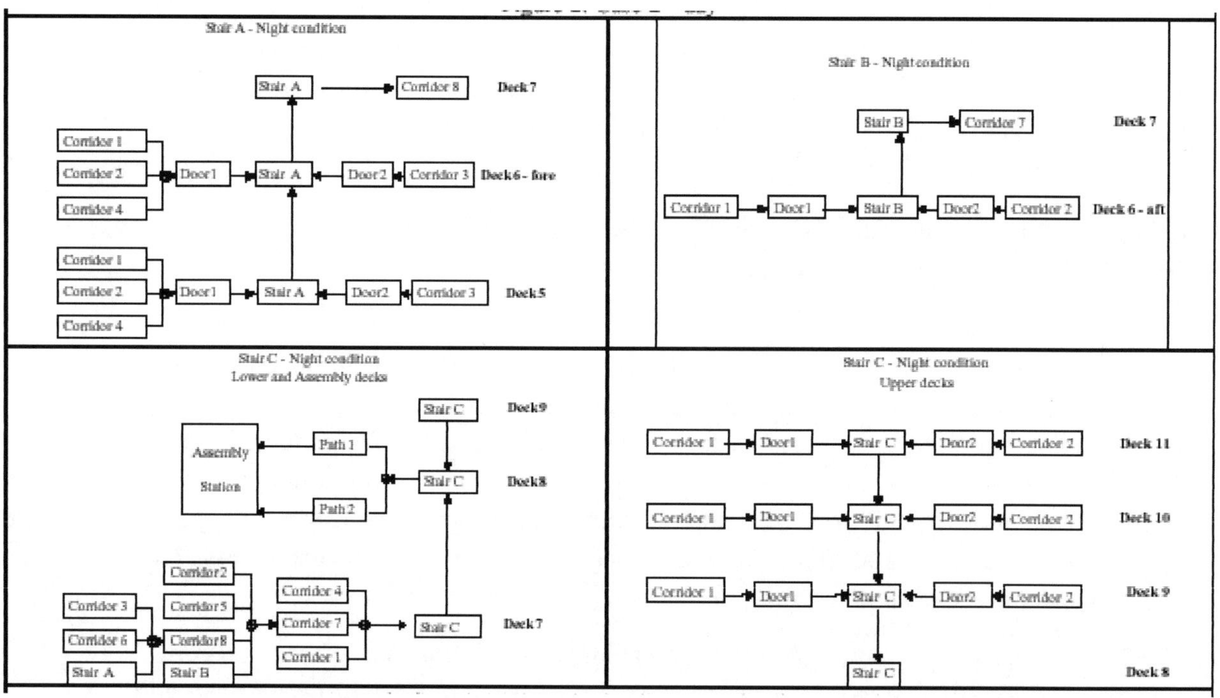

Figure 1 From MSC/1533 – Case 1 Night Case Hydraulic Representation

The MSC/1533 evacuation performance standard is defined as:

1.25 (R+T) + 2/3 (E+L) ≤ = n[7]; where "n" for passenger ships = 60 mins.

[7] Ibid 4

E+L is the embarkation and launching time = 30 mins.

R is response duration – for purposes of indicating a benchmark, R should be 10 mins for the night time scenarios and 5 mins for the day time scenarios.

T is the total travel time, calculated.

The example considers the overall ships geometry plus the factors and assumptions; the total evacuation time is calculated to be 41 minutes and 38 seconds; this meets the performance standard for the night; Case 1.

The focus is therefore the calculation of "T"; the other parameters are based on history/experience. The final calculation of "T" is stated as 437.5 seconds, that being the longest duration across the escape routes.

5. NETWORK VIEW

To display a enhanced perspective of the ships evacuation geometry, a node and edge (vector) notation is used to represent the geometry and the direction of flow; as required for a forward search graphical algorithm, as shown in

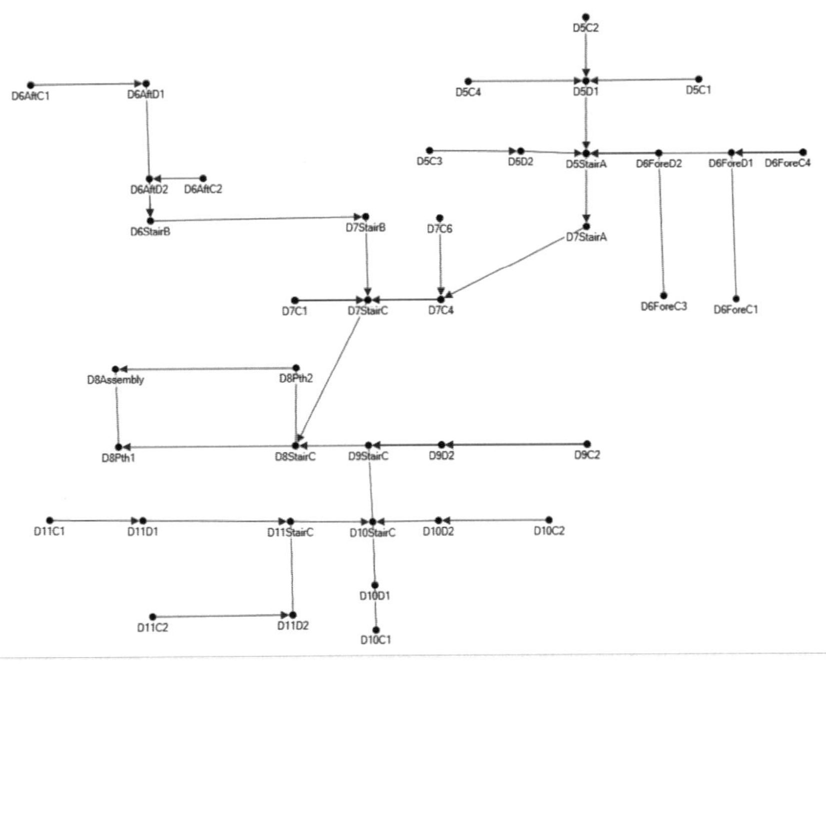

Figure 2 below.

The label refers to the location onboard ship:

"D"x refers to Deck x; "C"y refers to Corridor y; "D" In the body of label refers to Door. For example, D12D3 refers to Deck 12 and Door 3 the second "D" refers to Door).

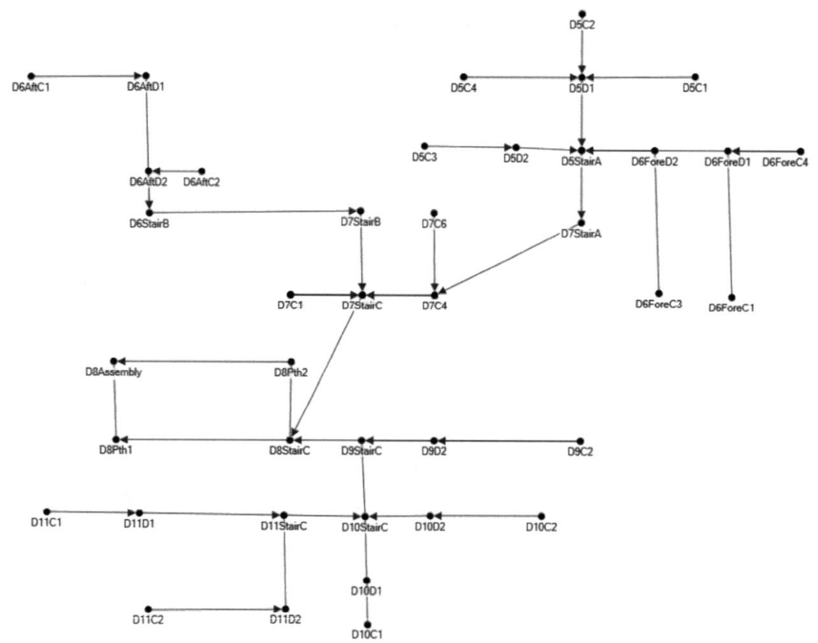

Figure 2 Edge-Nodes Representation.

The following observations are made:

a) A 2D view of the connected paths show pathways and their direction of traffic flow. In this case, flow originates from the source nodes and ends at a final destination on Deck 8 Assembly Area (D8Assembly). The 2D view also makes it easier to identify 3D aspects of location; and

b) For this example, not all pathways are modelled highlighting a limitation of MSC 1533, that is the paths are chosen a priori. In the real case, there are more connections/links and routes that passengers can take.

The network view also highlights in a clear manner the critical paths. For example, if Stair C between Deck 8 and 9 is closed, then the passengers from decks 9 to 11 are stranded and can't proceed to Deck 8 Assembly Area.

This network representation of escape routes is the basis of the evacuation analysis where passengers are located on edges and need to proceed to the Assembly Area in the shortest amount of time. This is a linear problem and can be solved using a multi-commodity approach.

The approach is suited to dynamic network reconfiguration; to simulate an emergency which involves transient phenomena such as fire, smoke or structural damage resulting in changes to pathway access. The model can be dynamically updated to reflect the changed network in real time.

This benefits design studies and can be used to audit existing designs for performance purposes; this is the "systems" view to evacuation and provides the basis of this research.

6. Multi-Commodity Modelling[8]

Multi-commodity flow modelling is a network flow representation of multi-source and sink identities that consider demand and capacity fluctuations.

A common scenario of a network-flow problem arising in industrial logistics concerns the distribution of a single homogeneous product from plants (origins) to consumer markets (destinations). The total number of units produced at each plant and the total number of units required at each market are assumed to be known. The product need not be sent directly from source to destination but may be routed through intermediary points reflecting warehouses or distribution centres. Further, there may be capacity restrictions that limit some of the shipping links. The objective is to minimize the variable cost of producing and shipping the products to meet the consumer demand.

Conceptually similar, passengers from locations in a ship (sources) are required to egress to an Assembly Area just prior to evacuation. The total number of passengers to assemble are known. Passengers do not need to travel directly to the Assembly Area but can take alternate routes to facilitate efficiency and minimise travel time. It is also known that routes have capacity restrictions. The objective is to minimise the travel time (Cost) and ensure all passengers reach their destination.

The modelling is characterised by:

a) flow capacity along edges and secondly, cost is assigned for each edge, i.e. the time duration to travel along edge;

---

[8] Applied Mathematical Programming, Bradley, Hax, and Magnanti (Addison-Wesley, 1977)

b) edge capacity, the sum of flows across an edge does not exceed its capacity requirements;
c) flow conservation at the source and sinks, i.e. Inflow demand equals outflow supply; and
d) flow conservation on transit nodes, flow entering a node is the same that exits.

A network comprising nodes and edges has many solutions and hence, one may optimise the flow to correspond to minimising cost or capacity. Therefore, for a minimum – time flow problem with n nodes:

$xi\,j$ = Number of passengers traversing from nodes i to j using edge *i–j*[9]

$$\sum_j x_{ij} - \sum_k x_{ki} = b_i \quad (i = 1, 2, \ldots, n), \qquad \text{[Flow balance]}$$

$$l_{ij} \leq x_{ij} \leq u_{ij}. \qquad \text{[Flow capacities]}$$

There are numerous publications in the field; multi-commodity modelling is a universal approach that can also be adaptable to many industrial problems[10].

The following observations are made:

a) Network models are special structures in linear programming, making them efficient and effective;
b) The approach presented is simply derived from specializing the rules of the simplex method to take advantage of the structure of network models;
c) The algorithms are extremely efficient and can permit the solution of a large (ship) network model using a special linear-programming procedure; ideal for the complexity of ship geometry and human dispersion; and
d) This technique can be easily modelled and implemented using MS Excel with the Solver add-on.

7. The Ship Model

The prime objective of this research is to confirm the use of multi-commodity analysis is a feasible method to solve evacuation cases at a high-level systems view.

Appendix 2 of MSC 1533 presents the IMO evacuation example[11]. Simplified and advanced analysis methods are listed therein. Using these techniques, personnel traversing the paths had no options to use alternative paths; the Guidelines do not allow for alternatives to be assessed; the prime focus is on crowding but assumes all identified paths are available.

[9] ibid 3
[10] Reference Applied Mathematics Programming, Chapter 8, Subparagraph 8.8 Reference MIT website; http://web.mit.edu/15.053/www/AMP-Chapter-08.pdf.
[11] Ibid 4

The Ro-Ro[12] ship representation of the model shown in Figure 1 for a night scenario case was created using the following techniques:

a) the approach is theoretical and mathematical focussed to develop realistic ship evacuation simulations at a system level for the needs and requirement determination in a design environment;
b) the methodologies are based on building evacuation studies;
c) the example cited herein is based on a benchmark scenario and is not specific to particular cases and detailed configuration or environmental issues found during an actual evacuation. In this case, the primary evacuation case is the night case in accordance with the FSS Code13 and includes a passenger manifest;
d) The prime aim of the exercise is to determine the egress duration for passengers, and to identify congestion points. If these points exist, then the design iteration is required to mitigate these points as they impact dramatically egress duration;
e) The approach in this example is similar to that found in MSC 1533 with respect to process and conventions; and
f) The assumptions included:
   i) Passenger travel velocity is based on building data and adapted to fit within the context of this analysis;
   ii) Passenger age, sex and skill levels have been averaged out for simplicity sake. The objective of this paper was not to better define travel metrics and parameters;
   iii) Damage, smoke, heat and toxic conditions and effects have not been included;
   iv) Group behaviour is not considered;
   v) Ship motion and trim have not been considered;
   vi) Capacity is defined as the number of passengers traversing an edge and is calculated and defined using people density. This capacity is used to define the congestion points when the density is equal to or greater than 3.5 persons/m^2;
   vii) Walking speed is in accordance to Table 3.4 of the MSC 1533 Circular[14]. It is assumed that for all ages and sex, there is a mean value and a standard deviation. The final velocity for each egress traverse is calculated using the inverse of the normal cumulative distribution for a specified mean and standard deviation using a random probability;
   viii) MS Excels Solver macro is used to determine the edge traverse flow and optimise (maximum or minimum) the final egress duration value subject to 'network flow constraints', i.e. calculate the number of people traversing specific edges to minimise egress duration. Figure 3 below presents the spreadsheet used in the calculations and a view of the Solver input parameters. Note that the non-linear GRC solver was used due to the random input quantities;

[12] Ro-Ro – Roll on, roll off.

[13] FSS Code - International Code for Fire Safety Systems (FSS Code) (resolution MSC.98(73))

[14] Ibid 4

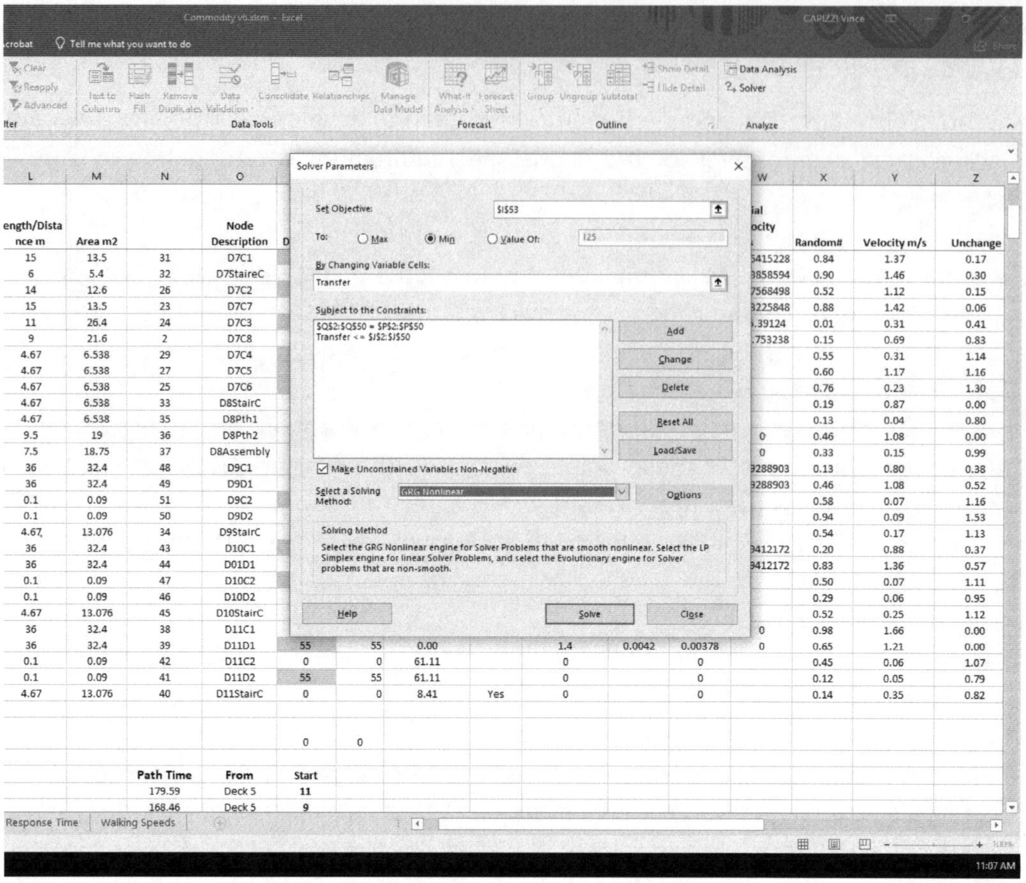

Figure 3 Excel Spreadsheet Solver View

ix) The routes are identified and an egress duration is determined for each path; from source to Assemble Area;

x) The path which exhibits the longest egress duration is chosen as the representative duration for the evacuation; and

xi) To express uncertainty in the solution, a Monte Carlo analysis is conducted to find the most likely time duration delivered (>500 iterations).

The model process is as follows:

i) Define nodes and edges using a hydraulic representation;

ii) Define edge properties, length, area and passage clear width;

iii) Define constraints to optimise (minimise travel duration); and

iv) Use Solver to determine the number of persons traversing each passage;

Graph edge properties can be changed to see how factors such as visibility, age or access affect the edge distance or time.

Multi Commodity (MC) network modelling provided was found to be a significantly better technique compared with traditional graph search algorithms. Moreover, it offers:

a) The ability to include enhanced features that cover most of the SOLAS requirements presented in MSC 1533; and
b) Demonstrated features that included multiple source and sink combinations, the advantage to add capacity/ and time constraints on edges and to consider the conservation of flow allowing identification of choke points.

This technique identifies design issues clearly and should be considered during the design phase; the MC tool is flexible and more adept to initial evacuation ship designs quickly.

To attend to the actual conditions during evacuation such as fire, smoke, visibility and ships motion, the MC network model can be adapted to represent individual passengers as required. This tool provides a resilient and realistic model of evacuation parameters.

7.1. Case 1 Result - Night Case Solution with no Capacity Restriction

In this solution, it is assumed that Capacity of flow is not a restriction, but however it highlights congestion areas.

The MSC1533 calculated result is 437.5 seconds. Table 1 presents results for this same case using MC modelling.

Egress duration sec	Number of Hits, Monte Carlo Analysis
Below 350 seconds	113
Between 350 and 450 seconds	369
Greater than 450 seconds	70

Table 1 Night Case Solution with no Capacity Restriction

The results indicate that the most likely result is between 350 and 450 seconds which aligns with the MSC 1533 result of 437.5 seconds.

Congestion was identified as identified in Table 2; generally, in accordance with MSC1533.

DECK	Location From	To
5	D5StairA	D6StairA
6	D6StairA	D7StairA
6	D6StairB	D7StairB
7	D7C7	D7StairC
7	D7C8	D7C7
7	D7StairA	D7C8
7	D7StairB	D7C7
7	D7StairC	D8StairC
8	D8StairC	D8Pth1
8	D8StairC	D8Pth2

8	D8Pth1	D8Assembly
8	D8Pth2	D8Assembly
9	D9StairC	D8StairC
10	D10StairC	D9StairC
11	D11StairC	D10StairC

Table 2 Congestion Areas

7.2. Case 2 Result - Night Case Solution with Capacity Restriction

The model was calculated using the Capacity restriction along the edge paths. The final transfer of persons to the Assembly Area from Path 1 and Path 2 was 208 and 241 respectively totalling 449 passengers with a total duration of 414 seconds.

To highlight egress difficulties due to structural damage and/or fire or smoke impediments, a 50-person Capacity restriction was applied to Path 1. Additionally, as a second capability example, Path 1 was completely shut off. The results of this analysis are shown in Table 3 below:

Capacity Restriction	Persons traversing Path 1	Egress Duration Impact Seconds
None	213	410
50 persons	50	444
Shut down – cannot traverse	0	557

Table 3 Night Case Solution with Capacity Restriction

The above result reflects the diminishing capacity of Path 1 to egress optimally with restrictions in place.

8. Conclusions

The following conclusions have been reached from this research:

a) Evacuation paths can be modelled that use cost functions that account for time, difficulty of cross compartment movement difficulty to move from compartment-to-compartment, additional constraints like smoke/fire, water, dark corridors, steam, oil/diesel, etc., and aids such as visual and audio signs and alarms, announcements and PPE. The "shortest" path is an optimised time value for egressing safely by minimising the cost and loss function based on local conditions and personnel status on attempting egress. But the egress time duration is not precise, as the probability comes into play to determine likely egress durations;
b) Multi-commodity graph modelling is an appropriate technique to calculate evacuation systems. It is suited to evacuation optimisation incorporating user friendly interactive communication with ships system configuration;
c) The tool is flexible and adept to conducting initial systems engineering needs, requirements and conceptual studies of ship evacuation systems. Various evacuation restrictions concerning fire, some toxic gas and structural damage can be played out quickly and efficiently by changing path capacity and managing velocity outcomes;
d) A solution cannot be found if persons have no alternative path to reach the destination. Perhaps this is its greatest contribution; and

e) With the capability to configure paths (edges), it can dynamically assess damage or fire fronts by incorporating environmental transients which can be used to configure person movements in real-time.

9. Recommendations

The research into this tool was not conducted to find replacement of specialist tools that have been developed over a long period of time. Rather, the aim of this research is to develop a tool that can be used simply and efficiently to provide a system view of evacuation and is aligned to the guidelines presented in MSC 1533.

It is recommended that the tool be further developed to cover transient phenomena during damage forced situations and to simulate the path dynamics to develop a time-based description of the damage scenario as it develops. This is the real potential of this work.

Lithium-ion Battery Fire Suppression in Submarine Battery Compartments

I. Burch[1], ian.burch@dst.defence.gov.au
M. Ghiji[2], mohammadmahdi.ghiji@vu.edu.au
G. Gamble[1], grant.gamble@dst.defence.gov.au
B. Suendermann[1], brigitta.suendermann@dst.defence.gov.au
P. Joseph[2], paul.joseph@vu.edu.au
K. Moinuddin[2], khalid.moinuddin@vu.edu.au
V. Novozhilov[2], vasily.novozhilov@vu.edu.au

[1] Defence Science and Technology Group
[2] Victoria University

ABSTRACT

Lithium-ion main storage batteries have the potential to improve the endurance of diesel-electric submarines through superior energy storage and charging capabilities when compared with traditional lead-acid batteries. A review of lithium-ion batteries has found that they pose a potential fire risk and conventional fire suppression systems may not provide an adequate level of safety in the event of a lithium-ion battery fire.

Studies have identified water-based fire suppression as effective for lithium-ion battery fires. Water extinguishes flames and cools the battery inhibiting exothermic reactions within the battery during a thermal runaway event. The application of liberal amounts of water or submerging the battery in water has been shown to achieve extinguishment of lithium-ion battery fires. However, these firefighting methods are not appropriate for submarines because the excessive volume and weight involved may result in submarine stability requirements not being met. Water mist fire suppression systems have the extinguishing and cooling capability of sprinklers or deluge systems whilst using a lower volume of water and are therefore potentially applicable for spaces such as submarine battery compartments.

In collaboration with Victoria University's Centre for Environmental Safety and Risk Engineering (CESARE), Defence Science and Technology Group is investigating the effectiveness of water mist fire suppression for lithium-ion battery fires. The aim of this work is to experimentally determine the extinguishing criteria for lithium-ion battery fires and develop Computational Fluid Dynamics models to predict the fire extinguishing capabilities of water mist on lithium-ion battery fires.

This paper describes why lithium-ion batteries are susceptible to fire, why water mist is seen as a promising fire suppression system for lithium-ion battery fires, and presents some of the preliminary experimental results of the collaborative research with CESARE.

INTRODUCTION

Lithium-ion batteries offer high energy and power density, light-weight and long lifespan [1, 2] and is the current preferred technology for mobile electronics, power tools, electric grid backup and aerospace applications [2]. Lithium-ion batteries are also gaining interest for use in diesel-electric submarine propulsion because their energy content could improve submarine range and endurance when compared to conventional lead-acid batteries [3]. In October 2018, the first of two lithium-ion battery equipped Soryu-class diesel-electric submarines was launched for the Japanese navy [4], these are the first submarines to be equipped with lithium-ion batteries. Lithium-ion batteries have also been developed for the second batch of three of the South Korean navy's next generation attack submarines [5].

A review [6] has found that, although occurrences are rare, lithium-ion batteries are a potential fire risk. Lithium-ion batteries are susceptible to thermal runaway, a complex combination of chemical reactions that are initiated by heat generated either internal or external to an individual cell. The reactions within the cells are exothermic resulting in a feedback loop that continues to produce heat until all reactive agents within the cell are consumed. The heat build-up can cause the combustible electrolyte within the cell to vapourise, vent and ignite resulting in heat transfer that could initiate thermal runaway in adjoining cells.

Studies [7, 8] have shown that water-based fire suppression is effective against lithium-ion battery fires as water extinguishes the combustion flames and cools the cells, inhibiting the exothermic reactions within the cell during thermal runaway. This has been achieved through the application of liberal amounts of water or by submerging the battery in water. This system of suppression is unlikely to be used on a submarine because of the restricted volume of water able to be carried, and stability requirements which may not be met upon the release of a large volume of water into a battery compartment. A system that has yet to be investigated for suppressing or extinguishing lithium-ion battery fires is water mist, a spray of fine water droplets that can provide both fire suppression and battery cooling. Water mist could extinguish burning electrolyte and potentially drive battery temperatures below the thermal runaway temperature and halt the reactions that generate heat. Water mist could be a viable extinguishing agent for submarine lithium-ion battery fire suppression because of its low water requirement and non-toxic nature.

Defence Science and Technology (DST) Group are currently collaborating with Victoria University's Centre for Environmental Safety and Risk Engineering (CESARE) to experimentally assess the thermal runaway and extinguishing processes and determine the criteria for lithium-ion battery fire suppression with water mist, criteria such as flow rate, orientation of the flow, water pressure and drop size distribution. This data will then be used to validate numerical analysis of lithium-ion battery fire suppression.

This paper describes lithium-ion battery chemistry, lithium-ion battery susceptibility to thermal runaway and why water mist could be the preferred choice for fire suppression. Also presented is an outline of the collaborative work being undertaken by DST Group and CESARE and some of the preliminary results.

LITHIUM-ION BATTERY COMPONENTS

Lithium-ion batteries are re-chargeable and comprise current collectors, electrodes (anode and cathode), flammable electrolyte, packaging materials, and separator. The term lithium-ion does not refer to singular cell chemistry but to a number of different chemistries. The choice of materials for the electrodes and electrolyte will affect the voltage, capacity, life, and safety of a lithium-ion battery. The following sections briefly describe lithium-ion battery components.

Current Collectors

The current collectors are the cell terminals and provide an even distribution of current across each of the electrodes. The anode and cathode materials are deposited onto copper and aluminium current collectors respectively so that the current collectors provide physical support to the electrode materials.

Electrodes

Typically, a metal oxide forms the cathode and graphite forms the anode. During discharge, lithium ions migrate from the anode and insert into voids in the cathode and upon charging, migrate from the cathode and insert into voids in the anode. The process is known as intercalation, the reversible insertion of ions into layered materials. The cell components and the intercalation processes are illustrated in *Figure 1*.

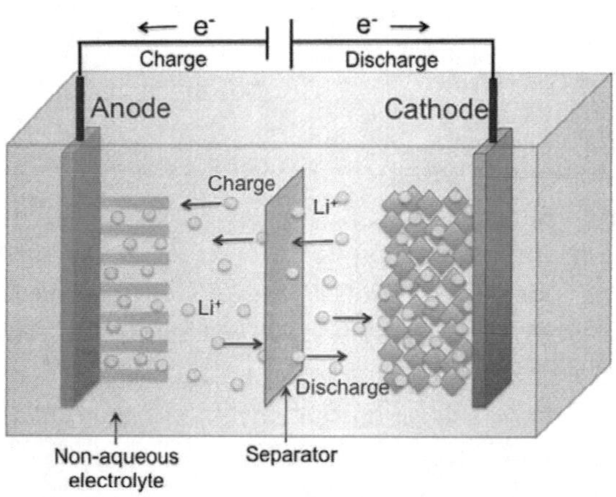

Figure 1. The insertion of lithium ions (yellow spheres) into the anode and cathode matrices during the charging and discharging processes [9].

Solid-Electrolyte Interface

During the initial battery charge, the intercalated lithium ions react with the electrolyte forming a passivating layer, the Solid-Electrolyte Interface (SEI), on the anode. The SEI allows lithium ion transport to the electrode but blocks electron flow to prevent electrolyte decomposition which consumes lithium ions and reduces the charge and discharge efficiency of the electrode material. The SEI forms mostly during the first charge, but continues gradually on further charging until the SEI layer is fully developed.

Separator

The separator is a thin porous membrane that physically separates the anode and cathode. The function of the separator is to prevent contact between the anode and cathode but allow lithium ion transport to and from the anode and cathode.

Electrolyte

The electrolyte is the medium that provides and facilitates the movement of lithium ions between the cathode and anode. The electrolyte in current technology lithium-ion batteries is flammable.

THERMAL RUNAWAY

If a lithium-ion battery temperature exceeds its safe operating temperature, the battery materials and electrolyte can lose their stability and generate heat which if not dissipated will generate more heat, a process known as thermal runaway.

Thermal runaway can be initiated by: internal defects producing an electrical short between the electrodes, elevated temperatures, over charging or discharging, or mechanical abuse. When a lithium-ion battery cell temperature exceeds the safe operating temperature of the cell, thermal runaway can take place in a multi stage process [10], typically

- the SEI layer begins to decompose exothermically at 90-120 °C (although this is dependent on the chemistry of the cell),
- exothermic reactions occur between the electrolyte and the anode above 120 °C and
- above 200 °C, thermal decomposition of the cathode evolves oxygen which can react with the electrolyte, followed by exothermic electrolyte decomposition.

The elevated temperatures produced during thermal runaway can induce electrolyte vapour production, pressurising the cell and triggering electrolyte venting or possibly rupturing the cell. Electrical equipment and wiring within the battery compartment can provide an electrical spark to ignite flammable liquids and gases however a fire can be started within lithium-ion batteries when the auto ignition temperature of the electrolyte is reached.

HAZARD AND RISK OF LITHIUM-ION BATTERY FIRES

Despite the extensive usage of lithium-ion batteries in electronic devices and vehicles, there is a fire risk associated with their use which is a concern in domestic electronic equipment [11], electric and hybrid electric vehicles [12, 13, 14, 15], and submarines [3, 16]. To achieve a desired voltage and energy content many small cells are closely packed in multi-cell batteries and the heat generated from a thermal runaway event in one cell can propagate to surrounding cells causing a cascading fire event.

Lithium-ion battery banks are generally safe under normal charge and discharge conditions. Battery Monitoring Systems (BMS) [17] are used to reduce the risk of thermal runaway by controlling lithium-ion battery characteristics such as current, temperature, and voltage within their safe operational window. However, due to the processes described above thermal runaway can still initiate.

To mitigate the risk of heat propagation to adjacent cells during a thermal runaway event, physical safety features such as an air gap between cells [18], thermal insulation [19] and heat absorbing phase-change and heat sensitive material between cells [20, 21] can be incorporated within the battery compartment. However these features may add weight and volume and negate the advantages of using lithium-ion batteries. The development of new chemistry lithium-ion batteries, focusing on fire resistant electrolytes, is ongoing and these are not yet commercially available [22].

LITHIUM-ION BATTERY FIRE SUPPRESSION

Lithium-ion battery fires are primarily considered to be Class B (flammable liquids) fires as defined in Australian Standard AS1850:2009 [23] as they contain flammable liquid electrolytes, and can be extinguished like other Class B fires. However, if the cell is not sufficiently cooled, thermal runaway may continue and the combustible materials within the battery may re-ignite.

The total energy available for combustion within a lithium-ion battery is dependent on the stored electrical energy and the combustion load, consisting of the electrolyte and the flammable battery construction materials. The quantity of electrolyte in a cell can maintain combustion for a limited time but adjacent cells may undergo thermal runaway if heat transfer to the adjacent cells is not prevented.

Lithium-ion battery fire tests [7, 8] have shown that applying liberal quantities of water or by submerging in water, a lithium-ion battery fire can be extinguished and the cell cooled inhibiting ongoing combustion or cascading to adjacent batteries. For lithium-ion battery fire suppression on a submarine, water mist may be a suitable extinguishing agent because of its ability to provide cooling and extinguishment and low water requirement compared with other water based suppression systems.

WATER MIST FIRE SUPPRESSION

The term water mist refers to fine water sprays, a continuum of droplets in the range between aerosol (droplet diameter $\simeq$ 5 μm) and fog (10 μm $\leq$ droplet diameter $\leq$ 1000 μm) [24]. The United States National Fire Protection Association (NFPA) has defined water mist as a spray with a range of particle sizes smaller than 1000 μm [25]. Compared to gaseous agents and conventional sprinkler systems, water mist as an agent has the following advantages:

- no toxicity or asphyxiation concerns,
- no environmental concerns, and
- low water volume requirements.

Extinguishing Mechanisms

There are three mechanisms associated with the extinguishment of fires by water mist:

- Cooling

  The cooling mechanisms provided by water mist for fire suppression comprise cooling the flame and cooling the fuel surface. Water droplets can penetrate the flame, reducing the flame temperature below adiabatic as well as reducing the temperature of the fuel below its flash point. Each of these mechanisms is capable of halting combustion [24].

- Oxygen depletion

During a fire, water droplets are converted to vapour and the total volume occupied by the droplets increases by over three orders of magnitude [26]. The water volume expansion results in the disruption in the entrainment of air into the flame as well as the dilution of the oxygen to below the limiting oxygen concentration, halting combustion.

- Radiation attenuation

The presence of water mist significantly decreases the radiant heat flux to materials in the proximity of the fire, through heat absorption, which will limit the spread of fire. Water mist will also reduce the radiation feedback to the fuel surface, reducing the pyrolysis rate of the fuel.

Spray Characteristics

The effectiveness of water mist in suppressing fires is related to three main spray characteristics; the drop size distribution, the flux density and the spray momentum.

- Drop size distribution

The drop size distribution refers to the range of droplet sizes within a spray volume. Water mist contains droplets of different sizes which vary with time and location as the droplets collide, vapourise, or impinge on surfaces and drop out.

A volume of water divided into small droplets has a larger total surface area than an equivalent volume divided into larger droplets. Smaller droplets are therefore more efficient at heat absorption leading to flame cooling and oxygen depletion. The smaller drop sizes also exhibit gas like qualities and can be carried around the compartment to obstructed areas and provide extinguishing conditions where the larger droplets would not reach. Larger droplets can penetrate the fire plume, cooling the fuel and slowing the pyrolysis rate.

- Flux density

Flux density refers to the amount of water spray present in a unit volume. An increase in flux density will provide more water into a compartment and provide greater cooling, that cooling can reduce the flame temperature and fuel temperature, suppressing the fire.

- Spray momentum

Spray momentum refers to the spray mass, velocity and direction. The spray momentum will determine whether the water droplets have sufficient energy to penetrate the fire plume or reach the fuel surface. Without sufficient energy to penetrate the fire plume, the flame temperature will remain above the adiabatic flame temperature and extinguishment will not occur.

EXTINGUISHMENT EXPERIMENTS AND MODELLING

DST Group are currently collaborating with CESARE to assess the combustion behaviour associated with lithium-ion battery fires and determine whether water mist will suppress the fire and prevent thermal runaway. The fire and extinguishing experimental data will be used to validate a Computational Fluid Dynamics (CFD) analysis of lithium-ion battery fire and suppression. CFD packages such as Fire Dynamic Simulator, developed by the National Institute of Standards and Technology [27], are used to simulate fire scenarios by fire safety engineering and research communities. CFD will be used to model lithium-ion battery combustion and water mist interaction and simulate fire scenarios with different geometries, mist parameters, fuels, obstructions and construction materials. Once validated, CFD could be used to assess the design parameters of water mist systems and to contribute to the generation of firefighting standard operating procedures.

In order to validate the CFD modelling, experiments will be undertaken that focus on lithium-ion battery fire behaviour and water mist/lithium-ion battery fire interaction. The following sections present some preliminary results.

Lithium-ion Battery Fire Characterisation Studies

Studies have been conducted on used, lithium-ion polymer batteries, from Motorola two-way radios (battery type: PMMN4066). The batteries contain two flat cells connected in series but were separated into single cells. The cells were mounted vertically with a side mounted propane gas burner to apply heat to the cell which also acted as an ignition source for vented gases, as seen in *Figure 2*. Cell surface temperature and mass loss were measured.

The images in *Figure 2* represent typical stages that occurred during lithium-ion battery fire tests. A small gas jet (A) has released as a result of electrolyte vapourisation and is easily ignited. This develops into a large rapidly venting gas jet (B) which extinguishes the flame. As the venting gas velocity decreases, re-ignition occurs (C). Eventually the electrolyte gases are consumed but the packaging and internal combustibles maintain the fire (D).

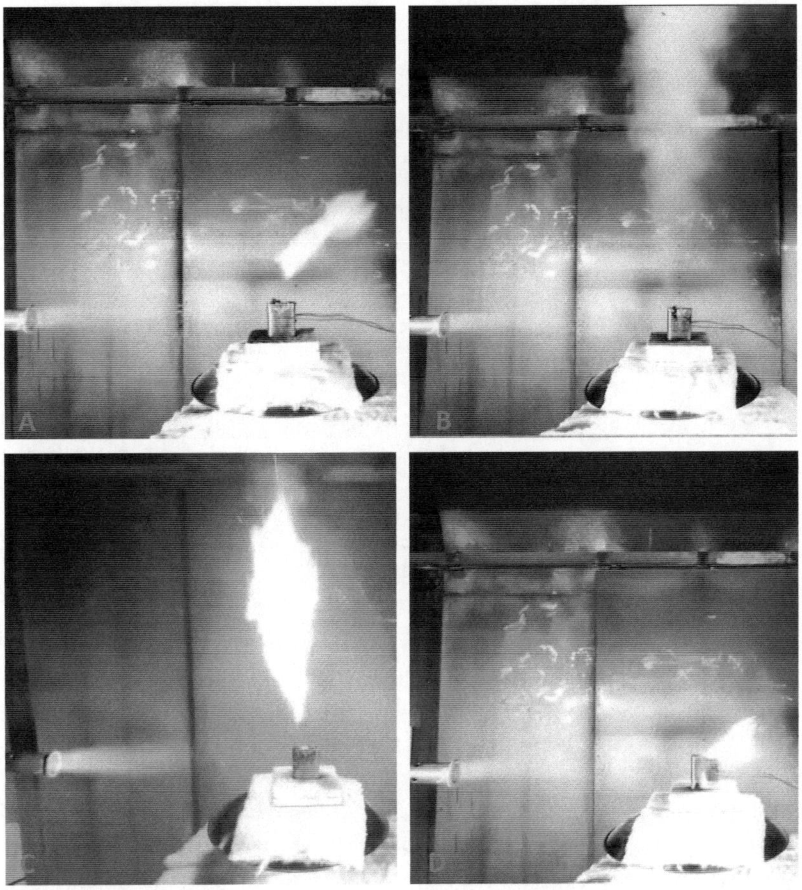

Figure 2. Typical stages of a lithium-ion polymer battery fire test. (A) A propane burner ignites a small vented gas jet. (B) The jet develops into a rapid venting prior to ignition that extinguishes the flame and (C) then re-ignites. (D) The combustible cell internal materials and packaging continue to burn after the electrolyte is consumed.

The stages of the fire in *Figure 2* correspond to distinct changes in the temperature and mass loss data shown in *Figure 3*. Firstly, there is a rise in cell surface temperature due to the action of the propane burner. Secondly, at 260 s (arrow A), a small jet venting through the pressure relief cap on the cell is seen as a reduction in mass. Ignition of the electrolyte gases by the propane burner occurred at this time. Thirdly, a marked increase in cell temperature transforms the jet into a rapid venting at 355 s (arrow B). This venting extinguishes the previous flame, and then is re-ignited about three seconds later by the burner flame, when the venting intensity has reduced. The rapid increase in temperature is indicative of thermal runaway which decreases after the internal reactions are complete. The battery continues to burn the combustible materials and packaging after the electrolyte is consumed. The spike increase in battery mass at arrow B in *Figure 3* is an idiosyncrasy of the load cell used to measure mass; the upward force of venting produces a transient downward reactive response, creating the illusion of greater mass.

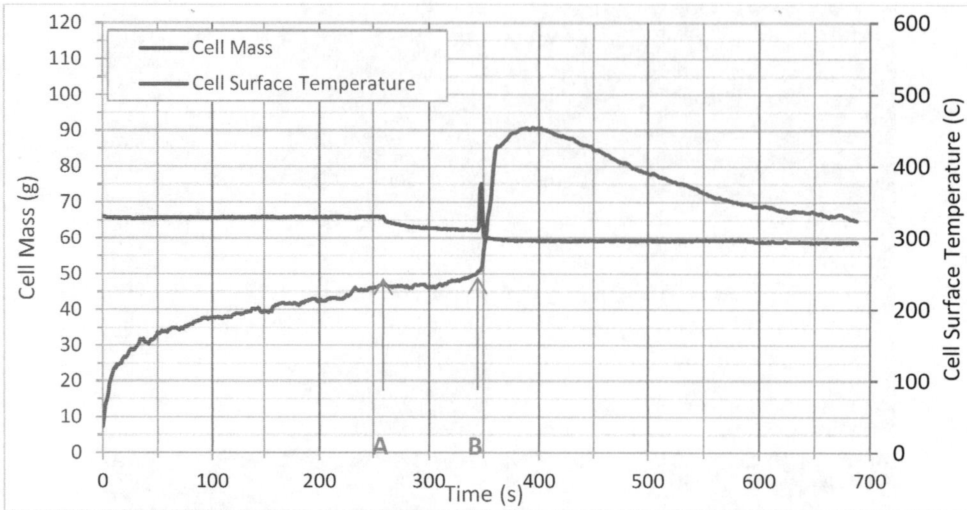

Figure 3. Mass and surface temperature behaviour resulting from heating a lithium-ion polymer cell using a side mounted propane gas burner.

A radiant panel was substituted for the gas burner to produce uniform cell heating to avoid localised heating produced by the propane burner. For these fire tests the cells were positioned with the large flat surface parallel to the radiant heater panel with a separate igniter used as an ignition source. Thermocouples are placed on the front and back surfaces of the cells.

Figure 4 shows the mass loss and the front and back surface temperatures of the cell when heated. Radiant heating results in an increase in cell temperature, and venting results in a loss of cell mass. The igniter was used at arrow A to ignite the gases and the panel heater was removed. This resulted in a drop in front surface temperature while the mass loss and back surface temperature remained stable. At arrow B major venting occurred extinguishing the flame. A short time later the temperature rises sharply - presumably thermal runaway has been initiated, and the mass drops (disregarding the load cell reactive spike). It was noted that the back surface temperature was higher than the front surface, possibly because the chemical reduction of electrolyte closer to the front surface had been more vigorous at an earlier time (100 s to 280 s) due to preferential heating of the front surface. After 380 s, there is no visible gassing.

It should be noted that not all tests resulted in two distinct outbursts of gas but the rapidly venting jet was always apparent.

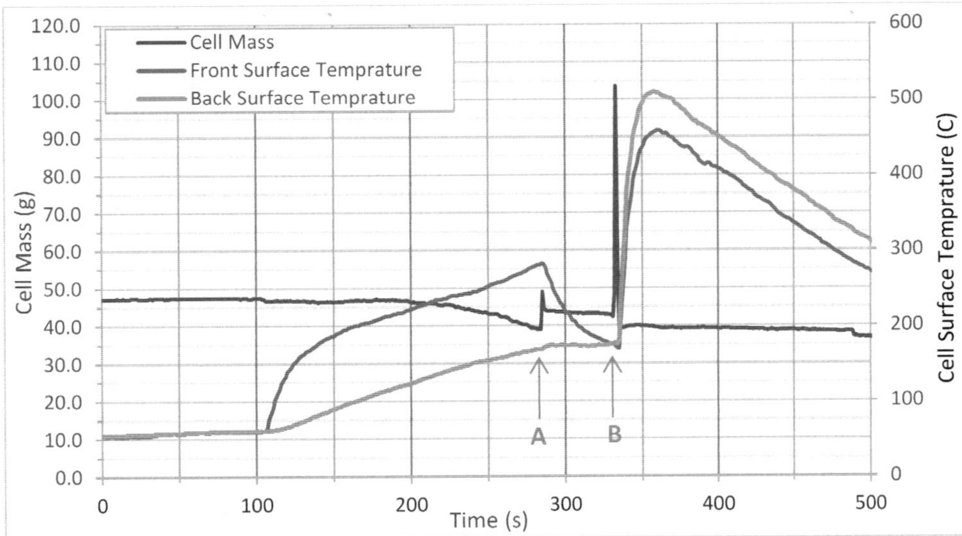

Figure 4. Mass and surface temperature behaviour resulting from heating the large surface of a lithium-ion polymer cell using a radiant burner.

Water Mist and Flame Interaction

To explore water mist fire suppression, tests were conducted with a water mist nozzle positioned above the flame of a propane fuelled Bunsen burner, a controllable fuel source compared to a battery venting event. The nozzle produced a hollow cone with droplet sizes ranging between 50 and 100 μm (NFPA defines water mist as having droplet diameters less than 1000 μm) and a flow rate of 0.8 ml/s. Different flame lengths were produced by adjusting either the air or propane flow rate and the extinguishing results are captured in *Table 1*. The results show that as the distance from the nozzle to the top of the Bunsen burner increased, or the flame length increased, the flame was more difficult to extinguish. The geometry of the experiment prescribes that greater distances between flame and nozzle result in lower mist flux density at the flame (due to both evaporation and greater cone volume), slowing the extinguishment process. The longer flame lengths require greater cooling to reduce the flame temperature.

Table 1. Effects of distance between the mist nozzle and Bunsen burner, and Bunsen burner's flame lengths and on the extinction time.

Distance (mm)	Extinction Time (s)		
	Flame Length = 80 (mm)	Flame Length = 140 (mm)	Flame Length = 200 (mm)
400	2.5	No	No
370	2	No	No
340	1	No	No
310	immediate	No	240
280	immediate	78	129
250	immediate	22.6	105
220	immediate	6	98
190	immediate	immediate	50
160	immediate	immediate	33

A Novel Water Mist Application Method

A different mist nozzle orientation has also been investigated for fire suppression. This mist system uses a flat spray positioned perpendicular to the flame of the propane fuelled Bunsen burner, rather than in the same plane as the flame. The nozzle was placed 150 mm horizontally and 20 mm above the top of the battery. On application of the water mist, extinguishment of the flame occurred within seconds. Video footage of the experiment shows the mist disrupting the connection between the fuel and the ignition source. If the water mist velocity at the burner is greater than the flame velocity, the flame and fuel are separated and extinguishment occurs. Similar to the extinguishing mechanism observed when a candle is extinguished by blowing it out.

Future Studies

To understand the interaction of water mist with a lithium-ion battery vent fire, further experiments will utilise a simplified battery comprising electrolyte as the only combustible material. Standard sized 18650 battery shells (65mm height x 18mm diameter) will be filled with a steel rod, to compensate volumetrically for the internal battery components, and electrolyte. These simplified batteries will be heated to determine electrolyte burning characteristics and the water mist parameters to produce extinguishment. Extinguishing experiments may extend to commercial 18650 batteries.

Future experiments involving the heating of an array of the simplified batteries will be used to investigate the onset of thermal runaway in adjoining batteries and the ability of water mist to provide cooling to prevent electrolyte venting, the subsequent fire and possibly thermal runaway. This research will incorporate larger scale batteries to assess scaling effects.

Additional experiments are planned to measure changes in battery mass and cell temperature to further understand progression to thermal runaway. This investigation will guide sensor usage to trigger the release of water mist.

Additional work will include ongoing development of Computational Fluid Dynamics models to predict the fire extinguishing capabilities of water mist on lithium-ion battery fires.

CONCLUSION

Lithium-ion main storage batteries have the capability to improve the endurance of submarines through superior energy storage. A review of the literature has shown that lithium-ion batteries pose a fire hazard; however the development of fire extinguishment systems for large banks of lithium-ion batteries in a marine environment is not yet sufficiently established to provide a satisfactory level of fire security.

Water mist could be a suitable method to extinguish lithium-ion battery fires on submarines because it offers suppression and cooling with a low water volume requirement. This collaborative research project is examining the ability and efficacy of water mist in protecting against lithium-ion battery fires. Initial extinguishing results, on gas flames and simplified batteries, using both vertical and horizontal nozzle orientations shows the potential of water mist as an extinguishing method.

The data gathered by this research will be used to further develop and validate CFD models and investigate design parameters of a water mist suppression system. Numerical modelling of the lithium-ion battery fire and water mist extinguishing will allow the design of a water mist system for any compartment configuration without the requirement of testing.

REFERENCES

1 Whittingham, M.S., 1976, *Electrical Energy Storage and Intercalation Chemistry,* Science, **192** (4244): p. 1126-1127.

2 Tarascon, J.-M. and Armand, M., 2001, *Issues And Challenges Facing Rechargeable Lithium Batteries,* Nature, **414** (6861): p. 359-367.

3 Depetro, A., 2016, *Future Submarine Fire Safety Study,* Master of Engineering Thesis VQT6062\S3914014, Victoria University.

4 Dominguez, G., 2018, https://www.janes.com/article/83625/japan-launches-first-soryu-class-submarine-equipped-with-lithium-ion-batteries , Jane's 360, 08 Oct 2018, accessed 30 May 2019.

5 Jeong, J., 2018, https://www.defensenews.com/industry/techwatch/2018/11/16/new-battery-can-double-the-operational-time-of-submarines-says-south-korea, TechWatch, Nov 16, 2018, accessed May 2019.

6 Depetro, A., 2017, *The Design & Safety Challenges of a Lithium-ion Main Storage Battery for Conventional Submarines.* Pacific 2017, Sydney.

7 Long, R., Kahn, M. and Mikolajczak, C., 2014, *Lithium-ion Battery Hazards,* Fire Protection Engineering, (4): p. 20-36.

8 Long, R.T., Sutula, J.A. and Kahn, M.J., 2013, *Lithium-ion Batteries Hazard and Use Assessment: Phase IIB*, National Fire Protection Research Foundation.

9 Roy, P. and Srivastava, S.K., 2015, *Nanostructured Anode Materials for Lithium-ion Batteries,* Journal of Materials Chemistry A, **3**(6): p. 2454-2484.

10 Spotnitz, R. and Franklin, J., 2002, *Abuse Behaviour of High-power, Lithium-ion Cells*, J. Power Sources, **113** p. 81-100.

11 Shilov, A, 2017, *Panasonic Recalls 280,000 Tablet Battery Packs Due to Fire Hazard*, accessed 25/06/2019, https://www.anandtech.com/show/11414/panasonic-recalls-280000-tablet-battery-packs-due-to-fire-hazard, AnandTech.

12 Ruiz, V., et al., 2018, *A Review of International Abuse Testing Standards and Regulations for Lithium-ion Batteries in Electric and Hybrid Electric Vehicles*, Renewable and Sustainable Energy Reviews, **81**(Part 1): p. 1427-1452.

13 Feng, X., et al., 2018, *Thermal Runaway Mechanism of Lithium-ion Battery for Electric Vehicles: A Review.* Energy Storage Materials, **10** (Supplement C): p. 246-267.

14 Larsson, F., Andersson, P. and Mellander, B.E., 2016, *Lithium-ion Battery Aspects on Fires in Electrified Vehicles on the Basis of Experimental Abuse Tests,* Batteries, **2**(2): p. 9.

15 Lecocq, A., et al., 2012, *Comparison of the Fire Consequences of an Electric Vehicle and an Internal Combustion Engine Vehicle,* 2nd International Conference on Fires in Vehicles-FIVE *2012,* SP Technical Research Institute of Sweden, Boras.

16 Buckingham, J., Hodge, C. and Hardy, T., 2008, *Submarine Power and Propulsion-Application of Technology to Deliver Customer Benefit*, UDT Europe, Glasgow.

17 Joachin, H., et al., 2008, *Electrochemical and Thermal Studies of LiFePO4 Cathode in Lithium-ion Cells,* ECS Transactions, **6**(25): p. 11-16.

18 Kizilel, R., et al., 2009, *An Alternative Cooling system to Enhance the Safety of Li-ion Battery Packs,* Journal of Power Sources, **194**(2): p. 1105-1112.

19 Hofmann, A., et al., 2017, *Preventing Li-ion Cell Explosion During Thermal Runaway with Reduced Pressure,* Applied Thermal Engineering, **124**(Supplement C): p. 539-544.

20 Hémery, C.-V., et al., 2014, *Experimental Performances of a Battery thermal Management System Using a Phase Change Material,* Journal of Power Sources, **270**(Supplement C): p. 349-358.

21 Duan, X. and Naterer, G.F., 2010, *Heat Transfer in Phase Change Materials for Thermal Management of Electric Vehicle Battery Module,* International Journal of Heat and Mass Transfer, **53**(23): p. 5176-5182.

22 Yim,T., Park,M.S., Woo,S.G., Kwon,H.K., Yoo,j.K., Jung,Y.S., Kim,K.J., Yu,J.S. and Kim,Y.J. (2015) *Self-Extinguishing Lithium Ion Batteries Based on Internally Embedded Fire-Extinguishing Microcapsules with Temperature Responsiveness*, Nano Letters, 2015, Aug 20, Vol. 15(8), pp. 5059-5067.

23 Standards Australia, 2009, *Portable Fire Extinguishers - Classification, Rating and Performance Testing*, AS/NZS 1850-2009.

24 Mawhinney, J.R. and Back, G.G., 2016, *Water Mist Fire Suppression Systems*, SFPE Handbook of Fire Protection Engineering, Springer. p. 1587-1645.

25 National Fire Protection Association, 2010, *NFPA 750 Standard for Water Mist Fire Suppression Systems.* NFPA, Quincy, MA, USA.

26 Liu, Z., A.K. Kim, and D. Carpenter, 2007, *A study of portable water mist fire extinguishers used for extinguishment of multiple fire types,* Fire Safety Journal, **42**(1): p. 25-42.

27 McGrattan, K., Hostikka, S., McDermott, R., Floyd, J., Weinschenk, C. and Overholt, K., 2013, *Fire Dynamics Simulator, User's Guide*, 6th Ed, NIST Special Publication, 1019: p. 262.

An Overview of the Integrated Platform Performance Modelling Approach used to Support the Future Submarine Program

Karl Slater, John Wharington and Mario Selvestrel
Defence Science and Technology Group, Department of Defence, Australia
karl.slater@dst.defence.gov.au

Abstract

An integrated platform performance modelling approach has been developed and is being used by Australia's Defence Science and Technology (DST) Group to help to inform and support defence decision makers in Australia's procurement of the future submarine.

In the context of analysing modern conventional submarines, integrated platform performance modelling is utilised to understand the whole of system performance of the platform. Of key importance are the interactions between the constraints placed on major subsystems within the submarine. These constraints need to be understood to develop a set of cohesive requirements for a future platform.

The novel integrated platform performance modelling approach has continued to be developed since 2010 in close collaboration with both Navy and the Future Submarine Project office. The modelling approach and tool have been used to provide objective quality evidence to key decision makers through the early design phases of this large capital project.

The integrated platform performance modelling approach has been able to analyse and assess platform performance whilst varying a number of parameters including; concept of operations, different technology options, and specific mission parameters. The results have provided evidence based advice to decision makers and capability development stakeholders to understand the implications and consequences of changes in the design or technology requirements.

This paper provides an overview of the modelling framework and tools developed at DST and its use in the Future Submarine Program along with the lessons learnt through adopting this approach. The future direction of the work program is also discussed.

1 Introduction

The intricate design, complex roles and criticality of many Defence assets make it essential to consider platform performance in an integrated and systematic manner. Otherwise, it would not be possible to assess whether complex platforms meet extensive technical and operational requirements.

The scene is eloquently set by the quote from [1] "The designer must fully understand the aspirations of the operator and equally the operator must understand the implications and consequences of their requirements." This quote provides a high level view of the important relationship between the submarine designer and the submarine operator.

Prior to 2009, the Maritime Division of the Defence Science and Technology (DST) Group – Department of Defence – conducted analysis of platform performance using a variety of methods, one of which was using empirical based equations in an Excel spreadsheet, aimed at supporting early stage submarine design trade-off studies. Whilst this spreadsheet was able to undertake various tasks, it was not a robust method for providing decision makers with objective quality evidence, nor was it repeatable or simple to integrate with other tools and as with all large spreadsheets was prone to user error.

Considering the complex and classified nature of submarine operations coupled with the extensive program requirements, it became apparent to DST that an analysis capability was required that could:

- Investigate top-level requirements and their impacts on submarine platform designs.
- Highlight key design and performance drivers.
- Integrate the capability of sub-systems/technologies to assess the impacts on platform performance parameters, enabling trade-off studies to be completed.
- Assess the overall performance of various platform concepts/designs whilst simulating the influences of different mission scenarios.
- Ensure any assessment was auditable and repeatable; and was conducted using a 'level playing field'.
- Validate manufacturer's claims.
- Support the maturation of performance and operational requirements.

The above challenges required a suite of system modelling capabilities; however at the time there were no suitable tools available to capture and integrate all the diverse requirements. Although the body of knowledge on using model driven techniques to support trade-off studies in design had begun to amass [2], there was very little literature on the use of these techniques in the submarine domain. The heavily integrated nature of submarines means that the design is tightly coupled with the desired mission scenarios, which is highly sensitive, and hence modelling complexity is increased and access to data is difficult. Compounding the above challenge was that the required modelling capabilities would be required to support the full life cycle of the program.

Figure 1 illustrates the key phases in delivering complex capabilities, including:

- Studies: where broad needs are defined.
- Concept Design: where a high level conceptual design of the capability is defined.
- Preliminary Design: where initial design parameters are defined.
- Detailed Design: where detailed design parameters needed for construction are defined.
- Build: the construction phase.
- In-Service: where ongoing support to the capability is provided.

An indicative view of where the IPSM can provide modelling support to these phases is shown in Figure 1, including:

- Requirement setting and analysis of alternate technology options.
- Undertaking trade-off studies to understand the implications of design decisions.
- Maturation of requirements as the capability needs and design options become better understood.
- Continuous evaluation and acceptance of designs from initial concept through to detailed design.
- Providing guidance on key aspects to be considered for test and evaluation.
- Analyse the Concept of Operations (CONOPS) as the capability is introduced into service.

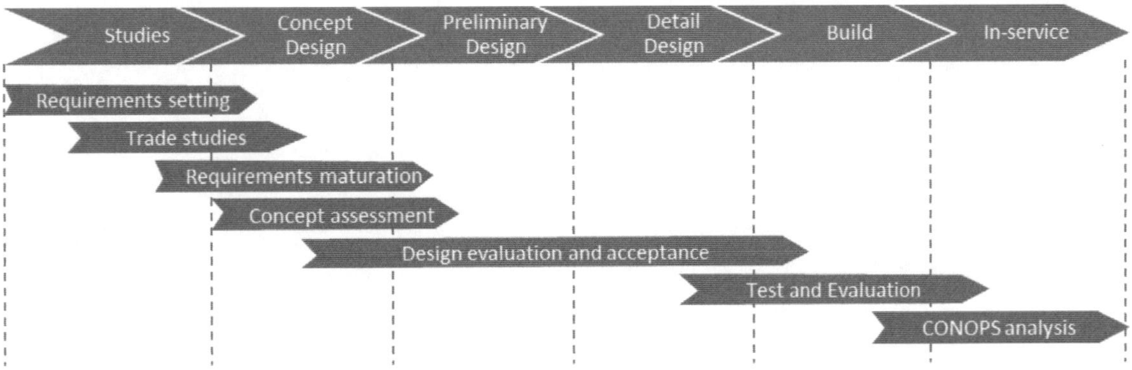

Figure 1 IPSM's Value in the Life-cycle

As such DST, in support of the Commonwealth of Australia, embarked on a task to develop an in-house software tool known as the 'Integrated Platform System Model' (IPSM). The IPSM is a platform systems analysis tool that has been and continues to be developed; and has become a critical modelling tool to support whole of platform performance studies. This paper provides an overview of DST's development of the IPSM tool.

For details on the genesis, theories and methodologies adopted in developing this tool, the reader is referred to [3]. Open literature details on the IPSM are available in [4], [5] and [6].

This paper provides a high level overview of the development of the IPSM covering aspects such as; software requirements, the phased approach to the model development, verification and validation undertaken, support and infrastructure required, example analysis, observations and lessons learnt, and lastly the future expansion plans for the tool.

2 Software and model overview

2.1 The Core of the IPSM

The Platform Systems Analysis team of DST's Maritime Division have applied the following principles in the development of the IPSM over the past decade:

- Continual software development with continual feedback from stakeholders to ascertain current and future needs.
- Continual improvement of processes for software development, numerical modelling and simulation.
- Adoption of modelling architecture according to systems engineering principles, including; use of classes and well defined types, engineering units, state machines, formal time domain simulation, reuse and encapsulation and repeatability.
- Platform system models are to be validated - validation efforts should be considered upfront.
- Quality control and pre/post configuration tools are essential.

A conceptual framework showing functional inputs, system models; and various infrastructure and support tools of the IPSM, is shown in Figure 2. The blocks inside the box labelled 'Modelica Environment' represent the core of the IPSM, consisting of various system models, operating environment models, systems requirements and design rules, all of which are written in an object oriented modelling code Modelica (discussed in more detail in Section 2.2).

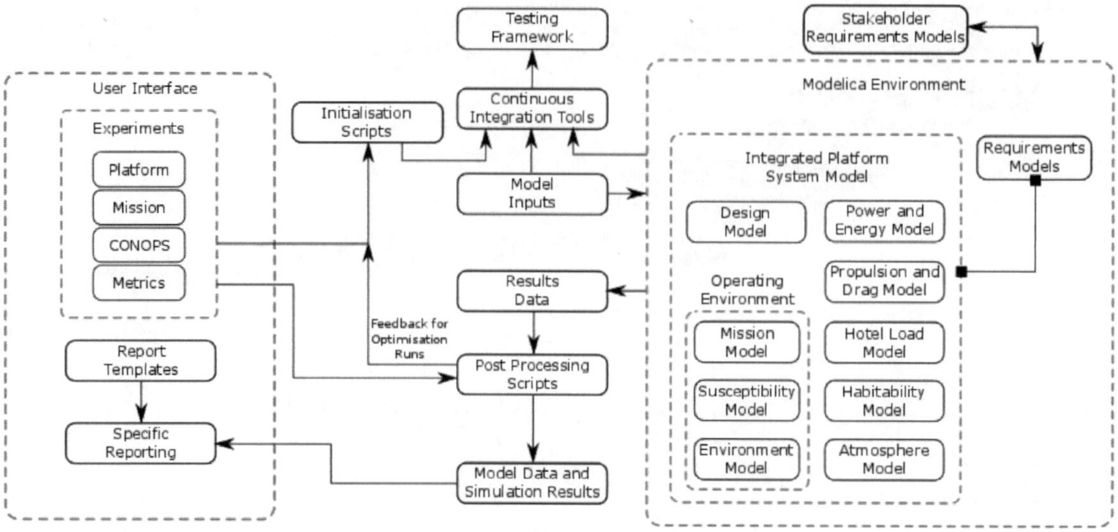

Figure 2 The IPSM Framework

The Modelica Environment is capable of modelling the integrated effects of the system models on mission performance, enabling an assessment on whether particular measures of success or requirements are being met. The algorithms and equations underpinning several of these platform system models are derived from research, development and analyses conducted by DST subject matter experts over several decades, in addition to external sources such as [1].

An idealised representation of an example submarine mission modelled for analysis in the IPSM is shown in Figure 3, where the submarines leaves home base and undertakes three open ocean transits interspersed with two choke points before arriving at the Operational Area (OA). An enlarged version of the Operational Area is also shown, highlighting the further detail in this area, which includes the patrol area and the snort box. The OA encompasses the movement of a submarine through a patrol area, transit to a safe area for battery charging using diesel engines (snort box), and returns to the patrol area. The ability to model this level of mission detail capability has become a feature of recent IPSM Phases, as will be highlighted in Section 3.

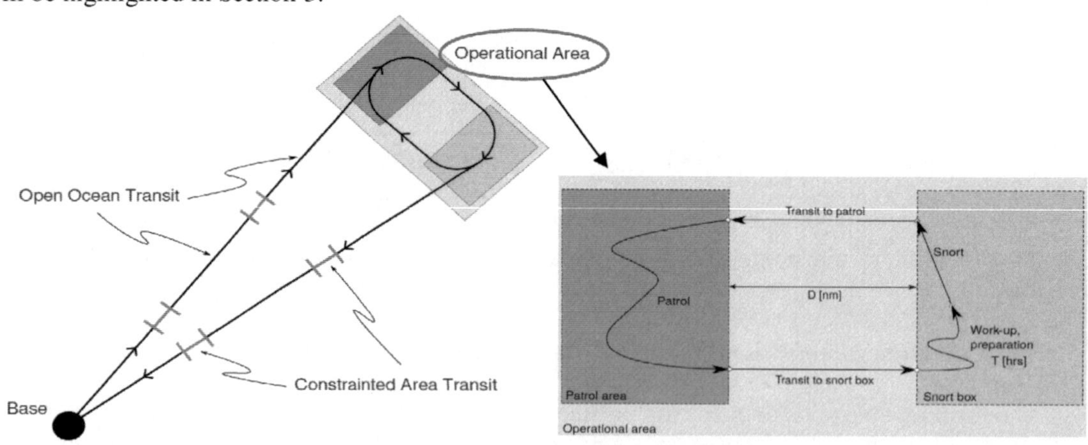

Figure 3 Idealised Submarine Mission Schematic Submarine Operational Area (OA)

The remaining blocks in the IPSM Framework shown in Figure 2 represents components and tools of the software infrastructure, as well links to the stakeholder requirements and data products. These products (visual, tabular and textual results) are derived from performing model simulations. IPSM simulations are conducted via single experiment, parametric studies (i.e. multiple simulations are performed across a range of values); and in versions since Phase Three - optimisation techniques have been used to determine the overall mission time and transit speed combination that maximises the time on task. Optimisation methods are currently implemented via the Dakota software package [7], which allows the flexibility of exercising different optimisation options.

Additional details of the software infrastructure will be discussed further in this paper.

2.2 Modelica Software Language

The IPSM tool has been developed using the Modelica modelling language, a systems-based modelling language with no upfront or ongoing licencing costs. The key features of Modelica include:

- Object-oriented language with replaceable classes.
- Equation-based modelling (also referred to as a 'system dynamic simulation').
- Capable of modelling complex physical systems such as those contained in a submarine.
- Widely used in the modelling community, as demonstrated recently in [8].

The main advantage with an equation based modelling approach is that it gives more flexibility to re-configure and optimise the system (Figure 4), in comparison to a traditional time-step approach. Figure 4 highlights that with minor reconfiguration two questions can be answered:

- Given Fuel Consumption Rate and Hotel Load, what is the Platform Speed.
- Given Platform Speed and Hotel Load, what is the Fuel Consumption Rate.

Using traditional approaches would require two question specific equation solvers to answer these.

Modelica implements system models internally as a nonlinear, constrained system of equations which are solved during simulation execution, given a set of initial parameters. This enables parametric analysis to be conducted, by altering these input parameters over several simulation runs, allowing the IPSM to determine the sensitivity of performance to the parameters of interest.

The flexibility of Modelica also enables the efficient application of external optimisation algorithms, which can automatically iterate through parameters without the need to recompile the model.

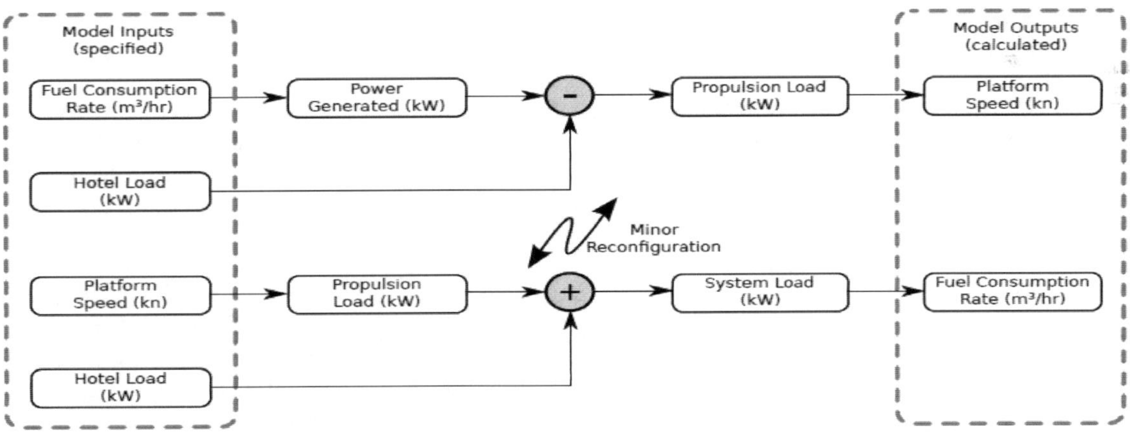

Figure 4 Modelica Offers Ease of Re-configuration

3 Phased development approach

The IPSM has been developed in a phased approach over several years as shown in Figure 5. The IPSM development team has generally comprised of three to five staff, with details of the various phases are provided in the following sections.

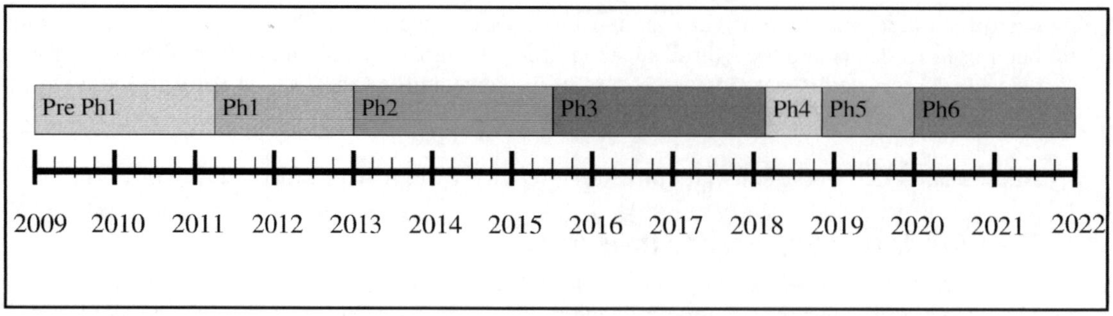

Figure 5 IPSM Development Timeline (future phases will be developed from 2019 onwards)

3.1 Pre-phase One

Prior to the development of the IPSM, modelling was undertaken in the group using custom software written by DST staff, mainly investigating the influence of scaling submarine displacement with performance. This approach was very limited in its capabilities and wasn't suited for overall systems analysis of different submarine options. Later, a spreadsheet using empirical based equations, as previously mentioned, was developed for analysing the alternate submarine options. This approach resulted in a large cumbersome spreadsheet that was difficult to use, manage and would not allow robust or repeatable results.

The experiences learned from these early attempts demonstrated that a different approach and method for analysing platforms systems and concepts were needed. Investigations were undertaken to find a more suitable method that was robust but flexible; repeatable but always auditable; and capable of system based modelling, hence Modelica was adopted.

3.2 Phase One

Development of the IPSM Phase One started in 2011 with a view of providing a platform systems analysis capability that could evaluate whole of platform performance. The initial focus was on developing power and energy modules, as power and energy mediums were considered to have a significant influence on the relative performance of concept submarine designs.

A visual-based version of Modelica – *ModelicaML* – was used to develop IPSM Phase One. However, upon release of the IPSM Phase One, it was deemed that any future enhancements would be done using the language-based version of Modelica. The visual based package wasn't powerful or flexible enough and as such modellers devoted significant effort editing the code generated by the in-built code generator.

As part of the exploratory process involved in developing the IPSM Phase One, a collaborative program was established between DST and QinetiQ GRC (UK based) to link the IPSM with Paramarine (a third party dedicated naval architectural software package), with a view to assess the trade-space between platform performance and naval architecture attributes [5]. The intent was not to create a submarine concept design tool, but to use the advantages of each tool to complement each other to assess potential design configurations. This exercise was deemed successful as not only did it demonstrate the ability of the IPSM to interrogate further aspects of the design space, it also showed the flexibility of the IPSM to interact with other modelling packages. As this activity was conducted as a demonstrator, no additional efforts were expended in linking Paramarine with IPSM.

Phase One also included a basic Graphical User Interface (GUI) to demonstrate its functionality, and was complete in mid-2012.

3.3 Phase Two

Phase Two of the IPSM added a parametric modelling capability (instead of only single experiment simulations), submarine atmosphere models; and waste and victualling models. It also included a significant redesign of the GUI to make it more intuitive and user friendly. Phase Two was delivered in late-2013.

As part of this phase, the development team was modelling the impact of habitability aspects of submarine design. New modules were developed to investigate aspects such as the use and storage of fresh water, grey and black water, solid waste; victualling parameters (cold, cool and dry stores) and atmosphere control consumables (CO_2 canisters and oxygen candles).

Phase Two was used to support analyses of data provided to DST in relation to initial military-off-the-shelf (MOTS) designs considered in 2012. The use of the IPSM and its outcomes were received very well by the stakeholders and hence further investment into the IPSM's development was made.

3.4 Phase Three

Phase Three was developed and used to support the competitor down select in 2015/16 and most recently, to provide support and evidence for developing and maturing effective requirements in the Preliminary Function and Performance Specification (PFPS). As part of the development of this phase, extensive validation activities were undertaken to validate the IPSM against 'real-world' Collins Class data. Upon successful validation, Phase Three was deployed to the client, to allow them to run their own simulations.

Phase Three added additional functionality that included an enhanced battery model and a mission management module with increased fidelity. This functionality allowed the user to model missions of any format so long as they could be decomposed for representation by discrete phases or legs. This enhancement has been especially useful as it allows for significantly more flexibility in the way in which missions are modelled, for example the ability to add any number of choke points and to alter mission leg lengths.

During the development of IPSM Phase Three the functionality of the IPSM grew significantly and as such maintaining the GUI to remain 'in-step' with the IPSM became a resource drain. Consequently, due to the prioritisation of work, the continued development of the GUI was dropped. Rather, the IPSM development team have developed fit for purpose training guides to assist any new users to access the IPSM via command lines in UNIX, providing greater functionality and flexibility to build platform models and run simulations. Additionally, using the command lines functions allows users to interrogate the tool, providing an insight on how the model has been developed.

3.5 Phase Four

Phase Four built on Phase Three by adding fidelity to components that were identified as significantly affecting mission performance in Phase Three. The areas addressed included higher fidelity battery (NREL [9] and Ceraolo [10]) and propulsion motor models.

Phase Four also made additional improvements to the command line user interface and underlying software build structure to make it easier to use by analysts who were less familiar with using the UNIX command line functionality.

3.6 Phase Five

Phase Five extended the mission modelling capability to consider interactions between:

- Mission State, or phase of the mission.
- Susceptibility to threats.
- Environment.

For example, Figure 6 illustrates that at the start of the mission, the platform is in an 'Open ocean transit' mission state, in a 'Temperate Southern' environment, with 'Nil' susceptibility to threats.

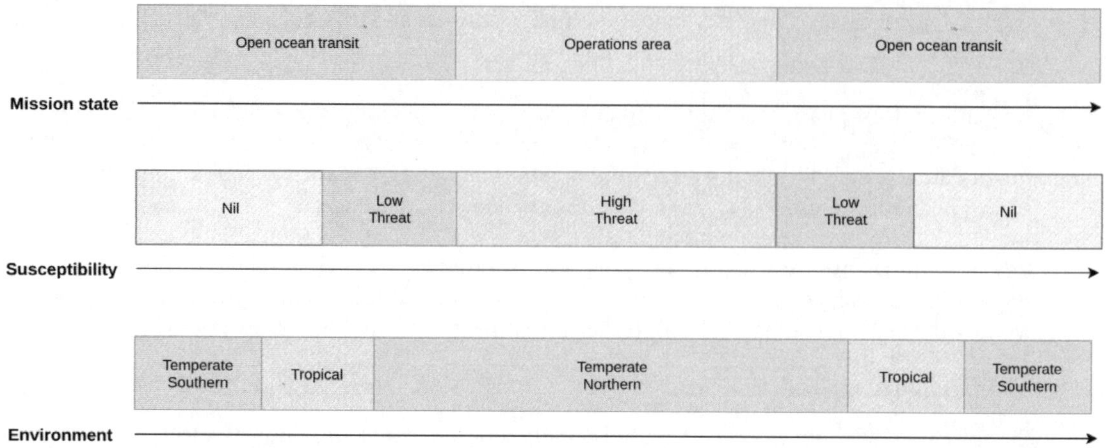

Figure 6 Experiment showing combined use of mission, susceptibility and environmental configurations

Phase Five also added detail modelling of thermal loads and hotel loads which allows the energy demands and operational performance of individual platform systems to be modelled. The detailed modelling takes into account:

- System response to platform states or operating variables, e.g. hotel loads can vary according to mission state.
- System-to-system and system-to-environment energy flows, e.g. energy loads are modelled at the system level (power train, combat system, heating and cooling, etc).
- Energy demands of waste heat handling systems, e.g. thermal loads and associated cooling loads are modelled at the system level (power train, combat system, heating and cooling, etc).

Phase Five also expanded the number of Measures of Effectiveness being evaluated to consider different CONOPS and threat profiles.

4 Verification and validation approach

The purpose of model verification is to verify that the model is consistent with its specification and fulfils its intention. This differs from model 'validation', which is primarily concerned with determining the degree to which the model is an accurate representation of the real world, from the perspective of the intended use of the model.

Software testing is conducted using Continuous Integration practices, which allows isolated IPSM code changes to be immediately tested and reported on when added to the main code base. Continuous Integration provides rapid feedback so that if a defect is introduced into the model, it can be identified and corrected as soon as possible. The IPSM development team have created a testing framework that has been integrated into the Jenkins software, which is used for implementing the Continuous Integration process.

For IPSM Phase One and Phase Two validation, unclassified open literature data sources were used. Additionally, 'brochure data' from prospective manufacturers were also used to support the validation of IPSM Phase Two.

For Phase Three and beyond, the validation activities were much more rigorous due to the availability of data. Firstly, the ISPM was cross-validated against various outputs from various power and energy models. Additional validation activities were then conducted using 'real-world' Collins Class submarine fleet data. The validation results showed that the outputs of the IPSM provided a close representation of the real world data, and subsequently the IPSM has been deployed for use by staff embedded in the client site.

5 Overview of the IPSM support and infrastructure

Supporting the IPSM requires hardware, software and supporting processes. Beyond model development and execution and technical support, it is essential to ensure the outputs of the model are accurate, robust, repeatable, and traceable; and in a form that serves the needs of the intended audience. As such, standards or best practice guides in systems engineering (e.g. IEEE 1220-2005), quality management (e.g. ISO9001) and software development (e.g. ISO/IEC 12207:2008) were used to guide the design and development of the IPSM; however the strict adoption of these standards would have introduced far too much overhead for the size of the IPSM and the relatively small development team.

5.1 Hardware Overview

Early in the development of the tool it was decided that it would be advantageous to have the tool available on a dedicated Defence server rather than on a stand-alone platform. This would allow multiple users to be able to use the tool from different locations across Defence and would allow the developers to keep tight control of the versions being used.

Subsequently, the IPSM is hosted on blade servers located within Defence premises. Multiple servers are used due to the different security classifications of data used in IPSM simulation runs. As the IPSM is hosted on multiple servers, an integration manager has been established that ensures the IPSM version that exist on each server are compatible, where possible – this is achieved via rigorous software testing, configuration management and version control.

5.2 Software Suite

The IPSM servers are configured with a Ubuntu Linux based operating system, however Putty or Xwin-32 is used for remote display of Unix from Windows workstations. The operating system uses several software and support tools to enable the model to be developed, compiled, tested, executed and managed, such as:

- *Modelica*: The IPSM is implemented in the Modelica modelling language and is compiled using the Open Modelica Compiler (OMC).
- *Dakota*: The IPSM uses Dakota to support state-of-the-art robust optimisation and sensitivity analysis. Broadly, the Dakota software's advanced parametric analyses enable design exploration, model calibration, risk analysis, and quantification of margins and uncertainty with computational models.
- *Lua*: a powerful, efficient, lightweight, embeddable scripting language. It supports procedural programming, object-oriented programming, functional programming, data-driven programming, and data description. It is used for building scripts in the IPSM for various purposes.
- *Perl*: used for building scripts for various purposes.
- *Gnuplot*: used to generate output plots.
- *TracWiki*: an open sourced, web-based project management and bug tracking system hosted on a Wiki page. Trac is used to manage and track all IPSM issues, from future enhancements to existing 'bugs' and, provides wiki facility to store supporting material for IPSM user and developers, including background information.
- *Git Distributed Version Control System*: Git is used for version control of model code, data, build scripts, test files and some documentation.
- *Jenkins*: an open sourced Continuous Integration Server that provides continuous integration services for software development. This includes building and testing (i.e. verification) software projects continuously, and monitoring the execution of externally-run jobs. In the IPSM, various Jenkins jobs are created to monitor the IPSM test suites. When the developer makes a commit to a code repository it can trigger a build of the IPSM code base, a build and execution of the IPSM unit test suite and a build and execution of the IPSM system test suite.

5.3 Configuration Management

The Configuration Management (CM) approach for the IPSM is as light weight as possible, whilst still retaining the ability to adequately control and manage any changes to the model code, the server software and hardware configuration, and any associated documentation. The goal is to give the IPSM developers the ability to identify (and return to) a known state at any point in time.

The IPSM consists of the model code itself, the model infrastructure, formal documentation, and numerous other project artefacts that are produced as part of the development process. Each of these items is placed under configuration control and baselined, as this information is critical to reproducing any previously known state of the IPSM. The following is a list of configuration items identified for control in the IPSM:

- Models focussed items, including:
  - Design documents.
  - Model source code.
  - Model input and data, including reports.
  - Test plans and procedures (unit tests, system tests and acceptance tests).
- Quality documentation, including:
  - Quality manual.
  - User guides.
  - Security documents.
  - Software release notes.
  - Version description documents.
  - V&V (verification and validation) focussed items including:
    - Review evidence (e.g. Validation data).
    - Verification and validation reports.

The configuration management tools used in the IPSM consist of the Git Distributed Version Control System, TracWiki and the internal Australian Defence archiving systems.

6 Example analyses

As demonstrated by the analysis conducted in [6] [11], the IPSM is capable of supporting a broad range of analysis tasks, ranging from evaluating design alternatives (to determine what capability is to be built), to developing concept of operations for the capability being introduced (to determine how the capability is to be deployed).

However, for demonstration purposes, two example analyses from using the IPSM are presented to provide a high level overview of the IPSM's capability. These examples also demonstrate how the model can be used to generate insights into the domain. The aim of this example analysis is to investigate the sensitivity of operational area (OA) performance to variation of operational parameters.

As discussed previously, a representation of a platform and of the scenario of interest is required. The platform representation at a minimum requires descriptions of the size of the platform (displacement) and parameters describing the power systems included in the design.

The subsystem used for representing scenarios or missions in the IPSM is flexible with the only requirement being that the desired scenario can be decomposed into discrete segments. Within the IPSM, the scenario is described as a state machine where each state represents a segment of the scenario. Each state has a configuration applied describing the submarine's behaviour for the present state, this includes: submarine speed, battery use profile, and relative priorities and availabilities of the installed power systems.

The first example scenario [6] used in this analysis is an idealised representation of a submarine area of operations and is shown in Figure 3. This scenario models the movement of a submarine through a patrol area, transit to a safe area for battery charging using diesel engines (snorting), and return to the patrol area.

In this analysis, the patrol box in Figure 3 represents the area of interest to the submarine; as such a measure of performance is defined as the OA Effectiveness: the proportion of time spent in the patrol box to the total time spent in the patrol area.

The ability of the IPSM to conduct parametric analysis allows the IPSM to determine the sensitivity of performance to the parameters of interest. This analysis will investigate the performance variation with regard to the distance between patrol and snorting boxes (in the OA), investigated by means of a parametric sequential search through a specified range. The resulting operational area effectiveness will be recorded for each run of the scenario.

Figure 7 shows the results of the analysis with respect to the OA effectiveness. The schematics on the figure highlight short standoff distance towards the left, and long standoff distance towards the right. The key points of the results are that the OA Effectiveness reduces linearly with the increasing distance between the patrol box and the snort box (see Figure 3). Where the distance between the two boxes is 0 nautical miles (nm) the OA effectiveness is large ~95%, however even at a modest 25 nm standoff distance the OA effectiveness has been significantly reduced to ~40%.

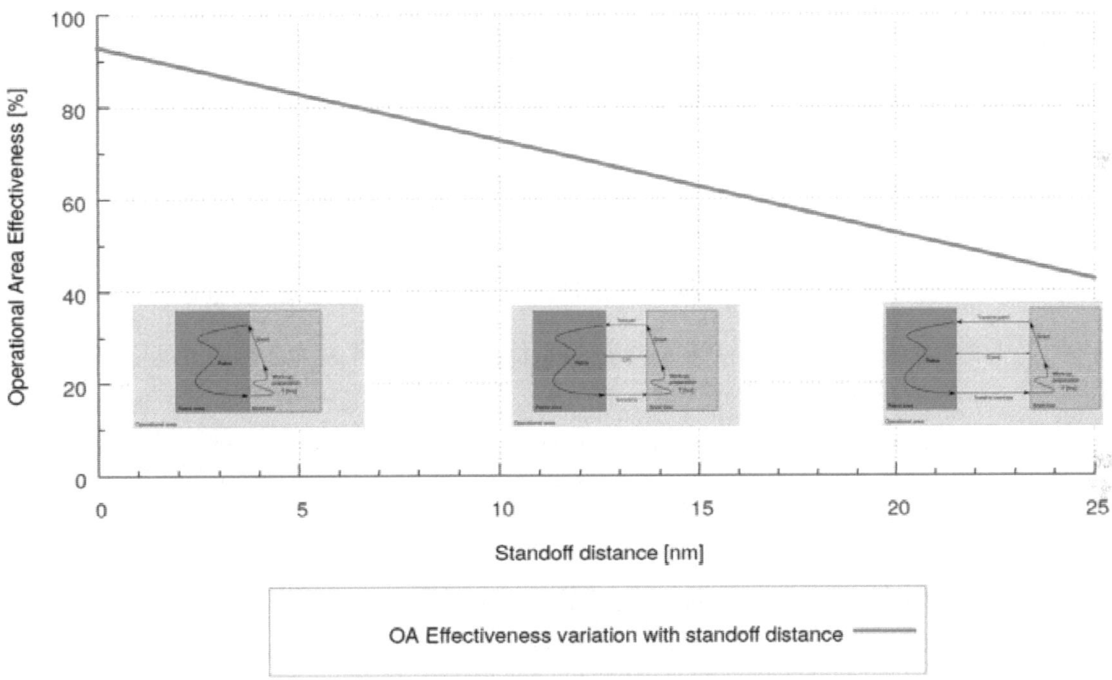

Figure 7 Operational Area (OA) Effectiveness Variation with Standoff to Snort Distance

In the context of a study conducted to support maturation of platform requirements, the outcomes of the example analysis could include:

- A requirement that specifies the minimum OA effectiveness for the submarine.
- Specification of a minimum distance between the patrol and snort boxes based on expected platform signature levels and expected threat environments.

The second example scenario [11] used in this analysis shows a method to calculate the maximum amount of time available to be spent in the Operational Area during a submarine mission. This analysis optimises the speed of advance (SoA) for the submarine undertaking the transit to the Operational Area. The time available to be spent in the Operation Area can be maximised if the submarine can transit at its maximum transit SoA.

The results of a parametric simulation are shown in Figure 8, which illustrates the variation of specific diesel consumption with the overall speed of advance. The speed of advance was varied by increasing the speed of the submarine when submerged (deep) whilst under battery power, whilst the speed during snorting was held at a constant 10 knots.

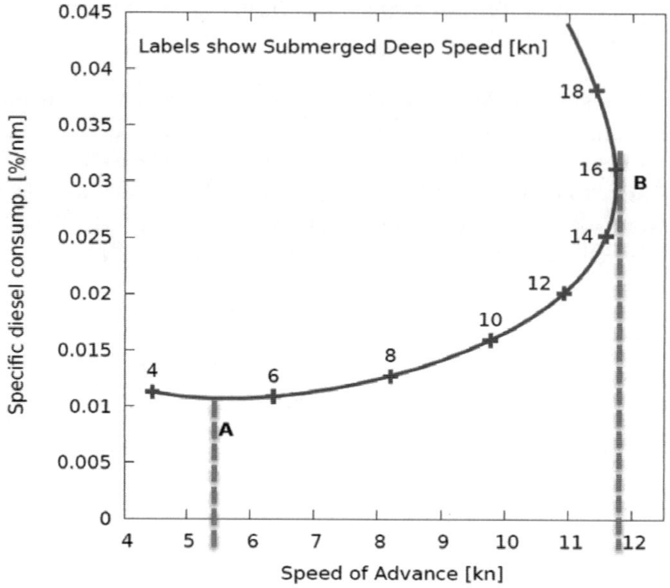

Figure 8 Max Range SoA and Max Time on Station SoA

The blue line on this plot shows the submerged speed whilst deep, the horizontal axis shows the overall speed of advance in the transit leg.

There are two points of interest, 'A' and 'B', marked on this plot. Point 'A' indicates the minimum with respect to specific diesel consumption and occurs at a SoA of approximately 5.5 knots. This SoA corresponds to the speed for achieving maximum range, but consequently less time in the Operational Area.

Point 'B' shows the maximum achievable SoA of approximately 11.8 knots. Increases to the speed whilst submerged deep beyond 16 knots will reduce the overall SoA. This result is somewhat counter intuitive as it shows that increased speed whilst submerged deep will result in a reduction of the SoA. This occurs due to the fact that as the speed of the submarine increases, the endurance of a single battery discharge decreases. Eventually the submarine spends a greater amount of time snorting (at a lower speed) than submerged deep (at a higher speed) thus resulting in the lower SoA. This SoA corresponds to the speed for achieving maximum time in the Operational Area, but consequently with less range.

It should be noted that a faster maximum SoA might be possible with variation of the snorting speed.

In the context of a study conducted to support the development of CONOPS for a platform, the outcomes of the example analysis could include:

- A range of speeds available to achieve either maximum range or maximum available time in the Operational Area.
- Detailed information for the submarine designer or the submarine operator with regards to variations to specific CONOPS and how they could impact on mission effectiveness.

7 Observations and lessons learnt

Throughout the development of the IPSM tool since 2011 a number of lessons have been learnt and observations made by the development team. A summary of these are listed below:

- Stakeholder support
  - Effective visualisation of results is critical for communicating findings to a broad audience.
  - Enabling access of the IPSM using a remote login ensures that all users are using the most updated software build of the tool.
  - GUIs designed for conducting analysis can be of limited utility as they tend to constrain the type of analysis that can be performed.
    - The IPSM development team spent a lot of time developing the GUI for use in Phase Two, but ultimately it became under-utilised as the IPSM's capability grew.
- Modelling
  - The end goal of the model cannot be forgotten i.e. it is not necessary to have a perfect or fully developed model to generate effective results.
  - Having multiple models with overlapping capability, that are independently developed, promotes trust in the development, use and validation of models.
- Analysis
  - Platform system modelling tools are highly effective and useful when representative mission scenarios and operational parameters are being considered.
    - Subsequently it is paramount that a continuous feedback loop is established between modellers and the submarine user community.
  - Reconstructing performance estimates provided by prospective tenders has enabled the impact of uncertainty in sub-system performance to be quantified; and all tenders evaluated fairly with a common set of assumptions.
  - The capability provided by the IPSM provides a useful test-bed for novel concepts of operation as demonstrated in [11].
- Software development
  - A software development team must be diverse in their skills and not all focussed on the technical modelling aspects.
  - The time required to undertake verification and validation tasks are often underestimated and not planned for sufficiently in the inception stages.
  - Invest effort in reducing the run-time of the model - a single experiment simulation of the IPSM is in the order of a few seconds, making it highly attractive from a usability perspective.
  - Analytical tools do not follow a linear environment where 'development' is followed by 'use'. For example:
    - Tools can highlight limitations in the analysis being conducted, requiring further development to address these limitations.
    - Introduction of new technologies need to be considered, requiring new system models to be developed and integrated into the overarching model.
    - Different mission scenarios or platform tactics need to be evaluated, requiring a different approach or new models to be developed for operational analysis.

8 Summary of key IPSM features

In summary the IPSM provides the following features and capabilities:

- Stakeholder support
  - A series of secure wiki pages used for information sharing, which provides up to date user guides; and other supporting material and requirements for users and developers.
- Modelling
  - Platform level performance parameters are calculated based on integrating multiple system models, as opposed to a coarse single-equipment approach.
  - Ease of re-configuration between inputs and outputs; and reuse of various parameters.
  - Effective management and segmentation of data, coupled with strict access control, to ensure varying levels of security classification restrictions are handled appropriately.
- Analysis

- The execution of single experiment simulations, parametric simulations and optimised simulations to generate performance and operational attributes
  - The ability to undertake platform related sensitivity analysis of performance for existing, alternative and new technologies
  - Post processing of simulations to produce graphical, tabular and textual results to analyse and compare results of multiple platform concepts against the same criteria.
  - Ability to accept user defined metrics to be applied in order to test a series of conditions on simulated results (i.e. for verification and validation activities).
  - An automated and flexible reporting function.
- Software development
  - Continuous and automated software testing (at unit level and system-level); and effective version control, quality management and configuration management.
  - Extensive logs of simulations and data products for tracking and auditing purposes.

9 IPSM expansion plans

The success in generating auditable, repeatable, clear, meaningful and reliable results using the IPSM has shaped and matured various platform requirements. Subsequently, the IPSM has become an exemplar modelling tool and in the short term the IPSM will continue to be deployed during various phases of the program as more platform details emerge. The IPSM will continue to be maintained, extended, verified and validated as more modelling and in-service Collins Class data becomes available.

The flexibility, adaptability and effectiveness of the IPSM, enables the tool to be used for in-service applications. As such, there is interest from the Collins Class submarine community in using the IPSM to support pre/post-mission planning analysis and activities.

DST is also interested in expanding the functionality of the IPSM to directly support assessments of measures of effectiveness. The stealthy nature of submarine operations means that platform performance parameters strongly interface with the threat domain. As such the IPSM development team are in the inception stages of developing frameworks to understand the trade-off space between platform performance, mission effectiveness and susceptibility to exposure; and are aiming to provide a demonstration of this capability in the future.

The case study provided in [11] highlighted how the IPSM can be adapted to assess the benefits of particular technology concepts. As such, the IPSM could be used to assess the capability impacts of various technology demonstrators. Additionally, this assessment could also consider any system and design integration impacts that this technology has on overall performance.

10 Concluding remarks

The IPSM has been developed iteratively over the past decade using DST's knowledge of submarine systems, coupled with their diverse experience in systems modelling and simulation. Subsequently, the IPSM has evolved from an Excel based exploratory tool to a highly powerful, integrated, sophisticated and validated integrated platform systems modelling tool.

The DST IPSM tool is a highly effective and respected tool for providing analysis and advice on platform performance parameters and trade-off studies.

Reflecting back on the quote mentioned at the start of the paper, "The designer must fully understand the aspirations of the operator and equally the operator must understand the implications and consequences of their requirements." [1] Understanding this relationship has been a key part of the successful development of the IPSM.

11 Acknowledgements

The authors acknowledge the following DST personnel that have both contributed to and supported the development of the IPSM: Mr Andrew Tynan and Mr Kevin Gaylor.

12 Bibliography

[1] R. Burcher and L. Rydill, Concepts in Submarine Design, Cambridge University Press, 1995.

[2] B. Kassel, S. Cooper and A. MacKenna, "Rebuilding the NAVSEA Early Stage Ship Design Environment," in *ASNE Day - Engineering the Affordable Global Navy Through Innovation*, USA - Virginia, 2010.

[3] J. Rodgers and et. al., "A Framework for Integrated Platform System Modelling," in *Maritime / Air Systems and Technologies (MAST) Conference*, Europe - Malmö, 2012.

[4] S. Harrison and et. al., "Analysis of Platform Configurations Using the Integrated Platform Systems Model," in *International Maritime Conference – Pacific 2012*, Australia - Sydney, 2012.

[5] K. Slater and et. al., "High Level Capability Analysis of Submarine Platforms," in *International Maritime Conference - Pacific 2013*, Australia - Sydney, 2013.

[6] S. Harrison, J. Wharington and K. Slater, "An Investigation into the Penalties Incurred due to Operational Constraints Imposed on an SSK-type Submarine in an Operational Area," in *International Maritime Conference – Pacific 2015*, Australia - Sydney, 2015.

[7] B. Adams, "Dakota, A Multilevel Parallel Object-Oriented Framework for Design Optimization, Parameter Estimation, Uncertainty Quantification, and Sensitivity Analysis: Version 6.5 User's Manual," Sandia Technical Report SAND2014-4633, 2016.

[8] J. Sutherland and et. al., "System-Level Design Trade Studies by Multi Objective Decision Analysis (MODA) utilizing Modelica," in *Proceedings of the 1st Japanese Modelica Conference*, Japan - Tokyo, 2016.

[9] N. DiOrio, A. Dobos, S. Janzou, A. Nelson and B. Lundstrom, "Technoeconomic modeling of battery energy storage in SAM," National Renewable Energy Laboratory (NREL), Golden, CO, USA, 2015.

[10] M. Ceraolo, "New dynamical models of lead-acid batteries," *IEEE Transactions on Power Systems,* vol. 15, no. 4, pp. 1184-1190, 2000.

[11] S. Harrison and et. al., "The Operational Benefits of Covertly Refuelling a Conventional Submarine whilst at Sea," in *Warship 2017: Naval Submarines & UUVs*, Bath, UK, 2017.

A Platform-Focused Power and Energy Modelling Capability for Submarines

Peter Dennis[1,2], Peter Hield[2], John Wharington[2] and Karl Slater[2]
1. The University of Melbourne, School of Engineering, Victoria, Australia
2. Defence Science and Technology Group, Department of Defence, Australia
peter.dennis@unimelb.edu.au

ABSTRACT

Maximising efficient use of the limited energy stores aboard a vessel is pivotal in achieving optimal operational performance. In the case of diesel-electric submarines, the complex nature of power systems and operating behaviours means that the most useful assessments of energy use are likely to be produced when accurate and adaptive models of power and energy systems (and the onboard systems they supply) are coupled with a robust and detailed model of the platform as a whole.

To provide this capability, a framework of granular, physics-based power and energy system models was developed and embedded into DST's mature and validated Integrated Platform Systems Model (IPSM). The new models apply an 'energy balance' approach to the platform, where the losses incurred at each phase of energy use are fully accounted for, including the energy consumed in dissipating waste heat flows.

Applications for the tool include performing objective comparisons of platform-level or component designs over specified missions and environmental conditions, as well as identifying suitable targets for recuperation of electrical energy from sources of heat. This capability is demonstrated via an analysis of the impact of seawater temperature and the characteristics of installed cooling systems on the energy consumption and mission performance of an example diesel-electric submarine.

INTRODUCTION

Maximising the efficient use of onboard energy stores is a critical part of optimising the operational performance of a platform. Energy conserved may then either be spent on improving performance characteristics like range, endurance, availability or transit speeds, or conserved to reduce fuelling costs, and potentially improve machinery life and maintenance intervals via lighter usage schedules. In addition, if design or technology changes are integrated sufficiently early in development, it may be possible for a smaller or cheaper design to achieve the same operational targets, or for capabilities to be extended on the same platform.

In the case of diesel-electric submarines, the complexity of power systems and operating behaviours often obscures the relationship between component-level changes and platform level performance impacts. For this reason, obtaining defensible assessments of the potential for emerging energy technologies to improve platform performance requires a model that integrates accurate and adaptive modes of power and energy systems with a robust and detailed model of the platform as a whole.

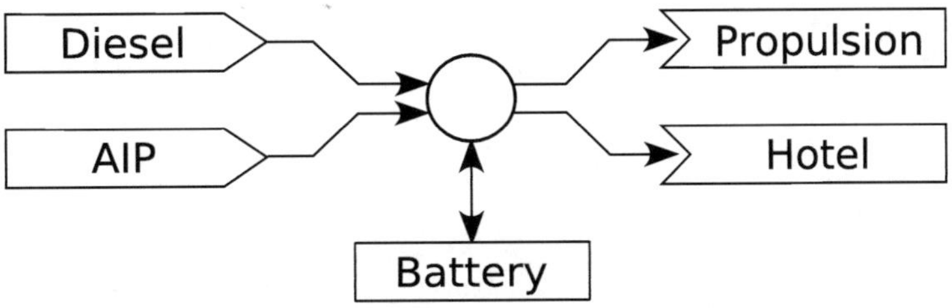

Figure 1: Top level electrical energy producers and consumers in an IPSM platform model.

DST's Integrated Platform System Model (IPSM) provides the required platform modelling capability, along with a framework for exercising these platforms via a library of mission templates and study designs. The IPSM also has a record of contribution to capability assessment studies spanning the life cycle of naval platforms, encompassing initial concept development and requirement setting, competitive evaluation processes, trade-off studies, and operational analyses [1,2].

The IPSM already includes a power and energy model that considers battery capacity, air-independent propulsion (AIP) reactant and diesel fuel reserves, energy conversion efficiencies, and variable propulsion and hotel loads. However, in this base model the systems interact only via electrical power delivery (waste heat flows are ignored), and the hotel load is modelled as a monolithic power requirement (Figure 1). This design is sufficient for assessing and comparing platforms where gross electrical loads and efficiencies are known (e.g. [3–5]), but detailed investigation of power and energy technologies requires more granularity in the model.

The present work therefore describes the development of a dynamic, modular hotel load model for use with the IPSM, incorporating a robust treatment of the costs and potential benefits associated with waste heat flows. The model applies an 'energy balance' approach to this problem, where inter-system energy transfers and losses incurred at each phase of energy use are fully accounted for, including the energy consumed in waste heat dissipation. In addition, the model is capable of identifying and quantifying opportunities for the recuperation of electrical energy from sources of heat.

Use of the updated tool is then demonstrated via a small study. This study examines the impact of seawater temperature and the characteristics of installed cooling systems on the energy consumption and mission performance of an example submarine, with the results used to suggest how the tool may be employed in future investigations.

THE NEW 'DYNAMIC HOTEL LOAD' MODEL

The intent of this new model is to decompose the monolithic hotel load, and to account for the influence of inter-system dependencies and waste heat management on the total energy consumption. As a result, each major system (e.g. combat, habitability, HVAC) is modelled separately, but interactions between the systems via electrical, thermal and other energy transfers are considered. The following sections detail the model structure for an individual system, and how these systems are then integrated into a complete model.

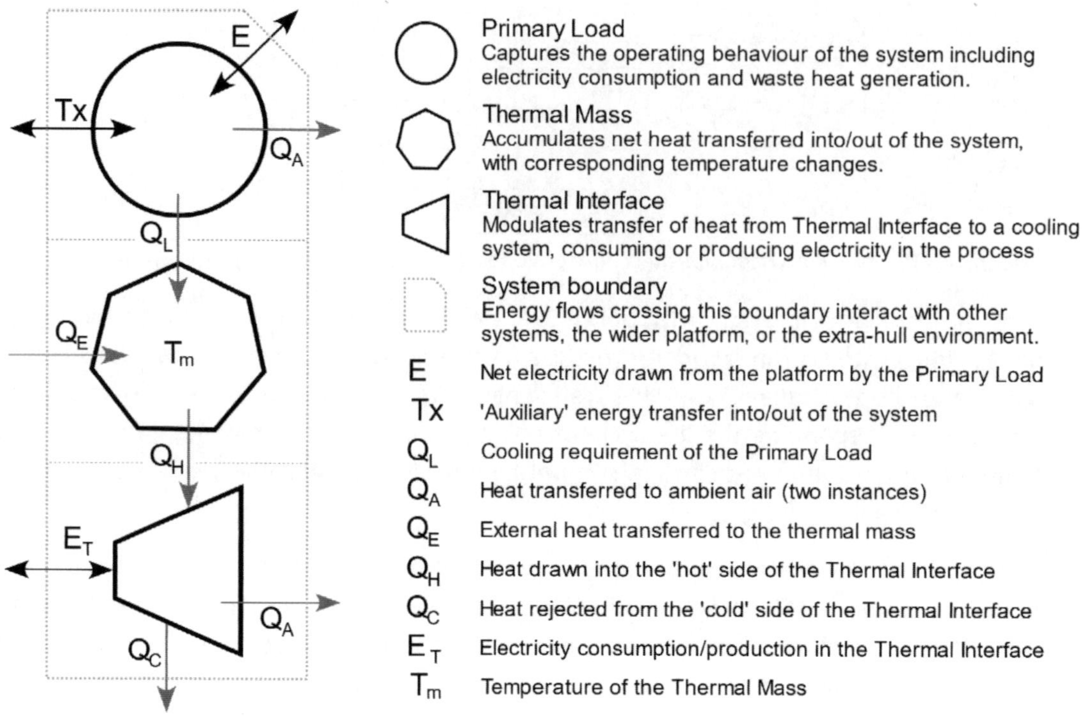

Figure 2: Schematic of energy flows for a single system

The Primary Load

The Primary Load (the circle in Figure 2) is responsible for determining the system's 'primary electrical demand' (E), and how this energy is further distributed. E represents the electricity consumed in accomplishing the systems primary task, and may be programmed to react to the present operating state of the platform (such as its speed, depth, or the nature of its current objective), or even to the activity of other systems in the network. For example, the HVAC system energy use may vary as a function of cabin temperature, the Combat system in response to the nature of current operations, or the Habitability system according to crew numbers and activity.

These primary electrical demands usually constitute the largest contribution to total hotel load, but the model is constructed to account for all energy flows to assess their individual contribution. This means that the net energy entering the Primary Load must be zero, i.e. $E + Tx + QA + QL = 0$ (Figure 2).

In this equation, Tx (the auxiliary transfer link) is used only rarely, but captures significant non-thermal exchanges between hotel systems (e.g. power supply relationships), or between a hotel system and the environment (like signal transmission or material ejection from the submarine). Finally, the net energy received via E and Tx is dissipated via Q_A (convection to ambient air) and Q_L (transfer to the coupled thermal mass).

The Thermal Mass and Heat Dissipation

Heat arriving in the Thermal Mass object (heptagon in Figure 2) via Q_L is combined with energy flows Q_E (heat transfer from externally-linked systems) and Q_H (heat removed from the mass by the Thermal Interface object) to determine net heat flux, causing T_{mass} to change accordingly. Q_E is particularly important for cooling systems, which must accept waste heat from other parts of the hotel load, while Q_H is controlled by the Thermal Interface object.

The Thermal Interface is responsible for mediating heat transfer between the Thermal Mass (connected via Q_H) and a coupled cooling system (connected via Q_C). The interface may be a simple heat exchanger, or an energy-conversion device such as a refrigeration system or a heat engine (like a thermo-electric generator or Rankine cycle). This flexibility allows straightforward comparison of the impact of a chosen technology on the energy demands of the system, and in turn the platform.

As heat passes through the interface, electrical energy may be produced or consumed (E_T), separate to the primary electrical demand above, with some heat potentially lost to ambient air (Q_A). The rate of energy transfer — as well as the efficiency of energy production or consumption — depends on the characteristics of the interface and the relative temperatures of the thermal mass and coupled cooling system.

Overall Model Structure

In the above sections, energy flows were discussed primarily in the context of a single system, however the Dynamic Hotel Load model links all hotel systems together in a network, where the key flows crossing system boundaries (E, Tx, Q_A, Q_E, E_T and Q_C, Figure 2) each relay energy to or from the next link in the chain. The overall design of this network is based upon shipboard system descriptions in well-known publications [6–8].

The electrical links E and E_T vary independently, but in combination determine how much electrical energy the system draws from the overall platform model (Figure 1). An additional hotel system (Supply Losses, not yet discussed) monitors these electrical flows and accounts for distribution losses such as power conversion inefficiencies and cable impedance. These losses are added to the total via Supply Losses' own E and E_T terms.

The thermal links Q_A, Q_C and Q_E are all inter-system links, forming the network of thermal flows shown in Figure 3. In key systems (HVAC and the cooling systems), Q_E is used to accept heat input from other systems.

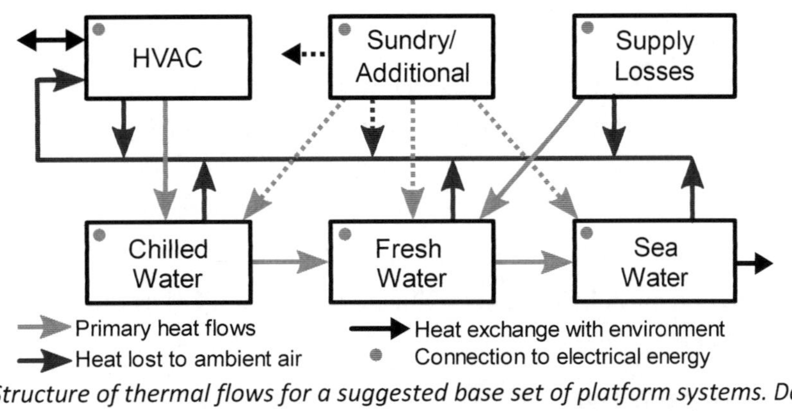

Figure 3: Structure of thermal flows for a suggested base set of platform systems. Dotted lines indicate the potential links for additional systems created using the 'Sundry/Additional' template.

In the case of HVAC, the energy accepted via Q_E is the total energy lost via Q_A across all systems (blue arrows in Figure 3). We also may note that the HVAC system is responsible for environmental interactions (black arrow) such as the induction of external air during snorts, and heat exchange with seawater across the hull.

In the case of cooling systems, Q_E constitutes the energy transferred into the system via the Q_C links of coupled systems (red arrows in Figure 3). Figure 3 also shows the cooling systems aligned in sequence in the bottom row, where each system accepts heat from zero or more non-cooling systems, and also passes energy down the chain from Chilled Water through Fresh Water to Sea Water (and ultimately out of the platform via seawater heat exchangers).

The Sundry/Additional system serves as a template for developers to account for all components of hotel load not yet captured by the other five systems. In the simplest case, all other behaviour could be aggregated into a single 'Sundry' system. However, the template may also be used to add as many hotel load systems as required to reach the desired level of resolution. Each such system may then be coupled to the appropriate cooling system (or systems), and set up with its own specific behaviours. For example, a model developer may choose to create separate models for habitability systems, ship services (like high pressure air, hydraulics or trim and balance), or to break a larger, complex system like the Combat system into smaller components.

Another important use of this template is to capture the impact of powertrain waste heat (e.g. heat generated by the diesel engines, propulsion motor, battery or AIP systems) that must be managed via the hotel HVAC and cooling systems. A hotel system may be created for each major powertrain component, absorbing waste heat generated from a component's inefficiencies as appropriate. If necessary, an electrical load (E) may also be modelled if electrically-driven equipment is required to remove heat from the component.

This waste heat is dissipated in the usual way via Q_A and Q_C, with the electricity consumed or produced in this process tracked as E_T. Thus, the modeller is not only able to investigate the relationship between powertrain efficiency and hotel loads, but also to assess the effectiveness of potential technologies for recuperation of powertrain waste heat (e.g. Olshima et al. [9]).

Summary of capabilities

Figure 4 shows the main energy flows typically observed in a diesel-electric submarine. The arrows representing these flows are coloured according to type (e.g. thermal or electrical), and sized in approximate proportion to the magnitude of each flow. This representation articulates that any energy conversion inevitably involves some degree of loss, and that these losses (and hence opportunities for improvement) are numerous and widespread.

The Dynamic Hotel Load model, in conjunction with the wider IPSM, is thus useful in providing the ability to investigate:

- how a specific system behaves within a particular scenario,
- how a system's behaviour changes across multiple scenarios,
- how a system is impacted by energy flows from the environment or other systems,

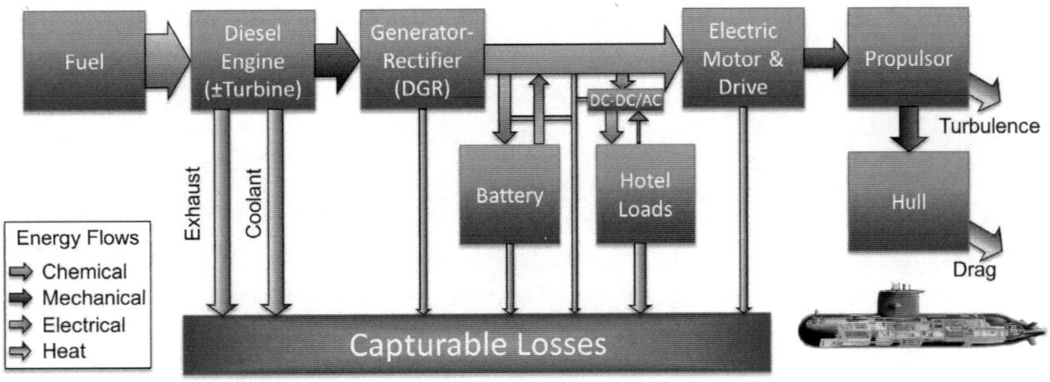

Figure 4: Sankey diagram of major energy flows in a diesel-electric submarine

- the energy demands of waste heat handling equipment,
- the impact of substituting system or component technologies,
- the impact of adding or removing complete subsystems,
- how and where energy may be recovered from waste heat flows.

The following section demonstrates one such investigation.

EXAMPLE APPLICATION

Definition of mission and metrics

Each individual IPSM simulation requires specification of both a platform and a mission. A mission defines a series of tasks that must be completed in order. Each segment has its own criteria for completion (e.g. distance travelled, fuel used or time on patrol) and associated operating constraints and CONOPS policies including: operating speeds, snorting strategy, environmental conditions, threat environment and required energy reserves at completion of the stage.

Figure 5 shows the mission considered in this paper, in which the platform must:

1. Transit from the base to the Operational Area (OA), including a 'Constrained Transit' (CT) segment characterised by increased detection threat,
2. Patrol in the OA, leaving to snort safely as required, until the allocated time or fuel is exhausted,
3. Return to base via Open Ocean Transit (OOT) once the patrol is complete.

The IPSM records time traces of all significant variables, and produces additional reports, graphs (like Figure 6) or summary metrics as requested. Two such metrics we consider in our example are Operational Area Effectiveness (OAE, the fraction of time in the OA leg spent in the patrol box), and Mission Effectiveness (ME, the fraction of total mission time spent in the patrol box).

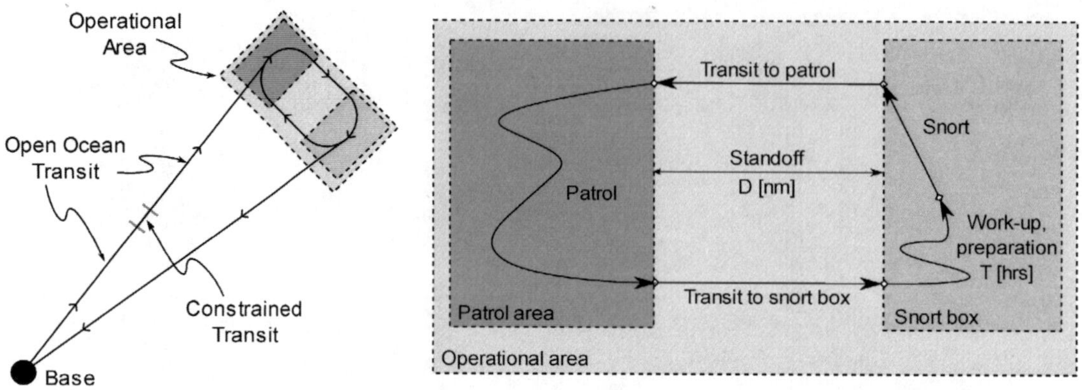

Figure 5: Example mission structure, with detailed view of the Operational Area.

Baseline scenario

Figure 6 was obtained by executing a simulation of the above mission with an example platform and a baseline parameter set. Some notable performance features are:

- During the transit legs at the start and end of the mission, battery discharge rate is high as a result of the higher speed and propulsion load, contributing to approximately daily snorting. No limitations on snorting apply in the open ocean.
- Just before the short 'constrained' transit, the mission requires the submarine to replenish the battery to a suitable level before proceeding into the area. Once the CT leg is commenced, a more conservative minimum battery level is applied in this leg (blue shading), as in the OA. Snorting is avoided where possible.

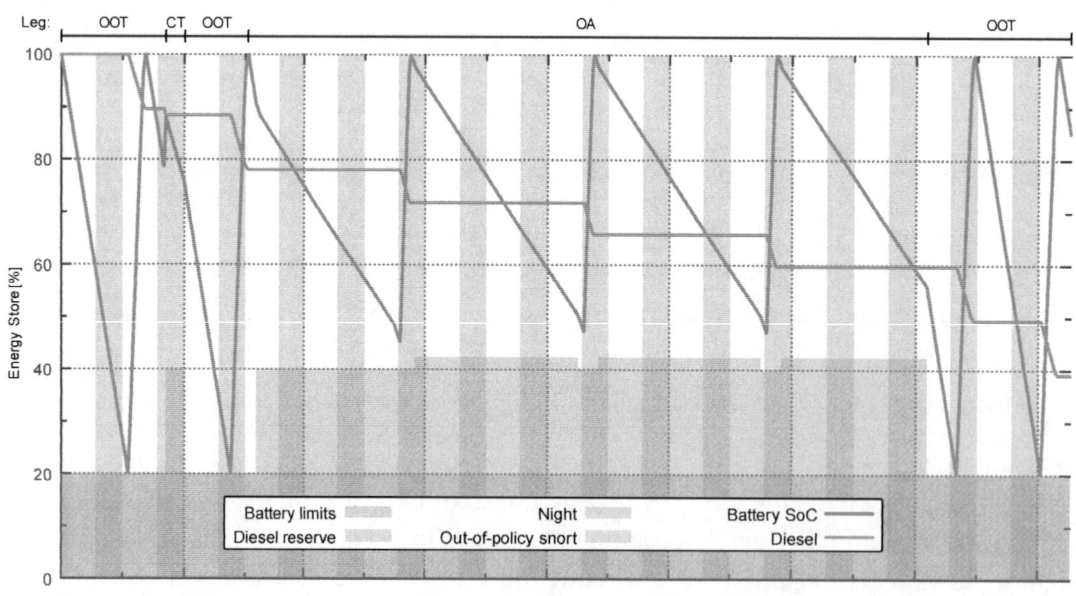

Figure 6: Evolution of energy store levels over the scenario illustrated in Figure 5, as generated by the IPSM. Mission segments depicted: Open Ocean Transit (OOT), Constrained Transit (CT) and Operational Area (OA).

- In the operational area, propulsion demands are lower, and snorting is only required once every three days. The submarine leaves the patrol box before the battery is fully depleted to ensure a) there is sufficient charge to transit to the snort box, and b) snorting will only occur at night as preferred (grey shading). No 'out-of-policy' snorts occurred in this simulation. Transits to and from the snort box are visible as short periods of increased battery discharge rate before and after each snort.

These results are repeated in the top-left plot of Figure 7, which provides similar summaries for two more scenarios, as well as some additional outputs: temperature traces for seawater, freshwater coolant and chilled water coolant in the second row, and the major thermal flows through the chilled water cooling system (with magnitudes normalised to unity) in the third. The two alternative scenarios are introduced in the sections below, and Table 1 at the end of this section summarises the inputs and key performance outcomes for all three cases.

In this first scenario, T_{sea} is relatively low at 5°C, resulting in favourable operating conditions for chilled water system (the chilled water is warmer than the fresh water, which is slightly warmer than T_{sea}). This results in a low electricity demand for cooling relative to the cooling load (bottom left). The spikes in cooling load visible during snort events are a result of the diesel engines pulling in warmer surface air, increasing the demand on the HVAC system, which subsequently dumps its waste heat to chilled water.

Modifying environmental conditions

In the second scenario (Figure 7, centre column), the platform specification is unchanged but T_{sea} is raised to 30°C. This imposes two detrimental effects on platform performance. First, there is a change in the rate of heat transfer between the submarine's hull and the seawater outside, worsening the load on the HVAC system. This is observable in the greater magnitude of the 'incoming heat load' relative to the first scenario (solid dark line). Second, it causes a higher freshwater coolant temperature, which in turn provides a large adverse temperature gradient for the chilled water cooling system's refrigeration equipment. This causes a reduction in chiller performance, resulting in electricity consumption that is a much greater fraction of the incoming heat load (solid grey line).

Overall, the impact of this higher electricity demand on total hotel load is significant, and requires the platform to switch from a 3-day to a 2-day snorting cycle to avoid exhausting the batteries between permissible snort periods. Diesel fuel consumption has also increased. The need to spend more time snorting, and in particular transiting to and from the snort box, has a measurable impact on the platform's overall performance (Table 1). Specifically, chiller energy use has more than doubled and the OAE and ME performance metrics have dropped by almost 5% in relative terms from Scenario 1 to Scenario 2.

Modifying system performance

While it is difficult to mitigate heat transfer between the hull and the ocean, it may be possible to upgrade the refrigeration plant within the chilled water system. In the third scenario (right-hand column of Figure 7), the environmental conditions are the same as in the second scenario, but the chiller is now operating with improved efficiency.

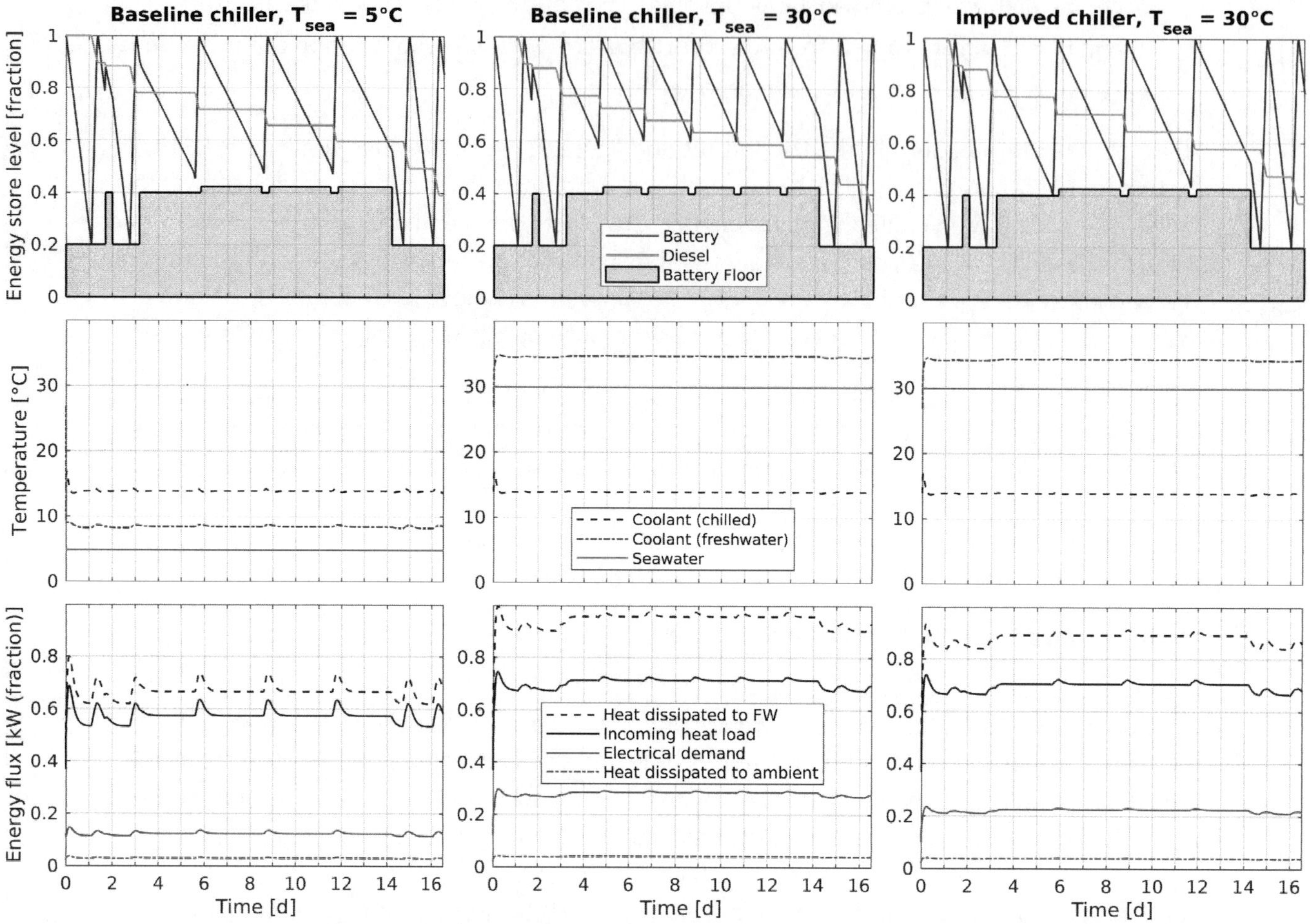

Figure 7: Simulation results for the three scenarios under examination.

Table 1: Summary of simulation inputs and results for the three scenarios considered

	Scenario 1	Scenario 2	Scenario 3
Sea temperature	5°C	30°C	30°C
Chiller performance	Standard	Standard	Improved
Mission duration	396.7 h	397.0 h	396.9 h
Chiller energy use (mean)	12.4%	28.1%	22.2%
Snort count (total/OA)	8/3	10/5	8/3
Diesel fraction remaining	39.0%	33.4%	37.1%
Time on station	237.8 h	226.8 h	237.1 h
OA Effectiveness (OAE)	90.1%	85.9%	89.8%
Mission Effectiveness (ME)	60.0%	57.1%	59.7%

This change has little effect on the incoming heat load (bottom-right), but reduces the penalty in electricity consumption of the chilled water system by around 38%, and in diesel consumption by around 52%. Significantly, the top-right plot of Figure 7 shows that this change is sufficient to return to a 3-day snorting pattern (although with lower battery reserve at the end of each cycle), and recover almost all of the penalty to OAE and ME metrics that was imposed by operating in warmer waters.

Modelling outcomes

This section presented an example of a simple, back-to-back comparison analysis achievable with the newly-upgraded IPSM. Similar analyses may of course be carried out on any number variety of platforms and scenarios. However, by leveraging IPSM's integration with Sandia National Labs' *Dakota* design of experiment tool [10], it is also possible to carry out parameter optimisations and other multi-variable analyses as required.

CONCLUSIONS

The intent and capabilities of a power and energy model developed for integration into DST's Integrated Platform System Model (IPSM) were described and demonstrated via a worked example. The new model provides a robust, granular and comprehensive framework for itemising and interrogating models of hotel systems and powertrain waste heat flows. Such studies may take the form of simple back-to-back simulations like those presented here, or more complex analyses by leveraging existing IPSM features and extensions.

As part of the IPSM, the new energy model provides DST with the capability to understand, assess and compare complete platform designs and technology insertion proposals in the power and energy domain. This is an important competency for the Department of Defence as a smart buyer, supporting rigorous decision-making and technical assurance in both new acquisition and life of type extension projects.

REFERENCES

[1] Karl Slater, Hamid Diab and Simon Harrison. 'Integrated Performance Modelling of Modern Conventional Submarines to Support Strategic Decision Making'. In: Warship 2017: Naval Submarines & UUVs, 14–15 June 2017, Bath, UK. 14th June 2017.

[2] Karl Slater, John Wharington and Mario Selvestrel. 'An Overview of the Integrated Platform Performance Modelling Approach used to Support the Future Submarine Program'. In: Pacific International Maritime Conference, Sydney, Australia (to appear). 6–8 Oct. 2019.

[3] Simon Harrison et al. 'Analysis of Platform Configurations Using the Integrated Platform Systems Model'. In: Pacific International Maritime Conference, Sydney, Australia. 2012.

[4] Simon Harrison, John Wharington and Karl Slater. 'An Investigation into the Penalties Incurred due to Operational Constraints Imposed on an SSK-type Submarine in an Operational Area'. In: Pacific International Maritime Conference, Sydney, Australia. 2015.

[5] Simon Harrison et al. 'The Operational Benefits of Covertly Refuelling a Conventional Submarine whilst at Sea'. In: Warship 2017: Naval Submarines & UUVs, 14–15 June 2017, Bath, UK. 14th June 2017.

[6] Ulrich Gabler. Submarine Design. Oxford, United Kingdom: Casemate UK Ltd, 2011.

[7] Roy Burcher and Louis Rydill. Concepts in Submarine Design. Cambridge, United Kingdom: Cambridge University Press, 1998.

[8] P.J. Gates. Surface Warships: an introduction to design principles. Brassey's Sea Power, Vol 3. Alexandria, Virginia, USA: Brassey's Defence Publishers, 1987. ISBN: 9780080347547.

[9] Noam Olshima et al. 'Waste Heat Recovery of Marine Diesel Engines'. In: Pacific International Maritime Conference, Sydney, Australia. 3rd Oct. 2017.

[10] Brian M. Adams et al. Dakota, A Multilevel Parallel Object-Oriented Framework for Design Optimization, Parameter Estimation, Uncertainty Quantification, and Sensitivity Analysis: Version 6.9 User's Manual. National Technology and Engineering Solutions of Sandia. 13th Nov. 2018. URL: https://dakota.sandia.gov/documentation.html.

A system-of-systems for at-sea evaluation of modern sonars

Piers Wren, Sonartech Atlas, piers.wren@sonartech.com.au

ABSTRACT

Submarine sonar systems are advancing in capability due to the developments in sensor technology and the application of modern algorithms. Measuring performance of a modern sonar system at sea is a complex problem that requires a multi-disciplined approach to ensure the aims of a trial are achieved. The underwater acoustic environment is complex and it is rare to experience conditions that support every aim of any trial. Decisions made through system development will impact the ability for a system to achieve successful sell-off via sea trial; the possibility for calibration and an ongoing measurement-based maintenance program.

This system-of-systems plays a pivotal role supporting the evaluation and transition-to-use of on-board systems, primarily through the conduct of at-sea trial activities. This support encompasses all phases of test including pre-trial planning, trial data collection and monitoring and post-trial data analysis. The monitoring of trials conduct and data collection must make efficient use of trials staff and post-analysis must support required turn-around times to allow for trials windows to be used effectively.

This paper considers the requirements on the systems supporting successful evaluation of sonar performance, particularly in a sea trials environment. The support systems described include physical items such as transducers, software systems used during a trial and systems engineering techniques that ensure stated requirements can be demonstrated in a wide variety of at-sea conditions. These same systems are also key enablers for continuous improvement via access to real-world recordings, and the introduction of machine learning algorithms.

BACKGROUND

Test and evaluation of a military sonar system in its operating environment is a difficult proposition to undertake. In order to meet the theoretical capabilities of a modern sonar, the ocean would need to be free from interfering noise sources and support propagation of sound out to ranges that are unlikely to be supported on any day; let alone the day chosen to run a sea trial. This presents sonar developers and their customers with the challenge of determining whether a sonar is able to demonstrate its required performance during a trial event. To answer this question the trial must capture far more than the performance of the sonar; it must also capture enough information about the environmental conditions on the day to allow experts to estimate theoretical performance in that environment to support comparison with the performance observed during the trial.

The qualification of a sonar systems begins long before the trial event. The initial systems engineering that sets the requirements against which the sonar will be compared must be cognisant of how theoretical performance translates into real-world performance and ensure that requirements have the flexibility to allow verification of the sonar in a variety of environments. Translation from theory to practice is highly dependent on the hardware and algorithms being verified. Capable sonar houses should have rules in place to do this translation based on previous delivery of similar product; or based on measurements taken from prototypes of the system under test.

Once development of a system that has appropriate real-world requirements is complete and that system is placed into an environment for qualification, the trials team can expect to spend significant effort measuring the environment in addition to evaluating the system under test. Significant pre-work is possible to ensure bathymetry maps of the area are up-to-date, weather predictions and history are available and the required people and equipment are prepared for the trials period. At a minimum, the trial must record the sound speed profile for the trials area so that this be later used to predict the propagation of sound during the trial. Additions to this information will be system-dependent but are likely to include:

- Alternate/additional measurements of acoustic emissions,
- Positioning information for all entities in the area; and
- Details of known interferences in the area.

TRIALS TOOLING

The tools required to execute a successful trial fall into three categories. Planning tools used prior to the trial, execution tools used during the trial and analysis tools used following the trial. An integrated trials suite enables these tools to interact with each other and simplifies reporting at various stages of the trial. It ensures trials operators are able to monitor their progress against the trial plan and allows trial analysts to compare the executed trial against the planned trial. Tooling to monitor the trial progress as the trial is still underway is crucial in order to detect and workaround issues as they occur. Live monitoring enables operators to correct for problems that might otherwise invalidate the results of the trial.

Figure 1: Integrated Sonar Trials Toolset

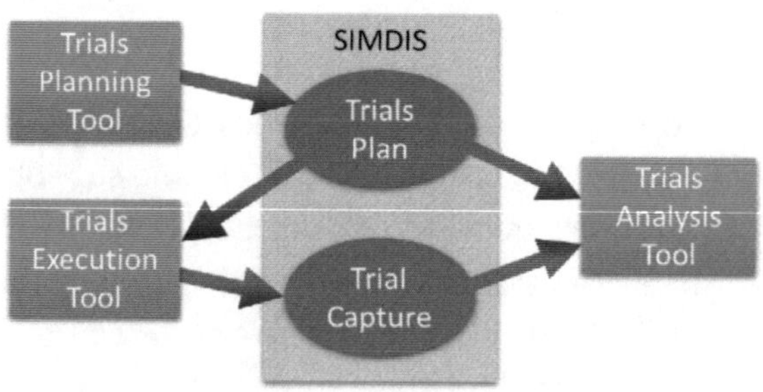

Sonartech is currently working towards a fully integrated sonar trial suite supporting customers through the planning, execution and analysis of a sonar trial event. The key to this integration is that each tool in the suite uses open formats to represent trials data. This enables sharing within the suite, and also allows the use of 3rd party tools on the same data. The enabling standards for this integration are the SIMDIS toolset available from the U.S. Navy Research Laboratory (NRL) and the Mission Data File (MDF) format available from Sonartech. The use of open, industry standards for scenario definition and data capture allows our sea trial tools to integrate with planning, visualisation and analysis tools available from 3rd parties, including the SIMDIS 3D display tool available from NRL. Example use cases of this integration could be:

- 3D planning and fly-through of the sea trial scenario.
- Execution of 'What-If' modifications to each scenario.
- Live comparison between the planned scenario and trial during execution.
- Automated control of trial equipment based on the planned scenario.
- Comparison between the recorded trial and planned trial.

The complexity of a sonar trial environment can cause results from multiple runs within the same trial to be significantly different from each other. To cater for this complexity, the system should be able to manage the results from multiple trial runs and multiple sensors within the trial and present a progressive view of the trial as each result is received. An intelligent system can also monitor the differences between measurements and recommend when enough data has been captured to achieve the goals of a trial; based on the remaining variances in the data.

Using open formats in this context allows each of the planning, execution and analysis tools to potentially be developed by different organisations. It also allows a single system to capture the truth data for a sea trial and share that data with each organisation that will analyse the trials data. The use of open formats will promote interaction between OEMs and allow each OEM to focus on developing the tooling that is required to test their own system rather that developing the entire suite required to plan, execute and analyse the trial.

SONAR PRODUCT DESIGN FOR TRIAL

Noting that sonar systems generally require interactions with one or more operators, it is important to understand the information presented to the operator during each event within a trial. To enable this, the sonar system should be able to record not only the data received, but also any operator interactions with the system during the trial. This information can then be used to recreate the trial experience from the operator's perspective and understand moments when an operator's response was not as expected.

Modern sonar systems will undergo calibration trials in addition to measurement/acceptance trials. Where a sonar has the ability to be adapted to its installed operating environment, the method by which it can receive calibration data should be considered during the initial specification. Ideally, the sonar system will be able to interact directly with the logs recorded during the trial, but at a minimum the system should be able to accept calibration data without the need for special equipment or significant operational downtime.

The next step for sonar trial and calibration is the addition of measurement and calibration features to the primary product. Such a system takes the integration presented above and merges it into the deployed system, enabling the operational crew to perform opportunistic trials when conditions are most suitable. This will allow the trial and calibration of the sonar system to be refined as the system is exposed to new environments and contact types; and to account for any drift in calibration as the system ages. Once deployed, it also reduces the need to execute specific trial events as much of the trial requirements can be met while the system is operational.

MACHINE LEARNING IN A SONAR TRIALS ENVIRONMENT

Modern sonars output an enormous amount of data that must be captured, in addition to capturing data from supporting sensors available in the trials environment. This presents a data management issue as the data must be captured, stored, catalogued and made available to downstream systems. Sonartech is very active in this field and offers a suite of recorders and data analysis systems which work with the open MDF format and have been adopted by the local and international customers.

This is an area where modern Machine Learning algorithms can help refine the performance of the sonar system. Libraries of sea trial data already exist in MDF and other formats that can be mined to compare the performance observed during a trial to an established ground truth of that event. This data forms training sets that can be used to train Machine Learning algorithms on the issues a sonar may experience while under trial so that those algorithms can search for patterns within the data. The aim of such an undertaking is for the algorithm to provide recommendations on best practices for the trial; and to add an interpretation of the results that considers influences which human analysts may not consider.

An adaptive sonar that is able to monitor the results generated from all sensors fitted to a platform in realtime would be able to systematically learn the biases inherent in each sensor, environment and signal type. Such a system could gradually adapt the processing results to be more correct, based on the biases it learns. The first generation of such an adaptive system may work with the tracker data provided by each sensor; however future generations are likely to interact directly with the sensor data to detect an localise contacts earlier than possible with the current, static systems.

CONCLUSION

This paper has discussed the difficulties of measuring performance of a sonar during a sea trials event, proposed a technical solution to enable integration across the sonar trials industry and discussed a future direction of opportunistically calibrated sonar systems. The concepts discussed in this paper are currently deployed, in development, or achievable using existing technology.

imc2019
PACIFIC INTERNATIONAL MARITIME CONFERENCE
INTERNATIONAL CONVENTION CENTRE SYDNEY, AUSTRALIA
8-10 OCTOBER 2019

Damage Assessment of Composites for Naval Vessels Subjected to Explosive Events

A.Gargano[a], M.E. Ibrahim[b], V. Pickerd[c]

Maritime Division, Defence Science and Technology Group, 506 Lorimer Street, Port Melbourne, 3207, Australia.

[a]alexander.gargano@dst.defence.gov.au, [b]matthew.ibrahim@dst.defence.gov.au
[c]vanessa.pickerd@dst.defence.gov.au.

ABSTRACT

Composite materials are used in a wide variety of naval applications due to their low weight, high stiffness and strength, corrosion resistance, as well as radar and acoustic absorbance, and non-magnetic properties. Despite these benefits, there remains concern over the resistance of composite materials used in naval operations to explosive loading. Large-scale damage has an immediate effect on operations, potentially resulting in water ingress and flooding. Small-scale damage can often go undetected and result in the reduction of the mechanical properties and structural integrity of the composite structure. This may impact on the operational capability and safety of the vessel. Hence, robust and reliable damage detection methods are required to assess the post-damage capability of composite platforms.

Current non-destructive testing (NDT) techniques used for composite structures include visual inspection, tap-testing, ultrasonics and radiography. However, these techniques have largely been proven effective only for thin, lightweight structures. In general, marine composites are thick (in the order of hundreds of millimetres) and of large lateral dimensions. This poses challenges for reliable in-situ NDT detection of damage in composite naval vessels, required to enable confident operation of platforms post-explosive events. In this paper, NDT techniques were used to evaluate the damage in small- and large-scale composite panels that were subjected to various explosive blast test conditions. Damage such as debonding, delamination and fibre rupture were detected using ultrasonic scanning, and resulted in large reductions in the residual flexural and compressive properties of the composite laminates. However, micro-sized damage

such as matrix and intertow cracking also resulted in substantial reductions to the post-blast residual stiffness, yet were undetected using ultrasonic scanning.
This work demonstrates the effectiveness and limitations of NDT methods in determining the extent and types of damage. Future NDT methodologies to overcome limitations of existing technologies are also discussed.

INTRODUCTION

Composite materials such as carbon, glass and Kevlar® fibre-reinforced composites are used in a wide range of naval vessels and ship structures as they are light-weight, with high stiffness and strength. Composites also have good seawater corrosion resistance, environmental durability and non-magnetic properties [1, 2]. Mine countermeasure vessels such as the RAN Huon Class Minehunter Coastal and some corvettes and patrol boats such as Visby-class corvette or Skjold-class patrol boat are largely built using composite materials. Composites are also used in the superstructure, advanced mast system, bulkheads, decks, propeller, propulsion shaft and rudder of some other naval ship classes. Navies are also exploring the feasibility of using composites for a variety of internal equipment and fittings, such as machinery, heat exchangers, equipment foundations, valves, pumps and ducts [1].

Despite the many benefits of using composite materials in naval ship structures, there is significant concern in their response to an explosive blast load from a weapons strike. Large overpressures and impulse from an explosive detonation can damage composite laminates in the form of matrix cracking, delamination between the plies and fibre fracture, which can ultimately lead to complete rupture [3-7]. Damage modes also associated with sandwich composites can include facesheet-core debonding, core crushing and core cracking [8-11]. Large-scale damage can have immediate effects such as water ingress and flooding. However even small scale (often undetected) damage can reduce the mechanical properties of the composite, thereby compromising the operational capability and safety of the vessel.

Nondestructive testing (NDT) techniques can be used to detect some types of non-visible (or barely visible) damage, but also provide information about the type, size and extent of the defect on a composite material. This information can be used to assist in the post-blast damage assessment of the structure. Thick structures (in the order of hundreds of millimetres) cause particular difficulty for both detection and sizing of defects for most NDT techniques. However, some techniques can be used to detect damage in thick composites resulting from explosive blast. X-radiography can be used to inspect composite materials, though has some challenges inspecting planar delaminations, microwave-wavelength radiation can be transmitted though glass-fibre based composites to detect the presence of moisture in particular, and low-frequency ultrasonic inspection can be applied to thick polymer-matrix composite sections [12].

In this paper, damage assessment using through-transmission and pulse-echo ultrasonic scanning, scanning electron microscopy (SEM) and visual inspection of small- and large-scale composite targets to far- and near-field and surface/contact explosive air blast loading is presented and evaluated. The materials assessed are representative composite structures commonly found in navy ships: (i) woven E-glass fibre vinyl ester, (ii) woven carbon fibre vinyl ester, and (ii) PVC foam stiffened glass fibre polyester.

The following results are separated into the explosive blast damage assessment of (i) small-scale and (ii) large-scale composite panels. The progression of failure modes in small-scale composite laminates was identified using a series of post-blast damage quantification techniques. Also, the effect of the different failure modes on the residual stiffness and strength was evaluated. From the large-scale tests, the capabilities of damage detection using NDT techniques were evaluated for thick section composite panels.

CASE STUDY 1 – SMALL-SCALE EXPLOSIVE BLAST TESTS

In this case study, small-scale composite laminate panels were subjected to a series of explosive air blast events. The aim of this work was to identify the progressive failure modes in composite materials when subjected to increasing blast impulses, achieved using three different loading conditions; (1) far field, (2) near field and (3) contact explosive tests, as depicted schematically in Figure 1. Damage assessment of the small-scale composite materials post-explosive blast event was conducted using visual inspection, ultrasonics, SEM and post-blast mechanical property testing.

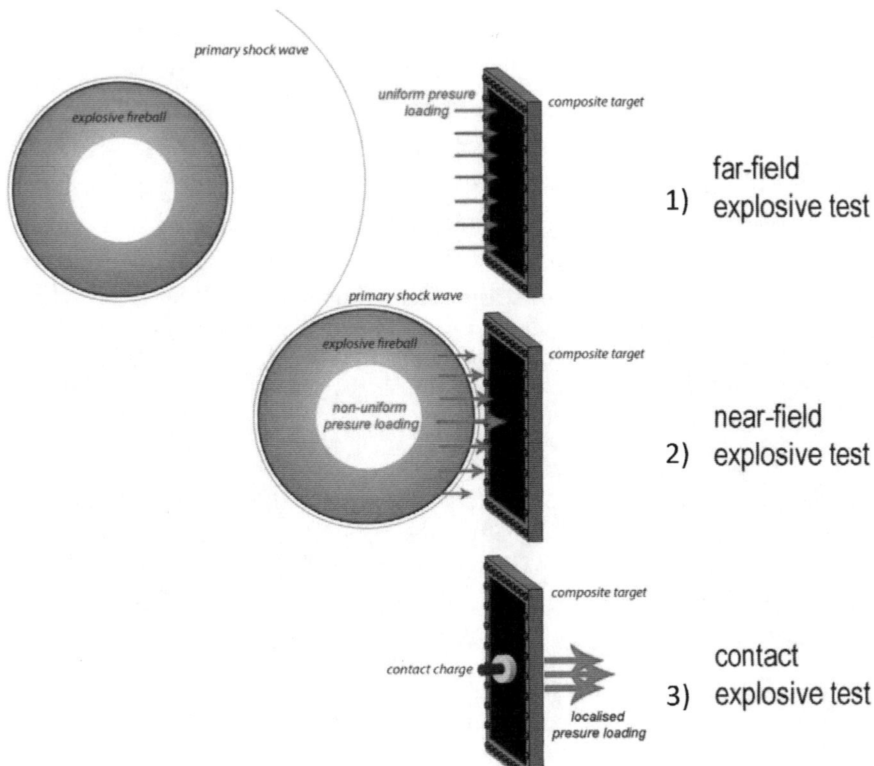

Figure 1: Far-field (top) near-field (middle) and contact (bottom) explosive air-blast conditions.

Materials and Methodology

Composite laminates comprising of either E-glass or carbon fibres with vinyl-ester resin were tested in the form of flat square plates (275 mm x 275 mm x 4 mm). The E-glass and carbon used in the laminates were both single ply plain woven fabric (areal density of 600 g/m^2). The fabrics were stacked into preforms with the warp tows aligned in the same direction, giving a cross-ply fibre [0/90] pattern. The fabric was then infused with liquid vinyl ester resin at room temperature using vacuum bag resin infusion (VBRI). To ensure the laminates had the same thickness, the E-glass and carbon laminates contained 10 and 7 plies respectively. The panels were held within a rigid steel window frame having a 250 mm x 250 mm aperture and placed inside an enclosed steel plate-lined concrete chamber. Far- and near-field explosive tests were conducted using bare spherical Type 4 plastic explosive charges that were hung from the ceiling of the chamber. The overpressure and impulse of the blast wave was controlled by varying the explosive charge mass and stand-off distance from the laminate plate, with near-field being a stand-off distance where the blast products and the primary shock wave interact with the target, typically less than 20 times the charge radius or a scaled distance of <1 m/kg$^{1/3}$. Contact explosive tests involved placing cylindrical PETN sheet explosive on the surface of the composite plates.

Damage Assessment

When subjected to a far-field explosion, the composite plates appeared undamaged by visual inspection, as shown in Figure 2. However, scanning electron microscopy revealed damage in the carbon fibre laminate in the form of intertow cracks between the fibres and polymer matrix together with transverse cracks within the polymer-rich regions (shown in Figure 3). When subjected to a near-field blast, the panels suffered delamination cracking at one or multiple ply interfaces and fibre rupture in the carbon laminate. This was revealed by through-transmission ultrasonics and is shown in Figure 4. In comparison, the glass fibre laminate was more resistant to damage and remained intact even when subjected to intense near-field blast load.

Figure 2: Front face images of the glass (left) and carbon (right) laminates subjected to a far-field explosive blast test.

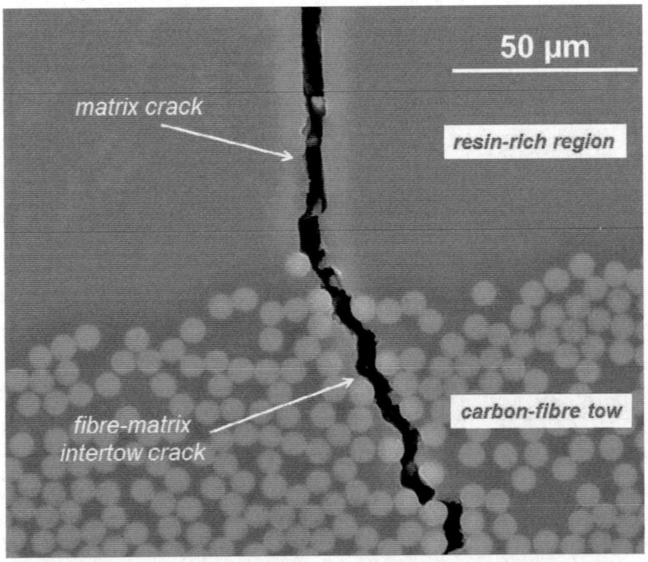

Figure 3: Scanning electron microscope images of the carbon fibre laminate showing cracking within the fibre tows and matrix cracking in the polymer-rich regions.

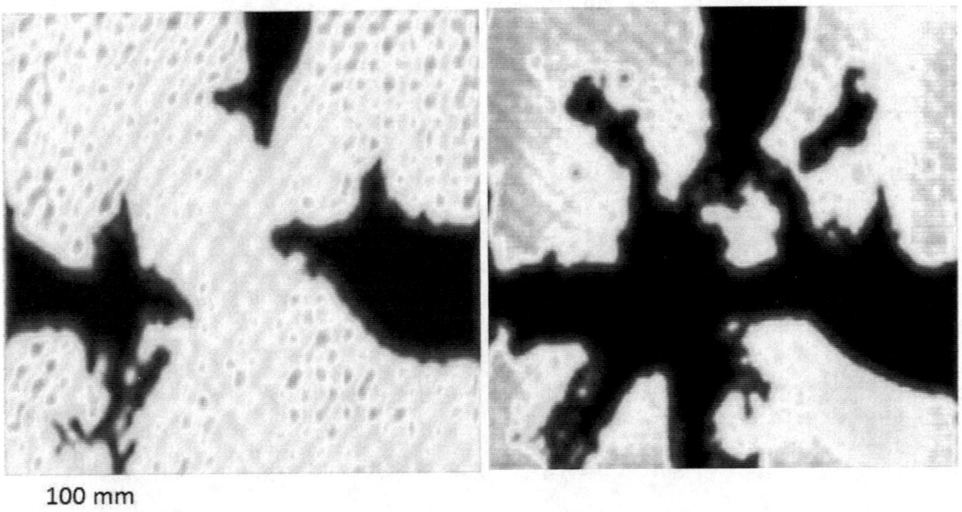

Figure 4: Through-transmission ultrasonic C-scan images of the carbon laminate when subjected to near-field blast test. The dark regions reveal internal delaminations. Note: the left image was subjected to ~30% lower blast load than the right image.

When subjected to a contact explosion, the laminates experienced localised damage around the charge location. Images of the front face of the carbon fibre-vinyl ester and glass fibre-vinyl ester laminates after explosive testing are shown in Figure 5. From visual inspection, the glass fibre laminates undergo burning and fibre rupture which resulted in a central hole, as well as surrounding delamination. Comparatively, burning and delamination was not visible in the carbon laminate. A slight discolouring was observed around the hole of the carbon laminate which was residual adhesive used to bond the charge to the plate.

Through-transmission ultrasonic scanning of the laminates revealed the extent of the surrounding delamination damage (shown as dark blue/black) around the central hole, as shown in Figure 6. The figure is shown with the C-scan images as an overlay to the strike face images shown in Figure 5. The C-scan images of both the glass and carbon fibre laminates show a substantial region (~20% of the sample area) which was delaminated by the explosive blast. In the case of the carbon laminate, this was undetected by visual inspection. It would be possible to identify delamination in thin laminates such as those shown here using the 'tap test' technique. However, when inspecting thick composites, deeply buried defects are inaudible to the human ear, making tap testing insufficient in providing an adequate evaluation for engineering certification [12].

Figure 5: Damage to strike face of glass (left) and carbon (right) fibre laminates subjected to 1.0 g of PETN contact explosive charge. Images are cropped and of higher magnification.

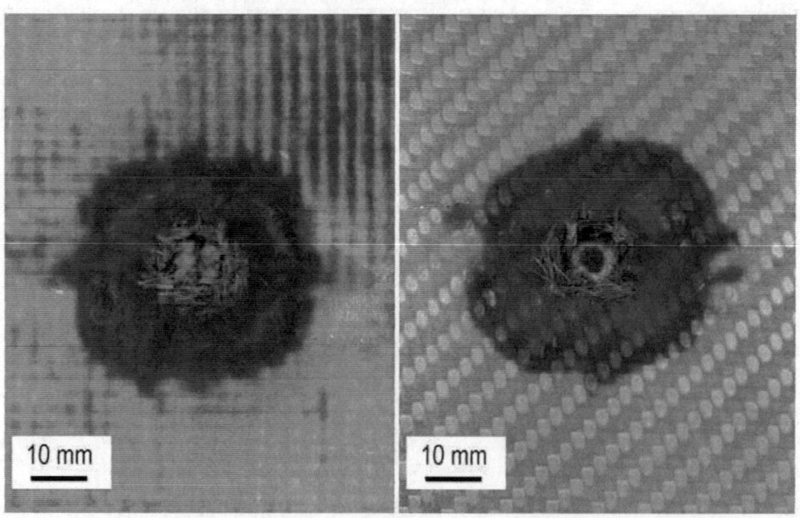

Figure 6: Through-transmission ultrasonic C-scan images showing delamination damage to the (left) glass fibre and (right) carbon fibre laminate subjected to a 1.0 g PETN contact explosive charge. Images are cropped and of higher magnification.

The effect of damage on the stiffness and strength properties of the composites was assessed by post-blast mechanical property testing. Table 1 shows the mechanical properties under bending (flexural) of the composites post far- and near-field explosive blast loading. The results show a decrease of ~17% in the flexural stiffness property of the carbon laminate for the far-field tests, due to the initiation of fibre-matrix intertow cracking and matrix cracking, despite the panel appearing visually undamaged. This raises concerns about the limitations of current visual damage detection techniques. The results show that damage induced by low-blast impulses can be detrimental to the mechanical properties, and are only visible using scanning electron or optical microscopy.

When subjected to an intense near-field blast explosion, both the flexural stiffness and strength of the carbon fibre laminate were substantially decreased (~70%) due to the initiation of large-scale delamination cracking and fibre rupture. Comparatively, the glass laminate only showed minor decreases in the flexural properties even when subjected to the extreme overpressures and impulses of the near field blast. The lesser extent of damage in the glass laminates provides evidence of greater resilience to far- and near-field explosive blast loading in glass laminates compared to carbon. This was believed to be due to the glass laminate having greater tensile and flexural strain energy density properties compared to the carbon laminate. The glass laminate had tensile and flexural strain energy density values of 11.5 and 9.3 MJ/mm^3 compared to the carbon laminate's 8.9 and 5.7 MJ/mm^3, respectively. The energy density is a combined measure of a materials modulus, ultimate strength and failure strain (area under the stress-strain curve) and is the amount of strain energy that can be absorbed before breaking.

Table 1: Post-blast flexural properties of the carbon and glass fibre laminates subjected to far- and near-field explosive blast tests. Numbers in brackets give standard deviation.

Blast impulse (Pa.s)	Carbon fibre laminate		Glass fibre laminate	
Far-field	**Modulus (GPa)**	**Ultimate strength (MPa)**	**Modulus (GPa)**	**Ultimate Strength (MPa)**
0*	42.5 (±1.0)	460 (±8)	23.1 (±1.8)	647 (±33)
109	41.8 (±1.1)	446 (±11)	#	#
154	42.5 (±0.3)	436 (±7)	*X*	*X*
219	35.1 (±0.2)	442 (±24)	23.6 (±0.4)	617 (±37)
Near-field	**Modulus (GPa)**	**Ultimate strength (MPa)**	**Modulus (GPa)**	**Ultimate Strength (MPa)**
353	14.0 (±11)	154 (±141)	21.9 (±0.3)	631 (±15)
472	#	#	19.1 (±0.2)	604 (±18)

Note: * represents pristine condition, '#' represents panels that were not post-blast mechanically tested due to either no or excessive damage, 'X' represents material was not blast tested at this impulse.

Table 2 presents the compressive properties of both laminates against contact explosive loading. Contact loads are presented in N.s as the pressure was not able to be measured during the tests and hence analytical calculations were used to determine loading using equations presented in Remennikov et al.[13]. The compressive modulus and strength of the carbon and glass laminates are reduced substantially (by up to ~60%) with increasing blast impulse. The compressive strength was more greatly affected, with maximum reductions of 52% and 63% to the glass and carbon fibre laminates, respectively. Two main damage types were identified from contact explosive loading including: (1) penetration/perforation approximately the size of the charge diameter, and (2) delamination surrounding the centre hole.

Table 2: Post-blast compressive properties of the carbon and glass fibre laminates subjected to contact explosive tests. Numbers in brackets give standard deviation.

Contact charge	Carbon fibre laminate		Glass fibre laminate	
Blast impulse (N.s)	**Modulus (GPa)**	**Ultimate strength (MPa)**	**Modulus (GPa)**	**Ultimate strength (MPa)**
0	56.9 (±1.2)	200 (±7)	33 (±1.2)	187 (±5)
0.65	42.3	101	28 (±0.9)	116 (±12)
1.46	43.2	103	27	108
1.97	36.1	103	22 (±1.7)	110 (±2)
2.43	36.9	74	20 (±1.8)	90 (±5)

A study was conducted to identify the effect of the two primary damage modes (central hole and surrounding delamination) in the laminates due to contact explosive loading. Hole sizes of 13 mm and 68 mm were cut from a pristine glass fibre laminate to represent the penetration area and delamination damaged area, respectively. The dimensions were based on the damage areas observed in the specimen subjected to the largest blast impulse tested, of 2.43 N.s. Table 3 presents the results for reduction in compression modulus and strength due to the central hole and the delamination. When subjected to the 2.43 N.s contact blast, the compressive modulus and strength of the glass fibre specimen was reduced by ~40% and ~50%, respectively. It was found that the total reduction in modulus (~40%) was equally due to both the central hole and surrounding delamination damage (~20% each). Comparatively, the central hole resulted in ~20% reduction in compression strength and therefore the remaining ~30% was due to the delamination damage. This finding is consistent with past studies in that delamination was the more detrimental failure mode for composite laminates loaded in compression [14, 15]. It is noted that the reduction in compressive strength from the 68 mm hole was ~10% more than the 2.43 N.s blast (68.7 MPa compared to 89.7 MPa). This suggests that a portion of the delaminated area maintained some of its residual strength post explosive blast. This demonstrates the importance of NDT to correctly identify the true extent and magnitude of damage.

Table 3: Effect of central hole and delamination area on the residual compression strength and modulus of the glass laminate.

	Pristine	**2.43 N.s blast impulse**	**13 mm hole**	**68 mm hole**
Compression modulus (GPa)	32.7 (±1.25)	19.7 (±1.80)	26.8 (±4.79)	20.6 (±3.30)
Ultimate compression strength (MPa)	187 (±5.29)	89.7 (±4.61)	147 (±4.43)	68.7 (±6.75)

CASE STUDY 2 – LARGE-SCALE EXPLOSIVE BLAST TESTS

In this case study, two representative HMAS Minehunter class deck and bulkhead panels were subjected to near-field explosive air blast tests. This work was also used for a feasibility study of in-situ ultrasonic inspection of full-scale stiffened thick-section GFRP panels.

Materials and Methodology

The bulkhead panel tested was 2500 mm x 2440 mm with a nominal thickness of 12.5 mm. Three box or 'top-hat' stiffeners were equally spaced across the panel and were nominally 100 mm proud of the surface. The decking panel was larger and thicker, measuring 3440 mm x 2440 mm with a nominal thickness of 21.5 mm. Two top-hat stiffeners were placed either side of the centreline, in-line with the longitudinal direction and were nominally 150 mm proud of the surface.

During testing, the four edges of the test panel were clamped using a series of plates and bolts and positioned upright. A bare PE4 charge was suspended from a supporting beam behind the test panel, and positioned in-line with the centre of the target. The bulkhead panel was subjected to a 0.5 kg charge at a stand-off distance of 0.4 m detonated on the stiffener side. The deck panel was subjected to a 1 kg explosive charge at a stand-off distance of 0.6 m detonated on the flat side. The tests were conducted in an outdoor testing range which eliminated shock reflections occurring from walls and ceilings (although not from the ground).

Damage Assessment

Post-blast damage assessment of the large-scale composite panels was conducted using phased-array pulse-echo ultrasonics and visual inspection. The inspection was performed to examine delamination and voids within the composite facesheet and discontinuity at the stiffener-facesheet interface. The C-scan images from pulse-echo ultrasonics pre- and post-blast of the bulkhead panel are shown in Figure 7. In the pre-blast scan, a highly reflective region marked 'C' was observed. This was due to sanding of the area for the application of the strain gauge. This created a parallel surface which resulted in excellent reflectivity of the incoming ultrasonic pressure wave. The post-blast C-scan image showed an increase of approximately 25-30% in the amount of ultrasonic reflection (shown by the orange colour) compared to the pristine sample. This was likely due to localised internal delamination and debonding at the stiffener-facesheet interface. This damage was more pronounced in the centre of the panel, closest to the charge location. Areas marked A and B in Figure 7 also show heavily damaged regions at the ends of the middle stiffener. Higher ultrasonic reflection in these regions was due to separation of the stiffener foam core from the surrounding glass fibre plies, and localised delamination in the glass fibre plies, as shown in Figure 8.

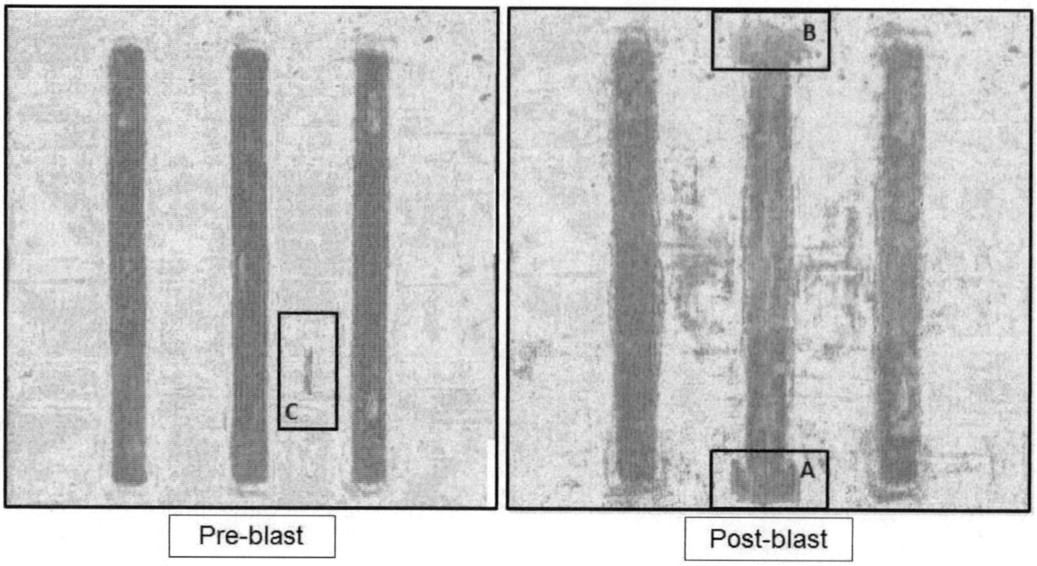

Figure 7: Pulse-echo ultrasonic C-scans of the bulkhead panel inspected pre- (left) and post- (right) explosive blast loading.

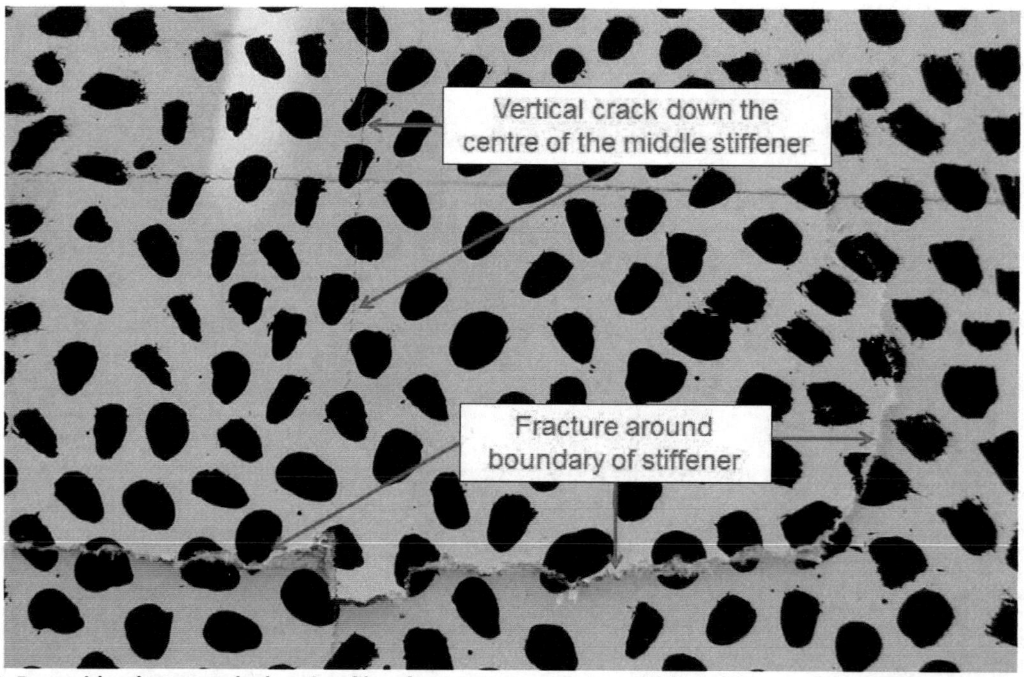

Figure 8: Rear-side photograph showing fibre fracture around the middle stiffener of the bulkhead panel. Note: the black and white speckle pattern was used for displacement measurement using Digital Image Correlation (DIC).

Visual inspection of the deck panel revealed several surface cracks at the boundaries of the stiffeners which were believed to be largely cracking of the gel-coat, as shown in Figure 9. The pulse-echo ultrasonic C-scan images pre- and post-blast of the deck panel are shown in Figure 10. The deck panel was almost double the thickness of the bulkhead panel which rendered it more challenging to penetrate ultrasonically at the selected 2.25 MHz frequency. There was much less discernible damage via ultrasonic inspection in the deck panel. Most notably, however, there was a loss of clarity along the boundaries of the two stiffeners which suggested discontinuity between the stiffener and facesheet, likely a result of debonding at the stiffener and facesheet interface. A slight increase in the variation of reflected sound along the length of the stiffeners was believed to be a result of debonding along the stiffener-facesheet interface. A greater amount of variation between the reflected ultrasonic signal in the centre of the panel (indicated by A), was likely due to distributed damage to the matrix and delamination of the stiffener.

Figure 9: Gel-coat cracks on exterior of deck panel.

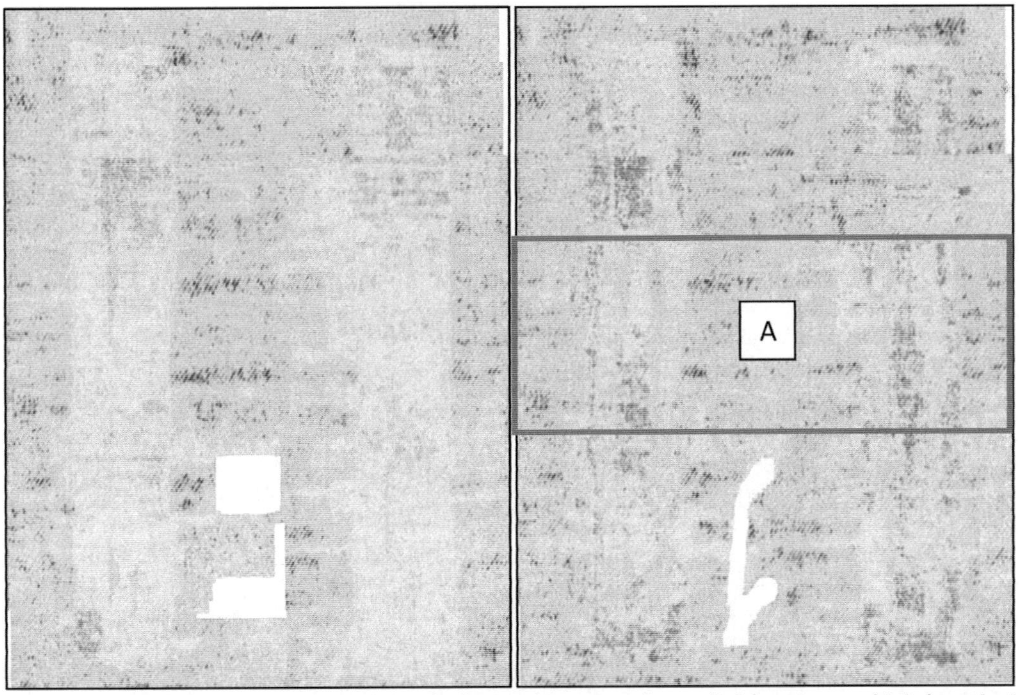

Figure 10: Pulse-echo ultrasonic C-scans of the deck panel inspected pre- (left) and post- (right) explosive blast loading.

SUMMARY OF NDT TECHNOLOGIES

Using the inspection method shown here, damage such as polymer cracking, delamination and stiffener-facesheet debonding were detected effectively and evaluated in terms of size and location. However, the ultrasound was not able to penetrate the air-filled foam due to its high porosity, and therefore this material could not be satisfactorily inspected. Cutting through the cross-section of stiffener revealed core longitudinal core cracking (shown in Figure 11), which was undetected using ultrasonics. This highlights both the effectiveness and limitations of ultrasonic inspection in large-scale thick composite panels containing fillers such as polyurethane foams.

The limitation of ultrasonic inspection for aerated foams could be best overcome via the use of x-radiography, or x-ray computed tomography, which has often been used for thick-foam inspections, and demonstrated on a 78 mm thick GFRP/foam sandwich composite material [16]. However, the increased overhead from health risks associated with ionising radiation currently renders the application of field-portable x-ray equipment unfeasible. Another limitation of ultrasonic inspection for thick structures is the phenomenon known as 'shadowing', in which almost all sound energy is reflected from the flaw nearest to the surface, leaving damage beneath the surface flaw undetectable - of particular relevance in thick structures. Furthermore, component geometry complicates the response from single-sided inspection methods, which are mandatory for more or less all marine-scale composite applications, creating some non-inspectable regions where a suitable reflection cannot be obtained due to simple geometry constraints.

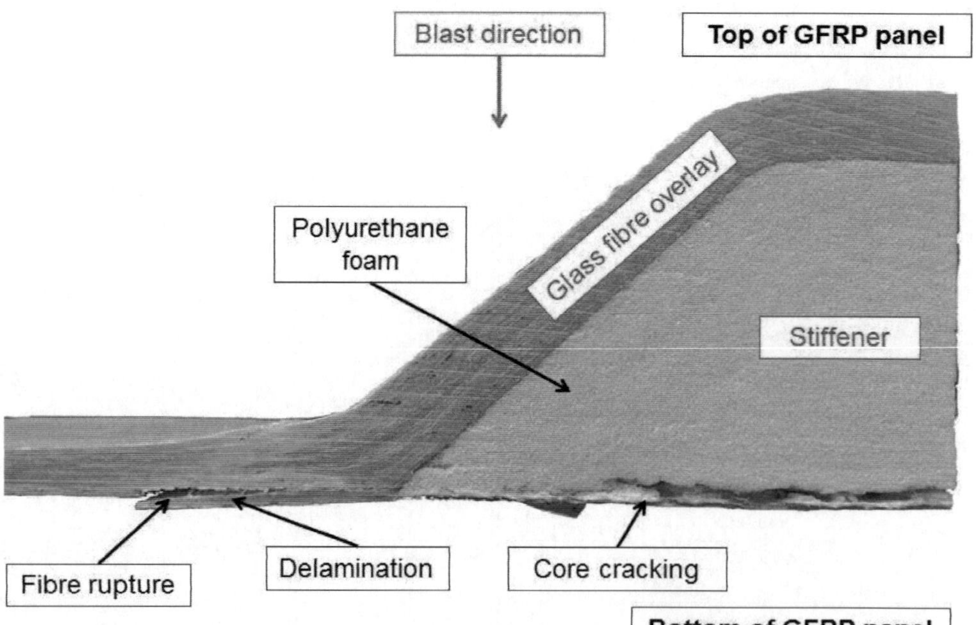

Figure 11: Cross-sectional photograph showing fibre rupture, delamination and core cracking in the bulkhead panel.

CONCLUSION

Nondestructive testing has been applied to the assessment of multiple failure modes that occur in common naval composite materials when subjected to explosive blast loads. The results showed the effectiveness of detecting, sizing and measuring the extent of delamination in the small-scale glass or carbon fibre laminates and stiffener-facesheet debonding in large scale composite panels. However, limitations arise with the detection of micron-sized damage such as fibre-matrix intertow cracking, damage within the foam core of a sandwich composite or stiffener and flaws hidden by the shadowing effect, particularly relevant in thick composite structures that are found on navy ships. It was found that delamination damage substantially reduced the compressive and flexural properties, although was largely dependent on the extent of the damage. However, the smaller damage modes such as fibre-matrix intertow cracking and matrix cracking that were undetected from the ultrasonic scanning also reduced the mechanical properties of the composite laminates. This showed that even small-scale and often overlooked damage can be detrimental to the composite structure which can limit operational capability and enhance the vulnerability of the naval vessel. This was proven for thin, small-scale composite laminates. Further work needs to be done analysing the degradation of the mechanical properties of large, full-scale stiffened composite structures due to explosive blast loading.

The increasing use of thick composite structures in load-bearing applications contributes significant challenges to detection, sizing and characterisation of flaws. Further development of NDT for thick-section composites is required to improve the structural integrity assessment and maintenance processes of current and future composite naval vessels. DST Group has an active research program into the detection, sizing and characterisation of thick composite structures suitable for maritime applications using acoustic non-destructive techniques. For example, recent advancements in NDE for the detection of internal delamination in thick-section composite propeller blades [17] and the susceptibility of ultrasonic inspection of delaminations to crack-tip closure [18] has been published by the DST research program.

ACKNOWLEDGEMENTS

The author would like to acknowledge the technical assistance of Mr. Trevor Delaney, Mr. Pat McCarthy, Mr. Alastair Douglas, Mr. Darren Wiese, Mr. Peter Vincent, Mr. Michael Brincat, Mr. Andrew McLean, Mr Chris Townsend and Mr. Darren Stanford in the preparation and execution of the experiments. The author would also like to acknowledge the support of the United States Office of Naval Research (ONR) under the direction of Dr. Y.S. Rajapakse and the assistance of Professor Adrian Mouritz and Associate Professor Raj Das from RMIT University.

REFERENCES

1. Mouritz, A. P., et al. (2001) Review of advanced composite structures for naval ships and submarines. *Composite Structures* **53** (1) 21-24
2. Chalmers, D. W. (1994) The Potential for the use of composite materials in marine structures. *Marine Structures* **7** (2) 1994/01/01/ 441-456
3. Comtois, J. L. R., Edwards, M. R. and Oakes, M. C. (1999) The effect of explosives on polymer matrix composite laminates. *Composites Part A: Applied Science and Manufacturing* **30** (3) 181-190
4. Hebert, M., Rousseau, C. E. and Shukla, A. (2008) Shock loading and drop weight impact response of glass reinforced polymer composites. *Composite Structures* **84** (3) 199-208
5. Tekalur, S. A., Shivakumar, K. and Shukla, A. (2008) Mechanical behavior and damage evolution in E-glass vinyl ester and carbon composites subjected to static and blast loads. *Composites Part B: Engineering* **39** (1) 57-65
6. Yahya, M. Y., et al. (2008) The blast behavior of fiber reinforced thermoplastic laminates. *Journal of composite materials* **42** (21) 2275-2297
7. Franz, T., Nurick, G. N. and Perry, M. J. (2002) Experimental investigation into the response of chopped-strand mat glassfibre laminates to blast loading. *International Journal of Impact Engineering* **27** (6) 639-667
8. Arora, H., Hooper, P. A. and Dear, J. P. (2011) Dynamic response of full-scale sandwich composite structures subject to air-blast loading. *Composites Part A: Applied Science and Manufacturing* **42** (11) 1651-1662
9. Gardner, N., Wang, E. and Shukla, A. (2012) Performance of functionally graded sandwich composite beams under shock wave loading. *Composite Structures* **94** (5) 1755-1770
10. Rolfe, E., et al. (2017) Failure analysis using X-ray computed tomography of composite sandwich panels subjected to full-scale blast loading. *Composites Part B: Engineering* **129** 26-40
11. Langdon, G. S., et al. (2013) The air-blast response of sandwich panels with composite face sheets and polymer foam cores: Experiments and predictions. *International Journal of Impact Engineering* **54** 64-82
12. Ibrahim, M. (2014) Nondestructive evaluation of thick-section composites and sandwich structures: A review. *Composites Part A: Applied Science and Manufacturing* **64** 36-48
13. Remennikov, A., Mentus, I. and Uy, B. (2015) Explosive breaching of walls with contact charges: theory and applications. *International Journal of Protective Structures* **6** (4) 629-647
14. Prichard, J. C. and Hogg, P. J. (1990) The role of impact damage in post-impact compression testing. *Composites* **21** (6) 1990/11/01/ 503-511
15. Sanchez-Saez, S., et al. (2005) Compression after impact of thin composite laminates. *Composites Science and Technology* **65** (13) 1911-1919
16. Lambrineas, P., et al. (1991) X-ray computed tomography examination of inshore minehunter hull composite material. *NDT & E International* **24** (4) 1991/08/01/ 207-213
17. Ibrahim, M. E., et al. (2014) *Advanced materials research*. Vol. 891, Trans Tech Publ
18. Ibrahim, M., Smith, R. and Wang, C. (2017) Ultrasonic detection and sizing of compressed cracks in glass- and carbon-fibre reinforced plastic composites. *NDT & E International* **92** 111-121

A GENERIC WARSHIP MODEL FOR SURVIVABILITY ANALYSIS

Anthony Woolley[+] and Noel Richards[++]
[+]Defence Science and Technology
[++]Deakin University
anthony.woolley@dst.defence.gov.au
noel.richards@dst.defence.gov.au

ABSTRACT

A generic warship model is proposed to provide a standardised capability for collaboration on naval platform survivability analysis within the scientific community. The generic warship model is based on a decommissioned Royal Australian Navy *River* Class destroyer escort, ex-HMAS *Derwent*. The model is free from security sensitivities, allowing for it to be easily distributed. This generic warship model will allow collaborators to test and compare output from sensitive vulnerability and recoverability computer algorithms without needing to share those algorithms. Data collected from blast, and fire and smoke experiments performed onboard ex-HMAS *Derwent* can also be utilised to verify vulnerability and recoverability algorithms using the *Derwent* model. Survivability experiments, using decommissioned naval platforms, are rare and expensive occurrences. These experiments are important for providing valuable data to generate, and verify and validate survivability algorithms; however, such experiments are limited in the scope of scenarios that can be examined. Conversely, modelling and simulation enables exploration of complex survivability scenarios that would be difficult to perform during experiments. A computer model based on ex-HMAS *Derwent* will provide a standardised, realistic warship simulation environment to analyse complex survivability scenarios, to promote collaboration, and to enhance warship survivability. Development of the three-dimensional generic warship model is discussed, highlighting problems encountered and overcome. Examples in vulnerability, fire, and crew movement studies are also presented.

1. INTRODUCTION

Survivability of a naval platform is dependent on the platform's capability to avoid a threat, resist threat effect and recover from resultant damage, referred to, respectively, as susceptibility, vulnerability and recoverability. Each of these three domains can be assessed individually or as an integrated combination of each domain. The combination of all three domains is known as Integrated Survivability [1]. It is believed that the ability to assess platform Integrated Survivability will enhance platform survivability; contribute to capability trade-off decision-making across the three survivability domains during all phases of the platform capability life-cycle; and assist with mission planning. An ongoing research program is examining requirements for naval platform Integrated Survivability analysis, with the aim of developing an Integrated Ship Survivability Assessment Capability (ISSAC) [1, 2].

A risk associated with naval platform survivability collaborations relates to security sensitivities. A failure to properly manage the risk may compromise the operational effectiveness of the platform and associated missions. Consequently, in managing the risk, the collaborations may be limited in their usefulness. Therefore, to alleviate security sensitivity concerns, and to maximise the scope and usefulness of the collaborations, a generic three-dimensional warship (platform) computer model was developed. This model will be available to those involved in the collaborations, to facilitate development of naval platform survivability models, and perform Modeling and Simulation (M&S) analysis in a standardised environment. That is, the warship model will be a common factor amongst the collaborators, thereby minimising (or, even, eliminating) the need to share intellectual property and security sensitive information. The collaborators can perform M&S using the warship model and share the output with other collaborators for analysis and comparative purposes, or for ongoing development and refinement of survivability algorithms and software tools.

The proposed warship model is derived from ex-HMAS *Derwent* (hereafter referred to as *Derwent*), a decommissioned Royal Australian Navy (RAN) *River* Class destroyer escort [3]. After decommissioning, *Derwent* was used in the Ship Survivability Enhancement Program (SSEP), where it was subjected to a series of live explosive and fire experiments [4]. Data collected from the SSEP was utilised to verify various survivability algorithms. The data may also be utilised by collaborators to verify survivability algorithms in conjunction with using the warship model.

This paper presents an overview of M&S for survivability analysis, including experiments such as the SSEP (Section 2); and then briefly describes *Derwent* (Section 3). Developing the warship model based on *Derwent*, identifying and overcoming challenges, and adapting it for use in the different M&S software tools is described (Section 4). An overview of applications in which the *Derwent* model has been used is also presented (Section 5).

2. NAVAL PLATFORM SURVIVABILITY MODELLING AND SIMULATION

A M&S environment makes use of algorithms to approximate the 'real world'. These algorithms can be generated from observations collected during experiments, either using full scale environments or scaled approximations of the real environments. Experiments provide opportunities for data collection to understand material response to various stimuli and facilitate development, and verification and validation, of algorithms. In the domain of naval platform survivability analysis, experiments to understand platform response to different threat scenarios using decommissioned naval platforms have occurred. For example, the SSEP consisted of a series of live explosive, fire and smoke events to measure the response of the platform [4]. The explosive events were designed to simulate missile detonations, with different explosive yields being tested. The fire trials were designed to analyse heat transfer and fire progression; and the smoke trials were designed to understand the progression of smoke through the platform, as well as smoke clearance rates. The data collected from the SSEP was used to verify vulnerability algorithms; and will be used to assist development of algorithms for a naval platform recoverability capability.

Naval Research Laboratory (NRL) also utilised a decommissioned naval platform for research, and test and evaluation purposes [5]. Ex-USS *Shadwell* (hereafter referred to as *Shadwell*) was a World War 2 era dock landing ship. After de-commissioning in 1970, *Shadwell* was later converted for use in damage control experiments, primarily fire and fire fighting analysis. For example, *Shadwell* was utilised to test halon replacement systems for fire suppression [6, 7]; to investigate fire spread and examine submarine ventilation doctrine for *Los Angeles* Class attack submarines [8]; and to verify output from the *Fire and Smoke Simulator* (*FSSIM*) [9]. NRL also developed a virtual environment based on *Shadwell* to examine the use of the virtual environments in training naval fire fighters [10].

Experimentation using scaled approximations of naval platform structures also occurs. For example, experiments involving explosive charges detonated inside 1m^3 steel containers were performed to understand structural deformations, and strain profiles, and to evaluate numerical modelling used in naval platform vulnerability analysis [11]. Scaled response experiments allows for an increased number of experiments and can be used to verify vulnerability algorithms to model platform compartments subjected to air explosion events.

Analysis of human movement transit time is another example of live experiments. For example, crew movement transit time relating to damage control activities was measured on board an active naval platform [12], and in structures analogous to naval platforms [13]. Measurements performed when *Derwent* was in active service were recorded for transiting a corridor, ladders, and door and hatch entry. These measurements contributed to the development of models for a fire management training software tool [12]. Later, a limited study, using a land-based facility, measured people movement transit times to examine the suitability of a commercially available human movement software tool [13].

Experiments, such as the afore-mentioned examples, provide data for specific situations and scenarios. If variations to those scenarios were required, more experiments would be necessary, requiring extra resources (time, funding, platforms on which to perform the trials, and specialist personnel). Alternatively, M&S tools can be utilised to duplicate those experiments and study new scenarios.

M&S is often utilised to understand complex, real world scenarios. In particular, M&S can facilitate analysis of situations that may be deemed high-risk, ethically prohibitive and/or financially prohibitive to perform in the real world. M&S software tools have been developed to analyse elements of naval platform susceptibility, vulnerability and recoverability [14, 15, 16, 17]. These tools can be used to examine aspects of survivability, and can be used to analysis variability within those scenarios. Analysis capability afforded by these tools can facilitate examination of configuration changes and trade-off studies; inform training and policy and enable analysis of actual naval damage incidents [18, 19]. The challenge associated with modelling survivability scenarios, though, is the need to accurately represent the real world within the M&S environment. In particular, one model that will be required is that of the naval platform. In some instances, this model will represent the actual platform being examined; however, there will be situations requiring the use of a generic warship model. A generic warship model can be used as a baseline within the M&S environment to enable analysis and comparison of new technologies or procedures. A generic warship model also has the advantage of enabling collaboration with national and international design authorities, research and development institutions, academia, and defence agencies. It will permit the sharing of results between collaborators without needing to share M&S capabilities, or the underlying algorithms. The *Derwent* model is proposed to be that generic warship model; to be a shared asset during collaborations; and to be utilised to further develop M&S capability for naval platform survivability analysis. Data collected during the SSEP means that *Derwent* model can also be used for the verification and validation of vulnerability and recoverability M&S software.

3. HMAS *DERWENT*

Derived from the Royal Navy (RN) Type 12M *Rothesay* Class frigate and commissioned in April 1964, *Derwent* was a *River Class* Destroyer Escort of the RAN [3]. Historically, the RN Type 12 frigate commenced with the *Whitby* Class frigate, and underwent several design class changes that included *Rothesay, Leander* and *River* Classes. *Derwent* was constructed at the Williamstown Dockyard in Victoria, Australia, with the hull being an all-welded design of light metals and alloys. Armament varied during *Derwent*'s operational life. Primary weapons consisted of two 4.5" guns in a twin turret; a *Seacat* anti-aircraft missile system (*Derwent* was the first RAN platform to launch a guided missile); the Australian designed *Ikara* anti-submarine missile system; and two triple mounted anti-submarine torpedo tubes. A triple barrel limbo anti-submarine mortar was removed during a modernisation program between 1981 and 1985. Crew complement consisted of 240 officers and enlisted personnel. *Derwent* was recognised with battle honours for operations during the Indonesian-Malaysian Confrontation (1964 to 1966). *Derwent* also served as escort for the troopship HMAS *Sydney* during the Vietnam War. *Derwent* was decommissioned in August 1994, and subsequently utilised as the trials platform in the SSEP. After the SSEP, on 21 December 1994, *Derwent* was scuttled in deep water off the coast of Western Australia.

4. A GENERIC WARSHIP COMPUTER MODEL

Development of any computer model needs to take into consideration the intended use of the model; resources available to assist with building the model; time constraints; and the required digital file format. These considerations can dictate, for instance, the software package used for the build; and whether to start with small details, or to build a broad shape and work down to the smaller details. The form of the 'building blocks' to construct the model is also governed by those considerations. The building blocks are the mathematical representations for the three-dimensional geometry, such as the use of polygons or Non-Uniform Rational B-Splines (NURBS). The building blocks affect, for example, computer memory space and translating the model to other formats for use in relevant M&S software. In the case of *Derwent*, the finished model had to be flexible because it would be used in different software tools, with each having different model requirements.

4.1 Polygons versus NURBS

The decision to use polygons or NURBS is generally governed by the end-use of the model and, to some extent, by the shape of the object being modelled. For the *Derwent* model, the hull was a large, simple and smooth shape. This suited NURBS modelling, which uses fewer control points and results in a smoother curve than a curve generated using polygons. However, some of the end use survivability analysis software required a polygon format. This meant the hull could be modelled using NURBS, but it would need to be converted to polygons when the final shape was achieved. Conversely, the decks primarily consisted of sharply defined shapes, which is better suited to polygon modelling. Consequently, even though it was valid to build the hull using NURBS and the decks in polygons, the decision was made to use polygons for the entire model.

4.2 Constructing the *Derwent* Computer Model

One method to construct the outer shape of an object is the use of photogrammetery [20], and this was the original method of choice to construct a model of *Derwent*'s outer hull. Photogrammetery is the process of obtaining measurement and details for an object using photographs showing the object in its entirety at varying angles, including aspects that would not normally be seen (such as, below the waterline). Unfortunately, there were not enough suitably consistent photographs of *Derwent* to enable the use of photogrammetry. Instead the printed design drawings of *Derwent* were available for scanning and these were used to construct the deck and infer the hull shape.

4.2.1 Constructing the Decks

The *Derwent* design drawings consisted of deck plans and a side profile. These were scanned, and the process of constructing computer models of the decks commenced. Due to the large physical size of the design drawings, the digitised images were distorted during the scanning process. Advanced image processing software, such as Autodesk *Maya* [21], has capability to make corrections to the images whereby warped lines can be straightened and compressed areas can be stretched. Correcting the distortions was made easier using a linear measurement scale, with evenly spaced distance markings, along the centerline of the ship in the original design drawings.

The next step was to create each digitised deck as a separate image layer superimposed over the other decks. A 'transparency' function within *Maya* allowed each deck to be aligned and scaled. Figure 1 shows the digitised scan of each deck aligned and in scale. The digital scans resulted in flat, non-interactive images that would be used to construct wireframes of the decks. To do this, the scans were imported, separately, into *Maya* and a method known as 'digital tracing' was utilised to manually trace, point-by-point, the outline of each deck.

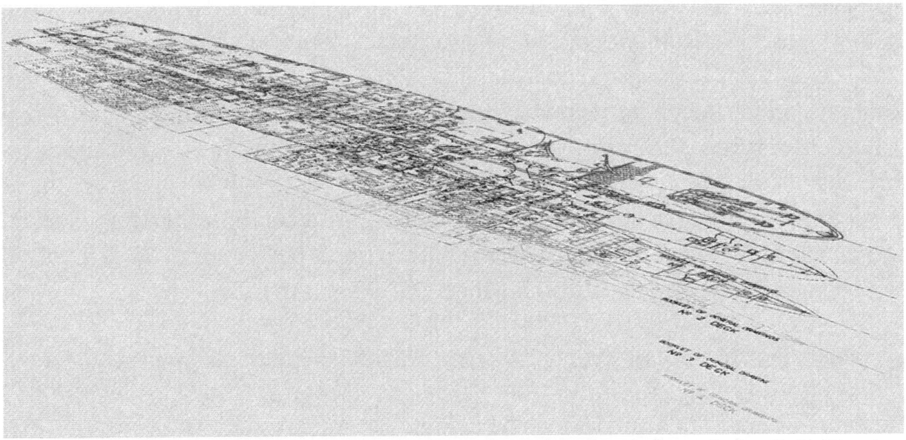

Figure 1: Pictorial layers of Derwent deck plans, aligned and scaled, ready for digital tracing

Examination of a profile view of *Derwent*'s design drawings revealed that the decks were not horizontal – there was a slight upward curve in Decks 2 and 3 on approach to the bow, as seen in Figure 2. To replicate this curvature, the wireframe decks were aligned with *Derwent's* profile view, and the digital tracing process was repeated. Figure 2 includes the wireframe overlay on the design drawings. The profile view was also used to measure the spacing between decks to enable the construction of bulkheads (walls) giving the model three dimensions and deck connectivity. Most of the M&S software, in which the *Derwent* model will be used, does not require the bulkheads and decks to have thickness; they could remain as two-dimensional planes (in three-dimensional space). Future development of the *Derwent* model, for use in training, will include that dimensionality and added texturing to give a semblance of visual realism.

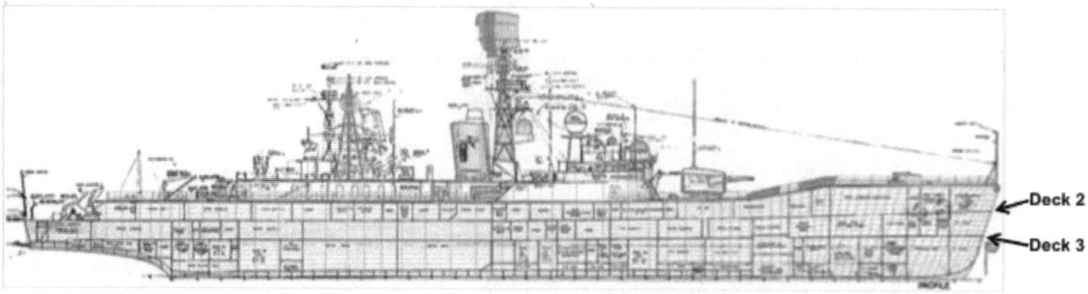

Figure 2: Derwent profile showing curves in Decks 2 and 3, with wireframe model overlay for digital tracing

4.2.2 Constructing the Hull

Imagery accurately portraying *Derwent*'s hull form was difficult to locate, and the design drawings only presented the cross sectional profile shown in Figure 2. To overcome the lack of information regarding the *Derwent* hull form, an existing computer model of a generalised hull form was acquired and fitted to the modelled decks. Each deck was used as a cut plane through *Derwent*'s hull and aligned in virtual space, along with a profile view of the ship, as shown in Figure 3. Vertices in the generalised hull form were then manually moved to the desired position fitting the shape inferred by the decks. A requirement of the generalised hull form was that it be very simple, since the technique of manually moving the vertices was a laborious task. Furthermore, moving too many vertices into position by hand increased the risk of error and could have resulted in diminished accuracy for the level effort involved. Conversely, polygon modelling algorithms have sophisticated methods to smooth surfaces created from very few vertices. Therefore, it was advantageous to manually move a few vertices to their exact positions, and use polygon algorithms to smooth the shape between the vertices. The result is shown in Figure 4. Since *Derwent*'s hull form was symmetrical, only half the hull was constructed via this method. Once completed, that half was copied and mirrored to create a three-dimensional hull form. The final step was to refer back to photos, and re-work the model, using digital tracing, until the new hull form fitted the evidence presented in photos that were available, while also accommodating the shapes of the decks. Figure 5 shows the completed hull form, along with completed superstructure details. Figure 6 presents a shaded exterior view of the finished *Derwent* hull form.

5. EXAMPLES USING THE DERWENT MODEL

The *Derwent* model is being utilised in several research programs to develop M&S capability contributing towards vulnerability and recoverability assessments for the fleet-in-being and to ensure future naval platform capability achieves, or exceeds, RAN survivability requirements. Currently, no susceptibility research programs are utilising the *Derwent* model; however, others have utilised a model of a *Leander* Class platform to test coupling between a radar cross-section simulation with *SURVIVE* [22]. The following sections provide examples of *Derwent*'s use in vulnerability and recoverability modelling.

5.1 Vulnerability Modelling

Simulation of warhead blast and fragmentation is achieved using the *CVAM* vulnerability modelling tool [23]. Figure 7 presents the *Derwent* model in the vulnerability modelling capability; and Figure 8 shows a simulated warhead detonation [24]. Ray tracing (white lines) represent the weapon's fragmentation distribution, while red shading identifies areas on the platform affected by the blast. In this example, the weapon detonation occurred in the operations room.

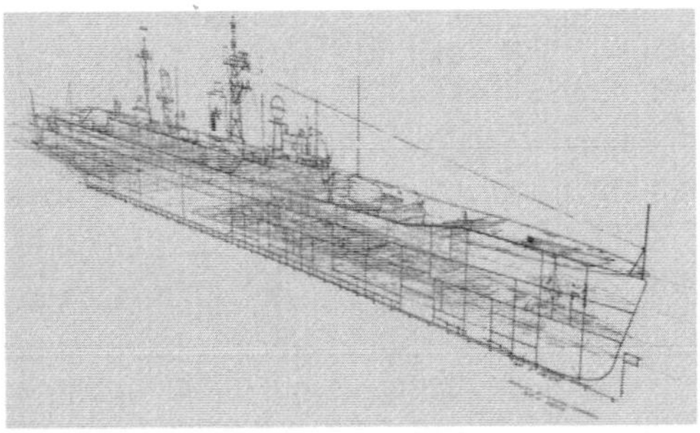

Figure 3: Decks and profile aligned

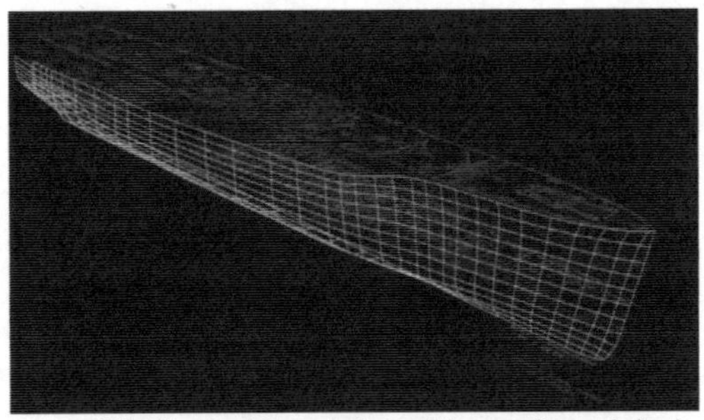

Figure 4: Half hull wireframe inferred from aligned decks

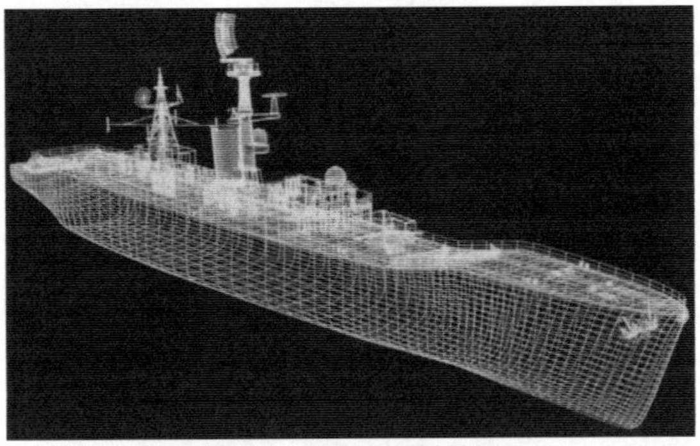

Figure 5: Completed wireframe hull form

Figure 6: Three-dimensional shaded representation

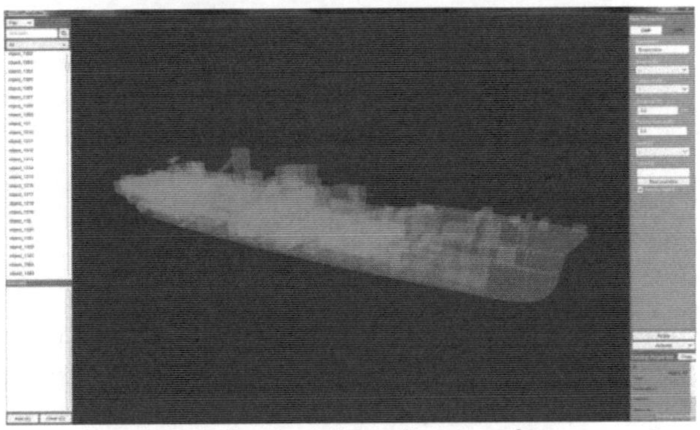

Figure 7: Vulnerability modeling editor interface

Figure 8: Simulated missile detonation [24]

5.2 Recoverability – Fire Progression and Heat Transfer

A research program involving DST Group and Fire Safety Engineering Group (FSEG), University of Greenwich, United Kingdom, has an aim of developing a recoverability capability linking the fire modeling software, *SMARTFIRE* (*SMF*), with the crew movement software, *maritimeEXODUS* (*mEX*) [2]. This capability is referred to as the *Naval Damage Incident Recoverability Toolset* (*NavDIRecT*). The *Derwent* model is ideal for this research program; it enables the developers of *SMF* to utilise a warship-like environment without needing access to sensitive fleet-in-being platform models and data. For example, Figure 10 presents part of *Derwent's* forward section for Decks 2 and 3. A hatch connects Decks 2 and 3; and holes in the hull of Deck 3 represent damage from a non-detonating missile that impacted the starboard hull and caused damage in the port hull. The open hatch, the weapon entry hole, and port hull damage act as vents for a fire that initiated from unspent missile fuel. Labels T1, T2, and T3 in Figure 10 represent three approximate locations where temperature recordings occurred during the SSEP. Figure 11 presents the temperature recorded at T1, T2 and T3 up to 500 seconds after fire initiation.

During development of *NavDIRecT*, a simplified version of the aforementioned fire scenario is being utilised. The simplified fire scenario commences at the location indicated in Figure 12 (similarly located to that in Figure 10 for the actual fire scenario); however, the fuel load and distribution (that is, items that will burn) is less than that used during the SSEP. Figure 12 presents a heat map output from the *SMF* simulation at 200 seconds after fire initiation [2]. During the SMF simulation, temperatures open hatch and at V1 and V2 along the weapon entry hole (shown in Figure 12) were measured. Figure 13 presents the temperatures for these locations up to 500 seconds after fire initiation.

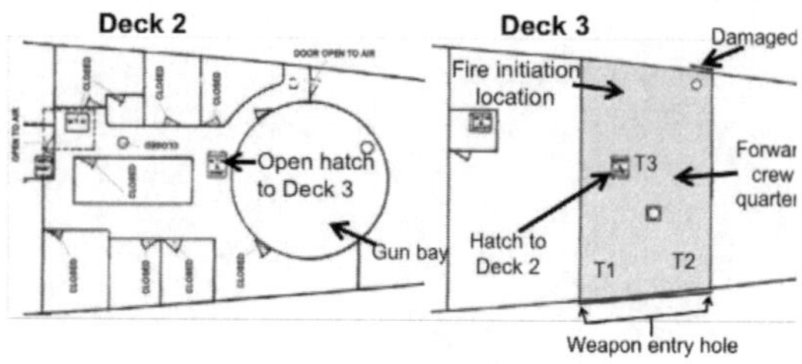

Figure 10: SSEP fire scenario configuration

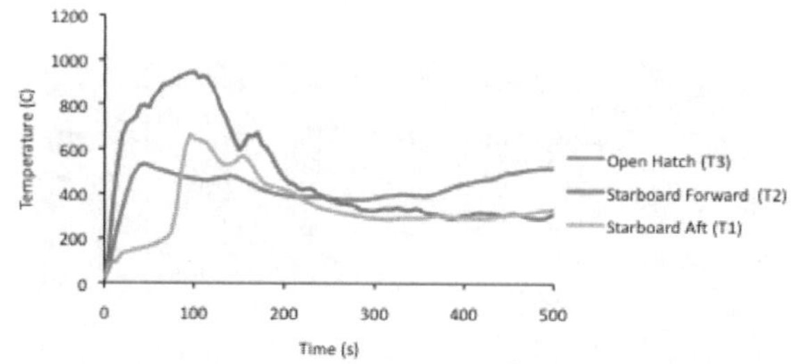

Figure 11: Measured temperatures (^0C) during the SSEP

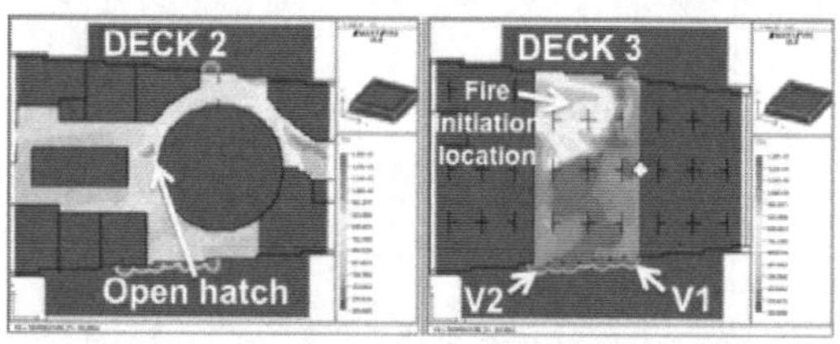

Figure 12: Simulated heat transfer through Deck 2 and Deck 3 at 200 seconds after fire initiation [2]

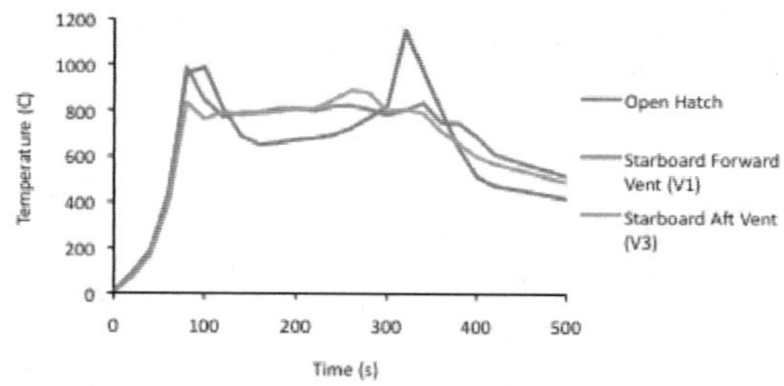

Figure 13: Simulated temperature (^0C) curves [2]

The differences between the *SMF* and SSEP experiment design has resulted in the obvious differences between the temperature curves presented Figures 11 and 13. Future refinement of the SMF scenario will align it with the SSEP experimental design, to enable verification of the fire modelling functionality within *NavDIRecT*. When verified, the combined fire modeling and crew movement (Section 5.3) functionality will provide the desired capability to assess naval platform recoverability against fire scenarios.

5.3 Recoverability - Crew Movement

Crew movement modelling for damage control activities is being developed using the *maritimeEXODUS* (*mEX*) software. *mEX* is a ship evacuation modelling tool; however, the research program with FSEG is enhancing *mEX* such that it will facilitate the movement of crew to simulate damage control activities [2]. The current modelling of *Derwent* within *mEX* has shown that artificial agents representing the crew can be assigned a search path to simulate a 'blanket search' to identify casualties and damage [2]. Verification of crew movement modelling on *Derwent* can be achieved using data collected during a crew movement trial while *Derwent* was in operational service [12]. Crew movement and fire modelling (Section 5.2) can then be combined to provide a recoverability capability from fire with verification using experimental data collected from trials performed on *Derwent*. Later work will expand the recoverability modelling capability to include flooding and toxic hazard response.

7. CONCLUSION

M&S provides insight into naval platform response to various threat effects, and enables repeated analysis of complex naval platform survivability scenarios. To perform that M&S, though, algorithms defining the behaviour of the objects being modeled are required. These algorithms can be generated, and verified and validated from experiments using full scale test environments or scaled representations of those environments. Consequently, there will be a continued need to perform experiments; however, M&S can be utilised to perform analysis of complex survivability scenarios not achievable via experimentation.

One model required in the survivability analysis is that of the naval platform. The proposed generic warship computer model, based on *Derwent*, is a valuable asset to enable further development of naval platform survivability M&S capability. Data collected from the experiments performed on *Derwent* during its operational life and during the SSEP will enable verification of naval platform vulnerability and recoverability algorithms. The research being performed to develop *SMF* and *mEX* for M&S of recoverability activities has demonstrated the value of the *Derwent* model as a collaborative tool. The model has facilitated the sharing of information relevant to naval platform survivability analysis without needing to share sensitive information; it has enabled the testing of enhancements to the M&S software using a naval platform environment; and it will be utilised to verify the M&S environment. Future collaborations in the domain of naval platform survivability will similarly benefit from the use of the *Derwent* generic warship model.

8. ACKNOWLEDGEMENTS

The authors acknowledge Director Navy Technical Bureau, RAN, as sponsor of this work. Thank you to Guido Papp, QinetiQ, for the initial development of the *Derwent* model. The authors thank staff of DST Group for contributing to the Integrated Survivability research program: Vanessa Pickerd and Alexander Gargano, vulnerability analysis; Serap Aksu, systems analysis; Grant Gamble, Brigitta Suendermann and Ian Birch, fire and smoke analysis; and Anthony Travers and Tom Whitehouse, Integrated Survivability. The authors also thank FSEG, University of Greenwich, for developing the *Derwent SMF* and *mEX* models.

9. REFERENCES

1. Brett, J. Gamble, G., Reid, W., Smith, S., Woolley, A., Yiannakopoulos, G., 2017. Integrated Survivability Analysis of Naval Platforms in High Threat Environments. Pacific 2017 International Maritime Conference, 3 – 5 October 2017, Sydney, Australia.
2. Woolley, A., Ewer, J., Lawrence, P., Travers, A., Deere, S., Whitehouse, T., Galea, E., 2016. A Modelling and Simulation Framework to Assess Naval Platform Integrated Survivability. MAST Europe, 21 – 23 June 2016, Amsterdam, Netherlands.
3. Royal Australian Navy, 2019. HMAS Derwent. http://www.navy.gov.au/hmas-derwent (accessed 19 June 2019).
4. Howe, J. S., 1996. Ship Survivability Enhancement Program: Management of the Program. DSTO-GD-0087. DSTO, Australia.
5. Carhart, H. W., Williams, F. W., 1987. The Ex-Shadwell – Full Scale Fire Research and Test Ship. NRL Memorandum Report 6074. Naval Research Laboratory, Washington, DC.
6. Peatross, M., Maranghides, A., Sheinson, R. S., Black, B., Smith, W. D., 1996. Conducting Real Scale Halon Replacement Tests Aboard the ex-USS Shadwell: Overall Test Design. Halon Options Technical Working Conference, 7 – 9 May 1996, Albuquerque, USA.

7. Darwin, R. L., Williams, F. W., 2000. The Development of Water Mist Fire Protection Systems for U.S. Navy Ships. Naval Engineers Journal **112**(6) pp. 49-57.
8. Hoover, J. B., Tatem, P. A., Williams, F. W, 1998. Meta-Analysis of Data from the Submarine Ventilation Doctrine Test Program. NRL/MR/6180-98-8168. Naval Research Laboratory, Washington, DC.
9. Floyd, J., Williams, F., Tatem, P., 2005. Validation of a Network Fire Model Using the Ex-Shadwell Submarine Ventillation Doctrine Tests. Fire Safety Science – Proceedings of the 8th International Symposium, 18 – 23 September 2005, Beijing, China.
10. Tate, D. L., Sibert, L., King, T., 1997. Virtual environments for shipboard firefighting training. Proceedings of IEEE 1997 Annual International Symposium on Virtual Reality. 1 – 5 March 1997, Albuquerque, USA. Sajdak, J. A. W., Karni, Z. H., 2006. Determination of a measure of total integrated system survivability. Warship 2006: Future Surface Warships, 20 – 21 June 2006, London, UK.
11. Pickerd, V., Bornstein. H, McCarthy, P. Buckland, M., 2016. Analysis of the structural response and failure of containers subjected to internal blast loading. International Journal of Impact Engineering 95, pp. 40-53.
12. Kennett, S. R., 1998. Estimation of Response Times for Fire Fighting Personnel on a Naval Vessel. DSTO-TN-0165. DSTO, Australia.
13. Grech, M., Drohan, D., Roberts, W., Gamble, G. I., 2012. Ship Recoverability: How Important is the Human Factors Input. Pacific 2012 International Maritime Conference, 30 January – 2 February 2012, Sydney, Australia.
14. Turner, S., Horstmann, P., Bain, G., 2006. Warship Survivability. Warship 2006: Future Surface Warships, 20 – 21 June 2006, London, UK.
15. Thornton, J. S., Coulthard, R., Hardy, J. N., 2006. Modelling Maritime Survivability. Warship 2006: Future Surface Warships, 20 – 21 June 2006, London, UK.
16. Test & Evaluation Solutions, 2019. Integrated Recoverability Model (IRM). http://www.irm-software.org/ (accessed 19 June 2019).
17. SURMA, 2016. SURMA Technical Description Software for Comprehensive Navy Combat Survivability Analysis. White paper, SURMA Technical Description, SURMA-M-101-G.
18. Kennett, S., Gamble, G., Li, J-D, 1998. Modelling of the HMAS Westralia Fire. DSTO-TR-0698. DSTO, Australia.
19. Buckland, M., Cannon, S. M., de Yong, L., Gamble, G. I., Jeremy, J. C., Lyon, T., McCarthy, P., Morris, B., Neill, R. A., Skeen, M. B., Suendermann, B. and Turner, T., 2009. HMAS Sydney II Commission of Inquiry Report on Technical Aspects of the Sinking of HMAS Sydney and HSK Kormoran. DSTO-GD-0559. DSTO, Australia.
20. Tumeliene, E., Nareiko, V., Suziedelyte Visockiene, H, 2017. Photogrammetric Measurements of Heritage Objects. Geospace 2017, 4-6 December 2017, Kyiv, Ukraine.
21. Autodesk, 2019. Maya. https://www.autodesk.com.au/products/maya/overview (accessed 19 June 2019).
22. Turner, S., Horstmann, P., Bain, G., 2006. Warship Survivability. Warship 2006: Future Surface Warships, 20 – 21 June 2006, London, UK.
23. Buckland, M., 2008. The use of warship vulnerability analysis tools as a measurement of blast and ballistic protection requirements. Pacific 2008 International Maritime Conference, 29 – 31 January 2008, Sydney, Australia.
24. Wallace, S., 2016. Maritime Platform Model for Vulnerability Analysis. (Short Poster) Australasian Simulation Congress 2016, 26 – 29 September 2016, Melbourne, Australia.

Addressing Military Survivability and Safety Aspects in Classification of Naval Design

Dr. Bhineka Kristanto; Corporate Technology, American Bureau of Shipping (ABS); ekristanto@eagle.org

CAPT Richard Delpizzo, USNR (ret.), F(SNAME); Global Government Services, ABS; rdelpizzo@eagle.org

ABSTRACT

It has become standard operating procedure for Classification Societies to provide services for naval ship designs around the world. This paper will briefly discuss the factors leading to this eventuality and describe some of the military survivability aspects of naval design currently being addressed by societies. Using the ABS International Naval Ship Guide (INSG) notations and NATO standard ANEP 77 as examples, the paper will describe some aspects of survivability that classification now addresses. Topics will include:

NATO ANEP 77: The Naval Ship Code is intended to address naval surface ship safety. It is a goal-based standard using IMO conventions and other sources as introduced through the International Naval Safety Association (INSA).

Hull Girder Ultimate Strength Assessment: Discusses the critical failure modes of all hull girder structural elements. All relevant failure modes for individual structural elements, such as plate buckling, torsional stiffener buckling, stiffener web buckling, lateral or global stiffener buckling, and their interactions are considered in identifying the weakest inter-frame failure mode.

Ballistic and Fragmentary Protection: From basic to enhanced levels of protection, this notation addresses ship arrangement details aimed at concealing crew and mission critical assets to protect from attack. Methods to slow ballistic projectiles and protect from structural penetration are used for higher notation levels.

Air Blast: Addresses the calculation of air blast loads and methodology for analyzing structure, using explosive amount and distance from blast to determine protection level desired.

Underwater Shock: Describes the assessment of a structure to withstand an underwater explosion.

Underway Replenishment: Addresses the general guidance to identify the fitness of principal equipment and the completeness of a ship's operational manual in order to perform Underway Replenishment safely.

Redundancy: Notations addressing redundancy arrangements, indicating a vessel is fitted with multiple propulsors and steering systems such that fire or flood in one space will not affect other propulsion systems.

Finally, the paper will consider evolving technologies being considered for naval designs that will need to be addressed by class societies in the future.

KEY WORDS:
Classification societies; naval classification; military notations; standards.

NOMENCLATURE:

ABS: American Bureau of Shipping

Other Classification Societies:

- BV - Bureau Veritas
- DNV GL - Det Norske Veritas Germanisher Lloyd
- IRS - Indian Register of Shipping
- LR - Lloyd's Register
- PRS - Polish Register of Shipping
- RINA - Registro Italiano Navale

ANEP: Allied Naval Engineering Publication

ARI: Acquisition Reform Initiatives

ASTM: American Society for Testing and Materials

CLF: Combat Logistics Force (US Navy)

ConOpS: Concept of Operations Statement

IMO SOLAS: Safety of Life at Sea

INSG: ABS International Naval Ship Guide

IT: Informational Technology

LOP: level of protection

LOS: line of sight

MVR: ABS Marine Vessel Rules

NATO: North Atlantic Treaty Organization

NSC: Naval Ship Code

OT: Operational Technology

RO: Recognized Organization

SDOF: single degree of freedom

UNREP: Underway Replenishment

FACTORS LEADING TO NAVAL CLASSIFICATION

> *"Change is with us. Change in what we do, how we do it, and sometimes it's even where we do it. It's inevitable that there will be change in how we procure our equipment for the future. The only message I'd like to give industry is that we're not afraid to adopt new ideas or new ways of doing business, or to seek the authority to reach out just a little bit further than we have in the past. It will be a lot better if we figure that out together, as the customer and the supplier." – ADM Boorda*

As the Cold War ended, Navies around the world began seeking new ways to continue building ships within ever tightening budgets. For example, in the United States, the US Government initiated Acquisition Reform Initiatives (ARI) in the 1990s and through to the early 21st century. Working with industry, they succeeded in transforming longstanding military acquisition practices into a more commercially relevant acquisition process. The American classification society American Bureau of Shipping (ABS) participated with the US Navy in converting many of their older military documents into commercial standards, as well as incorporating their design practices into rules and guides for their use. The American Society of Testing Materials (ASTM) Committee F25 on Shipbuilding also served as a conduit for converting obsolete military documents into relevant commercial performance standards with military unique aspects where needed.

As ASTM F25 was one of the preeminent standard bodies geared specifically towards US shipbuilding, it was the natural choice as one of the first organizations to gain increased government participation. The committee became an excellent forum for reviewing and comparing of military standards to existing industry standards. As government representatives became more familiar with the commercial counterparts to many military documents, they realized that they could perhaps be made suitable for naval applications with some changes. These changes were added as annexes to the standards known as "Supplementary Requirements" and can be found as part of many ASTM standards in Volume 01.07.

These are just two examples of the many ways in which navies throughout the world began working with their national industrial base, and the many classification societies around the world, to transform their acquisition practices from an insular model using government self-certification into a more commercial model using the long established commercial model of ship classification.

The commercial shipping industry has for many years employed the process of ship classification, using the independent third-party certification services of one of the recognized Classification Societies. As noted earlier, many navies (including the US Navy) gained familiarity with that commercial process through the building of many non-combatant auxiliary ships built to commercial standards. In America, the US Navy has participated as part of the ABS Technical Committee for many years, aiding in the development of rules for the design and certification of ships for the combat logistics force (CLF), sealift, and oceanographic service.

What is Classification?

First conducted over 250 years ago, 'Classification' has been performed by classification organizations called societies. The objective of ship classification is to verify the structural strength and integrity of essential parts of the ship's hull and its appendages, and the reliability and function of the propulsion and steering systems, power generation and those other features and auxiliary systems which have been built into the ship in order to maintain essential services on board. Classification societies (or the more common shortened term 'class societies') provide these services through the development of standards called Rules.

These rules form the basis for assessing the design and construction of new vessels and the integrity of existing vessels and marine structures. Implementing the published rules, the classification process typically consists of:

- A technical review of the design plans for a new vessel to verify compliance with the Rules
- Attendance by a class society Surveyor:
    - At the construction shipyard
    - At the relevant production facilities that provide key components for the vessel
    - At sea trials
- When these inspections or surveys are completed, satisfactorily verifying compliance with the rules, the class society issues a certificate of classification.

Once in service, the owner must submit the vessel to periodical onboard class surveys to verify that the ship continues to meet the relevant Rule requirements. An example of the process for American Bureau of Shipping (ABS) is provided below in Figure 1.

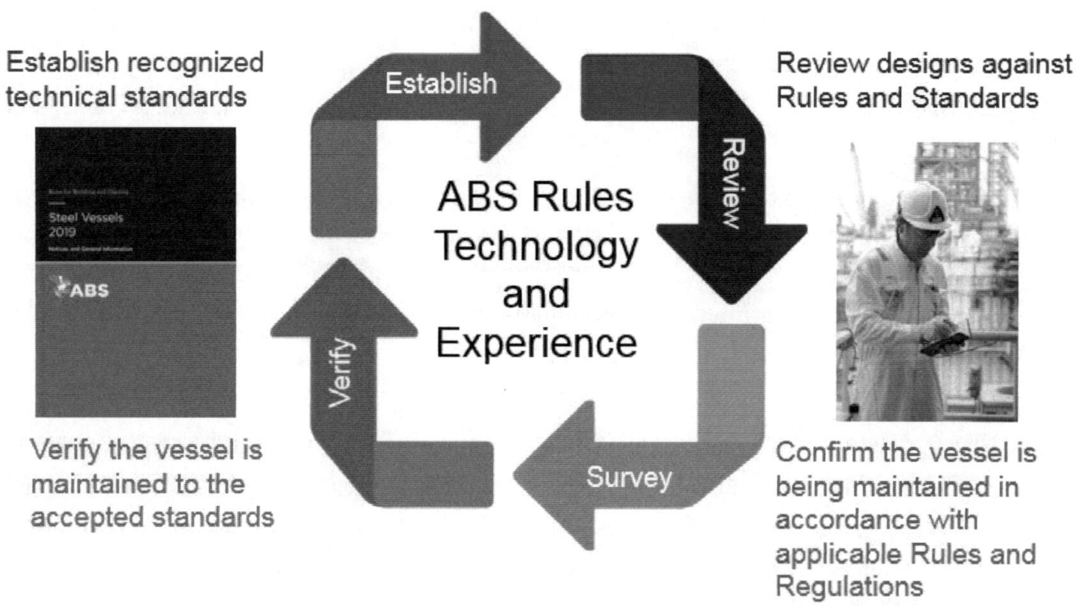

Figure 1: The ABS Classification Model

Use of Naval Classification Increases

As the navies worldwide began to incorporate more commercial processes and standards, they decided to build on their familiarity with class society rules and processes to move towards the next step: rules for combatants. Navies around the world, such as in the United Kingdom, Norway and Germany reached this conclusion, and employed their home class societies. Lloyd's Register worked with the British Navy and led this effort by publishing their

Naval Ship Rules in 1999. The UK Navy has since applied them to all British naval combatant acquisition programs.

Many societies have developed Rules to address a range of naval and naval auxiliary vessels, such as ABS, Bureau Veritas (BV), Det Norske Veritas Germanisher Lloyd (DNV GL), Indian Register (IRS), Lloyd's Register (LR), Polish Register (PRS), and Registro Italiano Navale (RINA). At least two dozen navies around the world have applied the process of classification to some degree for either the acquisition or in-service maintenance of their vessels. It has indeed become standard operating procedure for Classification Societies to provide services for naval ship designs around the world.

MILITARY SURVIVABILITY ASPECTS OF NAVAL DESIGN ADDRESSED BY CLASSIFICATION

Survivability is a critical consideration for a vessel. For commercial vessels, this entails the ability to withstand loads due to inclement environments and operate even if critical components or systems fail. Naval vessels have the additional burden of surviving in combat environments. Both offensive and defensive measures can be taken to improve the survivability of the vessel. Survivability for naval vessels is composed of three basic properties:

Susceptibility: This is principally a ship's ability to avoid detection from other offensive ships, which depends greatly on its Signature. Each naval ship has unique signature characteristics in various regions of the electromagnetic spectrum. The spectrums most commonly used for identification and detection purposes are visual, microwave and infrared. Reducing or modifying the various signatures such that they match the operating environment can enhance the survivability of the vessel. Of course, sound and visual identification still remain equally important to signature.

Vulnerability: This is a ship's ability to continue to conduct its mission, either fully or partially, and to protect its crew, after an offensive weapon has struck the ship. The design is to attempt to minimize damage as best as practicable; this can either be done via system redundancies; thwarting the ability of an enemy to successfully acquire targets on the ship (such as through providing protected accesses for crew transit onboard the ship to minimize open air movement); providing means to reduce the energy of the weapon effect; or structure hardening to prevent weapon penetration.

Lethality: This is the ability of a ship to successfully engage an enemy using its offensive weapons or other assets such as aircraft, helicopters or unmanned underwater, surface or aerial vehicles. This third aspect of survivability is typically outside of the scope of the Classification Society, which is principally concerned with hull, mechanical and electrical (HM&E) aspects of the ship design, as well as some non-offensive mission systems.

To better illustrate and describe some aspects of survivability that classification now addresses, the following are examples of military notations addressed in the ABS International Naval Ship Guide (INSG).

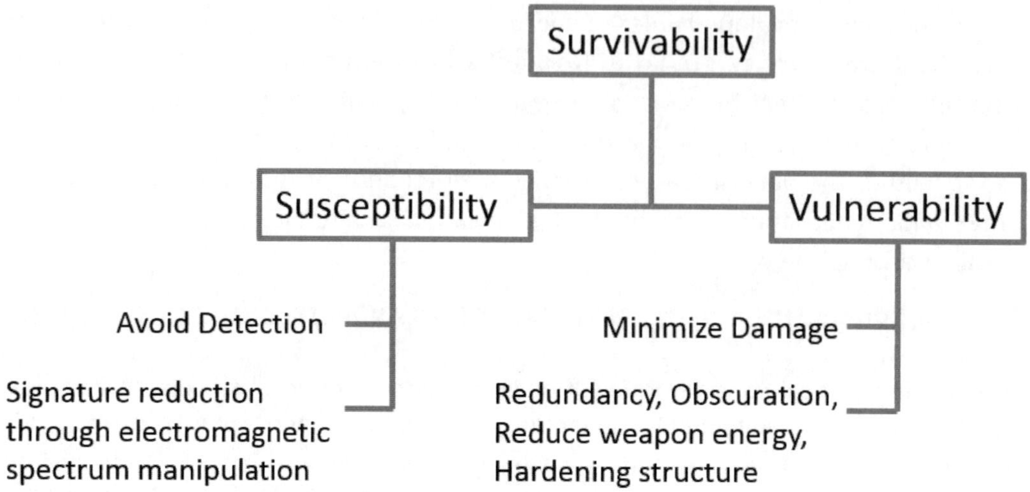

Figure 2: The Survivability Model for Classification

Notations from the ABS International Naval Ship Guide

The ABS Guide for Building and Classing International Naval Ships (INSG) is applicable to non-nuclear, displacement type, monohull surface ships. It addresses hull, mechanical and electrical systems specifically adapted to combatant and non-combatant vessel requirements in areas such as survivability, mission and system requirements and interfaces, and interpretations of statutory requirements. The structure and organization of the INSG is similar to commercial ABS Marine Vessel Rules (MVR), using a proven quality system process. In addition, naval shipbuilding standards from around the world were examined for reference and/or inclusion during development. Key military requirements were adopted to address redundancy and survivability. The INSG is a complete standard for building naval warships, including:

- Mission loads
- Vital space definition
- Damage control
- Fire/flooding requirements
- Revised finite element analysis (FEA) guidance and design pressure requirements
- Battle override features in control and automation

The INSG allows flexibility in tailoring the requirements as needed, typically to become a part of a comprehensive certification matrix. This matrix may include a combination with portions of other Rules (as needed), as well as in conjunction with Naval Administration[1] standards. The matrix is created to best accommodate project support as part of the acquisition process, which in many cases is very complex for a government program. In addition, INSG may

[1] The 'Naval Administration' is the agency within a Government or Nation responsible for the safe operation of government ships. The Naval Administration may be assisted or supported by other government departments, or it may delegate this duty to another agency within the Government.

support Statement of Compliance reviews/surveys, offered as third-party certification by ABS to criteria both originating from ABS Rules as well as other resources (e.g., builder's specifications).

NATO ANEP 77 – Addressing Naval Ship Safety

While it is clear that IMO SOLAS (Safety of Life at Sea, a convention of the International Maritime Organization) is not easily adapted to naval designs – indeed, ships of war are specifically exempted from application to the regulations – safety of life at sea is key for any ship. The Naval Ship Code (NATO Allied Naval Engineering Publication 77, or NSC) is intended to address naval surface ship safety. Using a goal based philosophy for naval ship safety, the Naval Ship Code provides the first high-level comprehensive safety standard for combatant and noncombatant military ships. Put simply, ANEP 77 is a sort of naval version of IMO SOLAS and is being applied to many NATO and non- NATO warships around the world.

ABS INSG offers the **NavalSafe** notation for military ship designs. ABS NavalSafe(x) is an optional notation assigned to the ship once the performance requirements of the defined chapter(s) of the NSC are met and the Naval Ship Safety Certificate is issued. The NSC is based on IMO conventions, resolutions and other sources that are applicable for the majority of naval ships.

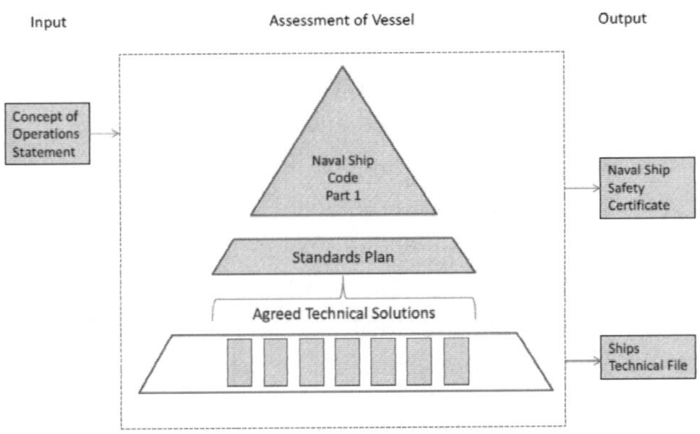

Figure 3 – Naval Safety Ship Certification Process. Source: NATO ANEP 77

The ABS NavalSafe(x) notation determines a minimum level of safety for government combatant and non-combatant vessels. The index 'x' in NavalSafe(x) notation represents: S (Structure), BSC (Buoyancy, Stability and Controllability, ES (Engineering Systems), SS (Seamanship Systems), FS (Fire Safety), EER (Escape, Evacuation and Rescue, C (Communication), N (Navigation), DG (Dangerous Goods), All (if all entries are applicable). The NSC is applicable to all surface craft used for government, non-commercial service, such as navy, coast guard, border patrol, customs etc. It applies principally to conventional powered vessels (non-nuclear) using conventional fuels such as diesel (for example NATO F76 fuel) or intermediate fuel oils.

The NSC requires that a Concept of Operations Statement (or ConOpS) be developed to compare the applicability of the criteria and standards chosen. Once this is determined, the Code can provide a path for a ship to be certified by a Naval Administration, along with recognized organizations (RO) such as classification societies, to establish that a vessel is safe to operate in accordance with the ConOpS provided (within the limits of those aspects addressed by the Code), as well as within the safety policies, and safety organization, of the government organization in which it will operate.

While the goal based nature of the Code allows the Naval Administration and ROs to consider alternatives to the typical safety requirements applied to commercial ships, it is important to emphasize two limitations: (1) It is not intended to be viewed as a complete and entire safety management system for a ship or fleet; and (2) While fully intended to apply to operating conditions and foreseeable damage scenarios applicable to peacetime and maritime security (as determined in the ConOpS), the Code is NOT intended to apply to combat operations, or its associated threat conditions; these are outside of the scope of the Code, and intended to be addressed separately by the appropriate departments within an Administration.

IMO SOLAS requirements may be applied as a solution option for a particular safety aspect of the design; arrangements other than that proscribed in IMO SOLAS may be evaluated using the Naval Ship Code.

Hull Girder Ultimate Strength

When hull girder structural members are compressed beyond their limit it reduces their load carrying capacity. It is therefore important to identify the weakest inter-frame failure mode by investigating all relevant failure modes for individual structural elements, such as plate buckling, torsional stiffener buckling, stiffener web buckling, lateral or global stiffener buckling and their interactions.

In addition to simple longitudinal strength calculations (where the calculated elastic stresses in the deck or bottom shell members are evaluated against prescribed allowable stresses), some navies seek this Ultimate Hull Girder Strength Assessment (**UHS**) notation because the ultimate hull girder strength analysis takes into account elasto-plastic properties of the material, nonlinear geometric behavior of the elements, and their buckling and post-buckling strength, in order to more accurately estimate the strength reserves of the ship's cross section, providing a rigorous, first-principles based alternative evaluation of the actual hull strength. This is particularly relevant for unconventional and highly optimized naval vessel designs. However, it should be stressed that the UHS notation applies to the intact strength of the vessel, it alone is not suitable for evaluation in a damaged condition to mitigate military damage.

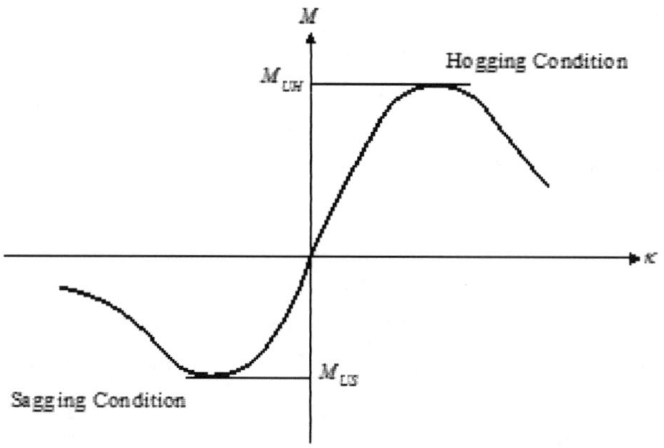

Figure 4: Vertical Bending Moment M versus Curvature κ

The hull girder ultimate strength is represented by hull girder ultimate bending capacity M_U (M_{UH} or M_{US}) (Figure 4) which is defined as the peak value of the curve with vertical bending moment M versus the curvature κ of the ship cross section.

Ballistic and Fragmentary Protection

From basic to enhanced levels of protection, this notation addresses ship arrangement details aimed at concealing crew and mission critical assets to protect from attack. Methods to slow ballistic projectiles and protect from structural penetration are used for higher notation levels. ABS uses the notations **BFP1, BFP2** and **BFP3** as optional notations that indicate the level of ballistic and fragment hazard protection. Ballistic and fragment hazards can be mitigated by various methods, which will determine the level of protection (LOP) the ships can attain. Selection of the desired LOP for ballistic and fragment hazard protective barriers should be based on the mission requirements, anticipated threats and results of vulnerability assessments, which should account for ship resiliency under combat conditions. The primary objective of ballistic and fragmentary protection is to provide technical guidance for achieving an LOP for naval ships against direct fire weapons by reducing the line of sight (LOS) and hardening the exposed surfaces. Hardened exposed surfaces may include both transparent (e.g., glazed) and opaque (e.g., steel, aluminum, composite, etc.) barriers comprised of single- or multi-layer construction.

BFP1 notation focuses on obscuration from open view of typical crew activities. An example where BFP1 might be employed regards the configuration of the access to the Bridge. Access doors on either side of a Bridge, that might allow clear visibility to the helm when opened, would allow the possibility of a straight line of sight (LOS) shot by a sniper at the watch officer, navigator or helmsman. Using a small right angle for the door entrance to the Bridge would eliminate this force protection weakness. Higher levels of this notation (BFP2, BFP3) would address items like protective glass around the Bridge, and enhanced protection of bulkheads and decks surrounding ammunition storage and crew gathering areas such as the mess and recreation areas.

Air Blast

ABS notation "**AB (Weight, Range)**" is an optional notation that indicates protection from an air explosion event. Ships complying with these requirements will be eligible for the optional notation AB (Weight, Range), where Weight is the equivalent TNT weight of the explosive material and Range is the safe distance from the source of explosion. Blast curve methodology for a hemispherical surface burst is used to predict the air blast loads acting on exposed surfaces of the ship resulting from open-air (unconfined) bursts of high explosive (HE) at the water surface.

Equivalent single degree of freedom (SDOF) systems is used for structural analysis. SDOF reduces the level of analytical effort and facilitates the use of parametric studies and the development of analysis. As shown in the figures below, the simplest method for solving the structural dynamic equilibrium equation is through a graphical representation solution. A dynamic response chart which provides the solution for a SDOF system having perfect elastic-plastic resistance and a triangular pulse loading is used. Given the applied loading and the basic SDOF system parameters, the ductility demand may be read directly from the response chart.

Analysis curves for hull plate thickness (Figure 5) and longitudinal stiffener (Figure 6) are also provided. These curves encompass all of the engineering steps from blast load determination through the nonlinear dynamic structural analyses and damage state predictions. Knowing the hull plate thickness (t) and longitudinal frame spacing (s) or the longitudinal stiffener particulars: length (L), cross sectional area (A), moment of inertia (I) and plastic modulus (z), an engineer can quickly assess the ability of the reviewed structure to meet the required level of protection.

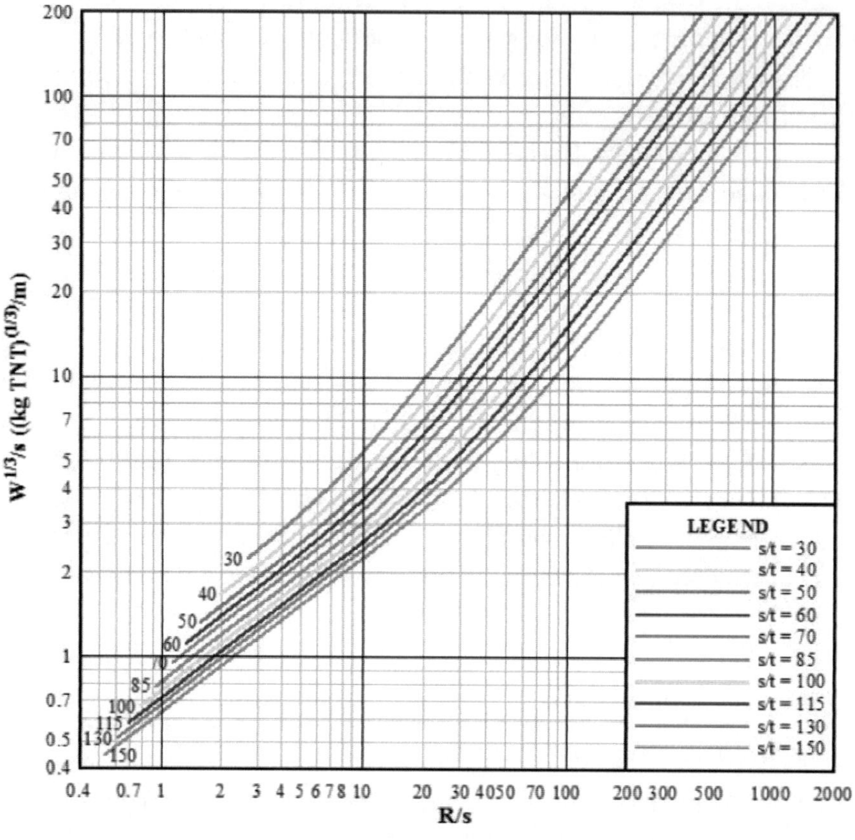

Figure 5: Scaled R-W for Hull Plates

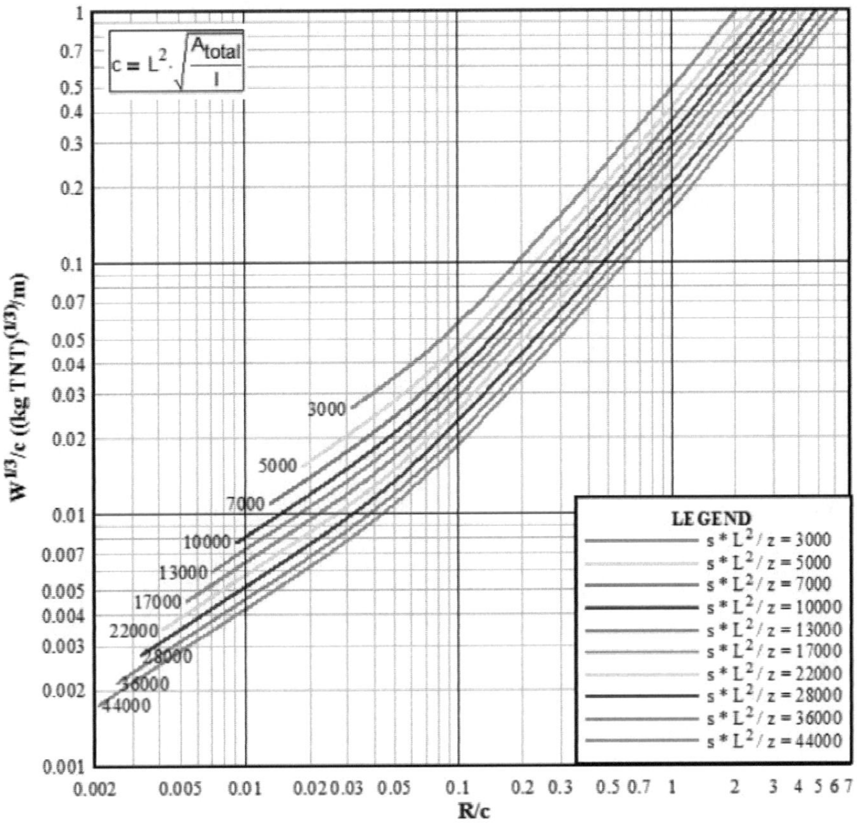

Figure 6: Scaled R-W for Longitudinal Stiffener

Underwater Shock

The ABS shock notation (described as **UNDEX (Weight, Range)**) provides steps needed to protect ship structures from underwater shock. The UNDEX (Weight, Range) notation covers the hull structure below the waterline, and shock effect on the equipment. The Naval Administration is to define the underwater shock threat in the form of charge weight and the expected safe distance from the source of the underwater blast. Shock loading can come from shock waves, gas bubble pulses and bulk cavitation. Only shock waves comprised of direct, surface reflected, and bottom reflected waves are considered.

Equivalent single degree of freedom (SDOF) analysis can be used to design the protective structures subjected to blast loads based on calculated underwater shock loading. Use of equivalent SDOF systems significantly reduces the level of analytical effort, and there are several numerical methods that can be used to solve the dynamic equation using SDOF. In the notation guidance, the linear acceleration method is used. In this method the response is evaluated at successive increments (Δt) of time. At the beginning of each interval, the condition of dynamic equilibrium is established. Then the response for a time increment (Δt) is evaluated approximately on the basis that the spring coefficient in the equation of motion remains constant during the interval Δt. Assessment of supporting members such as girders or transverses can either be done using a series of separate SDOF dynamic analysis for each of the supporting members or by more advanced structural analysis.

Underway Replenishment (UNREP)

Replenishment at Sea is crucial for any naval fleet engaged in power projection extending far from its nation's shores, as well as supporting allied navies in extended exercises. The notation **UNREP (Beam, Stern, Vert)** is applicable to both supplying and receiving ships, and covers:

- Ship's maneuvering
- Engagement
- Cargo transferring and its equipment
- Disengagement
- Breakaway

As guidance, the notation does not cover the following, as these items are typically addressed in fleetwide procedures for each navy:

- Cargo loading
- Cargo storing
- Equipment below the cargo hook (ex: personnel cage, cargo container, etc.)

The notation addresses general guidance to identify the fitness of principal equipment and the completeness of operational manuals in order to perform UNREP in a structured manner.

Redundancy

One property of vulnerability resistance addresses redundancy arrangements, indicating a vessel is equipped with propulsion and steering systems designed to provide enhanced reliability and availability through functional redundancy, such that a fire or flood in one space will not affect other propulsion systems. The objective of redundancy is to provide requirements which reduce the risk to personnel, the vessel or other vessels. The requirements for this notation are intended so that, following a single failure, the vessel is capable of either maintaining course and maneuverability at reduced speeds without intervention by other vessels, or maintaining position under adverse weather conditions to avoid uncontrolled drift and navigating back to safe harbor when weather conditions are suitable. In addition, it addresses aspects which would reduce the detrimental effects to the propulsion systems due to a localized fire in the machinery spaces.

Notations offered address multiple propulsion machines (**R1**), multiple propulsors and steering systems (**R2**), multiple propulsion machines in segregated spaces (**R1-S**)**, and** multiple propulsors and steering systems in segregated spaces (**R2-S**). Weather criteria may also be applied resulting in the addition of a plus (**+**) to the above notations. However, it should be stressed that while this notation has applications for military ships, it is intended principally for commercial vessels and is based upon a single failure concept only. Specifically, this concept accepts that failures may occur but that only one such failure is likely at any time. Of course, military ships intended to go in harms way may be expected to sustain more than a single failure point. This is determined based on the ship design's hazard and survivability analysis typically conducted by the Navy.

Evolving Technologies and the Future

Autonomous Operations are becoming more and more pervasive in naval operations, as navies operate successively larger and more operationally capable aerial, surface and undersea vehicles. This will result in additional platform space for air capable operations, control centers, mission space and deployment gear, hull accesses and ramps. On the command and control side, it will push navies to consider further defensive measures to preserve their networks and protect them from cyber-attacks. Cyber protection, both from an operational technology (OT) and informational technology (IT) perspective, has already become one of the preeminent properties for a superior warfighting ship, and this will continue for the foreseeable future. In some cases, class societies (such as ABS) are working with navies to publish guidance to help improve cybersecurity and mitigate related risk, such as through the ABS Guide for Cybersecurity Implementation for Government Vessels, Facilities and Assets (Volume 6 of the ABS CyberSafety© series of publications).

"Novel uses for increasingly sophisticated robotics, energy storage, 3-D printing, and networks of low-cost sensors, to name just a few examples, are changing almost every facet of how we work and live." – ADM Richardson

For at least the last 60 years, diesel engines, gas turbines and boiler/turbine plants delivered the power necessary - via a shaft and propeller - for ship propulsion for the world maritime industry. While these forms of propulsion, principally using diesel engines, is quite likely to continue into the next few decades, current technological advances offer a number of other options. These options typically center on transitioning from a conventional mechanical propulsion arrangement to a hybrid electric propulsion system. Modern electric propulsion systems may incorporate new technology, such as using fuel cells, as power generation, and batteries (principally lithium ion), super-capacitors and flywheels to provide energy storage to supply and/or supplement the electrical power needs of the vessel. The stored energy aspect augments energy needs to enhance safety in response to emergency scenarios, as well as the high power needs necessary to serve directed energy systems and related weapon technology. Fuel cells have already been used for supplemental electric power generation on naval platforms since the beginning of the 21st century, such as the air-independent propulsion system for the German Navy's Type 212 and 214 submarine classes, among others developed around the world.

SUMMARY

In the US Navy, Chief of Naval Operations Admiral Richardson has frequently stressed the need for experimentation, engineering and prototyping to ready the fleet for the next conflict and yield a relevant navy that is ready to defend the country from attack and protect US interests around the world. His focus on conceptual, geographic and technological agility has reverberated with navies throughout the world who are now considering a number of commercial processes and evolving technologies for naval designs. As a now established trusted partner in naval design and construction worldwide, class societies will continue to innovate and develop to meet the needs of the navies they serve.

REFERENCES

1) Admiral Boorda, J. M. (1996, March 15) [Interview]. Washington Watch Newsletter (see also www.navy.mil)

2) American Bureau of Shipping. Classification Services, from https://ww2.eagle.org/en/Products-and-Services/classification-services.html

3) American Bureau of Shipping. (2019, January) Guide for Building and Classing International Naval Ships.

4) Ashe, G. M., & Delpizzo, R. D. (2013). *Current Trends in Naval Ship Design*. Arlington, VA: American Society of Naval Engineers. ASNE Day 2013 Proceedings.

5) Delpizzo, Richard D. and Valluri, Sharat. (2017, March 17) An Introduction to NATO Standard ANEP (Allied Naval Engineering Publication) 77 and Its Application to Naval Ships. Colombia Mar 2017.

6) North Atlantic Treaty Organization (NATO). (2018). Naval Ship Code, Allied Naval Engineering Publication (ANEP) 77.

7) Richardson, J. M. (2016, January). A Design for Maintaining Maritime Superiority [CNO Directive]. US Navy, Washington, DC

8) Lopez, Todd (2019, June). Naval Engineers Must 'Lean In' to Advance Technological Agility. (2019, June). Retrieved from https://www.defense.gov/explore/story/Article/1882567/naval-engineers-must-lean-in-to-advance-technological-agility/

Subs in Schools: Engineering a Reliable Depth-Control System and Pressure-Hull

Cameron Harper[1], Ahmed Swidan[2] and Warren F. Smith[3]

[1] ASLT, UNSW Canberra at ADFA, Australia cameron.harper@defence.gov.au
[2] UNSW Canberra at ADFA, Australia a.swidan@adfa.edu.au
[3] UNSW Canberra at ADFA, Australia w.smith@adfa.edu.au

The "Subs in Schools" STEM project baseline remotely-controlled submarine kit suffers from several functional deficiencies; predominately its resistance to water ingress and effective depth-control. The aim of the project reported is determining whether the pressure-hull and depth-control system can be redesigned to significantly improve reliability, manufacturability and maintainability. Discussed is the intended approach for developing a new design and associated prototype. This includes construction and testing of the current design to identify limitations; conducting reliability analysis of the key components; presenting a proposed concept design for the improved version; and a summary of work completed to date. The tests on the existing model confirmed reports of leakage and limited depth-control functionality. Literature suggests that bulkhead penetrations are the primary weakness for water ingress; an argument supported by the experimental findings. The Failure Mode and Effect Analysis (FMEA) model in conjunction with the exponential random failure analysis model, is used to estimate the reliability of the pressure hull. A concept for a revised design is presented that will require proof of concept experiments and the subsequent manufacture of a testable prototype.

This paper reflects a work in progress as part of a 2019 undergraduate honours project by the lead author.

INTRODUCTION

Reengineering Australia's (REA) "Subs in Schools" (SiS) Technology Challenge invites school students to design, build and operate remotely-controlled submarines to compete in an annual national competition (soon to be international). The competitors are permitted to enter an REA-designed kit-model or alternatively, design their own submarine or variation of the kit-model. However, to date, no competitors have successfully manufactured a correctly functioning example of the REA entry-level submarine kit. Models invariably suffer from malfunctions associated with water ingress and a general inability to effect reliable depth-control [1]. The purpose of the kit is to provide the foundation to develop a functioning remote submersible fit to compete in the competition. The difficulty for it to perform basic submarine operations reliably, fundamentally undermines its intent and essentially renders it redundant. The overarching limitation of these aspects of the model appear to lie in a general lack of reliable functionality.

SCOPE/AIM

The reported study focussed on devising a solution to address the deficiencies in the kit-model's watertight integrity and depth-control system. While the baseline models may exhibit deficiencies in other areas; such as manoeuvrability, this project targets the design and assessment of an improved pressure-hull and depth-control system. The aim is to determine if a reliable pressure-hull and depth-control system can be designed that is straightforward to manufacture and simple to maintain for high school students. The scope involves assembling the existing kit design in accordance with the build manual [2] and testing for functionality. Identifying the deficiencies and determining the model's reliability, a revised design of the respective systems is to be conceived and manufactured. This prototype will then be tested, assessed for reliability and compared to the baseline design with recommendations to be submitted to REA.

CURRENT MODEL

The current SiS base kit-model required significant effort to complete construction. This was chiefly due to the ambiguous nature of the build manual [2]. Problems contributing to delays in the construction process included:

- Instructions often conflicted with images
- Several images were not relevant to the direction given
- Images were often unclear
- Terminology was vague, colloquial or incorrect
- Part numbers were not referenced in the instructions
- Lack of instructions/design drawings provided for manufacturing certain required parts
- Kit was incomplete
- Parts incompatible with 3D printed mounts
- Parts missing entirely
- No electrical drawings provided

DESIGN

The current Subs in Schools kit-model comprises 3D printed components, PVC piping and standard remote-controlled vehicle hardware as shown in Figure 1. The design has several main sub-assemblies. The first is a transparent double-sectioned watertight cylinder enclosed by o-ringed end caps and joined by a watertight cable conduit. This is known as the "pressure-hull" and houses all electronic components. The pressure-hull is mounted within an outer shell constructed from a section of PVC pipe. The forward section of the outer shell is capped with a nose cone which houses the bow plane assembly. At the stern, the tail cone houses the propeller, stern tube, and rudder and elevator assemblies. The conning tower is mounted on the outer shell and houses the main battery switch and connects to the antennae float.

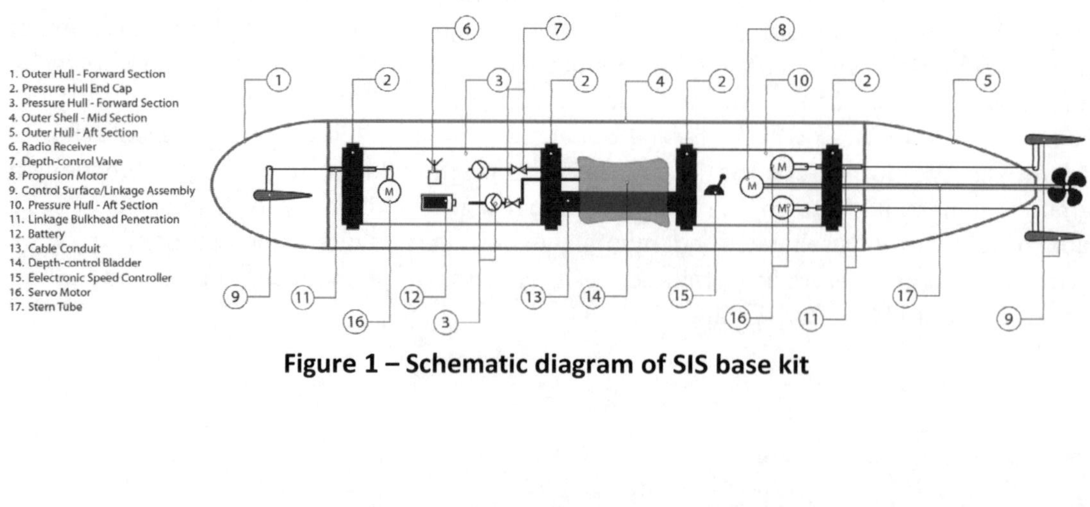

Figure 1 – Schematic diagram of SIS base kit

Depth-control is effected by a pump and bladder system. The bladder is located amidships within the void located between the forward and aft pressure-hull sections. To descend, air is transferred from the bladder via a dedicated pump and solenoid valve to the pressure-hull. As the bag deflates, water is able to occupy the void and the vessel's overall density increases. The vessel subsequently becomes negatively buoyant and consequently submerges. To resurface, the procedure is reversed by means of a separate dedicated pump-valve system.

Experiment

Upon successful completion of the kit, the model was subjected to basic function tests in a controlled environment. A test bath was filled with fresh water and the model was placed on the surface in order to establish the hydrostatic behaviour of the vessel (Figure 2). In order to keep the vessel transversely upright, it was necessary construct a weighted keel. To neutrally trim the vessel, ballast (in the form of stainless-steel washers) were added to the forward and aft sections as required.

Figure 2 – Completed base kit-model surfaced in test bath

With the vessel floating freely, a test of the propulsion and control surfaces was carried out. All systems functioned correctly on the surface and there was no evidence of water leakage into the pressure-hull.

The depth-control system was tested next. The submersion pump was activated, and the vessel began to dive. The hydrostatic behaviour of the vessel changed instantly. The vessel began to trim forward with constant redistribution of the ballast being necessary to keep it neutrally trimmed. The submersion process was noted as being particularly sluggish and due to the outer shell being entirely opaque, it was impossible to determine the status of the inflatable bladder. The time taken to fully submerge to a depth of 200mm was approximately five minutes.

Once submerged, a function test of propulsion and control surfaces was re-conducted. All systems functioned correctly. However, when the surfacing pump was activated, the vessel would only fully surface by the head (or stern), depending on how the ballast was distributed. This process was also sluggish with the vessel not able to fully surface at any point. Upon completion of the test, the model was dismantled for diagnosis. It was found that the water-tightness of the pressure-hull had not held and, although they were still functioning, the electronic components were partially submerged in water in both sections.

The water was drained, and the model dried before conducting a secondary test. This test was conducted as in the previously described manner. However, during this test the bladder was pre-pressurised to determine the ability of a positively pressurised hull to resist water ingress. During the second descent, air was observed to escape from the bow-plane linkage seal (see Figure 3). It was determined that these seals were the most likely source of leakage in the forward and after sections of the pressure-hull.

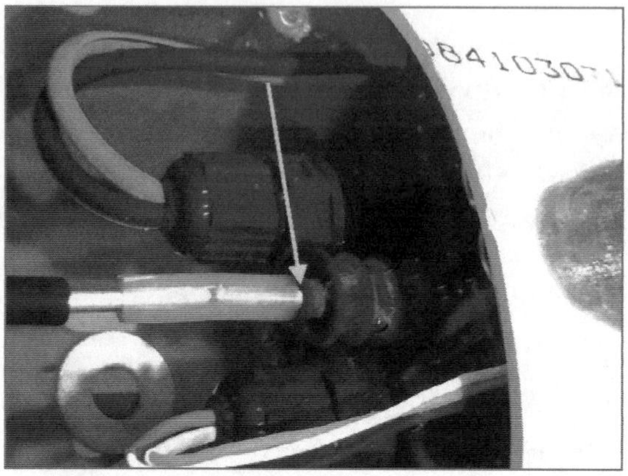

Figure 3 – Image of the bow-plane linkage seal (leakage point indicated)

Experimental Results/Identified Problems

The preliminary tests showed that the pressure-hull suffered from significant water ingress even at shallow depths. This is a considerable problem as it prevents the model from being properly tested. Until a solution to the problem can be found, the model is restricted to surface experiments in order to prevent damage to the sensitive electronic components. In addition, the depth-control system failed to operate correctly. The time taken to fully submerge the vessel was considerable and the model failed to surface with a satisfactory trim.

Interestingly, these two problems appear to be coupled. As predicted in the literature, pressure-hull water ingress appears to be isolated to the bulkhead penetrations; specifically, the bellow seals that fit over the bow-plane control linkages (Figure 3). It was hypothesised that design of the depth-control system may be the primary source of this defect. Because the depth-control system displaces air from the pressure-hull in order to surface, the pressure-hull enters a vacuum state as the submarine ascends. The design of the bellows seals is such that they are insufficient to support this negative pressure and, combined with the positive external pressure forcing water in, are failing.

The embarked water subsequently increases the overall displacement of the vessel beyond the buoyant capability of the depth control system, and leaves it partially stranded at depth. At this stage, the stern tube and pressure-hull end cap seals appear to be functioning correctly. However, with the inherent vacuum effect in the current depth-control system, reinforcing the linkage seals may successively induce failure in these systems.

Assuming these water tight integrity problems can be solved, the performance of the current depth-control system remains unsatisfactory. Even functioning with no leaks, responsiveness is low and the time to reach depth, unacceptable. The source of the problem appears to also be with the fundamental design of the depth-control system. The pumps currently transfer air (sourced from the pressure-hull) in and out of the bladder. The compressive property of air is most likely reducing the efficiency of the pump as the vacuum pressure in the hull increases. Theoretically, this means that the responsiveness of the depth-control system degrades as the depth increases.

RELIABILITY ANALYSIS

A crucial factor influencing further functional testing of the submarine is the pressure-hull's considerable vulnerability to water ingress. In this section, determining the reliability of the various components that potentially contribute to failure of the pressure-hull are discussed. In order to analyse the reliability of these components, the method of exponential analysis will be applied in conjunction with Fault-Tree Analysis (FTA) and Failure Mode Effect Analysis (FMEA) techniques [3-8]. The exponential analysis method is employed primarily for failures that occur randomly as a result of fatigue or overload. It should be noted that for this analysis, it was necessary to estimate several key parameters as the circumstances of the experiments did not permit the repeated collection of reliability data.

The pressure-hull assembly is composed of several components that effect its ability to maintain watertight integrity. These components and their functional relationship with respect to failure, are presented in Figure 4. From the components identified, a subsequent failure mode effect analysis was carried out as presented in Table 1.

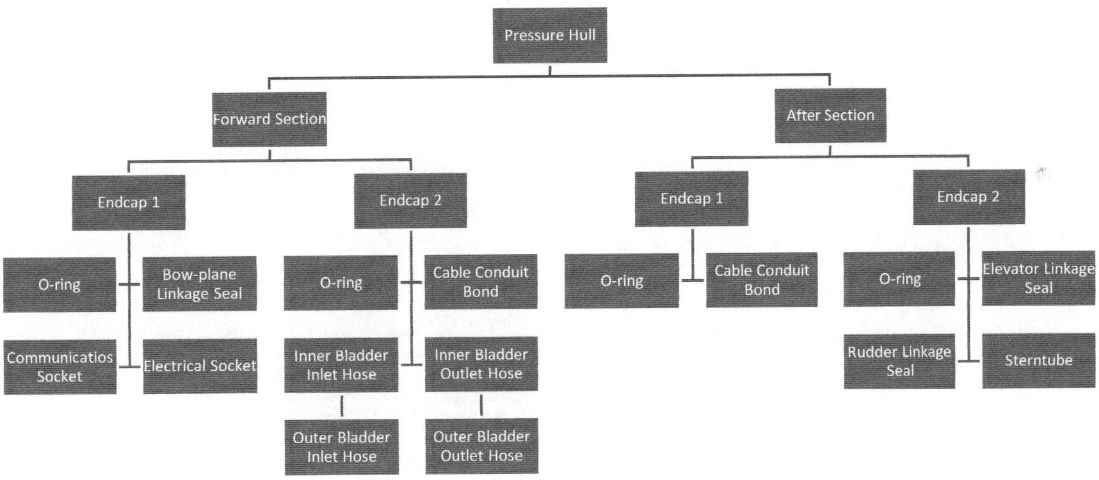

Figure 4 - Functional Component Layout: SiS Kit Model Pressure-Hull

By considering the estimated likelihood of failure along with observations from the experiment, the Mean Time to Failure (MTTF) was approximated. These values were determined by discussing their feasibility with a competition subject matter expert and a conferring with a reliability-engineering specialist. The constant failure rate was calculated by the relationship:

$$\lambda = \frac{1}{MTTF} \quad (1)$$

The resulting failure rates can be viewed in Table 2.

Table 1 – Failure Mode Effect Analysis of SiS Kit Model Pressure-Hull

Component	Failure Mode	Failure Cause	Consequence	Likelihood	Reliability Analysis Model
Bow-plane Linkage Seal	Water Ingression	Pressure Overload	Electronics System Failure /Reduction of Buoyancy	High	Random Failure
Elevator Linkage Seal	Water Ingression	Pressure Overload	Electronics System Failure /Reduction of Buoyancy	High	Random Failure
Rudder Linkage Seal	Water Ingression	Pressure Overload	Electronics System Failure /Reduction of Buoyancy	High	Random Failure
Stern Tube Seal	Water Ingression	Degradation of Lubricant/ Pressure Overload	Electronics System Failure /Buoyancy Failure	Possible	Random Failure
O-ring - Forward Section Endcap 1	Water Ingression	Pressure Overload	Electronics System Failure /Buoyancy Failure	Low	Random Failure
O-ring - Forward Section Endcap 2	Water Ingression	Pressure Overload	Electronics System Failure /Buoyancy Failure	Low	Random Failure
O-ring - After Section Endcap 1	Water Ingression	Pressure Overload	Electronics System Failure /Buoyancy Failure	Low	Random Failure
O-ring - After Section Endcap 2	Water Ingression	Pressure Overload	Electronics System Failure /Buoyancy Failure	Low	Random Failure
Electrical Connector Socket	Water Ingression	Shock Fracture	Electronics System Failure /Reduction of Buoyancy	Low	Random Failure
Aerial Connector Socket	Water Ingression	Shock Fracture	Electronics System Failure /Reduction of Buoyancy	Low	Random Failure
Cable Conduit Bonds	Water Ingression	Shock Fracture	Electronics System Failure /Buoyancy Failure	Low	Random Failure
Outer Bladder Inlet Hose	Water Ingression	Pressure Overload	Electronics System Failure /Buoyancy Failure	Low	Random Failure
Inner Bladder Inlet Hose	Water Ingression	Pressure Overload	Electronics System Failure /Buoyancy Failure	Low	Random Failure
Outer Bladder Outlet Hose	Water Ingression	Pressure Overload	Electronics System Failure /Buoyancy Failure	Low	Random Failure
Inner Bladder Outlet Hose	Water Ingression	Pressure Overload	Electronics System Failure /Buoyancy Failure	Low	Random Failure

Table 2 - Constant Failure Rates of Pressure Hull Components

Component	Approximated Mean Time to Failure (MTTF)	Calculated Constant Failure Rate Per Hour, λ
Bow-plane Linkage Seal	1.25h	0.80
Elevator Linkage Seal	1.25h	0.80
Rudder Linkage Seal	1.25h	0.80
Stern Tube Seal	4.00h	0.25
O-ring - Forward Section Endcap 1	24.00h	0.04
O-ring - Forward Section Endcap 2	24.00h	0.04
O-ring - After Section Endcap 1	24.00h	0.04
O-ring - After Section Endcap 2	24.00h	0.04
Electrical Connector Socket	36.00h	0.03
Aerial Connector Socket	36.00h	0.03
Cable Conduit Bonds	48.00h	0.02
Outer Bladder Inlet Hose	72.00h	0.01
Inner Bladder Inlet Hose	72.00h	0.01
Outer Bladder Outlet Hose	72.00h	0.01
Inner Bladder Outlet Hose	72.00h	0.01

The subsequent reliability of each component is calculated using the exponential relationship:

$$R(t) = e^{-\int_0^t \lambda dt} = e^{-\lambda t}, t > 0 \quad (2)$$

This yielded the following failure rates:

$R_{link.seal}(t) = e^{-0.80t}$

$R_{o-ring}(t) = e^{-0.04t}$

$R_{socket}(t) = e^{-0.03t}$

$R_{cndt.bnd}(t) = e^{-0.02t}$

$R_{stn.tbe}(t) = e^{-0.25t}$

$R_{hoses}(t) = e^{-0.01t}$

A summary of the reliability parameters is presented in Table 3. By analysing the resulting data, it is observed that the reliability of the pressure-hull is dominated by the reliability of the linkage seals. This suggests that by improving the reliability of these components, the functionality of the pressure-hull will be significantly improved. The comparative reliability of each component was then plotted over a submerged operating period of twenty hours (Figure 5).

Table 3 - Summary Results from Exponential Reliability Analysis: Pressure Hull Components

Component	R(t)	λ(t)	Approximate (MTTF)
Bow-plane Linkage Seal	$e^{-0.80t}$	0.80	1.25h
Elevator Linkage Seal	$e^{-0.80t}$	0.80	1.25h
Rudder Linkage Seal	$e^{-0.80t}$	0.80	1.25h
Stern Tube Seal	$e^{-0.25t}$	0.25	4.00h
O-ring - Forward Section Endcap 1	$e^{-0.04t}$	0.04	24.00h
O-ring - Forward Section Endcap 2	$e^{-0.04t}$	0.04	24.00h
O-ring - After Section Endcap 1	$e^{-0.04t}$	0.04	24.00h
O-ring - After Section Endcap 2	$e^{-0.04t}$	0.04	24.00h
Electrical Connector Socket	$e^{-0.03t}$	0.03	36.00h
Aerial Connector Socket	$e^{-0.03t}$	0.03	36.00h
Cable Conduit Bonds	$e^{-0.02t}$	0.02	48.00h
Outer Bladder Inlet Hose	$e^{-0.25t}$	0.01	72.00h
Inner Bladder Inlet Hose	$e^{-0.25t}$	0.01	72.00h
Outer Bladder Outlet Hose	$e^{-0.25t}$	0.01	72.00h
Inner Bladder Outlet Hose	$e^{-0.25t}$	0.01	72.00h

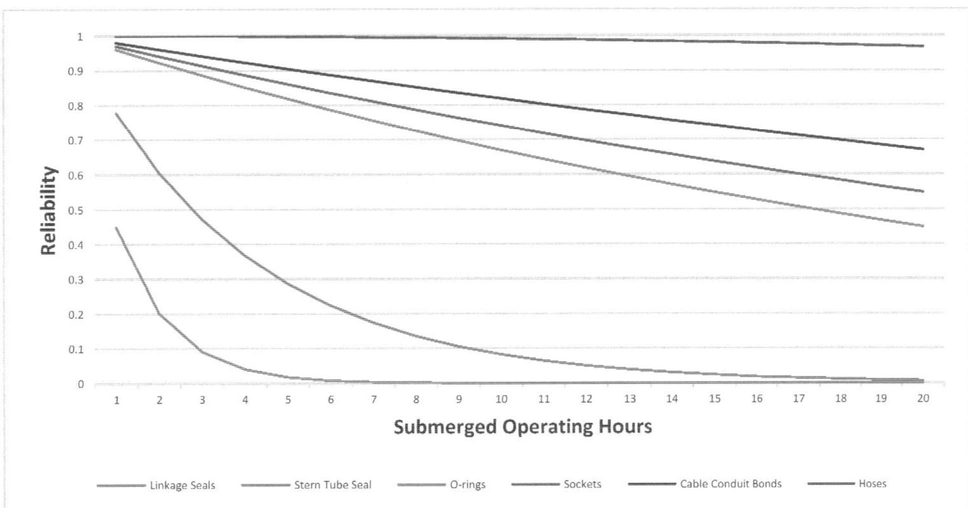

Figure 5 - Reliability of Pressure-hull Components as a Function of Time

PROPOSED DESIGN

Design Requirements

Based on the results from preliminary experiments, relevant literature and the intent of the SiS competition, several requirements were formulated for the redesign. The requirements are presented in Table 4 and reflect a significant improvement in reliability.

Table 4 – Design Requirements

Mandatory	Desirable
Pressure hull must reliably resist water ingression	The design should incorporate as many existing components as possible
Depth-control system must function correctly and reliably	The design should allow for ease of maintenance/access
Desired depth must be capable of being reached in a timely manner	The design should be as simple as reasonably possible
The redesign must be capable of being manufactured by SiS applicants	
The depth-control system must be of a VBS type	
The depth-control system must be capable of functioning while stationary	
The vessel must be able to adjust trim without the use of control planes	

Concept Design

The revised design attempts to draw on Burcher and Rydill's philosophy of simplicity in the pressure-carrying envelope [9]. In addition, a two-tank system has been adopted in light of the prevalence of this configuration found in the literature [10-13]. The proposed general arrangement and new components have been designed with special consideration given to ease of manufacturing, maintainability and robustness.

To combat the water ingress problem, the control surface servo motors are to be mounted external to the pressure- hull (see Figure 6-5, 13). Water resistant servos will be supplemented for the existing servos. This relocation will remove the requirement for moving components to penetrate the pressure-hull and reduce all bulkhead penetrations to cable runs, water pipes and the stern tube (Figure 6-15). The static nature of these remaining penetrations will make sealing and consequently manufacturing, much easier.

The revised depth-control system includes forward and aft ballast tanks which operate independently to effect pitch/trim control without the use of the control surfaces (see Figure 5-3, 11). The new system transfers water from the external environment into sealed ballast tanks via a pump and valve arrangement designated to each tank. While the pumps and valves will be located within the pressure-hull, there no longer exists an interface between the pressure-hull and the ballast tanks. This should aid in reducing the load on the pressure-hull seals and stern tube as pressure will only be a function of the vessel's depth. In addition, the positive displacement of water instead of air should increase the responsiveness of the system due to water's more efficient displacement. The intention is to utilise the pumps only for filling the tanks and to employ the pressurised air within as a means to jettison the water when surfacing.

This will reduce the need for additional hardware and added complexity. To aid with alignment, maintenance and robustness, a series of structural components have also been designed. The redesign includes a forward and aft separable self-aligning collar assembly. The forward and aft ballast tanks can be separated from the main pressure-hull by means of a bayonet fitting included in the collars. The collars also include a void space for fitting electrical connectors with a slot where wires can penetrate and be loomed along the inside of the outer shell. The new design is intended to be capable of being integrated with the existing outer shell components and has attempted to minimise substitutions for fundamental parts in the current kit where possible.

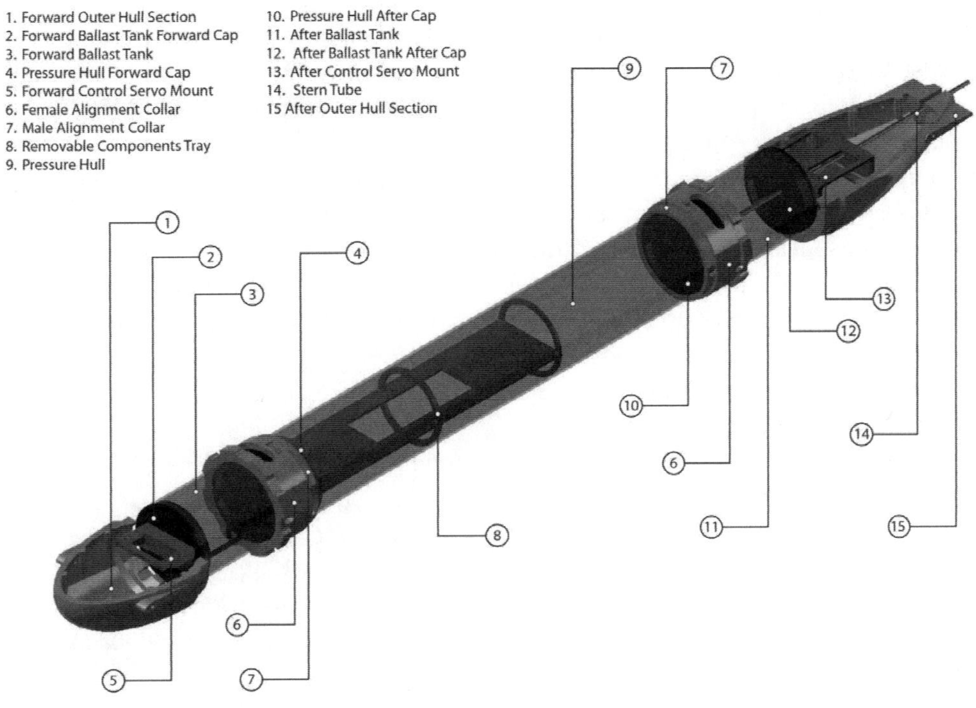

Figure 6 – The proposed redesign for the pressure hull and depth-control system

CONCLUSION

At this stage of writing, the project to develop a reliable pressure-hull and depth-control system that is straightforward to manufacture and simple to maintain is well underway with half of the milestones already completed. The base-model has been completed and tested, with the applicable faults identified, and a design solution conceptualised. However, there is still work to be achieved in the coming months. A detailed design is needed with associated proof of concepts experiments carried out. In addition, reliability modelling; suitable for accurately analysing both the existing system, and the revised prototype, needs to be implemented.

Acknowledging the work is ongoing, as an undergraduate thesis, it is anticipated that complete detailed results will be available when the paper is presented and updates will be available from the authors.

REFERENCES

[1] Baily, N., Nathan, R., Zhang, S., & Khoo, L. (2017). *Aegir Group 2017 Portfolio*. Carlingford High School, NSW, pp. 1-19

[2] REA. (2017). 2017 *Submarine Build Manual - Level 4: SUB Class*. pp. 1-63

[3] Tronskar, D. J. (Unknown). *Root Cause Analysis - An essential element of Asset Integrity Management and Reliability Centered Maintenance Procedures*, pp. 1-69

[4] Todinov, M. (2005). *Reliability and Risk Models - Setting Reliability Requirements* (Vol. 1). West Sussex, England: John Wiley & Sons Ltd. (305), pp. 27-28

[5] Siemens. (2016). "How to conduct a failure modes and effects analysis (FMEA)". In *Siemens White Paper* (pp. 1-8): Siemens PLM Software.

[6] Goldberg, F., Vessely, W. E., F., Roberts, N. H., & Haasal, D. F. (1981). *Fault Tree Handbook*. Washington: Systems and Reliability Research Office of Nuclear Regulatory Research U.S. Nuclear Regulatory Commission, pp. 1-45

[7] Singh, M., M.D. Jaybhaye, Lathkar, G. S., & S.K. Basu. (2012). "Reliability Analysis of Electro- Hydraulic Based Auto-Leveling System for Mobile Platform". *International Journal of Engineering and Innovative Technology (IJEIT), 1*(2), 6.

[8] Kang, H.G. and Polley, J. (2014). "Reliability Engineering Design Project". Australian Maritime College, pp. 1-17

[9] Burcher, R., & Rydill, L. (1994). *Concepts in Submarine Design* (Vol. 2). Cambridge, New York: Press Syndicate of the University of Cambridge, pp. 74-75, 86-97

[10] Jong-Yong Parka, & Kim, N. (2018). "Design of a safety operational envelope protection system for a submarine". *Ocean Engineering* (148), pp. 602-611.

[11] Chen, Y., Wang, G., Xu, G., Zhang, W., & Wang, W. (2013). "Hovering control of submarine based on L1 adaptive theory via ballast tanks". *International Journal of Advanced Robotic Systems (July- August 2017),* pp. 1-10.

[12] Tangirala, S., & Dzielski, J. (2007). "A Variable Buoyancy Control System for a Large AUV". *IEEE JOURNAL OF OCEANIC ENGINEERING, 32* (4), pp 762-781.

[13] Ye, Y. (2014). "Buoyancy Control for Low-speed Under-actuated AUVs". *Applied Mechanics and Materials,* pp. 644-650

Model Submarine Sail Geometry Optimisation – A Numerical and Experimental Investigation

Jeremy Wotherspoon[1], Ahmed Swidan[2], Warren Smith[3], Alaa Osman[4]
1 SBLT, UNSW Canberra at ADFA, Australia j.wotherspoon@student.unsw.edu.au
2 UNSW Canberra at ADFA, Australia a.swidan@unsw.edu.au
3 UNSW Canberra at ADFA, Australia w.smith@adfa.edu.au
4 UNSW Canberra at ADFA, Australia a.osman@adfa.edu.au

Abstract

This paper aims to address improvements to the current model of the Subs in Schools (SiS) Technology Challenge, with a focus on the redesign and improvement of a model submarine's sail. The sail is an important feature with regards to resistance as it contributes disproportionately to total resistance, thus affecting required power and subs' endurance. A series of wind tunnel experiments were conducted at the University of New South Wales Canberra, in order to benchmark a range of model sails. This benchmarking data was utilised to validate RANS (Reynolds-averaged Navier–Stokes equations) Computational Fluid Dynamics (CFD) simulations using ANSYS™. The validated numerical simulations were conducted to investigate the influence of different design variables being; slenderness ratio, bluntness ratio and the maximum frontal cross-section area on sail resistance in order to optimise the geometry within the dimensional constraints provided by SiS technology challenge. The computed results were in good agreement with the measured data and provided insight to the effect of submarine sail geometry design variables on resistance. An educational package on the effect of these design variables on the drag were provided to the SiS students. This would allow students to perform further investigations to enhance the hydrodynamic performance of the SiS model.

This paper reflects a work in progress as part of a 2019 undergraduate honours project conducted by the lead author.

Nomenclature

P	Density (kg/m^3)	L	Total length of model (m)
μ	Kinematic viscosity on model (m^2/s)	R	Total resistance (N)
A_W	Wetted area surface (m^2)	R_F	Frictional resistance (N)
C_D	Total resistance coefficient	R_{VP}	Viscous pressure resistance (N)
C_F	Frictional resistance coefficient	Re	Reynolds number
C_{VP}	Viscous pressure resistance coefficient	V	Speed of model (m/s)

Introduction

Small scale submarine models exist primarily for the purpose of research with the goal to provide advancements to full scale submarines. The common methods of testing the submarine models are as follows: seakeeping experimental tests using towing tank, wind tunnel or numerical simulations. The most noteworthy example being the first substantial submarine model tests, the Gertler streamline bodies experiments from 1950 (Gertler, 1950), which resulted in the development of "tear drop" shaped hulls first introduced on the USS Albacore in 1958. This research greatly improved the understanding of flow around submerged bodies and resistances in submarine design.

However, Re-Engineering Australia (REA) produced a program piloted in 2014, namely: the Subs in Schools (SiS) Technology Challenge that exists to increase interest in STEM fields in general and to introduce students to the engineering challenges related to submarines and underwater vehicles. The current model of SiS has teams assemble a submarine model, with more focus on providing coding, producing engineering drawings and debugging issues that occur throughout the process. The concept within this paper is to address a previously un-addressed field of potential within the SiS technology challenge being resistance. This investigation into resistance will provide the students an understanding of fluid flow, addressing areas for improvement, such that they can re-design parts to improve their model's performance. Overall increasing their engineering knowledge and the experiences gained within the competition.

This research uses a model submarine sail as a case study such to optimise a component of the model within the guidelines of the competition. This is to demonstrate potential areas of improvement and add knowledge to the teams on how to approach hydrodynamic improvements of the model. In this paper, CFD simulations are utilised with the purpose of estimating the hydrodynamic performance characteristics, e.g. drag and lift coefficients, velocity vectors and the pressure contours around the model submarine' sail. The methods reported in this paper are the pre-processing RANS based CFD simulations including detailed grid-independence study to provide confidence in the computed results The CFD simulations were employed to analyse the influence of 3 design parameters being; slenderness ratio, bluntness ratio and the maximum frontal cross-section area between 3 varying geometries, such to find the optimal values of these design variables in order to achieve a minimal drag force for a given sail volume.

This case study and the processes involved were performed to understand what is required to perform hydrodynamic improvements on submarine model components. The process, results, summary and knowledge gained will provide useful and detailed information to the students on factors affecting resistance and hydrodynamic performance which they can use to perform improvements within SiS.

Sail Design

The largest appendage and a significant contributor to overall drag with estimates of up to 30% of total resistance is the sail (Friedman, 1984). The sail has many key features when it comes to submarine operations such as "stowage and support for the masts when raised, a conning position when in harbour for berthing, safe transfer to the open deck at sea without swamping, underwater handling stability, and an ability to operate covertly while submerged at shallow depth" (Joubert, 2004). Therefore, the sail can be seen as a key feature affecting both operational factors and performance. Submarine sail design is a fine balance with regards to height, too short and submarine speed will be restricted when using masts. However, if it is too tall or too thin the snap roll will cause excessive forces on the sail and risk damage to the hull, masts and other sensitive components housed within the sail (Hervey and Till, 1994).

Flow Resistance of the Submarine Sail

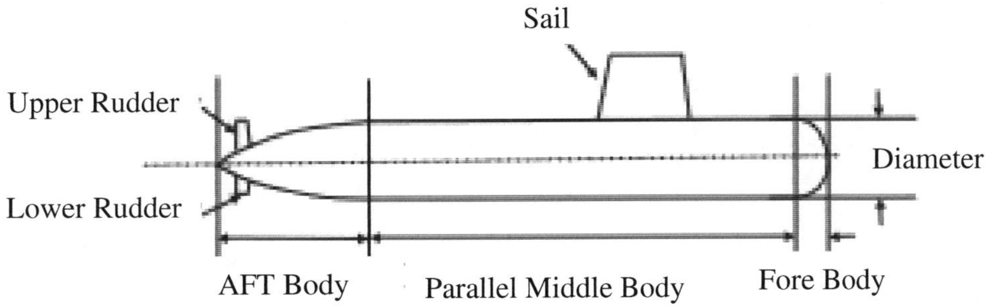

Figure 1. Side view of a generic submarine hull (Renilson, 2015).

When discussing resistance of any component of a deeply submerged submarine it can be thought of as two components; friction resistance and form resistance, wave resistance can be neglected due to the deep submersion (Bertram, 2012). Submarines are designed to operate in submerged conditions as that is their primary area of operation, at depth there is no wave making and also for snorkeling at low speed wave resistance is also negligible.

$$C_D = \frac{R}{0.5\rho A_W V^2}$$ ... Equation 1. Total resistance coefficient (Moonesun, *et al.* 2014).

$$C_F = \frac{R_F}{0.5\rho A_W V^2}$$ ... Equation 2. Frictional resistance coefficient (Moonesun, *et al.* 2014).

$$C_{VP} = \frac{R_{VP}}{0.5\rho A_W V^2}$$ ... Equation 3. Viscous pressure resistance coefficient (Moonesun, *et al.* 2014).

Despite the not ideal nature of a sail on a submarine as can be seen by the non-streamline nature of the sail as seen in Figure 1. However, the large appendage in one form or another is necessary for any modern conventional submarine, as it acts as the bridge conning position and the housing for various periscopes, masts and ducting (Bertram, 2019). Idealised hydrodynamics would suggest that if the sail must exist it should be as thin as possible to reduce drag as described in equation 3. This being the viscous pressure equation for a bare hulled submarine however the theory still applies to just the sail (Moonesun, *et al.* 2014).

Viscous pressure resistance is due to the sail inducing local flow fields with different velocities than the average, this results in an increase in shear stresses and increased energy losses in the boundary layer. This form of resistance is minimised by having a length where the sections vary by small increments along the sail. Also, when considering frictional resistance as described in equation 2, this still is valid when solely considering the sail as this form of resistance is cause by viscous shear of water over the surface (Bertram, 2019). This is fundamentally reduced by causing a reduction in wetted surface, therefore for a given volume a sphere would be the ideal shape to achieve this, however a spherical sail produces its own problems, this being an increase in the form resistance. Therefore, the two forms of resistance for submarines and their optimal solutions contradict one another, as demonstrated in Figure 2.

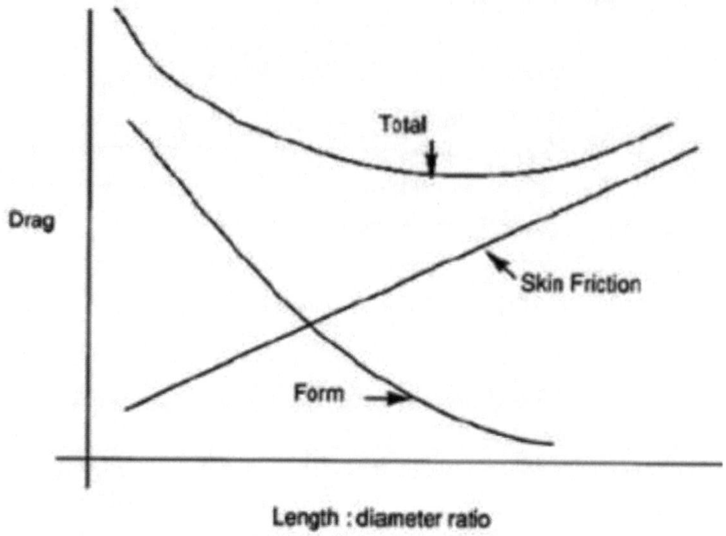

Figure 2. Graph displaying the contradicting nature of form and frictional drag (Moonesun *et al.*, 2013).

Thus, the main areas of focus in the study of sail optimisation is with the interference drag between the hull and the sail, and location of the sail along the length of the hull as these interactions have the largest part to play. However, the flow around the geometry of the sail itself is still of notable effect due to the sail's large contribution to total resistance, a flow simulation of a maneuvering submarine can be seen in Figure 3, demonstrating flow interference caused by the sail. Recommendations for sail design exist, in order to minimise skin friction drag, the sail should be kept as small and smooth as possible (Mises, 2012). Pressure drag is minimised by streamlining of the shape, with maximum thickness being approximately 45% of length and thickness to length ratio should not exceed 12.5%.

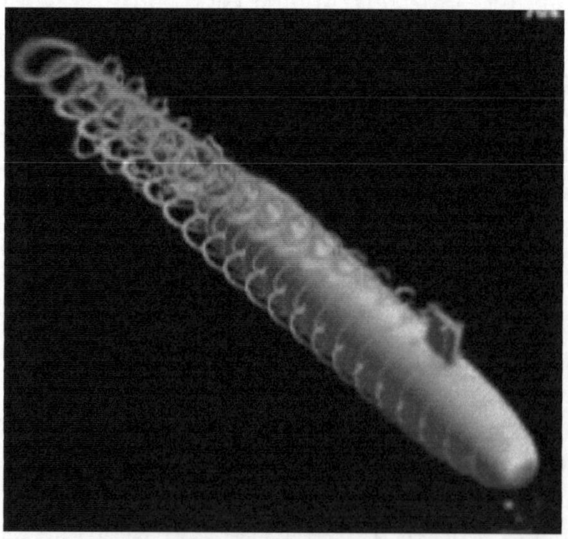

Figure 3. Flow vortices around a maneuvering submarine (Renilson, 2015).

Design of Experiments (DOE): Geometries

For the both CFD and experimental processes various simplified sail geometries were generated within the constraints of the SiS technology challenge, being the sail must be longer than 100mm, must be wider than 30mm and must have a height of at least 50mm (REA, 2019). A simplified sail geometry was utilised such to ignore the transition region addressed in other optimisation studies. These geometries were formed and evaluated using 3 design variables, these being a length to width ratio (slenderness ratio), a width to frontal radius ratio (bluntness ratio) and a height by width (maximum frontal cross-section area). Geometries are detailed in table 1 and 2 and displayed in Appendix A.

Table 1: Primary Sail Geometric Parameters

Sails: Volume 'V'(mm^3)	Length 'L'(mm)	Width 'W'(mm)	Frontal Radius 'FR'(mm)	Height 'H'(mm)
1: 2.1027e+05	200	30	20	60
2: 2.1027e+05	127	50	30	56
3: 2.1027e+05	159	40	70	51

Table 2: Sail Design Variables

Sails: Surface Area 'SA'(mm^2)	Slenderness Ratio 'SR'	Bluntness Ratio 'BR'	Maximum frontal cross-section area 'FCSA'(mm^2)
1: 32118	6.66666	1.5	1800
2: 23656	2.54	1.66666	2800
3: 25408	3.975	0.57143	2040

CFD Simulations

Modelling for CFD Simulations

The modelling of the geometries was done using the inbuilt geometry modeler within ANSYS-FLUENT. The geometry itself was created through the use of a sketch of an ellipse as the front of the sail, a semi-circle as the tail and two connecting lines. This sketch was then extruded to form a geometry. Which was removed from an enclosure being the wind tunnel of overall geometry with a length of 900mm, with a width and height of 460mm.

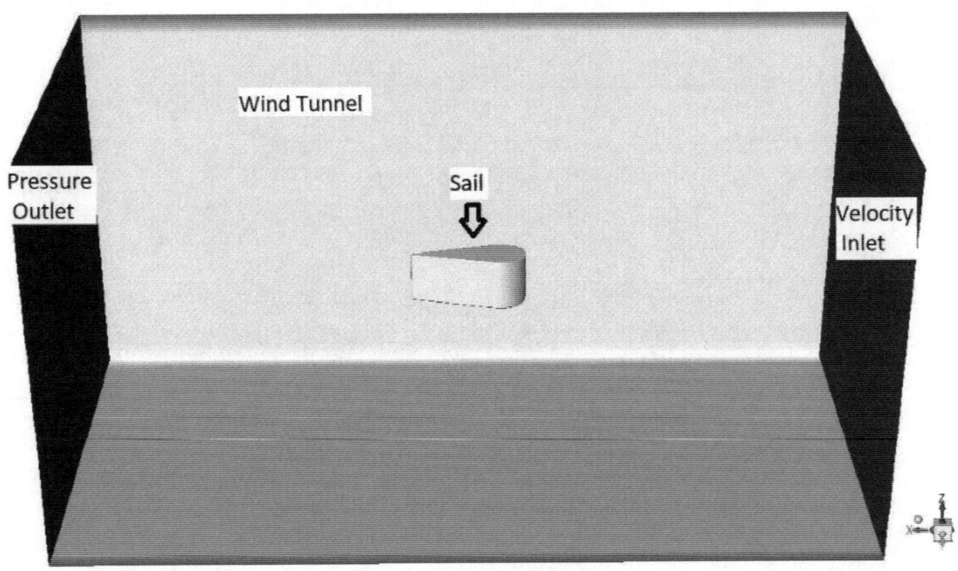

Figure 4: Display of basic setup (Front walls of wind tunnel removed for visualisation)

This removal of the sail was done using a Boolean subtraction operation, such that the sail itself was not analysed, but the flow around the surface was. This subtracted geometry was meshed within the ANSYS-FLUENT in-built meshing tool. The modelling physics were setup in ANSYS-FLUENT setup, the input information is based on an approximate movement velocity of the submarine modelled at three different velocities being, 0.75m/s, 1m/s and 1.25m/s displayed in figure 4. The speed induces a laminar flow around the sail geometry, with Reynolds number calculations detailed in Appendix B.

Independent Mesh Investigation

The independent mesh investigation undertaken involved the comparison of 3 different mesh sizes, meshed under the same conditions in order to verify an appropriate size mesh to use for the simulations. This was done with respect to the variation in drag, the number of elements, therefore computational time. The analysed geometry for this mesh investigation was geometry 2 as it has the smallest surface area and therefore with the primary focus of the mesh being on retaining the geometry and analysis on the geometry itself, a smaller surface area plays the largest role in reducing the number of elements. Proximity and curvature sizing functions were used so the sail curves maintained their curvature and smoothness and the proximity allowed the flow directly around the sail to be more accurately captured. The meshing method applied to the geometry was a multi zone mesh due to the massive variation in the geometry surface and the far field. A prism mapped mesh type with a pave surface mesh method was used for the semi-structured mesh on the sail such to retain the sail edges and curvature at the front and rear of the sail. A tetra/pyramid free mesh type was used so that the fine prism inflation layer could be applied. The inflation layer consists of 10 layers with a growth rate of 1.1 and a first layer thickness of 1E-05. The various number of elements for the different mesh qualities are detailed in table 3, with their effect on the total drag for the same continuity.

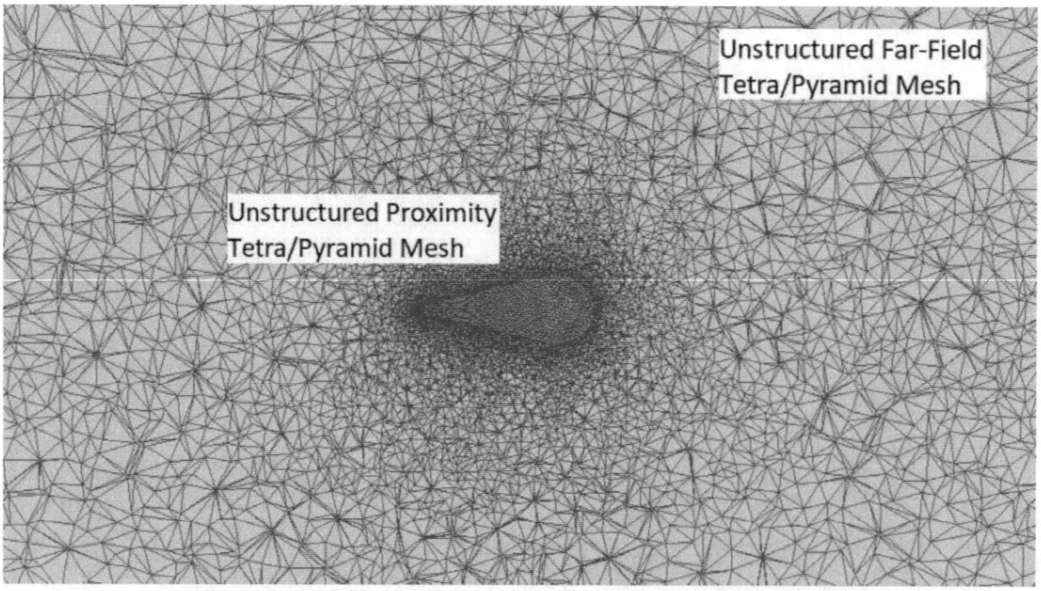

Figure 5: (a) Top view of whole mesh (Medium Mesh)

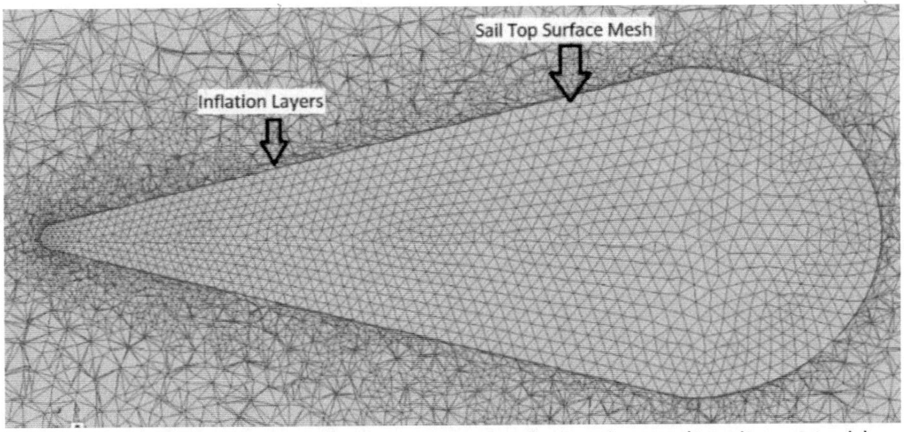

Figure 5: (b) Top view of sail mesh and inflation layers (Medium Mesh)

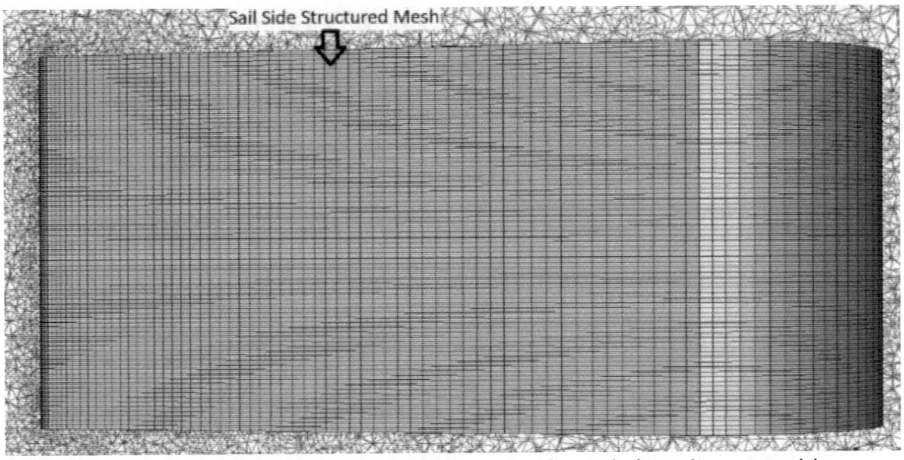

Figure 5: (c) Side view of sail structured mesh (Medium Mesh)

Table 3: Independent Mesh Analysis Results

Fineness of Mesh 2	Elements	Nodes	Comparison Continuity	Total Drag (N)	Variation (%)
Course	617825	206876	4.93E-04	0.70103	
Medium	797387	267492	4.89E-04	0.45686	-34.83
Fine	1405492	450138	4.95E-04	0.44586	-2.41

The results of the independent mesh investigation as detailed in table 2, displayed the variation from the course to the medium mesh provides a variation of the drag value of -34.83% for an element increase of 179562. However, the increase from medium to fine provides a drag variation of -2.41% for an element increase of 608105. For maintaining the same meshing format and respective mesh quality the medium mesh was deemed feasible with respect to the variation in drag for the number of elements present.

Procedure for CFD Analysis

The CFD analysis module provided within ANSYS-FLUENT is a simple set up solver, however in order to achieve accurate and stable solutions is more difficult. This module provides the ability to control relaxation factors to aid with convergence rates and stability of solutions. The simulations were performed using laminar flow and energy equations, with the simple scheme method of pressure-velocity coupling.

The gradient method used was least squares cell based, pressure equation was second order, momentum equation first order upwind and energy equation second order upwind. With the under-relaxation factor controls primarily set to default conditions.

The required convergence was 1e-04, being loosely converged for the simulations as this allows the difference in results to be clearly illustrated between the geometries and can be easily achieved for all geometries either using default controls or through minor manipulations of relaxations factors. To achieve this level of convergence it typically required approximately 220 iterations being for geometry 2 and 3 whereas geometry 1 required manipulation of relaxation factors. All simulations were run for at least 500 iterations passed required convergence to ensure the stability of solution.

CFD Results

The numerical values from the CFD solution are listed in tables 4, 5 and 6. Table 4 details the pressure including pressure drag, table 5 details the velocity including x-directional velocity and table 6 provides an overall drag summary.

Table 4: (a) Pressure Results for Sail 1

Velocity (m/s)	Minimum Pressure (Pa)	Maximum Pressure (Pa)	Pressure Difference (Pa)	Pressure Drag (N)	Pressure Drag Variation (%)
0.75	-929.46	291.26	1220.7	0.10200	
1	-1655.0	516.80	2171.8	0.18086	77.301
1.25	-2592.0	806.53	3398.5	0.28267	56.291

Table 4: (b) Pressure Results for Sail 2

Velocity (m/s)	Minimum Pressure (Pa)	Maximum Pressure (Pa)	Pressure Difference (Pa)	Pressure Drag (N)	Pressure Drag Variation (%)
0.75	-825.45	286.982	1112.4	0.23088	
1	-1463.6	509.23	1972.9	0.40920	77.230
1.25	-2288.1	794.77	3082.9	0.63860	56.061

Table 4: (c) Pressure Results for Sail 3

Velocity (m/s)	Minimum Pressure (Pa)	Maximum Pressure (Pa)	Pressure Difference (Pa)	Pressure Drag (N)	Pressure Drag Variation (%)
0.75	-862.13	289.54	1151.6	0.094154	
1	-1537.4	513.57	2051.0	0.16665	77.001
1.25	-2409.3	801.28	3210.6	0.26000	56.015

Figure 6: (a) Contours of static pressure geometry 1 at 1.25m/s, side view

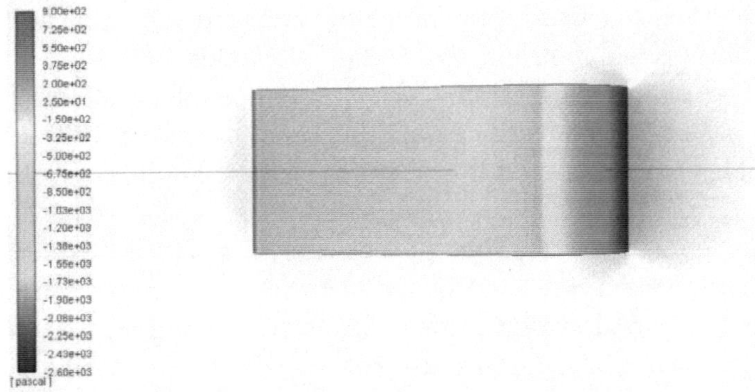

Figure 6: (b) Contours of static pressure geometry 2 at 1.25m/s, side view

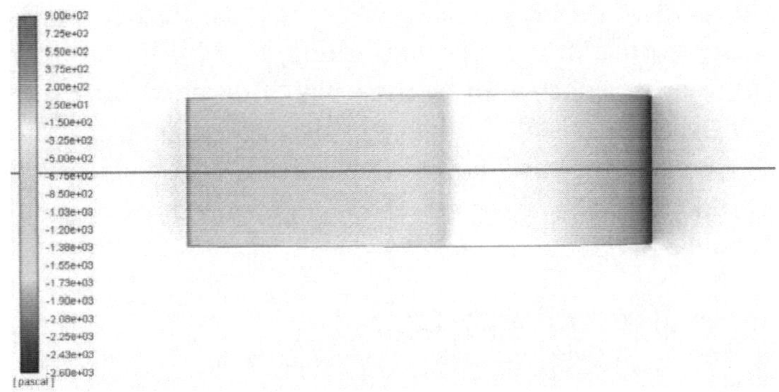

Figure 6: (c) Contours of static pressure geometry 3 at 1.25m/s, side view

The contours of static pressure around the 3 geometries, are provided for the inlet velocity of 1.25m/s, in figure 6. The colour maps of the 3 geometries in figure 6 are all scaled from -2600Pa (dark blue) to 900Pa (Red) for the 1.25m/s, this allows for effective comparison between the different geometries.

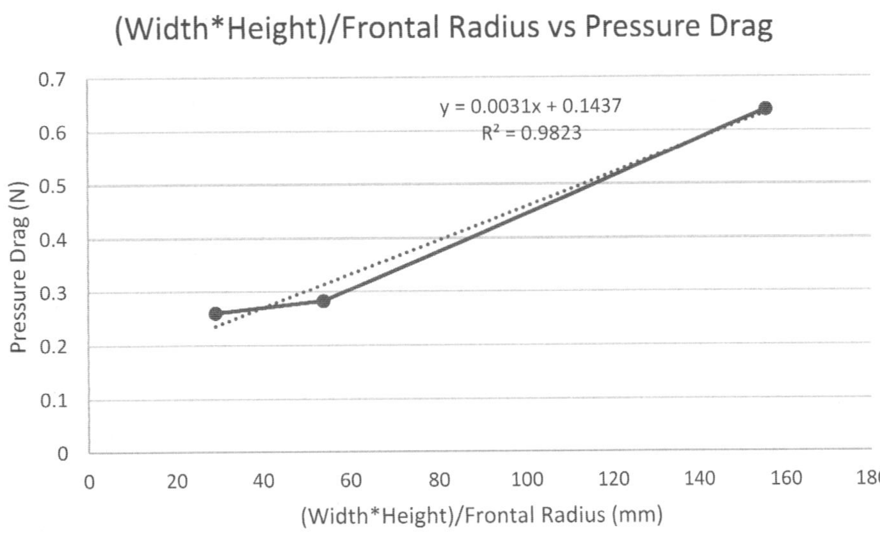

Figure 7: Graph of primary features of the geometries and the influence on the pressure drag at 1.25m/s.

When comparing the maximum and minimum pressures as well as the pressure drag between the various geometries, the design variables and their influence must be considered as well as the response to velocity variations. Figure 6 displays that the static pressure is mainly influenced by the frontal region of the sail. This is dictated by primarily by design variables of bluntness ratio and the maximum frontal cross section.

All other velocities and corresponding pressure drag values, exhibit similar R^2 when approximated with a linear fit. This demonstrates the direct relationship between a high width and height and an increased pressure drag value, however this effect can be greatly reduced through having a larger frontal radius, as despite having the second largest maximum frontal cross-sectional area it has the lowest pressure drag value. This demonstrated in figure 6 (c), as there is a sharp pressure increase at the very front of the sail, however the variation in pressure is spread across the length of the frontal radius thus reducing the pressure drag. In contrast figure 6 (a) and (b) the pressure changes rapidly over a short distant as demonstrated by their high bluntness ratio, causing much greater pressure changes especially visible in figure (b) with the highest bluntness ratio. Also, as the inlet velocity increases the variation in pressure drag reduced significantly, this suggests that if a faster velocity could be achieved the magnitude of the drag would increase but the relative effect would decrease.

Table 5: (a) Summary of velocity results for sail 1

Velocity (m/s)	Maximum Velocity (m/s)	Maximum x- axis Velocity (m/s)	Minimum x- axis Velocity (m/s)	Viscous Drag (N)	Viscous Drag Variation (%)
0.75	1.0653	0.99757	-0.82608	0.062760	
1	1.4738	1.3345	-1.2336	0.090060	33.775
1.25	1.8907	1.6713	-1.6799	0.11853	25.237

Table 5: (b) Summary of velocity results for sail 2

Velocity (m/s)	Maximum Velocity (m/s)	Maximum x- axis Velocity (m/s)	Minimum x- axis Velocity (m/s)	Viscous Drag (N)	Viscous Drag Variation (%)
0.75	1.1668	1.1405	-0.58954	0.033373	
1	1.6326	1.5332	-0.83681	0.047497	34.425
1.25	2.1548	1.9255	-1.1137	0.062566	25.58

Table 5: (c) Summary of velocity results for sail 3

Velocity (m/s)	Maximum Velocity (m/s)	Maximum x-axis Velocity (m/s)	Minimum x-axis Velocity (m/s)	Viscous Drag (N)	Viscous Drag Variation (%)
0.75	1.4853	0.94838	-1.4330	0.050828	
1	1.6659	1.2875	-1.5572	0.072715	35.76
1.25	2.2330	1.6262	-1.7257	0.096143	26.31

Table 6: Summary of Drag forces of three sail configurations, namely 1, 2 and 3, at 1.25m/s

Sail	Viscous Drag (N)	Pressure Drag (N)	Total Drag (N)
1	0.11853	0.28267	0.4012
2	0.062566	0.63860	0.701166
3	0.096143	0.26000	0.356143

Figure 8: (a) Velocity path lines around geometry 1 at 1.25m/s, top view

Figure 8: (b) Velocity path lines around geometry 2 at 1.25m/s, top view

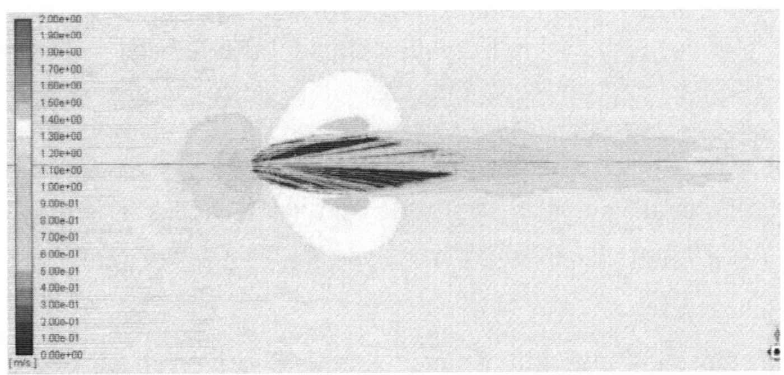

Figure 8: (c) Velocity path lines around geometry 3 at 1.25m/s, top view

The velocity path lines around the 3 geometries, are provided for the inlet velocity of 1.25m/s, in figure 8. The colour maps of the 3 geometries in figure 8 are all scaled from -0m/s (dark blue) to 2m/s (Red), this allows for effective comparison between the different geometries. Figure 8 a, b and c illustrate the region of maximum velocity for all sails is slightly away from the surface itself and just downwind of the widest point of the sail. This is what can be expected with reference to figure 6 also as this is the region of lowest pressure on the sail, allowing the flow to move more easily from high to low pressure.

A key region within figure 8 is downwind of the tail, with the darker inconsistent region due to the effects of wake. Wake plays a large effect in the calculation and the effect of drag on an object. It can be seen from the size of the wake fields, that they interact as a type of pressure, which is also represented in table 5 along with the x-axis velocities, the negative x-axis velocity indicates there are significant wake vortices. As with any form of pressure drag as previously discussed it can be reduced by minimising the changes in geometry over a given length, being a traditional aerodynamic shape, with a long frontal radius, small width and height and a length such that it has the required volume for its purpose. Viscous drag, as can be seen in table 5 and 6 is only a small contributor to total drag however on this scale and at these velocities it is still significant and can easily become more so as surface roughness and external features come into play. However, viscous drag is primarily a feature of surface area as described in equation 2 previously.

The highest performing sail as described within table 6 was sail 3, as it had the lowest drag for the given volume. Sail 3 did not have the smallest width, length or surface area; however, it did have the smallest bluntness ratio due to its large frontal radius. As has been examined through the static pressure contours and velocity path lines the effect of the large frontal radius in slowly dissipating the pressure is of significance to a sails performance.

Conclusion

In this paper, a series of submarine sails have been numerically modelled, and analysed using ANSYS-FLUENT. The drag characteristics of these isolated sail geometries were compared with respect to the individual static pressure contours and velocity path lines of the geometries. The pressure drag, viscous drag and total drag measurements were taken at 3 different velocity points on 3 geometries with variations in chosen design variables such to achieve a minimal drag for a constant volume. The findings detail the significance of a frontal radius in minimising pressure drag. Where possible width and height should be reduced such to reduce maximum frontal cross-section and bluntness ratio and thus increasing the slenderness ratio, however these have constraints both within the SiS technology challenge and in real life operational considerations. Surface area also contributes with respect to the viscous drag, although only a small component of the total drag it is still of concern in terms of optimisation. Although, these series of simulations and corresponding results are currently without validation, however they stand in agreement with the theory expected and accurately represent the physics of the sail.

Future Work

This is an ongoing project and still requires further work on the topics of validation which will be achieved through conducting series of experiments utilizing a sub-sonic wind tunnel facility using three dimensional models of the presented geometries. The knowledge gained from the current simulations will go into further work to extend to an optimisation of this geometry within the predefined constraints. This extension will in further detail explain the effect of the design variables on the performance of the geometry and provide a more detailed evaluation on the improvements possible utilising CFD simulations and whether this knowledge can be effectively represented in an educational package for use within the SiS technology challenge.

References

- Bertram, V. (2019). Submarine Hull Design. [online] Ntnu.edu. <https://www.ntnu.edu/documents/20587845/1266707380/2012Chennai_SubmarineDesign.pdf/9bb180be-a08d-48ea-af4e-7e5e19210d3b>
- Bertram, V. (2012). Practical ship hydrodynamics, second edition. 2nd ed. Amsterdam: Elsevier.
- Friedman, N. (1984). Submarine design and development. Annapolis, Maryland: Naval Institute Press.
- Gertler, M. (1950). Resistance Experiments on a Systematic Series of Streamlined Bodies of Revolution-For Application to the Design of High-Speed Submarines. <Apps.dtic.mil. https://apps.dtic.mil/dtic/tr/fulltext/u2/a800144.pdf>
- Hervey, J. and Till, G. (1994). Submarines. London: Brassey.
- Joubert, P. (2004). Some Aspects of Submarine Design Part 1. Hydrodynamics. Archive.org. <https://archive.org/stream/DTIC_ADA428039/DTIC_ADA428039_djvu.txt>
- Mises, R. (2012). Theory of Flight. Dover Publications. <https://www.britannica.com/technology/submarine-naval-vessel>
- Moonesun, M., Mikhailovich, K., Tahvildarzade, D. and Javadi, M. (2014). Practical scaling method for underwater hydrodynamic model test of submarine. Journal of the Korean Society of Marine Engineering, 38(10), pp.1217-1224. <http://jkosme.or.kr/_PR/view/?aidx=3471&bidx=285>
- Moonesun, Mohammad & Javadi, Mehran & Charmdooz, Pejman & , Karol & Mikhailovich, Uri. (2013). Evaluation of submarine model test in towing tank and comparison with CFD and experimental formulas for fully submerged resistance. Indian Journal of Geo-Marine Sciences. 42. 1049-1056.
- REA (Re-Engineering Australia). (2019). 2019 Level 4 – Submarine Build Manual. < https://3ixa2917tenz3fyiar4esjdf-wpengine.netdna-ssl.com/wp-content/uploads/2019-SiS-Build-Manual.pdf>
- Renilson, M. (2015). Submarine Hydrodynamics. Launceston: Springer.

Appendix A: Details of the simulated three geometries

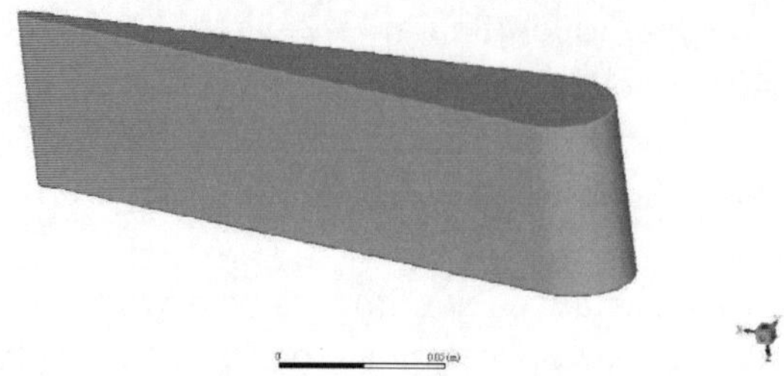

Figure 9: (a) Isometric view of geometry 1

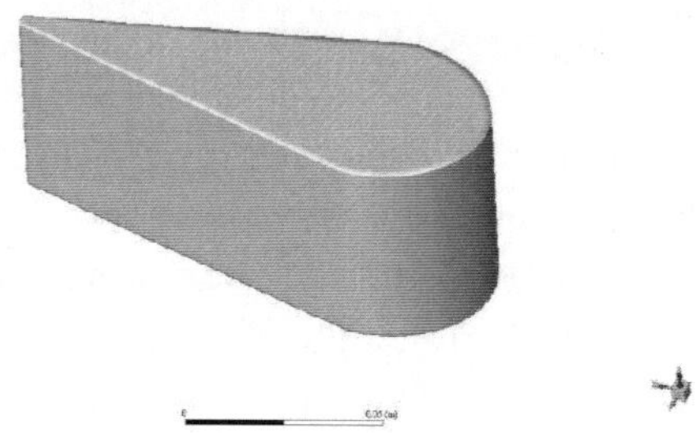

Figure 9: (b) Isometric view of geometry 2

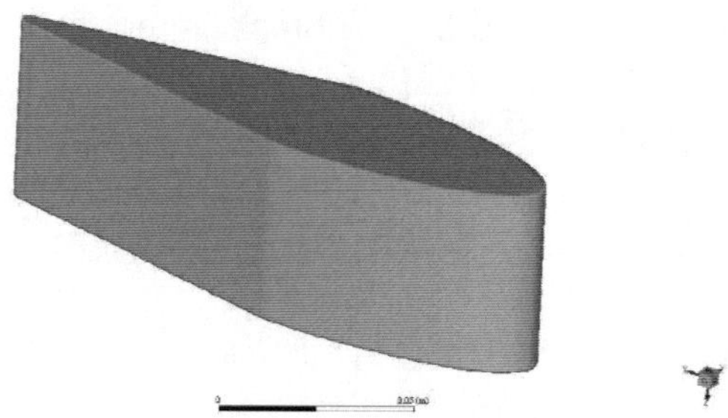

Figure 9: (c) Isometric view of geometry 3

Appendix B: Reynold's Number Calculations

$$Re = \rho VL/\mu$$

Equation 4. Reynold's number (Bertram, 2019).

Sail	Velocity (m/s)	Reynold's Number ()	Corresponding Air Speed (m/s)
1	0.75	149282.1535	10.90308104
1	1	199042.8714	14.53744139
1	1.25	248803.5892	18.17180174
2	0.75	94794.1675	10.90308104
2	1	126392.2233	14.53744139
2	1.25	157990.2792	18.17180174
3	0.75	118679.3121	10.90308104
3	1	158239.0828	14.53744139
3	1.25	197798.8534	18.17180174

Taking Sub in School to the next level – Acoustic Testing

Sean Buckland[1], Ahmed Swidan[2], Warren Smith[3]
UNSW Canberra at ADFA, Australia, s.buckland@student.unsw.edu.au
2 UNSW Canberra at ADFA, Australia a.swidan@unsw.edu.au
3 UNSW Canberra at ADFA, Australia w.smith@adfa.edu.au

Re-Engineering Australia (REA), Defence Science & Technology (DST), Australian Submarine Corporation (ASC) and industry partners developed the Subs in Schools (SiS) program to engage Australian students in STEM subjects. The program has been running in secondary schools for over five years and requires students to design, manufacture and test submarine hull model forms. The program is aligned with generating interest in the growing Australian shipbuilding industry and preparing young Australians for a career in engineering to contribute to the Royal Australian Navy (RAN) and innovative industries alike. This work aims to expand the current testing criteria through implementing an acoustic testing regime in the SiS competition. The competition seeks to simulate applicable and 'real life' tests, whilst exploring the theory of sound within the context of SiS, thus the inclusion of acoustic testing in the sea trials aspect of SiS would align with the reality of the operational requirements of submarines. The acoustic testing included underwater measurements with hydrophones and a stationary model submarine, tests conducted in an isolated, insulated air medium (acoustic testing box) and resonance analysis and mitigation of the submarine as a structure. For testing and analysis purposes, comparison of the underwater and air data was undertaken to analysis the effect of mass loading and damping of the propeller and structure as a whole (stationary tests).

Outcomes of this research are expected to be, the recorded acoustic data on the model submarine for further benchmarking, extrapolation of underwater acoustic data to in-air data, an educational module on underwater noise and development of an acoustic testing system for the SiS competition.

INTRODUCTION

The question, 'Can additional trials be included in the Subs in Schools competition to test realism?', has inspired the need for this research. Research and initial testing were conducted to establish the feasibility of underwater acoustic testing. Due to the small scale of the submarine and surface models measured the source noise was not distinguished against the ambient underwater noise level, and thus raw data could not be processed in a meaningful way. Testing was then moved to the air medium, where the effect of damping from mass loading on the propeller and structural vibrations was expected to increase to the source noise, in conjunction with the propagation characteristics of noise in air when compared to water.

This paper is limited to the application of acoustic testing underwater and in air, at a model scale, and primarily within the SiS program. Though it does seek to imitate real engineering design tests. There has been considerable research in the field of underwater acoustics, for communication and military purposes, but not so of acoustic testing of model submarines at this scale, in both the underwater domain and in air, hence this paper explores the phenomena of the existing literature at model scale and its applicability to SiS and actual scale testing.

This paper seeks to determine if stealth can be measured in the Subs in Schools competition. It seeks to do this through conducting acoustic tests in both air and water domains. The output noise of a machine is a key indicator of the quality of build and efficiency of the system. Thus, through acoustic testing, inefficiencies can be found and mitigated. Through testing in two distinct mediums, the difference and similarities of the results in each medium will be analysed and comparisons drawn where applicable.

METHODOLOGY

The primary method of transmitting and receiving underwater signal are piezo-electric transducers. Whilst processes such as magnetostriction, hydraulics (mechanics) and electrodynamics, have been used in specific applications successfully, this research will employ piezoelectricity in a passive sonar function to measure signals.

Optics in underwater testing have been used successfully, mainly testing for the presence for turbulence, through the phenomena of refraction, but this approach is too limited for the purpose of this research. In the underwater testing, a Bruel & Kjaer hydrophone was used as the probe in conjunction with PicoScope PC oscilloscope and software. Test were conducted in a stationary tank, with the model submarine prevented from moving forward by foam stops.

In order to test for stealth, there must be a consistent baseline, which different models can be measured relative to, in order to be compared to each other. A portable acoustic testing box was manufactured for this purpose. The tests conducted in air were done with a Bruel & Kjaer Microphone used with supporting apparatus including an internal pre-amplifier and input/output measuring amplifier.

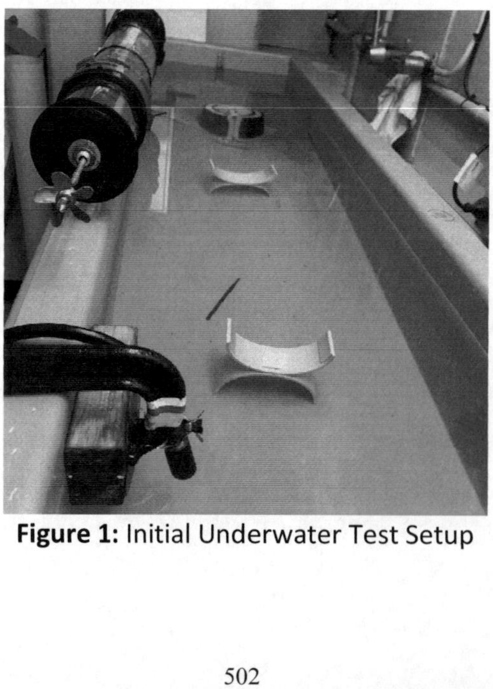

Figure 1: Initial Underwater Test Setup

EXPERIMENTAL SETUP

Measuring Acoustics

The basic premise of measuring sound waves in both mediums discussed thus far, is similar. Using a hydrophone/microphone, an transducer employed in a passive mode, to detect fluctuations in pressure, through the piezo-electric effect, thus giving a voltage signal through an oscilloscope, which can be read through appropriate software on a PC, and interpreted by applying known scaling factors of the respective probe.

Water as a propagation medium has many distinct differences from air as a propagation medium, some beneficial and others detrimental to wave propagation. Electromagnetic radiation suffers little attenuation in the air and can travel great distances, whilst sound lacks penetration and suffers high levels of attenuation. Underwater the roles are reversed. Acoustic waves travel at 1482 m/s in water at 20°C (4.32 times the speed of sound in air at 20°C), thus they can travel large distances with less attenuation. The dB scale will be used to measure pressure level of the sound wave. The log scale is used because such a vast range in sound intensities are encountered, in air and water.

$$SPL = 20 \times log_{10}\left(\frac{P}{P_o}\right)$$

Where SPL = level of sound pressure
P = measured sound pressure
Po = reference sound pressure (taken to be 1 μPa underwater and 20 μPa in air)

The differences in reference pressure and impedance of the medium and the defining factors in analysing the differences in the measured signals, but for the purpose of this report an arbitrary power level will be used to easily compare signals.

$$Voltage\ Signal\ in\ dB\ scale = V_{dB} = 10 \times log_{10}(V)$$

Impedance describes a measure of opposition to alternating current (Butler & Sherman, 2016). The impedance of water is 3500 times greater than air thus, a sound in water with a given pressure, is 3500 less intense in air (Lurton, 2002). Reciprocally, if a sound in water and another in air have the same intensity, then the pressure is much smaller in air.

$$c = \sqrt{\frac{E}{\rho}} = \sqrt{\frac{1}{\chi\rho}}$$

Where

$c = bulk\ speed\ of\ sound\ in\ medium\ (\frac{m}{s})$

$E = Elastic\ Modulus\ (Pa)$

$\rho = density\ of\ medium\ \left(\frac{kg}{m^3}\right)$

$\chi = compressibility\ coefficient\ (\frac{m^2}{N})$

Through use of the wave equation,

$$\lambda \times f = c$$

Where
$\lambda = wavelength\ (m)$
$f = frequency\ (Hz)$
$c = bulk\ speed\ of\ sound\ in\ medium\ \left(\frac{m}{s}\right)$

Comparisons can be drawn between tests underwater and in air, within the frequency domain.

Instrumentation

The below tables detail the equipment used in testing and some key characteristics of the respective equipment.

Table 1: Amplifier 2610 Specifications

Model	2610
Gain	0.003 - 50
Frequency Range	2 Hz – 200 kHz
FSD	10 µV – 30 V
Mass	5.2 kg

Table 2: Microphone Specifications

Model	4147 (4193 new)
Voltage sensitivity	12.5 mV/Pa
Frequency Range	70 mHz to 20 kHz
Polarization	200 V
Preamplifier	Type 2639

Table 3: Oscilloscope Specifications

Model	2205A
Bandwidth	25 MHz
Memory	16 kS
Sample Rate	200 MS/s
Input Sensitivity	4 mV/div to 4 V/div
Voltage Resolution	12 bit

Table 4: Transducer Specifications

Model	8104
Voltage sensitivity	56 uV/Pa ± 15 uV
Frequency Range	0.1 Hz to 10 kHz
Length	120 mm
Mass (inc. cable)	1.6 kg

Figure 2: Measuring Amplifier used in Air Tests

TEST CONDITIONS

Although both the microphone and hydrophone have designated sensitivities, these are more nominal values than absolute values. The microphone was able to be calibrated using a designated calibrator as a known sound source, but the hydrophone was unable to be calibrated, thus the nominal sensitivity was used.

Fundamental Frequency (RPM)

For comparisons to be drawn between the two mediums, the source of sound must be operating under the same parameters in both mediums. Rudimentary methods can estimate the RPM using the max RPM of the brushless DC motor and the power setting of the controller, but to accurately determine the RPM of the shaft, hence fundamental frequency, a digital tachometer was used in later tests.

Underwater

The underwater testing was conducted in a $1520\ mm \times 350\ mm$ test tank with water at a depth of 180mm. The model submarine running at minimum RPM (approximately 15% of full power at 3000 RPM). The model was prevented from moving forward by foam stops held down by a 10kg weight. The probe was then placed 100mm adjacent to the propeller at the same depth level, xx mm underwater. The probe support was clamped to the side of the test tank.

Air

An acoustic testing box was purpose designed for gathering air data for the model submarine operating under the same conditions as the water tests. The dimensions of the acoustic testing box are 1200 mm x 400 mm x 300 mm (less the sound absorbent material thickness for internal space). Ambient conditions in the anechoic chamber were compared to the acoustic testing box to validate the use of the box as an isolated and insulated testing medium and illustrate its limitations.

Three calibration tests using a Bruel & Kjaer ½ inch microphone calibrator yielded the following average microphone sensitivity, known sound source of 94dB read 241.2 mV on PicoScope with input gain set to 1 and output gain set to 0.3. Thus 94 dB correlates to 72.40 mV, therefore for testing purposes a microphone sensitivity of 1298.34 dB/V will be used for air measurements.

Processing

Acoustic analysis was processed using PC Picoscope 6 as the interface. MatLab R2018a was then employed to further analyse the raw data through domain transformations and relevant scaling.

Gain

The gain of the pre-amplifiers using in underwater testing were both set to 20 dB, which correlates to a factor of 10 in voltage gain, through use of the logarithmic relationship between power and dB. Therefore, the output voltage of the underwater signal was divided by 100 ($\frac{1}{10\times10}$) to determine the actual voltage signal. Due to the differing signal intensities measured when testing different propellers, a non-constant voltage multiplier had to be used to obtain the most suitable signal. An input gain of 1 was used across all air tests and the output voltage multiplier was 0.3 for the three bladed propeller and 0.1 for the four and five bladed propellers.

Results and Discussion

The Influence of the Testing Environment and Propeller Configuration

For the underwater tests above, the noisy signal depicts no significant change in spectral tones with a change in number of propeller blades. The tones seen are relatively consistent among the three different tests and would seem independent of the number of blades. This suggests the fundamental frequency of the propeller, shaft and model is driven by something other the Blade Passing Frequency (BPF).

Varied Propeller Data Underwater

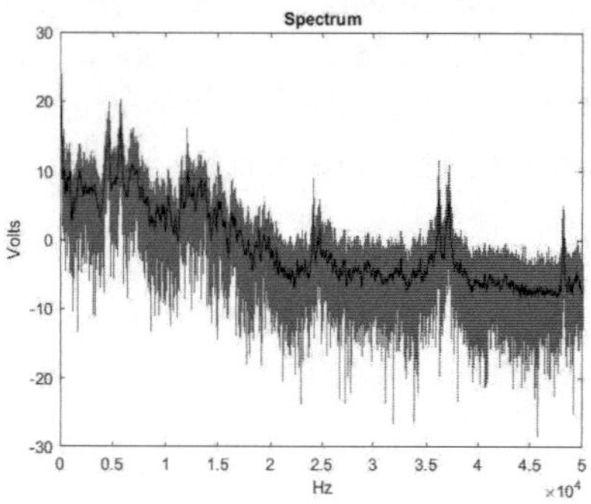

Figure 3:Three Blade Propeller

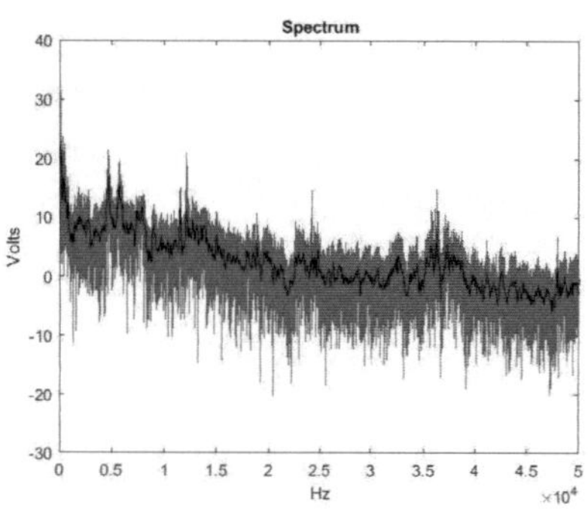

Figure 5:Four Blade Propeller

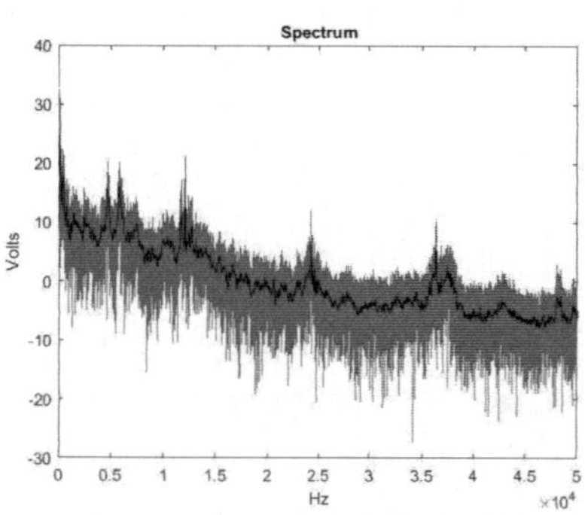

Figure 7: Five Blade Propeller

Varied Propeller Data in Air

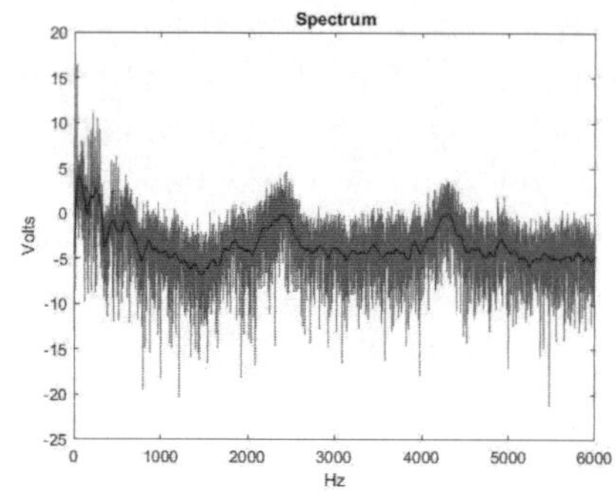

Figure 4: Three Blade Propeller

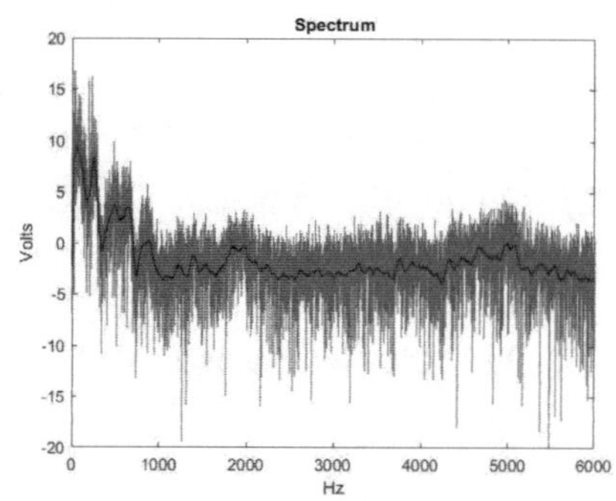

Figure 6: Four Blade Propeller

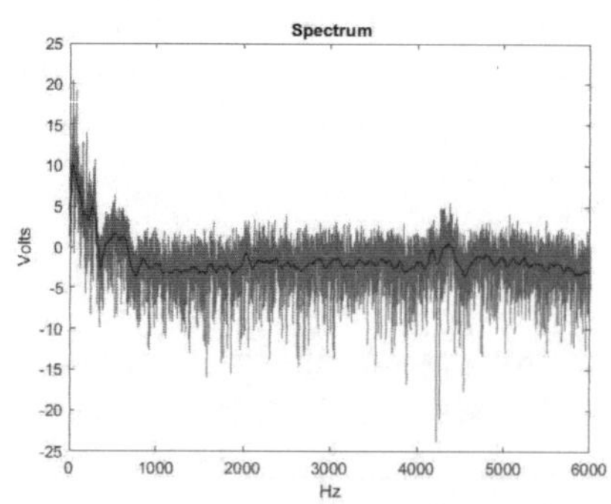

Figure 8: Five Blade Propeller

The influence of Propeller velocity

The below data was measured in the acoustic testing box with the five bladed propeller attached and shaft RPM was recorded at the time of testing with use the digital tachometer. The lack of variance in tones between 600 RPM and 2000 RPM suggests that the shaft speed is not the cause of the fundamental frequency and again the BPF is not the cause either. The spectrum of the 3500 RPM tests shows higher frequency tones, likely caused by structural vibrations of the whole model.

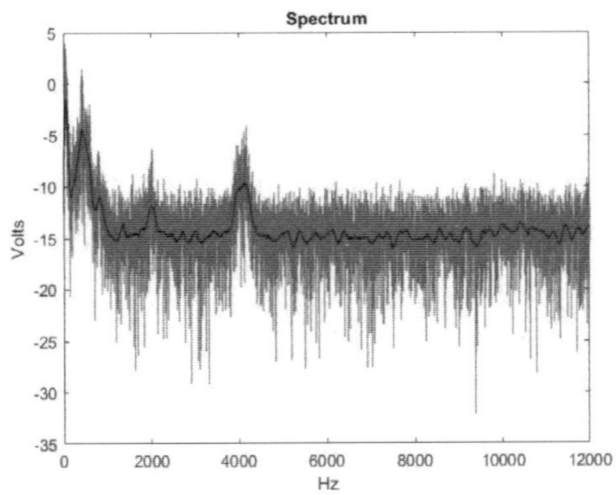

Figure 9: 600 RPM

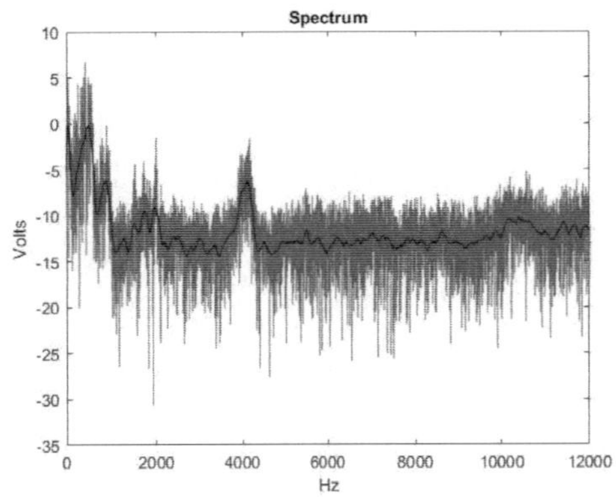

Figure 10: 2000 RPM

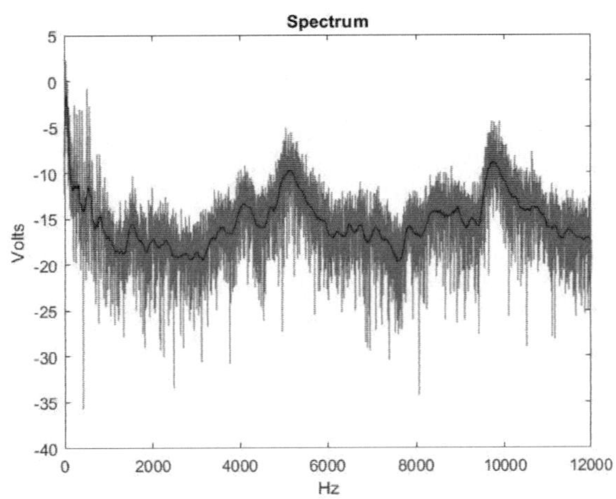

Figure 11: 3500 RPM

Discussion

The power axis in the above spectrums are given as $10 \times log_{10}(V)$ on the Y axis so the volts can only be interpreted qualitatively as of yet, as the dB scale has not been applied with respect to a reference pressure. The signals are relatively noisy, but this is a side effect of having to use pre-amplifiers. This will be mitigated in future testing and processing through use of appropriate filtering.

Following on from the above discussion of the spectrums, data thus far has not led to conclusive evidence of the source the fundamental frequency. The spectrums present a relatively noisy signal, but demonstrate the repeatability of the data.

Repeatability of data is crucial to obtaining unique signatures of each hull/propeller. Collection of unique data for individual models will provide confidence in the certainty of the measured data.

Future Works

Further testing will be conducted to reach conclusive evidence of the cause of the fundamental frequency. Data comparisons such as Figure 13 shown below, will be used to process and analyse future tests.

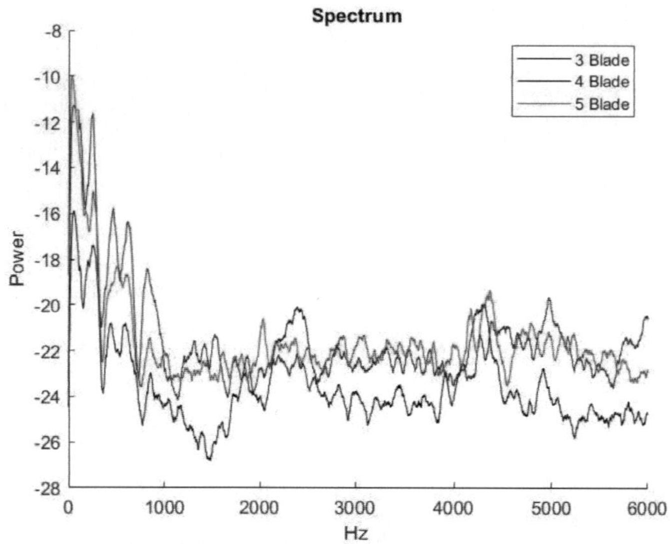

Figure 12: An overlay of the filtered spectrum of the varied propellers

References

Acoustics and Vibrations Unit UNSW ADFA, 2006. *Short Course on Underwater Acoustics,* Canberra: Department of Defence.

Anon., n.d. *ONDA.* [Online] Available at: http://www.ondacorp.com/tecref_tutorialhydrophone_theory.shtml [Accessed 17 05 2019].

Bruel & Kjaer, 1986. *Sound Intensity.* Denmark: K Larsen & Sons.

Butler, J. & Sherman, C., 2016. *Transducers and Arrays for Underwater Sound.* 2nd ed. USA: Springer.

Coates, R., 1990. *Underwater Acoustic Systems.* London: Macmillan .

Department of Planning, Transport and Infrastructure, 2012. *Underwater Piling Noise Guidelines.* [Online] Available at: https://www.dpti.sa.gov.au/__data/assets/pdf_file/0004/88591/DOCS_AND_FILES-7139711-v2-Environment_-_Noise_-_DPTI_Final_word_editing_version_Underwater_Piling_Noise_Guide.pdf [Accessed May 2019].

Dylejko, P., 2007. *Optimum Resonance Changer for Submerged Vessel Signature Reduction,* Sydney: UNSW.

Etter, P., 2018. *Underwater Acoustic Modelling and Simulation.* 5th ed. Boca Raton: CRC Press.

Ji-Sung, J., Hyung-Taek, K. & Won-Ho, J., 2014. *Numerical Study on Non-Cavitating Noise of Marine Propeller.* Korea, inter.noise.

Lecoffres, Y., 1999. *Cavitation: Bubble Trackers.* Rotterdam: Balkema.

Listewnik, K., 2013. Sound Silencing Problem of Underwater Vehicles. *Solid State Phenomena,* Volume 196, pp. 212-219.

Lurton, X., 2002. *An Introduction to Underwater Acoustics: Principles and Applications.* 2nd ed. New York: Springer.

REA, 2019. *Re-Engineering Australia.* [Online]
Available at: https://rea.org.au/
[Accessed 04 March 2019].

Ross, D., 1976. *Mechanics of Underwater Noise.* New York: Pergamon Press.

Seol, H., Jung, B., Suh, J.-C. & Lee, S., 2002. PREDICTION OF NON-CAVITATING UNDERWATER PROPELLER NOISE. *Journal of Sound and Vibration.*

Smith, S., 1999. *The Scientist and Engineer's Guide to Digital Signal Processing.* 2nd ed. San Diego: California Technical Publishing.

Task Force 72, 2019. *Task Force 72: Scale Model Ship Association Inc.* [Online]
Available at: http://taskforce72.org/forum/index.php
[Accessed 27 March 2019].

UNSW School of Physics, n.d. *Acoustic impedance, intensity and power.* [Online]
Available at: http://www.animations.physics.unsw.edu.au/jw/sound-impedance-intensity.htm
[Accessed 05 09 2019].

Wang, Y., Ding, K. & Fu, J., 2012. *SUBMARINE UNDERWATER STRUCTURE-BORNE NOISE AND FLOW NOISE DUE TO PROPELLER EXCITATION,* Wuhan, China: Acoustics Australia.

Design and Build of a Device to Measure Resistance and Power of Model Submarines in the Submarines in Schools Technology Challenge

Simon Von-Limont[1], Ahmed Swidan[2], Alaa Osman[3], and Warren Smith[4]

ASLT, RAN, UNSW Canberra at ADFA, Australia simon.von-limont@defence.gov.au
[2] UNSW Canberra at ADFA, Australia a.swidan@unsw.edu.au
[3] UNSW Canberra at ADFA, Australia a.osman@unsw.edu.au
[4] UNSW Canberra at ADFA, Australia w.smith@adfa.edu.au

The objective of the Submarines in School Technology Challenge (Subs in Schools) is to stimulate Australian school students to pursue a science, technology, engineering and mathematics (STEM) related career. It has an obvious connection and motivation to grow interest in industries being developed by the SEA1000 Future Submarine Program. One of the principal improvements envisioned to the program is the introduction of more real-world engineering tests into the competition including the measurement of the hull resistance and power. The work reported in this paper aims to enhance student learning experiences through developing a test rig to measure the hull resistance and required power. The instrumentation will be selected based on analytical results and computational simulations to determine the key design elements, e.g. motor requirements and the tension in the towing cables. Having these tests conducted during the competition will facilitate and encourage students engaging in relevant engineering theory to enable them to incorporate resistance and powering aspects more appropriately into their designs.

This paper reflects a work in progress as part of a 2019 undergraduate honours project by the lead author.

INTRODUCTION

Subs in Schools (SiS) was developed by Re-engineering Australia (REA), in association with the Department of Defence and a variety of industry stakeholders, including SAAB Australia and the Australian Submarine Corporation (ASC) [1]. It was established in 2014 as a result of the Australian Government announcing the SEA 1000 Future Submarine Program to replace the Collins class submarines [2]. This is Australia's largest ever Defence procurement and Subs in Schools is intended to develop employability skills in students and interest them in careers in Defence industries to support this program [3].

As presented in Table 1, Subs in Schools offers four project level programs of increasing complexity. This report relates to the most advanced and comprehensive level 4 submarine challenge which allows students to engage with marine engineering and naval architecture theory through experiential learning, including 3D printing, electronics, Computer Aided Design (CAD) and coding. It involves a number of team deliverables, including a project portfolio, a working scale model, a trade booth and a verbal presentation. Students must also engage with industry to seek sponsorship/support, and through this process they learn valuable soft skills such as teamwork, communication and project management [1].

Table 1 – Subs in Schools Project Levels [1]

Level	Suggested School years	Project
1	7-8	Build a mini underwater remotely operated vehicle (ROV) from a kit
2	8-9	Design and build a large underwater ROV
3	9-11	Design an internal accommodation space in CAD
4	10-11	In year 10 construct the REA submarine kit. In year 11 design and build a model submarine either improving the existing design or creating a new design.

The competition requires a practical demonstration comprising trials, where teams operate their submarines to complete surface maneuvering, flotation and ballasting, submerged maneuvering trials and a "timed way-point voyage" [4]. The trials test a number of design aspects including watertight integrity, buoyancy, propulsion and control. Feedback from previous competitions indicates that the model submarines being developed are overpowered, and they do not emulate realistic performance of full size submarines.

A previous UNSW Canberra student, Ryan Cavenagh, attempted to address the overpowering issue [5]. The underlying philosophy was to explore means for limiting the available propulsion power. The strategy was based on developing techniques to measure the resistance of model submarines on the surface, and tools typically used for surface ships were employed. For example, matching of the Froude number (Fr) and modelling in Maxsurf were utilised to determine benchmarking data including resistance and velocity. From this data, two variations of drop tower towing mechansims were designed to measure the resistance: one using multiple pulleys, and the other using a wheel system, as depicted in Figure 1.

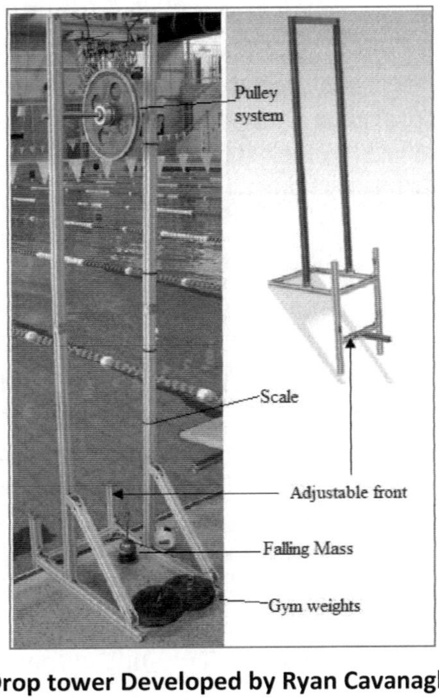

Figure 1 – Drop tower Developed by Ryan Cavanagh in 2018 [5]

The drop towers were used to tow the model submarines (an example is described in [6]) on the surface by applying a known tow force by means of a falling mass. The resulting velocity can be measured. This is the reverse of a standard towing tank where resistance is measured for a given velocity. In the competition the team with the highest velocity (hence the lowest resistance) would win that aspect of the competition.

Several limitations with the falling mass design as configured were observed. Firstly, it was only capable of towing on the surface and for a test distance of 6.4m. This limit was a function of available proof of concept materials (drop height, pulley wheel) but having a much longer test length to operate in a standard 25m pool is highly desirable. Secondly, the frame is heavy, it requires significant weights to counterbalance and it needs to be constructed and deconstructed each time it is used, which is not conducive to portability. Thirdly, the velocity is measured manually which introduces uncertainty and offers poor sensitivity or via video which leads to significant data processing.

There are three primary techniques for measuring resistance on a submarine. These include physical scale modelling using a towing tank or wind tunnel and applying a modified surface vessel resistance method; using CFD modelling to predict resistance; and, using analytical approximation techniques. To design a device capable of simultaneously measuring total resistance and speed to the Subs in Schools program and to enhance students learning experiences, analytical solutions were compared with CFD results to make sure that the towing tank device components are adequate and capable of measuring the hull resistance accurately.

METHODOLOGY

Depicted in Figure 2 is a basic overview of the methodology used to design an alternate system/device for SiS. The first phase was to estimate the resistance to obtain an approximate value that the system must be able to both tow and measure. For example, electronic components have specified operating limits and analysis is required to ensure each component integrates within the system appropriately. The second phase is to verify that the approximation results are accurate by comparing them to numerical results using CFD simulations. This is to provide a level of certainty. The third phase is to design and build the system/device, and the fourth phase is to test the model and compare the experimental data with the computed results.

Analytical Method

There is a direct relationship between resistance and propulsion. For example, theoretical power is the power needed to overcome the resistance of a naked hull at a defined speed i.e. $Power = force \times velocity$. Thus, if the resistance at a constant velocity is known, then the theoretical power required to overcome that resistance can be determined. Resistance is a measure of the horizontal component of opposing force on a vessel as it traverses through water. The primary components of resistance on a submerged submarine are skin friction and hydrostatic pressure [7].

To calculate the resistance on the REA model submarine, an analytical method was employed using a recommended procedure in chapter 4 of reference [7]. The analytical results are compared with corresponding CFD results in the coming section.

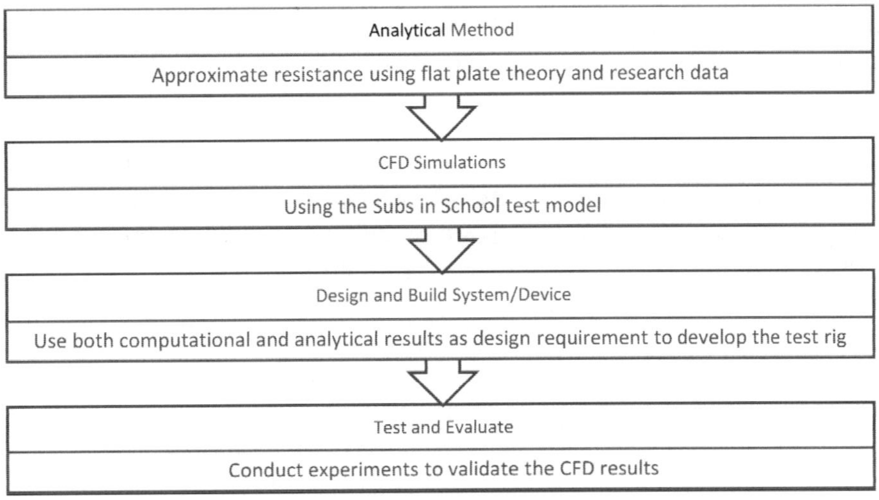

Figure 2 – Summary of the methodology

Numerical Method

Over the past decade, CFD has been developing significantly. As such, CFD is now a powerful tool that can accurately predict total resistance components of a submarine model moving in 1 Degree of Freedom undersea. To build a simple towing device facility, CFD is employed to predict the corresponding drag force at a range of operational speeds. Although Fr is irrelevant in submerged towing tank testing, it may be used to calculate a realistic towing speed. The selected speed range was primarily chosen based on Froude scaling of full-scale submarines, e.g. HMAS Collins. This submarine has an unclassified submerged speed v of 20 kn (10.29m/s)) and a length L of 77.4m [7]. Thus, it was decided to run simulations at a range of water inlet velocities from 1 to 2.5 m/s in 0.5 m/s increments, e.g. Fr scaling.

The numerical simulations were executed using the Reynolds Averaged Navier Stokes (RANS) solver Fluent. The code resolves the incompressible RANS equation in integral form utilising the finite volume method. The motions of the model were restricted in all degrees of freedom, i.e. applying an inlet flow at a given velocity with the body remaining stationary using the traditional fixed grid method. The simulations were performed using 10 nodes of the high-performance computing cluster (consisting of 322 cores), at the National Computational Infrastructure, Canberra, Australia.

Illustrated in Figure 3 is the REA model submarine geometry. The main particulars of the model are as follows: length L/D = 1/0.11, the forward and aft control surface have a chord of 0.061 m. The sail has a chord of 0.25 m and span to 0.104 m from the body centerline.

A slow growth rate was selected for grid generation to ensure smooth transition between neighbouring structured cells, as illustrated in Figure 4 (a). The grid was generated with at least seven density levels in order to capture pressure distribution around the submarine model. The finite volume meshing method was implemented for all simulations to solve the differential equations for a viscous 3D monophase flow represented by RANS equations in which the effect of turbulence is solved using two equation Shear Stress Transport (SST) k-ω eddy viscosity model. To resolve the near wall turbulence quantities all y+ wall treatment was selected that attempts to combine the high y+ wall treatment for coarse grids and low y+ treatment for fine grids, where y+ is a non-dimensional wall distance.

The computational domain is shown in Figure 4(b). The boundary conditions used were velocity inlet at the right-handed side of the domain, pressure outlet at the left-handed side of the computational domain, non-slip wall at the hull surface of the model and slip-wall at the domain wall, while the plane that bisects the submarine model vertically was treated as a symmetry plane, as shown in Figure 4(b).

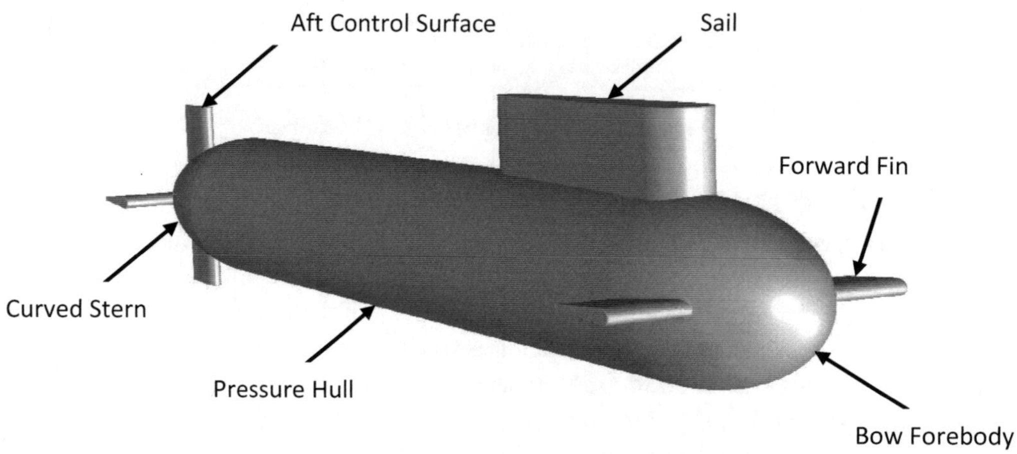

Figure 3 – The CAD REA model submarine

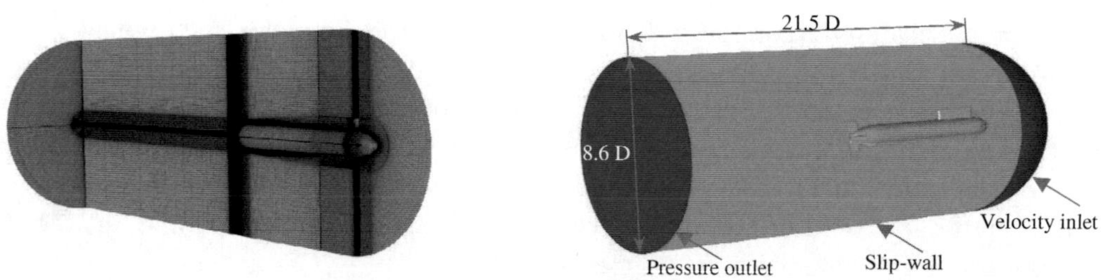

Figure 4 – (a) the generated fine grid and (b) the boundary conditions and the numerical SiS model submarine

The applied physical model in all simulations uses the Eulerian multiphase segregated iterative method to solve the conservation equations for mass, momentum and energy (RANS equations). This model solves the flow equations for the velocity components and pressure in an un-coupled manner. First, the linearised components of the momentum equations are the prevailing pressure and mass fluxes through the control volume faces (inner-iterations), followed by a Semi-Implicit Method for Pressure-Linked Equations (SIMPLE) to resolve the pressure-velocity coupling, while the linkage between the momentum and continuity equations is achieved through predictor and corrector stages.

Grid Independence Study

To study the sensitivity of computed results against the used sizes of cells to discretise the domain and the 3D geometry of the submarine hull model, grid-density independence studies were conducted, as shown in Table 2. Provided is a summary of the independence grid studies and associated uncertainties against the measured data. The second column presents the grid size and the third column compares between the results utilising three applied grid densities (first column). The total number of generated cells varies from around 6 million, 12 million and 14 million cells for coarse, medium and fine fixed grids, respectively.

Table 2 – Summary of Numerical Independence Study

Grid level	Grid Size (millions)	Drag force difference to fine grid (%)
Coarse	6	1.7
Medium	12	0.6
Fine	14	-

Results presented in Table 2 shows that the uncertainty error bounds did not enhance significantly despite required higher computational cost. Thus, it was decided to use the medium sized grids for the presented results on the basis of the slight change in timing of drag force magnitudes with respect to the applied coarse stationary grid, as presented in Table 2. The time-step varies in order to satisfy a chosen steady state solution.

Results and Discussion

Presented in Figure 5 is an image of the pressure distributions around the submarine hull, with a demonstrated stagnation pressure at the bow and the fins facing the flow direction. The pressure distribution is of importance during the design stage of the REA model as it provides users with the locations that are under higher stress levels on the hull. These locations are with higher risk of failure, e.g. the stern fins. Insights on the flow behavior around the submarine model and areas of improvement to reduce drag force on the hull and to achieve higher speeds can be obtained by studying Figure 6.

Good agreement is illustrated in Figure 7 between the analytical and the computed results with a deviation of less than 5% between both results. This comparison could provide the designer of the towing device with important information for efficient facility design, e.g. regarding the load cell's range.

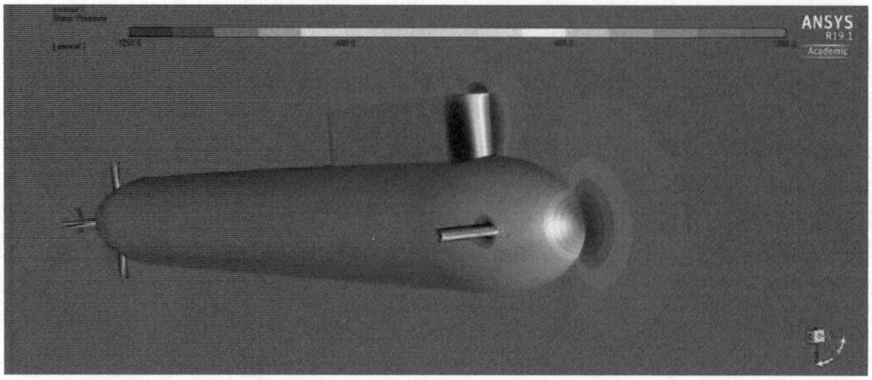

Figure 5 – Pressure contours around the submarine model

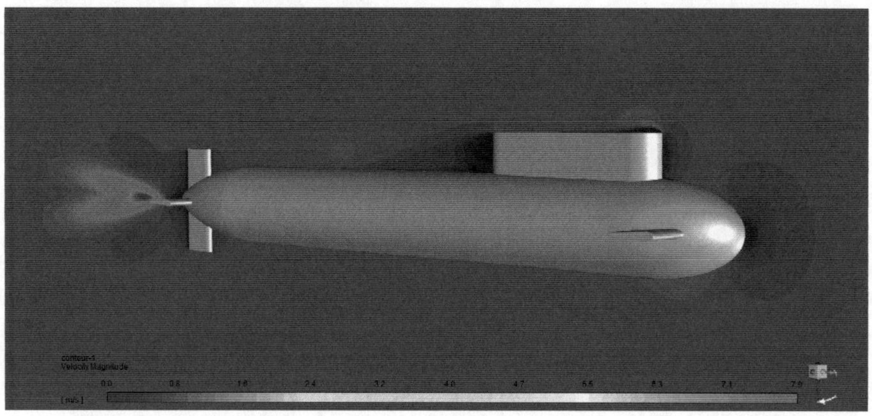

Figure 6 – Velocity flow behavior around the submarine model

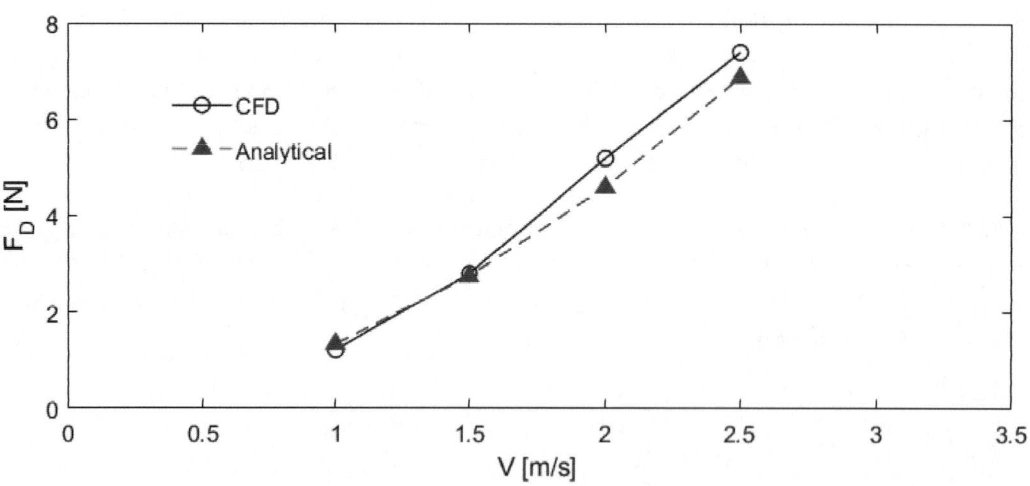

Figure 7 – Comparison of the total drag force CFD results against the analytical results

SYSTEM REQUIREMENTS

Analysis of the analytical and computed resistance results was undertaken to select the design specifications for the new experimental system/device. This involved first approximating the resistance on a built REA model submarine by using approximation theory. This yielded an approximate resistance value of 7 N at a maximum velocity of 2.5 ms^{-1}, providing a baseline level of resistance for which the device must be able to overcome.

The design requirements for the device were then established. The primary design requirements include:

- The device must be able to measure both resistance and power of model submarines;
- It must be lightweight and portable to meet airline check-in luggage requirements;
- It must be powered and controlled by electronics and be a self-contained unit i.e. no external power source or counterweights;
- Critical components must be water resistant; and
- Must be user friendly i.e. requiring very little setup and effort on behalf of the user.

CONCEPT DESIGN

Based on the design requirements a concept study was conducted to determine the most suitable concept on which to conduct further study. The concept chosen was a portable device that incorporates a DC motor driven fishing reel and a load cell. To measure resistance, the device will have a main towing component at one end of the pool and another guide rope retaining unit at the other end of the pool. Model submarines will be trimmed to a depth of 0.5m through a calibration lap of the pool then a towing harness will be attached to the nose cone. Another guide harness will be attached near the tail and this harness will be connected to two ropes/lines that run the length of the pool.

The maximum velocity value was used to determine the torque and RPM requirements for a DC motor to couple to a tailored fishing reel. This was achieved by approximating the net mass of the model submarine and using this mass, the approximated force and velocity in dynamics analysis. This yielded a minimum torque of 1.23Nm and a minimum motor RPM of 186, and a DC motor was purchased based on this analysis. The change in velocity due to the spooling effect of the reel was also analysed, and it was determined that there was only a 4.2% change in velocity over a 25m pool course and this was deemed acceptable. The DC motor chosen has a built-in rotary encoder to measure the angular velocity, and this will be used to code the motor to wind the fishing reel at the characteristic velocity.

The fishing line between the reel and model submarine passes over a grooved bearing that is mounted atop a load cell that will be coded to record a peak and average resistance reading for the duration the model submarine is at a constant velocity. A conceptual representation of the system is shown in Figure 8. The proposed device/system components are listed and depicted in Figure 9.

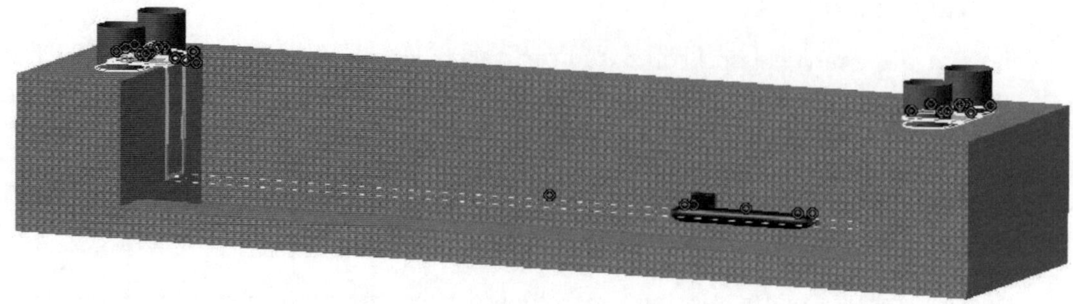

Figure 8 – Conceptual Representation of Model Tow Configuration

Number	Component
1	Aluminium frame and telescopic legs (variable depth adjustment)
2	Grooved roller bearings
3	Bearing to Load cell mount
4	90 degree folding brackets for legs
5	Collapsible buckets (for anchoring sub-system)
6	Motor to Reel coupling
7	Gear DC motor with encoder
8	Arduino Uno
9	12V Li-ion battery
10	Fishing Reel
11	Removable bucket mounts (rubber base for anchoring sub-system)
12	Arduino shield for resistance and power readouts
13	10kg rated load cell and driver

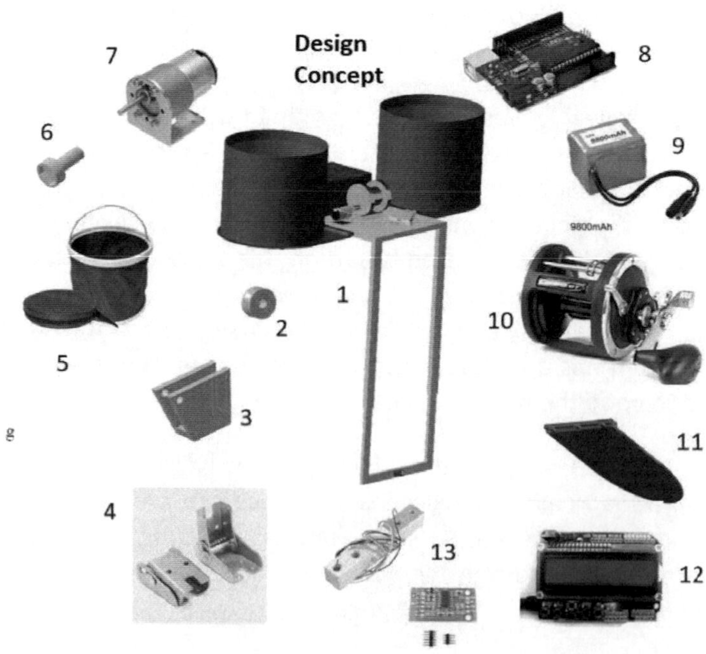

Figure 9 – Proposed Model Towing System Components

To measure power, the inbuilt drag system on the fishing reel will be set to a value that all model submarines within the competition are capable of overcoming. This will be achieved by pulling on the fishing line whilst reading the force measured on the load cell and adjusting the drag system until it begins unwinding line at the desired force value. The submarine will be connected via a harness designed to keep the line away from the propeller and the submarine will be ordered to proceed at full speed ahead. The time taken for the submarine to traverse through a given displacement will be recorded in order to measure the maximum power output. An additional "bollard pull" test could be conducted by turning off the drag system and measuring the maximum thrust of the model submarines at zero speed ahead.

FURTHER WORK

Work to build and test the device is scheduled to commence in the third quarter of 2019 and results will be presented at the conference.

Once the device is built, experiments will be conducted to validate that the device is able to accurately measure resistance and power. This will be achieved by first towing the respective harnesses through the water and measuring the resistance. Experiments will then be conducted using the model submarine, for example a parametric study of resistance while varying velocity. The model will be towed at the characteristic velocity and the harness resistance value will be deducted from the gross resistance value to give the net resistance on the submarine. This value will be compared to analytical and CFD approximation results to determine the accuracy and levels of uncertainty with the device.

CONCLUSION

A review of theory and a previous thesis have yielded foundation theory and system design requirements necessary to achieve the aim of this project. A planned methodology has been established, design work and procurement are in progress and construction of the concept design is scheduled to commence in July 2019. CFD results were in good agreement with analytical results, and CFD was able to predict the pressure contours and the velocity distributions around the 3D submarine model.

This design project aligns well with Australia's strategic interest to stimulate interest in its youth to pursue STEM and submarine related careers.

Acknowledging the work is ongoing, as an undergraduate thesis, it is anticipated that complete detailed results will be available when the paper is presented and updates will be available from the authors.

REFERENCES

1. (REA), R.E.A.L., *Subs in Schools Overview Levels 1 to 4 2017.* 2017.
2. CASG, *Future Submarine Program.*
3. Australia, S. *Chapioning subs in schools STEM program*. 2017 01/04/2019]; Available from: https://saab.com/region/saab-australia/about-saab-australia/latest-news/stories/stories---australia/2017/championing-subs-in-schools-stem-program/.
4. REA. *2019 Competition Regulations Level 2, 3 & 4: ROV, Spatial and SUB Classes*. 2019; Available from: https://3ixa2917tenz3fyiar4esjdf-wpengine.netdna-ssl.com/wp-content/uploads/2019-ROV-SD-SUB-Competition-Regulations-V1.0.pdf
5. Cavanagh, R., *Measuring Resistance of Hull Forms in the Subs in Schools Competition*. 2018, UNSW Canberra.
6. ENVIZAGE, *Subs in Schools (Submarine Kit).* 2019.
7. Renilson, M., *Submarine Hydrodynamics*. 2018, Springer International Publishing AG.

INTEGRATING MARITIME UNMANNED SYSTEMS ONBOARD MINE HUNTERS AT HEART OF STAND-OFF MINE CLEARANCE

Antony PENN, ECA GROUP, penn.a@ecagroup.com

Abstract

Developed by ECA GROUP, UMIS™ is the new generation Mine Counter Measure (MCM) unmanned system able to carry out any Mine Counter Measure (MCM) mission, using a collaborative system of surface and underwater robots even in the most challenging environments. In use within several navies since 2016, and being chosen in 2019 by leading NATO navies for MCM missions, the Belgian and Netherlands Navies, UMIS is a game changing solution for mine hunting at sea. It has proved to be faster in fulfilling operations, more efficient, cost-effective and safe, as the crew is kept away from minefield. Comprehensive and modular, UMIS™ includes a wide range of interoperable unmanned vehicles such as USVs, UUVs (AUVs, ROV, MIDS ROVs), towed sonars, UAVs, etc. UMIS™ also integrates UMISOFT™, a C2 MCM comprehensive software suite allowing easy and complete management of the entire unmanned mission: preparation, run and supervision, data processing, analysis and management. UMISOFT™ enables to manage parallel drones missions, robot-from/on-robot deployment (such as USV deploying and retrieving a UUV) as well as automatic launch and recovery systems (LARS from USVs) and can be integrated into the combat management system (CMS) of the vessel (MCMV).
We believe that next generation mine counter-measure vessels will be shaped by unmanned systems onboard.

ASSESSMENT ON RISKS AND THREATS OF MARITIME MINES

Naval mines can be used in offensive purpose, to ensure destruction of enemy's assets or to deny access into a harbor, to block a choke point... Mines represent the majority of these threats since their objective is to at least damage ships, at most to sink or destroy them completely. Easy to produce and deploy, this relatively cheap mean with a great destruction potential attracts also terrorist organizations.

Used in defensive way, mines are used to protect own or allied force's assets in order to control a "safe" zone.

Yet there are also remnant bombs dropped during World War II with some 400 000 mines estimated lying in the Baltic Sea, North Sea and the Channel. Not to mention those dropped during more recent conflicts and some hundreds of millions of them still remaining in stock all over the world.

Maritime mines have become a threat when found at the bottom of a harbor or in a fishing area since their explosive material can still be active despite decades spent in seawater; they may also have become unstable over time, which makes them potentially even more dangerous. They represent a threat for economy, since over 80% of world's trade is carried by sea.

In the last decades several incidents causing material damages and human lives losses at sea have been reported (e.g. USS Samuel B Roberts in 1988, USS Princeton in 1991, then more recently in 2018 s ships, including minehunters, have been damaged by mines used in Bab-el-Mandeb Detroit area).

PASSIVE AND ACTIVE MINE COUNTER MEASURES – TWO DIFFERENT BUT COMPLIMENTARY APPROACHES

There are two approaches for Mine Counter Measures: passive and active. Passive approach implies a certain number of preventive measures to design and equip the vessel in order to minimize the risk for a ship to detonate the mine and/or to resist to mine detonations while the active measures deal with mine hunting and their neutralization or destruction.

Magnetic Risk Management as A Passive Protection Against Mines: From Platform Design to Maintenance Treatment

The first implies to design, build, equip and maintain the platform in the way to protect it from mine activation and detonation by controlling its magnetic and acoustic signatures. Therefore, the design of the platform is specific: the hull built of fiber glass or wood and equipped with a low acoustic signature propulsion systems. Yet these kind of ships are expensive, can be slow and therefore vulnerable to enemy's attacks. More affordable and stealthy at sea, steel-hulled ships used by most navies can preventively be fitted with degaussing systems and/or depermed using underwater deperming stations. The magnetic signature of the ship is a critical point when it comes to mine threat. Degaussing systems installed onboard vessels as well as deperming of ships reduce the risk of magnetic sensors detecting the steel hulls of both surface ships and submarines. Degaussing reduces magnetic signature by passing an electric current through coils arranged inside the ship depending on the ship's heading, position on the globe...whereas deperming is an operation performed from time to time in order to reduce permanent magnetization of a vessel (surface ship or submarine). Both aim at altering the magnetic signature of the vessel in order to protect vessels by acting as a form of camouflage from magnetic detection, and as a precaution against magnetic mines.

The traditional deperming process sees the entire ship or submarine wrapped with a massive wire coil. Electricity is then passed through the coil to reduce the magnetic 'signature' of the vessel - a process that can take up to 10 days.

An innovative approach uses wire coils laid down on the sea bed through which an alternating current is passed as the vessel moves over the coils. This deperming 'over-run' process takes less than one day and therefore provides a significant operational and economical advantage when compared to the traditional wrapping.

The new-generation deperming system developed by ECA GROUP and proposed to the RAN by Advanced Magnetic Ranges Australia Pty Limited (AMRA - a French-Australian Joint Venture bringing together Amog, Tycon and ECA GROUP) simplifies this laborious process significantly and reduces the time required from as much as ten days to about one day. It is an 'over-run' treatment range for submarines and surface ships with:

- Magnetic treatment of both submarines and frigates
- Fast treatment time (low down time)
- High precision of measurement and treatment
- On-board Degaussing system calibration and checking
- Low maintenance needs
- Overrun station that operate without any cable-wrapping operations – Reduced manual handling
- Treatment loop and sensors that are fixed so displacement-related damages are avoided and maintenance is reduced.

Ship trajectory over the underwater coils and magnetic sensors is determined accurately through a DGPS system. The over-run treatment range can be combined with the RAN's existing multi-influence range that is designed to detect and measure a variety of other 'signatures': acoustic, magnetic, electric, and pressure. ECA GROUP manufactures ranging systems for the French Navy and is installing a similar over-run range for another South East Asian Navy.

Active Mine Counter Measures: Threat Detection And Neutralization - Two Distinct And Complex Operations

Though it is a complex matter to detect remnant underwater mines and ammunition[1] lying on the bottom of an ocean, it is equally complex to have sea divers destroy them due to environmental constraints (e.g. sea currents, water depths, etc.).

This led naval forces and the industry worldwide to develop systems designed to reduce the risks associated with neutralizing such threats while optimizing the effectiveness of such disposal operations, i.e. destroying the largest possible number of threats in the shortest possible time.

One of the first effective systems built to eliminate underwater threats stemmed from France's ambition to develop its own nuclear deterrent weapons delivered by submarines. With its nuclear powered ballistic missile submarines (SSBN) based in the Brest roadstead, the French Navy had to be able to detect and destroy any underwater threats on the access routes of these submarines to make its deterrence concept credible. This was one of the main reasons behind the development of the French Navy mine-hunting fleet.

Minehunter Ships As Main Response to Mine Threats Since 1970

Maritime mine clearance has conventionally been conducted since the 1970s by dedicated ships called minehunters. Mines are detected and classified using mine-hunting sonar systems installed under the ship's hull or towed by the ship. Sonars retrieve seabed images and the analysis is made by specialized operators onboard.
When detected, objects that are likely to be mines, a mine identification and neutralization vehicle, such as a PAP or mine killer (such as K-STER), is deployed to identify and neutralize the mine.
The cost of these ships can be astronomically high as they need to be able to enter minefields and detect mines without being detected by the mine, which would trigger its explosion.
Were any mines to explode, the ships also need to be able to withstand or at least survive such a blast.

In addition to a non-magnetic hull, resiliency of these ships and their equipment is paramount, today's military standards require a large number of protection systems such as degaussing, shock resistance, low radar and acoustic signatures or CBRN protection.

The growing and evolving threat (more and more sophisticated mines fitted with acoustic sensors to detect, localize and recognize particular a target such as military ships, asymmetric threats such as underwater IED and new conflicts) generated the need to operate more safely, keep crews out of danger zone, protect costly equipment onboard and reduce operations' costs and time, has led navies to specify and evaluate different automated solutions through several R&D programs (SeaKeeper, ESPADON, etc.) challenging industry to develop new operational solutions.

New solutions are now available, naval drones also called unmanned systems are integrated within naval forces and especially Mine Clearance units. These drones can be autonomous underwater vehicles (AUV) or unmanned surface vehicles (USV). French mine clearance divers (GPD), Latvian and Romanian Navies use A9 AUVs, Kazakstan Navy uses INSPECTOR class USVs, Singapore Navy uses USVs equipped with mine-killers. They can be remotely deployed in the minefield by specialist mothership, non-dedicated ships or from land. In order to detect and classify mines, they carry or tow acoustic sensors. In order to identify and neutralize mines, these naval drones (especially USVs) deploy other drones that are remotely controlled from the support vessel or from land.

One example is the Unmanned Mine Counter Measures Integrated System (UMIS™) developed by ECA GROUP. This system is a result of decades of developments in robotics, hydrodynamics and IT systems in order to address the growing need of operational autonomy at sea.

Mine Countermeasures: Legacy vs. Stand-off concept

Several decades of mine hunting using dedicated ships show some shortfalls of traditional and current mine hunting concepts.

Existing high frequency sonars are designed to detect ground and in-water mines and are ill-suited to detection of mines near the interfaces (e.g. drifting and buried mines)... Detection of buried mines is known to be feasible with low frequency sonars (e.g. parametric sonars or synthetic aperture sonars, possibly even combinations thereof) operated close to the seabed, mounted on towfish or AUVs. The other steps, such as buried mine classification, identification, relocation and neutralization remain very challenging today.
A near surface sonar is not suited to the detection of drifting mines due to lack of "height of eye". An airborne LIDAR is a more effective solution.

The second is the operations time since the ship's hull mounted sonar cannot function properly at high speed and the operations are made in sequence by the minehunter: detection/classification then identification/neutralization.

The third is the risk for personnel especially when it comes to enter the minefield or to dive for a mine neutralization operation.

Delivering the UMIS™ system to the first customers in 2016 and being evaluated then chosen by Belgian and Netherlands Navies to provide a set of unmanned systems as a 'toolbox' for the next-generation mine counter measures operations for NATO missions, ECA GROUP's solution has proven to be faster in fulfilling operations, more efficient, cost-effective and safe, as the crew is kept away from minefield. Using conventional mine hunters on a minefield have many drawbacks.

First, the high cost to reduce the ship's magnetic signature to a very low level. Secondly, the ship detects mines using a Hull Mounted Sonar (HMS) of which mine classification depends, on its range and resolution. ECA GROUP sonar expert Dr. Marc Pinto explains that "*Physics of sonar dictates that you can't have both range and resolution: Long range sonars have poor resolution. High resolution sonars have very limited range. With conventional mine hunters, it is either required to approach and manoeuver close to the possible mine for a good classification, which is risky and time-consuming, or, the minehunter stays far from the targets, leading to a low resolution and a high probability to miss some mines, which is also a high risk*".

In comparison, ECA GROUP's UMIS™ deploys Autonomous Underwater vehicles (AUVs) carrying the sonar close to the seabed and Unmanned Surface Vessels (USVs) which bring identification and disposal vehicles in the minefield, leaving the mothership away in a safe zone. The UMIS manages a set of drones which cooperate autonomously into the dangerous area. "It is proved that, operating in parallel, the UMIS system can divide by 3 at least, the time of missions, and comes with a higher efficiency and clearance rate." reports

VADM (Ret.) Christian CANOVA, FRN, Undersea Warfare Expert and former NATO Maritime Command Deputy Commander.

TECHNOLOGICAL ADVANCES ADDRESSING NEW OPERATIONAL CONCEPTS: THE BELGIUM AND NETHERLANDS NAVIES "TOOLBOX" CASE

In 2017 Belgium and Netherlands Navies specified their requirements for the replacement of their MCM fleet (tripartite mine hunters in operations since 1985 built by Naval Group and equipped by ECA GROUP with PAP mine identification and disposal ROV) after two unmanned MCM trials conducted in August-September 2016 and Mid-2017. The two navies adopted a completely new approach, the stand-off mine hunting concept requiring a new technological solution for the unmanned system as well as the ship for its deployment.

In May 2019 the Belgium Naval & Robotics consortium, composed of Naval Group and ECA GROUP, was notified the contract to supply twelve mine-hunting vessels to the Belgian and Netherlands navies, leading navies for Mine Counter Measures in NATO. Six ships are destined to the Belgian Navy, while the other six will be delivered to the Netherlands Navy. As a part of the contract around 100 drones, constituting approximately ten drone systems (toolboxes) will be delivered by ECA GROUP.

The contract, worth nearly two billion euros, will last ten years. After a design phase of the ship and adaptations for the tools, Belgium Naval & Robotics will launch the production phase of the ships and drone systems with a first delivery in 2024. This contract includes a vast cooperation plan with the Belgian industry in order to carry out a significant part of the contract in Belgium, in particular the production of ship equipment and naval drones composing the toolbox as well as their maintenance. The winning offer includes also training solutions using virtual reality to be implemented at NATO MCM training center in Ostende as well as an implementation of two R&D labs dedicated to cybersecurity and mine counter measures unmanned systems. Over 39 Belgian industry and scientific partners are involved in this program.

An Innovative Solution For Robotic Mine Warfare As A Response To An Operational Need

Belgium Naval & Robotics, thanks to Naval Group's recognized expertise, will deliver a militarized ship with military characteristics (acoustic and electromagnetic discretion, shock resistance) adapted to the operational use targeted by the Belgian and Dutch navies designed for mine warfare using an ECA GROUP system of robots. In particular, this vessel incorporates a launch and recovery system for ECA GROUP's INSPECTOR 125 unmanned surface vehicles (USV). This innovative, robust and reliable system ensures the safety of operators and enables to protect the ship, the USV and its payloads, while offering a high level of mission availability.

The drone systems on board these new-generation vessels are composed of latest generation of drones developed by ECA GROUP. They are integrated into the ECA GROUP command and control MCM UMISOFT™ system that is connected to the Naval Group's I4drones® system to form the mine-warfare mission system itself interfaced with the ship's combat management system.

The solution includes: A18-M autonomous underwater vehicles (AUVs), T18-M towed sonars and Mine Identification & Destruction Systems (MIDS) composed of SEASCAN and K-STER C remotely operated vehicles (ROV). All these drones can be operated autonomously from the USV INSPECTOR 125. The drone system also includes unmanned aerial vehicles (UAVs) and influence sweeps.

The program provides for the supply of drone systems that can also be used for expeditionary mine warfare. Containerized and equipped with handling and communication systems, the mine-hunting drone systems can be airlifted and deployed directly from the coast without the need of a ship.

Toolbox To Configure According The Mission: UMIS™ as a modular and mobile MCM system

The UMIS™ toolbox is particularly interesting as a flexible and stand-alone unmanned solution. To meet navies' requirements and their logistic constraints, ECA GROUP designed specific storage and transportation equipment for each possible configuration. The containerized configuration can be deployed in coastal areas, ports and harbors.

A typical UMIS™ configuration is its containerized version that is composed of a dedicated C2 container and its set of containers for the "tools" themselves. Only ISO containers are used in order to facilitate their transportation and deployment.

The C2 container will be deployed on land or installed on ship of opportunity, while the drones will be able to be deployed from beaches, river banks, port docks, etc. or non-dedicated ships. The main functions of the MCM C2 system is to operate the drones like USV INSPECTOR 125, mid-size AUV A18-M, towed sonar T18-M, MIDS configuration made of inspection ROV SEASCAN and the "mine killer" K-STER and men portable AUV A9-M and its equipment such as LARS (Launch and Recovery System), but also to prepare the missions and tasks, process and manage the data collected.

A complete UMIS™ in its containerized configuration is air-transportable and can be deployed very quickly anywhere in the world. It is fully compliant with safety and transportation standards.

Reducing the duration of the operations with drones missions "parallelization"

The Toolbox, consisting of stand-off (remote) sub-systems corresponds to the gradual change of paradigm in MCM, providing high-level flexibility in the planning and completion of operations that can be carried out in parallel and at a distance from the platform and its crew. This innovative solution increases the speed of mine clearance in a given area and above all, prevents exposing human life to the risks inherent to this type of operation.

The stand-off concept inherent to the project requires full control of durable, autonomous vehicles with a wide variety of payloads (including sonars).

Commonality For Operations Efficiency, Logistic Flexibility and Reduced Cost of Ownership

An important commonality of tools (eg: common software, AUV-derived towed sonar, common launching and recovery systems) allows an optimization of the logistics with an undeniable operational impact.

As an example, the T18-M towed sonar is derived from the A18-M autonomous underwater vehicle, which reduces the logistic footprint required by maximizing the use of common subsets between the two systems. In addition, since the tow fish is battey powered (such as the A18-M AUV), it is no longer necessary to feed it by its towing cable, which makes it possible to use a small diameter towing cable and consequently a smaller towing winch and a more compact electric generator aboard the USV the INSPECTOR 125 deploying it.

The reduction of the logistics footprint also extends to accessories, tools, software, spare parts that can be shared by several drones, for example battery chargers for A18-M / T18-M, battery replacement tools etc. .

The use of energy autonomous vehicles as for the T18-M towed sonar, but also for the ROVs SEASCAN for inspection and K-STER C for neutralization, allows the increase of their areas of use (maximum immersion and deployment distance from the surface drone) and their use in demanding environment conditions (up to 4 knots). This is made possible by the significant reduction in the diameter of the towing or connecting cables between the vehicles and the surface drone (USV), which reduces the hydrodynamic drag and therefore the efforts on the cables enabling higher performances (speed and or depth) that are impossible to achieve with vehicles powered by the surface drone whose size and mass are constrained by the ships they are deploy from.

These technological breakthroughs integrated into the robot systems that will be delivered to the Belgian and Dutch navies allow great flexibility in carrying out MCM operations.

DRONES SYSTEMS "TOOLBOXES" SHAPING NEW GENERATION MINE COUNTER-MEASURE VESSELS

Integrated approach – the system dimension central to MCMV missions

In the context of the Belgian/Dutch contract, the toolbox own mothership (MCMV) benefits from Naval Group's expertise in the complete chain of the Combat System (design - integration - maintenance) and provides an operational solution with a reduced risk. That expertise includes a consistent choice of off-the-shelf Sea Proven equipment, integrated and implemented via the Combat Management System (CMS). Rapid data processing, sharing of the tactical situation (interoperability of the ship in joint or international operations) maximizes mission efficiency.

Naval Group and ECA GROUP offer an integrated global solution, the Integrated Mission Management System (IMMS), which brings together all the operational information systems of the MCMV and the MCM Mission System which provides end-to-end management of the Tools' mission (Supervision, Planning, Evaluation, Data Management and Analysis, Command / Control of vehicles).

A combination of the Mine Warfare (MW) and Planning and Evaluation (PE) components, the MCM Mission System is a solution based on existing products SMMD® and UMISOFT™ which has been the subject of technical cooperation between Naval Group and ECA Robotics since 2016. Within the ship's Command Information Center (CIC), operators coordinate all of the operations from their Multi-Function Consoles.

Which architecture for optimal deployment of UMIS™ drones systems?

In the Belgian-Netherlands MCMV replacement program, aspects related to the deployment, storage, maintenance and reconfiguration of the USV are crucial. This is for the deployment and recovery of drones, to minimize the constraints related to swell and movement of the platform, especially in high sea-state.

Naval Group has been working on this issue for many years and while different architectures have been imagined, such as stern ramps or gantry systems, Naval Group engineers have finally opted for a shipside launch and recovery of the USV. The major problem is the pitch which, unlike the roll, cannot be controlled by a stabilization system.
Of all the comparative and very exhaustive studies carried out, it appears that the best location is close the quiet point of the ship. In addition, the safest way is to adapt a preferential route relative to the swell and to be positioned at the right of the quiet point, which because of the forms of hull, is located between the middle and the back third of the ship. Analyzes and simulations have been multiplied and they have shown that this is the optimal location to deploy drones and in particular USVs in high sea states.

Beyond the calculations and simulations, this is a solution that is put into practice by the offshore oil industry, whose ships equipped with drones work in very harsh seas and also launch and recover their ROVs from the ship side.
Note that the ships will also, in addition to the side LARS, have the possibility of deploying their AUVs from the rear, using a hydraulic crane. The robots will then be launched and recovered using specific cages.

Consistent with the choices made for deployment and positioning, launch and recovery devices (LARS) were positioned at the quiet point. There are two of them, one on each side, in the form of tilting A-frames, associated with floating docks capable of handling the USV, who's mass can reach 18 tons.

A particularly innovative floating docks system has been conceived in a system approach and allows in particular to limit the pendulum phenomena of the assembly by working in a combined and automated way with the kinematics of the gantry combining guide arm, winches and a synchronized damping system as well as constant tensioning to limit the overall effort in ensuring the road stability of the floating dock.

Tilting LARS fit into the superstructure when folded. With this design, surface drones that are large and heavy equipment do not need to be moved for reconfiguration or maintenance. This is a real asset because it simplifies and secures operations. Not only for the personnel, but also for the mission because each handling operation presents a risk and it is thus necessary to limit them as much as possible in order to avoid damaging the USV and injuring the personnel.

It is also for the reason of redundancy that Naval Group has opted for two identical, symmetrical and independent LARS, which makes it possible, in case of possible failure of one device, to be able to continue the mission with the other one.

The LARS are located on each side of a large "garage" bringing together all the marine gear of the toolbox. Curtains closes this space, so sailors can proceed to maintenance or reconfiguration operations sheltered from the weather.

Conclusion

Since the beginning of the 20th century, ships and their crews had to enter the minefields in order clear the area. This was firstly made using mine sweeping systems towed by minesweepers, where the aim was to trigger mine fuses, in order to make the mine explode, or to cut the mooring line of moored mines. Then in the 50's with the technology evolving, sonars and remotely operated vehicles appeared giving birth to mine hunting and minehunters. But in both cases the ships have to enter the mine fields in order to detect or neutralize the mines.

The progress made in the past decade, in batteries, computing technology, low consumption electronics, sensors, materials, communications have enabled the development of a complete range of unmanned systems (AUVs, USV,s) that are now available to be used to perform mine detection and clearance without having the need for the ship and her crew to enter the minefield. This is the route taken by the Belgian and Dutch navies to replace their current mine-counter measure fleet that selected ECA GROUP solution for the system of robots (toolbox) that will be provided as part of the program.

The contract is for ECA GROUP the result of two decades of innovation in underwater robotics and mine warfare as well as the recognition by two major NATO navies in the field of underwater mine warfare of ECA's expertise in this area.

The toolbox solution that will be delivered to the Belgian and Netherlands navies can easily be adapted to fulfil any other navy requirement.

The regulatory approach for vessels capable of autonomous and remote controlled operation

Grant Judson
Principal Advisor Navigation
Standards Division
Australian Maritime Safety Authority
Email: grant.judson@amsa.gov.au

Rachel Horne
Senior Advisor Maritime Regulation
Standards Division
Australian Maritime Safety Authority
Email: rachel.horne@amsa.gov.au

ABSTRACT

The Australian Maritime Safety Authority (AMSA) is a Commonwealth statutory authority responsible for maritime safety, protection of the marine environment from pollution, and search and rescue. As part of these responsibilities, AMSA regulates vessels operating within Australia's Exclusive Economic Zone (EEZ), including vessels capable of autonomous and remote controlled operation. AMSA's challenge is to put in place appropriate domestic regulatory arrangements for these vessels, which are flexible enough to support current as well as emerging technology in a rapidly evolving industry.

Our knowledge of this technology is in its formative stages, as is our regulatory approach. Until a defined whole of Government policy on autonomous and remotely operated vessels is confirmed, AMSA will pursue regulatory solutions in line with our policy settings and statement of regulatory approach. These principles will guide AMSA in mapping out a fit-for-purpose regulatory treatment based on the risks presented, including considerations such as vessel size, area and concept of operational use, the ability to comply with collision regulations, environmental impacts and differing levels of autonomy.

Internationally, AMSA influences the regulation of the systems and technology that will support shipping of the future, through our work at the International Maritime Organization (IMO). Our interests include ensuring that these systems and infrastructure are safe and effective when operated in Australian waters. AMSA continues to play an influential role on the international regulation of Maritime Autonomous Surface Ships (MASS).

This paper explores the regulatory challenges posed by autonomous and remotely operated vessels, together with the work AMSA is undertaking in response, at a domestic and an international level.

1. INTRODUCTION

Major advances in technology in recent years have now made autonomous and remotely operated vessels a reality. In the past, law makers conceptualised vessels in traditional terms of a physical structure, physical machinery and experienced seafarers on board to operate it. The maritime laws currently in place at the domestic and international levels are well suited to such vessels – they balance the safety of vessels, their crews, and environmental protection, while allowing such vessels to operate with a level of intervention commensurate with the risks posed by the operation. However, these laws do not easily accommodate the sorts of technological advances being seen today, much less those that will be reality in coming years.

As Australia's maritime safety regulator, AMSA's challenge is to adapt longstanding regulatory and operational arrangements to provide for the safe operation of autonomous and remotely operated vessels, in a way that will be effective in the short, medium and long term. AMSA cannot achieve this in isolation, and is actively engaging with leaders and key stakeholders in the fields of autonomous and remotely operated vessel design, technology, operation and regulation. AMSA will leverage its experience and that of its stakeholders, to identify the best way forward, in order to provide effective regulation and continue to ensure the safety of people and vessels, and the protection of the marine environment in Australia.

AMSA's role

AMSA is the Commonwealth statutory authority responsible for maritime safety, protection of the marine environment, and search and rescue. AMSA regulates commercial vessels operating within Australia's Exclusive Economic Zone (EEZ), including vessels capable of autonomous and remote controlled operation. AMSA is committed to safety and the protection of the marine environment; it encourages and supports innovation and new technology, where possible.

AMSA is established by the *Australian Maritime Safety Authority Act 1990,* and is responsible for administering the following:

- *Marine safety (domestic commercial vessel) National Law Act 2012* (National Law)
- *Navigation Act 2012* (Navigation Act)
- *Shipping Registration Act 1981* (Shipping Registration Act)
- *Protection of the Sea (Prevention of pollution from ships) Act 1983* (PoTS Act)

Internally, AMSA has an Autonomous Vessel Team, which is made up of subject matter experts in regulatory and operational matters with expertise at both a domestic and an international level. The team is responsible for managing enquiries and applications received in relation to autonomous and remotely operated vessels. This ensures consistency in our treatment of these vessel operations, and provides AMSA with crucial data and intelligence sources to inform our future approach.

Domestic trends

<u>What is operating today</u>

New technology continues to be developed and used on vessels operating in Australian waters in commercial, research and government settings. To date, most autonomous applications have related to data gathering and hydrographic survey. However, trends indicate that the systems and infrastructure that support autonomous and remote operation will play a large part in the broader maritime industry in the future. A range of sub-surface and surface autonomous and remotely operated vessels are already in operation. Some examples include:

- *Sub-surface:* Undersea gliders, autonomous underwater vehicles and under ice exploration systems; and
- *Surface:* OCIUS Bluebottle, IX Blue Unmanned Survey Vessel (DriX), Saildrones, C-Worker 5 and 8, and WAM-V.

Each of these vessels has different physical and operational characteristics, which in turn changes the way AMSA has approached their regulation.

Future domestic industry insight

The technology currently being developed will soon be implemented in sectors of Australia's maritime industry. This does not mean that autonomous and remotely operated vessels will replace all conventional vessels, but that Australia's maritime industry will increasingly feature more connected, intelligent and automated systems. The systems and infrastructure that supports high levels of automation and remote operation of vessels in Australia has the potential to significantly improve efficiency and safety for operators and seafarers alike.

While these changes bring with it opportunity, AMSA is cognisant of the need to ensure that the migration to such systems and infrastructure does not negatively impact on safety and the protection of the environment in Australia. AMSA will develop an appropriate regulatory approach which is consistent with our statement of regulatory approach, and is fit for purpose.

International trends

International trends indicate that the advent of highly automated and technologically advanced vessels is a reality today. A wide range of technology, systems and infrastructure are being developed to support these vessels. It is becoming increasingly evident that the commercial shipping industry is not working towards completely unmanned vessels, but instead to increasingly 'smarter' ships, which are designed and equipped to be able to operate across a range of autonomous operation.

Some significant international projects include:

- *One Sea Autonomous Maritime Ecosystem* – The primary aim of Finland's One Sea project is to lead the way towards an operating autonomous maritime ecosystem. The project brings together leading maritime experts, technology companies, businesses and government in a collaborative approach to creating a maritime environment suitable for autonomous ships by 2025.
- *Yara Birkeland* – The vessel "Yara Birkeland" is intended to be the world's first fully electric and autonomous container ship with zero emissions. The 80m x 15m ship will operate in Norwegian waters, and be monitored from a shore control station, with a human in the loop, whilst the ships systems and supporting infrastructure are

responsible for a large number of the ship functions. The ship will be delivered in 2020, and will gradually transition to fully autonomous mode by 2022.

- *The AutoFerry project in Trondheim, Norway* – The primary aim of this project, led by the Norwegian University of Science and Technology, is to develop ground breaking new concepts and methods which will enable the development of all-electric autonomous passenger ferries for transport of people in urban water channels.

Many of the international trends relate to the use of systems and infrastructure to support decentralised control, monitoring and support arrangements for vessels; and provide enhancements to safety and efficiency in the maritime industry.

2. AMSA'S INTERNATIONAL INFLUENCE

IMO progress with addressing autonomous vessels

The International Maritime Organization (IMO) is a specialized agency of the United Nations, and is the global standard-setting authority for the safety, security and environmental performance of international shipping.

In 2017, the IMO's Maritime Safety Committee (MSC) agreed that there was a need to address the regulation of autonomous shipping. The committee also agreed to the use of a universal term, Maritime Autonomous Surface Ships or MASS, to describe a ship which, to a varying degree, can operate independently of human interaction. MASS are generally considered to be large ships which undertake international voyages, rather than smaller domestic vessels. In 2017, the IMO initiated a Regulatory Scoping Exercise (RSE), to examine IMO regulatory instruments with an aim of determining:

a. what regulations apply to MASS, do not apply to MASS, prevent MASS operation, or require amendment to allow for MASS operation, and
b. the most appropriate way of addressing the issues identified during the course of the exercise, for example where a specific convention prevents MASS operation.

The RSE is being conducted by a number of volunteering member states, and is due for completion at the 102nd session of the MSC in 2020. Only after the IMO is satisfied with the results of the RSE will it consider the next steps to properly addressing MASS, and/ or making regulatory change if considered necessary.

AMSA's role at the IMO

AMSA considers it to be in Australia's best interests to remain engaged in the IMO's work on MASS. Key characteristics of our maritime environment, including Australia's remoteness, sensitive marine environment, challenging sea conditions, and a significant dependence on shipping, means that Australia brings a different perspective to this work than some other IMO member states. For example, while we have a focus on environmental protection, and retaining our "blue economy" and our raw mineral export industry, other member states have a focus on ship building, the production of ship technology, and the maintenance of a large ship register. Our interests can be summed up as wanting to ensure that the on board systems and the infrastructure that will support shipping are safe, effective, and fit for purpose when operated in our region of the world.

Whilst Australia is not directly engaging in the design, build or operation of MASS today, it is important that future regulation is appropriate for such vessels operating in Australia in the future.

It is for these reasons that Australia takes an active role in the work on MASS, championing its progress, and focusing on ensuring that changes to international regulation is modern, practical, safe and will be effective for Australia.

3. REGULATORY REQUIREMENTS THAT APPLY DOMESTICALLY

AMSA regulates domestic commercial vessels (DCVs), regulated Australian vessels (RAVs) and foreign vessels. The majority of autonomous and remotely operated vessels operating in Australia are DCVs.

Domestic commercial vessels

DCVs are Australian vessels for use in connection with a commercial, governmental or research activity that operate within Australia's Exclusive Economic Zone (EEZ). Australia's DCV fleet is made up of approximately 35,000 vessels, in a broad range of categories such as fishing vessels, ferries, charter vessels, kayaks, jet skis, and barges. All DCVs over 24m are required to be registered under the Shipping Registration Act, and will receive an official number if registered.

Under the National Law, DCVs are required to have a unique vessel identifier, certificate of survey, a certificate of operation, and must be crewed by persons holding the required certificate of competency. There are also general safety duty requirements that must be complied with. The owner and master must also develop and implement a safety management system (SMS), that ensures the safe operation of the vessel so far as reasonably practicable. DCVs are inspected by AMSA Port Marine Surveyors or accredited Marine Safety Inspectors.

AMSA is responsible for administering the National Law regulatory framework, which was not developed to cater for autonomous and remotely operated vessels. Despite this, the broad definitions of 'vessel' and 'domestic commercial vessel' mean that autonomous and remotely operated vessels are subject to the same requirements as other domestic commercial vessels. Given that autonomous and remotely operated vessels do not usually have crew on-board (or at all, in the case of fully autonomous vessels), and have complex operating systems not provided for in established technical standards, this causes significant challenges from a regulatory perspective.

AMSA has introduced a range of risk-based *general exemptions* which provide a degree of flexibility for designated kinds of operations and vessels. Operators may also apply for a *specific exemption*, which allows AMSA to assess individual operations and vessels on a case by case basis and potentially use administrative conditions to create a tailored regulatory approach. AMSA may only grant exemptions when satisfied that doing so will not jeopardise the safety of a vessel or a person on board a vessel, and the exemption does not contradict an international Convention. While using the exemption mechanism may be appropriate through the formative stages of AMSA's approach to regulating autonomous and remotely operated vessels, it does not necessarily provide the foundations of a longer term framework.

Regulated Australian Vessels (RAV)

RAVs are commercial vessels that are required to be registered under the Shipping Registration Act and are either proceeding on an overseas voyage or surveyed and certified under the Navigation Act. There are several hundred RAVs and the composition of the fleet varies from small 6.5m vessels to fishing vessels to large 272m LNG carriers.

Vessels registered under the Shipping Registration Act receive an official number. RAVs require safety certificates, which are issued based on their compliance with international Conventions, (such as SOLAS, ILLC, and COLREGS), as determined by the regulations (Marine Orders), based on the class of the vessel. RAVs may either be surveyed by a recognised organisation (Classification Societies) on behalf of AMSA or by AMSA itself. RAVs are subject to flag State control inspections by AMSA Port Marine Surveyors. An autonomous RAV would need to comply with relevant regulations and international conventions depending on its size and class.

Provisions for flexibility through exemption or alternative means of compliance are provided within international Conventions, but these are limited by a variety of strict parameters and conditions.

One of the common scenarios AMSA sees is where a small autonomous or remotely operated vessels is operated off a mothership. AMSA treats these vessels as *"ships equipment"*. The smaller vessel is considered to be part of the ship for the purposes of survey, and is included and operated in compliance with the ship's safety management system. This has been common practice in industries such as passenger vessel and offshore oil and gas for many years.

The more complex scenario is where the autonomous or remotely operated vessel is a standalone, independently operating vessel. To a degree, for RAVs, AMSA is reliant on the IMO making changes to the international framework so that they can then be implemented in the domestic framework. In the meantime, AMSA would be looking for appropriate safety assurance and risk management processes to be in place, and for support from a recognised organisation, before it could consider using a flexibility arrangement to allow the vessel to operate, for example through an exemption, equivalent means of compliance, or alternative design arrangements.

Foreign vessels

Foreign vessels entering Australian ports are subject to AMSA's port State control regime and may receive an inspection based on the eligibility criteria. Vessels are expected to be certified and in compliance with the relevant conventions ratified by the vessel's flag State. Vessels sailing under a flag not party to a Convention receive "no more favourable treatment" and are expected to meet the same standard as any other vessel. Vessels may be detained under the

Navigation Act if considered unseaworthy or substandard (in line with SOLAS CH1 Reg 19, implemented by section 23 and 24 of the Navigation Act).

Australia is a member of two regional port State control regimes, in cooperation with other countries: the Tokyo Memorandum of Understanding (TMoU) and the Indian Ocean Memorandum of Understanding (IOMoU). Australia implements IMO guidance for port State control as well as regional guidance in collaboration with our memorandums of understanding. These guidelines are based on the relevant Conventions.

Vessels operating within Australia's EEZ are also subject to the Protection of the Sea (Prevention of pollution from ships) Act and can be subject to inspection and possible prosecution if there are reasonable grounds to believe reckless or negligent actions have caused pollution.

Autonomous and remotely operated vessels operating in Australia will need to comply with all relevant Conventions and regulations applicable to the size and type of vessel, including the *International Regulations for Preventing Collisions at Sea, 1972, as amended*.

Defence vessels

Vessels that fit the definition of "defence vessel" under the National Law, or "naval vessels etc." under the Navigation Act, are not regulated by AMSA. This means vessels that are either:

- a warship or other vessel that:
    (i) is operated for naval or military purposes by Australia or a foreign country; and

    (ii) is under the command of a member of the Australian Defence Force or of a member of the armed forces of the foreign country; and

    (iii) bears external marks of nationality; and

    (iv) is manned by seafarers under armed forces discipline (however described)

  or
- a Government vessel that is used only on government non-commercial service as a naval auxiliary; or
- a vessel used by a foreign country for customs or law enforcement purposes.

AMSA can provide guidance on whether a vessel is likely to be a defence vessel or a naval vessel on a case by case basis, but ultimately it is a question of fact to be determined by the Department of Defence.

4. FUTURE REGULATORY DEVELOPMENT

Vessels capable of autonomous and remote operation present regulatory challenges under the current international and domestic frameworks.

Internationally, the RSE process currently underway at the IMO will assist to determine how safe, secure and environmentally sound MASS operations might be addressed in IMO instruments. At the 101st session of the MSC, the IMO also approved interim guidance for the trials of MASS, which will be released as an MSC circular in due course. The interim guidance sets out how the IMO expects relevant authorities and stakeholders to trial MASS technology in international waters. Until the IMO has progressed its work on addressing MASS through

IMO instruments, there is little that safety regulators can do regarding MASS technology on internationally trading ships.

Domestically, it is possible to address some of the challenges presented by the systems and infrastructure that support autonomous and remotely operated vessels, initially by using established flexibility mechanisms, and then by looking towards regulatory change. As our knowledge of this technology is in its formative stages, and because of the need to ensure any regulatory changes are appropriate for current, emerging, and unknown technology, AMSA is taking a staggered approach to regulatory change.

First steps - short term

AMSA's priority has been, and remains, to obtain more practical, 'hands on' experience in how autonomous and remotely operated vessels operate, what they are being used for, and in what environments; and then to work through how they actually sit inside the current domestic regulatory framework –and what changes may be required to the framework

In order to guide AMSA's everyday work in relation to autonomous and remotely operated vessels, which includes providing regulatory advice, processing applications and conducting compliance and enforcement activities, AMSA has put in place a set of initial policy principles, which outline the way in which decisions will be made under the current regulatory framework. Some of the key principles are:

- autonomous vessels and remotely operated vessels must be as safe as manned vessels;
- risks to the safety of people, other vessels and the environment must be appropriately addressed;
- the owner and/or master is responsible for the safety standards and operations of the vessel; and
- for lower risk operators who have a strong safety compliance history, a 'lighter touch' regulatory approach that focuses on the ultimate safety risk, rather than on certification, may be appropriate.

With the benefit of some experience, and with policy principles in place, AMSA is able to undertake the next step of beginning to identify sub-categories of autonomous and remotely operated vessels. These sub-categories are intended to assist in identifying an appropriate regulatory approach, and include considerations such as vessel size, area and concept of operational use, the ability to comply with collision regulations, environmental impacts and differing levels of autonomy. For example, AMSA is exploring the sub-category of "ultra-small marine equipment", which would apply where a vessel or system is <3m, <100kg, operates <5 knots, and which remains in theoretical line of sight and under the positive control of another vessel.

Second step - medium term

AMSA will continue to explore sub-categories of autonomous and remotely operated vessels, to guide an appropriate regulatory approach within the current framework. Where necessary and appropriate, specific exemptions may be granted to enact the preferred approach.

Currently, there is little flexibility in the National Law to determine whether a 'thing' is a vessel or not, and if so, what requirements should apply. AMSA will continue to work with the Department of Infrastructure, Transport, Cities and Regional Development, to identify potential changes to the National Law which could remove some of the issues currently being faced, and which could pave the way for a better suited, more flexible regulatory scheme.

AMSA is using industry partnerships to help form our approach. For example, a collaborative relationship has been established with the Trusted Autonomous Systems Defence Collaborative Research Centre (TAS CRC). The TAS CRC is a federal government funded organisation, whose mandate includes providing support to safety regulators who need to gain trust and assurance in the systems, infrastructure and technology that support autonomy in the maritime industry. TAS CRC and AMSA will work collaboratively to explore assurance and trust in automated systems. The TAS CRC assists AMSA by providing subject matter expertise in regulating for a rapidly developing industry.

Third step – long term

AMSA expects that in the longer term, there will be a coherent and appropriate international regulatory framework in place that supports safe operation of MASS. AMSA will work to translate changes occurring at an international level, into the domestic maritime framework where necessary and appropriate, which will help to ensure that all vessels operating commercially in Australia's waters are meeting the same consistent and appropriate standards.

AMSA will continue to monitor emerging technology and trends in the Australian domestic maritime industry, and internationally.

5. AMSA'S FUTURE PLANS

AMSA has identified a number of actions to progress, in order to ensure the regulatory approach to autonomous and remotely operated vessels is appropriate now and in the future.

Firstly, AMSA will continue to contribute to the work of the IMO. AMSA is proactive in influencing the work of the IMO to ensure that international developments are in line with Australia's interests. Proactive engagement internationally provides the opportunity to learn from international developments, and to ensure that potential changes and amendments to the international regulatory framework are reflective of the way Australia's industry will develop and position Australia well for the future.

Secondly, AMSA is committed to consulting and collaborating with industry, including face to face engagement where possible. This consultation and collaboration is a key opportunity to gather information, test proposals for change, and ensure any proposed regulatory solutions are fit for purpose now and into the future.

Thirdly, AMSA is prioritising identifying ways to increase flexibility within the regulatory scheme, for example through amendments to the National Law. These potential changes would allow AMSA to determine what is and is not a vessel, and what requirements should apply at the outset, rather than via exemption down the track.

Finally, AMSA is committed to building internal expertise to respond to enquiries and assess applications for regulating efficiently and effectively. There is a focus on making decisions on a case by case basis, and looking closely at the risk that particular vessels and their operations present to people, the environment, and other vessels. AMSA has a dedicated team who work on these issues, which proactively liaises across different areas of the organisation, steering progress in the right direction, and ensuring proactive progress in some of the more challenging areas.

The Australian maritime industry can expect to see increased engagement from AMSA on these issues in the future.

6. CONCLUSION

AMSA's challenge in the context of autonomous and remotely operated vessels, is to provide safe, effective, and flexible regulation, which appropriately balances safety and protection of the marine environment, and meets the needs of a complex, evolving industry.

The first step, of gaining real experience, and establishing policy principles, has already occurred. The second step, of proactively consulting and collaborating with industry, and exploring potential changes to domestic maritime regulation, is underway, and is positioning AMSA well to successfully undertake the third step, which is designing and enacting regulatory change.

As AMSA looks to the future, we see the growing wave of technological advancements and the changes they will enable, and we see the opportunities for Australia's maritime industry expanding. Whether our stakeholders are involved in scientific research, surveying, or the movement of people or goods, they can be assured that the domestic maritime regulatory framework will be suitable and appropriate to enable a minimum level of safety assurance for them, and the Australian community.

CBRNe Enabled Drones – The State of Play and Problematic Future

H J Slack
Frazer-Nash Consultancy, Level 11, 385 Bourke Street, Melbourne, Australia, VIC 3000, Australia; h.slack@fncaustralia.com.au

Keywords: UAV, CBRNe, Maritime

ABSTRACT
Unmanned Aerial Vehicles (UAVs) from both foreign military and non-state actors are becoming an increasing threat to maritime vessels. The small size and manoeuvrability of UAVs has broadened their appeal, causing significant disruption to a variety of locations such as airports and aircraft carriers, as well as potentially posing direct threat to life. UAVs have ushered in an era of low-cost and easily accessible weapons that can deliver a Chemical, Biological, Radiological, Nuclear and Explosive (CBRNe) payload from a remote location. This paper discusses this present and future threat to maritime vessels, and the mitigation strategies available. A CBRNe UAV attack can have a significant impact on a maritime vessel. Without adequate sensors to detect a UAV or the subsequent CBRNe payload, the attack could result in; loss of life, loss of the vessel and a time-consuming and costly decontamination effort. The threat posed by UAVs will only increase in the future as the technology advances and the expense declines. A safe future for military and commercial maritime vessels increasingly depends on an organisation's ability to remove or react to a CBRNe threat. The placement of suitable detectors is therefore a key element of protection, allowing for a quick response time once the threat is detected.

1. INTRODUCTION – THE PRESENT THREAT

Unmanned Aerial Vehicles (UAVs) have become commonplace in both military and commercial applications. UAVs are already extensively used by a number of military forces including the ADF. Systems previously used include the ScanEagle, Wasp, Skylark, Shadow and Heron UAV. [5]. The reduction in cost of buying or building an UAV has narrowed the capability gap between non-state actors and conventional military forces by providing an accessible airborne capability. For non-state actors, UAVs are an extremely attractive option to deliver a payload in a maritime environment. They can be purchased at a relatively low cost and can therefore be seen as expendable. UAVs can be modified or custom made with open-source information, and specialised training is not required for operation.

UAVs are already being using by state and non-state groups due to their accessibility. In Ukraine drones have been used by government forces and separatists for intelligence and reconnaissance gathering efforts. Similar trends have been seen in Iran and Syria, with sightings of UAVs dropping what have reported to be explosive payloads. In October 2016, the Islamic State skilled two Kurdish soldiers using an armed UAV. [9]

In 2016 and 2017 the US reported a hostile interaction with a Sadegh-1 drone in close range of an F/A-18E Super Hornet trying to land on USS Nimitz. This was of particular concern as the

UAV has weapon capability. [10] In 2017, Islamic State documents were obtained which detailed the network's UAV programme. The devices used were commercially available but reconfigured to carry small bombs and munitions. [11] In August 2018, an alleged assassination attempt on President Nicolas Maduro of Venezuela took place in Caracas. Two DJI M600 UAVs each carrying 1kg of C-4 explosive were used. President Maduro was unharmed. [1]. The first suspected UAV swarm attacks were witnessed in 2018 at an airbase in Syria and a naval station on the Mediterranean Sea. [10, 11]

The Islamic State have previously released videos that show UAV footage from above the battlefield demonstrating their Intelligence, Surveillance and Reconnaissance (ISR) capabilities. The Lebanese militant group Hezbollah and Hamas, the Palestinian group that operates in the Gaza strip, are confirmed to have military UAVs in their arsenal supplied by Iran [11]. However, a whole host of other non-state groups have used UAVs in a military use through the rise and popularity of commercial systems. New America's World of Drones database details non-state groups which have deployed UAVs including: Isis, Hezbollah, Hamas, Houthi rebels, Libyan rebels, various Syrian rebel groups, the Revolutionary Armed Forces of Colombia (Farc) and Colombian and Mexican drug cartels. [1, 11]

Therefore it is essential to appreciate that weaponised UAVs are not a distant or possible threat, but a clear and present one, and a threat that is likely to grow swiftly as ever-more sophisticated technology becomes ever-more accessible. While terrorist groups cannot match armed forces in conventional weapons particularly in the air domain, UAVs have changed the battlespace. They provide a strategic benefit by providing an airborne capability. They are easy to adapt and can provide a range of capabilities including but not limited to; surveillance, offensive, disruptive, and smuggling goods. As far as the military is concerned, UAVs can be equipped with explosives or weapons which are remotely piloted creating an ever changing threat for armed forces. The risk that a UAV will be used for terrorist attacks is of high concern for those in charge of counter-terrorism measures.

2. THE MARITIME THREAT – CBRNE PAYLOADS

CBRN attacks provide a huge threat to maritime vessels. A payload could be dropped onto or near a vessel, or fitted with explosives and detonated nearby. Depending on the weapon chosen, the resulting impact could result in an expensive and time consuming decontamination effort, to the loss of the vessel and the loss of life. A UAV carrying a potential CBRNe threat situated downwind or above the vessel would pose a significant threat to the vessel and crew if the payload was released. Ultimately the damage incurred could result is mission failure. The resulting impact of an attack could all be achieved be a single low cost UAV operated from a remote location from a single non-state actor.

The present threat is an example of symmetric warfare where low level technology is capable of causing significant damage to large scale maritime vessels. The risk posed is comparable to the current crisis faced by governments and institutions with regards to hacking.

While the purchase of UAVs in not controlled and training is a simple matter, the CBRNe element is under much stricter regulation. There are a number of international treaties and organisations that manage the proliferation and control of such substances. A complete list of

CBRN treaties, organisation and regimes relating to disarmament, arms control and non-proliferation of weapons of mass destruction are detailed by the Nuclear Threat Initiative. [2]

Radiological and nuclear threats can both cause mass destruction but are highly unlikely to be delivered by UAV in a maritime setting, due to their difficultly to obtain and control. Two particular conventions to highlight are the Chemical Weapons Convention (CWC) and the Biological and Toxins Weapons Convention (BTWC).

The Chemical Weapons Convention (CWC), enforced by the Organisation for the Prohibition of Chemical Weapons (OPCW), details specific substances that are controlled in terms of development, production, stockpiling and the use of Chemical Weapons. [3] A number of chemicals that can be used in an attack are in common use in the manufacturing industry and can also be produced by non-state actors. The Biological and Toxins Weapons Convention (BTWC) enforced by the United Nations prohibits the development, production and stockpiling of bacteriological and toxin weapons. [4]

3. DETECTION OF THE THREAT

Maritime vessels in the Australian Defence Force (ADF) and other military forces are currently equipped with CBRN detectors that warn crew when a threat is detected. CBRN detectors are commonplace on maritime vessels, substantially predating the recent rise in the use of UAVs. The placement of CBRN detectors on-board a vessel can be key to rapidly detecting a threat in order to protect the crew member and to contain the attack. The quicker the agent can be detected, the more chance that no crew will be harmed and a decontamination effort will be effectively and efficiently carried out. CBRN detectors are needed in conjunction with technologies to remove the threat of a UAV in case a different method is used or the attack is not detected by the vessel. The risk previously foreseen would be the vessel travelling through an area in which a CBRN agent had been released. UAVs add a new element to this space allowing for the threat to be dropped onto or near the vessel with potentially very little warning.

The detection methods currently available to detect drones are:

- Radio-frequency;
- Infrared;
- Acoustic; and
- Visual.

Military vessels are generally equipped with radar and a form of electro-optical device for tracking or scanning. Civilian ships are generally equipped with navigational radar which will be less likely to detect an UAV.

4. ELIMATING THE THREAT

Once a UAV has been detected, the threat must be removed. RC Protocol Manipulation is a series of steps that can be used to remove the threat posed by a UAV. [6].

Soft kill level 1: detection and monitoring – watching to ensure the UAV is not a threat. In a maritime environment the monitoring period may be drastically reduced to that of an urban environment such as when the vessel is alongside due to its perceived risk.

Soft kill level 2: taking control of the UAV – if technology is available on-board to achieve this, the UAV can be taken away from where it will cause the most disruption. Technology to achieve this is unlikely to be available at sea and also unlikely it could would be at an acceptable state of readiness.

Hard kill level 1: projectiles and kinetics – this includes nets dropped by other UAVs and bullets. Nets have recently been trailed, however this solution is unlikely to be made available on-board due to the readiness level required. Bullets are a realistic and available method of UAV neutralisation depending on the distance from the vessel.

Hard kill level 2: Lasers, magnetic, and other means. Semi-active laser targeting has been considered as a potential solution. This is a considered a good solution for surface-to-air strikes.

One popularly considered option is jamming the signals for the device. For operations occurring in line of site, radio-frequency is used, and beyond this it is maintained using satellite.

The Dutch police have trained birds of prey to take down UAVs in an urban setting. This technique allegedly causes no harm to the bird and could be a potential solution for vessels alongside, although their applicability at sea is dubious. [7]

The Hard Kill Challenge in New Mexico enables a range of Defence contractors to test novel technology during the course of a competition. Systems involved in the 2017 competition included high-energy weapons and an attack UAV utilising a net to drop onto hostile drones. [8]

5. CURRENT PROTECTION METHODS

During a demonstration the US Navy shot down a UAV with a laser over the Persian Gulf. Raytheon is pitching is C-RAM air defence system as a UAV countermeasure. Northrop Grumman is altering its G/ATOR air defence system to monitor UACs, as is Lockheed Martin and its An/TPQ-53 radar. [6]

6. FUTURE THREAT

The popularity of UAV use in military and commercial settings has led to a number of debates on what has been called 'killer robots'. [12, 13, 14] UAVs controlled by AI could be capable of undertaking missions without human direction which has caused much debate on the ethics of such systems for state use. Ethical concerns are highly unlikely to be considered by non-state actors.

7. ADVANCED DRONE TECHNOLOGY

UAV swarms have already been used as an offensive tactic, however coupled with a CBRN or explosive payload this could produce devastating effects and prove difficult to protect against. It is already possible to build a UAV with 3D printing however this space has not been fully utilised. Advances in autonomy, optical navigation, AI and mounted weapons all create opportunities and problems for protecting against this threat.

8. FUTURE OF THE PROTECTION SPACE

The protection space of the future is critical in protecting life and assets. The wide availability of UAVs make it difficult to develop ways to counter against them. Military and commercial efforts are trying to develop technologies to spot unmanned aircraft.

Radar is currently been used to detect UAVs however as previously discussed it is difficult to distinguish smaller devices from birds, therefore calibration of detection needs to be altered. It is important that fine calibration does not produce excessive false positives which may encourage poor behaviour and lead to ignoring future signals and alarms. Radar may be able to be used in conjunction with other technologies in order to effectively detect these systems.

Black Dart is an annual event held in California where anti-UAV specialists gather to test weapons against pilotless aircraft. [15] This event highlights the growing need for a protective solution against such devices.

The acquisition systems of most developed countries produce large, expensive weapons, which provide an easy target for UAVs. In order to produce new technology, governments and militaries undergo an onerous process to bring novel systems into service. The regulation of this process limits rapid and reactive development in order to bring new technology into service. Non-state actors have fewer barriers of this nature and can take greater risks with technology although have a much lower financial backing.

The Royal Navy is currently looking into 'laser beam riding' where a low-power beam carrying encoded information is directed to the specified target. A missile is launched and it then 'rides' the beam to the target. The low power and narrow frequency of the beam make it difficult for the target to detect, inhibiting counter measures. This technology has been proposed as a solution to fast in-shore attack craft, drones and shore installations. [10]

Much recent effort has been focused on how to protect against UAVs however importance still needs to be placed on the detection of CBRN agents. It is entirely possible a UAV may be able to release its payload close enough to a vessel, or even onto the vessel while undetected or unmanaged. CBRN detectors while present on most maritime vessels must be able to detect the threat covering its area. The placement of such sensors has become of increasing importance as a number of military forces have looked into the best combination of sensor placement to be able to detect a threat in the vast majority of instances.

9. CONCLUSIONS

UAVs are now commonplace in the battlefield for surveillance and intelligence gathering, and are increasingly being adapted to carry weapons. A number of non-state actors have taken advantage of this technology, using UAVs to deliver a CBRNe payload.

The combined effect of a CBRNe weapon and an UAV as a delivery method have already been used by non-state actors. UAV technology is becoming more advanced and more readily available.

Countering UAV technology requires solutions that will be effective over a wide range of circumstances in order to be effective on-board a vessel.

Advances in technology have been seen in the soft and hard kill categories over the past few years, with the US military organising events to test solutions to the ever expanding problem. UAVs armed with a CBRNe payload may require a new tactical angle to remove the threat as the dispersion of the payload too close to the vessel would have the intended outcome of the assailant.

As well as the development of soft and hard kill technologies it is important that the placement of sensors on vessels are intelligently chosen if a CBRNe threat has been released with or without the knowledge of the crew. Ensuring these sensors cover the area of the boat will help to save lives and the cost of a decontamination operation.

REFERENCES

[1] In-Depth: World of Drone, Bergen P, Sterman D et al. 'https://www.newamerica.org/in-depth/world-of-drones/5-non-state-actors-drone-capabilities/'

[2] Nuclear Threat Initiative 'https://www.nti.org/learn/treaties-and-regimes/treaties/'

[3] Organisation for the Prohibition of Chemical Weapons, Convention on the Prohibition of the Development, Production, Stockpiling and Use of Chemical Weapons and on their Destruction 'https://www.opcw.org/sites/default/files/documents/CWC/CWC_en.pdf'

[4] United Nations Office for Disarmament Affairs, The Biological Weapons Convention 'https://www.un.org/disarmament/wmd/bio/'

[5] Legal and Practical Challenges Associated with the use of Unmanned Aerial Vehicles in the Maritime Environment, Soundings, Issue 11, May 2016, Moloney, S. 'http://www.navy.gov.au/sites/default/files/documents/Soundings_No_11_0.pdf'

[6] Defence One, Against the Drones, Watson, B. 'https://www.defenseone.com/feature/against-the-drones/'

[7] The Guardian, Eagle-eyed: Dutch police to train birds to take down unauthorised drones, Thielman, S. 'https://www.theguardian.com/world/2016/feb/01/dutch-netherlands-police-birds-unauthorized-drones'

[8] Pentagon Tests Lasers and Nets to Combat a Vexing Foe: ISIS Drones, The New York Times, 23 September 2017, Schmitt, E. 'https://www.nytimes.com/2017/09/23/world/middleeast/isis-drones-pentagon-experiments.html'

[9] ISIS used an armed drone to kill two Kurdish fighters and wound French troops, report says, The Washington Post, 11 October 2016, Gibbons-Neff, T. 'https://www.washingtonpost.com/news/checkpoint/wp/2016/10/11/isis-used-an-armed-drone-to-kill-two-kurdish-fighters-and-wound-french-troops-report-says/?noredirect=on&utm_term=.f8ea89b88e43'

[10] Answering the Asymmetric Maritime Threat, 39 March 2019 'https://www.thalesgroup.com/en/united-kingdom/news/answering-asymmetric-maritime-threat'

[11] How do we thwart the latest terrorist threat: swarms of weaponised drones?, The Guardian, 20 Jan 2018, Sims, A. 'https://www.theguardian.com/commentisfree/2018/jan/19/terrorists-threat-weaponised-drones-swarm-civilian-military-syria'

[12] From CNN: The Rise of the Killer Robots, 5 November 2015, Conn, A. 'https://futureoflife.org/2015/11/05/from-cnn-the-rise-of-the-killer-robots/?cn-reloaded=1'

[13] https://www.hrw.org/topic/armas/killer-robots

[14] Human Rights Watch, Killer Robots and the Concept of Meaningful Human Control. 'https://www.hrw.org/news/2016/04/11/killer-robots-and-concept-meaningful-human-control'

[15] Army Air Defenders participate in Black Dart 2018, 25 September 2018 Rader, J. 'https://www.army.mil/article/211603/army_air_defenders_participate_in_black_dart_2018'

Reduced Project Duration via Intelligent Scheduling for Submarine Production and Maintenance

Jordan Ciricillo
General Dynamics Electric Boat
jciricil@gdeb.com

Jorge Payne
General Dynamics Electric Boat
jpayne@gdeb.com

Robert Richards
Stottler Henke Associates, Inc.
richards@shai.com

Submarine production has benefited significantly from more intelligently allocating resources and managing other constraints, thereby increasing efficiency and reducing overall project duration. Intelligent scheduling is being leveraged by General Dynamics Electric Boat (EB) for the scheduling of various aspects of submarine construction, to increase the speed of production. Scheduling is an NP-complete problem that is the size of the solution space grows exponential as the size of the project grows linearly and therefore problems of any reasonable size cannot be solved simply mathematically. Unfortunately, most commercial project management software does not benefit from intelligent scheduling technology. Most 'solutions' such as resource leveling greatly simplify the problem and thus result in far suboptimal results. Intelligent scheduling includes a strategy that leverages scheduling heuristics learned from many of the world's best human schedulers, including those involved in submarine scheduling, in order to solve complex scheduling challenges in reasonable amounts of time.

The goal of scheduling is to optimize the synchronization of resources and other constraints to minimize the duration of the project. For submarine production and maintenance, resources include human resources, equipment resources, and physical-space resources. This paper reviews some of the literature on this topic showing different techniques and results showing the major difference in schedule duration due to the scheduling engine. Real-world experience from EB is provided to further illustrate the real-world impact, and the lessons learned. Thus, without adding one extra resource, submarines are being produced by EB more rapidly just by utilizing better scheduling technology.

INTRODUCTION

Scheduling, at its most basic, is the process of assigning tasks to resources over time, with the goal of optimizing the result according to one or more objectives [1]. Scheduling is heavily used in ship and submarine production to minimize the time and cost associated with the completion or production of small to large, simple to complex projects. The Aurora scheduling framework is one example of a general-purpose intelligent scheduler that has been successfully applied to a variety of domains [2], [3], including submarine production. Intelligent scheduling combines graph analysis techniques with heuristic scheduling techniques to quickly produce an effective schedule based on a defined set of tasks and constraints [4]. This typically includes the following:

Temporal: Tasks must be scheduled between the project start and end dates; each task has duration and an optional start date and an optional end date.

- Calendar: Tasks can only be scheduled during working shifts; tasks cannot be scheduled on holidays.

- Ordering: Tasks can optionally be assigned to follow either immediately after/before another task or sometime after/before another task; optionally with a specific offset time in between.
- Resource: Each task can require that resources be available for the task to be scheduled.

The framework distills the various operations involved in creating a schedule that respects all of these constraints into reconfigurable modules that can be exchanged, substituted, adapted, and extended. This framework is used as a foundation to create domain-specific scheduling tools that respect the constraints specific to domains. Additionally, heuristics are tuned on a domain-specific basis to ensure a high-quality schedule for a given domain.

The scheduling framework consists of two primary components: the engine and the user interface. Both components may be customized to create a domain-specific scheduling tool.

This paper describes lessons learned from working on some of the world's most complex scheduling challenges, specifically submarine production.

Heuristics: Importance Of

Scheduling is an NP-complete problem, that is the size of the solution space grows exponential time and therefore problems of any reasonable size cannot be solved simply mathematically. Most 'solutions' such as resource leveling greatly simplify the problem and thus result in far suboptimal results. Stottler Henke has employed a strategy that includes leveraging scheduling heuristics learned from many of the world's best human schedulers in order to solve complex scheduling challenges in reasonable amounts of time.

Consider the following extremely simple example (which is ~~therefore~~ easier to use to illustrate this point) where:

- Three activities, called Activity 1, 2, and 3, from three different orders are all competing for time on similar machines in a particular work center.
- The priority is highest (or the due date is soonest) for Activity 1 and lowest for Activity 3.
- Two different machines exist, A which is expensive and precise and B, which costs less and has higher throughput.
- Machine A is required for Activity 3, but it can also process activities 1 and 2, though it is not efficient to do so.

Let's look at a solution from a simple scheduler: Activity 1 is chosen first for assignment, since it has the highest priority, and it so happens that at the moment Activity 1 can begin, only Machine A is available, so Machine A is assigned to Activity 1. Activity 2 is assigned to Machine B, which has become available soon after Machine A. Activity 2 is soon completed, owing to Machine B's fast production rate. When Activity 3 is finally examined, its required machine, Machine A, is busy, and worse, busy on an activity that it wasn't essential for. Meanwhile, Machine B is idle.

Obviously, this is a suboptimal solution since a different assignment would have prevented Machine B from being idle and prevented expensive Machine A from being assigned to a task that didn't need it. Of course, a more complex scheduler could "look ahead" to see if the cheaper machine might be soon available, but for any such workaround there's a corresponding example that still causes problems. And each of these rules has to be anticipated and created by the scheduling system software developer.

Perhaps a scheduling system could be written that systematically tried every possible solution and selected the best, and therefore optimal, one. In the example above, the number of possible solutions is 2 choices for Activity 1 times 2 choices for Activity 2 times 2 choices for Activity 3 = only 4 possible solutions. However, consider an activity list consisting of only 30 simple resource assignments where (for simplicity's sake) only one resource is required for each activity. Assume on average 4 meaningfully distinct choices (e.g. different machines) for each activity. This means that there are 30 distinct decisions with 4 choices each, so the number of solutions is 4 x 4 x 4 .... x 4 =

4^{30} = over a million trillion possible solutions,

which are clearly impractical to systematically search. This calculation was based on an extreme over simplification, the more realistic, and complicated planning problem is much more difficult. This is the essence of NP-Complete problems. The widely recognized and clearly applicable NP-Completeness Theorem states that to guarantee an optimal solution to an NP-Complete problem requires exponential time (e.g. M^N where M is the average number of options per choice and N is the number choices) which is clearly impractical in this case, since N is typically in the thousands. An optimal solution can simply not be guaranteed for this application.

Therefore, to determine near-optimal solutions in reasonable timeframes requires good heuristics learned from actual human experts on a large number of situations. We have developed both general heuristics for producing good solutions and the techniques and architecture to incorporate domain specific knowledge and heuristics into the planning system. Our expertise includes substantial experience eliciting the required knowledge and cognitive processes from expert planners, then mimicking those processes in software to create advanced intelligent planning and scheduling systems. To wit, Aurora mimics the *decision-making process of expert schedulers*.

FLEXIBLE / RECONFIGURABLE ARCHITECTURE

To achieve maximum flexibility, we designed Aurora to have a number of components that could be plugged in and matched to gain varied results. The scheduling system permits arbitrary flexibility by allowing a developer to specify what code libraries to use for different parts of scheduling. Each of the pluggable components must extend the corresponding general base class that defines the entry-point methods. This allows the objects that are integral to Aurora to interact with them successfully. The libraries may make use of any of the Aurora objects (such as activities and resources) that pass through the interface. These objects provide support for additional attribute caching, permitting domains to make use of custom properties in the scheduling heuristics. The primary pluggable components include a preprocessor; a scheduling queue prioritizer; the actual scheduler, which usually applies several scheduling methods; a conflict solution manager; and a postprocessor. See Figure 1 for a more detailed breakdown of configurable operations.

From this reconfigurable Aurora architecture, we have been able to build quite varied complex and successful scheduling systems; accomplishments range from scheduling the downlinks of US Air Force satellites [5] [7] & scheduling related to space debris tracking [6], to scheduling medical residents during their education at Harvard's Medical School, to scheduling the final assembly of the Boeing 787 jetliner and various other aircraft for Boeing, as well as similar operations for Bombardier and Learjet, to combining intelligent scheduling with Critical Chain Project Management (CCPM), to scheduling the manufacturing facilities of pharmaceutical production.

Further details regarding some of these accomplishments and lessons learned from the experience are provided in the sections below.

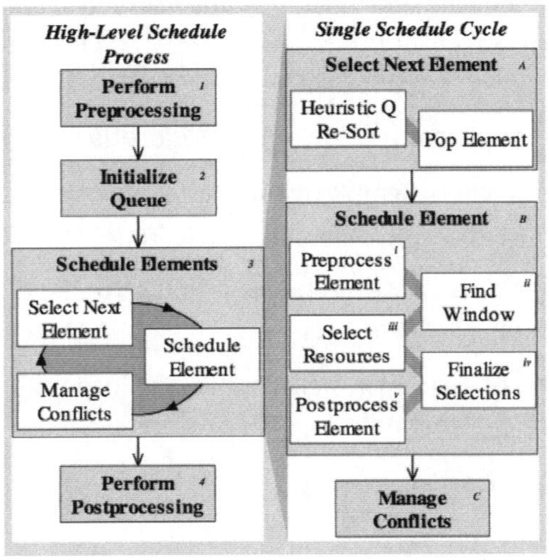

Figure 1. Aurora's reconfigurable scheduling system process breakdown.

General Dynamics Electric Boat

Aurora is being leveraged by General Dynamics Electric Boat (EB) for the scheduling of various aspects of submarine construction, to increase the speed of production. To help maximize efficiency, customizations have been provided to further benefit EB and to provide greater efficiency to the users of Aurora, that is, the user interface has been adapted to make the EB specific use cases even more streamlined.

Electric Boat needed a project management and scheduling tool that could not only support the large models and complex constraints and resource requirements found in submarine construction, but additionally wanted the option of performing of Critical Chain Project Management (CCPM). Aurora's critical chain capability was originally developed for Boeing for aircraft manufacturing which faces many similar scheduling complexities found in submarine production. Due to the complex project management and scheduling challenges found at Boeing, the resulting product CCPM implemented in Aurora has resulted in Aurora being the world's most capable critical chain software solution.

Here are some of the powerful and many times unique capabilities of Aurora that Electric Boat leverages:

- Multiple-pass intelligent resource-constrained scheduling – This intelligent resource leveling tool results in shorter project schedules than the single-pass resource leveling option provided in Microsoft Project and Primavera P6.
- Mixed-mode scheduling – Aurora provides both forward and backward scheduling, available on a task-by-task basis.
- Schedule Rationale – For each task, Aurora provides the rationale/explanation on why it was scheduled where it was scheduled, so it is easy to determine what changes could be made for a task to occur earlier.
- Interface with other scheduling tools – Since the benefits of scheduling are only a mouse click away, Aurora is designed to interface with Microsoft Project, Primavera P6, Artemis, TeamCenter and others. You do not have to change the way you work to receive the benefits of Aurora's scheduling engine.
- 64-bit version can handle projects into the hundreds of thousands of tasks.
- Ability to run how the client wants to run. Aurora can be run as a web-based application or a standalone application under Windows, Mac and Linux.
- Supports more types of constraints beyond finish-to-start, start-to-start, finish-to-finish & start-to-finish, including:
    - physical space constraints, including taking into account the creation and elimination of the space during the project,
    - ergonomic constraints,
    - shift-based constraints,
    - resource Links — graphically depict when resource availability is driving the start date.
- Supports complex human classifications
    - E.g., occupation plus various skills and or certifications, and Aurora optimizes taking these into consideration.
- The ability to leverage knowledge about resource constrained task placement during execution. That is, during execution tasks many times start and complete at different times than calculated during scheduling, therefore resources may become available for a task that was originally scheduled later but could be done now, Aurora understands the details of the schedule and finds these opportunities, thus shortenkng the execution and utilizing resources that otherwise would lie idle. Thus, Aurora determines in real time what is best to work on to minimize project execution time.
- When performing Critical Chain:
    - Ability to take variability of tasks in a chain into account in buffer consumption. That is, if a chain consists of a series of low variability tasks at the beginning then a few high variability tasks at the end of the chain,

standard buffer consumption reports could give an overly optimistic view of the situation.

- Ability to handle short-duration tasks, and update buffer reports on any timeframe (e.g., once every hour).

Thanks to Aurora's modeling capabilities, GDEB now has a tool that allows them to see the level of detail necessary, so the model of reality reacts correctly to the actual reality, resulting in a level of trust, so GDEB is providing the best path forward. An example of the need for modeling human resources with details beyond just an occupation, such as occupation plus a set of specializations and/or certifications, includes specializations that certain welders have. For example, there may be a resource set of welders, all of whom can perform Shielded Metal Arc Welding, then there may be subsets that can also perform Gas Tungsten Arc Welding, there can also be different levels such as apprentice or master. One welder may fall into many different subsets and to make a different resource set by hand for each and maintain this is overly complicated. It is better to have a dataset with the welders and the skills and let Aurora manage the details and allocate the welders optimally.

One of the unique and powerful capabilities in Aurora is the explanation facility. Aurora provides an explanation capability that shows the rationale for why every task is scheduled where it is, that is, each task includes the reasons why it is scheduled at its current time. This is a powerful capability that provides transparency into why the schedule is scheduled the way it is and builds trust by the users. Figure 2 shows a sample explanation. What is usually seen is that the start date may be affected by a start-no-earlier than constraint, then the start date may be later due to one or more predecessors not completing until later, and then finally the actual scheduled start date may be further delayed due to a resource not becoming available until after all the predecessors have completed.

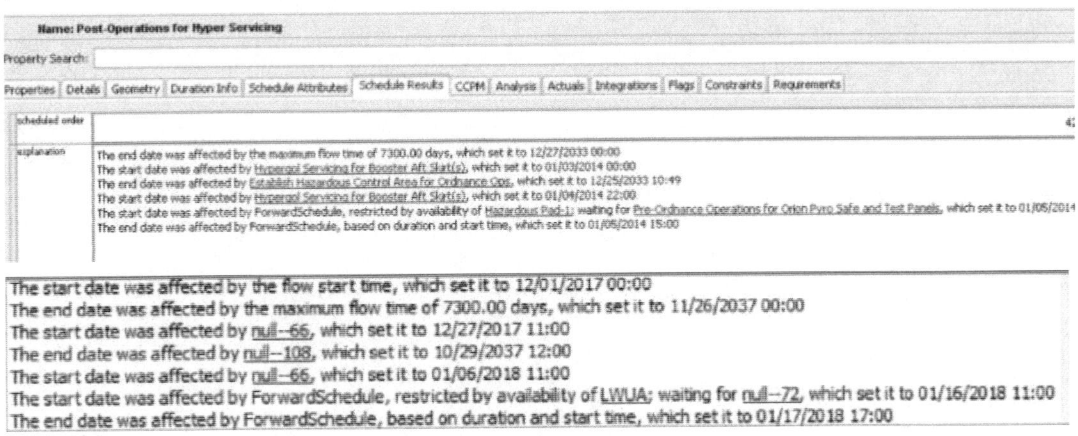

Figure 2. Automatically generated explanation

EB has some of the most sophisticated fabrication capabilities in the world, however, to increase efficiency sometimes it is best to outsource/farm out less specialized work. Aurora already provided many of the graphical and tabular reports to help the user determine what is best to outsource. Aurora has been modified to provide a convenient interface for visualizing which tasks can be outsourced and providing a one-click option to outsource a task that adjusts the actual model appropriately, see Figure 3.

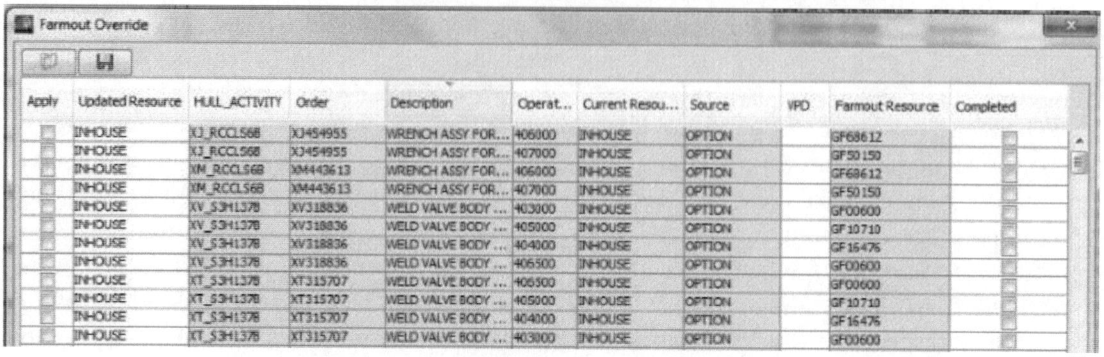

Apply	Updated Resource	HULL_ACTIVITY	Order	Description	Operat...	Current Resou...	Source	VPD	Farmout Resource	Completed
	INHOUSE	XJ_RCCL568	XJ454955	WRENCH ASSY FOR...	406000	INHOUSE	OPTION		GF68612	
	INHOUSE	XJ_RCCL568	XJ454955	WRENCH ASSY FOR...	407000	INHOUSE	OPTION		GF50150	
	INHOUSE	XM_RCCL568	XM443613	WRENCH ASSY FOR...	406000	INHOUSE	OPTION		GF68612	
	INHOUSE	XM_RCCL568	XM443613	WRENCH ASSY FOR...	407000	INHOUSE	OPTION		GF50150	
	INHOUSE	XV_S3H1378	XV318836	WELD VALVE BODY ...	403000	INHOUSE	OPTION		GF00600	
	INHOUSE	XV_S3H1378	XV318836	WELD VALVE BODY ...	405000	INHOUSE	OPTION		GF10710	
	INHOUSE	XV_S3H1378	XV318836	WELD VALVE BODY ...	404000	INHOUSE	OPTION		GF16476	
	INHOUSE	XV_S3H1378	XV318836	WELD VALVE BODY ...	406500	INHOUSE	OPTION		GF00600	
	INHOUSE	XT_S3H1378	XT315707	WELD VALVE BODY ...	406300	INHOUSE	OPTION		GF00600	
	INHOUSE	XT_S3H1378	XT315707	WELD VALVE BODY ...	405000	INHOUSE	OPTION		GF10710	
	INHOUSE	XT_S3H1378	XT315707	WELD VALVE BODY ...	404000	INHOUSE	OPTION		GF16476	
	INHOUSE	XT_S3H1378	XT315707	WELD VALVE BODY ...	403000	INHOUSE	OPTION		GF00600	

Figure 3. Farmout /Outsource interface

Aurora for EB has been enhanced to provide the ability to handle less than perfect data sources, such as having an override for the status of work-in-progress tasks, so schedulers can easily override data from external sources to match reality when those external sources have not yet been fully updated. For example, the latest data may include information about open tasks that actually have zero (0) duration remaining. This may occur if an operation which has an initial estimate of 10 hours, experiences unforeseen circumstances that cause the operation to actually need more than 10 hours to complete. However, the current external system that data is read from simply calculates the remaining duration from the original duration minus the hours worked. Once the hours exceed the original duration, Aurora will see the remaining duration as zero (0). Therefore, a dialog is provided, see Figure 4, that shows all the open operations and their currently calculated remaining durations. The user has the option to change any of the remaining durations or to mark an operation complete. This information can also be saved out separately and later read back in if desired.

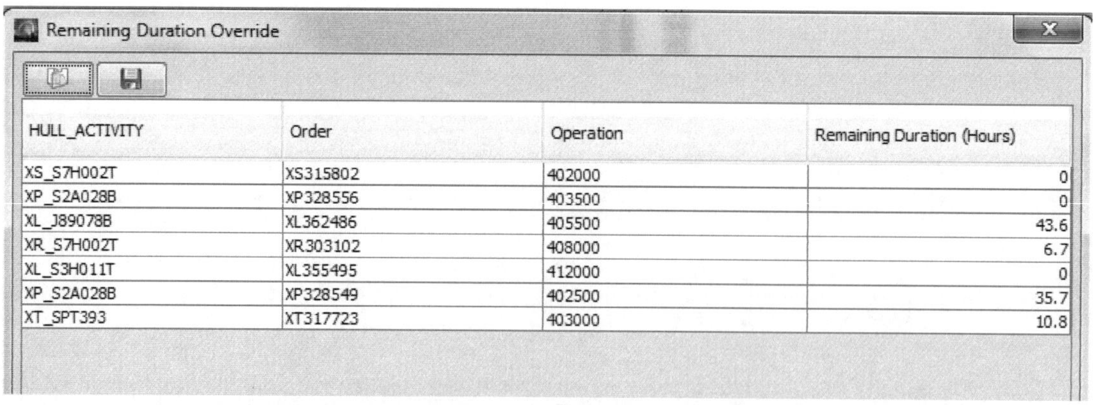

HULL_ACTIVITY	Order	Operation	Remaining Duration (Hours)
XS_S7H002T	XS315802	402000	0
XP_S2A028B	XP328556	403500	0
XL_J89078B	XL362486	405500	43.6
XR_S7H002T	XR303102	408000	6.7
XL_S3H011T	XL355495	412000	0
XP_S2A028B	XP328549	402500	35.7
XT_SPT393	XT317723	403000	10.8

Figure 4. Remaining duration override interface

Overall EB mostly needed enhancements related to ways to increase the efficiency of the user experience. That is, certain data that is read into Aurora from external systems is not updated in a way that Aurora needs for various reasons. For example, when an operation is outsourced in the external system it means the actually outsource process steps will be commenced. This is not appropriate for situations where long-term scheduling is occurring, and outsourcing is used to meet deadlines that may occur months or years in the future. The ability to easily outsource items to test long-term schedules is useful and necessary, but it is not desired to start the outsourcing process since more changes may occur during the interim and the actual outsourcing specifics may change.

Concurrent and Non-concurrent Constraints

Many domains benefit from the concept non-concurrent constraints, due to the fact that workspace is limited and there are many situations where tasks should not be performed too close to each other at the same time. Figure 8 shows non-concurrent constraint for tasks A, B and Figure 9 shows concurrent constraints for task B, A, & C.

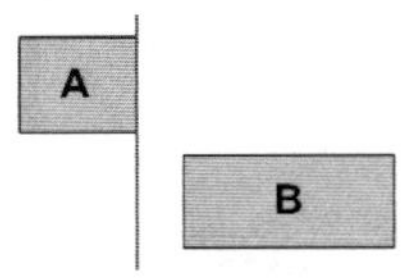

Figure 8. Non-concurrent tasks

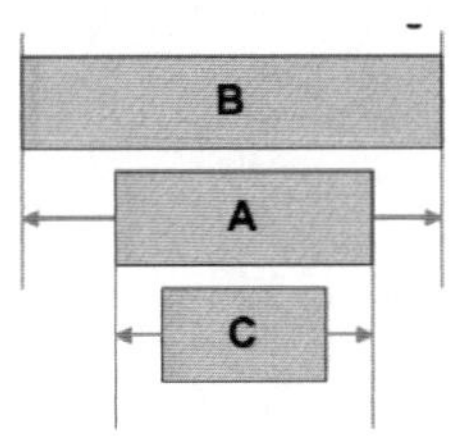

Figure 9. Concurrent tasks

A variation of the non-concurrent constraint is the ability to mark activities as being 'hazardous' to other activities. The result of such a hazardous marking means that Aurora will never schedule the hazardous activities to occur simultaneously with any of the activities it is hazardous to. Graphical enhancements now allow for hazard activities to be denoted in the PERT Chart, with special arrows emanating from the activity causing the hazard and pointing to the activities affected. Figure 10 illustrates hazardous constraints.

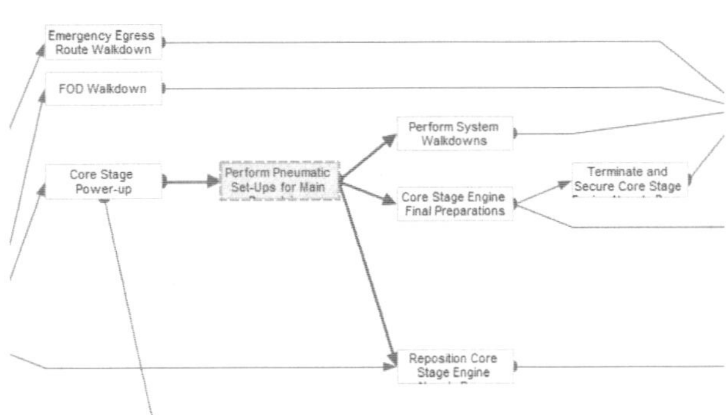

Figure 10. Hazardous constraints shown with red arrows

INTELLIGENT SCHEDULING BENEFITS

The use of Aurora for scheduling has typically meant that 10%+ more tasks can be accomplished with the same resources in the same amount of time (or the same tasks accomplished in 10%+ less time) when compared with other scheduling methods.

One real-world example considers the analysis of a refinery turnaround project. Note that no Microsoft Project results are provided because the MS Project software could not successfully resource-level this project.

The project network consists of over 2,500 activities. A view of the network is shown in Figure 11. Note the red lines link tasks with Finish to Start constraints, this network also has some start-to-start constraints that are shown with yellow lines, some may be seen in the upper-left portion of the network shown in Figure 11.

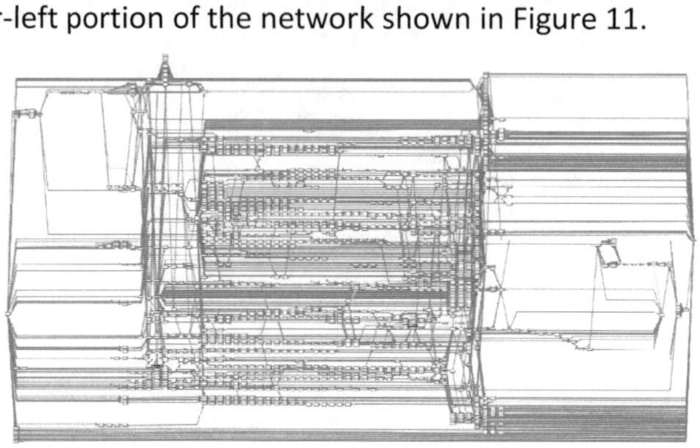

Figure. 11. Turnaround Project Network

The results of the analyses are shown in Figure 12.

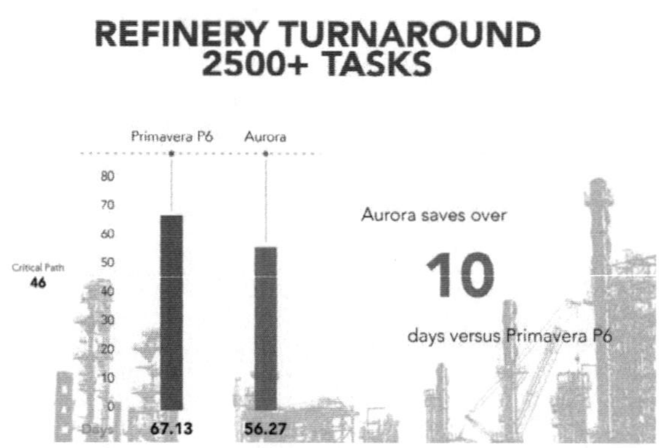

Figure 12. Scheduling Results – Refinery Project

The difference in absolute terms is over 10 days. There are a few ways to compare these results; the simplest is to simply compare overall durations, using Aurora's intelligent scheduling results as the basis: Primavera P6 resource-leveling is over 19% longer than intelligent scheduling. Using the Primavera P6 resource-leveling as the bases: Intelligent scheduling is over 16% shorter than Primavera P6 resource-leveling.

Another valuable perspective lies in comparing the resource-constrained result with the

Critical Path, that is, the situation assuming unlimited resources. Why is this perspective valuable? Because the Critical Path is the best-case scenario, and the valid schedule when considering resources must always be longer than the Critical Path, so the length longer than the Critical Path is the only portion of the total project duration that the resource-leveling or intelligent scheduling can affect.

The Critical Path for the refinery turnaround project is 46 days.

Primavera P6 resource-leveling results longer than Critical Path:	21.125 days
Percent longer than Critical Path:	45.9 %
Aurora results longer than Critical Path:	10.27 days
Percent longer than Critical Path:	22 %

The percent difference between days more than Critical Path for Primavera P6 versus Aurora is over 100%.

These results demonstrate the significant benefit of leveraging Aurora's intelligent scheduling. Recall that everything besides the method for scheduling is the same in both cases. Leveraging Aurora saved over 10.5 days, and all of the associated costs with all the resources that are needed, as well as the lost revenue from the refinery being unavailable.

Of course, the cost savings and other benefits of leveraging Aurora are huge for the initial plan, but even more potential benefit comes in the *execution phase* of the project, where unexpected circumstances need to be dealt with. By leveraging intelligent scheduling, updating the schedule can be done quickly, and the updated schedule will be shorter than if one used resource-leveling only. Therefore, every time a schedule update is performed, the overall benefit of leveraging Aurora's intelligent scheduling increases.

Conclusions

General Dynamics Electric Boat's submarine production has benefited significantly from intelligently scheduling and modeling to the detail necessary so that the schedule itself reacts correctly to real-world changes during execution, thereby increasing efficiency and reducing overall project duration. Intelligent scheduling includes a strategy that leverages scheduling heuristics learned from many of the world's best human schedulers, including those involved in submarine scheduling, in order to solve complex scheduling challenges in reasonable amounts of time.

The capabilities that have benefited GDEB includes:

- Large multi-project support, able to handle 100,000+ tasks per project.
- Multiple-pass intelligent resource-constrained scheduling, resulting in shorter projects and greater transparency.
- Mixed-mode scheduling, supporting both forward and backward scheduling, available on a task-by-task basis.

- Schedule explanations for each task providing greater understanding and transparency.
- Scheduling and re-scheduling occur wall clock time fast, so many what-ifs / scenarios can be performed rapidly.
- Support for various constraint types, which allow for the correct modeling of GDEB realities.

So now the user has the ability to model their situation to the level of detail required, can find optimal schedules, and finally perform the scheduling in such a short amount of time that various other what-if scenarios can be performed as desired.

References

[1] M. L. Pinedo, Scheduling. Cham: Springer International Publishing, 2016.

[2] A. Kalton, "Applying an Intelligent Reconfigurable Scheduling System to Large-Scale Production Scheduling," presented at the International Conference on Automated Planning & Scheduling (ICAPS) 2006, Ambleside, The English Lake District, U.K., 2006.

[3] R. Richards, "Critical Chain: Short-Duration Tasks & Intelligent Scheduling in e.g., Medical, Manufacturing & Maintenance," presented at the 2010 Continuous Process Improvement (CPI) Symposium, Cal State University, Channel Islands, 2010.

[4] Kalton, A., R. Richards. (2008) Advanced Scheduling Technology for Shorter Resource Constrained Project Durations. AACE International's 52nd Annual Meeting & ICEC's 6th World Congress on Cost Engineering, Project Management and Quantity Surveying. Toronto, Ontario, Canada. June 29 – July 2, 2008.

[5] Mohammed, J., Stottler, R., "Rapid Scheduling of Multi-tracking Sensors for a Responsive Satellite Surveillance Network," Proceedings of the Infotech@Aerospace 2010 Conference, Vol. 1, AIAA, Reston, VA, 2010.

[6] Stottler, R., Thompson, R., "Globally Optimized Scheduling for Space Object Tracking," Proceedings of the Infotech@Aerospace 2011 Conference, Vol. 1, AIAA, Reston, VA, 2011.

[7] D. Stottler and K. Mahan, "Automatic, Rapid Replanning of Satellite Operations for Space Situational Awareness (SSA)," presented at the Advanced Maui Optical and Space Surveillance Technologies Conference, 2015.

Paul Houghton
Stirling Labs
paul.houghton@stirlinglabs.com

Accelerating Familiarisation Using Virtual Reality and Self-Directed Learning

Introduction

Familiarisation is a process in organisations that is frequently used to come to understand the nature and context of a process or a thing. A person will gain familiarity not only with workflow, processes and tasks in a traditional sense but now includes familiarity with systems, digital tools and concepts in addition to their traditional job. The need for familiarisation has increased but the resources to support it have not.

SHIPBUILDING TRENDS IN AUSTRALIA

The Naval Shipbuilding Strategy has highlighted the potential for growth in shipbuilding industries across Australia in terms of already known new builds, sustainment activity and newer builds coming on line. With this comes an opportunity to establish an advanced manufacturing and export capability suited to the 21st century.

New approaches, tools and processes are all a part of this and building on Industry 4.0 changes that are strong in Europe comes the concept of Shipyard 4.0 foreshadowing the digitalisation of the industry and other advanced manufacturing that Australia will tackle. This will entail riding a wave of change that is sweeping shipbuilding worldwide with changes such as:

- Distributed design where the design function can be done entirely separate from the production, even in another State,
- Modularity where blocks of a new ship can be built by highly efficient specialists in factories, not necessarily in the shipyard,
- Industry 4.0 combining additive manufacturing, automation and data sharing in logistics networks,

- Lifecycle Planning where consideration of the evolution of a ship is built in from the concept design on,
- Digital twin where the initial CAD model evolves with the ship and is used throughout every stage of it's life, and
- Vessel specific systems that allow for unique capabilities across a fleet.

In addition to new ships being built and sustained across multiple locations across Australia, other influences will shape the nature of Australian Shipbuilding over time. Naval technologies will change faster than the shipbuilding process and so upgrades will be constant which could also mean structural changes such as the new towers on the Anzac frigates to accommodate upgraded electronic systems.

There will be regulation changes too which will reflect the changes in Defence, Government and Society around environmental concerns, energy efficiency, security including cyber security and safety and public liability. This changing landscape of requirements will have continuing impact on the operation and even the design of ships.

During this time the shipyards will adapt and refine their processes and workflows to build an export capability that can be maintained after the current crop of ships. This will only be possible if a workforce with appropriate skills and talent are available to not only do the work of the day but re-skill for tomorrow. This in the face of resource industry booms and other sector demands such as manufacturing and construction.

Thus the role of learning will be crucial in achieving the goals of the National Shipbuilding Strategy, improving productivity, reducing re-work and lowering costs. In amongst this learning mix is the process of familiarisation, often performed in a physical setting and delivered face-to-face in groups by experienced staff.

DIGITALISATION

The key new driver for the bulk of the changes to the Shipbuilding industry, in line with big changes in many other industries, is digitalisation. This is already underway in some aspects of the workflow of a ship build, predominantly beginning in the design office. There is much to come and fast if the examples of Australia's overseas partners are to be a guide. Other industries such as automobile manufacturing and aerospace are considered to be years ahead.

The new tools can be adapted from other industries and with them, the possibility of importing the skills and talent to use them but this is not always possible and competition for these new skills is becoming fierce.

The emergence of product lifecycle management (PLM) technologies that are tailored to the unique challenges of developing advanced ships has been partially responsible for Integrated Product Development Environments which allows for parallel production across different and diverse teams. This results in productivity gains, improved shipbuilding performance and ultimately gains in the total cost of ownership of a ship.

The PLM combines tools for product modelling using 3D computer-aided design (CAD), project management and coordination tools, 'database' features to manage parts, suppliers, and production information in a completely integrated way. This approach scales across multiple teams and multiple shipyards.

Another idea across this new, integrated shipbuilding landscape is that of a digital twin or virtual twin. The concept here is that the digital model developed as part of concept design, carries through to detailed design and engineering, production and ultimately to the operation of the vessel. The design is developed as a 3D model, same with the engineering phase although not always in the same systems. Bringing the actual build data into a digital model is challenging but possible. Then the operational data and live data from sensors etc can become part of this evolving digital twin model.

Sensor data for the operational ship is gathered from large numbers of sensors and analysed for anomalies which can be used to tune performance or predict possible failures. Currently this data is largely brought together in the form of data lakes but increasingly this data is analysed at the edges using machine learning on the sensors themselves.

A more recent addition to the suit of digital tools in shipbuilding is that of virtual reality. The digital models at concept or detailed design stage can finally be experienced by the non-expert as if they were real - allowing inspection, review and problem solving as if the ship were already built. Parts of the same model can provide the detail of compartments or blocks for production. The addition of collaboration tools inside of VR now allows for multiple people to undertake these reviews which can occur across the world involving experts for issue resolution, annotation of models for change by designers and recording of joint decisions for contract variation.

TALENT, SKILLS AND LEARNING

There is a need for talent and skills but not only to bring onboard new skilled staff but to up-skill existing staff in a continuous manner. This aspect of Australia's shipbuilding capability and skill needs was noted in a report by the Rand Corporation [1] and was also published in the more recent Naval Shipbuilding Plan [2].

There are initiatives underway to address this skills gap, pre-eminent among these is the National Shipbuilding College working with existing providers such as TAFE and Universities initially. The nature of continuous change in the industry challenges the premise of existing providers whose current operations largely focus on long-cycle, certified training.

There is certainly an important role for certified training just as there is for other types of learning in the enterprise and across the shipbuilding industry. The new digital tools promote learning in many ways as a natural consequence of their capacity for integration, coordination, collaboration and knowledge sharing. These are capabilities common to the needs of a future shipbuilding industry and a learning environment.

Given the nature of the future changes to the industry and the speed with which these changes will have to come, an examination of non-certified learning should occur to both speed up the adaptation to change and provide a means through which the skill portfolios of individual workers can better reflect their actual talents rather than their job descriptions.

There is an option to explore newer, emerging learning modes that can be supported by digital technology. One of these is the need to contextualise the application of work skills and knowledge in the industry. For example, a contractor arrives to perform a work package on a ship refit. They have the right qualifications and have worked on other ships but the unique nature of the ships described earlier and the unique situation of any one task makes it difficult to quickly come 'up to speed' to do the task at hand. In this case the contractor is given a familiarisation with the specific context in that yard, on that ship, in the compartment he will be working on.

FAMILIARISATION

Familiarisation often currently takes the form of job shadowing where someone new to the job or area of the workplace is asked to follow a more experienced worker around [3]. Trickle-down-familiarisation (eg one crew member training another) is not no longer acceptable in terms of the speed of uptake. It is unstructured, leads to incomplete knowledge of the equipment capabilities, especially the lesser used functions [4]. Alternately, this can also done in groups with a leader which compounds the issues with the way an individual can learn what is needed to successfully do the job since this familiarisation is group based and inherently lowest common denominator for expediency.

Familiarisation is a form of learning in an organisational setting, be it a Shipyard, a design office or a clerical area. It is not standardised, is very difficult to assess, cannot be measured as to quality and therefore cannot be part of a program of continuous improvement [3]. Yet, it is occurring to differing levels of success and with little consideration of the learning nature of the task in that there is little evidence of the conscious application of learning theory to the various familiarisation efforts.

Taking the drivers for familiarisation and the constraints that work against it such as the resources needed for one-to-one familiarisation, the need to transport contractors to a site to become familiarised and the increasing complexity and speed of change in the workplace - a new, more considered approach to familiarisation is needed. Familiarisation is becoming more frequent, more complex and in many ways, more urgent in modern shipbuilding industry.

Any new approach should also take advantage of the digitalisation of the shipbuilding industry to extend the availability and access to familiarisation and improve the richness of it in the context of the job itself. In this way it could be possible to accelerate the process of familiarisation and also increase the number of workers who could participate because it would lessen the need for existing workers to be involved.

While Familiarisation isn't the same as Certified training, it is sometimes called Familiarisation Training which can cause confusion and it is useful to differentiate the two and also to place other forms of organisational learning into this context because the terms can be used interchangeably by some whereas there are clear differences.

In work done by Stirling Labs, this point became clear and a workshop was undertaken to differentiate these types of learning with a client organisation who was seeking to establish an approach to Familiarisation that would fit into their business context and avoid

Organisational Learning

Certified Training	Induction	Orientation	Familiarisation	Drill
Theory and concepts	Less formal	More formal	More frequent	People
Application	Organisation focus	Job focus	Workspace focus	Simulation
Generic	Welcome	Expectations	Location, tools	Communication
			Meaning in context	
			Ownership	

the confusion. These differences can be summarised by the following:

LEARNING MODELS

If we consider how familiarisation occurs currently, it appears that there has been very little thought given to the process other than it being the transfer of information from one to many or optimally, one to one. If we think of familiarisation as a form of learning though, there are some considerations that would make the process more effective and also take advantage of the use of digital technology.

Firstly, the participants are essentially adults who have already undertaken at least 10 years of school and formal learning, are able to think in terms of concepts and in the context of their job, are motivated to learn. This differentiates familiarisation from the sort of learning that occurs in school which is called pedagogy or child learning. Adult learning is called andragogy, a term used by Malcolm Shepherd Knowles and he defined it as the art and science of adult learning. It now refers to any form of adult learning [5].

Knowles defined 5 assumptions about adult learning that are different from those that apply to child learners, they are:

- Self-Concept - As a person matures his/her self concept moves from one of being a dependent personality toward one of being a self-directed human being.
- Adult Learner Experience - As a person matures he/she accumulates a growing reservoir of experience that becomes an increasing resource for learning.
- Readiness to Learn - As a person matures his/her readiness to learn becomes oriented increasingly to the developmental tasks of his/her social roles.
- Orientation to Learning - As a person matures his/her time perspective changes from one of postponed application of knowledge to immediacy of application. As a

result his/her orientation toward learning shifts from one of subject centredness to one of problem centredness.

- Motivation to Learn - As a person matures the motivation to learn is internal [6].

Knowles went on to develop ideas around another characteristic of adult learners and that is that they are self directed in that learners are motivated to assume personal responsibility and collaborative control of the cognitive (self monitoring) and contextual (self management) processes in contracting and confirming meaningful and worthwhile learning outcomes [7].

FAMILIARISATION+VR+COLLABORATION

Virtual Reality has long been on the periphery of technology (since the 1950's) for a variety of reasons but recently the decreasing cost of the computing power, increasing capability of headsets and quality of software have combined to make the VR experience (VRE) affordable and useable. A key factor are the head-mounted displays (HMD) which is worn on the face and all head movement is captured, when used with appropriate software, the user's perspective of the VRE will always be correct regardless of where virtual objects are or where the user is looking. This is in contrast to other technologies that are sometimes referred to as VR, such as "caves" or 3-degree-of-freedom, video based HMDs. [8]

The VR hardware provides a computer-generated visual experience and the software can use this capability to present a VRE that creates what Slater, M. [9] calls the place illusion (PI) and plausibility illusion (Psl). Essentially the VRE causes the sensation of being in a real place (PI) and that the scenarios witnessed, such as colleagues conversing, is actually occurring in this place (Psl).

A central advantage of using VR for visualisation is the elimination of abstraction such as translating a 2D plan or a flat screen 3D representation or projection. This means that the immediacy of the model being viewed is instantly natural because that is how we see the world and have done since birth.

From a learning standpoint, there are even more benefits that can be derived from VR. Research now suggests that immediacy and 'realness' of VR make it suited to support two powerful dimensions of learning:

- Experiential Learning - experiential learning engages most of the senses, builds social-emotional skills, creates a context for memorisation, expands critical thinking and is unquestionably more relevant to real life applications of what's being studied. [10]
- Personalised learning - a diverse variety of programs, learning experiences, instructional approaches, and strategies that address the distinct learning preferences, interests, aspirations, weaknesses, or cultural backgrounds of individual students. The result of this is an educational experience that's more fitting to you as an individual and maximizes what you can get out of each interaction. [10]

This also allows for the an individual learner to engage and re-engage with the model and annotations which will take advantage of the self-directed nature of the familiarisation model. This means the combined impact of experiencing a ship as if it were real, doing this as many times as you need to learn (become familiarised). It allows for the an individual learner to engage and re-engage with the model and annotations which will take advantage of the self-directed nature of the familiarisation model.

If we can then add the capability to collaborate in VR, then we not only allow interpersonal dynamics to be part of the experience but we open up the possibility for interaction with experts, mentors or other learners inside the space. These interactions are not limited to the physical availability of others since these interactions can occur from anywhere in the world via the internet.

This additional capability of new VR environments supports a third type of learning relevant to our familiarisation model

- Mastery Learning - where a learner moves forward when you truly master a subject, without feeling humiliated for taking your time. The end result is actual knowledge that can spread interconnect with other subjects in a number of unexpected ways, both creatively and intellectually. [10]

The need for familiarisation in organisations is currently recognised and resources are being allocated to it but this paper is suggesting a new approach based on treating familiarisation as an adult learning activity that can take advantage of new technologies such as VR. The costs associated with the more traditional approach of physical familiarisation in groups means that is not done as regularly as needed or effectively as possible for any one individual.

The benefits of treating familiarisation as a form of adult learning that promotes self-directed learning appear appropriate and can be tested. The use of VR offers many advantages including familiarisation learning on ships that are still in the design or construction stage. Add to this the capacity for learners to solve problems together using collaboration tools and it becomes an approach that offers a powerful logic for exploration.

More obvious amongst the benefits are:

- Effective use of resources and time because the individual drives the learning and accesses help when needed for clarification or additional learning,
- Use of VR for experiential, personalised and mastery learning for familiarisation supports the nature of self-directed adult learning,
- Collaborative problem solving as a context for familiarisation,
- Communities of practice can be built on the back of these activities to continue learning and
- Knowledge sharing becomes possible across the distributed digital shipyard.

Stirling labs will be undertaking a number of projects in the next year with clients who are already interested in exploring this approach to familiarisation and through that time, the models will be tested and refined.

References

[1] RAND Corporation, Australia's Naval Shipbuilding Enterprise: Preparing for the 21st Century, 2015

[2] Department of Defence, Naval Shipbuilding Plan, 16 May 2017

[3] Murray Goldberg https://www.maritimeprofessional.com/blogs/post/familiarization-training--the-ugly-stepsister-of-certification-13568

[4] Australian Maritime Safety Authority (AMSA) https://www.amsa.gov.au/print/1924

[5] Kearsley, G (2010) Andragogy (M Knowles) The Theory into practice database. Retrieved from http://tip.psychology.org

[6] Knowles, M. (1984). Andragogy in Action. San Francisco: Jossey-Bass

[7] D. Randy Garrison, Self-Directed Learning: Toward a Comprehensive Model. November 1997 Adult Education Quarterly 48(1):18-33

[9] Robert Spencer, Jeremy Byrne and Paul Houghton (2019). The Future of Ship Design: Collaboration in Virtual Reality, COMPIT 2019 Tullamore, Ireland 25 March - 27 March 2019

[9] SLATER, M. (2009), Place illusion and plausibility can lead to realistic behaviour in immersive virtual environments, Philosophical Trans. Royal Society of London, Series B, Biological Sciences 364.

[10] Lucas Rizzotto, How VR, AR, & AI Can Change Education Forever, https://www.roadtovr.com/vr-ar-ai-can-change-education-forever-part-1-todays-problems/

The Artificial Intelligence applied to *CAD*/CAM Systems for designing the future Naval Ships

Rodrigo Perez Fernandez, Universidad Politécnica de Madrid, rodrigo.perez.fernandez@upm.es

ABSTRACT

At this time all industrial designs are made by *Information Technology* (IT) tools, as *Computer Aided Designs and Manufacturing* (*CAD*/CAM), maintained and sustained with suitable databases and with lifecycle management of the information. To address the challenge of the *Artificial Intelligence* (AI) for designing better Naval Ships, it is necessary the *CAD*/CAM tools to adapt in different manners.

The difference items of a Naval Ship should not be autonomous objects that decide by themselves, but they should be supervised by a central system that takes into account the product as a unique entity. A Naval Ship is a product with a huge value, but also has a vast importance for what it contains, so it cannot be left to the edge of technological autonomy. It is necessary to have a higher responsibility and this responsibility must be of a human intelligence.

Future Naval Ships must be connected to be Smart thought *Internet of Things* (IoT) technology. The connection of smart devices within a Warship must be human controlled. The control should start from the design tools because they control the shipbuilding process, from the early stages of the design up to operation management through detail and production stages. The collection of design systems, *Product Lifecycle Management* tools and other devices must be inter connected among them and will be part of the *AI* platform for the Naval Ships. The information shared in the scope of the *AI* must be managed by the human along the whole lifecycle of the Warship, starting from the beginning of the initial design. This need requires the *CAD* tools to be prepared with specific characteristics to handle that information.

1. INTRODUCTION

The *Industry 4.0* revolution involves a big bunch of interconnected and interlaced technologies, which can be studied separately, but should be implemented as a whole integrated development in each industrial field, moreover, in the *Computer Aided Design*, *Manufacturing* & *Engineering* (*CAD*/CAM/CAE) System, from now on referred just as *CAD*, development industry.

Each technology reveal a set of limits which cannot be distinguish clearly from its neighboring technology.

Augmented reality, virtual reality and mixed reality are closely related to the digital twin and interlaced with the *Big Data* which is generated by the *CAD* tools and all surrounding solutions, which applies some *cloud/edge/fog* computing to this data in a merged technology between finite state machines and *Artificial Intelligence* (AI) cognitive processes.

To perform in an agile manner all these computing, it requires a network which support different connection ways to add special devices, i.e. *Internet of Things* (IoT) which can access to the data, creating and modifying it, in a different layer which affects to the basic information layer created by the *CAD* System in the shipyard.

This network should be secured (*cybersecurity*) but open to allow distributed work, which must be step controlled in a manner that records any modification of each working step done in an open, transparent, trusted and non-modifiable working method for all actors involved in process, like: shipyard, engineering offices, classification society and ship owner (*blockchain*).

Results of the design should be easy integrated with future building ways like 3D printing, generating printing orders directly from the *CAD* model.

Shipbuilding phases involve design and manufacturing, but an integrated *Industry 4.0 CAD* System should also be involved in operation and maintenance phases.

When a warship comes for a reparation, sometimes the model is not available for this operation of maintenance.

In reparation and maintenance steps, replicating the full engine room or any other local in the *CAD* System could be a nightmare, unless *CAD* system has an *AI* processing toolkit which, from a cloud of points, can recreate *CAD* equivalent items which can be converted with a minor user intervention, and lesser as the *AI* learns, in a *CAD* integrated and full modifiable design.

At the end, this is a brief sum up of the *Industry 4.0* technologies which can be applied to a *CAD* System (as illustrated in *Figure 1*), included in, or as an integrated surrounding solution or as information generators for the evolutive design process.

Figure 1. Related *Industry 4.0* Technologies in a Shipbuilding *CAD* environment [7]

2. THE ORIGIN OF THE AI

To be able to understand how *Industry 4.0* became today's buzzword, it is necessary to understand that the *Fourth Industrial Revolution* takes the automation of manufacturing processes to a new level by introducing customized and flexible mass production technologies. This means that machines will operate independently, or cooperate with humans in creating a customer-oriented production field that constantly works on maintaining itself. The machine rather becomes an independent entity that is able to collect data, analyse it, and advise upon it. This becomes possible by introducing self-optimization, self-cognition, and self-customization into the industry. The manufacturers will be able to communicate with computers rather than operate them. All of this will be possible through *AI* example.

As regards *AI,* it can be said that there is no single origin of the concept known as *AI* and therefore there is no consensus to define that concept, but to be able to understand any of the definitions that apply, it is convenient to know at least briefly the most relevant facts and some of the milestones in its history.

It can be considered that *AI* was born as a philosophical study on human intelligence based on the concern of man to imitate the behaviour of other beings with capabilities beyond the reach of human beings (such as flying or diving), reaching the point of trying to imitate itself. In this sense, it can be said that *AI* is the search to imitate human intelligence. It is clear that it has not yet been completely achieved, but it is also increasingly true that we are closer.

The first man who became aware of his own existence and was able to think, surely wondered how his thought would work and would conclude the idea superior creator, an intelligent being capable of creating another one. In this sense, the idea of a virtual design for intelligence is as old as human thought.

In 1920, the Czech writer Karel Čapek [6], published a science fiction stage play called *Rossum's Universal Robots*. The play is about a company that builds artificial organic humans in order to lighten the workload of other people. Although in the play these artificial men are called robots, they have more to do with the modern concept of android or clone. They are creatures that can be passed as humans and have the gift of being able to think.

The English mathematician Turing publishes an article entitled *Computing Machinery and Intelligence* that opens the doors to *AI* [5]. The article itself began with the simple question: Can machines think?.

Later Turing proposed a method to evaluate if the machines can think, which got to know itself like the Turing test. The test, or Imitation game, as it was called in the document, was presented as a simple test to judge if machines could think.

1956 Dartmouth conference convened by McCarthy and where the term *AI* was coined. The conference was attended by researchers from Carnegie Mellon University and IBM, including: Minsky, Newell and Simon. In this conference extremely optimistic forecasts for the next ten years were made that were never fulfilled, which caused the almost total abandonment of the investigations during fifteen years, known today as *AI* winter [4].

3. ARTIFICIAL INTELLIGENCE AS A FUTURE SOLUTION

Control system in the ship can include some *AI* predictive processes integrated in the bridge overall control system, which helps to deduce the consequences of maintenance operations, from doing in the correct time as well as delaying or skipping them [2]

This procedure applied is including in the edge/fog computing methods, due to only involves the ship inner communication network, delaying the massive download of operation data upon arrival at port. During navigation data transfer should be only applied to critical operations.

Navigation data can be also useful in design and production phases, to correct some processes in order to obtain more efficient systems, and more efficient designs. This is only possible applying some *AI* process to this data, classifying, processing and getting some results.

This working methodology, to be deeply profitable for both actors, requires a joint venture between *Ministry of Defence* (MoD) and shipbuilder.

AI processes based on navigation data, in the multi-boat paradigm, can obtain information to improve design and production processes, which can be applied to the current series, or an evolved variant of this naval ship type, or other ones.

Ship operation phase is not the only one which produces a set of *Big Data* to be processed by an *AI* system, in the production phase, some calculus can be done in the workshops or even delegated in a cloud system, to be distributed.

This data, in *CAD AI* tools, can be classified generating working sequences, design automatic checks, and automatic design processes.

Now, setting the focus on the Designer, sometimes for the tight schedule or for the start-stop working requirements of the design office, like meetings, designer does not have much time for its principal task: create full effective designs, which requires concentration and calm.
This problem can be solved by *AI*, applying some design rules, and exposing a bunch of effective solutions, delegating the final decision on designer but helping discarding a big group of previous solutions which have shown some problems (learned lessons).
At this point, some sceptic people can think: *AI* is going to substitute designers work, but *AI* is going to augment the capabilities of this designer, making work less stressing and more efficient.
All available *AI* processes which can be applied in a *CAD* system to help designer are represented in *Figure 2*.

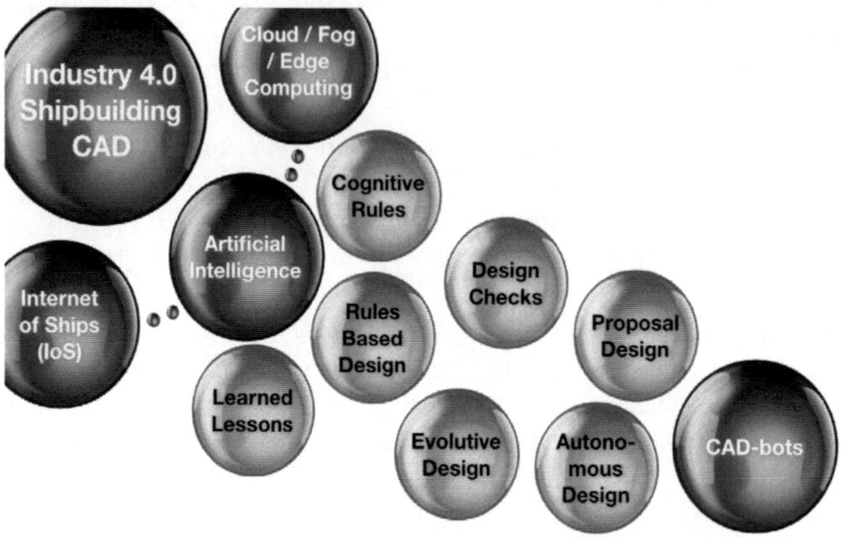

Figure 2. *AI* Evolution in the Shipbuilding *CAD* World [7]

3.1 First level AI CAD tools

AI applied to a *CAD* system should be based on the standard ways to control a design:

- Rules Based Design.
- Learned Lessons.
- Cognitive Rules.

First on the list, rules based design, it is the shipyard standards book of rules. These rules are the base of any dockyard design, and are the first one to be learned by our *AI* system.
To improve it, some other technologies can be applied, like Cognitive *AI,* which is the base of natural language processing in an *AI* solution.
This cognitive *AI* can be run over the shipyard standards book of rules, and with some user help, try to improve understanding in application of the rules from the *AI* tool.
All *AI* solution requires a specific time to learn the correct ways to apply the set of rules.
Next step, also based on natural language spelling, can be add the Learned Lessons to our *AI* solution, creating a mixed ecosystem of rules to be applied to the current design process.
And last step, but not less important, Cognitive Rules, these rules are deduced from the current design and also from production error input in the *CAD* system, like design incidences.

3.2 Second level AI CAD tools

With all first levels fulfill, second level steps can be applied:

- Design checks.
- Evolutive design.

First improvement applied in the *CAD AI* is the capability, in a non-highly intrusive manner, help user with the design, offering a list of design rule application failure to the current work.
This list, if is based on the second part of this tools, evolutive design, can not only the design issues, also offer an improved solution to user.
The bunch of solutions offered is based on the experience accumulated to similar issues found in the historic of the design improvements applied.
These steps generate Learned Lessons input in the *AI* system.
Second level tools requires a high level design knowledge to be trained faster, like a design checker role.

3.3 level AI CAD tools

Once the previous two levels have been accomplish their training, a third level of tools arise:

- Proposal design.
- Autonomous design.

First third level tool, proposal design, is one of the most required tool, in a natural language interchange: "Please, which are the best ways to connect this equipment and that pump?".
Based on the cognitive learning process of the *CAD AI* tools, the answer can be interpreted in machine language as route some pipe lines which connects both items and let me select the better one on my design criteria.
This last step also generate Learned Lessons input in the *AI* system.
Second third level tool, autonomous design, require a big bunch of approvals of the previous question, and based on them and in the *Piping & Instrumentation* (P&I) diagrams, can look for the best location for equipment items in the naval ship, as well as, create a piping solution of the system.
This step generate *Cognitive Rules* input in the *AI* system. These third level tools require months or years of training in the *CAD AI* tools, creating which can be named as *CAD*-bots, autonomous designers which, based on a simple input, P&I diagrams, can recreate a 3D accurate design, which accomplices which the most when not all design rules applied.

4. AI CAD-BOTS

These *CAD*-bots are *AI* augmented cognitive bots which can perform automatically some design parts based on the lessons-learned and cognitive rules, but always fulfilling the complete list of the shipyard standards book of rules.
There is one exception, in the case of contradictory ones. At this point, *AI* tools, represented by the *CAD*-bots can offer designer which rules have a conflict and reapply approved rule to the previous design with reduced effort.
These *CAD*-bots can design and offer user a set of the most efficient solutions, helping in some complex design processes, like the study of the impact of deep design modifications, or even classification rule changes, augmenting user capabilities.

CAD-bots cannot solve all design problems, but their orientations is to do the most repetitive tasks as well as help in some design corrections.

5. AI APPLICATIONS IN MARINE INDUSTRY

The naval industry has always been very traditional and seems to be always in the tail of the implementation of the improvements that in other industries have matured previously. This fact has a reasonable explanation in the difficulty that the naval industry has to convert investments into profits. The naval industry is not friend of risks, especially when the simple fact of making warships is a huge risk in itself. On the other hand, the naval industry collects, to a greater or lesser extent, all the other industries and it may be thought that what is good for the others must also be good for it.
Focused on the shipbuilding industry, *AI* shall address important limitations among which are the lack of data or the confidentiality of the same. The marine industry has focused on immediate results, it is looking very quickly for a solution and does not store data and results in a systematic way that allows it to be used again in similar scenarios. The development of powerful algorithms requires that they can be applied in similar conditions with some recurrence. The data must be correctly structured and reasonably clean so that they can be used with advantage [10]. In the most successful cases there may be limited series of warships that have the same characteristics. It is difficult to find a systematic use of the data, but still there are interesting contributions that occur in the different phases of the life cycle of a naval project. However, it is possible to find a recurrent pattern in the characteristics of the steel parts of the ship. So *AI* systems can assess if they are correctly defined.

5.1 AI in marine design

The first stage of the lifecycle of naval artefacts is designing. We can find interesting approaches to the use of *AI* in this phase. And it must be said that some of them are quite old and date back to the time of the first explosion of *AI* interest. In 1989, the *Defense Advanced Research Projects Agency* (DARPA) of the United States promoted a workshop held at *Rutgers University*, New Brunswick, NJ to support research initiatives of hydrodynamic designs of naval ships. One of the objectives was to clarify the relationships between the hydrodynamic design problems of ships and the areas of *AI* research related to the design and analysis of complex systems [4]. Note that the results cannot be said to be very promising, since they concluded the need to acquire computational fluid mechanics analysis tools *CFD* and integrate them into the design processes and effective control of design processes, focusing on concurrent design and including approaches to explore feasible design space configurations and systematically recording and storing results. However, the expectations remained open and unspecified. Later approaches have been made to apply *AI* to the resolution of complex design problems through expert systems and the appropriate selection of them for certain problems such as structure dynamics or vibrations.
Where it is possible to find a greater variety of proposals is in the task of optimizing the designs using *AI* algorithms that analyse the design space of certain naval ships in which it has been made a systematic parameterization of the variables that allow to define different design alternatives. One of the examples can be found in the article, [16], for the application of *AI* to the design of cargo ships. However, it is not easy to find applications of *AI* to really systematic processes that are in real use, beyond purely academic or research attempts that have not finally materialized in the field of design. The reason can be found in the difficulty to develop these tools and the low return that companies and organizations derive from them.

5.2 AI in naval products and companies

It is possible to find some more tangible applications with a certain validity in the field of the operation of the ships, that is to say, the operation and the management of the transport. In this sense, the initiatives are much more numerous, although some are in phases of research or prototypes, others are in more advanced stages of implementation.

Thus, it is possible to find prototypes of unmanned vehicles that are used in very hostile environments and that require the support of *AI* This is well described by the MIT professor, Henrik Schmidt, in his course *Unmanned Marine Vehicle Autonomy, Sensing and Communications*. These types of artefacts in difficult environments, such as ice water, where communication is an impossible limitation, the role of *AI* is crucial, [11].

Another interesting field of application of *AI* is described in an article about the use of *AI* techniques for the detection of small warships, [15]. The approach is interesting because it raises a situation in which the naval ship is the subject of the observation, but can also be the owner of the application. Having on board systems with detection techniques of other warships, based on *AI,* opens the horizon of unmanned naval ships and their possibilities of realization.

AI also is being integrated into the combat systems of modern ships as essential to identify threats. Thus, the software STARTLE® of the company Dstl was selected by the Royal Navy for the management of threats and is described as a software that continuously monitors the ship's environment in a cut and medium range, processes the data it receives and through techniques of *AI* helps crews to make decisions. "It is inspired by the way the human brain works, emulating the conditioned fear response mechanism of mammals. It quickly detects and evaluates potential threats, the software significantly increases the situational awareness of the human operator in complex environments," [9]. More recently the company Rolls-Royce has signed an association agreement with Google to use the latter´s machine learning engine to improve the company's intelligent awareness systems, [13].

It is also possible to locate *AI* applications in management systems for the exploitation of energy at sea or proposals of companies dedicated to energy in ships. Recently the company Eco Marine Power announced that it would start using the Neural Network Console provided by Sony Network Communications Inc., as part of a strategy to incorporate *AI* in various technological projects related to the ship, including the further development of the patented system of Aquarius MRE® propulsion (Marine Renewable Energy) and EnergySail®, [14].

One of the great references in the naval field is the marine area of Rolls Royce that is trying to promote the application of *AI* in ships in two lines: the intelligent management of assets that covers energy, health, data and fleet management and a second line of business of remote and autonomous operations. The latter includes intelligent detection or recognition, remote operations, autonomous navigation systems and connection with ships, [12]. As we can see on their website, not all lines of work are in operation, but some are in development.

5.3 The future of AI in marine business and industry

The exploitation of the marine business has an undeniable field of growth for *AI* There are many computer solutions that, based on the operating data of the different ship systems, can help manage assets in a more optimal way. The application of *IoT* to ships provides both data collection and the ability to act on assets to obtain their best performance, [10]. For this, it is necessary to have some essential elements. In the first place, we must have comprehensive solutions that cover all aspects of connectivity and integrate them in a coherent manner. It is necessary to have the signalling, connectivity and appropriate representation model based to provide interactivity with the end users. IBM with its program MAXIMO®

and SENER with FORAN® are developing a proposal that integrates reality model made during the initial design stage, with the solution of IBM to merge asset management with the power of data *IoT* obtained of sensors, devices and people to have visibility of them in real time. Have a model of the database in a single database, it allows obtaining virtual reality and augmented reality images on which the obtained data can be superimposed and compared with the technical performance measures expected for each element of the monitored warship. In this way it will be possible to act in the way that each situation advises.

Another interesting field of work for the future of *AI* is image recognition. In this sense and placing the focus on the marine industry, two fields of application appear. The identification of images in autonomous vehicles that can help the mission and the operation of them. Although it does not only have application in ships and unmanned devices, it can also be used in surveillance systems and detection of possible threats or risks in manned ships. Part of this is what one of the projects that have been mentioned above covers, [9].

The recognition of images through *AI* is also of interest in design stages. The need to have virtual models of the objects that are part of a project makes real models can be scanned and then try to be recognized to create the virtual model. This is of particular interest in ship revamping and retrofit. The need to have a virtual model from a real one in order to evaluate the possibilities of retrofitting, including the processes and manoeuvres necessary to carry out such operations. While cloud applications are able to work with that amount of visual information, they have not yet passed the threshold of identifying the elements that appear in the scene and converting them into analytical geometric representations or not, on which can obtain measurements or manipulate as a whole. An extension of this, can be applied to the component models that are used in the design stages by the *CAD* applications. Currently, it is increasingly common for components to be modelled in *CAD* that are obtained through external files that for the most part have been obtained for marketing purposes. These format are superficial representations of many faces that do not have a geometric and parametric representation. This, which is useful simply to see a model makes it useless or even a problem to carry out projects, since it is necessary to have metadata that only exist when the models have a formal geometric representation. That is to say, it is not the same to handle the six faces of a cube, that the cube in its totality. This limitation opens a field of action to *AI* programs that are able to recognize that certain faces form a determined surface, and in turn that certain surfaces form a certain solid. By the moment the available programs help the user, but it is finally the user who validates the conversion. However, *AI* programs can go on igniting this type of recognition that makes a human being to be more decisive each time.

The realization of a naval project is something certainly complex, and not only by the transversality of it but also by the number of tools that must be handled and the limitations imposed by design rules or standards of various kinds: construction, security, etc. *CAD* systems provide more and more tools, but they are also increasingly complex to be used optimally. Along with this scenario, the marine engineering companies and the shipyards are faced with very demanding deadlines and staff or very young who, although they are familiar with new technologies, do not know the art of naval architecting and marine engineering or with very old people who are more reluctant to work with the *CAD* and with new capabilities. Therefore, it would be interesting to have a virtual assistant who can provide all the information necessary to do the job correctly. SENER and IBM are developing a project that integrates the cognitive abilities of Watson® with the functionalities of the *CAD* in the different stages of design.

The platform Watson® will have a corpus of information and data that will integrate everything needed to use correctly and optimally the FORAN® system Furthermore it also will include all regulations applicable to different types of naval ships, regulation of *IMO*, fighting against the pollution, safety regulations, etc. It can even integrate the design and work rules of the shipyards themselves in such a way that *CAD* users can make designs according to all the regulations and applicable standards at the different levels: administration, construction and ship-owners. The system will be trained to learn in order to give the correct information to each type of intention and will be able to learn and trained in different and new things that may affect the design. The integration can be done at different levels, completely decoupled from the *CAD* or coupled to it to perform certain operations in the *CAD*. It launches events that are captured by the listener system and this in turn links with the virtual assistant to provide the information or data that are linked to those operations. The interaction with the system may also be on demand or taking advantage of the natural language processing capabilities of the Watson® system. This project, see *Figure 3*, allows the capabilities of the FORAN® system to be increased to provide the cognitive and analytical capabilities of the Watson® system and make them available to the marine industry to put it in the field of *digitalisation 4.0*.

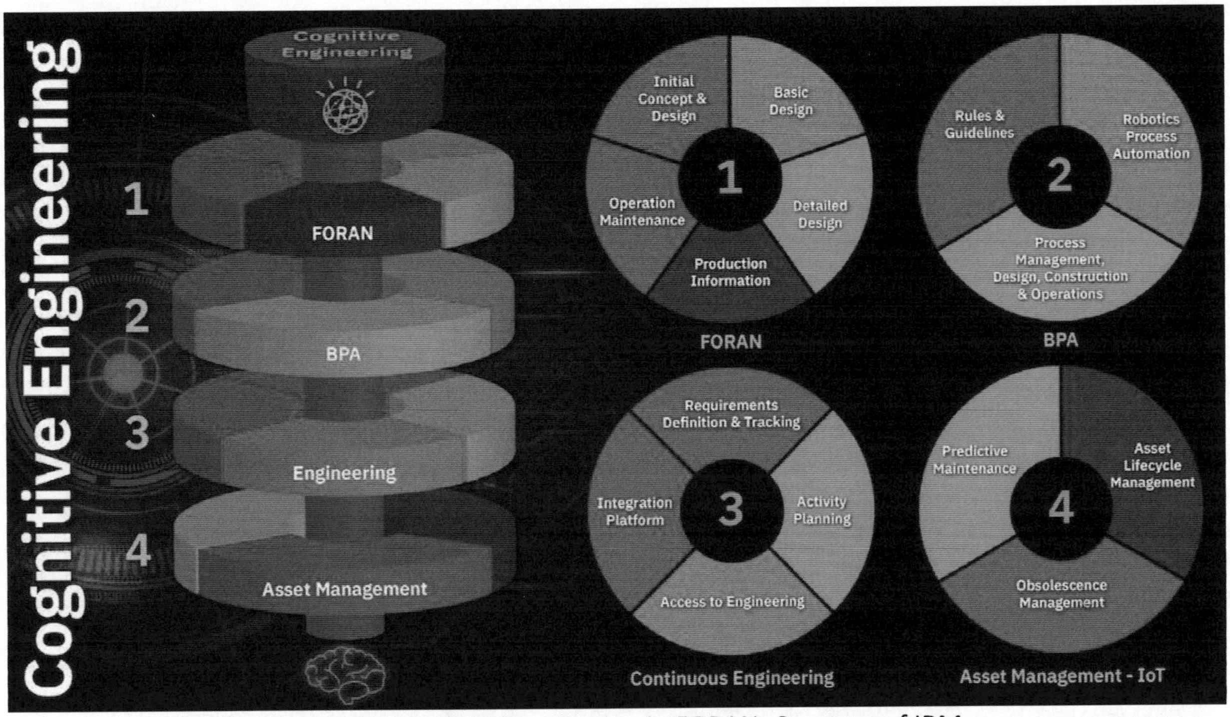

Figure 3. IBM Cognitive Engine in FORAN. Courtesy of IBM

As we have found possibilities of application of *AI* in the field of design, it will also be possible to see the realities of *AI* in manufacturing processes. Perhaps the first steps we will see in the ability of the machines to select the material of a manufacture. Machines that select clipping plates to take advantage of the material according to the remaining pieces or robots capable of organizing the movements of intermediate products in the workshop. The novelty of *AI,* will be that these machines and robots will not have to be programmed, but they will receive the data or the objects with which they have to work and they will know how to act. "In the future, robots will no longer have to be expensively programmed in a time-consuming way with code pages that provide them with a fixed procedure for assembling parts, we just have to specify the task and the system will automatically translate these specifications into a program," [8].

6. FINAL CONSIDERATIONS

AI is one of the enabling technologies of digital transformation that has the greatest potential among those that make up the fourth industrial revolution. Knowing their characteristics and possibilities is essential to decide their application to certain processes and products, especially industrial ones and very particularly those related to the marine sector. It is important to identify the value that *AI* can contribute to the use cases where it can be applied.

AI automates the learning of repetitive tasks and the discovery of relationships through data. It is necessary to populate accurate and reliable data to *AI* systems as well as to provide sufficient information that is well structured and correctly tagged. The *AI* highlights the importance of the data.

It is necessary that those who use *AI* know how to make themselves understood and ask the right questions. The *AI* must be correctly fed, with questions and answers. An *AI* system is as intelligent as the individuals who prepare it.

AI adds intelligence to products, which means that it makes those technologies that incorporate them better, but we must not forget that it is these technologies that provide the core of value to work processes and methodologies.

The use of *AI* in industrial environments such as the naval one is just beginning. There is still a long way to go, in the field of design, optimization of projects, maintenance of data and results. Fields such as the recognition of images, for their conversion into models, the automatic intervention in the validation of the requirements, the optimal exploitation of the processes inherent to naval engineering, are still practically unexplored.

With or without debate, the truth is that *AI* is present day by day in our environment, and its development will be growing. We will see *AI* in applications that we would not have imagined weeks ago. Musical compositions that are the most listened to can be created with *AI* or the most curious and tasty cooking recipes, intelligent cars without a driver or the best doctor capable of correcting diagnosis and treatment. But in spite of all these advances, we will continue to need intelligent people, people who are smarter than machines and who are ahead of them, because there are capacities of the human being who can never be embedded in an artificial intelligence.

As a summary of this paper, the author would like to leave a final reflection for the reader consideration: if the brain can think, a *CAD* software could do it with AI.

REFERENCES

[1] Benayas-Ayuso, Arturo & Pérez Fernández, Rodrigo. (2018). Automated/controlled storage for an efficient MBOM process in the shipbuilding managing the *IoT* technology. SMART SHIP, 23 and 24 January.

[2] Pérez Fernández, Rodrigo & Benayas-Ayuso, Arturo & Perez-Arribas, Francisco. (2018). Data management for smart ship or how to reduce machine learning cost in IoS applications. SMART SHIP, 23 and 24 January.

[3] Nakamoto, Satoshi. (2009). Bitcoin: A Peer-to-Peer Electronic Cash System. Cryptography Mailing list at https://metzdowd.com.

[4] Aggarwal, A. (2018). "The Birth of *AI* and the first *AI* Hype Cycle".

[5] Turing, A.M. (1950). "Computing Machinery and Intelligence". Mind 49, pp.433-460.

[6] Capek, K. (2017). "R.U.R. (Robots Universales Rossum)". Books Mablaz.

[7] Benayas-Ayuso, A. and Pérez Fernández, R. (2019). "What should shipbuilding expect from the *CAD*/CAM systems of the future?". The Naval Architect magazine. April. pp. 28-31.

[8] Wurm, K. (2017), Prototype Robot Solves Problems without Programming, Siemens, https:www.siemens.com/innovation/en/home/pictures-of-the-future/industry-and-automation/the-future-of-manufacturing-autonomous-assembly.html

[9] Mathews, A. (2016), Artificial intelligence to play key role in maritime combat, https://aerospacedefence.electronicspecifier.com/marine/artificial-intelligence-to-play-key-role-in-maritime-combat

[10] Muñoz, J.A. and Pérez Fernandez, R. (2017). "*CAD* tools for designing smart ships in the world of the Internet of Things" SMART SHIP 2017, 24 and 25 January.

[11] O'Leary, M.B. (2017), MIT News, http://news.mit.edu/2017/unlocking-marine-mysteries-artificial-intelligence-1215

[12] Rolls-Royce (2018), Ship Intelligence, https://www.rolls-royce.com/products-and-services/marine/ship-intelligence.aspx#

[13] Kingsland, P. (2018), Ship Technology, https://www.ship-technology.com/features/rolls-royce-teams-google-ai-driven-ship-awareness

[14] MI News Network (2017), Marine insight, https://www.marineinsight.com/shipping-news/eco-marine-power-study-use-artificial-intelligence-research-projects

[15] Del-Rey, N; Mata, D.; Jarabo, M.P. (2017), Artificial intelligence techniques for small boats detection in radar clutter. Real data validation, https://www.sciencedirect.com/science/article/pii/S0952197617302610

[16] Abramowsky, T. (2013), Optimization, Application of Artificial Intelligence Methods to Preliminary Design of Ships and Ship Performance, https://www.researchgate.net/publication/259361068_Application_of_Artificial_Intelligence_Methods_to_Preliminary_Design_of_Ships_and_Ship_Performance_Optimization

L3HARRIS™
FAST. FORWARD.

Integrating Autonomy – Maintain, Launch, Execute and Recover

Emma Parkin MEng (Hons)

L3Harris Technologies, UK. Emma.Parkin@asvglobal.com

Joe Chilcott BEng (Hons) MSc CEng MIET

L3Harris Technologies, UK. Joe.Chilcott@L3Harris.com

TABLE OF CONTENTS

LIST OF FIGURES

Integrating Autonomy – Maintain, Launch, Execute and Recover

SYNOPSIS

In recent years, autonomy has been subject to significant interest and investment within the maritime industry. Several high-profile trials, such as Unmanned Warrior 2016 and Autonomous Warrior 2018, have drawn attention to the rapidly emerging and advancing technology. Following the successful operation of 7 autonomous vessels during Unmanned Warrior 2016, L3Harris completed a series of demonstrations at Autonomous Warrior 2018. The continually evolving technology has enabled increasingly complex autonomous operations to be trialled. This accessibility and the enhanced capabilities have resulted in the increased adoption of autonomous vessel technology. Navies around the world have shown much interest in the enhanced military capability that autonomous vessels bring to the operational theatre. It is clear that the technology is available and the demand exists, but does integration into a modern operation warship pose an indisputable challenge? The operating cycle of an USV when hosted onboard a warship can be summarised as 'Maintain, Launch, Execute and Recover'. Understanding how the USV integrates into the mothership at each of these stages will increase the effectiveness and efficiency of operating the USV. To fully understand this, all aspects of integration should be considered—people, processes and technical interfaces. Two key onboard systems that an autonomous vessel will need to interact with are the combat system and the IPMS. As autonomous technologies become more established and proven, the confidence gained will have implications for its possible implementation on larger vessels, potentially leading to fully autonomous cargo ships and cruise ships.

KEYWORDS: Autonomy, Unmanned, Integration, IPMS, Combat System, Launch and Recovery

1. Introduction

In recent years, autonomy has been subject to significant interest and investment within the maritime industry. Several high-profile trials, such as Unmanned Warrior 2016 and Autonomous Warrior 2018, have promoted massive interest in the rapidly emerging and advancing technology.

Following the successful operation of 7 autonomous vessels during Unmanned Warrior 2016, L3Harris then held a series of more advanced capability demonstrations at Autonomous Warrior 2018. Most recently, in support of the Defence Science and Technology Laboratory (Dstl), L3Harris operated a 9-metre (30ft) vessel outfitted with advanced, fully autonomous navigation capability for reconnaissance, interdiction and patrol tasks. The vessel, 'MAST 9', seen below in Figure 1, demonstrated COLREG aware collision avoidance. While navigating waterways at speeds of up to 40 knots for over 80 hours, this vessel successfully executed seven different task types. ASView, L3Harris' proprietary autonomous control system, enabled remote mission commanders to track and follow target vessels for interdiction tasks. The vessel, designed and built by L3Harris, also demonstrated collision avoidance through the use of radar to provide situational awareness.

"The exercise successfully showcased an integrated system of systems approach to executing autonomous defence tasks with little or no human intervention," added Ian Campbell, Defensive Surface Warfare, Platform Systems Division, Dstl.

Figure 1. An image of MAST 9 from Autonomous Warrior 2018

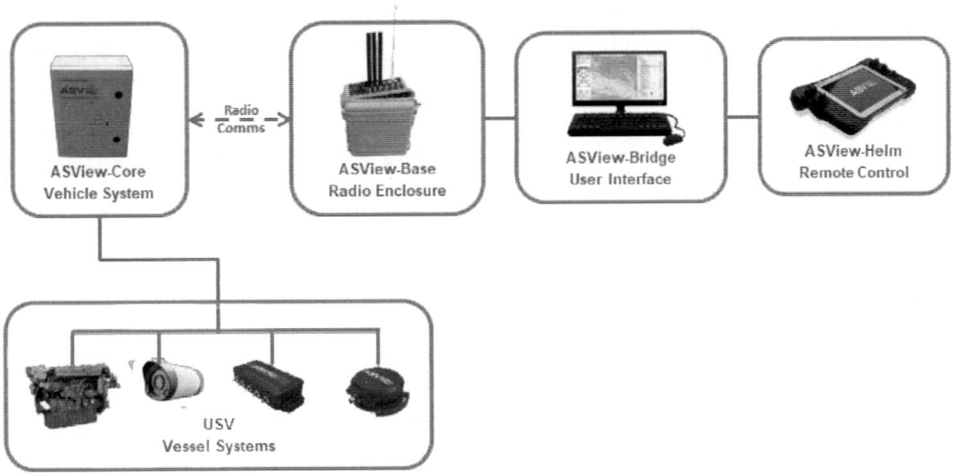

Figure 2. ASView System Components

The current operating philosophy of an Unmanned Surface Vessel (USV) is to operate as a standalone system for ship's crew. This is feasible for small USVs and short deployments. However, for larger USVs and enduring deployments, an integrated solution will increase efficiency of operations. The aim of this paper is to consider how a medium sized USV could integrate into a mothership. This paper will review typical USV architectures, autonomy and current operating philosophy. The importance of considering the entire operating cycle and the challenge of fusing the Maintain, Launch, Execute and Recover stages onboard an operational warship will be reviewed.

2. Autonomy

It is important to define the difference between automation and autonomy. An automated system is a system that performs repeatable tasks without human assistance, whereas an autonomous system performs in an uncertain environment and can resolve system failures without human intervention. For decades there have been automated systems available to aid and support operators. Only in recent years has the technology evolved, allowing computing power to become available to make autonomous maritime systems a reality. The increased pace of change towards autonomous technology is due to multiple reasons:

- Personnel – A combination of the shortage of skilled people in the maritime sector and their associated high through-life costs.
- Safety of personnel – Society's views on the value of personnel has changed significantly, as well as the behaviours of our adversaries. This shift has put greater pressure on reducing loss of life and maximising the safety of personnel.
- Cheaper/smaller vessels – Autonomous vessels do not need to support human life. There are no requirements for accommodation, domestic facilities, HMI operating positions and so on. This reduces the space and weight requirements resulting in a significantly smaller vessel than traditional manned vessels.
- Reduction in human error and fatigue – Despite the high calibre of operators, human errors are inevitable. This is exacerbated by combat stress and long operational hours, leading to fatigue. A machine cannot get tired or stressed and therefore has greater longevity.

Navies and industries throughout the world are investing in autonomous technologies to enable maintaining and/or increasing capability while reducing associated costs. Examples of where USVs can realise this benefit is through the ability to operate up to 24 hours a day with zero or minimal human intervention. The continually evolving technology has allowed for increasingly complex autonomous operations to be trialled, which is enabled by L3Harris' autonomous vessel systems being supplied to end-users with comprehensive training packages. The increasing accessibility and capability has resulted in increased adoption of autonomous vessel technology. Governments' worldwide and industry-leading commercial companies, such as Ocean Infinity and Fugro, are just some of the many companies investing in autonomous technology.

Globally, Navies have shown significant interest in the enhanced military capabilities that autonomous vessels bring to the operational theatre. As the maritime environment is complex and dynamic, when integrating a new system such as an USV, considerations such as personnel, safety and policies are paramount. As stated by the Department for Transport (2019), "People remain at the heart of the maritime sector's journey towards maritime autonomy", putting more emphasis on the need to understand how autonomous systems will integrate within a ship, its operators and its processes.

The technology is available, and the demand exists, but integration into a modern operational warship poses an indisputable challenge.

The Royal Navy's announcement of 'NavyX', which is its new Autonomy and Lethality Accelerator programme [1], demonstrates the vision and commitment the Royal Navy is making to the development of autonomy and ensuring it is leveraged as a force multiplier. This programme emphasises the need for a step change in the pace of development and procurement of these systems. At the heart of this accelerator is the desire to create multi-disciplined collaborative teams from across the industry and academia. Furthermore, this emphasizes the need to look to the future to see how the Royal Navy will employ these new capabilities and how they will be fully integrated in future capability solutions. Consideration is needed to determine how current processes and doctrines will be adapted to accommodate autonomous capabilities. From a technical perspective, how will these systems be developed in such a way as to be interoperable and scalable for future enhancements? Potentially pan industry and academia. Finally, how will training for these systems be provided in a practical and cost-effective manner? It is time to consider how unmanned and autonomous systems will be integrated into larger motherships.

3. Unmanned Surface Vessels

An Unmanned Surface Vessel (USV), sometimes also known as an Autonomous Surface Vessel (ASV), is a vessel which can operate on the water without the need for a human to be onboard. As a crew is not necessary, USVs offer many advantages. For an USV to operate correctly in autonomous mode, there are several systems which need to work harmoniously together. The key systems within an USV are:

- The situational awareness system
- The communications system
- The autonomous control system
- The physical hardware.

The situational awareness system is comprised of 'vision' sensors fitted onto the USV and can include optical cameras, infra-red and radar capabilities, amongst others. A comprehensive and reliable situational awareness picture is critical to successful and safe conduct and is required to develop a truly COLREG compliant collision avoidance system. The communications system provides the link between the base station and the USV. Communications bearers can include satellite, UHF, VHF, Wi-Fi and 4G. Both the range of communications and the rate of data transfer are dependent on the location of operation and the choice of communications bearer.

The autonomous control system receives input from all sensors and systems to inform decisions. This means both the situational awareness and communication systems feed into the autonomous control system, and vice versa. Principally, the autonomous control system is responsible for the decision-making. This consists of the Last Response Engine, which makes the vessel safe if higher levels of autonomy have failed to do so already, the Collision Avoidance System, which strives to determine a safe path for the vessel to traverse, and the Control Plan, which is a high-level mission system used to define mission objectives.

4. Current RN Operating Philosophy

The Royal Navy currently operates vessels with varying levels of automation and capabilities. Historically, each function of the ship had been built and operated as a discrete system with specific SQEP to operate these systems. Numerous discrete systems cause the ship to be manpower intensive. Lean manning has continued to drive the requirement for more automation, as depicted in Figure 3 [2]. To continue on this path of growing automation and autonomy, the availability and fusion of data on a ship is paramount. Therefore, integration of these systems is essential to ensure the availability of required data throughout all of the ship's functions. Project Nelson is "using Artificial Intelligence and data science to build a 'Ship's Mind'" [3], which is an example of the paradigm shift towards integrated data platforms supporting data fusion.

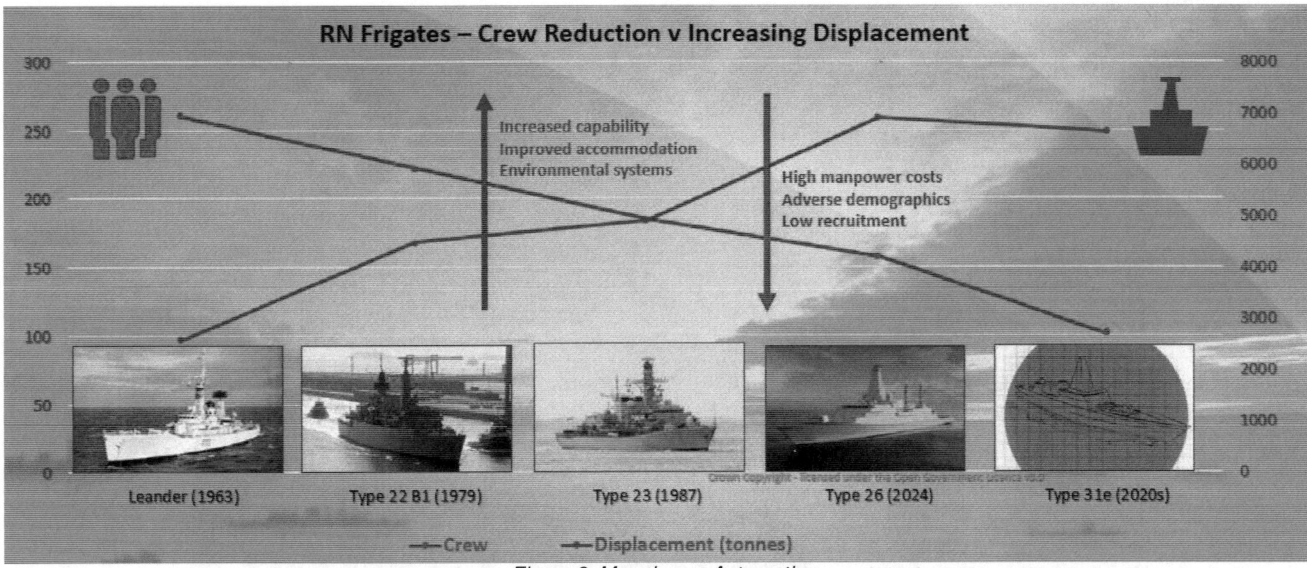

Figure 3. Manning vs Automation

Recently, there has been a desire to deliver versatile, capable and scalable vessels on tight budgets and schedules, personified perhaps in the T31e Programme. The question is, therefore, 'How do we provision for these cutting-edge technologies for future enhancements?' To do this, we must have an appreciation of how these systems will integrate and operate.

The Queen Elizabeth Class aircraft carriers have demonstrated what can be achieved through collaboration across industry and a high degree of integration. Primarily this collaboration has supported lean manning requirements, enabled by a high degree of system integration. This has been achieved through the use of a shared network environment entitled the Integrated Network Equipment (INE). The INE is a number of routers, access switches and network nodes interconnecting all of a ship's networks, thereby making the sharing of data between different systems far easier. This has enabled a number of holistic, command focused features within the Integrated Platform Management System (IPMS), such as capability reporting which enables the ship's capabilities, both combat systems and auxiliaries, and power and propulsion systems, to be assessed and prioritised in accordance with the current Command Aim. The capability reports are collated into the top 3 from each department, eventually allowing the Command Advisor to focus the ship's efforts to support the Command Aim. This is a powerful feature which highlights the benefits of integrating data throughout the ship to provide a holistic picture for the command hierarchy. This pattern of integration and enhanced functionality must continue as is emphasised with the additions of new technologies such as USVs.

There are still a great number of improvements to be made such as data availability and integration. Historically systems onboard have been developed in a 'stove pipe' approach, making data integration and holistic analysis difficult. Recent developments in shared computing environments show great promise as a common IT infrastructure throughout a ship, providing not just a shared network, but multifunctional shared glass, virtualised PCs and Servers, network storage and associated services. There are two main stove pipes that must be considered in the context of this paper, focussing on the integration of an USV—the combat systems/weapons engineering team and the marine engineering team. The integration of an USV onto a ship requires input from both organisations because each will have a role in the operating cycle of the USV, covering maintenance, deployment, mission execution and recovery of the USV.

5. The Operating Cycle – Maintain, Launch, Execute and Recover

Maintain, Launch (Deploy), Execute and Recover are the four stages that make up the operating cycle that an autonomous vessel will go through whilst being hosted onboard a large mothership (Figure 4). Ensuring successful integration into an operational warship requires consideration of the people and processes interacting with the autonomous vessels at each stage.

Figure 4. The Maintain, Launch, Execute and Recover Concept Render

The two key onboard systems that an autonomous vessel will need to interact with are the Combat Management System (CMS) and the IPMS. These interfaces will be a mixture of live remote data links and hardwired links whilst stowed.

The integration of offboard systems, such as USVs, into both the IPMS and the CMS is a complex matter. As the USV moves through the different stages of its operating cycle, the system exercises primary control and changes in monitoring, as indicated below:

1. Maintain Phase – IPMS is likely to have overall control and monitoring capability, allowing the engineering department to ensure the vessel is prepared to perform for the complete duration of its forthcoming deployment.
2. Launch Phase – At this stage, the control of the vessel transfers from the IPMS to the CMS. The vessel becomes an 'active' asset which the CMS will need to manage.
3. Execute Phase – During this phase, the CMS will be overseeing the USV, providing Command and Control (C2) and setting goals to achieve. However, basic information on the state of the vessel will still be communicated to the IPMS to allow the engineering team to monitor and store operational health data for further analysis.
4. Recovery – Just like the launch phase, the CMS will have control of the vessel. However, once onboard, when the system is no longer 'active', the IPMS will re-gain control to prepare the vessel for the next mission.

Standardizing this operating model will create a standard interface for an USV allowing a mothership to interchange USVs, depending on operational requirements.

5.1. Maintain

The maintain phase focuses on ensuring the vessel is ready for operational deployment, which will involve running diagnostics and trending data from the autonomous systems. The larger USVs are currently around 12m, as shown in Figure 5, and can be stowed on a mother ship, which host relatively large equipment. Considering the equipment that the USV will likely be hosting, the maintenance process, and the potential impact to the ship's trim and stability, it could be argued that the Marine Engineering department is best to take responsibility for the USV whilst stowed on the mother ship, supported by Mission System personnel. This is due to the USV having similar systems, propulsion, steering, and hydrostatic characteristics and supporting systems, all of which the Marine Engineer is most experienced with.

Figure 5. L3Harris Mine Countermeasures (MCM) Autonomous Vessel for Thales UK

To support the Marine Engineers maintaining the USV, there must be a data interface from the USV's onboard systems to IPMS. This data interface will provide health and diagnostic data for processing and analysis. Analytics, such as that provided by an Enhanced Health Monitoring system, can be used to detect alarm conditions and compare live data to benchmark data, as shown in Figure 6.

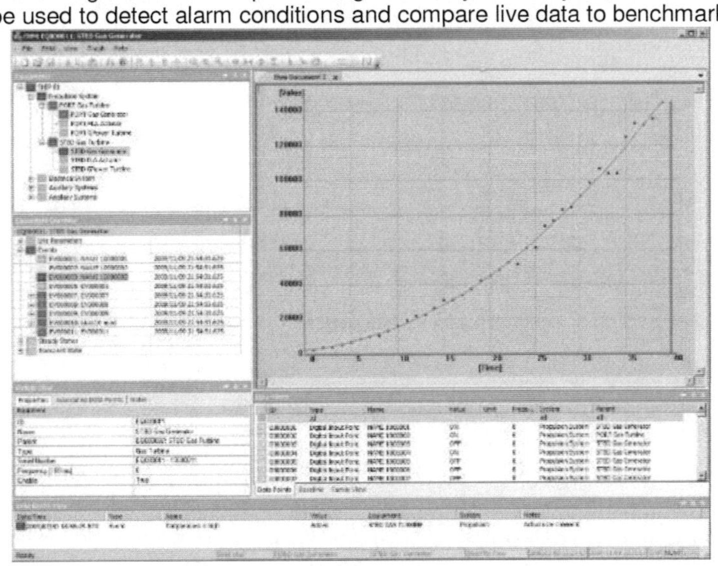

Figure 6. L3Harris Enhanced Health Monitoring

AI techniques can be used to identify 'unknown unknowns', determining patterns and trends which would be almost impossible for a human to spot. The Enhanced Health Monitoring system could incorporate such trends for predictive maintenance, enabling equipment defects to be rectified prior to occurring and thereby maximising the availability of the USV. To achieve predictive maintenance, a great deal of data needs to be collected and analysed, with emphasis on the importance of integration and accessibility of data for the development of future capabilities.

There are numerous other tools and features within IPMS that add value and capability if a simple interface exists between the stowed USV and IPMS. For example, a recent feature of IPMS is the development of auto prioritisation of ship defects to support the mission intent/command aims. With the USV providing health data to IPMS, the potential defects that occur on the USV can be prioritised to reflect and observe the ship's context and current command aim, respectively.

5.2. Launch

Launching an USV from a mothership is a complex operation which requires a number of considerations. First, and most significant within the defence sector, is the challenge of fitting a launch system into a typically small and cramped mission bay. Second, there is the logistical challenge of having so many individual moving parts in one system, potentially operating in hostile conditions. The unpredictability of wave patterns, particularly in high sea states, makes it difficult to securely latch the USV with the launch and recovery system without causing any damage. There are then further constraints on the design of the system imposed by the restricted space in the mission bay. Finally, there is the concern of control of the system moving from the IPMS to the CMS as the USV moves from the maintain phase to the launch phase.

In the near future, there will be multiple offboard capabilities hosted onboard a mothership, spanning across air, surface and sub-surface. It will be difficult to manage numerous separate, disparate systems and ensure the correct information is fed into both the IPMS and the CMS, allowing them to make decisions and plan missions. The challenge here is ensuring that all involved parties receive the same real time information simultaneously, avoiding the possibility of poor decisions being made based on outdated information. With the vast amount of data being collected and distributed on an USV, it is important to ensure that the user is not overloaded with unimportant information. A potential solution to this is to set boundaries on the information sent by the USV, meaning that unless information is outside the safe boundaries, the operator does not need to know what the readings are.

It is proposed that upon completion of the Maintain phase, when the USV is ready for deployment, the engineering department can 'activate' the vessel so that it appears on the CMS as an available asset to use for planning and executing missions. This will rely on a connection between the IPMS and the CMS.

5.3. Execute

During the execution stage, both the IPMS and the CMS will be fed information from the USV. The vessel will report back to the operations centre, via the capability reporting feed, on the progress of the mission and any required/useful mission specific information to the CMS to inform future command decisions. Additionally, the USV will report health monitoring information to IPMS allowing the engineering team to prepare for necessary preventative or corrective maintenance on completion of the mission.

To reduce the amount of data being transferred back to IPMS, the system can set safe operating parameters. This means only data which falls outside of these parameters will be passed over the data link, thereby reducing the amount of data and bandwidth required.

5.3.1. Controlling a single USV through the CMS

The control system supplied with the L3Harris USV is based on a Service- Orientated Architecture. The interaction between the CMS and the USV is outlined in the diagram below:

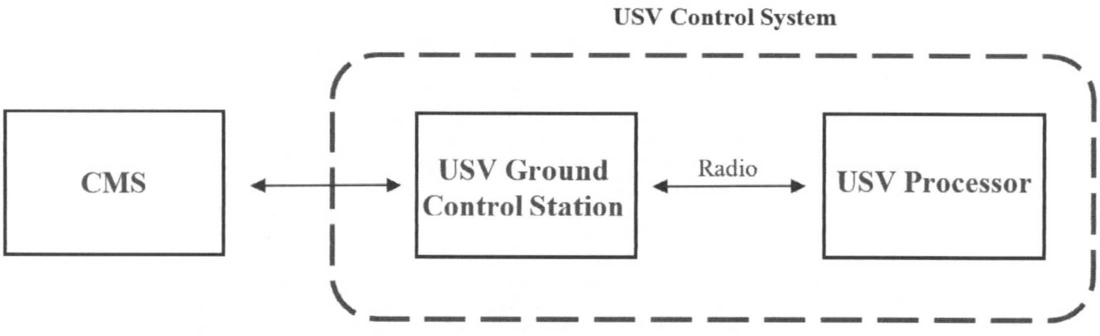

This architecture allows the CMS to send commands to the USV's ground control station, which then converts the message to a format readable by the USV processor.

5.3.2. Controlling a squad of USVs through the CMS

Controlling a squad of USVs could be done in a number of different ways. First, the CMS could communicate with the USV individually with the squad planning element being hosted on the CMS:

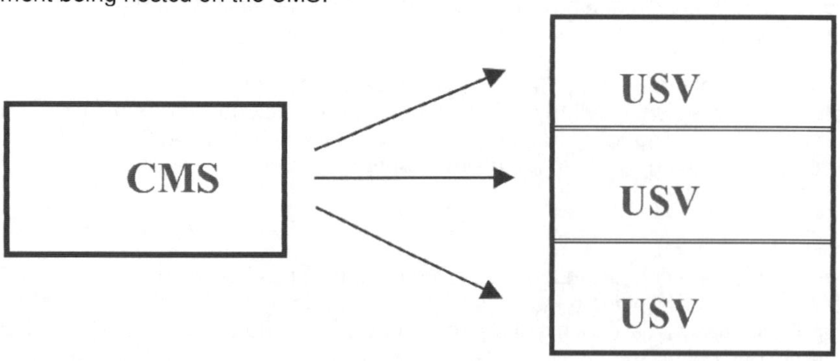

The above scenario would work well when the offboard systems are within a good communication range from the mothership. Once out of the communication range, the behaviour of the USV will be dependent on the autonomous control software. An example behaviour could be that the USV returns to the last known location where it was receiving communications, known as the Last Response Engine.

The next two proposed architectures would suit a situation where the offboard systems were further afield from the mothership. This is because there would be less information to be sent between the mothership and the offboard systems.

One of the offboard systems could be nominated as the Lead USV, which would be 'smarter' than the others. This USV would receive the aim of the mission from the CMS and would then issue individual missions to the squad of Follower USVs. In this scenario, the planning would be carried out on the Lead USV.

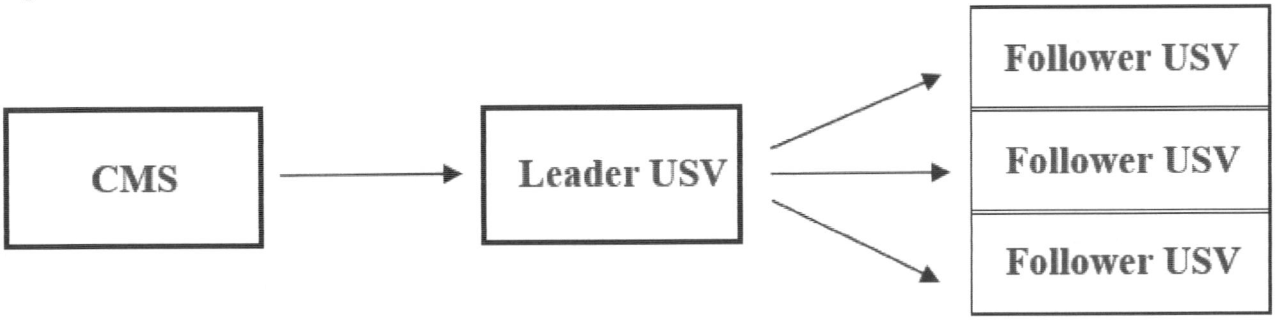

Alternatively, there could be negotiation tasking between USVs where a mission is sent to the whole squad and they negotiate between themselves about how to achieve the mission in the most efficient way.

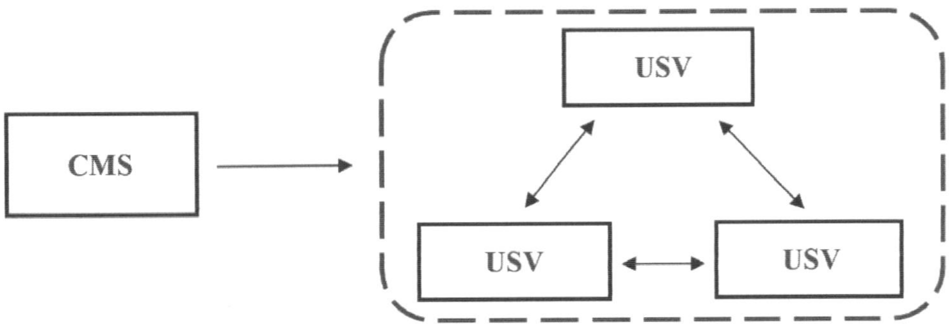

Throughout the above scenarios, there will also have to be health information fed back to the IPMS – this could be done either via the USV directly or fed from the CMS to the IPMS using the communications links already established.

5.4. Recover

Recovery of an autonomous vessel is one of the biggest challenges for maritime autonomy across industry and is preventing a wider uptake of this technology. Innovative autonomous launch and recovery solutions are currently subject to significant investment. This makes achieving the full operating cycle of an autonomous vessel being completely unmanned a real possibility in the near future.

During the recovery stage, much like the launch stage, the control of the process will fall under the CMS remit. However, it would be very advantageous to maintain a link with the IPMS during this phase of the operating cycle.

Some advantages of maintaining a link with the IPMS include:
- Utilising the functionality of the mission bay cameras to aid in the recovery process
- A wireless link to IPMS when the USV is being launched/recovered to update the operator on the status of the USV during that phase of the operating cycle.

6. Next Steps

For successful integration of offboard systems across all domains and manufacturers, there is a need for a standard USV to

CMS/IPMS interface. Not only is this in keeping with the current trend towards open architectures, but it will also allow for the inevitable rapid pace of technological change and associated capability upgrades. The lifespan of an USV is around 10 years, much shorter than the lifespan of a mothership which is between 30 and 50 years. Having a defined standard interface will allow different systems to be modular, interchanged and updated multiple times during the lifespan of a mothership.

As the uptake of autonomous and unmanned offboard systems increase in the near future, the integration of these systems needs to be carefully considered. The solution will be based on previously independent and unconnected systems that worked together seamlessly and simultaneously.

7. Future Trends

The future of autonomy is far broader than the topics covered within this paper. The technologies being developed to optimise autonomous vessels will have other applications for the operations of the support vessel. As autonomous technologies become more established and proven, the confidence gained will result in larger vessels and potentially lead to fully autonomous cargo ships and cruise ships. The path to this vision has been set out in the Maritime 2050 Route Map [4]. As this vision becomes a reality, USVs will naturally grow in size, complexity and capability, raising a question of how we can integrate people and processes into/surrounding the autonomous vessel. This is the beginning of a journey with many challenges ahead.

8. Conclusion

Successful integration of an offboard autonomous vessel into a mothership's platform management system will rapidly increase the effectiveness and uptake of this technology across defence and commercial sectors. Currently, autonomous vessels are treated as a separate component, operating independently from the mothership. However, platform-wide integration will allow the USV to be incorporated into the operating doctrine, just as any other system in the mothership. This will ensure that the USV capability can be kept in a high readiness state for use by the ship's command.

As discussed throughout this paper, the successful integration of an USV into a mothership is not just a technical challenge; the personnel, processes and policies must be considered and developed in parallel. The operating cycle of an USV is captured in four phases—Maintain, Launch, Execute and Recover. By considering the stages that an USV goes through, all of the stakeholders can be engaged, and interfaces can be refined and enhanced to optimise the solution.

A Royal Navy vessel has two main information components that the USV operating cycle will interface with—the CMS and the IPMS. To ensure a smooth operation between the two, a common computing environment and infrastructure is essential. An integrated approach will allow for much more scalability, such as the handling of squads of autonomous vessels operated from a central control point on the mothership or even interchanging multiple USV depending on mission objectives.

References

1. Gov.UK, (2019). £75m investment set to revolutionise Royal Navy Operations. Gov.UK. [Viewed 1 May 2019]. Available from: https://www.gov.uk/government/news/75m-investment-set-to- revolutionise-royal-navy-operations

2. Chilcott J. and Kennedy N. (2018). Enabling Lean Manning through Automation. iSCSS Symposium Proceedings 2018. p.g. 199-211.

3. CTI Digital, (2019). Project Nelson [online]. CTI Digital. [Viewed: 24 May 2019]. Available from: https://www.ctidigital.com/our-clients/nelson-royal-navy

4. Department for Transport., (2019). Technology and Innovation in UK Maritime: The case of Autonomy [online]. Department for Transport. [Viewed: 5 April 2019]. Available from: https://www.gov.uk/government/publications/maritime-2050-navigating-the-future

5. L3 ASV (2018). L3 ASV and Dstl Complete 1,380km of Autonomous Reconnaissance Missions at Autonomous Warrior [Viewed 21st April 2019]. Available from: https://www.asvglobal.com/l3-asv- and-dstl-complete-1380-km-of-autonomous-reconnaissance-missions-at-autonomous-warrior/

6. L3 ASV (2016). MAST Goes into Action at Unmanned Warrior [Viewed 22nd April 2019]. Available from: https://www.asvglobal.com/mast-goes-action-unmanned-warrior/

L3HARRIS™
FAST. FORWARD.

The Detection and Tracking of Anthropogenic Acoustic Sources Utilizing a Hybrid Autonomous Underwater & Surface Vehicle (AUSV)

(Detecting, Locating and Tracking Anthropogenic Acoustic Targets)

Neil Trenaman, Ocean Aero, Inc., ntrenaman@oceaenaero.us

INTRODUCTION:

The automated detection and classification of maritime traffic is a challenging problem and is of great importance to many organizations. For maritime domain awareness an automated boat detection system could alert authorities of potential illegal or immanent national security threats by unidentified vessel traffic. However, not all vessel traffic and recreational or small boats, for example commercial snorkeling and diving boats would be considered as a threat. For this reason, a classification system is needed to discriminate these different types of boats. The need for similar systems arises in the monitoring of harbor traffic for national security. There are many different methods for boat detection, examples including radar, electro-optic (EO) and infrared (IR) cameras, and both active and passive sonar. Active sonar and radar provide little additional information beyond detection. Radar and optical methods are limited by line of sight for detection, and optical systems can be obscured by rain, fog, or may require daylight. Active sonar can be used or detection of quiet targets, but the high level of reverberation in shallow water environments often causes many false detections, which limits its utility.

As an alternative, passive sonar has been proven to be an efficient tool for the detection and identification of self-emitting targets. There is significantly more work to be done the on the classification of the acoustic signature(s) of small boats (1).

Detection and extraction of underwater acoustic signals play an extremely important role in marine vehicle tracking and is one of the key technologies in underwater source detection, location, tracking, recognition and as well as acoustic communication. Energy detection, feature detection and matched filter detection are commonly used range from high detection performance to low computation complexity, which can usually work well to some extent. Among these researches of the underwater passive detection methods, energy detection is mostly discussed and used in practical for its merits of lowest computational cost and easy to realize, while shows a poor performance under low SNR marine environment. However, over the past several decades, the marine environment has been more complexed in both natural and anthropogenic influences that ambient noise revealed. And as a consequence, the existing underwater passive detection methods are facing serious challenges, for which advanced detection scheme with better performance is still worthy investigating, especially under low signal-to-noise ratio (SNR) region (2).

SOUND IN THE OCEAN

Anthropogenic sound is created in the ocean both purposefully and unintentionally. The result is noise pollution that is high-intensity and acute, as well as lower-level and chronic. The locations of noise pollution are along well-traveled paths in the sea and particularly encompass coastal and continental shelf waters. Increased use of the sea for commercial shipping, geophysical exploration, and advanced warfare has resulted in a higher level of noise pollution over the past few decades.

NATURAL SOUND SOURCES IN THE OCEAN

The ambient acoustic environment of the ocean is highly variable. At a given time and place, a broad range of sources may be combined. In addition, conditions at a particular location may affect how well ambient sounds are received (e.g., sound propagation, water depth, bathymetry, and depth). Natural phenomena known to contribute to oceanic ambient noise include: (a) wind, sea state and swell patterns, (b) bubble distributions, (c) currents and turbulence, (d) seismic activity, (e) precipitation, (f) ice cover and activity, and (g) marine life.

ANTHROPOGENIC SOUND SOURCES IN THE OCEAN

Human activity in the marine environment is an important component of the total oceanic acoustic background. Sound is used both as a tool for probing the ocean and as a byproduct of other activities. Anthropogenic noise sources vary in space and time, but may be grouped into general categories: (a) large commercial ships, (b) air guns and other seismic exploration devices, (c) military sonars, (d) ship-mounted sonars, (e) pingers, (f) acoustic harassment devices (AHDs), (g) polar ice-breakers, (h) offshore drilling implements, (i) research sound sources, and (j) small ships (3).

NEED FOR DETECTION:

The global threat of illegal or unfriendly foreign incursions into sovereign national martime domains is a growing concern for border security and national defense agencies. Those agencies tasked with the detection, identification and tracking of these threats are generally under resourced to deal with the increase in events both in the spatial and temporal coverages needed to provide a comprehensive deterrent and early warning response.

Passive acoustic systems are the most promising sensor technology available to fill several gaps in maritime domain awareness. They offer the ability to detect, track and classify any vessel generating noise on, or under the water. One of the largest sensor gaps for maritime law enforcement is the inability to monitor small low visibility boats or small vessels not reliably detected by radar, nor are they required to transmit Automatic Identification System (AIS) messages.

TEHCNOLOGY PLATFORM – OCEAN AERO AUSV

Ocean Aero has developed an environmentally powered autonomous marine vehicle that has the unique ability to operate as both an Autonomous Surface Vehicle (ASV) and an Autonomous Underwater Vehicle (AUV) resulting in a hybrid, dual-modality Autonomous Underwater and Surface Vehicle (AUSV) that offers a single platform that can self-deploy to a remote operational area and dive underwater to perform a variety of mission or simply avoid detection. The AUSV uses two modes of propulsion on the surface. The primary mode is wind power. The vehicle uses a two-element rigid wing (similar to an airplane) to efficiently sail at speeds over 5 knots. The wing and flap are controlled by the CPU using environmental data from an anemometer, an ultrasonic boat speed sensor, an IMU (inertial measurement unit) and a GNSS antenna. Based on the mission parameters, the vehicle uses proprietary algorithms to determine the fastest approach to target waypoints. This mode of propulsion uses very little power and with the solar panels recharging the batteries, the AUSV can operate for extended period of time at sea.

The second mode of propulsion is thruster power. The AUSV uses a SeaBotix thruster to motor at speeds over 2 knots. This capability is useful for low wind conditions, launching and recovering the AUSV from a ship or pier, and flooding the vehicle for quick dives. The flooding feature is due to the thruster's innovative rotating mount. While sailing, the thruster is stored inside the hull to reduce drag. It rotates outside the hull for motoring and can be positioned at 90 degrees to the hull for rapidly flooding or dewatering the vehicle.

For operations on the surface, the user can remotely create, upload and execute missions using a web-browser interface and satellite communications. The mission control dashboard uses waypoints to define operational parameters (sailing, motoring, station keeping, diving, decks-awash loitering, etc.) and user-defined zones to restrict operations.

Underwater, the AUSV 'Navigator' model can dive to 200 meters using a robust ballast system similar to a submarine. The system can navigate underwater using predefined missions and use two thrusters to achieve speeds over 2 knots.

For command and control, the AUSV uses Satellite and Wi-Fi radios for communications and optional 900 MHz and cellular antennas for expanded capabilities.

Utilizing a hybrid autonomous underwater & surface vehicle (AUSV) developed by Ocean Aero, Inc. initial tests and evaluation for a three-dimensional passive acoustic hydrophone array coupled with on-board real-time processing for the detection, tracking and reporting of anthropogenic acoustic targets will be illustrated. Ocean Aero AUSVs equipped to detect, recognize and track particular classes of surface or subsurface vessels that are deemed to be of high interest as potential threats to maritime border security is discussed.

SENSOR TECHNOLOGY

The Ocean Aero AUSV was equipped with a custom designed three dimensional passive acoustic array consisting of twelve hydrophones spaced in a cluster and located underneath the AUSV hull (Figures 1& 2).

Figure 1: The 3-D multi-hydrophone array during bench tests

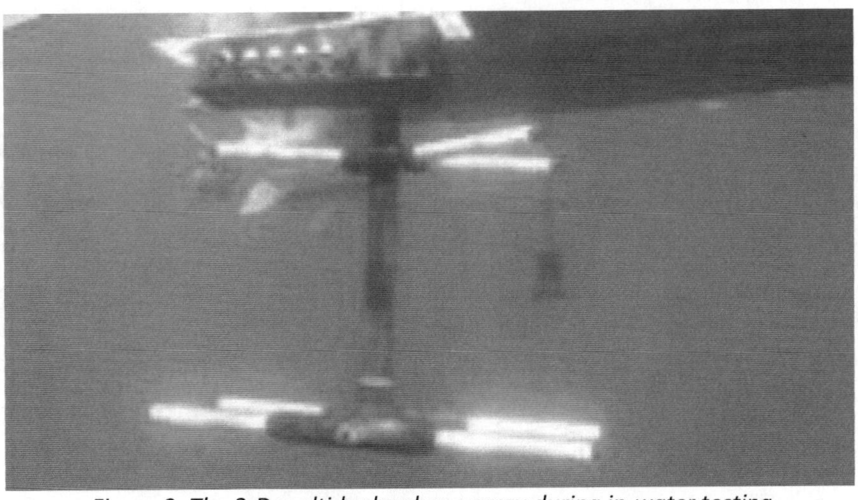

Figure 2: The 3-D multi-hydrophone array during in-water testing

FIELD TRIALS

Results of initial sea-trials are very promising and the graphical display of the target of interest from initial detection to tracking in real time are presented below.

With reference to the target frequency of 500Hz indicates that the type of vessel of interest would be smaller than a 12m fishing vessel traveling at a speed of 7 knots.

FIELD TEST CRITERIA:

- Preliminary Goal: AUSV shall have a Hydrophone array capable of detecting other sea vessels within 5 nautical miles generating acoustic signatures above 500 Hz.
- Verification: Visual, RADAR & AIS
- Test Equipment: AUSV, GPS, 3-D Passive Acoustic Hydrophone array and signal processing computer, support boat and target boat.
- Expected Result: Identify target signature with an acoustic signature above 500Hz.
- Confirm range to target, bearing and compute latitude and longitude position.
- Track the identified target and update range, bearing and new latitude and longitude position.

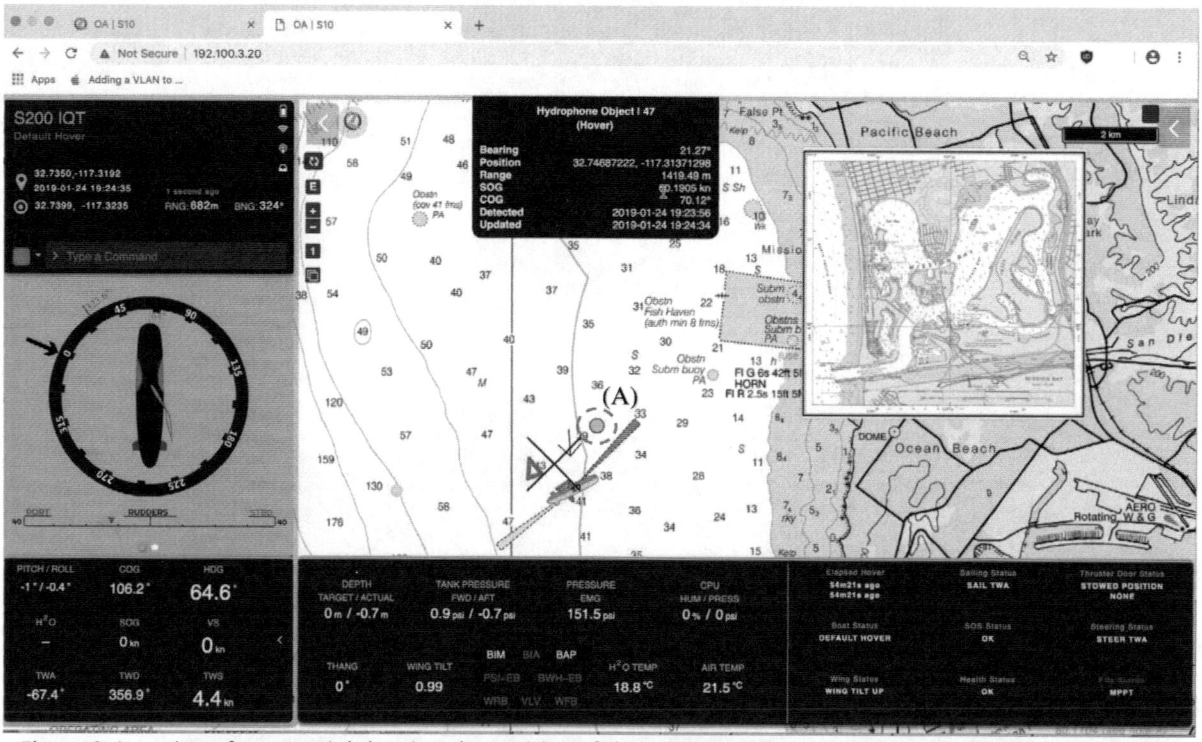

Figure 3. Location of at-sea trial showing the position of the AUSV, nearby AIS targets and the initial detection of the 500Hz target of interest (A)

The passive acoustic array initially detected the target boat and after a continuous set of sixteen positive returns, the target boat's latitude and longitude position was confirmed and continually updated as the target boat moved through the detection area.

The sequence of depicted in Figures 4a,b,c indicates the initial detection and location of the target of interest denoted by the marker(●) at point A (Figure 4a) and the progress of the target as it transected to point B (Figure 4b) and then point C (Figure 4c).

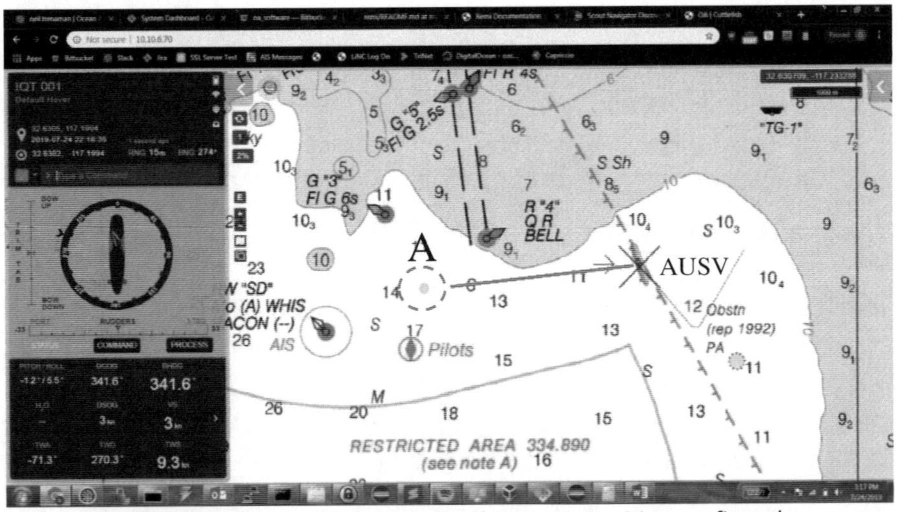

Figure 4a: Target (A) detected and Lat Long position confirmed

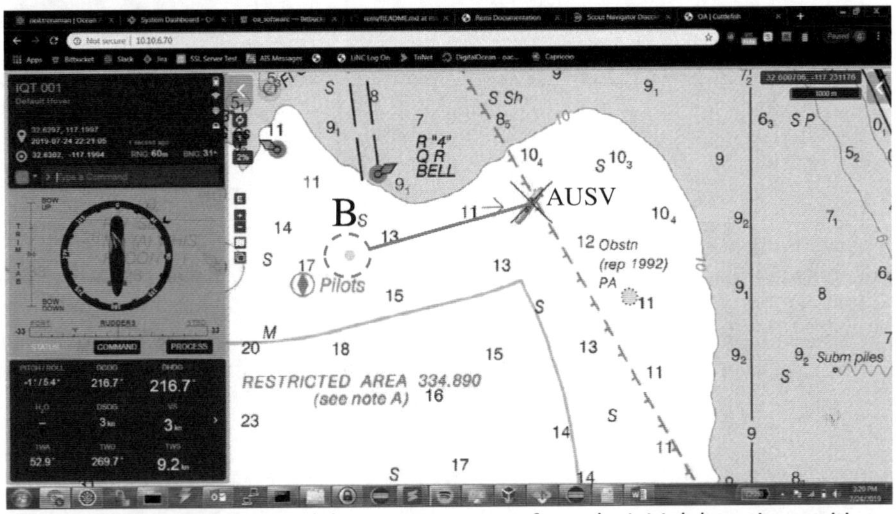

Figure 4b: Tracking Target (B) as it moves away from the initial detection position

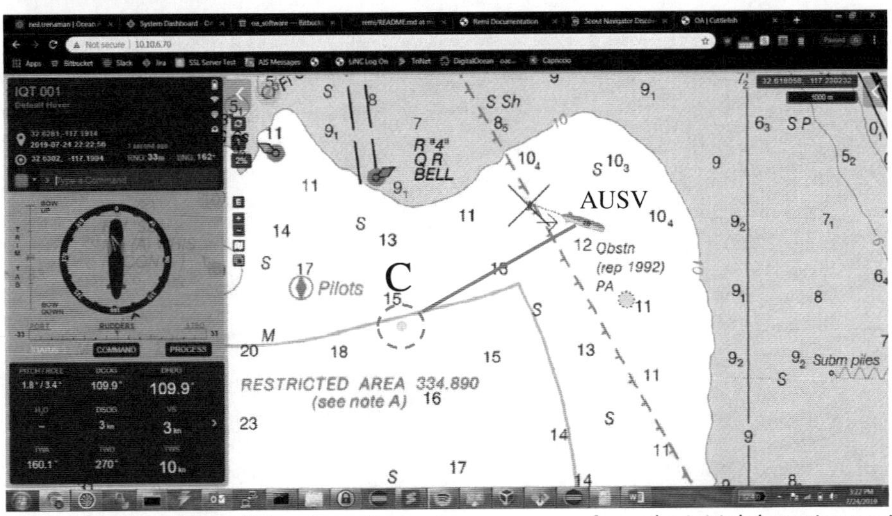

Figure 4c: Tracking Target (C) as it continues to move away from the initial detection position

CONCLUSION

The Ocean Aero AUSV equipped with a three-dimensional passive acoustic hydrophone array successfully demonstrated the feasibility of using an autonomous underwater and surface vehicle (AUSV) for the detecting, locating and tracking of anthropogenic acoustic targets in a pre-specified acoustic frequency range, which for this test was >500Hz.

The following goals were achieved :

1. AUSV equipped with a Hydrophone array capable of detecting other sea vessels within 5 nautical miles generating acoustic signatures above 500 Hz.
2. Verification via visual, RADAR & AIS methods were realized
3. Identify target signature with an acoustic signature above 500Hz.
4. Confirming range to target, bearing and computing latitude and longitude position.
5. Tracking the identified target and updating range, bearing and new latitude and longitude position.

Further test and evaluation in a wider range of ocean environments is a desirable next step as well as assessing options for integrating the hydrophone array directly into the hull of the AUSV or alternatively having the hydrophones imbedded in a thin line towed array.

References:
1. Ogden, GL., Zurk, LM., Jones., ME. Peterson, ME. 2011. Extraction of small boat harmonic signatures from passive sonar. The Journal of the Acoustical Society of America 129, 3768 [DOI: 10.1121/1.3583500]

2. Dahai Cheng, Huigang Xu and Ruiliang Gong, Huan Huang, 2018. Ships Matching Based on an adaptive acoustic spectrum signature detection algorithm. Signal & Image Processing : An International Journal (SIPIJ) Vol.9, No.4, August 2018

3.Hildebrand, JA. 2009. Anthropogenic and natural sources of ambient noise in the ocean. Marine Ecology-Progress Series. 395:5-20.

Mitigation of Slamming of Catamarans in Waves

Ahmed Swidan[1, 2*], Dane Christophersen[3], and Tristan Bauer[4]

[1]UNSW Canberra at ADFA, Australia. A.Swidan@unsw.edu.au
[2]Arab Academy for Science, Technology and Maritime Transport, Egypt.
[3]Royal New Zealand Navy, New Zealand. danechristophersen@gmail.com
[4]Austal, Australia. tristan.bauer@utas.edu.au

High-speed catamarans have, over the past three decades, extended their service areas from protected waters to the open ocean where impacts with waves can result in structural damage. The work detailed in this paper investigates the wetdeck slamming loads experienced by an amended NPL model that is equipped with a centrebow between the two demihulls while encountering regular waves. The investigation centred on the influence of early flow separation through utilising three interchangeable centrebow configurations and assessing the slam force magnitudes. The systematic and random uncertainties associated with the seakeeping test results are quantified in detail. This experimental investigation therefore provides a new dataset for the slam forces on an arched wetdeck structure of catamaran vessels in head seas. It was found that the early water separation when employing two wedge sections would reduce the slam forces by approximately 44% when compared with the parent centrebow configuration. The proposed new passive technique presents a step forward towards lighter and faster catamarans.

INTRODUCTION

A catamaran experiences wetdeck slamming when operating in large waves as the wetdeck, the exposed deck area between the two demi hulls of the catamaran, impacts the water surface with a high relative vertical velocity. Wetdeck slamming is a significant design issue for catamarans since it can decrease catamarans' transport efficiency and can cause major structural damage and avoiding its occurrence is one of the main reasons a vessel's master will reduce speed or change course in heavy weather, adversely affecting the vessel's operation and schedule.

During operation within the normal limits of a typical catamaran only the demihulls interact with the water. Operating in sufficiently large sea states, pitch and heave motions may cause the lower most exposed surface of the cross-deck structure connecting the hulls, known as the wetdeck, to impact the water surface. With repeated loading and unloading, fatigue failure becomes increasingly likely, and previous catastrophic failures of catamarans have occurred, despite being designed to classification society rules (Swidan, 2016).

Figure 1: Saint John Paul II, wave piercing catamaran, length overall = 110m (Incat, 2019).

A centrebow is intended to enhance catamaran performance in rough seas by damping the heave and pitch motions, as commonly included in INCAT's Wave Piercing Catamaran (WPC) designs, as shown in Figure 1. In comparison to conventional catamaran demihulls the high slenderness ratio of WPCs provides little reserve buoyancy to dampen pitch and heave motions, increasing the likelihood of the wetdeck contacting the water surface and a corresponding increase in the number of slamming occurrences. A centrebow can be described as a partial arched bow form which is truncated at typically 0.2-0.33 LOA and then a flat wetdeck continues aft towards the stern.

The central keel of the centrebow is located slightly above the calm water surface at full load displacement and static trim to avoid unnecessary resistance while operating in slight seas, as shown in Figure 1. Under rough sea conditions which would induce sufficient vessel motions, the centrebow enters the water at relatively high vertical velocities, consequently producing severe slam loads. The addition of a centrebow is beneficial to the catamaran design, although it increases the overall complexity and light ship weight of the vessel. An investigation into the magnitude of the slamming forces and induced loads on a WPC model demonstrated that slam induced bending moments were found to reach 11% of the product of hull weight x length, and the slam loads were found to reach as high magnitude as 132% of the hull weight (Davis, et al., 2017). A range of centrebow geometries, e.g. relatively longer and shorter centrebows, have been tested to explore whether there are significant improvements available to reduce the severity of the slamming loads. It was found that the larger centrebow volume induces larger slam loads due to an increase in the volume of trapped water (Shahraki, 2013). Relatively shorter centrebows were determined to displace water forward of the demihull bows, thereby minimising the developed slamming loads (Shabani, 2017).

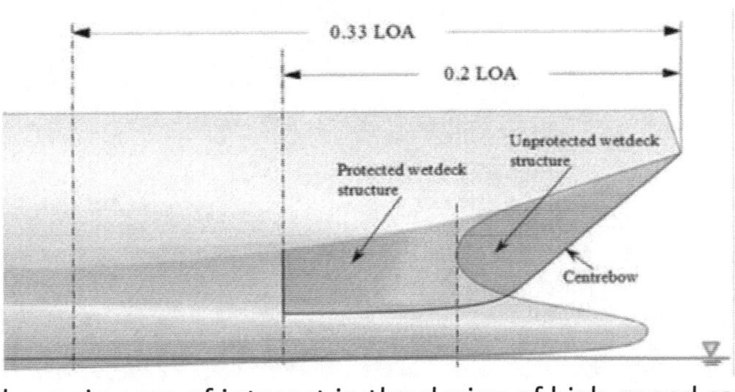

Figure 2: Showing the main area of interest in the design of high-speed catamarans, which is the impact loading in the vicinity of the centrebow (Swidan, 2017).

Controlling the vessel motions have also been explored, where an active T-foil was included at the bow in conjunction with stern tabs. The study showed a potential in reducing the motions when the system was passive, although not as effective (AlaviMehr, 2016). Modifications to the geometry which intentionally influences the flow can potentially offer reduction in the slamming loads (Swidan, 2016).

It is desirable to explore potential methods to minimise wetdeck slamming while maintaining the benefits that the centrebow offers. The purpose of this research was to investigate the effect of a passive device on reducing the total slamming load experienced by a generic catamaran hull model fitted with three centrebow configurations.

MODEL AND EXPERIMENTAL SETUP

The Test System

The model tests were performed in the towing tank facility of the Australian Maritime College (AMC), see Figure 3. The AMC's towing tank is 100 m long by 3.55 m wide. Experiments are generally carried out at a water depth of 1.5 m depending on the size and type of model being tested. In this study, a water depth of 1.4 m was necessary to avoid model to carriage possible collisions due to the expected magnitude of model motions. Two disadvantages were recognized utilizing this set-up, 1. the wave-damping mechanism along the side of the tank was ineffective, and 2) the wave maker is calibrated for a water depth of 1.5 m, as such there were slight differences between the target wave heights and the achieved ones during the experimental tests. The hydraulically operated single flap paddle type wave generator produces waves with wavelengths of 0.4-6.5 m and 0.4 m high. A PC based DAQ and processing system is provided to which a variety of analogue and digital instruments can be connected.

The Test Model

The test model was constructed using the National Physics Laboratory (NPL) demihulls and fitted with a generic centrebow, that looks like those used by INCAT, see Figure 4. Figure 4 shows the body plan of the developed catamaran model with the parent centrebow. The model's main particulars are presented in Table 2.

Figure 3: Showing the test model attached to the carriage of the towing tank facility at the Australian Maritime College, encountering a wave height of 72.7 mm at a speed of 1.53 m/s.

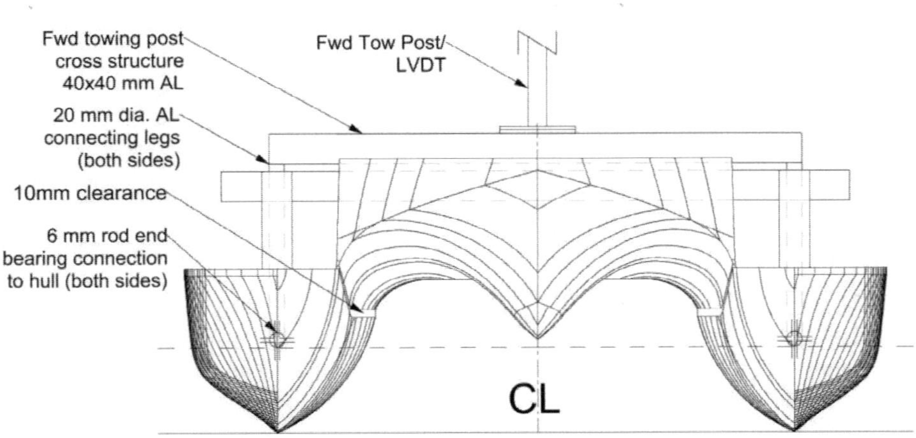

Figure 4: Body plan of the NPL's demihulls fitted with a generic centrebow and the forward towing configuration.

Table 1: Main particulars of the test model

Length Overall	2661 mm
Length Waterline	2502 mm
Displacement	52.16 kg
Draft	115 mm
Beam at waterline	918 mm
Beam Centres	690 mm
LCG (model) from transom	1162 mm
Pitch radius of gyration (about LCG)	893

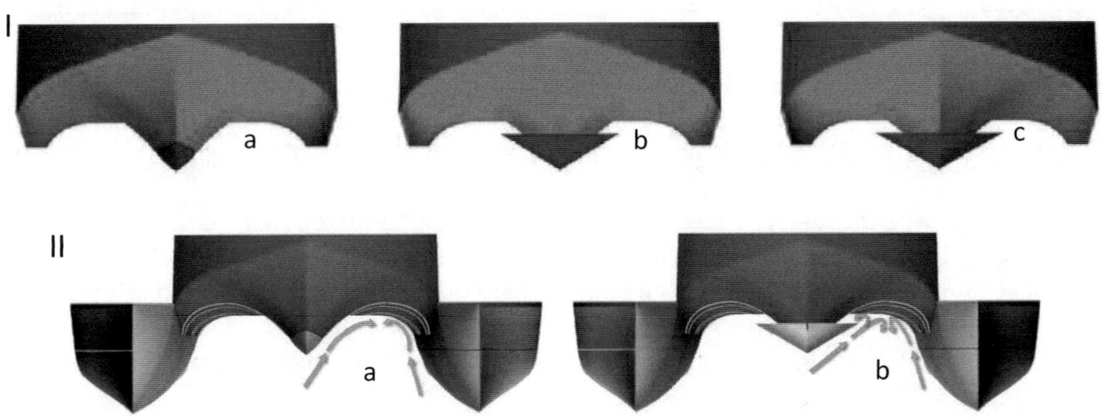

Figure 5: I-(a) Showing the parent centrebow; (b) 35 mm wedge fitted to parent centrebow; (c) 45 mm wedge. (a), (b) and (c) are viewed from forward and omit the demihulls. Figure 3 II-a showing the semi-entrapped flow behaviour between the parent centrebow and the demihull, (b) illustrates an early flow separation around the tip of the 35 mm wedge.

To study the influence of an early flow separation on the wetdeck slam force magnitudes, two interchangeable wedge sections were designed, 3D printed and fitted on the parent centrebow, see Figure 5-I and 5-II. These three model configurations, namely: parent, 35 mm wedge and 45 mm wedge, are shown in Figures 5-I and 6. Figure 6-a shows the width of the horizontal face of the wedges were set to 35 mm and 45 mm, which occupied approximately 20% and 25% of the 180 mm tunnel width at the centrebow truncation, respectively.

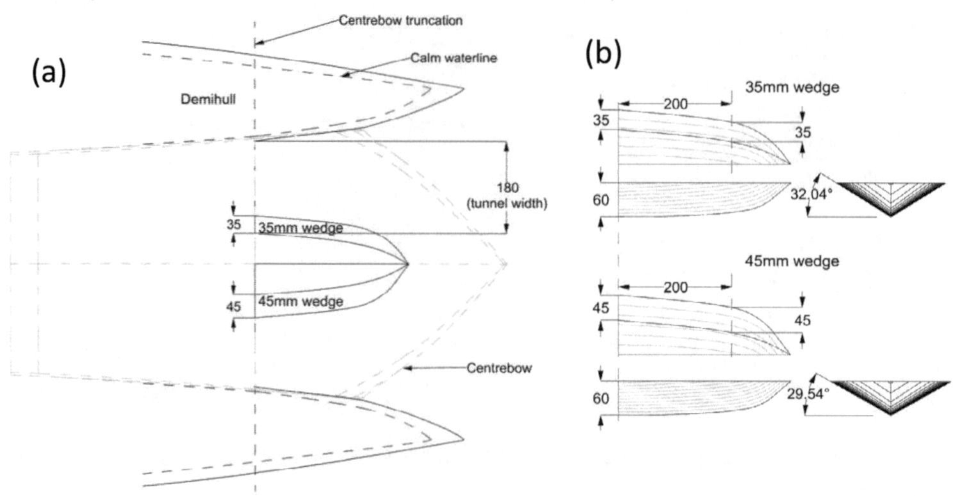

Figure 6: Showing (a) the horizontal faces that occupy 20% and 25% of the tunnel width on a plane 70 mm above the waterline for the 35 mm and 45mm wedge respectively, and (b) the lines and principal geometry of the 35mm and 45mm wedge sections.

Figure 6-b shows the overall height of the wedge section (applicable to both wedges) at the aft extent was 60 mm which reduced toward the bow to follow the keel line of the parent centrebow. The wedges are henceforth referred to as 35 mm and 45 mm which relates to the width of the horizontal surfaces. The increase in volume over the parent centrebow due to the wedges was 0.4946x10-3 m^3 and 0.6371x10-3 m^3 for the 35 mm and 45mm wedge respectively.

Instrumentation

To characterise the model's response during impact, the key parameters measured in these model tests were; the vertical force, wave profile, heave, pitch and the carriage speed. A summary of the instruments and signal conditioning hardware is given in Table 2.

Three load cells were considered to be the minimum number sufficient to avoid moments during the impact phase based on recent drop-test experiments, such as those of Swidan et al. (2016). All the total impact force measurements presented in this paper are the sum of the three load cell outputs, as shown in Figure. 7. The model was attached to the carriage by two tow posts, allowed to move freely in the vertical direction within linear slides, to which DC-EC series LVDT sensors were attached to capture pitch and heave data.

Table 2: Details of Gauges

Gauge	No. of Channels	Manufacturer	Model	Maximum Range
Carriage speed	0	SICK	DME5000	150 m
Load cell	3	Applied Measurement	S1W	350 N
LVDT	3	Hampton	DC-EC-5000	
Static wave probe	2	Static wave probe		0.5 m
Moving wave probe	2	General Acoustics	USS 20130	0.5 m

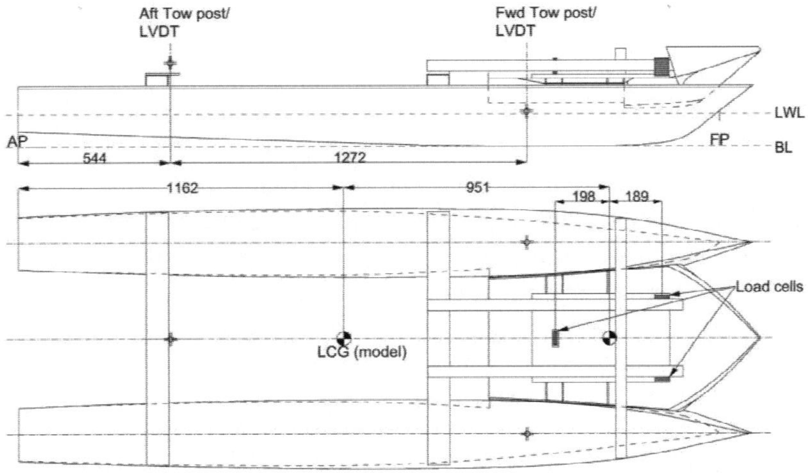

Figure 7: Plan and profile views of the model setup showing locations of load cells and tow posts

Wave heights were recorded primarily by a static wave probe located 9 m forward of the wave-maker. Supplementary to this were two moving wave sensors to capture the instantaneous wave profile the model was encountering. These were attached to the carriage with one abreast the LCG of the centrebow and the other abreast the LCG of the model, set as far outboard as possible to avoid any bow wave or wake while avoiding wall effects of the towing tank. This was achieved by conducting several calm water runs with a rigidly fixed reference set very near the water surface and observing any disturbance to the water surface. Minor seiching was observed with a long standing wave with a period of approximately 45 s which affected this process, therefore the best compromise was used. It was assumed the seiching was a consequence of the reduced water depth used and would not influence the final results due to the very long wavelength and elimination through post processing methods. Settling time of the tank between runs was maintained at approximately 45 min between runs due to the wave dampening mechanisms on the side of the towing tank being unavailable at a water depth of 1.4 m.

The load cells were calibrated off the model by suspending each from a rigid point and applying known masses. This was checked both when applying and removing the masses to ensure the output was a linear relationship and was repeatable. Once the load cells were applied to the model, the loads were zeroed by the associated wheatstone bridges and a check was performed to ensure the total load output from all three load cells were producing an accurate load output. A known weight was placed in three arbitrary positions along the centrebow and the 'record zeros' function in the DAQ system captured the load values from each load cell. This also provided confidence that there was no noticeable binding in the model setup and that the load cells would carry the entirety of the applied load.

A video camera recording at 1920 x 1090 resolution at 25 frames per second was set up immediately forward of the vessel as close to the water surface as possible to view the flow within the centrebow tunnels. The signals from all the instruments were acquired using a modular National InstrumentsTM compact data acquisition system (cDAQ 9174) with National Instruments LabVIEW software used to record the signals. A sampling rate of 10 kHz was used. The measured data was automatically converted into its corresponding unit of measured data using the calibration factors that were input into the DAQ software prior to the tests.

Test Conditions

Tests in irregular waves and oblique seas are recommended for testing catamaran operational limits and for determining extreme phenomena, however this presents an increase in the number of variables and associated complexity in analysis (ITTC, 1999). This investigation was limited to regular head seas to provide high-quality data that could be used as benchmark for validating numerical methods by designers and researchers. The three centrebow configurations were all subject to the same test conditions. Overall, there were 18 conditions in total, with 3 test runs each to ensure repeatability of the recorded data. Prior to the testing starting, 20 runs were performed to assess the model and instrumentation quality. The two model speeds of 1.53 m/s and 2.89 m/s, and initial wave frequency of 0.8Hz were common with previous studies, e.g. by AlaviMehr (2016).

The equivalent Froude number of the speeds of 1.53 m/s and 2.89 m/s are of 0.31 and 0.58 respectively, which would represent 20 and 38 knots respectively for a full-scale vessel assuming an approximate scaling factor of 45. Each tested wave height of 75, 90 and 105 mm was non-dimensionalised against the wavelength to give corresponding wave frequencies of 0.876, 0.8 and 0.741Hz, respectively. To present the measured data, the frequencies are shown in non-dimensional circular encounter frequency, w_e^*, which factors in the models heading and speed with respect to the waves. The formula is presented in equation (1).

$$w_e^* = 2\pi f_e \sqrt{\frac{Lwl}{g}} \qquad (1)$$

Table 3: Summary of Test Conditions

Configuration	Speed [m/s]	Wave height [mm]	Corresponding wave frequency [Hz]
Parent model hull	1.53	75, 90 and 105	0.876, 0.8 and 0.741 respectively
	2.89	75, 90 and 105	0.876, 0.8 and 0.741 respectively
With 35mm Wedge	1.53	75, 90 and 105	0.876, 0.8 and 0.741 respectively
	2.89	75, 90 and 105	0.876, 0.8 and 0.741 respectively
With 45mm Wedge	1.53	75, 90 and 105	0.876, 0.8 and 0.741 respectively
	2.89	75, 90 and 105	0.876, 0.8 and 0.741 respectively

The average water temperature was 18°C, the average air temperature was 20°C and the density of the water was 998.85kg/m^3 for the duration of the testing (ITTC, 2011). The test conditions are summarised in Table 3.

UNCERTAINTY ANALYSIS

To ensure the high quality of the measured data, systematic and random uncertainties were determined. Systematic uncertainties stem from the instrument themselves, and the accuracy of the reading is usually published by the manufacturer. Random uncertainties are caused by factors which are uncontrolled, e.g. environmental.

Systematic Uncertainties

The uncertainties based on the instrumentation used were sourced from the technical data provided by the manufacturers.

Table 4: Summary of the systematic uncertainty of the instruments

	Carriage position/speed	Load cell	LVDT	Moving wave sensor
Uncertainty source	Waldkirch DME5000-212	VA DC- EC-5000(MC)	XSENS MTI-10 series	General Acoustics 20130
Linearity error (%)	0.3	0.03	0.25	0.18

Table 4 summarises the systematic uncertainties of each instrument used during the experiment. The linearity error is the residual difference between the sensor output curve and line of best fit.

Filtering of measured slam forces has been avoided to obtain the peak of the transient slam load magnitudes that occurs within a very short duration of time. Butterworth low pass filter was utilized to attain velocities from positional data by taking the gradient between incremental data, filtering was necessary to eliminate noise in the source data.

Random Uncertainties

Each condition, see Table 3, was repeated for a minimum of three runs to confirm that the recorded data was repeatable. Additionally, it helped to provide confidence in the instrumentation being used including the recording DAQ system.

Figure 8 shows a good agreement between the repeated three tests of all measured data using two LVDTs, static wave sensor, carriage speed sensor, and three identical load cells. Figure 8 illustrates three runs at 1.53 m/s set carriage speed and 90 mm set wave heights at a set wave frequency of 0.8 Hz.

The root mean square error (RMSE) for the peaks of the forward LVDT was 1.43 mm and the aft LVDT RMSE was 1.45 mm over 10 cycles, as shown in Table 5. The steps in the wave data is due to the 50 Hz output of the wave sensor being sampled at 10 kHz as previously described.

The carriage speed showed good repeatability across all runs. The static wave probe showed some uncertainty in the wave height produced by the wave generator, as shown in Table 5, with the greatest error found for the wave height of 90 mm at ±6.27%. The total wave height did not achieve the desired magnitude, and this was likely caused by the lowering of the towing tank depth to 1.4m. The amplitude heave motions were compared to assess the repeatability of the forward and aft LVDTs. The results showed the peaks to be uncertain within 5%. Figure 8 presents the achieved wave heights (H_w) of 84 and 101 mm with a maximum deviation of 6.25% with the target wave heights of 90 and 105 mm, respectively.

The typical cyclical loading on each of the load cells for three repeatability runs using the parent configuration in 105 mm waves and 1.53 m/s carriage speed, is presented in Fig. 8.

Table 5: Summary of recorded wave heights and their respective RMSE and percentage error

Average Wave Height [mm]	RMSE ± [mm]	Error ± [%]
72.74	0.6	1.65
83.71	2.6	6.27
101.19	0.36	0.72

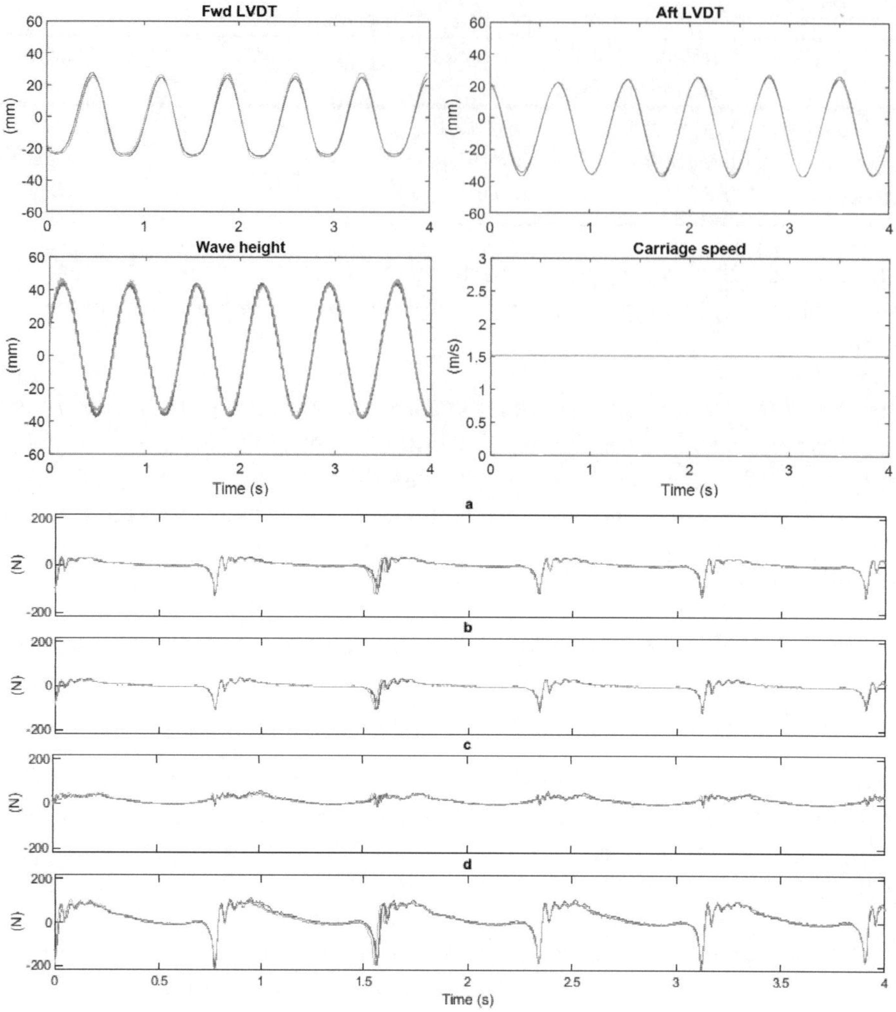

Figure 8: Showing 3 run results of condition 1. The results include the aft and Fwd LVDTs, static wave height sensor, carriage speed, and the measured forces of (a) the port side load cell, (b) starboard load cell, (c) Aft load cell, and (d) The total load, e.g. instantaneous summation of the three load cell measurements.

The asymmetrical loading was approximately 9.95% greater on the starboard load cell. This was likely due to the asymmetrical mass of the centrebow due to the plugs where pressure sensors have previously been fitted. These results ensured high-quality experimental data that provides a benchmark for researchers and industry performing numerical and computational simulations.

RESULTS AND DISCUSSION

The presence of wetdeck slam events, where water impacted the archways, was limited to 9 conditions, are summarised in Table 6 with the achieved wave height. The wave length, λ, was calculated from $\lambda = g/\omega^2 2\pi$, with an initial wave frequency, f, selected as 0.8Hz based on previous experiments of a similarly sized model (AlaviMehr, 2016). The non-dimensional wave height, h_w, was constant for all conditions and calculated from $h_w = kHw/2$, where k was the wave number calculated by $k = 2\pi/\lambda$.

Non dimensional wave circular encounter frequency, ω_e, was calculated by $\omega_e = \omega + \omega^2 V_m/g$, where V_m is the model velocity. The wavelength, λ with an initial wave frequency, f, is selected as 0.8 Hz based on recent tests of a similarly sized model by AlaviMehr (2016). Table 6 presents, the non-dimensional wave height, h_w, k that is the wave number calculated, w_e^*, which is the non-dimensional wave circular encounter frequency and V_m equal to the model velocity.

Figure 9 illustrates a corresponding load cycle for both the parent centrebow and the 45 mm wedge configuration in condition '1'. The major characteristics of the trend for the parent centrebow are labelled in Figure 9-a. Figure 9-b shows the introduction of a small peak in the total load prior to the maximum slam load, when the wedge impacts with the water surface. This minor depression was likely to be significantly reducing the subsequent major slam load. An overlay of the two data sets is shown in Figure 9-c to demonstrate there is a clear difference in the magnitude of the maximum slam load and subsequent vibration, and similar load throughout the remainder of the cycle.

Figure 10 illustrates that the wedge is forcing water transversely outwards during water entry, disrupting the flow of the jet rising up the sides of the demihulls, and the highly turbulent flow within the tunnels is also a possible contributing factor for reducing slam load magnitudes due to the higher air content. It was probable that there was a region of trapped air above the horizontal surfaces of the wedges during submersion, as there was visible flow separation occurring during the relative descent of the centrebow.

Figure 11 demonstrates consistency of the previously identified reduction in slam loads across the 10 s sample data for each configuration operating in a condition 'a'. Only the points of maximum load are shown to avoid confusion. Visual inspection of these plots indicates that the parent centrebow repeatedly experiences the highest slamming loads, while the 45mm wedge offers the greatest reduction. Table 7 presents the mean load and percent reduction for each configuration over the 10 s period.

Table 6: Summary of Test Conditions when Slam Occurred

	Condition	V_m [m/s]	H_W [mm]	h_w	f [Hz]	λ [m]	k	w_e [rad/S]	w_e^*
All hull forms	1	1.53	84	0.12	0.8	2.44	2.5	9.01	4.53
	2	1.53	101	0.11	0.74	2.85	2.2	8.04	4.06
	3	2.89	101	0.11	0.74	2.85	2.2	11.04	5.57

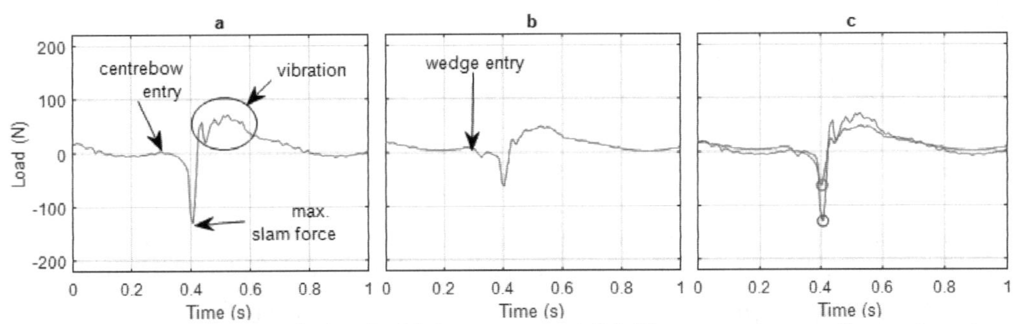

Figure 9: Showing a slam cycle load of (a) parent and (b) 45mm wedge configurations in condition '1', and (c) compares between the two configurations.

Table 7: Mean peak load amplitude and percentage comparison per each configuration

		Mean peak force magnitude [N]			Mean peak force [%]		
	Condition	(1)	(2)	(3)	(1)	(2)	(3)
Configuration	Parent	-125	-186	-129	100	100	100
	35mm wedge	-87	-135	-105	69.6	72.6	81.4
	45mm wedge	-55	-112	-90	44	60.2	69.8

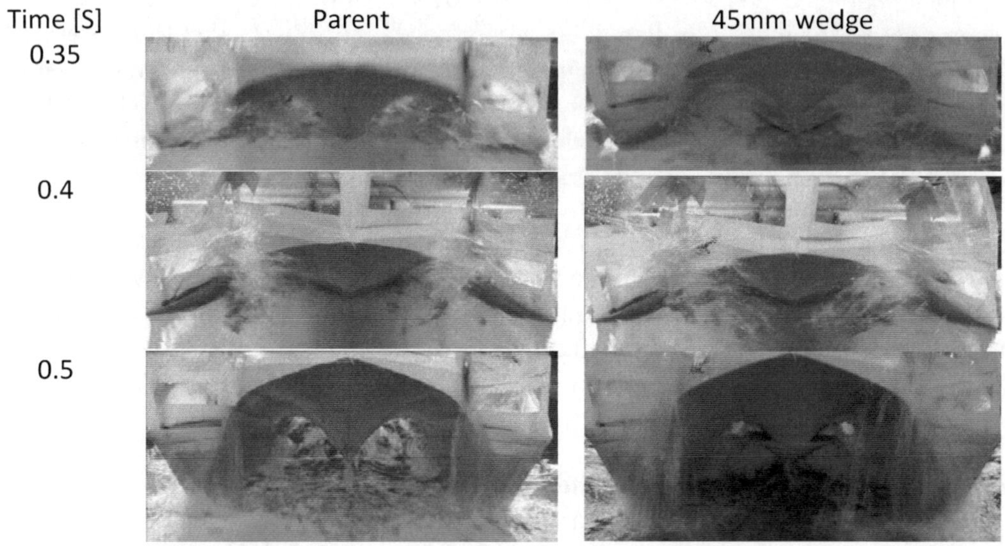

Figure 10: Photos taken of the parent and 45mm wedge configurations during a slam cycle.

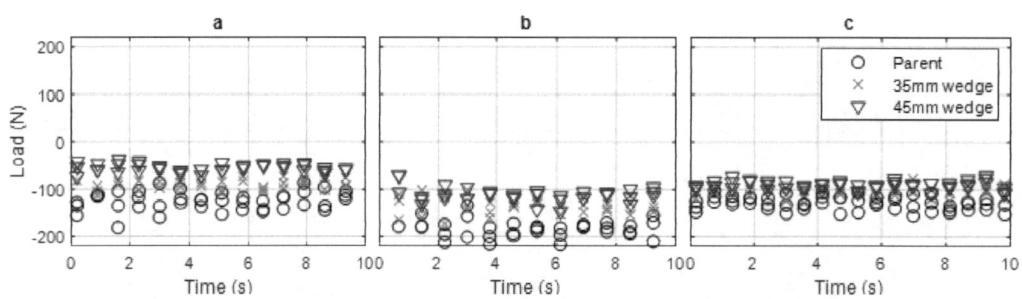

Figure 11: Showing all measured slam peak force magnitudes of conditions 1, 2 and 3.

CONCLUSIONS

This paper reported on a series of seakeeping tests to investigate the influence of early flow separation on the wetdeck slam force magnitudes by utilizing a generic catamaran hull model that is fitted with three centrebow configurations during wave impacts. In contrast to previous seakeeping tests these experiments focuses on reducing wetdeck slamming loads by allowing the flow to separate prior to slam event rather than modifying using larger or shorter centrebow configurations.

Since full details of the generic hull form are presented the results provide a comprehensive set of benchmarking data for use in the validation of numerical techniques to predict slam impact magnitudes of catamarans encountering regular waves.

The systematic and random uncertainties associated with the towing tank test results were quantified in detail and demonstrated the excellent repeatability of the tests; for example, the uncertainty of the measured peak slam loads was found to be less than 10%.

The developed wedge centrebow geometry in this paper could be considered as a novel finding towards minimising wetdeck slam severity that would affect passenger comfort, hull structure, cargo lashing and machinery operations of catamarans. The largest reduction in total impact loads, observed when utilizing the 45 mm wedge to be approximately 44% of the parent centrebow in a test condition. Thus, the addition of a wedge to the centrebow of a WPC offers a potential passive device to minimise slam loads and there is benefit in pursuing the concept.

ACKNOWLEDGEMENTS

The Authors would like to acknowledge Dr Jason Lavroff for assistance in the experimental design as well as Mr Tim Lilienthal and Mr Kirk Myer for assistance in the model setup.

REFERENCES

AlaviMehr, J., 2016. The Influence of Ride Control Systems on the Motion and Load Response of a Hydroelastic Segmented Catamaran Model. PhD thesis, University of Tasmania, Hobart, Australia.

Davis, M. R., French, B. J. & Thomas, G. A., 2017. Wave slam on wave piercing catamarans in random head seas. Ocean Engineering, pp. 84-97.

Incat Tasmania (2019). 112m high-speed wave-piercing catamaran. Retrieved 18 Feb, 2019, from: http://www.incat.com.au.

ITTC. (2011). Fresh water and sea water properties. Hjortekaersvej 99: 26th ITTC Specialist Committee on Uncertainty Analysis.

Shabani, B. (2017). *The effects of tunnel height and centre bow length on motions and slam loads in large wave piercing catamarans.* PhD thesis, University of Tasmania, Hobart, Australia.

Shahraki, J. R. (2014). *The Influence of Hull Form on the Slamming Behaviour of Large.* PhD thesis, University of Tasmania, Launceston, Australia.

Swidan, A., Thomas, G., Ranmuthugala, D., Penesis, I., Amin, W., Allen, T. & Battley, M. Experimental drop test investigation into wetdeck slamming loads on a generic catamaran hullform. *Ocean Engineering*, vol. 117C, pp. 143-153. 2016.

Swidan, A. (2016). Wetdeck Slamming Loads– Experimental and Numerical Investigation. PhD thesis, University of Tasmania, Launceston, Australia.

Swidan, A., Thomas, G., Penesis, I., Ranmuthugala, D., Amin, W., Allen, T. & Battley, M. Wetdeck slamming loads on a developed catamaran hullform - Experimental investigation. *Ships and Offshore Structures*, vol. 1, pp.1-9. 2017.

Ice Belt Extents for Protection in Waves

Andrew Marsland MEng CEng MRINA: Principal Naval Architect, RNZN Naval Engineering Authority, andrew.marsland@nzdf.mil.nz

Ed Ballard PhD CEng FRINA: Associate Director, MARICO Marine NZ Ltd, ed.ballard@marico.co.nz

INTRODUCTION

Floating ice is a significant hazard to ships operating in polar regions. In addition to ice fields that can be observed and monitored from satellites, ships must also contend with ice drifting away from ice fields, as well as ice calved from glaciers and ice shelves. This can lead to a proliferation of icebergs, bergy bits and growlers in waters that are otherwise 'ice free'. Bergy bits[1] and growlers[2] are of particular concern. Due to their smaller size they are more difficult to detect visually or using radar, especially amongst waves, in darkness or poor visibility, yet they have the potential to breach watertight integrity. Ice in polar waters is often multi-year sea ice, or ice of land origin, both of which are stronger and more damaging than first year sea ice.

Incidents highlighting the risk posed by floating ice in polar waters include:

- BCM Atlantic (2000): Trawler built to DNV Ice Class A, sank after striking ice at 6-7 knots;
- MV Explorer (2007): Adventure tourism passenger vessel built to DNV Ice Class A, sank after striking ice of land origin at a speed of 4-5 knots
- Snow Dragon (2019): Chinese research icebreaker, struck an iceberg that was not detected due to poor visibility at around 3 knots. Damage was sustained to the ships mast and bulwarks.

To mitigate the risk posed by floating ice, polar vessels incorporate structural ice protection, in the form of thicker plating and stronger frames around the waterline. Internationally agreed minimum requirements for polar vessel design are embodied in the IMO Polar Code (IMO, 2014), which references IACS Polar Class Rules (IACS, 2016) for structural ice protection requirements.

The risk of ice damage may be increased by waves causing the instantaneous waterline to be above or below the extents of structural ice protection, whether transiting an apparently 'ice free' area, or an observed ice field. Data from one wave buoy deployed in the Ross Sea as part of a programme to characterise the Southern Ocean and Ross Sea wave environment (Garrett & Durrant, 2019) indicated a mean significant wave height of 3.4m (Sea State 5) from the time of deployment in 3/10th ice coverage, until it became fast in 10/10th ice. The greatest significant wave height over this period was 10.8m (Figure 1). Waves in this marginal ice zone are driven by the ocean swells circumnavigating Antarctica, with no shielding land mass.

[1] Typically 100 – 300m^2 in area, protruding 1-5m above the water.
[2] Typically 20m^2 in area, and protruding less than 1m above the water.

Therefore, it must be assumed that vessels operating in Antarctic marginal ice zones will encounter waves together with floating ice.

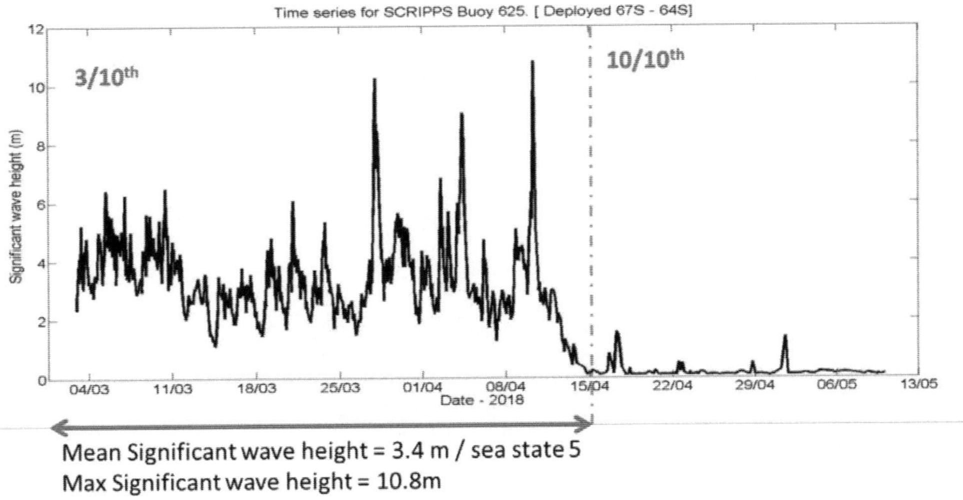

Figure 1: Significant Wave Height Data from Wave Buoy Deployed in Ross Sea Marginal Ice Zone, March / April 2018

The Royal New Zealand Navy (RNZN) conducts summer patrols to the Ross Sea, to support the Ministry of Primary Industries with its fisheries monitoring and compliance activities under the Convention for the Conservation of Antarctic Marine Living Resources (CCAMLR). From the early 2020s, the RNZN will transport fuel and supplies to McMurdo Sound using the new Polar Class tanker HMNZS AOTEAROA, and a dedicated Southern Ocean Patrol Vessel is also planned to enter service from 2027 onwards. Noting the current and future level of polar operations, together with the ice hazard described, the RNZN commissioned seakeeping analysis on a number of representative hullforms, to understand the level of protection that IACS Polar Class ice belt extents provide to vessels operating in waves. This paper presents some of the findings and conclusions from this analysis, building on earlier work by the NZ Defence Technology Agency (DTA) and MARICO Marine NZ Ltd (DTA, 2018).

STRUCTURAL ICE PROTECTION – POLAR CLASS RULES

IACS Polar Class Rules (the Rules) are appropriate for ships operating in areas where multi-year ice may be expected. For ships other than icebreakers, the Rules allow for varying levels of ice protection: from PC1 to PC7, where PC1 is the most ice capable. The design case for all Polar Classes is a glancing ice impact at the bow; this impact load is scaled to account for the expected loads on different areas of the hull. Unless the ship is stern acting, the highest loads are expected at the bow, and in the 'ice belt' region around the waterline. For example, in the midbody region of a PC5 vessel, in the ice belt the ice load is 50% of the bow impact load, below the ice belt the load is 30% and on the bottom the load is 0%. Ice belt extents are defined by the location of the Upper and Lower Ice Waterlines (UIWL, LIWL), as shown in Figure 2. This is a simplified version of the hull areas defined in the Rules, which divide the hull further into different longitudinal regions. For the purpose of this investigation, Polar Class 'ice belt extents' were taken as including both the bow region and the ice belt aft of the bow region, together representing the areas of greatest ice reinforcement.

The design load corresponds to the single highest load that is expected to occur once per year by a ship operating in ice infested waters, in ice conditions commensurate with the PC rating of the vessel (IACS, n.d.). The effect of ship speed is combined with ice strength and thickness in a single 'Class Factor', and therefore the design ice load could equally be achieved at low speeds / stronger ice, or higher speed (e.g. transit speeds) / weaker ice. The Rules allow for a degree of permanent set in the plating and framing, under the design load.

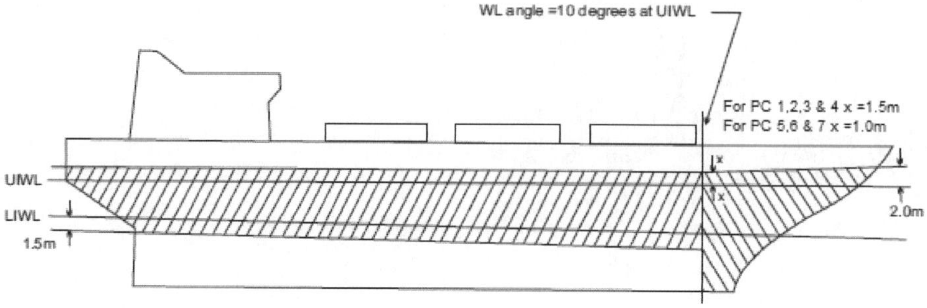

Figure 2: Polar Class Ice Belt Extents (Simplified)

ICE BELT EXTENTS ANALYSIS

Vessel Characteristics

Four vessels have been investigated, a conventional Offshore Patrol Vessel (OPV), two versions of a Polar Patrol Vessel (PPV) and a naval tanker design. The characteristics of the vessels are summarised in Table 1.

Table 1 Vessel Dimensions and Loading Conditions

Characteristic	Units	Offshore Patrol Vessel	Polar Patrol Vessel		Naval Tanker	
			PPV-A	PPV-B	Polar Light	Polar Deep
Description	-	Conventional hullform, bilge keels, fin stabilisers, twin screw, twin rudder	Ice bow, fin stabilisers, twin skeg, twin screw, twin rudder	As for PPB-A, but with 8m midship stretch	Twin skeg, twin screw, twin rudder, bilge keels	
Polar Class	-	No	Yes	Yes	Yes	Yes
Length Between Perpendiculars	m	73.5	86.4	94.4	168.7	
Beam	m	14.0	19.0		24.5	
Mean Draught	m	3.8	5.7		7.0	8.4
Displacement	t	1900	5700	6500	21000	26100
Block Coefficient, C_B	-	0.49	0.61	0.64	0.73	0.75
Prismatic Coefficient, C_P	-	0.67	0.69	0.72	0.75	0.77

Analyses were performed for a range of vessel headings from head seas to stern seas at 15-degree intervals. The vessel speeds ranged between zero and a maximum of either 16 knots (OPV) or 15 knots (PPVs and Naval Tanker).

Analytical Methods

Hydrodynamic analyses for the vessels were performed using the 3D diffraction analysis software HydroStar, developed by Bureau Veritas, to determine the motions and relative motions characteristics of the vessels.

A mesh for each vessel at the selected loading condition draught was used to compute the vessel motions and relative motions Response Amplitude Operators (RAOs). The motions of the vessel incorporate non-linear damping effects via the use of linear and quadratic damping components. As part of the 3D potential flow analysis, HydroStar calculates the velocity potential throughout the fluid domain surrounding the vessel, accounting for the radiated and diffracted wave fields. The velocity potential at the surface around the vessel allows the fluid pressure to be determined, from which amplitude and phasing of the wave elevation above the mean water level can be calculated at selected locations (e.g. Figure 3). Combining the absolute vertical motions at points on the vessel with the adjacent wave surface elevation allows the motions (amplitude and phasing) of the water surface relative to the hull to be determined.

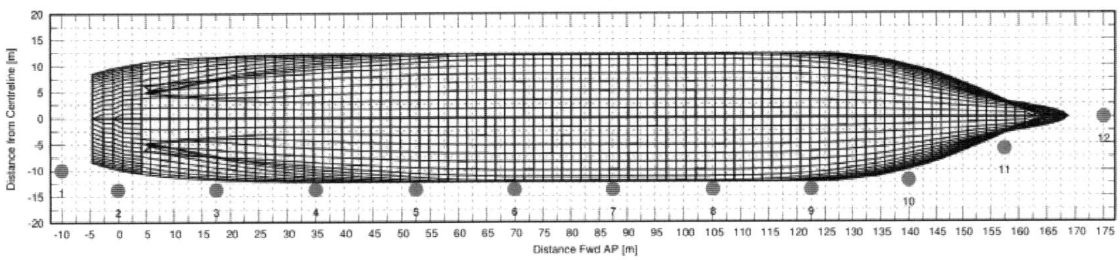

Figure 3: Example of Locations of Relative Wave Elevation Calculations, Plotted Against Hull Mesh to Waterline

Statistical Methods – Ice Belt Effectiveness

Assuming that the wave spectral density $S_w(\omega)$ represents the distribution of the wave energy and that RAO(ω) represents the transfer function of any first order quantity (motions, accelerations, relative wave elevation etc.), the spectral density of the first order responses can be defined as:

$$S_R(\omega) = RAO^2(\omega) \times S_w(\omega) \quad [1]$$

The spectral moments of the responses can then be defined as:

$$m_n = \int_0^\infty \omega^n S_R(\omega) d\omega \quad [2]$$

From the spectral moments, statistical properties of the responses such as the variance, m_0, are able to be calculated.

Assuming that the response is a narrow-banded Gaussian process, the distribution of instantaneous response amplitudes is given by the Normal distribution. Assuming zero mean response, the cumulative distribution of the instantaneous response *x* is given by:

$$P(x) = \frac{1}{2}\left[1 + erf\left(\frac{x}{\sqrt{2m_0}}\right)\right] \quad [3]$$

The distribution of instantaneous response amplitudes can be used to determine the proportion of time for which the sea surface remains within the vertical extents of the ice belt. If the upper and lower extents of the ice belt relative to the mean waterline of the vessel at a particular loading condition are given by $Z_{IB,Upper}$ and $Z_{IB,Lower}$ respectively, the probability of the water surface lying between those values is given by:

$$P(Z_{IB,Lower} < x < Z_{IB,Upper}) = \frac{1}{2}\left[1 + erf\left(\frac{Z_{IB,Upper}}{\sqrt{2m_0}}\right)\right] - \frac{1}{2}\left[1 + erf\left(\frac{Z_{IB,Lower}}{\sqrt{2m_0}}\right)\right] \quad [4]$$

Figure 4 illustrates this method, showing how the relative wave elevations are normally distributed about the mean waterline. Depending on the heights relative to the waterline of the upper and lower limits of the ice belt at a particular longitudinal position, the probability of the sea surface being within the ice belt is found from the area under the normal distribution between these upper and lower bounds (shown shaded red).

This probability (Equation [4]) was used as an appropriate metric of ice belt effectiveness. The implicit assumption is that ice impacts will occur at the instantaneous waterline; this is imperfect, however it is thought to be generally representative of the coverage provided by the ice belt.

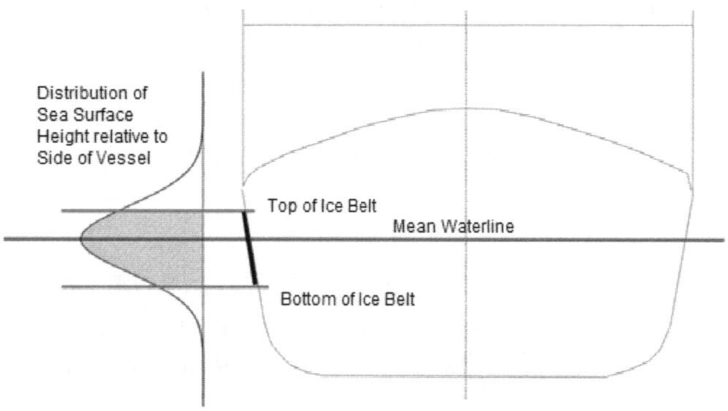

Figure 4: Normally Distributed Relative Wave Elevations

Wave Environment in Southern Ocean and Ross Sea

Wave spectra were developed from wave measurements recorded by wave buoys deployed in the Southern Ocean and Ross Sea (Garrett & Durrant, 2019), using a clustering technique designed to provide a limited number of spectra that were most representative of the entire recorded population.

Wave height was calculated for each wave measurement sample using spectral moments, allowing grouping based on the band in the Douglas Sea Scale they fell into, with sea states 3, 5 and 7 being of interest[3]. The waves in each sea state band were then used as input to a clustering of curves analysis using the CLARA algorithm, set up using the methods described by Hamilton (2010). Ten output medoids (the spectral curves which best describes each cluster of data) for each sea state grouping and location were used as representative spectra. Figure 5 and Figure 6 present the 10 medoid spectra for the Ross Sea and Southern Ocean wave conditions respectively.

[3] No sea state 3 conditions were recorded in the Southern Ocean.

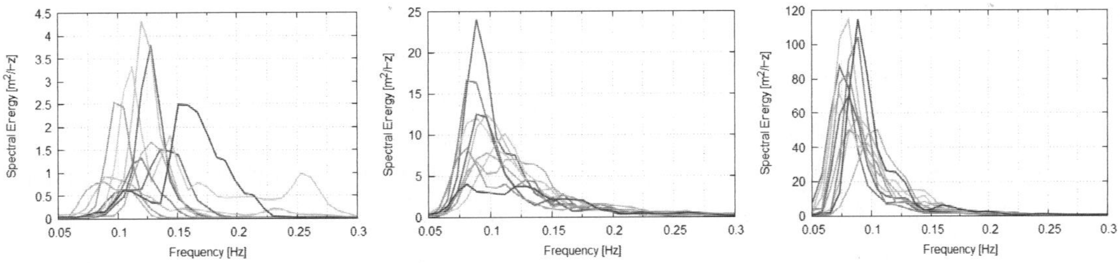

Figure 5: Ross Sea Medoid Spectra, Sea States 3 (left), 5 (centre) and 7 (right)

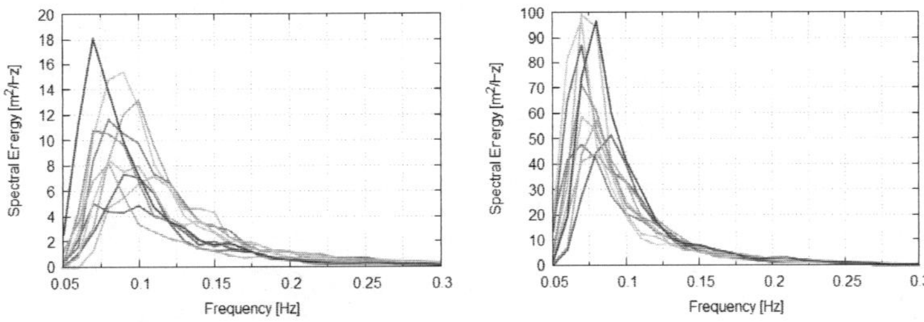

Figure 6: Southern Ocean Medoid Spectra, Sea States 5 (left) and 7 (right)

Analytical ITTC Wave Spectra

To provide a basis for comparison, analyses were also performed using the ITTC two parameter spectrum, which is defined in terms of significant wave height H_{sig} and peak period T_p as follows. This spectrum is commonly used to define open ocean conditions:

$$S_{ITTC}(\omega) = \frac{5}{16} H_s^2 \omega_p^2 \omega^{-5} exp\left(-\frac{5}{4}\left(\frac{\omega}{\omega_p}\right)^{-1}\right) \qquad [5]$$

Wave Spreading

Wave spreading was accounted for by using a *cos n* type function, whereby the directional short-crested wave spectra S(ω,θ) can be expressed in terms of the unidirectional wave spectra as follows:

$$S(\omega,\theta) = S(\omega)D(\omega,\theta) = S(\omega)D(\theta), \qquad [6]$$

where,

$$D(\theta) = \frac{\Gamma\left(1+\frac{n}{2}\right)}{\sqrt{\pi}\Gamma\left(\frac{1}{2}+\frac{n}{2}\right)} cos^n(\theta - \theta_p). \qquad [7]$$

The primary wave direction is given by θ_p and typical values for n for wind seas are between 2 and 4. For the present analyses a value of 2 was used.

RESULTS

Ice Belt Effectiveness in Ross Sea and Southern Ocean Spectra

Polar plots were developed indicating ice belt effectiveness in Southern Ocean and Ross Sea conditions. The results represent P_{min}, the lowest probability (worst-performance) across all of the points calculated along the length of each vessel.

Figure 7 illustrates the relative performance of the PPV-A, PPV-B and Naval Tanker (Polar Light and Deep) vessels with ice belt extents corresponding to PC1-4 and PC5-7 requirements[4]. The sea conditions represented are the weighted average of the 10 Ross Sea, Sea State 7 spectra. For each vessel, it is clear that the larger PC1-4 ice belt extents result in an increased probability of the sea surface lying within the ice belt.

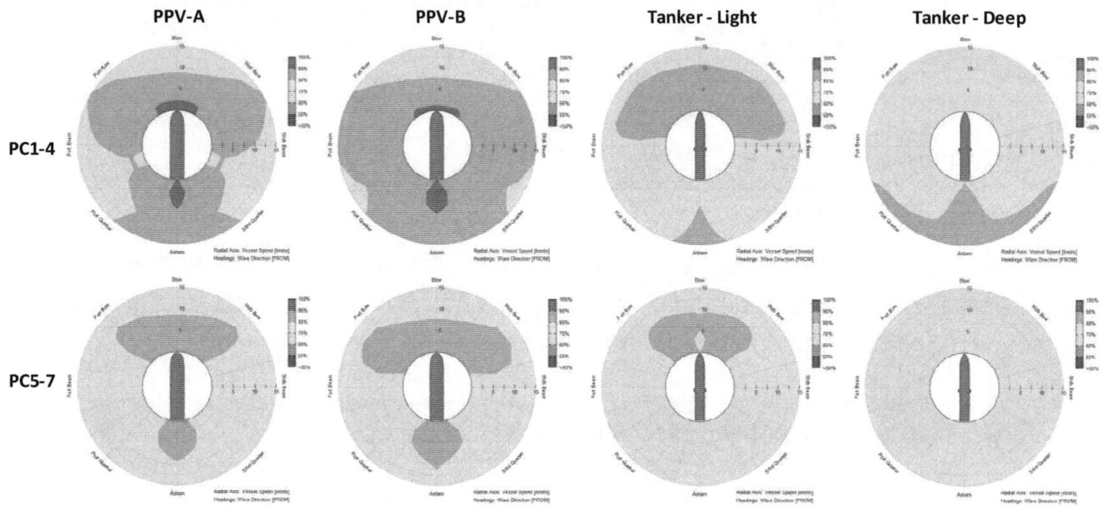

Figure 7: Comparison of Effectiveness of Different Ice Belt Extents, Ross Sea, Sea State 7

For Sea States 3 and 5, the ice belt effectiveness P_{min} was high, with vessels generally having greater than 90% ice belt effectiveness at speeds of 0 to 15 knots[5].

Ice Belt Effectiveness Results in ITTC Spectra

The spectra found via the clustering process exhibit characteristics which in many cases are quite dissimilar from common analytical spectra. Consider as an example Spectrum 1 for the Ross Sea, Sea State 7 conditions. Figure 8 compares this spectrum against an ITTC analytical spectrum, which has the same nominal H_{Sig} and T_p. It can be seen that the ITTC spectrum has a broader energy spread than the cluster spectrum, with a much lower peak.

[4] Note that Polar Class ice belt extents are grouped into PC1-4, and PC5-7. See Figure 2.

[5] The exception was the naval tanker in Polar Deep condition, with PC5-7 extents, for which P_{min} was found to be 80-90% throughout the same speed range.

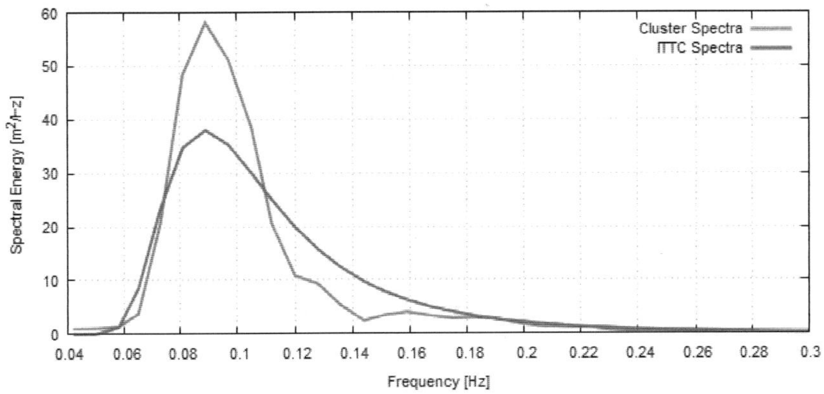

Figure 8: Comparison of Ross Sea, Sea State 7, Spectrum 1 with corresponding ITTC Spectra

Figure 9 compares the ice belt effectiveness of the ice belts on each vessel in the selected wave conditions for cluster and ITTC spectra. It can be seen in this particular case that the cluster spectra typically result in better ice belt effectiveness, with a wider range of conditions having green or dark green colouring.

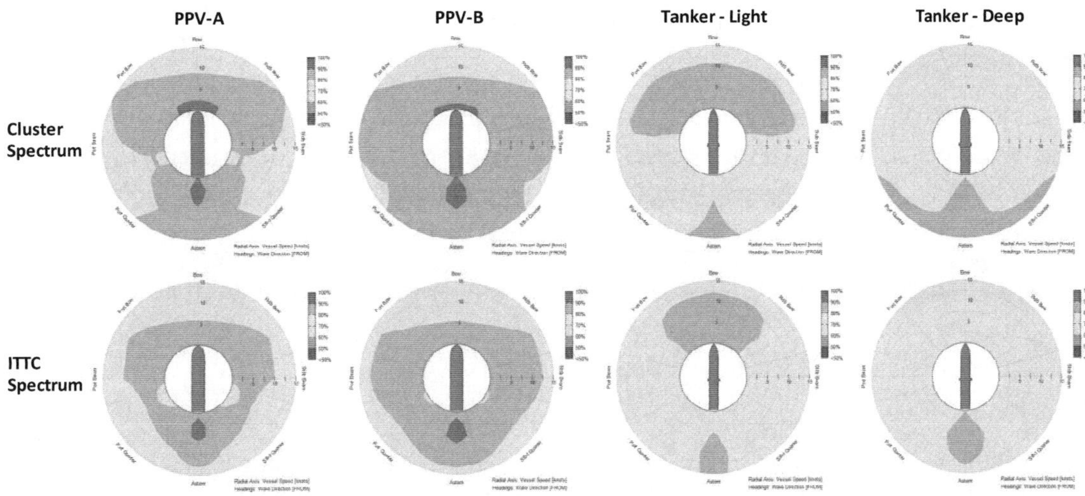

Figure 9: Comparison of Effectiveness of Different Ice Belt Extents, Ross Sea, Sea State 7 versus equivalent ITTC Spectra

The improved ice belt performance in the cluster spectra can be explained by the relationship between the distributions of wave energy in the spectra and the relative wave elevation (RWE) responses. Consider the relative wave elevation RAOs at the bow of the vessels in bow quartering seas at a speed of 10 knots. Figure 10 compares the cluster and ITTC spectra against the relative wave elevations for each of the vessels. It can be seen that the peak RWE responses for the tanker are much closer to the peak period of the waves and so are more affected by the concentration of wave energy in that region of the cluster spectra. The PPV vessels have their peak RWE response at a higher frequency, away from the majority of the wave energy for both spectra.

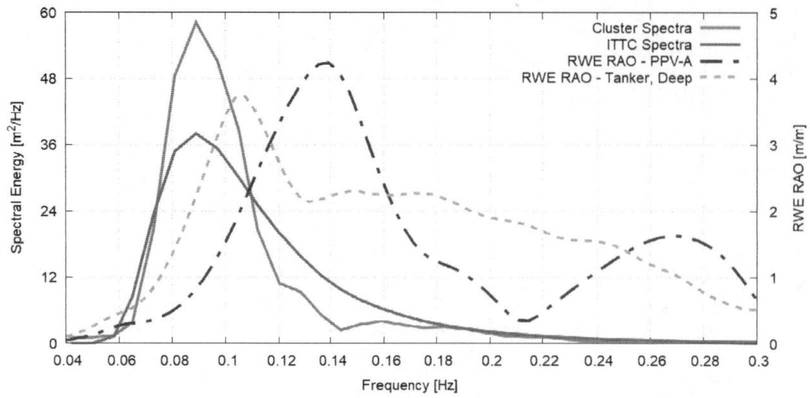

Figure 10: Comparison of Cluster and ITTC Spectra Against Relative Wave Elevation RAOs

This example illustrates the importance of correctly specifying the form of the wave spectra, as even for the same nominal H_{Sig} and T_p, the vessel performance may differ significantly. Whether that difference results in an under- or over-prediction of vessel performance will depend on the frequency dependent characteristics of the particular vessel response.

Ice Belt Effectiveness – General Performance Curves

Whilst the ice belt effectiveness results for the representative Polar Class vessels in general demonstrate a high degree of coverage, the results are significantly influenced by the location of the UIWL and LIWL (Figure 2). For example, the tanker has a large difference between the two waterlines due to its operating both at full cargo deadweight and empty in polar waters, which results in generous ice belt extents.

In order to understand required ice belt extents in a more general sense, performance curves were generated from the response variance results m_0, using Equation [4]. The specific m_0 results used were the weighted average results across the Southern Ocean spectra, at all headings.

Two examples corresponding to Sea State 5, Southern Ocean at 10 knots are presented in Figure 11. The variation of P (Equation [4]) is plotted with respect to H / T_{mean}, where H is the total ice belt height outside of the bow region, and T_{mean} is the mean draught for the considered loading condition. With reference to Figure 2:

$$H = 1.5\text{m} + (UIWL - LIWL) + \begin{cases} 1.0\text{m (PC5-7)} \\ 1.5\text{m (PC1-4)} \end{cases} \quad [8]$$

To produce these curves, the arrangement of ice protection was assumed to be as per Polar Class[6] (Figure 2), but UIWL was nominally raised or lowered to generate results for a wider range of H, whilst keeping T_{mean} fixed. The minimum possible H is 2.5m, when UIWL = LIWL

[6] It should be noted that results for OPV are somewhat artificial, because the basis vessel could not sustain the weight of Polar Class ice protection. However, the inclusion of this vessel helps to demonstrate the effect of different hullforms on ice belt performance.

and PC5-7 extents are applied, although this is clearly only a theoretical case. Parallel sinkage due to additional structural weight was neglected.

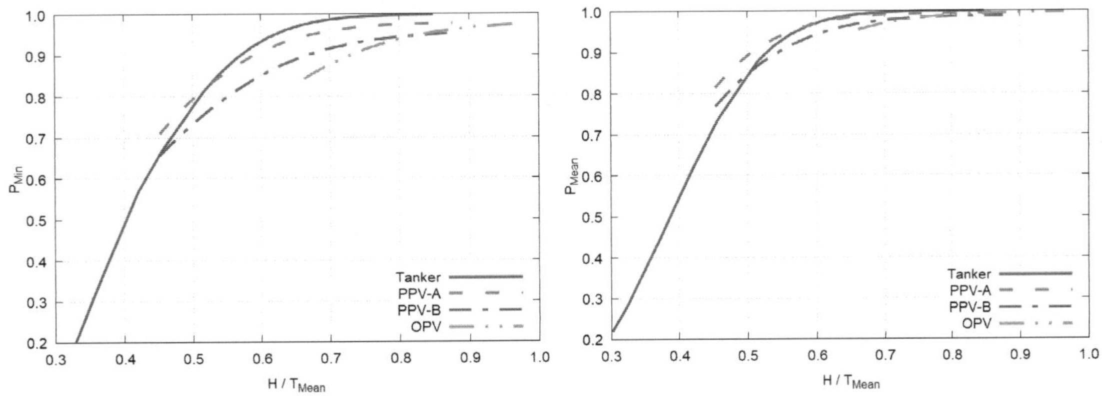

Figure 11: Ice Belt Effectiveness (Minimum (left) and Mean (right) Across All Waterline Locations). Sea State 5, 10 knots, Southern Ocean.

Determining ice belt effectiveness P by considering only the worst-case location along the waterline (P_{min}) is conservative, implicitly assuming that ice impact will occur at the worst-case location. A more realistic assumption would be that impact is equally likely to occur anywhere along the waterline (as represented by P_{mean}, the average performance of all locations along the waterline). When P_{mean} is taken as the measure of ice belt effectiveness, the results collapse down approximately onto a common performance curve. This suggests that, for this sea state and speed, ice belt effectiveness is determined largely by H / T_{mean}, and is not otherwise sensitive to hullform characteristics or principal particulars.

Performance curves for a range of sea states and ship speeds confirmed this insensitivity amongst the basis vessels, with the slight exception being sea state 5 / 3 knots (Figure 12), in which ice belt effectiveness of the OPV hullform is stepped below that of the other vessels. This is caused by significant roll at low speeds due to a hullform with low natural roll damping.

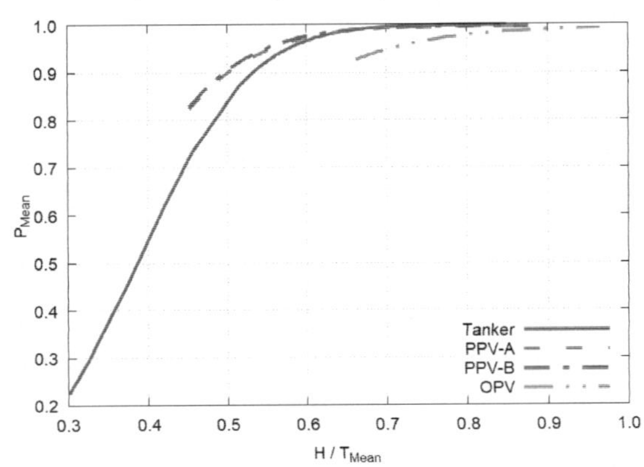

Figure 12: Ice Belt Effectiveness (Mean Across All Waterline Locations). Sea State 5, 3 knots, Southern Ocean.

P_{mean} results for all four hullforms combined[7] are presented in Figure 13. It was found that, for any given value of H, results for 'PC5-7' extents were nearly identical to those for 'PC1-4' extents. This is because, in fixing H the only difference between the two was the absolute height (above baseline) of the upper ice belt extent at the stem.

Unsurprisingly, ice belt effectiveness is reduced in higher sea states, and at higher speeds. Increasing H / T_{mean} leads to an increase in ice belt effectiveness, however beyond H / T_{mean} = 0.6, diminishing returns apply.

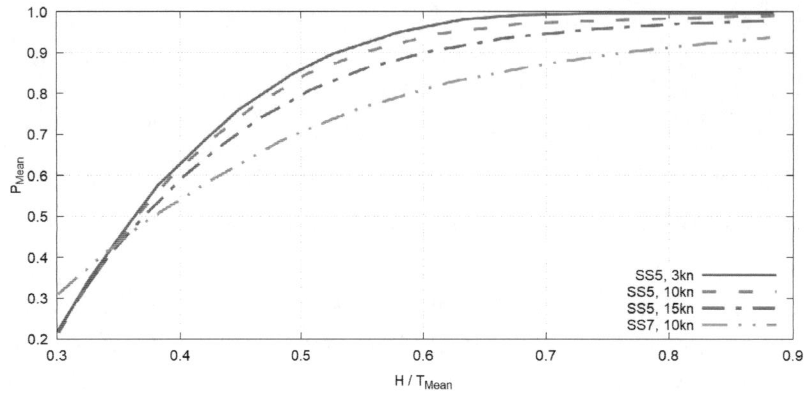

Figure 13: Ice Belt Effectiveness, in Southern Ocean Wave Conditions.

POLAR VESSEL DESIGN IMPLICATIONS

It would be difficult to set an 'acceptable' value of ice belt effectiveness, as this is only one component in determining the overall risk of structural damage due to collision with floating ice. Other factors include the prevalence of ice, the size, thickness and physical properties of ice and weather conditions, all of which are uncertain. However, high values of effectiveness must reduce the overall risk. For the basis Polar Class vessels H / T_{mean}, the ratio between total ice belt height and mean draught, is approximately 0.5 – 0.6 in the deepest loading condition. This provides typical ice belt effectiveness (Figure 13) of at least 85% at SS5 / 10 kn and 70% or greater even at SS7 / 10 kn.

While acknowledging the limited number of vessels contributing to this study, the results indicate that there may be little value in augmenting ice belt extents beyond Polar Class, to improve protection in waves, provided that:

- H / T_{mean} is greater than 0.5; and
- The vessel has good seakeeping performance, and in particular does not suffer from excessive roll and pitch in the operating wave conditions. This may be particularly problematic at low speeds, where fin stabilisers, if fitted, will be ineffective. Polar vessels need to operate at low speeds around ice and in waves, therefore seakeeping performance requirements should capture this. Analysis has shown that it is important

[7] Produced using the lower bound results. The SS5 / 3 knot curve excludes OPV results due to this vessel's roll low speed roll issue. The curve as plotted therefore represents vessels in which exhibit good seakeeping performance at low speeds.

to use the specific wave conditions for the area of operation, since ship motions are sensitive to the spectral energy distribution.

Statistical analysis of relative wave elevation data, as described in this paper, could be undertaken during initial design to further improve confidence in the protection that Polar Class ice belt extents provide in waves, for the specific hullform and operating parameters including wave environment, ship speeds and draughts.

REFERENCES

Defence Technology Agency (2018). Assessment of OPV Ice-Belt Effectiveness. DTA Technical Memorandum No. C1405-2, 2 October 2018

Garrett, S. & Durrant, T (2019). Characterising the Southern Ocean and Ross Sea Wave Climate. In *Proceedings of Pacific 2019 International Maritime Conference*.

Hamilton, L. J. (2010). Characterising spectral sea wave conditions with statistical clustering of actual spectra. Applied Ocean Research, Volume 32, Issue 3, July 2010, Pages 332-342.

International Association of Classification Societies (IACS) (n. d.). History Files and Technical Background documents for Unified Requirements. Available online at www.iacs.org.uk [accessed 8/7/2019].

International Association of Classification Societies (IACS) (2016). Unified Requirement I: Requirements Concerning Polar Class.

International Maritime Organisation (IMO) (2014). International Code for Ships Operating in Polar Waters. Adopted November 2014; in force 1 January 2017.

Characterisation of Miniature Piezoresistive Pressure Sensors for the Measurement of Slamming Loads.

Peter Graham, Dr. Daniel Sgarioto
Defence Science and Technology
peter.graham@dst.defence.gov.au

ABSTRACT

Understanding the effect of highly non-linear wave impact loads on structural integrity is important during Life-of-Type and Life-of-Type extension assessments. Towing tank testing utilising pressure sensors provides a cost effective method of examining these non-linear wave loads within a repeatable, controlled environment.

A survey of the relevant literature reveals many challenges and intricacies associated with the use of pressure sensors for this purpose. In the selection of a sensor, many factors need to be considered that affect the quality of the results obtained such as the sensor type, pressure range, sensitivity, sensor frequency bandwidth, output type, physical construction, size and weight.

This study examines the selection of a suitable pressure sensor to be used during a towing tank trial to measure hydrodynamic loading on a hydro-elastic generic patrol boat model. Prospective sensors were selected and tested for signal-to-noise ratio, dynamic response, photoflash sensitivity and thermal transient response in order to select the most appropriate sensor. Comparisons between results obtained for each sensor revealed some undesirable responses due to environmental influences. The information presented in this paper provides the basis for the selection of pressure sensors for use in scale towing tank or tank sloshing model test programs.

INTRODUCTION

The selection of a pressure sensor for scale model towing tank testing requires many factors to be considered. These include the type or construction of sensor, physical size and weight, the output type, media being sensed, pressure range, frequency range and sampling frequency of the output. This study was undertaken as there are currently no relevant standards or procedures for the selection of pressure sensors to measure non-linear wave impact loads in scale model tow tank testing. Previous studies, as outlined below, have used pressure sensors to measure tank sloshing, water impact pressure using wedges, cylinders and scale models which have alluded to issues with acquired pressure data and subsequent follow-on investigations were performed predominantly with Piezoelectric (PE) or Integrated Electronic Piezoelectric (IEPE) sensors.

Rosen et al. [1] found that due to the large number of sensors required, noise caused by coupling of signals between cables highlighted the importance of cable shielding. Thornhill et al. [2] stated that there were relatively high levels of noise in their acquired data and that several sensors failed. Choi et al. [3], Van Nuffel et al. [4] and Pistani et al. [5] found mounting issues related to whether the sensor face was flush with the model surface. If the sensor sits proud of the surface, turbulence will be generated causing incorrect pressure measurements. A sensor sitting recessed will trap air causing variability in the peak pressures that are measured. This is also the case if the measurement face of the sensor is not flush and sits recessed in a duct. The authors from [3, 4, 5] along with Judge et al. [6] and Kim et al. [7] also found that the acquired data contained spikes and level shifts when the sensors were moved between air and water, and vice versa, due to a difference between the water and air temperature. While the majority of studies used piezoelectric pressure sensors, which are known to have temperature sensitivity, Pistani et al. [5] found the same sensitivity in the piezoresistive pressure sensor when the ambient air temperature was the same as the water temperature. The conclusion was that strain gauges in the Wheatstone bridge self-heated due to the applied excitation voltage which locally raised the temperature. When water contacted the sensor it caused the contraction of the metal diaphragm resulting in the spike on the output. Judge et al. [6] experimented with coating the face of the sensor with a thin layer of dielectric grease to isolate the sensor from any temperature variation while still transferring dynamic pressure.

The previous studies show that a wide range of sampling frequencies have been used to measure the pressure caused by wave impacts. Rosen et al. [1] found that the sampling frequency of 2.5 kHz was sufficient for most situations, but on the low side for more extreme impacts. Van Nuffel et al. [4] compared peak pressure measurements and found that a sampling frequency of at least 200 kHz was required to reproduce the signal with approximately 1% error. Repalle et al. [8] found in their tank sloshing experiment that a minimum sampling frequency of 20 kHz was required to measure impact pressure but that 40 kHz is more appropriate.

As there is often a requirement in towing tank experiments to measure both hydrostatic and wave impact pressure, this study was limited to strain gauge or micro-electro-mechanical system (MEMs) based piezoresistive sensors rather than PE or IEPE sensors. PE and IEPE sensors are not suitable for use in measuring static pressure due to the decay in the output signal caused by charge leakage.

PIEZORESISTIVE PRESSURE SENSORS

A survey of commercially available piezoresistive pressure sensors was undertaken and two prospective sensors were selected due to their small physical dimensions, low weight, frequency bandwidth and diaphragm flush with the sensor face. The sensors selected were the TE Connectivity (Measurement Specialties) XP5 [10] and Keller 2MI [11]. The Endevco 8510C pressure sensor [9] was an existing item in our inventory which was included as a comparison to the XP5 and 2MI sensors.

Endevco 8510C

The Endevco 8510C piezoresistive pressure sensor uses a micro-machined diaphragm with four small piezoresistors implanted on the surface in a Wheatstone bridge configuration. The

8510C pressure sensor is a vented gauge type and is not sealed on either side of the diaphragm. On the rear of the sensor there is a small tube to allow the air pressure to push against the back of the diaphragm, equalising the force from the front of the diaphragm at ambient pressure. This results in the output of the sensor being centred about the ambient pressure. The sensor face has a diameter of 3.86 mm and an overall length of 19.1 mm. The front of the sensor has mesh covering the face as can be seen in Figure 1a.

There are two reasons for this mesh. It covers the inlet to the sensor and offers mechanical protection while also reducing the amount of high intensity light directly hitting the silicon diaphragm which causes a transient response on the sensor output. The diaphragm is mounted behind the mesh and held in place by epoxy adhesive. As a result, the manufacturer states that while the 8510C pressure sensor can be used in water it should be 'only for short durations' and 'it is best to use de-ionised water or water without normal contaminants found in tap water, which is quite corrosive' [14]. The manufacturer therefore advised that the sensor then needs to be dried out thoroughly following each immersion. The sensors will be used in a tow tank and it is impractical to fill the tank with purified water for the duration of the testing. It is also impractical to test for short periods and dry each of the pressure sensors out at the end of each session due to the amount of time required to setup and calibrate the sensor suite. Also as a typical tow tank test program can span over many days progressive damage to these sensors may occur which could compromise the acquired data.

TE Connectivity (Measurement Specialties) XP5

The TE (Measurement Specialties) XP5 pressure sensor, pictured in Figure 1b, was selected as one of the sensors for further investigation because it is the same physical size and thread pattern as the 8510C. The XP5 is also totally sealed with a flush diaphragm at the sensing face and the range of the sensor can be selected closer to the expected maximum pressure typically observed in scale model testing in towing tanks. The sensor range selected was 104.42 kPa, which was the lowest pressure range at the time. The transducer is completely sealed within a titanium body with a laser welded diaphragm on the face of the sensor and the sensing element is a temperature compensated Wheatstone bridge constructed using micro-machined silicon strain gauges.

(a) (b)

Figure 1. (a) 8510C pressure sensor; (b) XP5 pressure sensor.

Keller 2MI

The Keller 2MI pressure sensitive element, as shown in Figure 2a, is a micro-machined piezoresistive sensor that is mounted in a stainless steel case and covered in a thin coating of silicone elastomer. The element dimensions are 4.5 mm diameter and 3 mm in height. The small dimensions allow the element to be used in applications where space and weight are important factors and can be used in all gases and liquids that are compatible with silicone elastomer and stainless steel. The 2MI can be used in either constant current or voltage configurations. The end user is to install configuration and external temperature compensation resistors as specified on the supplied calibration sheet.

To enable the sensor element to be easily installed, moved and replaced, the sensor element needs to be mounted in a secondary housing tailored to the particular measurement application under consideration. A secondary housing with locking nut was constructed of aluminium to reduce weight and encloses the temperature compensation and zero adjustment resistors close to the element and allow the transition to a more robust cable. The housing was filled with potting compound to provide mechanical strength and ensure it is fully waterproof. In comparison with the 8510C and XP5 sensors, the finished sensor is larger in diameter at 7 mm and similar in length at 23 mm. The 2MI pressure sensor mounted in the secondary housing that was developed for specific towing tank measurement applications is shown in Figure 2b. The maximum pressure range of the selected sensor was 100 kPa.

(a) (b)

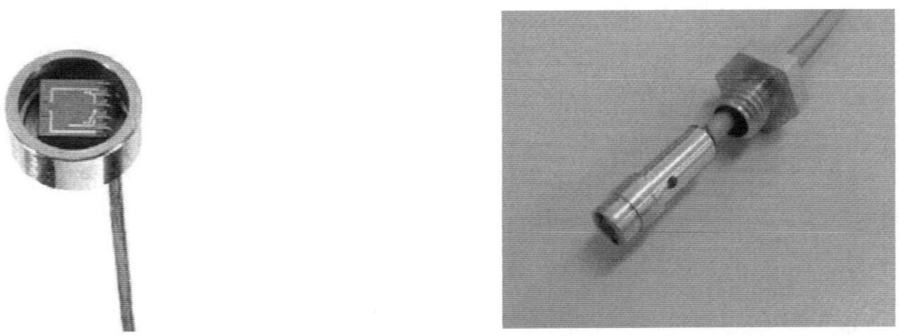

Figure 2. (a) Keller 2MI sensor element without silicone covering; (b) in housing with locking nut.

TESTING METHODOLOGY

There are currently no standards or procedures for the selection of pressure sensors for use in measuring non-linear wave impact loads in scale model tow tank testing. As a result, a suite of tests was developed to characterise the noise, dynamic, photoflash and thermal transient responses which are relevant to scale model tow tank testing. These responses may be defined as various parameters in the sensor data sheet, but it can be difficult to gain a full understanding on their effect on the acquired data. The testing regime was designed to compare the Endevco 8510C, against the Measurement Specialties XP5 and Keller 2MI sensors for their performance and response in a laboratory setting.

All of the sensors were connected to an Endevco Model 136 bridge amplifier and the analog signal was acquired using a Liberty data acquisition system made by HBM Inc. The GP-8 analog input module has 16-bit resolution and a maximum sampling rate of 100 kilo-samples per second (kS/s) per channel. While there is a large variation in the sampling rates of the previous studies mentioned earlier, a conservative sampling rate of 20 kS/s was selected for this study.

Noise

Noise is always present in data acquired from sensors to varying extents. It can be generated by many sources including switching power supplies, motors and computers. The noise can be transmitted through the air and picked up with cables or conducted through the power supply. The data sheet for the 8510C sensor states that electrical noise is 5×10^{-6} Vrms and the Model 136 bridge amplifier has electrical noise of 20×10^{-6} Vrms relative to the input plus $1\text{x}10^{-6}$ Vrms relative to the output. Given an average sensor amplifier gain of 42.33 the highest Signal to Noise Ratio (SNR) should be approximately 74 dB. The XP5 and 2MI sensors do not state a noise parameter in the data sheets.

As the 8510C sensor cable is 1.5 m long an 8 m extension cable was used to connect the sensor to the bridge amplifier. This extension cable is unshielded and has unterminated connections to the sensor outputs. Potentially both of these issues could increase the amount of noise, reducing the SNR at the output of the amplifier. A shielded extension cable was constructed and compared with the unshielded variant to examine the difference in the amount of noise exhibited on the output of the bridge amplifier. The XP5 sensor has a cable length of 2 m and therefore also required an extension cable to connect to the amplifier. The effect of using both types of extension cables will be compared. As the 2MI pressure sensor has been constructed with a 10 m cable, only the sensor and bridge amplifier will be compared to the results from the other sensors.

Dynamic Response

The dynamic performance of each of the sensors was tested using a PCB Dynamic Pressure Pulse Calibrator. The calibrator generates a step-function pressure pulse of precisely known static air pressure from a large volume reservoir. The pressure of the pulse is set by an input control valve and is displayed on a large dial gauge. The pulse actuation and duration are controlled by a press button which actuates a solenoid valve connected to the test housing.

The 8510C, XP5 and 2MI pressure sensors were tested with pressure pulses at 80 % of the maximum pressure range to gain a better understanding of the frequency response for each of the sensors.

Photoflash and Light Sensitivity

High intensity light from a camera flash or strobe light can cause a transient pulse on the output of piezoresistive transducers. This is due to the silicon diaphragm absorbing photons emitted from the flash or high intensity light. These photons excite the electrons of the material that makes up the bridge elements causing a change in the current flow through the piezoresistors. This change in current flow exhibits itself as a positive or negative voltage spike on the output.

As flash photography is often used as a part of the documenting process for towing tank testing, the sensors were tested for photoflash response. The response from the three pressure sensors was measured with the flash at a distance of 1.5 m and rotating the sensor from zero degrees (flash directly facing the diaphragm) through to ninety degrees, in fifteen degree increments. A Canon Speedlite 580 MkII flash module configured in manual trigger mode and a 1/64 second flash duration was used for this test program.

The three sensors were also tested to examine if there is any response due to changes in indoor lighting. The simple test involved placing the sensor perpendicular to a light source for duration of five seconds and then rotating the sensor so that it was parallel to the light for a further five seconds. This was repeated six times to create a test duration of one minute. The distance between the light and sensor was kept at a constant 1.5 m. The light source was a twin tube fluorescent light typical in laboratory lighting.

Thermal Transient Response

Thermal transient response is the output of the sensor when subjected to a step-function temperature change [13]. This sudden change in temperature can cause a change in the physical stress of the diaphragm and therefore a change in the value of the piezoresistors or microstrain gauges. A change in temperature will also cause a change in the resistance of individual piezoresistors or gauges. An imbalance in the Wheatstone bridge can occur if only part of the bridge is exposed to the thermal transient, resulting in a transient response at the output. Sensors are typically designed with thermal compensation which works only for equilibrium temperature and not thermal transients. As the pressure sensors will be moving between air and water when measuring slamming loads, therefore a minimal response should be exhibited on the sensor output due to the change in media and temperature of the media.

The thermal response was tested by immersing the pressure sensors mounted in the side of an open top enclosure in water to a set level, resulting in an expected hydrostatic pressure change. Various water temperatures were tested to vary the difference between the ambient air and water temperatures and therefore gain an understanding of the thermal response of each sensor. Two mounting positions in the enclosure produce a measured output of approximately 0.12 kPa and 0.31 kPa due to hydrostatic pressure changes alone.

RESULTS

The following results were measured during the testing program in response to noise, dynamic response, photoflash and light sensitivity and thermal transient response for the 8510C, XP5 and 2MI pressure sensors.

Noise

The SNR for each of the sensors was calculated using the peak-to-peak electrical noise and the maximum peak-to-peak voltage for the full range of the sensor output as shown in Equation 1.

$$SNR(dB) = 20\ log_{10} \frac{Max.Pressure\ _{V\ p-p}}{Measured\ Noise_{V\ p-p}} \tag{1}$$

In the laboratory bench tests for noise the 8510C, XP5 and 2MI sensors all performed similarly with an average SNR of 65.74, 65.46 and 65.46 dB respectively. There was no increase in noise due to the inclusion of the extension cable or between the original unshielded and shielded version of the cables for both the 8510C and XP5 sensors. As the 2MI was constructed with a 10 m long cable it was not tested with the extension cables. Figure 3 shows the electrical noise on the measured output of the sensors at ambient pressure. While laboratory testing in this case demonstrated no change in noise levels between shielded and unshielded extension cables, shielded cables should be used in noisier environments.

Data acquired during previous test programs that utilised the 8510C sensors found a SNR of between 42.4 and 63.8 dB with an average of 49 dB. When comparing the previously acquired data to the results from the laboratory testing there is reduction of 16 dB in SNR which is due to a 6.3 times increase in the measured noise.

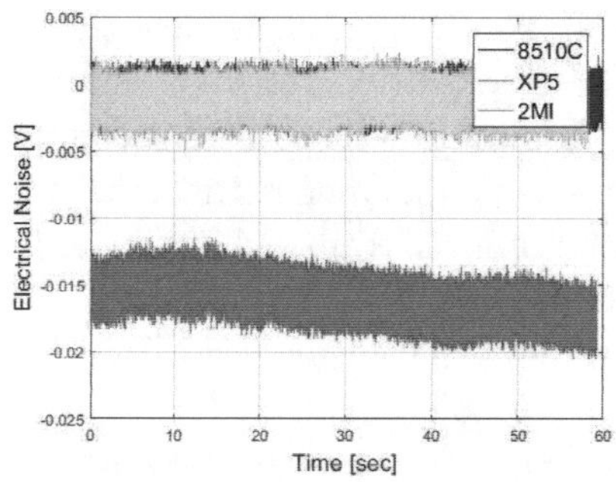

Figure 3. Pressure sensor noise response of all sensors.

Dynamic Response

The dynamic response of the 8510C, XP5 and 2MI sensors were tested with a set pressure pulse corresponding to approximately 80 % of the sensors range. The average measured steady state magnitude of the pressure pulse of the sensors including the peak overshoot is shown in Table 1. A pressure pulse for the 8510C, XP5 and 2MI sensors is shown in Figures 4a, 5a and 6a respectively. A magnified view of the pressure pulse peak for the 8510C, XP5 and 2MI sensors is shown in Figures 4b, 5b and 6b respectively. The overshoot for the 8510C and 2MI pressure sensors had a decaying oscillation with a measured frequency of 9.7 and 9.1 Hz respectively, which decayed within 0.7 of a second. The XP5 decayed from the overshoot with no oscillation and was still trending down at the end of the pressure pulse. It is estimated that the decay would reach a steady state in approximately 1.2 seconds. The measured response of the XP5 sensor after the negative going edge of the pulse also overshot the original zero pressure by 1 kPa returning to zero 1.2 seconds later. The spikes at the start and end of the measured pressure pulses for all sensors were caused by the actuation and release of the PCB pulse calibrator solenoid.

Sensor	Pressure Pulse (kPa)	Av. Steady State (kPa)	Peak Overshoot (kPa)
8510C	275.8	274.42	0.4
XP5	82.74	81	1.35
2MI	82.74	83.2	0.16

Table 1. Dynamic Pressure Pulse measurements for all sensors.

As shown, the 8510C and 2MI sensors matched the pressure pulse more closely and were within approximately 0.5 % of the set pressure for the pulse when compared to the XP5 which was 2.1 %.

(a) (b)

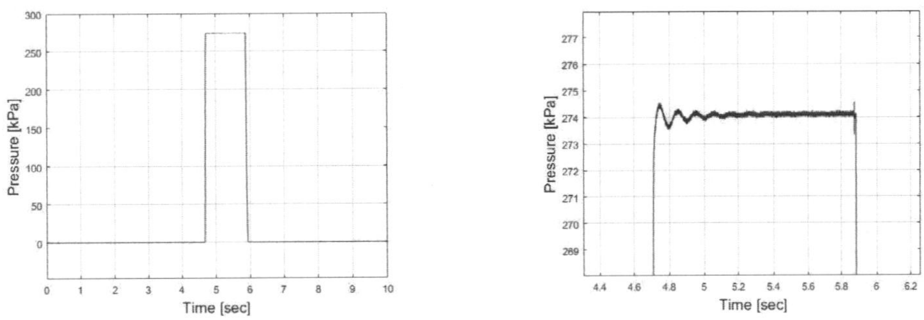

Figure 4. (a) Pressure Pulse response for 8510C sensor; (b) magnification of the peak.

(a) (b)

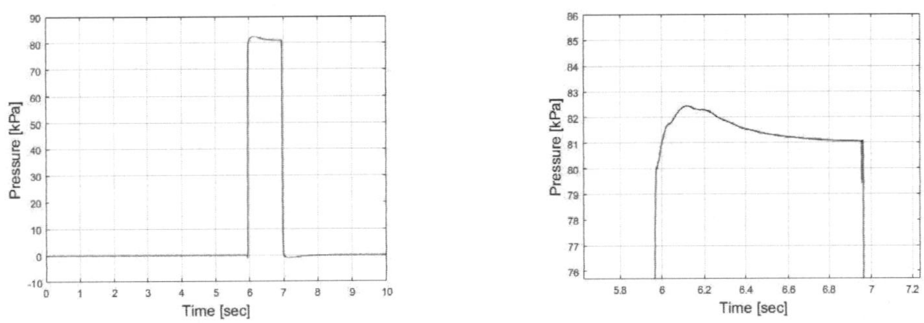

Figure 5. (a) Pressure Pulse response for XP5 sensor; (b) magnification of the peak.

(a) (b)

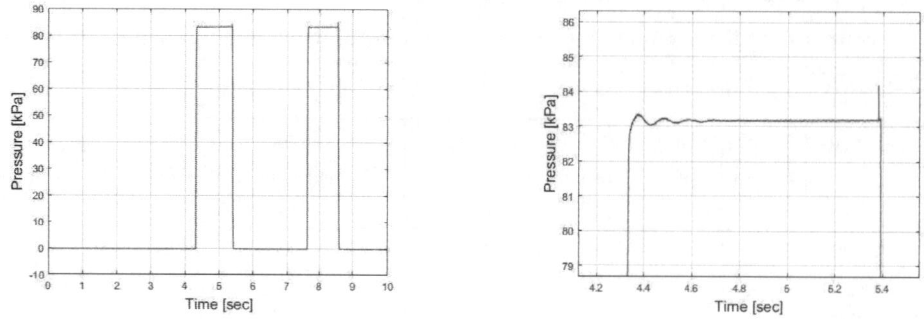

Figure 6. (a) Pressure Pulse response for 2MI sensor; (b) magnification of the peak.

The response from the three sensors was underdamped, with the XP5 sensor being more heavily damped than the 8510C and 2MI sensors. The higher damping factor of the XP5 sensor may be due to the increased mass of the diaphragm and may contribute to the lower dynamic frequency when compared to the 8510C and 2MI sensors. The useable dynamic frequency bandwidths for the 8510C, XP5 and 2MI sensors are calculated as 64 kHz, 21.6 kHz and 60 kHz respectively, using a general industry 'rule of thumb'. This 'rule of thumb' states that, 20% of the natural frequency is the useable bandwidth of piezoresistive and piezoelectric pressure sensors due to the change in gain of the sensor exceeding 5% as the dynamic frequency approaches the natural frequency [10, 12 & 13]. As a result, the signal from each sensor should be filtered to remove high frequency components of the signal before the data acquisition system and folding due to the Nyquist theorem. Typically manufacturers will only define the natural or resonant frequency in datasheets, so care needs to be taken to ensure that the maximum frequency expected during testing is lower than 20% of the natural or resonant frequency.

Photoflash and Light Sensitivity

The results for the 8510C sensor in Table 2 show that the high intensity light from the flash was measured at all angles except for 90 degrees. At 90 degrees the light output is parallel to the diaphragm and was blocked from hitting the diaphragm by the construction of the sensor. The level of response varied, as the angle between the flash and the sensor increased due to the mesh covering the duct leading to the sensor diaphragm blocking the light more effectively at some angles. The magnitude of the measured spikes varied from -0.77 to -23.13 kPa, which is 0.22 – 6.7 % of the total range of the sensor but in the negative direction.

The 2MI pressure sensor diaphragm is sealed using a thin layer of silicone elastomer to protect it from the media being sensed which allows some of the light through to the sensor diaphragm. The measured photoflash response, as shown in Table 2 and Figures 7a and 7b, was constant at -4.59 kPa across all angles except for 90 degrees, which produced a response of -0.4 kPa. The constant response is because the silicon attenuates the light rather than blocking it.

The XP5 sensor has a diaphragm that is covered by a thin layer of titanium, which blocks light and is therefore not affected by photoflash as shown in Table 2 and Figures 7a and 7b.

Flash Angle (deg)	8510C50 (kPa)	XP5 (kPa)	2MI (kPa)
0	-21.54	0	-4.59
15	-13.8	0	-4.59
30	-23.13	0	-4.59
45	-19.32	0	-4.59
60	-18.34	0	-4.59
75	-0.77	0	-4.59
90	0	0	-0.4

Table 2. Pressure sensor flash response of all sensors.

(a) (b)

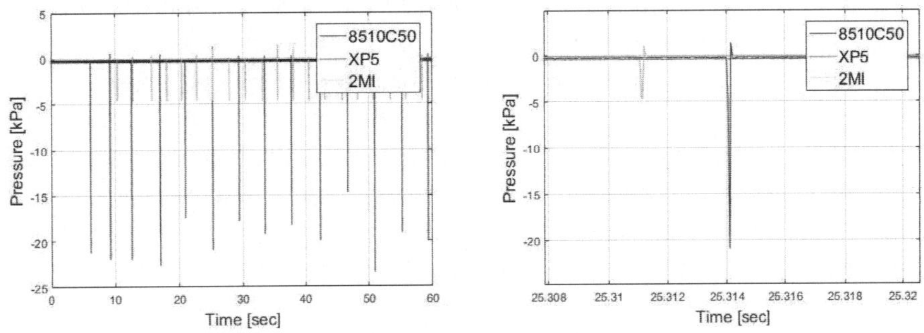

Figure 7. (a) Pressure sensor flash response of all sensors; (b) zoom of single flash.

When testing the response of the 8510C sensor to fluorescent lights the negative peaks shown in Figure 8a was due to weight of the sensor diaphragm causing deformation during rotation. A more consistent result showing level shifts due to an increase in light hitting the diaphragm would be expected if the 8510C was sensitive to fluorescent light. As the titanium diaphragm of the XP5 blocks all light, the negative peaks shown in Figure 8a are also due to the movement of the sensor. A more accurate test for these two sensors would be to leave them in position and block the light with an opaque cover.

There was a small level shift of 13 Pa from the 2MI sensor when it was rotated towards or away from the fluorescent light at 1.5 m as shown in Figure 8b. This variation on the output of the sensor is the equivalent of a 1.3 mm change in hydrostatic pressure.

(a) (b)

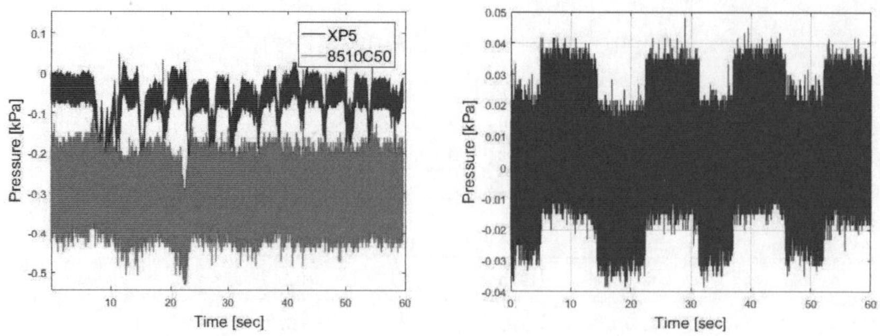

Figure 8. Fluorescent Light response at a distance of 1.5 m of (a) 8510C and XP5 pressure sensors; (b) 2MI pressure sensor.

Thermal Transient Response

The 8510C and XP5 pressure sensors were mounted in the side of an enclosure which was placed in the water causing the sensors to be submerged to a depth of 31.5 mm and 11.5 mm. The increase in water pressure should produce an expected measured pressure of 0.31 kPa and 0.12 kPa respectively. The test involved placing the sensors in the water and allowing them to stabilise for approximately 50 seconds before stopping the acquisition system. A second test involved removing the sensors from the water and allowing them to stabilise for a further 50 seconds. The two tests were repeated three times with the average response for the 8510C sensor shown in Table 3. Figures 8a and 8b show the measured response of the 8510C when it is placed in and removed from water.

At an ambient air temperature of 22°C and water temperature of 21°C the response from the 8510C sensor when it was placed in the water showed a step change with an average of 0.25 kPa. When the enclosure was removed from the water the sensor showed an average negative step change of 0.43 kPa. This response is 0.18 kPa larger than when the enclosure was placed in the water showing the effect of a change in the thermal compensation on the output of the sensor. The test was repeated in air and water temperatures of 21°C and 3°C respectively at the same water depth. The 8510C sensor showed a step change when the sensor was placed in the water with an average pressure response of 0.3 kPa. When the sensor was removed from the water it showed an average negative step change of 0.15 kPa. The step change is 0.15 kPa smaller than when the sensor was placed in the water which again shows the effect of a change in the thermal compensation.

When the XP5 sensor was placed in the water the output showed an average initial change of -18.35 kPa which trended up towards an average value of -10.5 kPa over 40 sec. On removal from the water the XP5 sensor had a large negative change of -10.2 kPa before a large positive change of 20.1 kPa. The output of the sensor stabilised at an average -0.18 kPa upon removal from the water. When the test was repeated at air and water temperatures of 21°C and 3°C respectively the response was similar to the previous test with a smaller magnitude on immersion and a larger magnitude on removal from the water. The average response for the XP5 sensor is shown in Table 3. The measured responses when placed in and removed from the water are shown in Figures 9a and 9b.

The large change in the output of the XP5 pressure sensor is due to the effect of the difference in the thermal conductivity between water and air on the sensor. The micro-machined strain gauges in the Wheatstone bridge self-heat due to the current passing through the gauges. The water conducts the heat away from the diaphragm more efficiently than air causing contraction of the metal diaphragm and also a change in resistance of the strain gauges. If the sensors were used in one media or if the thermal conductivity of the two media was more closely matched, the large difference in the response due to a change in media would not be experienced.

Temperature difference Air to Water (deg)	Average Sensor response into water (kPa)		Average Sensor response out of water (kPa)	
	8510C	XP5	8510C	XP5
1	0.19	-10.88	-0.47	3.64
-1	0.25	-18.35	-0.43	9.92
-18	0.3	-8.68	-0.15	5.47

Table 3. 8510C and XP5 average measured output when placed in and removed from water at different temperatures.

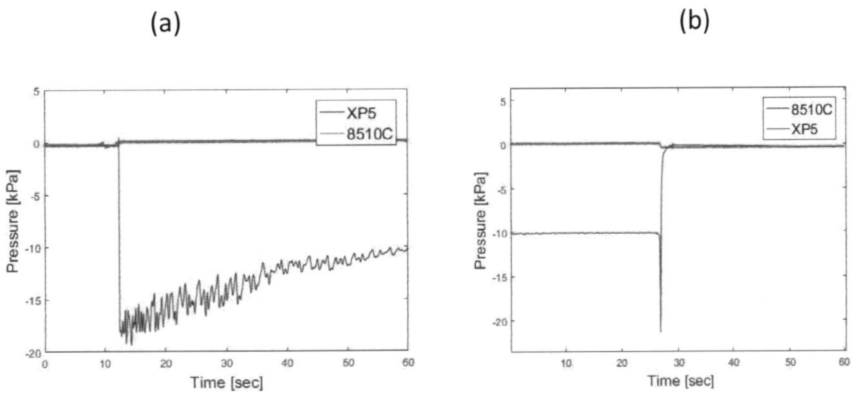

Figure 9. Hydrostatic response of 8510C and XP5 for (a) immersion in water; (b) removal from water.

The thermal response test for the 2MI pressure sensors was carried out at a different time as they were not available when the 8510C and XP5 sensors were tested. The testing for the 2MI sensor was carried out using similar methodology as detailed below.

The 2MI pressure sensor was mounted in the side of an enclosure which was placed in water causing the sensor to be submerged to a depth of 30 mm resulting in an expected measured pressure change of 0.29 kPa. At an ambient air temperature of 22.5°C and water temperature of 20.4°C, the measured response from the 2MI when placed in the water showed an average step change of 0.35 kPa as shown in Figure 10a. When the enclosure was removed from the water, the measured response showed a step change of -0.31 kPa, as shown in Figure 10b. The difference between the two results was due to a change in the thermal compensation of the sensor. The test was repeated with air and water temperatures of 22.1°C and 21.3°C respectively and a water depth of 26 mm with an expected measured pressure of 0.26 kPa.

The average response of the 2MI sensor is shown in Table 3 which indicates it is the least affected due to thermal effects.

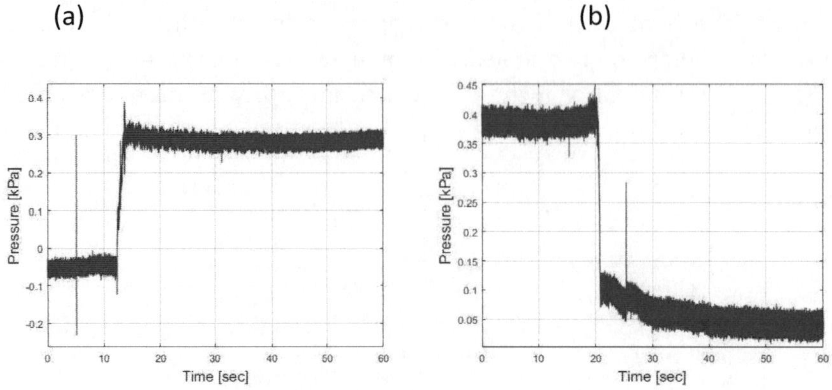

Figure 10. Hydrostatic response of 2MI for (a) immersion in water; (b) removal from water.

Temperature difference Air to Water (°C)	Water Depth (mm)	Sensor response into water (kPa)	Sensor response out of water (kPa)
0.8	26	0.28	-0.24
2.1	30	0.35	-0.31

Table 3. Hydrostatic response of 2MI pressure sensor.

During discussions with the supplier of the XP5 sensors about the thermal transient response, a suggestion was made to coat the sensor surface with a compound that would isolate the face of the sensor but still allow transmission of the pressure force. Judge et al [6] mentioned that this may be a viable technique for removing the spikes in the acquired data caused by thermal transients. Coating the gauges with PVC tape, dielectric grease and RTV silicone were considered as ways of isolating the sensor that could be applied during the towing tank testing program. As the sensors are intended to test slamming pressures at speeds of up to 2.5 m/s, coating the sensors only in grease was dismissed because of the chance of being removed during a run. It was considered that the application of PVC tape and RTV silicone could damage the sensors due to adhesion to the face of the sensor. Hence a thin layer of grease was applied to the face of the sensor before applying the tape or silicone. To reduce the effect of self-heating due to the amount of current flowing through the micro-machined strain gauges in the Wheatstone bridge, the sensor excitation voltage was reduced from 10 V to 5 V. This reduction in the excitation voltage halved the current flowing through the bridge and therefore reducing the amount of heat produced by the micro-machined strain gauges.

The testing demonstrated that the reduced excitation voltage and coating the XP5 sensor with PVC tape reduced the transient response but it also reduced the sensitivity of the sensor so that no hydrostatic response was measured. The response from the sensors with the reduced excitation voltage and coated with RTV silicone was very inconsistent. Some results demonstrated a thermal response similar to the uncoated sensor, while other tests demonstrated a response more closely matching the expected output. There was no discernible correlation between actual output and RTV silicone thickness. In addition the RTV silicone requires 24 hours to fully cure which would cause long delays in setting up the sensors and therefore this treatment option was not considered viable for towing tank measurement applications.

Dynamic air pressure pulse testing was undertaken with a sensor coated with grease and RTV silicone. The results showed that the response was more critically damped most likely due to the increase in the mass of the diaphragm causing a reduction in the frequency bandwidth of the sensor.

CONCLUSIONS

This paper describes the study examining the selection of a suitable pressure sensor to be used during a towing tank trial to measure hydrodynamic loading on a hydro-elastic generic patrol boat model. A survey was conducted of relevant literature which revealed many challenges and issues when using pressure sensors to measure highly non-linear wave impact loads in scale model towing tank testing. As there are no standards or procedures available to guide selection of pressure sensors for this purpose, a test program was developed to characterise the noise, dynamic, thermal transient and photoflash responses which are relevant to scale model tow tank testing.

The results of the study of three piezoresistive sensors for noise, dynamic, photoflash and light sensitivity and thermal transient responses showed that while some pressure sensors have the benefits of being completely isolated from any light effects, they may demonstrate a large transient thermal response when immersed or removed from water rendering them unsuitable for the intended application. For pressure sensors that do not have an opaque covering over the diaphragm to block light, care needs to be taken due to the effects of photoflash response and light sensitivity on the piezoresistors in the sensors Wheatstone bridge. As a result it is important to have a thorough understanding of the sensor design and responses to ensure that the most applicable sensor is chosen in relation to the particular testing application under consideration. The results of the study show that the Keller 2MI pressure sensor is the most appropriate of the sensors examined for the measurement of non-linear wave impact loads in scale model tow tank testing.

Information in this report can be used as guidance for selection of piezoresistive pressure sensors for scale model tow tank testing where wave impact loads are to be measured.

REFERENCES

1. Rosen, A. and Garme, K. (2000, Nov, 7 – 8). Experimental investigation on a high-speed craft in waves. Hydrodynamics of High Speed Craft: Wake Wash & Motions Control, London, UK.

2. Thornhill, E., Oldford, D., Bose, N., Veitch, B. and Liu, P. (2003, Feb). Planing hull performance from model tests. International Shipbuilding Progress, 50(1&2) pp. 5 – 18.
3. Choi, N., Choi, Y., Kim, H., Kwon, S., Park, J. and Lee, K. (2010, June, 20 – 25). A Study on the Characteristics of Piezoelectric Sensor in Sloshing Experiment. Proceedings of twentieth (2010) International Offshore and Polar Engineering Conference, Beijing, China.
4. Van Nuffel, D., Vepa, S., De Baere, I., Degrieck J., De Rouck, J. and Van Paepegem, W. (2011, June, 19 – 24). Pressure Measurement on the Surface of a Rigid Cylindrical Body During Slamming Wave Impact. Proceedings of the ASME 2011 30th International Conference on Ocean, Offshore and Arctic Engineering, Rotterdam, The Netherlands.
5. Pistani, F. and Thiagarajan K. (2012, June, 2). Experimental measurements and data analysis of the impact pressures in a sloshing experiment. Ocean Engineering, 52, pp. 60 – 74.
6. Judge, C. and Ikeda, C. (2014, November, 2 – 7). An Experimental Study of Planing Hull Wave Slam Events. 30th Symposium on Naval Hydrodynamics, Hobart, Tasmania, Australia.
7. Kim, S-Y., Kim, K-H. and Kim, Y. (2014, November, 30). Comparative study on pressure sensors for sloshing experiment. Ocean Engineering, 94, pp. 199 – 212.
8. Repalle, N., Truong, T., Thiagarajan, K., Roddier, D., Seah, R. and Finnigan, T. (2010, June, 6 – 11). The Effect of Sampling Rate on the Statistics of Impact Pressure. Proceedings of the ASME 2010 29th International Conference on Ocean, Offshore and Arctic Engineering, Shanghai, China.
9. 8510C Piezoresistive Pressure Transducer, Meggitt PLC (Endevco Corporation), https://buy.endevco.com/
10. XP5 Miniature Pressure Sensor Datasheet, TE Connectivity (Measurement Specialties Inc.), https://www.te.com/usa-en/products/sensors/pressure-sensors.html
11. 2MI Piezoresistive Miniature Pressure Transducer, Keller AG für Druckmesstechnik, http://www.keller-druck.com/home_e/paprod_e/2mi_e.asp
12. Selecting Piezoresistive vs. Piezoelectric Pressure Transducers, Kulite Semiconductor Products Inc., http://www.kulite.com/technology/reference-library/
13. Piezoresistive Pressure Transducers Instruction Manual, Endevco (Meggitt Inc), 24 February 1997
14. Private Communication with XP5 Supplier and Manufacturer. 27 February 2015.

Research thrusts and recent advancements in the modelling of helicopters during launch and Recovery Operations to Ships

Hurmat Ain, Bradley MacPherson, Robert Toffoletto and Rhys Lehmann

Defence Science and Technology Group, Melbourne, Victoria, Australia

Correspondence author: Bradley.MacPherson@dst.defence.gov.au

1. INTRODUCTION

The Defence Science and Technology (DST) Group within Australia's Department of Defence is devoted to applying science and technology to protect and defend Australia and its national interests. It delivers expert, impartial advice and innovative solutions for Defence and other elements of national security.

The Helicopter Systems Effectiveness (HSE) area within DST Group is dedicated to maximising helicopter safety and operational effectiveness in land and maritime operations. The core capabilities of the HSE group are:

- Development, provision, verification and maintenance of a range of rotary winged aircraft performance and flight behaviour models, for manned and unmanned platforms.
- Computational Fluid dynamic (CFD) modelling of ship airwake for various naval platforms and modelling helicopter aerodynamic characteristics.
- World-leading night vision lab to research aviation vision enhancement devices.

The development of flight models and CFD modelling are fundamental to the work performed within HSE group. Depending on the use of the flight model, different levels of model fidelity are produced. The flight models are used to support accident investigation, human-in-the-loop (HIL) simulation, slung load limit investigation and operations research activities carried out within DST Group. The airwake of the ship together with helicopter flight models has been extensively used to support First of Class Flight Trials (FoCFT) and investigate flight dynamic issues that are of major concern to Royal Australian Navy (RAN) when operating helicopters from ships.

To develop these models an extensive and detailed level of data is often required. Where data is lacking, HSE group generates data from a variety of sources and techniques that involves flight and technical documentation, DST Group low speed wind tunnel experiments, analytical modelling and estimation techniques, CFD modelling on DST Group advanced High Performance Computing (HPC) systems, flight and sea trials. With the increasing adoption of Unmanned Aerial Systems (UAS), HSE group is now incorporating UAS models into our core research activities.

This paper gives an overview of HSE group's capabilities and research undertaken for launch and recovery operations from naval platforms for rotary wing aircraft. The research outcomes can also be extended to support unmanned (rotary or fixed wing) applications. All research and development gains directly benefit the Defence's vertical lift capability through application of our improved modelling and simulation techniques.

2. LAUNCH AND RECOVERY RESEARCH

The provision of helicopter landing deck on naval ships significantly enhances the capabilities of maritime operations. However, operating a helicopter on the deck of a ship poses many challenges to its operators due to several factors and requires rigorous analysis. Prime amongst these is launch and recovery, which is the take-off and landing phases for aircraft operating from naval platforms. The complexity of helicopter-ship operation arises due to the coupling between ship airwake and helicopter downwash, including the effect of the atmospheric boundary layer (ABL), ship motion at various sea states and operations in degraded visual environments (DVE). To date, a FoCFT is still the best practice to determine ship-helicopter operating limits (SHOLs). To minimise the cost and practicality of covering every wind condition and sea state during a FoCFT, it is imperative to develop new experimental and numerical methodologies to simulate ship-helicopter operation accurately prior to the trial.

Figure 1: Helicopter launch and recovery on the Deck of LHD ship

With the advancements in computational and experimental techniques, several countries (including the US and UK) have attempted to develop SHOLs using a combination of computed airwake databases, often obtained using CFD codes; high-fidelity helicopter dynamic models; and high-fidelity simulators. It has been proposed by a number of researchers that high-fidelity piloted simulation might be one possible alternative [1], [2] and [3] - this method is known colloquially as a virtual SHOL. DST Group has attempted to use these same tools without a Human in the Loop (HIL) simulator to predict the potentially high risk conditions during launch and recovery operations from ships. In effect, the helicopter flight model is immersed in the ship airwake at various Wind Over Deck (WOD) angles and speeds to investigate what affect this has on the helicopter's control and power margins. Simulation generated candidate SHOLs are now regularly used to optimise flight trials by providing information on critical operating conditions to target during the trial at sea. The candidate SHOLs are also used for the prediction of conditions not experienced during the sea trial. In DST Group, the modelling framework used to generate candidate SHOLs is called the CSHOL – Candidate SHOL.

There is currently a NATO led investigation underway looking at the differences between the various methods employed by different nations. DST Group, together with a number of other partners within NATO, are currently conducting research to find ways to improve upon the advice for FoCFTs and investigating how modelling and simulation could reduce the dependence on sea trials for the generation of a SHOL. However, it has yet to be demonstrated that this combination is able to adequately reflect the results obtained during a FOCFT. Some of these techniques are listed in Table 2. The green column in Table 2 refers to the DST virtual SHOL tool – CSHOL.

Table 2: Classification of integrated modelling and simulation methods used by different nations (reproduced from the NATO AVT-315 report)

	Flight simulation methods				Experimental methods	
	Real-time	**Desktop (Piloted)**	**Desktop (Non-Piloted)**	**Desktop (UAV)**	**Wind/water tunnel**	**Flight test + wind tunnel**
Aircraft	Flight dynamic model	Flight dynamic model	Flight dynamic model	Flight dynamic model	Physical model	Aircraft
Airwake	CFD model	CFD model	CFD model	CFD model	Physical model	CFD/Physical model
Controller	Human pilot	Software controlled pilot model	Prescribed trajectory or fixed in space	UAV Controller	Prescribed trajectory or fixed in space	Human pilot
Deck motion	Simulated	None[1]	None	Simulated	Physical model	Physical model
Measured quantities	• Pilot ratings • Control margins	• Control margins • Control activity • Landing accuracy	• Forces • Moments	• Control margins • Control activity • Landing accuracy	• Forces • Moments	• Land based control margins
Derived quantities		Pilot ratings	Pilot ratings		Pilot ratings	Ship based control margins
Example tools	VirtualSHOL (UK,US)	DST CSHOL				SHOL-X (Aeromath)

3. DST GROUP VIRTUAL SHOL TOOL - CSHOL

Prior to the introduction into service of a new aerial system (helicopter or UAS) or ship type, the RAN has a requirement to certify the specific helicopter-ship combination for operations. Part of this certification includes the determination of the SHOLs via a FoCFT. This testing is the responsibility of the Aircraft Maintenance and Flight Trials Unit (AMAFTU). The SHOLs indicate the safe operating limits in the conduct of naval aviation operations. One of the many risks associated with helicopter-ship operations is the interaction between the rotor system and the turbulent airwake in the vicinity of the ship. This is especially true during launch and recovery, when aerodynamic loading of the rotor system can be considerably altered by the passage through an eddy or a region of cross-flow. The changes in rotor loading can affect the aircraft control and the power margin for a safe landing. To reduce the risks associated with the turbulent ship airwake, prior to a FoCFT, AMAFTU, generally, requires an indication of the airwake characteristics in the vicinity of the helicopter deck. This information is used to help determine the WOD angles and wind speeds that may cause aircraft controllability problems during launch and recovery.

In the past, DST Group has provided the ship airwake information to AMAFTU derived from wind-tunnel tests [4]. These tests would usually take the form of smoke flow visualisation, which give indications of turbulent or disturbed air around the ship in question. In more

[1] DST Group currently does not have the capability to include the effect of ship motion in the CSHOL simulation; however, there is a plan to include this in the future.

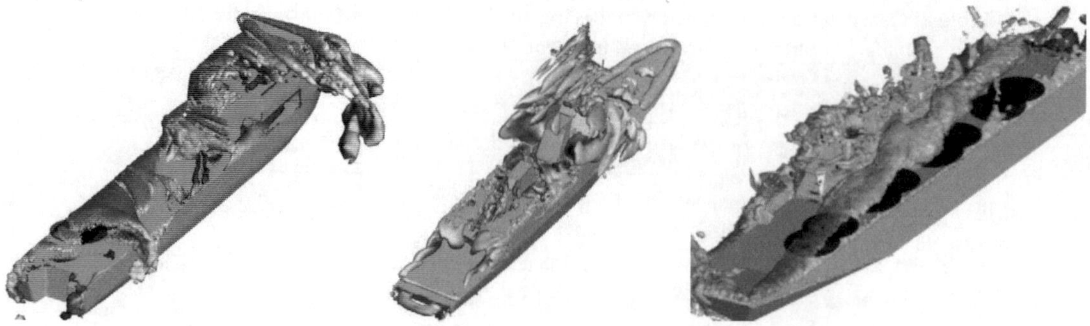

Figure 2: Example of few CFD ship airwake from DST Group database

recent years, DST Group has been using a combination of Computational Fluid Dynamics (CFD) as well as wind-tunnel tests [5], [6] and [7]. This led to the development of an extensive database of ship airwake for various naval vessels (see Figure 2). On these occasions, the CFD results have been able to provide more detailed information, such as wind speeds in the airwake (including the variation with time). The wind-tunnel tests are now being used more for the validation of the CFD results.

The acquisition of a database for ship airwake as well as significant experience in the development of flight models and control systems played a vital role towards the development of CSHOL. CSHOL now is DST Group's key modelling capability to produce candidate SHOLs for the RAN. The architecture of the CSHOL tool presented in Figure 3 indicates that a high fidelity flight model of a candidate helicopter (or UAS) is embedded into the time-dependant wake of a given ship. The tool then uses a virtual pilot to conduct an approach and landing to the given ship. From the simulation a number of parameters can be used to help determine the candidate SHOL. These parameters include: pilot controls (limits and variability), accuracy of flight path and landing position. Unlike a piloted simulation, the virtual pilot has the advantage of repeatability, which, in turn, allows this method to properly compare numerous parameters that may affect the SHOL. These parameters can include: variation in wind direction and speed, landing position, approach path, hover height and duration. The virtual pilot can also be 'tuned' to follow the desired path as aggressively as required.

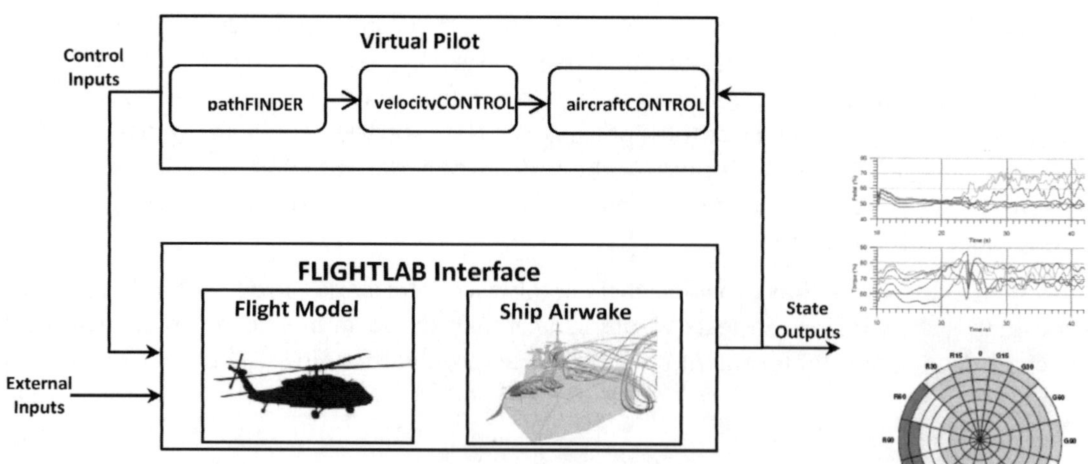

Figure 3: Simulation architecture of DST CSHOL

The superposition of the airwake and simplification of ship motion and sea state effects on airwake has served the DST and RAN reasonably well, guiding and optimising at sea flight trials to problematic areas of the SHOL. The current simplifications offer some scope to greatly reduce the sea based flight trials. Over the coming years, RAN will commission new vessels, many that will have manned helicopters operating from them, along with a range of UAS platforms. The need for reliable modelling and simulation based launch and recovery limit production is greater than ever.

3.1. CSHOL Applications and Challenges

ADF expects to operate every helicopter at sea including Army helicopters. This is unique to Australia. Operating Army helicopters that are not built for maritime environment poses several other challenges. CSHOL has been successfully used to provide advice to RAN by performing risk reduction studies for various Army and Navy helicopters operating on a variety of naval platforms. Some of the areas investigated include different approach and flight pathways to landing on a ship using an autopilot, on-spot engagement and disengagement simulation for issues such as blade sailing [8] and mast bending [9]. Blade sailing is a phenomenon where a helicopter blade experiences excessive flapping motion, which can lead to undesirable consequences and limit the operational capability of the helicopter. Mast bending moment is a result of a rotor system that has less lag in control response because of the large hub moment typically generated due to flexing of blades in the absence of hinges.

A key challenge has been the accurate representation of the ship-helicopter dynamic interface, in particular the fluid dynamics of the ship air wake and the interaction with the helicopter. To date, Defence along with other Naval Aviation operating countries, have modelled the ship air wake, obtained through trials, CFD or wind tunnels, with no coupled interaction of the helicopter's air wake. Typically, the combined air wake is a superposition of the two independent air wakes at any point of interest. Another key factor commonly left out of the ships airwake modelling is the ship motion and sea state. The ship's airwake is commonly anchored at the ships centre of gravity and ship roll and pitch set to zero. DST group has started research programs internally and with our university partners to begin to address the key deficiencies of the SHOL production problem. These are discussed in the following section.

3.2. Enhancement of Existing Capability via Collaborations

3.2.1 Coupled CFD ship and helicopter

One of the more difficult aspects of simulating the helicopter-ship interface is the accurate modelling of the interaction of the rotor aerodynamics into the environment. In the past, researchers have tackled this problem in one of two ways. The first is to simplify the model by de-coupling the interaction between the rotor down-wash and the ship airwake. This increases the speed of the simulation to the point where simulations can be conducted in real time; however, it is done at the expense of model fidelity. The second is to model the helicopter and ship wake interaction using high-fidelity CFD simulation typically requiring multiple days to run on a supercomputer.

The University of Sydney, in collaboration with DST Group, has developed a computationally efficient tool for the simulation of ship-helicopter dynamic interface. The numerical

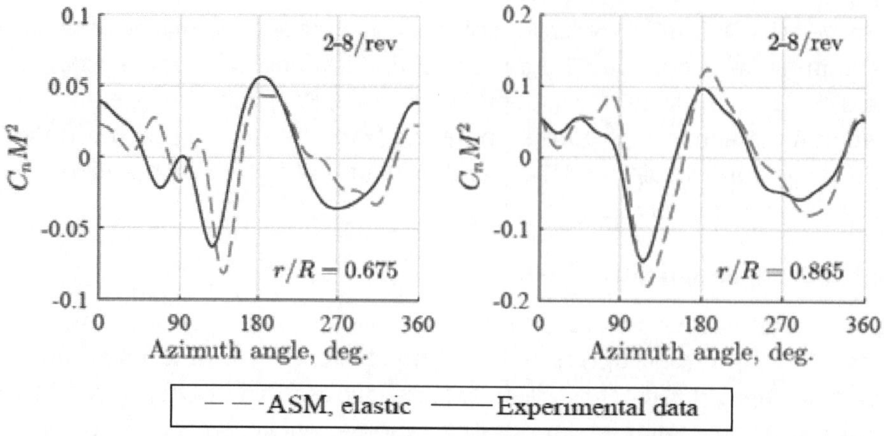

Figure 4: 2-8/rev vibratory components of normal force coefficient as a function of azimuth angle at two spanwise stations on the UH-60A rotor in forward flight at an advance ratio of 0.368. ASM simulation results compared to experimental data.

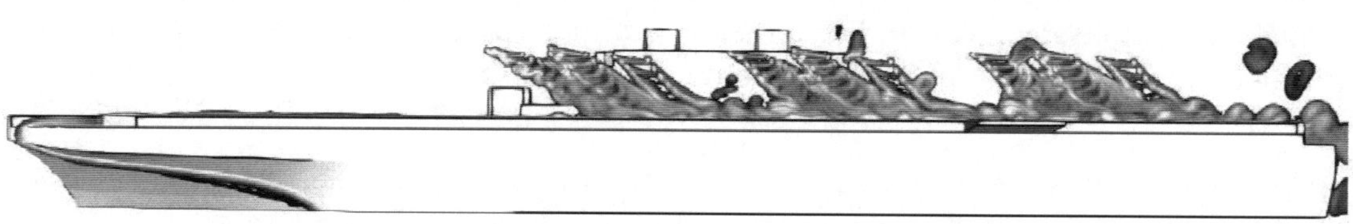

Figure 5: Vorticity magnitude on a plane through the port landing spots on the LHA ship with three tandem rotor helicopters

simulation uses a hybrid approach by utilising detached eddy simulation (DES) for ship airwake and actuator surface method (ASM) to model helicopter rotor. The savings from this hybrid method derive from the substitution of the rotor blade geometry for distributions of momentum source terms added to the Navier-Stokes equations in the vicinity of the blades. With such a hybrid method there is no need for the computational grid to resolve the geometry of the rotor, thus saving grid points and removing the need for overset grids, sliding planes, and grid deformation. Setup and configuration of ship-helicopter dynamic interface simulations are also significantly simpler when using a hybrid method, again due to the removal of the rotor blade resolving grid. Further details on the implementation of ship airwake and rotor model can be found in [10]. The Hybrid ASM has been validated against a number of scenarios ([3], [11]) and shows great promise going forward. Examples of some of the studies include comparisons with the X2, S-76, UH-60, and CH-47B. An excellent agreement was found between the ASM simulation and experimental data for the case of UH-60A rotor in forward flight (see Figure 4).

The hybrid ASM simulation tool has been developed with the objective of making studies of concurrent or adjacent rotorcraft operations over a range of operating conditions accessible, while still retaining the ability to assess unsteady rotor loads which may influence pilot workload. The simulation has shown a good potential of investigating concurrent operations of helicopters by capturing promising aerodynamic interactions for the case of three tandem rotor helicopters operating simultaneously on the flight landing deck of the Landing Helicopter Assault (LHA) ship. The partially developed vorticity field in Figure 5 shows the flow field at a particular WOD condition dominated by the rotor wakes. The tool can be further used to investigate other factors such as sea states, ship motion and atmospheric boundary layer. Additionally, this concept can also be extended to the helicopter formation flight issues which are heavily dependent on rotor-rotor coupled airwake interaction. While the preliminary simulation of dynamic interface served as the proof of concept, full potential of the tool is yet to be explored for the Defence specific naval platforms with various helicopter combinations.

3.2.2 Characterisation of Ship Motion and Sea State of Ship Airwake

Another major limitation of current ship-helicopter dynamic interface analysis comes with the assumption (typically due to reduce complexity) that the ship airwake is not affected by ship motion. However, the research on ship hull design and its influence on seakeeping characteristics anticipate that the ship motion and sea state impart energy into airwake which could affect seakeeping performance of a ship, especially for rough sea states. This could result in a time lag between ship motion and airwake response. Therefore, it is important to consider motion-induced effects on the airwake for dynamic interface modelling [12].

DST Group in collaboration with the University of Melbourne is working on developing a unique experimental capability to investigate the effect of ship motion and sea state on ship airwake. An extreme air sea interaction (EASI) facility (see Figure 6 for schematic and actual facility) located at the Department of Mechanical and Manufacturing Engineering, previously used as a hybrid tow tank and/or wind tunnel is now being utilised to investigate the influence of ship motion, ABL and waterline on ship airwake and its effects on helicopter landings. The facility has dimensions of 60 x 2 x 2 m (length x width x height) and the water is filled to a depth of 1m. The freestream velocity is controlled and monitored remotely and the pressure gradient of the facility can be controlled using an adjustable ceiling. Underwater optical access is available through a viewing window measuring 1.5 x 1.0 m (width x length) located at 20 m away from the upstream end of the tank for the purposes of optical technique based experiment apparatus such as Particle Image velocimetry (PIV) or Laser Doppler Velocimetry (LDV) systems. Current activity involves investigation of air-sea interaction, ABL and ship motion on generic platforms such as simple frigate ship (SFS2) and NATO generic destroyer (GD) ship. This work is in initial stages and results of this research outcome will be part of a NATO report.

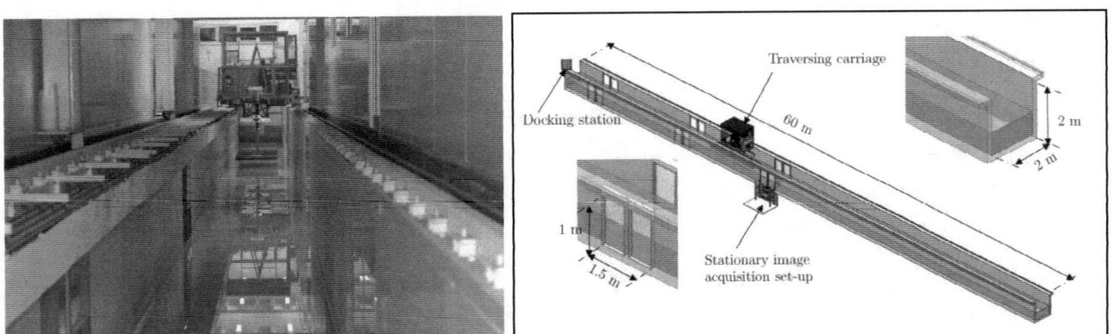

Figure 6: Extreme Air Sea Interaction (EASI) facility at the University of Melbourne

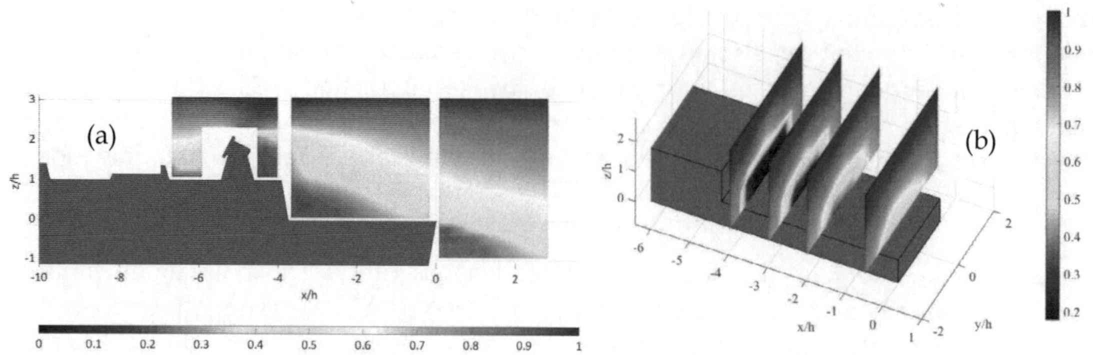

Figure 7: Normalised velocity contours over (a) NATO-GD and (b) SFS2

Some preliminary measurements using cobra probe and PIV were completed recently to validate ABL and turbulence intensities in a wind tunnel. Figure 7 presents the averaged velocity contours normalised over free stream velocity on the flight landing deck of SFS2 and NATO-GD. Next phase of the work will measure airwake on the landing deck due to prescribed motions, ABL and waterline. The results of this activity can be used to validate CFD simulations and can also be extended to Defence specific naval platforms.

4. SUMMARY

DST Group has been involved in launch and recovery research for 25 + years. The research has included investigation of alternative landing systems [13], maritime helicopter handling qualities standards [14], generation of candidate SHOLs with and without human in the loop simulation, mast bending, blade sailing, on deck modelling, Night Vision Goggles operations and crash investigation. Of the research activities listed above, the major activity has been production of candidate SHOLs. Over many years of collaboration with military research partners through The Technical Cooperation Program (TTCP) and NATO working groups, the augmentation of flight trials with modelling and simulation has greatly assisted the generation of SHOLs [15]. Our research has clearly shown that the airflow generated around a ship, from the hull and superstructure, can generate significant eddies and volumes of cross flow in the region of the flight deck. These flow regions can reduce the performance margin of a loaded helicopter or a UAS system. These factors, coupled with others such as pilot visual cues can lead to significant operational constraints that limit the ability to utilise the helicopter as part of the weapon system. Not being able to have a ship borne aerial system in the air significantly reduces the offensive and defensive capabilities of the naval system. Our research and advanced models ensure DST Group capability to support ADF safe and efficient ship-helicopter operations for current and future naval platforms.

5. FUTURE THRUSTS

With adoption of UAS platforms, we are increasing research to ensure full exploitation of the capabilities from naval platforms. The UAS platforms allow an increased ability to upgrade systems to meet changing threat environments at a far shorter development and acquisition cycle than traditional manned systems.

Application of the techniques applied in this paper to UAS platforms introduces a range of nuances, particularly when considering small scale UAS vehicles. As a vehicle becomes smaller relative to the dominant turbulent airwake structures, the demands on controllability and control power become increasingly significant. This influences the requirements for the ship airwake CFD solution and the fidelity constraints of the flight dynamic model.

Additionally, as human piloting demands reduce from high bandwidth attitude and trajectory control to basic navigation control or autonomous operations, the demands on the vehicle flight control system increase in consequence – the most relevant factor being disturbance rejection. Disturbance Rejection refers to the ability of the vehicle to reject external disturbances, either through dynamics inherent in the vehicle design or through active control from the vehicle flight control system. In order to simulate the vehicle performance in the CSHOL scenario, it is essential to develop a good understanding of its disturbance rejection characteristics and how these interact with the turbulence environment (this is akin to tuning the virtual pilot model in the manned helicopter scenario). This will be conducted through a combination of linear analysis of flight test results and non-linear studies involving the coupled CFD and flight dynamic model solution. As an extension of this work, DST Group is embarking on a resilient flight work program, focussing on establishing methods for robust control of autonomous unmanned vehicles in challenging aerodynamic environments including maritime operations. This study aims to facilitate comparisons between classical control techniques, non-linear adaptive control, and modern 'AI' inspired control strategies, to inform the development of future UAS platforms with superior performance in dynamic environments such as helicopter ship operations.

ACKNOWLEDGEMENTS

The authors acknowledge the financial contribution from our Defence clients for being a strong supporter of our research and the contribution of our collaborative partners Advanced VTOL, Prism Defence, University of Sydney and University of Melbourne for bringing forward their expertise and facilities to advance our research. We are also thankful to our partners in TTCP and NATO for sharing their research outcomes and experiences.

REFERENCES

[1] R. Rajagopalan and S. Mathur, "Three Dimensional Analysis of a Rotor in Forward Flight," in *20th Fluid Dynamics, Plasma Dynamics and Lasers Conference*, Buffalo, NY, USA, 1989.

[2] W. Z. Shen, J. H. Zhang and J. N. Sorensen, "The Actuator Surface Model: A New Navier-Stokes Based Model for Rotor Computations," *Journal of Solar Energy Engineering*, vol. 131, no. 1, 2009.

[3] D. Linton, G. Barakos, R. Widjaja and B. Thornber, "Coupling of an Unsteady Aerodynamics Model with a Computational Fluid Dynamics Solver," *AIAA Journal*, vol. 56, no. 8, pp. 1-14, 2018.

[4] C. D. Edward, "Flow Visualisation about the Helicopter Deck of the Hydrographic Ship," DSTO-TR-0762, 1999.

[5] G. Clarke and R. Toffoletto, "CFD investigation into the impact of an AMSD Upgrade on the Flow Field about an ANZAC Class Frigate Flight Deck," DSTO-TR-2254, 2009.

[6] R. Toffoletto and N. Blacker, *Computational Fluid Dynamics Investigation into the Effect of a Ski Ramp on the Spanish NAVANTIA Ship,* DSTO-TN-0748, 2007.

[7] R. Toffoletto, *Analysis of LHD Anemometer Locations,* DSTO-CR-2009-0219, 2009.

[8] M. Groom, *Simulations of Rotor Engagement and Disengagement in Ship Airwake,* Master Thesis: University of Sydney, 2016.

[9] H. Ain, *Bending Moment Analysis for ARH Tiger Rotor Mast on Canberra Class Ship Flight Deck,* DST-Group-TN-1863, 2019.

[10] D. Linton, R. Widjaja and M. Thornber, "Validation of an Actuator Surface Model with CFD-convected Wake Model for Hover and Forward Flight," in *Proceedings of the 7th Asian/Australian Rotorcraft Forum*, 2018.

[11] D. Linton, R. Widjaja and B. Thornber, "Simulations of Tandem and Coaxial Rotors using a CFD-coupled Rotor Model," in *Proceedings of the 21st Australasian Fluid Mechanics Conference*, 2018.

[12] S. Shukla, S. Sinha and S. N. Singh, "Ship-helo coupled airwake aerodynamics: A comprehensive review," *Progress in Aerospace Sciences,* vol. 106, pp. 71-107, 2019.

[13] I. Ho and R. Toffoletto, "Evaluation of Alternative Helicopter-Ship Landing Limit Systems," in *7th Australian Pacific Vertiflite Conference on Helicopter Technology*, 2009.

[14] I. Ho and R. Toffoletto, *Comparison of Various Methodologies for Helicopter Landing Limits,* DSTO-TN-0786, 2007.

[15] A. Hoencamp, "Helicopter-Ship Qualification Testing: SHOL-X Methodology," 2016.

FLOATING PONTOON MOTIONS, OPERATIONAL SAFE MOTION LIMITS

Elizabeth Freeman, Water Research Laboratory, UNSW Sydney, elizabeth.freeman@student.unsw.edu.au.

Ron Cox, Water Research Laboratory, UNSW Sydney, r.cox@unsw.edu.au.

Kristen Splinter, Water Research Laboratory, UNSW Sydney, k.splinter@unsw.edu.au.

Francois Flocard, Water Research Laboratory, UNSW Sydney, f.flocard@unsw.edu.au

ABSTRACT

Increasing Navy Fleet size requires an increase and diversity of maritime facilities.

By 2035, around half of the world's submarines will be operating in the IndoPacifc region (Defence 2016). Submarines offer a unique defence capability and play an essential role in supporting Australia's maritime security. When berthed, submarines sit lower in the water than other large military vessels and experience more excessive motions in operational wave conditions. As such, shore connected floating pontoons are an important option for developing and operating submarine facilities and are particularly appropriate for both the docking and loading of submarines. The safety of Navy personnel during loading of munitions and general operational tasks require that safe working limits be placed on the motions of these piled floating pontoons. At present, there are no standards assigning suitable limits of motion for floating pontoons in order to maintain the safety/comfort of users undertaking operational activities.

This research proposes a set of operational safe motion limits in the form of accelerations to ensure the comfort and safety of floating pontoon users. Experimental work has been undertaken and data collected on the motions of a piled floating pontoon subjected to regular waves, representative of a harbour environment. Pontoon widths, drafts and skirts have been examined under varying wave conditions.

The research has found that in many circumstances, the pontoon motions exceed proposed safe working limits and that the application of different pontoon design features can be used to reduce the accelerations of motion to within operational safe motion limits.

1 INTRODUCTION

Piled Floating pontoons are public access structures that provide a point of access and egress to vessels. They undergo dynamic motions resulting from wind, current, wave and berthing forces. Piled floating pontoons are an important option for developing and operating submarine facilities and are particularly appropriate for both the docking and loading of submarines. The safety of Navy personnel during loading of munitions and general operational tasks require that safe working limits be placed on these motions as excessive motions have the potential to cause a standing person to become unstable. Currently, no design standards exist defining these allowable limits for floating pontoons. Nor do any standards exist defining how postural stability should be considered when designing floating pontoons.

This paper firstly provides a set of motion limit criteria to aid floating pontoon design to ensure the comfort and stability of users (Section 2). The limits have been determined based on an extensive review of literature specific to how humans respond to motion. Methods for calculating the natural frequency of heave and roll along with heave accelerations are presented (Section 3). Section 5 summarises the results from a set of laboratory experiments (Section 4) on the dynamic motions of a box type piled floating pontoon subjected to boat wake conditions, modelled as regular waves with a height of 330mm (prototype) and period ranging from 2-7 seconds (prototype), with emphasis given on what happens when the forcing frequency corresponds with the calculated natural frequency of the pontoon. The effect of draft on recorded accelerations is examined and discussed for each of the tested scenarios. Testing has shown that the motions (accelerations) and displacement tend to peak when the natural frequency and forcing frequency correspond. Equations for calculating heave acceleration whilst providing simplistic estimates of heave, underestimate the acceleration when resonance occurs.

2 SAFE MOTION LIMITS (SML)

For this research, the Safe Motion Limits (SML) relating postural stability of a standing person with respect to dynamic motions of a floating pontoon will be associated to those motions originating from the moving environments described by Freeman et al. (Freeman et al. 2017). Dynamic motions exceeding those identified in the literature have the potential to result in motion sickness, postural instability, fatigue and discomfort. Defining these safe motion limits can be classified into age related groups, as well as allowable limits for both comfort and operation. The level of training of person's standing on the floating pontoon should also be considered (civilian or naval). Table 1 stipulates the SML to be adopted for this research based on a minimum level of training (civilian), for age bracket older children and adults (ages 7 to 65 years). It should be noted that the complex multidirectional behaviour of a floating pontoon is expected to create more instability than if the criteria identified in Table 1 were to act in isolation.

Table 1. Safe Motion Limits (SML) for Older Children and Adults (ages 7 – 65 years)

CRITERIA	LIMIT	REFERENCE
OPERATION (PEAK VALUES)		
Peak Vertical Acceleration	0.1g	(NSW Maritime 2005)
Peak Lateral Acceleration	0.1g	(NSW Maritime 2005) Powell and Palachin (2015)
Peak angle of tilt	6°	Rosen et al. (1984)
COMFORT (RMS VALUES)		
RMS Vertical Acceleration	0.02g	NORDFORSK (1987) Stevens and Parson (2003)
RMS Lateral Acceleration	0.03g	NORDFORSK (1987)
RMS Roll	2°	NORDFORSK (1987) Stevens and Parsons (2002)

3 STRUCTURAL RESPONSE DUE TO WAVE LOADING

Correctly analysing wave loading and the response of a floating structure to wind, wave and current requires advanced numerical or physical modelling techniques. Simplified desktop methods do exist and can be used in cases where suitable and required.

A brief description of a proposed desktop method for undertaking a simplistic analysis of the motion response of a floating pontoon is detailed below. Further discussion on how the results correlate with this method is provided in Section 5.

3.1 Analysis Method

3.1.1 Natural Frequency of Structure

In order to understand the motion response of a floating structure the natural frequency of the various axis needs to be determined. A number of methods are available including Gaythwaite (2016); BSi (2000) and DNV (2012). Rigid maritime structures, such as those supported by piles, tend to have high natural frequencies due to the stiffness of the moorings. This high natural frequency means that the amplitude of motion is large for shorter wave periods such as those produced by boat wake, however in typical design situations consideration of the effect of boat wake on a floating structure is often neglected. It is important that the natural frequency of the structure does not coincide with the natural frequency of the predominate wave climate. When the pontoon natural frequency and forcing frequency coincide resonance results in dynamic amplification.

The relative motion response of a floating structure depends upon the relative water-depth (d)-to-draft (D) ratio (d/D), the structure-beam (B)-to-draft (D) ratio (B/D) as it affects the virtual mass of the structure, and the beam (B)-to-wavelength (L) ratio (B/L), as well as the wave direction and the degree of mooring restraint (Gaythwaite 2016). Gaythwaite (2016) states that in general, the heave response is negligible for $L < 0.75B$ and near unity for $L > 4B$.

The natural frequencies, horizontal and vertical, have been calculated for the pontoon tested. The results will be discussed relative to the natural frequencies calculated to determine if peak accelerations occur when the natural frequency of the structure coincides with the forcing frequency.

3.1.2 Accelerations of Structure

This research is aimed at highlighting the importance of understanding the accelerations a pontoon might experience resulting from small amplitude boat wake in order to assess and better understand the potential impact on postural stability. There are several methods available to determine the acceleration of a floating body. For desktop assessment purposes we have calculated the heave acceleration with two alternative methods. The first method, (DNV 2012) includes consideration of draft but no inclusion of wave period. The second method (Gaythwaite 2016), includes wave period but no draft component. Results will be discussed relative to these estimates.

From a design viewpoint if approximate accelerations are cacluated at the design stage it is possible to make amendments if the safe motion limits (Table 1) are exceeded enabling better design. These caclulated accelerations will be discussed relative to the recorded data.

3.1.3 Response Amplitude Operator

The response of a floating structure or vessel is usually summarized in terms of response amplitude operators (RAOs). The response amplitude operator defines the response to a wave of unit amplitude (BSi 2000). In sheltered locations, for vessels or pontoons, the stiffness of a flexible mooring system can generally be neglected. If we consider the structure as freely floating, we can obtain conservative estimates of response by assuming RAOs of unity. Surge and sway amplitudes equal horizontal wave particle amplitudes and heave amplitude equals vertical wave particle amplitude. Roll and pitch amplitude equal the maximum wave slope. This assumption cannot be used when the natural frequency/period of the structure is near the forcing frequency/period. For stiff restraining systems such as pontoons held by piles, the horizontal motions are theoretically governed by the stiffness of the restraining system. However, the deflections of structures such as piles are generally small, and it will normally be adequate to assume that a floating structure will move within the tolerances of any fendering and guide system.

Summarising the response of a floating structure in this way does not provide the designer with acceleration motion data. The designer will not know how a person's stability will be affected by the motions. However, an analysis of RAOs has been included to better understand how RAOs and accelerations can be related.

4 METHODS

A testing program was developed to obtain the dynamic motions (accelerations and RAO) of two varying width floating pontoons when subjected to regular waves, typical of boat wake. The following sections detail the testing undertaken and the results obtained. This paper presents the results for the narrow pontoon. Full analysis of results including those obtained from the wider pontoon and skirt testing is to be presented as part of PhD thesis.

4.1 Experimental Setup

Laboratory testing was undertaken to investigate the dynamic motions of two piled box pontoons. The narrow pontoon (Pontoon 1) was approximately half the beam width of the wide pontoon (Pontoon 2) in the direction of wave propagation. The main aim of the investigations was to record the accelerations experienced as a result of various tested wave scenarios. The work investigated how altering the beam and draft impacted on the recorded accelerations and then compared these accelerations to the safe motion limits nominated above (Section 2).

4.1.1 Model Construction

Both piled floating pontoon models were constructed of grey PVC sheet at a scale of 1:10 prototype. PVC sheet was also used for internal ballast while iron and steel weights were used to alter the draft of the pontoons. Delrin, a highly crystalline engineering thermoplastic specified for high load mechanical applications, was used to construct wear/impact buffers at the pontoon/pile interface. They provided a low friction sliding connection between restraining piles and the pontoon with a 42mm (prototype) annular clearance in order to better reproduce the typical pile to HDPE fender typically used on floating pontoon projects. The pontoon/pile connections were such to allow free vertical movement under the tested wave conditions as would be experienced in real world settings.

Both pontoon models were constructed to have uniformly distributed internal ballast over pontoon plan area in order to achieve desired buoyancy and draft based on the prototype being tested. Figure 1 details the model floating pontoon dimensions as tested.

4.1.2 Setup

The physical model testing was conducted in the 1.2 m wide wave flume at UNSW Water Research Laboratory, with flume main dimensions as follows, 44 m long, 1.2 m wide and 1.6 m deep. The floating pontoons tested were rectangular prisms with six degrees of freedom:

surge (in the direction of wave propagation, x_b), sway (perpendicular to the direction of wave propagation, y_b) and heave (vertical, z_b), as well as the three rotations around the centre of gravity (roll, pitch and yaw). On each pontoon, five Inertial Measurement Units (IMU) were used to measure triple-axis accelerations and triple-axis gyrations of each floating pontoon. The IMUs were positioned on each corner of the pontoon, as well as in the centre of the top face. The accelerations recorded were in units of g (gravity, m/s2).

Figure 1. Floating Pontoon Design (model) for Narrow (Pontoon 1) and Wide (Pontoon 2)

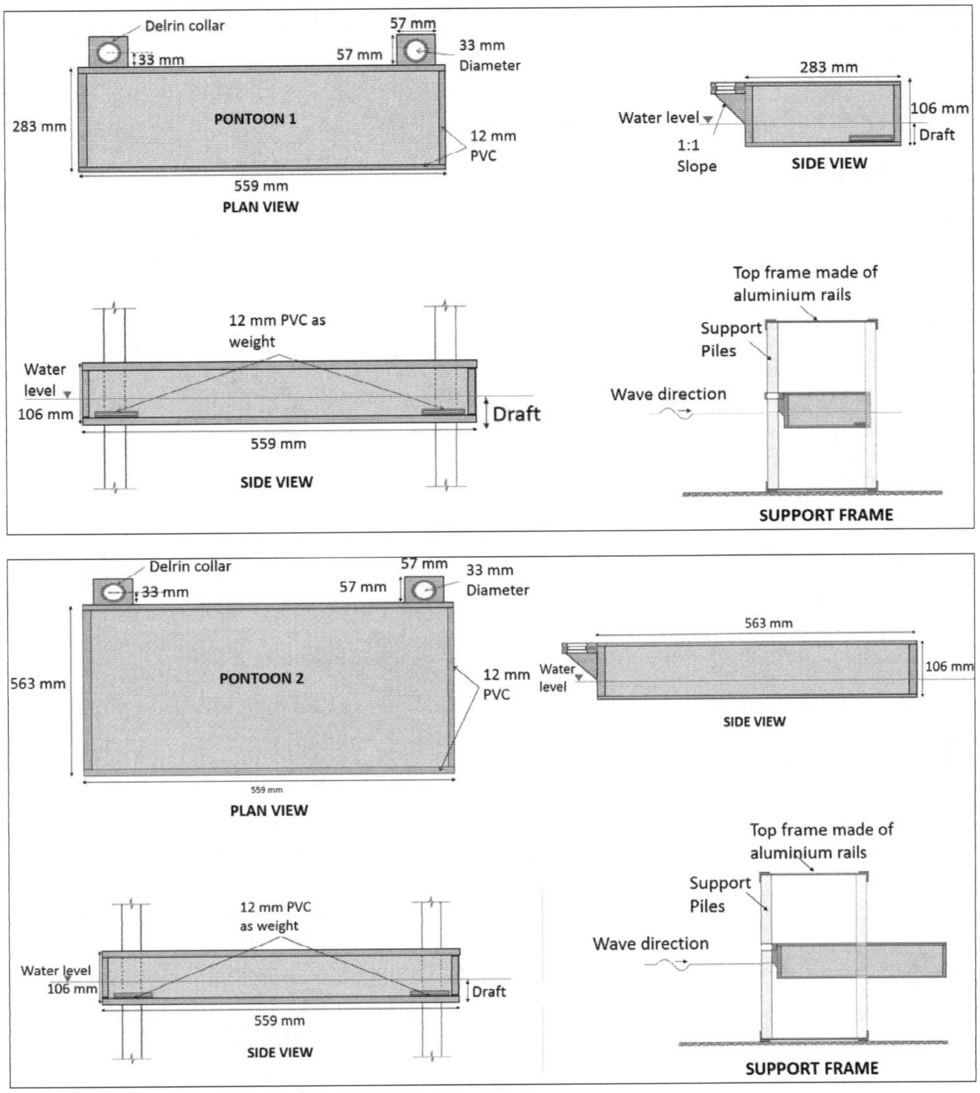

Pontoons (Diagram not to scale).

The draft of each pontoon was altered to four different depths (560mm, 585mm, 635mm and 680mm prototype) by adding lead and steel weights. Accelerations were recorded for each wave period/draft combination. The results presented are at prototype scale.

4.1.3 Wave Environment

Wave paddle control software developed at UNSW WRL allowed for generation of monochromatic waves for each wave period against each draft. Waves were generated by a vertical wave paddle situated at one end of the flume. Two probes were set up between the paddle and the structure to measure the incident and reflected waves. A third probe was positioned in the lee of the structure to measure the transmitted wave characteristics. In this study, only monochromatic waves representative of boat wake conditions were tested. The wave heights and periods adopted were based on those typically found in Sydney Harbour (Patterson Britton and Partners, 1987), where more than 137 public access points (wharves, jetties and pontoons) for boat users (Transport NSW 2014) are currently in use. The corresponding prototype wave period and wave height are presented in Table 2. Three runs, of approximately 189 seconds (prototype), were conducted for each of the wave periods to ensure similarity. All tests were completed in a water depth of 3.6 m (prototype).

Table 2. Monochromatic Wave Testing Parameters (Prototype Scale)

WAVE PERIOD (s)	2	3	5	7
MEAN INCIDENT WAVE HEIGHT (mm)	330	380	330	320

5 RESULTS AND DISCUSSION

The results presented in this paper are for the narrow pontoon (Pontoon 1). Exceedance of the safe motion limits in the vertical (heave) and horizontal (surge) planes is presented and discussed for each of the drafts tested. The effect of equivalence between natural frequency and forcing frequency is analysed relative to exceedance of the SML. The RAO in heave is presented along with discussion on the dynamic amplification resulting from natural frequency resonance.

5.1 Operational Criteria: Peak Vertical and Lateral Acceleration

5.1.1 Natural Frequency

The natural frequency in heave and roll (narrow pontoon) has been calculated for each draft tested, using methods presented in Gaythwaite (2016) and is presented in Table 3.

Table 3. Natural Frequency for Each Draft Tested – heave and roll (Prototype)

NATURAL FREQUENCY AND PERIOD	**DRAFT (m)**			
	0.560	**0.585**	**0.635**	**0.680**
$F_{N\text{ - heave}}$ (Hz)	0.37	0.36	0.34	0.33
$T_{N\text{ - heave}}$ (s)	2.7	2.8	2.9	3.0
$F_{N\text{ - roll}}$ (Hz)	0.43	0.42	0.39	0.36
$T_{N\text{ - roll}}$ (s)	2.319	2.404	2.585	2.764

As detailed in Table 3, the natural frequency in heave for each draft tested is approximately 3 seconds. We would expect dynamic amplification and resonance when the wave period coincides with the calculated natural frequency. Figure 2 shows the cumulative probability plots for heave accelerations for each draft tested. The figure shows the probability of exceeding our peak vertical safe motion limit of 0.1g for each of the tested wave periods. In heave, exceedance of our safe motion limit occurs a higher percentage of the time for the 3 second period wave for each of the drafts tested, corresponding well with what we would expect to happen as a result of correspondence between the natural frequency and forcing frequency. The plots show the probability of exceeding our safe motion limit for the 3 second period wave peaks at a draft of 585mm (20.64%) (top right). All other wave periods recorded very similar percentage exceedance for each draft and wave period. As discussed in Section 3.1.1, heave acceleration is predominantly dependent on the beam to wavelength ratio (B/L) and as such altering draft has minimal impact.

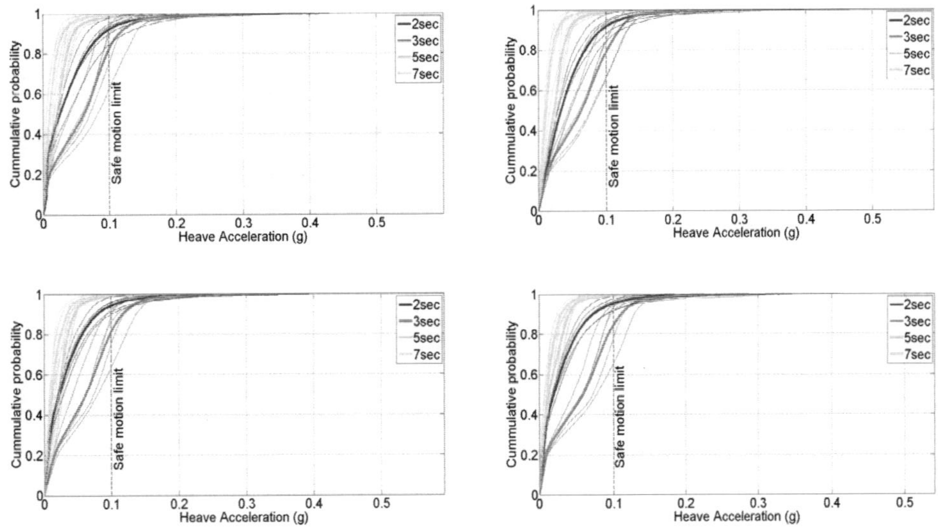

Figure 2. Cumulative Distribution Functions for the Narrow Pontoon in heave for each draft: (top left) 560mm, (top right) 585mm, (bottom left) 635mm and (bottom right) 685mm. Thin lines indicate individual sensors (1-5) and thick lines represent the mean of all 5 sensors.

Surge acceleration occurs in the direction of wave propagation (x-axis). It can be reduced by altering the amount of play between the pile and pontoon, however, there needs to be enough space to allow free vertical movement as the pontoon passes over the wave. Table 3 indicates how our natural period in roll changes with draft. The natural period in roll relates to the surge acceleration. The natural period in roll changes from approximately 2.3s (560mm) up to 2.76s (685mm). The change in roll period is due to the change in metacentric height resultant from changing draft. As our natural period in roll changes from approximately 2s up to approximately 3s we would expect our exceedance curves (surge) to change accordingly.

Figure 3 shows the cumulative probability in surge of exceeding the safe motion limit for each of the tested wave periods and each of the tested drafts. The highest exceedance occurred for a draft of 585mm (prototype) (top right) with the 2 second period wave exceeding the SML 36.31% of the time, followed by 5 second (27.19%), 3 second (26.78%) and 7 second (26.26%). A draft of 635mm (bottom left) then recorded the highest exceedance however for this situation the 3 second period wave had the highest probability (27.79%) followed by 5 second (24.62%), 2 second (14.97%) and 7 second (13.42%). 560mm (top left) was next in terms of exceedance with the 2 second period wave recording the highest exceedance followed lastly by 685mm (bottom right) where the 3 second period wave recorded the highest exceedance. The natural roll frequencies presented in Table 3 correspond well with the exceedance curves, whereby we see our predominant exceedance change from being the 2s period (560mm) up to the 3s period (685mm).

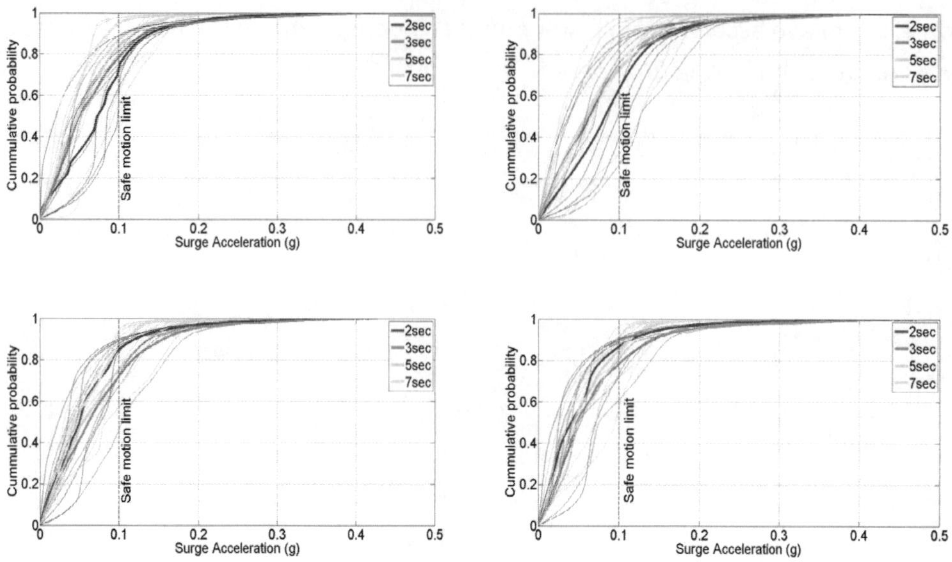

Figure 3. Cumulative Distribution Functions for the Narrow Pontoon in surge for each draft: (top left) 560mm, (top right) 585mm, (bottom left) 635mm and (bottom right) 685mm. Thin lines indicate individual sensors (1-5) and thick lines represent the mean of all 5 sensors.

These curves show significant exceedance even in the very mild tested wave climate (330mm prototype). The short period waves resulting from boat wake have the potential to cause dynamic amplification of a floating pontoon if the natural frequency is not considered.

5.1.2 Comparison Between Calculated and Recorded Acceleration

An effective method for calculating the projected accelerations of a floating pontoon at the design stage would be useful to minimise exceedance of the SML. Prior to this work there have been minimal guidelines on suitable acceleration limits of floating pontoon motions. Estimates of the heave accelerations have been calculated (Table 4 and Table 5) for each wave period/draft combination using the two methods mentioned in Section 3.1.2.

Table 4. Calculated Heave Acceleration and Effect of Altering Draft Using DNV (2012) (Prototype)

CALCULATED	DRAFT (mm)			
	560	585	635	685
HEAVE ACCELERATION (g)	0.29	0.286	0.275	0.266

Table 5. Calculated Heave Acceleration and Effect of Wave Period Using Gaythwaite (2016) (Prototype)

CALCULATED	WAVE PERIOD (s)			
	2	3	5	7
HEAVE ACCELERATION (g)	0.4	0.17	0.06	0.03

The calculated accelerations presented in Table 4 and Table 5 highlight that increasing draft reduces heave acceleration and as wave period increases magnitude of heave acceleration reduces. The recorded peak heave accelerations obtained during testing are presented in Table 6 for each draft and wave period combination.

Table 6. Maximum Recorded Heave Acceleration for Draft and Wave Period

		Draft (mm)	Wave Period (s)			
			2	**3**	**5**	**7**
Peak	**Heave (g)**	**560**	0.60	0.53	0.34	0.22
		585	0.59	0.57	0.30	0.22
		635	0.59	0.50	0.33	0.21
		685	0.48	0.54	0.26	0.22

Comparing the calculated accelerations (Table 4 and Table 5) with Table 6 the method proposed by DNV (2012) shows the closest correspondence with the recorded results. For a draft of 685mm using DNV (2010) method the calculated heave acceleration was 0.266g (Table 4) compared with the recorded peak acceleration, ranging from 0.48g to 0.22g (685mm) (Table 6). Similarly, for the other tested drafts DNV (2012) underestimates maximum heave acceleration at the resonance frequency however provided good agreement outside of this.

5.1.3 Response Amplitude Operator

Dynamic amplification occurs when the natural frequency of the structure corresponds with the forcing frequency. When dynamic amplification occurs heave response will be larger than unity and it cannot be assumed that the heave amplitude of the structure will equal the wave amplitude. Figure 4 shows the response amplitude operator for each of the drafts tested. RAOs for drafts of 560mm (top left) and 685mm (bottom right) indicate dynamic amplification at the 3 second period wave with amplitude of motion above the wave amplitude. This corresponds with the natural period in heave presented in Table 3. For a draft of 585mm peaks in the RAO occur at 3 and 5 second however amplitudes of motion are approximately equal to the wave amplitude. Dynamic amplification occurs at a period of 5 seconds for a draft of 635mm (bottom left) which is relatively different from the calculated natural frequency.

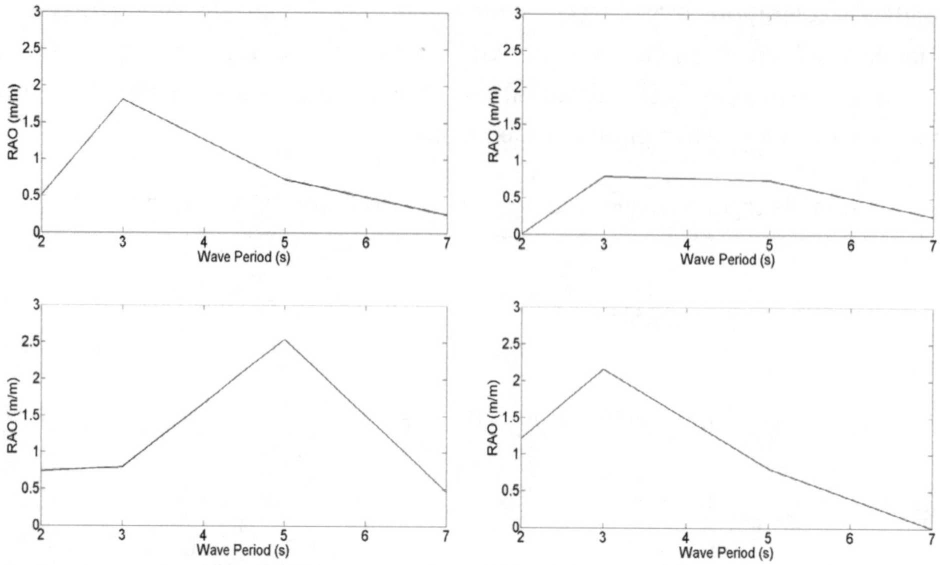

Figure 4. Response Amplitude Operator in heave for each draft: (top left) 560mm, (top right) 585mm, (bottom left) 635mm and (bottom right) 685mm.

6 CONCLUSION

A series of physical laboratory experiments were conducted to examine the dynamic motions of a piled floating box pontoon of varied draft under boat wake conditions (prototype = 330 mm) in order to compare these to safe motion limit criteria as defined in the literature for personal safety and comfort. The paper has highlighted how the dynamic nature of the pontoon relates to the natural frequency with exceedance of the safe motion limit occurring when the natural frequency corresponds with the forcing frequency. The DNV (2012) method for estimating heave acceleration underestimates heave acceleration when the natural frequency and forcing frequency correspond however provided good approximate values outside of resonance. Floating pontoon design can be bettered by considering the nominated SML (Table 1), calculating both the natural frequency and heave acceleration and ensuring the forcing frequency is far removed so dynamic amplification does not occur. The research has highlighted a gap in design guidelines/standards specific to allowable motions of floating pontoons. Although no specific guidelines exist, the paper proposes safe motion limits (what level of motion accelerations might be acceptable) and how to undertake preliminary design in order to minimise motions of floating pontoons. For important pontoon structures physical wave flume testing to measure response motions and accelerations is recommended.

7 REFERENCES

BSi, 2000. BS6349-6 Maritime Stuctures — Part 6: Design of Inshore Moorings and Floating Structures. In *Bsi*.

Defence, A.G.D. of, 2016. *2016 Defence White Paper*,

DNV, 2012. *DNV Rules for Classification of Ships Hull Structural Design Ships With Length 100 Metres and Above*,

Freeman, E.L., Cox, R.J. & Splinter, K.D., 2017. Suitable Criteria for Safe Motion Limits of a

Floating Pontoon Relative to the Postural Stability of a Stationary Standing Person. In *Coasts and Ports Conference*. pp. 21–23.

Gaythwaite, J.W., 2016. Design of Marine Facilities. Available at: http://ascelibrary.org/doi/book/10.1061/9780784414309.

NSW Maritime, 2005. *NSW Maritime Engineering Standards and Guidelines for Maritime Structures*, NSW Maritime.

Rosen, D.S., ASCE, M. & Kit, E., 1984. A Simulation Method for Small Craft Harbour Models. *Coastal Engineering*, pp.2985–3000.

imc2019
PACIFIC INTERNATIONAL MARITIME CONFERENCE
INTERNATIONAL CONVENTION CENTRE SYDNEY, AUSTRALIA
8-10 OCTOBER 2019

IMC 2019 CONFERENCE FULL PAPER

Enabling Instant Situational Awareness in Naval Bridge Operations

Robert Ventresca
Vice President, Worldwide Marketing
Thinklogical, A Belden Brand
robert.ventresca@thinklogical.com

EXECUTIVE SUMMARY

The proliferation of intelligence, surveillance and reconnaissance (ISR) information has changed the landscape of naval strategic thinking. Real-time analysis of ISR information has influenced the form and function of naval bridge design, transforming the modern bridge into the nerve center for all onboard operations.

STRATEGY OF INFORMATION SUPERIORITY

Defense and intelligence organizations have long been pursuing a strategy of information superiority to defend against a broad range of threats. However, the nature of these threats has changed. Today we need to respond to threats in hours, if not minutes – often without having any advanced warning of what the threat will be

The unpredictability of these threats requires a responsive command and control information gathering workflow that is robust, yet flexible and nimble to meet dynamic mission requirements. Ship commanders are therefore demanding better tools to collect, process, analyze and disseminate this information, leading to instant situational awareness and faster, better-informed decision making

KEY CHALLENGES IN NAVAL SHIP'S BRIDGE DESIGN

But giving that access demanded by commanders is functionally difficult. It requires a highly sophisticated IT architecture and audio-visual infrastructure. Providing access also opens the system to cyber hackers and the insider threat – the threat of accidental or intentional hacking or data breach by an employee or contractor

To manage access and mitigate the insider threat, organizations create information assurance directives to secure the information. These two concepts are in conflict. On the top we want any to any access to improve situational awareness and fulfill the mission – on the bottom, we want to lock down and secure our information – losing it through an insider or cyber event wipes out our strategic advantage.

LEGACY SHIP'S BRIDGE ENVIRONMENT

The other key driver is enhanced operational flexibility. Traditional ship's bridge infrastructures have "siloed" hard-wired workstations, each designed for specific functions. This limits collaborative workflows and increases staffing requirements to support the various dedicated operations. Because these systems are purpose-specific, a fault or failure (or combat battle damage) will render the station inactive and it's function unavailable.

These legacy systems typically use copper-based cabling, which adds weight and is susceptible to electrical emanations that can be "sniffed" or eavesdropped upon. If access to multi-classification networks is required, legacy systems often can't support this capability, or if they do, they either violate standard Information Assurance best practices for data protection due to space limitations or require clumsy manual desktop KVM (keyboard, video, mouse) switches to connect to and move between secure networks.

USS COLE INCIDENT

On 12 October 2000, a suicide bombing by a surface vessel significantly damaged USN Guided Missile Destroyer USS Cole while it was in port in Yemen for refueling. The attack was attributed to Al Qaeda. 17 sailors were killed and 39 injured. The bombing destroyed a mechanical room located below the galley, rendering the ship unseaworthy. [i]

One analyst noted later, "The U.S.S. Cole was a wake-up call for the Navy."[ii]

The incident was a key driver in the U.S. Navy's move towards "mirror-image" redundant fiber-optic based operations, communications and network backbones on new ships and retrofits. Fiber provides speed, bandwidth, reliability, immunity to EMI, and less weight compared to copper cabling systems.

NEXT-GENERATION NAVAL SHIP'S BRIDGE REQUIREMENTS

This is what is driving the need for a next-generation ship's bridge and combat information center infrastructure.

- Achieve information superiority and instant situational awareness through immediate access to critical ISR resources via "any-to-any" switching
- Enable mirror-image redundancy and rapid reconfiguration of command/CIC resources to quickly adapt to dynamic mission requirements
- Simplify management of multiple classifications through a single information assurance (IA) approved AV/IT infrastructure
- Lower total cost of ownership (TCO)

DESIRED END STATE – NEXT GENERATION SHIP'S BRIDGE

Next-generation, fiber-based naval bridges combine a video-enhanced cockpit area for navigation with mirror-image multi-function consoles incorporating an integrated command center and combat information center (CIC). This co-location of multi-purpose systems allows for greater situational awareness and collaboration between those who are driving the ship and those who are operating weapons and sensors.

The new design delivers unparalleled flexibility in how commanders station people on the bridge, providing for as few as two people standing watch on the bridge, replacing up to twelve watch standers typical for naval vessels. Because the integrated command center/CIC and the bridge are co-located, moving personnel takes literarily just a few seconds. With the open architecture of the next-gen design, commanders can re-assign any task to any console available on the bridge, providing a rapid response to the changing tactical needs of the ship.

NEXT-GENERATION SHIP'S BRIDGE INFRASTRUCTURE

In addition to fibre-optic cabling, the underlying infrastructure for secure video and data distribution plays an important role in enabling next-generation ship's bridge and combat information center capabilities. This enabling technology is comprised of two parts:

- Signal Extension (extend KVM - keyboard, video, mouse, USB, peripherals)
    - Moves computers and data sources out of the workspace. This allows for back-racking PCs and data sources in a location away from users
    - All processing and data remain on the source (computer, etc.) for security
- Secure Switching (matrix switching of multiple sources with multiple destinations)
    - Allows for switching of sources from multiple sources (and multiple classifications) on a single infrastructure

SIGNAL EXTENSION

The flexibility of a typical computer system is limited by the length of the cables that connect the computer to the display and peripheral devices, such as the keyboard and mouse. Display and peripheral signaling standards were not designed for lengths of greater than 10-15 feet. Therefore, displays and peripherals need to be located near their computers. And just installing longer cables does not work – the signal degrades and displays just flicker and keyboard and mouse controls don't work reliably.

In order to achieve greater cable lengths, you can use a secure signal extension technology, which allows a user access to a computer from up to 80 kilometers, or 50 miles. This involves installing a Transmitter, which connects to the computer just as the display, keyboard and mouse would in a typical computer system. The Transmitter then sends the video and data over *fiber optic cable* to a Receiver, installed at the user's station. The Receiver connects to the display and peripheral devices, just as they would connect to the computer – but over a significantly greater distance.

Typical commercial audio-visual products are not designed for use in mission critical secure facilities. In a secure infrastructure, the extension system does not modify or touch the signal or information in any way, to ensure the integrity of the data. It is a stateless device, with no buffers, memory or storage. It strictly transports only the signal from the computer, sending pixels for the display, keyboard strokes and mouse movements. All the data, applications and computing power remain in the computer.

Secure extension technology allows organizations to remove the computers from the user environment. The computers are now moved to a more secure IT friendly environment like a server room. This single step creates all kinds of benefits: reduced heat, reduced noise, better use of space, higher availability, and lower total cost of ownership because they are stored in a more IT friendly environment.

SECURE SWITCHING

Remember that the computers are in a server room, each connected to a Transmitter. These transmitters are now connected to a secure Matrix Switch. On the other side, all the destination workstation desks are connected to the switch via a receiver. The video display, keyboard and mouse at each desk are connected to the receiver. So now, any source can be connected to any destination whenever we want them to be connected.

If a user wants to send what he sees on his display to a video wall, he can do so with a couple of keystrokes. If a user wants to switch from one source to another, he can do so with a couple of keystrokes. If a presenter in a conference room wants to change what is on the video wall, he can do so with a drag and drop on a touch panel. Everything can go everywhere – unless the administrator wants to restrict it – which he can do with a couple of keystrokes. So now, even the classification of a control room can change in just seconds – the administrator simply restricts which sources can be switched to the control room with a couple of keystrokes.

To enable this next-generation C2 infrastructure it required to use a KVM switch with Isolation technology, which reduces or eliminates the need to air gap multiple separate systems to eliminate the chance of crosstalk or data misdirection. Using this isolation technology, a secure matrix switch can ensure that no stream of traffic is ever compromised by another stream; that is, top secret traffic is always isolated from secret, which is isolated from unclassified.

Allowing multiple classifications to flow through a single switch is the only way to achieve any-to-any switching capability in a multiple classification environment, creating a flexible and responsive collaborative workflow and leading to instant situational awareness. It is also typically required that the matrix switch be approved and certified to information assurance accreditations such as NATO NIAPC, Common Criteria EAL4, US DOD DISA JTIC UCR and TEMPEST.

SUMMARY

Through this secure fiber-optic enabling infrastructure technology, the next-generation ship's bridge and CIC can deliver on the intended requirements:

- Achieve information superiority and instant situational awareness through immediate access to critical ISR resources via "any-to-any" switching
    - Access to any potential source of information that might be needed, for an event that is not predictable
    - Flexibility to reconfigure desks, consoles and conference rooms when and as the mission changes
    - Ability to design robust, collaborative and efficient workflows for faster decision making
    - Enable mirror-image redundancy and rapid reconfiguration of command/CIC resources to quickly adapt to dynamic mission requirements
- Simplify management of multiple classifications through a single information assurance (IA) approved AV/IT infrastructure.
    - Dedicated stateless hardware at desk supporting multi-class information sources
    - PCs/sources back-racked and cabling physically separated/air-gapped in IT room
    - USB ports, network cables and hard drives physically isolated from the users for security
    - Reduced staffing
- Lower total cost of ownership
    - Pooling/consolidation of IT and AV infrastructure – less equipment to purchase and maintain
    - Less space, power, cabling, and cooling needed
    - Fiber-optic infrastructure; higher bandwidth, less EMF emanations
    - Future-proof architecture results in longer time between technology refreshes

---

[i] CNN Online article: "USS Cole Bombing Fast Facts" updated March 27, 2019 https://www.cnn.com/2013/09/18/world/meast/uss-cole-bombing-fast-facts/index.html

[ii] Cabling Installation & Maintenance Magazine, August 1, 2001 article: "Fiber-loaded SmartShips testing the waters" https://www.cablinginstall.com/home/article/16466028/fiberloaded-smartships-testing-the-waters

Development of a hardware-in-the-loop test rig for testing turbogenerator waste heat recovery systems

Noam Olshina[1], Chris Manzie[1], Michael Brear[1], Peter Hield[2], Owen Tregenza[2]
1 The University of Melbourne, Vic, Australia
2 Defence Science and Technology, Vic, Australia
nolshina@student.unimelb.edu.au

ABSTRACT

The overall energy efficiency of maritime platforms is increasingly important, for both commercial and military applications. This research project focuses on recovering some of the energy lost through the exhaust of a diesel engine using a turbocompounding system consisting of a radial flow turbine coupled to an electrical generator (turbogenerator). This system can be retrofitted to existing diesel engines and only marginally increases the engine weight, volume and complexity. Additionally, the reduced exhaust gas temperature reduces the vessel's thermal signature.

To gain a better understanding of the expected performance of the turbogenerator on a diesel engine, and the effectiveness of Model Predictive Control (MPC), the turbogenerator and MPC controller is tested in a test rig using Hardware-in-the-Loop (HIL) simulations. As opposed to simulation only based development, HIL simulations enable the testing and development of control systems under more realistic operating conditions. In this work, the test rig is used to subject the turbogenererator to realistic surface vessel and submarine operating conditions. This paper details the test rig used in this project, the performance mapping of the turbogenerator, the methodology used to reproduce engine operating conditions within the test rig, and the interface between the test rig and the simulation environment.

INTRODUCTION

Improving the energy efficiency of maritime vessels results in environmental, economic and for the military, strategic benefits [1].

WHR is a class of technologies for improving the energy efficiency of engines by targeting the energy contained in the exhaust gasses which would otherwise be wasted. Of the total energy contained in the fuel, approximately 40% is converted into useful work, 30% lost through the hot exhaust gasses and the remainder lost through other systems.

There are several Waste Heat Recovery (WHR) technologies in use and the subject of ongoing research [2]. The research presented in this paper focuses on turbocompounding, which involves the addition of a secondary power turbine (turbogenerator) in the exhaust stream of an engine. Turbocompounding provides competitive fuel efficiency gains compared to other WHR technologies with minimal increase in engine weight and complexity [2].

Diesel-electric submarines are required to operate at constant speed to minimise acoustic emissions, as well as maximise engine reliability and longevity. This has been difficult to achieve in practice. Air is drawn through a snorkel mast and exhausted underwater through the hull. This results in a low engine intake air pressure and a high and varying exhaust back pressure as ocean waves pass over the submarine (Figure 1). Consequently, the engine experiences high and varying exhaust gas temperatures and engine speed fluctuations, increasing engine wear and reduced reliability, resulting in a demanding maintenance schedule [3].

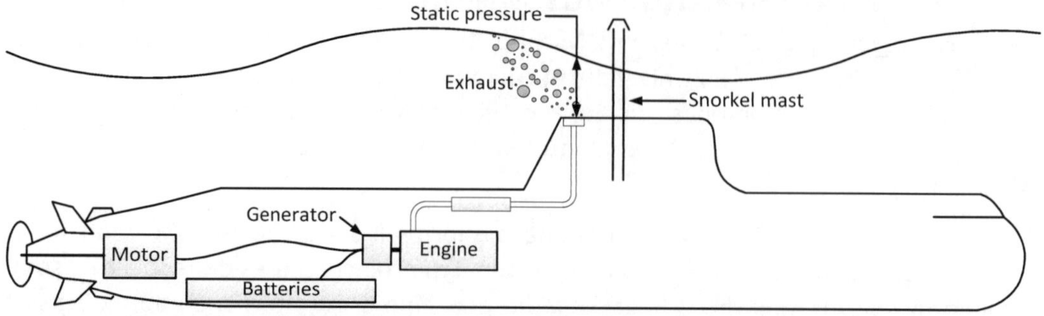

Figure 1: Schematic diagram illustrating how ocean waves cause engine back-pressure fluctuations.

Naval surface vessel engines are required to provide a variable power output consisting of the hotel load and propulsive power demand. They must also respond to changes in demand quickly to meet mission objectives. In doing so, the engine operates over a large range of set points with a significant amount of time transitioning between these.

Unlike naval applications, commercial vessel powertrains are required to mostly operate at steady state, and without the challenges of the unique operating environment of submarine engines. Therefore, the expected performance and idiosyncrasies of commercially available marine turbocompounding systems, such as that developed by MAN Diesel & Turbo [4], operating in challenging naval environments are unknown. However, this commercial experience provides additional motivation for and guidance in exploring this form of WHR for naval applications, and ultimately the development of a suitable naval turbocompounding system.

To control the complex engine-turbogenerator system, model predictive control (MPC) offers attractive advantages compared to other control techniques. MPC allows for the fulfilment of multiple objectives simultaneously, while satisfying system constraints. These objectives may include maintaining engine speed, maximising power or minimising fuel use, all while ensuring specified engine operating limits are not exceeded. Constraints may also include emission constraints, which have regulated limits [1]. MPC is a novel control strategy for maritime engine control.

Previous work undertaken by Defence Science and Technology and the University of Melbourne explored the use of MPC to control submarine engines subjected to varying exhaust back pressure [5] as well as for fixed diesel engines [6]. These studies demonstrated the potential of MPC as a control solution for these applications, with distinct advantages over traditional PID control.

The current focus of research efforts is to experimentally determine the effectiveness of turbocompounding on naval submarines and surface vessels using an MPC control strategy. To accomplish this, the MPC controller will be tested by controlling a simulated engine coupled to the real turbogenerator and subjected to realistic operating conditions. This type of experiment is referred to as Hardware-in-the-Loop (HIL) testing.

HIL testing offers a number of advantages over alternative control system testing methodologies. Most importantly, since the engine is simulated, testing can be done safely without risk to personnel or damage to a real engine, especially when testing on the limits of the engine's operational range. For the application here, the simulated engine, its configuration with the turbogenerator and the design of the MPC controller can all be modified with relative ease. This enables relatively inexpensive studies to determine the most suitable engine-turbogenerator configuration for a given application (eg: surface vessel vs. submarine), or to tune the various design parameters of the MPC formulation to improve the closed loop system performance.

For successful HIL testing, the turbogenerator test rig must be capable of reproducing realistic operating conditions and compensating for differences between the test rig and simulation environment. For MPC, an accurate model of the turbogenerator (and wastegate valve) is needed.

This paper begins by describing DST's test rig and a brief discussion of the MPC formulation. The main sections detail the development of the test rig control systems used to reproduce realistic operating conditions, the identification of the turbogenerator and wastegate valve performance maps needed for the MPC controller's internal system model, and the methodology used to interface the test rig with the simulation environment. The paper concludes by outlining the main results of this paper and the future work for this project.

TEST RIG

The DST test facility includes a test cell, air processing plant and an adjoining control room. The air processing plant can deliver air at pressures and mass flow rates comparable to the exhaust gas produced by the marine diesel engines. The test rig shown in Figure 2 is located in the test cell and is comprised of the turbogenerator system, hot air pipework, actuated valves, sensor and control network and control and data acquisition systems.

Figure 2: Photograph of test rig including: (1) turbogenerator system in dark frame, (2) actuated wastegate valve, (3) actuated bypass valve, (4) actuated back pressure valve (behind bypass valve), (5) pressure relief valve, (6) supply air pipe, (7) exhaust air pipe, (8) coolant radiator.

The turbogenerator system has been provided by Bluebox Energy Ltd. and includes the turbogenerator, actuated wastegate valve, water and oil cooling systems, power electronics, sensors and proprietary control system. The turbogenerator itself was manufactured by Bowman Power Group. It consists of a radial flow turbine directly coupled to a high speed 3 phase, 4 pole permanent magnet synchronous generator. The generator is rated for 60kW AC power output, with rotational speeds up to 55kRPM. Generated AC power is converted to DC and absorbed by the test facility's load bank. A proprietary control system is used to control the speed of the turbogenerator, either at a fixed speed setpoint or in 'maximum power' or 'maximum efficiency' modes.

In addition to the wastegate valve at the turbogenerator inlet, there are three actuated butterfly valves used to control the airflow throughout the rig. By controlling these valves appropriately, the operating conditions that the turbogenerator would be exposed to during service for both surface vessel and submarine applications can be reproduced. These valves include an inlet valve, turbogenerator bypass valve and back-pressure valve.

The inlet valve is used to reduce the pressure supplied by the air processing plant, so that in pressure in the volume upstream of the wastegate valve is approximately that of a marine diesel engine's exhaust. The bypass valve is controlled so that its upstream pressure follows an exhaust manifold reference pressure provided by the engine simulation. The back-pressure valve is used to reproduce the fluctuating exhaust back pressure typical of submarine operating conditions.

The test rig is fitted with a comprehensive sensor network used to measure the pressure and temperature at critical points within the test rig, as well as the mass flow rates through the turbogenerator and bypass lines. In addition to the sensors provided by Bluebox Energy Ltd. with the turbogenerator system, which transmit data over a CANBus, additional sensors have been added to independently measure the turbogenerator speed, generated DC voltage and DC current.

A LabVIEW system is used for data acquisition as well as to control of the valves and set the turbogenerator speed setpoint. A real-time dSPACE system is used as a platform to run the simulations of the engine in parallel with the MPC controller. Where required, data is transmitted between these two platforms using their analog inputs and outputs.

MODEL PREDICTIVE CONTROL

Model predictive control (MPC) techniques encompass a family of advanced control schemes which share defining characteristics. A mathematical model of a system to be controlled is used to determine the optimal sequence of inputs (and corresponding system states and outputs) along a fixed prediction horizon, while ensuring that system constraints are satisfied. The control input is determined using constrained optimisation techniques to find the minimum of a predefined cost function, which encapsulates the control objective. Interested readers are directed to [1] for a brief introduction into MPC and a selection of instructive references.

The MPC controller used in this work is a variant of standard MPC formulations known as Economic MPC (EMPC) and is based on the formulation presented in [7]. The main difference between standard MPC and EMPC is that the cost function explicitly penalises economic criteria such as fuel consumption. For example, for submarine operation, the cost function is defined as

$$J_{\text{sub}} := \sum_{i=0}^{N-1} \alpha_1 \left(\omega_{\text{e}} - \omega_{\text{e,ref}}\right)^2 + \alpha_2 \left(T_{\text{ex}} - T_{\text{ex,ref}}\right)^2 + \alpha_3 \left(P_{\text{gen}} - P_{\text{gen,ref}}\right)^2 + \alpha_4 \dot{m}_{\text{f}}, \qquad (1)$$

where ω_e is the engine speed, T_{ex} is the exhaust gas temperature, P_{gen} is the sum of the power generated by the engine $P_{gen,e}$ and the turbogenerator $P_{gen,tg}$, and $\dot{m}_f$ is the fuel mass flow rate. Subscript $(\cdot)_{ref}$ is for a used defined reference values which may be set by an operator, $\alpha_1,...,\alpha_4$ are weights and N is the MPC prediction horizon length.

For surface vessel gensets where meeting power demand and maximising fuel efficiency are of primary interest, we can set the weights in (1) to be $\alpha_1 = \alpha_2 = 0$. Similarly for propulsion engines where the engine is directly coupled to the ship's propellers and engine speed and fuel efficiency are of importance, we set $\alpha_2 = \alpha_3 = 0$ in (1). In both cases, by explicitly penalising the fuel consumption, the fuel efficiency of the engine/turbogenerator system is maximised.

In designing the EMPC controller, there are numerous design parameters which affect both closed loop performance and the computational complexity of the controller. Selection of these design parameters to achieve the specified closed loop performance without over specifying the EMPC implementation hardware is a challenging problem [8]. Some of these parameters include EMPC controller's prediction horizon time, horizon length (N) and the fidelity of the mathematical model of the engine-turbogenerator system.

HARDWARE IN THE LOOP CONFIGURATION AND CONTROL

To include the turbogenerator within a HIL simulation, three aspect of the test rig control are considered:

1. Representation of the simulated engine operating conditions within the test rig, by using the test rig's inlet, bypass and back pressure valves.
2. Development of the turbogenerator and wastegate valve models for use within the EMPC controller's mathematical plant model.
3. The hot air supply system was originally designed to replicate compressor exit conditions for gas turbine engine and thus cannot replicate the temperatures typically seen in a diesel engine exhaust system. Additionally, this system is not capable of rapid changes in temperature typical of diesel exhaust systems during engine transients. Therefore, the difference in temperature between the simulated exhaust manifold temperature and the actual test rig gas temperature upstream of the wastegate valve must be considered and compensated for in the interaction between the test rig, the EMPC controller and the simulated engine. These aspects of the HIL simulation are described in the sections below.

Experimental techniques often make use of dimensional analysis to distill the seemingly complex relationship observed between variables into simple relationships between dimensionless groups of these variables. The methodology of finding suitable dimensionless groups which can be used to characterise complex system behaviour is well established and known as the Buckingham-Pi principle [9, Chapter 5], and is a technique used extensively in this work. For example, to derive a mathematical model of the turbogenerator for use within the EMPC controller, which can be used for varying turbogenerator inlet pressures and temperatures.

Relevant dimensionless quantities are first derived using the Buckingham-Pi principle with the dimensioned variables of mass flow $\dot{m}$, speed ω, power P, pressure p and temperature T at the inlet $(\cdot)_{\text{in}}$ and outlet $(\cdot)_{\text{out}}$, turbine diameter d_{tg} and specific heat capacities c_p and c_v. This results in the following dimensionless groups relevant for the turbogenerator and control valve mapping:

$$\Pi_{\text{tg}} = \frac{p_{\text{in}}}{p_{\text{out}}}, \qquad \text{M}_{\text{tg}} = \frac{\dot{m}_{\text{tg}}\sqrt{c_p T_{\text{in}}}}{d_{\text{tg}}^2 p_{\text{in}}}, \qquad \Omega_{tg} = \frac{d_{\text{tg}}\omega_{\text{tg}}}{\sqrt{c_p T_{\text{in}}}},$$
$$\text{T}_{\text{tg}} = \frac{T_{\text{in}}}{T_{\text{out}}}, \qquad BSR_{\text{tg}} = \frac{d_{\text{tg}}\omega_{\text{tg}}}{2\sqrt{2c_p T_{\text{in}}\left(1-\Pi_{\text{tg}}^{1-1/\gamma}\right)}}, \qquad \Upsilon_{tg} = \frac{P_{\text{tg}}}{d_{\text{tg}}^2 p_{\text{in}}\sqrt{c_p T_{\text{in}}}}. \tag{2}$$

where $\gamma = c_p/c_v$.

VALVE CHARACTERISATION AND CONTROL

The simulated engine operating conditions that must be reproduced within the test rig include the exhaust manifold pressure and, for the submarine operating condition, the wave induced back-pressure.

For the exhaust manifold pressure, the test rig's inlet valve is set to a fixed position to reduce the pressure supplied by the air processing plant to a pressure closer to the simulated exhaust manifold pressure. The bypass valve is then actively controlled so that its upstream pressure follows a reference exhaust manifold pressure provided by the engine simulation. As the control architecture for the bypass valve and the back-pressure valve are the same, the architecture will be illustrated using the back-pressure valve.

The back-pressure due to waves passing over the exhaust port of a submarine is reproduced by controlling the test rig's back pressure valve so that it's upstream pressure, p_{ex}, tracks a reference exhaust back-pressure, $p_{\text{ex,ref}}$. The reference back-pressure can be any suitable timeseries representative of ocean conditions that a submarine may encounter while snorting. These timeseries could be obtained from direct measurement on board a submarine, generated from wave spectra, or simply consist of a single sinusoidal waveform representative of a given sea state.

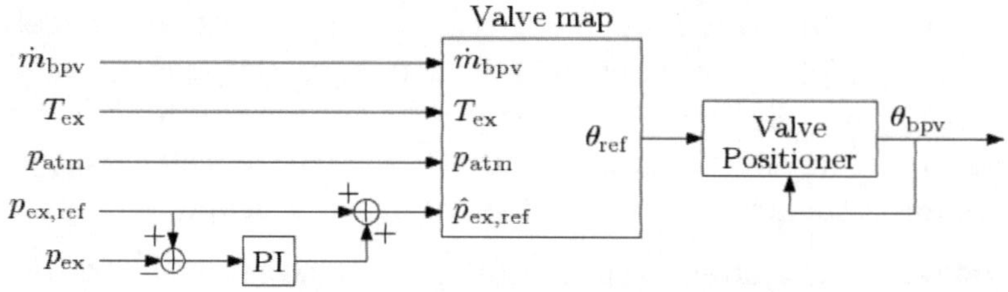

Figure 3: Back pressure valve control architecture.

Since the pressure ratio across a butterfly valve is a nonlinear function of valve angle and mass flow rate, a typical PID controller would struggle to reproduce the reference upstream pressure over the required operating range. Therefore, a model-based controller shown in

Figure 3 has been developed using a PI / feedforward architecture. A valve map is used to convert the back-pressure timeseries into a reference valve angle, which is then tracked by the valve's actuator. In addition, to the feedforward valve map, a PI controller is used to further reduce any error between p_{ex} and $p_{\mathrm{ex,ref}}$ and provides an augmented reference back-pressure, $\hat{p}_{\mathrm{ex,ref}}$ to the feedforward map. The valve positioner, which is integrated within the valve's actuator tracks θ_{ref}.

To develop the valve map, we model the valve as an isothermal orifice, considering the gas as a compressible fluid [10, Chapter 2]. Firstly, define the pressure ratio across the valve as $\Pi_{\mathrm{bpv}} = p_{\mathrm{atm}}/p_{\mathrm{ex}}$, where p_{atm} is the atmospheric. The mass flow rate through the valve is given by

$$\dot{m}_{\mathrm{bpv}} = A_{\mathrm{eff}}(\theta_{\mathrm{bpv}})\frac{p_{\mathrm{ex}}}{\sqrt{RT_{\mathrm{ex}}}}\Psi(\Pi_{\mathrm{bpv}}), \tag{3}$$

where T_{ex} is the valve's upstream temperature,

$$\Psi(\Pi) := \begin{cases} 1/\sqrt{2}, & 0 < \Pi \leq 1/2 \\ \sqrt{2\Pi(1-\Pi)}, & 1/2 < \Pi < 1, \end{cases} \tag{4}$$

and $A_{\mathrm{eff}}(\theta_{\mathrm{bpv}})$ combines the valve's coefficient of discharge and effective throat area into a single function of valve angle, θ_{bpv}. The map used in the feedforward controller is an inversion of (3) in the form $\theta_{\mathrm{ref}} = f(\dot{m}_{\mathrm{bpv}}, \hat{p}_{\mathrm{ex,ref}}, p_{\mathrm{atm}}, T_{\mathrm{ex}})$.

A series of experiments was conducted to identify the form of $A_{\mathrm{eff}}(\theta_{\mathrm{bpv}})$, by measuring $\dot{m}_{\mathrm{bpv}}, p_{\mathrm{atm}}, p_{\mathrm{ex}}, T_{\mathrm{ex}}$ and θ_{bpv} at operating points spanning the operating range of the valve. It was found that $A_{\mathrm{eff}}(\theta_{\mathrm{bpv}})$ was well fitted by a quadratic function as seen in Figure 4.

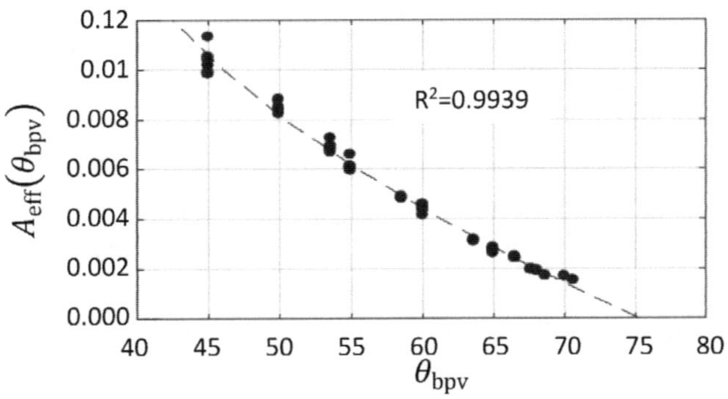

Figure 4: Back pressure valve characterisation

To validate the performance of the control architecture, and to tune the gains of the PI controller, the closed loop system was tested using a number of $p_{\mathrm{ex,ref}}$ signals, which were generated using a sinusoidal signal with amplitude and frequency corresponding to sea states 2 to 5. Figure 5 shows an 60 second extract from the generated $p_{\mathrm{ex,ref}}$ and measured p_{ex} for sea state 5. This extract indicates that the back pressure is well reproduced.

TURBOGENERATOR PERFORMANCE MAPPING

To determine the performance of the turbogenerator, experimental data was collected covering its entire operating range. Each datum represents data collected at 10kHz over 10 second intervals, and then averaged to reduce the impact of noise. Between operating points, sufficient time was allowed for any transient dynamics to settle. In total, 301 data points were collected.

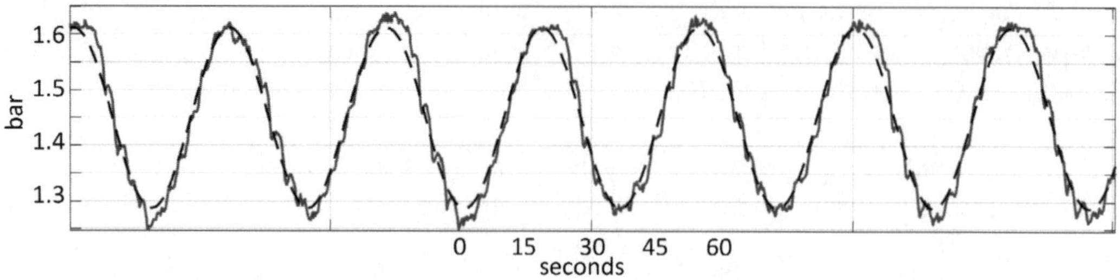

Figure 5: Time series of back pressure for sea state 5. Black dashed line indicated the reference pressure-. Solid grey line indicated the measured pressure.

For turbogenerator mapping, it is common practice to use quasi-nondimensionless variables by omitting constants c_p and d_{tg} from those in (2) and scaling other variables (for example, to use kPa instead of Pa or °C instead of K) [11]. These variables, which are used in this work include

$$\Pi_{tg} = \frac{p_{in}}{p_{out}}, \qquad \dot{m}_{tg,co} = \frac{\dot{m}_{tg}\sqrt{T_{in}}}{p_{in}/1000}, \qquad \omega_{tg,co} = \frac{\omega_{tg}}{\sqrt{T_{in}}},$$
$$\mathrm{T}_{tg} = \frac{T_{in}}{T_{out}}, \qquad BSR_{tg} = \frac{d_{tg}\omega_{tg}}{2\sqrt{2c_pT_{in}\left(1-\Pi_{tg}^{1-1/\gamma}\right)}}, \qquad \upsilon_{tg,co} = \frac{P_{tg}}{p_{in}/1000\sqrt{T_{in}}}. \tag{5}$$

By rearranging these equations, it is easy to calculate the real mass flow rate and power output based on the turbogenerator speed and the inlet pressure and temperature.

Quasi-nondimensional mass flow map

To model the turbine's quasi-nondimensional mass flow map, various physics based models promulgated in the literature and compiled in [11] were trialled, but none found to produce satisfactory accuracy. One factor explaining this is the larger size of the turbine used in this experiment compared to typical automotive turbines, which are the subject of the majority of modelling efforts described in the literature. Typically, automotive turbine mass flow models are only dependent on pressure ratio. In large radial flow turbines, the centrifugal forces imparted onto the gas by the rotating turbine blades act to retard mass flow. Therefore, the turbine's rotational velocity becomes a significant factor contributing to quasi-nondimensional mass flow model accuracy.

A data-driven model consisting of a 3rd order polynomial of the form $\dot{m}_{tg,co} = f_{\dot{m}}\left(\Pi_{tg}, \omega_{tg,co}\right)$ was developed which includes dependence on pressure ratio and quasi-nondimensional speed. Although this model cannot be used to predict turbine choking, since the polynomial fit diverges for Π_{tg} and $\omega_{tg,co}$ outside of the experimental range, the safety constraints imposed by the on-board controller restrict operation in the choked region. Figure 6 shows the turbogenerator's quasi-nondimensional mass flow map with Π_{tg} and $\omega_{tg,co}$ vs. $\dot{m}_{tg,co}$.

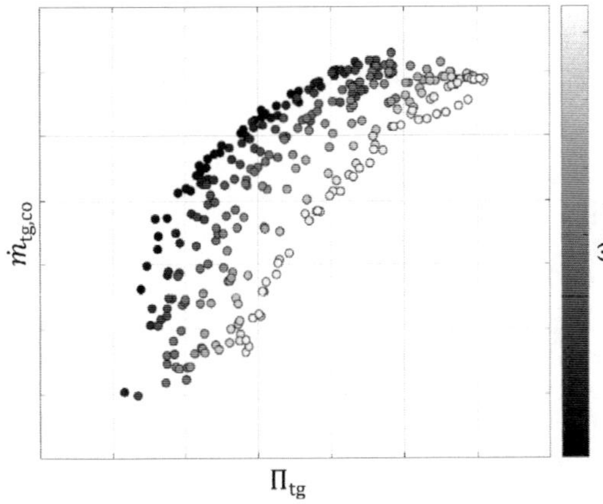

Figure 6: Quasi-nondimensional turbogenerator mass flow map. Correlation coefficient of 3rd order polynomial fit = 0.9866. Scales have been omitted to retain turbogenerator vendor's intellectual property.

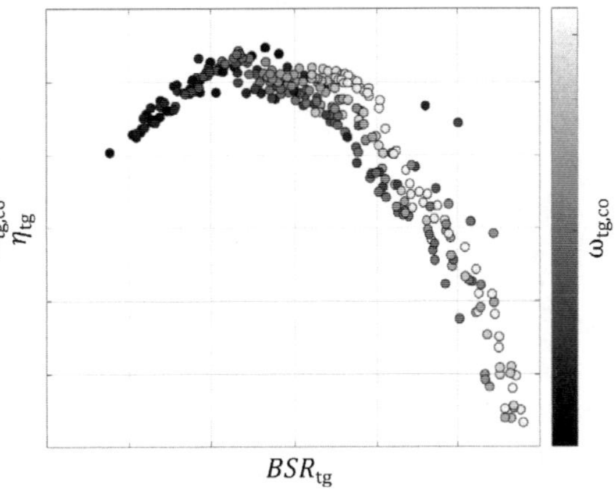

Figure 7: Turbogenerator turbine isentropic efficiency. Correlation coefficient of quadratic fit = 0.9538. Scales have been omitted to retain turbogenerator vendor's intellectual property.

Turbogenerator DC power output

The power produced by the turbogenerator's turbine and delivered to the generator is given by

$$P_{tg} = \eta_{tg} P_{tg,ideal},$$

where η_{tg} represents the isentropic efficiency of the turbine. $P_{tg,ideal}$ is the power that the turbine would produce in an ideal isentropic expansion given by

$$P_{tg,ideal} = \dot{m}_{tg} c_p T_{tg,in} \left(1 - \Pi_{tg}^{1-1/\gamma}\right).$$

Since the power extracted from the expanding exhaust gas can be calculated from experimental data using

$$P_{tg} = \dot{m}_{tg} c_p \left(T_{tg,in} - T_{tg,out}\right),$$

the isentropic efficiency is calculated from experimental data by

$$\eta_{tg} = \frac{\left(T_{tg,in} - T_{tg,out}\right)}{T_{tg,in}\left(1 - \Pi_{tg}^{1-1/\gamma}\right)}.$$

To model the turbine efficiency, we follow the approach in [11] which uses a quadratic function of BSR_{tg}. Here we have used a quadratic function as well, but once which includes the dependence on $\omega_{tg,co}$, as this was found to provide a superior fit. This model takes the form $\eta_{tg} = f_\eta\left(BSR_{tg}, \omega_{tg,co}\right)$. Figure 7 shows the turbogenerator's isentropic efficiency map, with BSR_{tg} and $\omega_{tg,co}$ vs. η_{tg}.

Having calculated P_{tg}, along with measurements of the generated DC power, $P_{tg,gen}$, we can determine the losses due to mechanical friction within the turbogenerator and due to the combined electrical conversion efficiency of the generator and power electronics. Figure 8 shows the plot of P_{tg} vs. $P_{tg,gen}$, with a line of best fit given by

$$P_{tg,gen} = \eta_{gen}\left(P_{tg} - P_{tg,0}\right),$$

where η_{gen} is the combined mechanical and electrical efficiency and $P_{\mathrm{tg,0}}$ represents the minimum power delivered to the generator before power begins to be exported. This point coincides with the minimum rectified voltage required by the power electronics before becoming active.

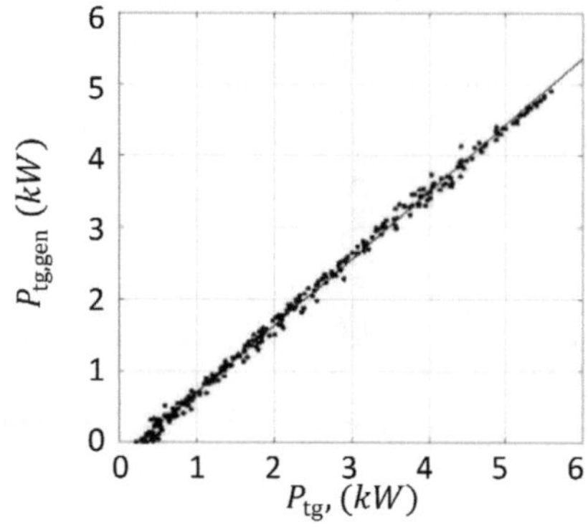

Figure 8: Turbogenerator power map

Wastegate valve mass flow

The model for the wastegate valve's mass flow rate is the same as the back pressure valve's, but uses different coefficients in $A_{\mathrm{eff}}(\theta_{\mathrm{wgv}})$.

INTERACTION BETWEEN EMPC CONTROLLER, ENGINE SIMULATION AND TEST RIG

There are numerous interactions between the simulated engine, the test rig and the EMPC controller. The EMPC controller provides control inputs for the simulated engine's fuelling rate $\dot{m}_{\mathrm{f}}$, the load on the engine τ_{l} (collected into the vector $\mathrm{u_e} = [\dot{m}_{\mathrm{f}}, \tau_{\mathrm{l}}]$), and the turbogenerator speed setpoint u_{tg} and the wastegate valve position u_{wgv} (collected into the vector $\mathrm{u_{tg}} = [u_{\mathrm{tg}}, u_{\mathrm{wgv}}]$).

The EMPC controller also requires state feedback from the simulated engine and test rig. For the engine, the states include the engine speed ω_{e}, turbocharger speed ω_{t}, and the pressures $p_{(\cdot)}$ and temperatures $T_{(\cdot)}$ in the intercooler $(\cdot)_{\mathrm{ic}}$ and cylinder block $(\cdot)_{\mathrm{cc}}$ inlet manifold $(\cdot)_{\mathrm{im}}$ and exhaust manifolds $(\cdot)_{\mathrm{em}}$. These are collected into the vector $\mathrm{x_e} = [\omega_{\mathrm{e}}, \omega_{\mathrm{t}}, T_{\mathrm{ic}}, T_{\mathrm{cc}}, T_{\mathrm{em}}, p_{\mathrm{ic}}, p_{\mathrm{cc}}, p_{\mathrm{em}}]$. For the turbogenerator and wastegate valve, the states include turbogenerator speed ω_{tg}, turbogenerator inlet pressure p_{tg} and temperature T_{tg} and the wastegate valve position x_{wgv}. These are collected into the vector $\mathrm{x_{tg}} = [\omega_{\mathrm{tg}}, x_{\mathrm{egv}}, T_{\mathrm{tg}}, p_{\mathrm{tg}}]$. In addition, the EMPC controller also requires the generated back-pressure, and for the cost function, the turbogenerator exported DC power and the exhaust temperature. The engine simulation also requires the mass flow rate through the wastegate valve to calculate the exhaust manifold pressure and temperature.

Since the air supplied by the air processing plant is substantially less than typical exhaust manifold temperatures, the control inputs to the turbogenerator, as well as the feedback to the EMPC controller and the simulated engine must be scaled accordingly. This scaling is accomplished using quasi-nondimensional variables, since these variables remain invariant between the EMPC - engine simulation and the test rig. In the following, we denote simulated and scaled measured quantities with a hat ($\widehat{\ }$) and unscaled measured quantities without.

Exhaust temperature

The simulated exhaust temperature is calculated using both the simulated turbocharger turbine and adjusted turbogenerator's mass flow rate and outlet temperature. The simulated exhaust temperature, $\hat{T}_{\mathrm{ex}}$ is given by

$$\hat{T}_{\mathrm{ex}} = \frac{\hat{\dot{m}}_{\mathrm{t}}\hat{T}_t + \hat{\dot{m}}_{\mathrm{tg}}\hat{T}_{\mathrm{tg,out}}}{\hat{\dot{m}}_{\mathrm{t}} + \hat{\dot{m}}_{\mathrm{tg}}}.$$

To convert the measured $\dot{m}_{\mathrm{tg}}$ to the scaled $\hat{\dot{m}}_{\mathrm{tg}}$ for use within the simulation and control environments, the quasi-nondimensional mass flow rates are equated, and then solved for $\hat{\dot{m}}_{\mathrm{tg}}$. This yields

$$\hat{\dot{m}}_{\mathrm{tg,co}} = \dot{m}_{\mathrm{tg,co}} \Rightarrow \frac{\dot{m}_{\mathrm{tg}}\sqrt{T_{\mathrm{in}}}}{p_{\mathrm{in}}/1000} = \frac{\hat{\dot{m}}_{\mathrm{tg}}\sqrt{\hat{T}_{\mathrm{in}}}}{\hat{p}_{\mathrm{in}}/1000} \Rightarrow \hat{\dot{m}}_{\mathrm{tg}} = \frac{\hat{p}_{\mathrm{in}}\sqrt{T_{\mathrm{in}}}}{p_{\mathrm{in}}\sqrt{\hat{T}_{\mathrm{in}}}}\dot{m}_{\mathrm{tg}},$$

where $(\cdot)_{\mathrm{in}}$ indicates the turbogenerator inlet conditions. To convert the measured $T_{\mathrm{tg,out}}$ to the scaled $\hat{T}_{\mathrm{tg,out}}$ for use within the simulation and control environments, the temperature ratios are equated, and then solved for $\hat{T}_{\mathrm{tg,out}}$. This yields

$$\mathrm{T}_{\mathrm{tg}} = \hat{\mathrm{T}}_{\mathrm{tg}} \Rightarrow \frac{T_{\mathrm{in,tg}}}{T_{\mathrm{out,tg}}} = \frac{\hat{T}_{\mathrm{in,tg}}}{\hat{T}_{\mathrm{out,tg}}} \Longrightarrow \hat{T}_{\mathrm{out,tg}} = \frac{\hat{T}_{\mathrm{in,tg}}}{T_{\mathrm{in,tg}}}T_{out,tg},$$

Turbogenerator speed set point and feedback

The EMPC controller provides a set point to the turbogenerator, $u_{\mathrm{tg,mpc}}$. This is then converted to a quasi-nondimensional speed setpoint, $u_{\mathrm{tg,co}}$ using the MPC's internal model of the turbogenerator inlet temperature, and then back into an actual speed setpoint for the turbogenerator using the rig's turbogenerator inlet temperature as

$$u_{\mathrm{tg,co}} = \frac{u_{\mathrm{tg,mpc}}}{\sqrt{\hat{T}_{\mathrm{in}}}} = \frac{u_{\mathrm{tg}}}{\sqrt{T_{\mathrm{in}}}} \Longrightarrow u_{\mathrm{tg}} = \sqrt{\frac{T_{\mathrm{in}}}{\hat{T}_{\mathrm{in}}}}u_{\mathrm{tg,mpc}}.$$

Similarly, the actual turbogenerator speed is fed back into the MPC controller and simulated engine model using

$$\omega_{\mathrm{tg,mpc}} = \sqrt{\frac{\hat{T}_{\mathrm{in}}}{T_{\mathrm{in}}}}\omega_{\mathrm{tg}}.$$

Turbogenerator power

The DC power generated by the turbogenerator, $P_{\mathrm{tg,gen}}$ is scaled for use by the EMPC controller by equating the quasi-nondimensional power of the test rig, $\upsilon_{\mathrm{tg,co}}$, to the EMPC quasi-nondimensional power in the controller and simulation environment, $\hat{\upsilon}_{\mathrm{tg,co}}$, and solving for $\hat{P}_{\mathrm{tg,gen}}$. This yields

$$\upsilon_{\mathrm{tg,co}} = \hat{\upsilon}_{\mathrm{tg,co}} \Rightarrow \frac{P_{\mathrm{tg,gen}}}{p_{\mathrm{in}}/1000\sqrt{T_{\mathrm{in}}}} = \frac{\hat{P}_{\mathrm{tg,gen}}}{\hat{p}_{\mathrm{in}}/1000\sqrt{\hat{T}_{\mathrm{in}}}} \Rightarrow \hat{P}_{\mathrm{tg,gen}} = \frac{\hat{p}_{\mathrm{in}}\sqrt{\hat{T}_{\mathrm{in}}}}{p_{\mathrm{in}}\sqrt{T_{\mathrm{in}}}}P_{\mathrm{tg,gen}}.$$

CONCLUSION & FUTURE WORK

This paper describes the turbogenerator test rig used to map the performance of the turbogenerator, as well as serving as a platform to test an EMPC controller using hardware-in-the-loop simulations. The control of the air flow through the rig to reproduce wave induced back-pressure fluctuations and engine operating conditions is described. Furthermore, the methodology of integrating the test rig into the simulation environment is outlined.

The future work enabled by the HIL simulations on the test rig includes verifying the efficacy of the EMPC controller for both submarine and surface vessel applications, determining the engine/turbogenerator configurations most suitable to these applications, and exploring the trade-off between the controller performance and computational complexity. These studies all aim to find a combination of system and controller design parameters which maximise the energy efficiency of diesel engines while ensuring all other system performance objectives are met.

REFERENCES

[1] N. Olshina, C. Manzie, M. Brear, P. Hield, O. Tregenza, "Waste Heat Recovery of Marine Diesel Engines," *in Pacific 2017, International Maritime Conference*, 2012.

[2] D. V. Singh and E. Pedersen, "A review of waste heat recovery technologies for maritime applications," *Energy Convers. Manag.*, vol. 111, pp. 315–328, 2016.

[3] P. Hield, "The effect of back pressure on the operation of a diesel engine," Australian Government Department of Defence - Defence Science and Technology Organisation, Technical Report DSTO-TR-2531, 2011.

[4] MAN Diesel & Turbo, "Waste Heat Recovery System (WHRS) for Reduction of Fuel Consumption, Emissions and EEDI".

[5] T.J. Broomhead, C. Manzie, P. Hield, and M.J. Brear, "An experimental investigation of additional actuators on a submarine diesel generator," *Control Eng. Pract.*, vol. 55, pp. 26–37, Oct. 2016.

[6] T.J. Broomhead, C. Manzie, P. Hield, R.C. Shekhar, and M.J. Brear, "Economic model predictive control and applications for diesel generators," *IEEE Trans. Control Syst. Technol.*, vol. 25, no. 2, pp. 388–400, 2016.

[7] N. Olshina, C. Manzie, P. Hield and M. Brear, "A Framework for Robust Nonlinear Economic MPC Without Terminal Constraints for a Class of Time Varying Systems," *2018 15th International Conference on Control, Automation, Robotics and Vision (ICARCV)*, Singapore, 2018, pp. 668-673.

[8] V. Bachtiar, C. Manzie, W.H. Moase, E.C. Kerrigan,"Analytical results for the multi-objective design of model-predictive control," *Control Engineering Practice,* 56:1-12, 2016.

[9] F.M. White, "*Fluid Mechanics*", 8th ed., McGraw Hill, 2016

[10] L. Guzzella, C.H. Onder, "*Introduction to Modeling and Control of Internal Combustion Engine Systems,*" 2nd ed., Springer, 2010.

[11] L. Eriksson and L. Nielsen, *Modeling and Control of Engine and Drivelines*, 1st ed., Wiley, 2014.

Hardware-in-the-Loop for Naval Platform Power and Energy Test and Evaluation

Wang Y. Kong and Nathan D. Marks
Department of Defence Science and Technology Group, Australia
wang.kong@dst.defence.gov.au, nathan.marks@dst.defence.gov.au

ABSTRACT

Traditional test and evaluation of naval platforms utilises a combination of dedicated physical platforms, and modelling and simulation tools. The former provides a high level of assurance, but is expensive, late in the process, and problems can be costly to mitigate. The latter is flexible and economical, but is ultimately constrained by the accuracy of the model.

Hardware-in-the-Loop (HIL) is a hybrid technique that leverages the benefits of both approaches by linking physical equipment to a real-time simulation. One or more real components are replaced with an equivalent model, where the real and the virtual interact as they would in the full system.

Through life management of a platform can be improved by the HIL technique. Key components can be tested as if they were part of the full system earlier in the design cycle, without committing to a full complement of equipment, or a specific system design. Operating conditions can be evaluated in a more repeatable and controlled environment, improving the verification process and support for in-service platforms. New control or equipment technologies can be evaluated with minimal cost, risk, and modification to the test platform, reducing the effort required for technology insertion.

This paper discusses HIL applications to support naval power system test and evaluation, and through life sustainment. Details of a HIL experiment where the main storage battery of a submarine power system is emulated via simulation will be presented to demonstrate the value of including HIL in the repertoire of test and evaluation tools.

INTRODUCTION

Traditional procurement processes for power and energy systems in maritime platforms involve detailed studies which determine the specifications for the various components. Original equipment manufacturers (OEMs) are then selected to supply the equipment based on these specifications. These items are verified via factory acceptance testing before delivery. It is then up to the shipbuilder to integrate the equipment into the platform and perform final integration testing. This method is well suited for mass produced commercial platforms where functionality, configuration, and requirements are generally straight forward, which leads to an established design procedure with known risk mitigation strategies.

Naval platforms differ from commercial platforms in a number of ways. Functional requirements are more complicated as they must satisfy a wide range of design objectives. Advanced bespoke equipment is often necessary to achieve stringent performance requirements, and this creates a number of risks. These technical risks include integration of new equipment that can lead to unforeseeable problems which might not be fully realised until late in the program; as well as certainty in design assurance, since designers are often working with developmental products which they may have little or no prior experience. Ultimately these risks affect the schedule and cost of the program, which can delay or derail successful delivery of a platform. An additional aspect presenting risk for naval platforms is the requirement for upgrading or replacing equipment throughout their service life to maintain a capability edge. This creates a number of challenges for through life design and verification of these platforms.

Effective design testing and evaluation, conducted as early as possible, is a strategy that can alleviate these challenges. This would likely involve a dedicated land based test facility which allows these activities to be conducted well in advance of the final design. The scale of this facility will depend on the risk appetite and budget. The most expensive solution is to replicate the entire power and energy train. This minimises the integration risk for the system prior to installation by allowing the largest range of tests, and should speed up the production process. The alternative is a reduced scale system where only selected equipment is tested. As a reference system for the platform, this test facility also provides a mechanism for live troubleshooting and verification of any changes to the design. Due to the considerable cost in setting up and running such a facility, there is a strong imperative to maximise its utilisation. Furthermore, as the design of the platform evolves throughout its life, the impacts of upgrading and redesigning this facility can be significant.

Hardware-in-the-Loop (HIL) is an experimental technique which enables a real item of equipment to seamlessly interact with a virtual real-time simulation, thus enabling real world testing to occur with a reduced set of equipment. Electric ships and power systems are particular key areas for the utilisation of HIL, as they offer reduced cost and risk for integration testing, and improved flexibility compared to full experimental validation [1] – [7]. This technique can help to mitigate some of the costs and risks of a land based test system, and maximise its utilisation. By providing options for testing that are appropriate for the risk appetite, HIL allows designers and decision makers to trade-off between budget and effort required. It can also be a valuable tool for workshopping faults in a controlled manner, maintaining uptime during maintenance periods, and investigating the integration of new technologies early in the design cycle.

This paper will detail the concept of HIL, and discuss how it can be of benefit to testing and evaluation of a large scale naval system. A number of examples will be presented to support this discussion. The role this technique can play in supporting long term research, development, troubleshooting and upgrade of a naval platform will be outlined. Finally, results from an experimental HIL based battery emulator will be presented to demonstrate the requirements for implementing HIL systems, and the performance they can provide.

HARDWARE-IN-THE-LOOP

The fundamental concept of Hardware-in-the Loop is to break a traditional simulation configuration at the terminals of the system or subsystem that will be replaced (tear point) with its physical counterpart, known as the Item-under-Test (IuT). A hardware interface converts low power signals representing the characteristics at these terminals in the simulation to the appropriate stimulation for the IuT, while sensors measure the response from the IuT to feed back to the simulation and close the loop. This technique has strong similarities with how a player interacts with a virtual-reality game, where the user (IuT) interacts with the game (real-time simulation) through a head-set (interface), and the user's response is measured and fed back through controllers (sensors).

The nature of the signals exchanged with the IuT defines the type of HIL, which can be divided into Control Hardware-in-the-Loop (CHIL) and Power Hardware-in-the-Loop (PHIL). Diagrams outlining these two types of HIL are shown Figure 1 below.

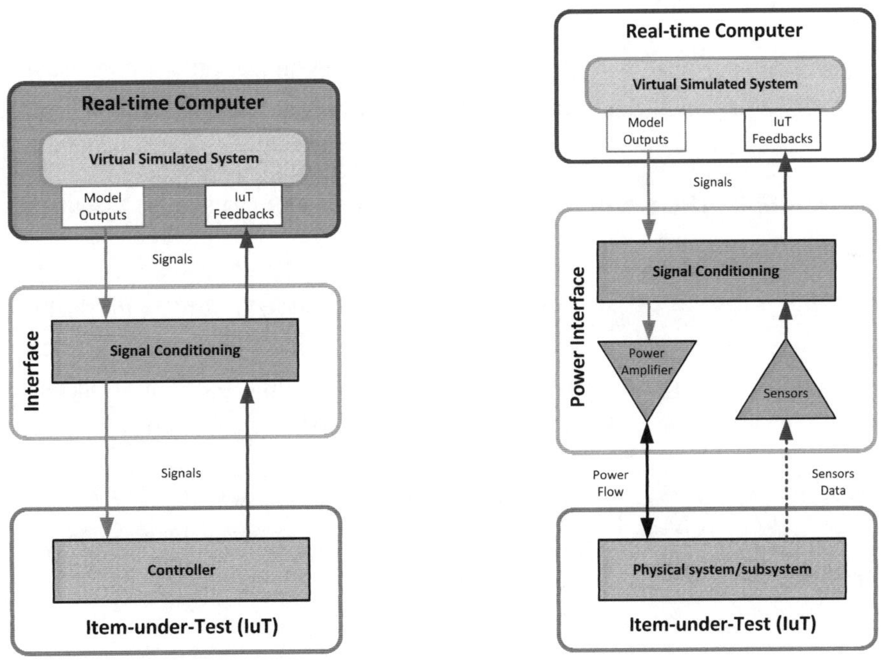

(a) (b)

Figure 1: Conceptual forms of Hardware-in-the-Loop (HIL) – (a) Control Hardware-in-the-Loop (CHIL) and (b) Power Hardware-in-the-Loop (PHIL).

Control Hardware-in-the-Loop as shown in Figure 1(a) is focused on the testing of controllers, where a physical controller is the IuT. This controller interacts with simulated versions of the equipment it is intended to control. No real power is exchanged between the interface and the IuT. The interface merely acts as a conduit for control and sensor signals that are exchanged between the IuT and the simulation. CHIL can be used to test items such as software and control algorithms in an environment that is easily controlled, with no risk to physical hardware.

In a Power Hardware-in-the-Loop system as shown in Figure 1(b), the IuT is a physical system or subsystem such as a motor, engine or power distribution system of a ship. The roles of the power interface are to amplify the control signal from the virtual simulation to the power level required to stimulate the IuT, and to condition measurements from the IuT for feedback to the simulation. For electrical systems, the power amplifier facilitates the bi-directional exchange of electrical power by controlling the voltage or current at the interface, and is generally some type of power electronic converter. For mechanical systems, assertion of speed or torque is required, which can be achieved with an electrical motor controlled via a power electronic converter.

For design assurance and prototype testing, both PHIL and CHIL provide a means to emulate the behaviour of a system at both sides of the interface. This enables the system and the integrated platform to be tested in a more realistic environment which can help to expose faults that are difficult to discover via traditional testing.

Some key application examples of HIL, and a discussion of the benefits it can provide as an additional tool for test and evaluation are presented in the following section.

EXAMPLE HARDWARE-IN-THE-LOOP APPLICATIONS FOR TEST AND EVALUATION OF MARITIME PLATFORMS

Hardware-in-the-Loop can be utilised to replace any desired system or subsystem with its physical counterpart. In this section, examples based on a simplified submarine power and energy system, as shown in Figure 2, will be used to illustrate and discuss how HIL can support test and evaluation of maritime platforms, in conjunction with or in place of a full test system.

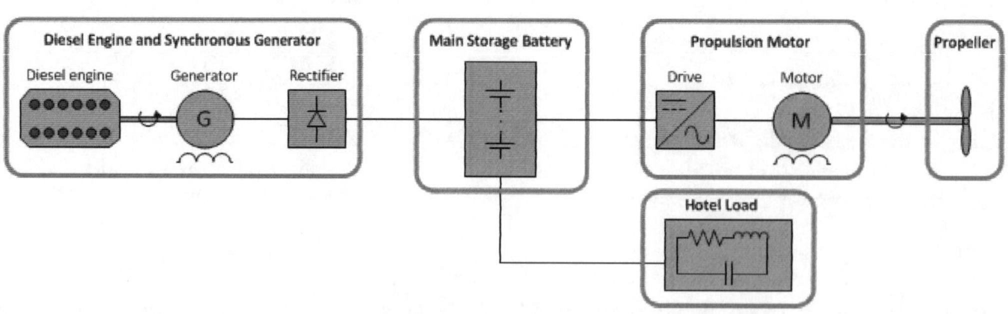

Figure 2: A simplified submarine power and energy system.

Hardware-in-the-Loop Emulation of a Propeller Load

An excellent candidate for PHIL is the emulation of the propeller with an electrical machine controlled by a power electronics converter as shown in Figure 3. While more expensive and complicated compared to a traditional water brake or other dynamometer load, this system should provide better replication of the propeller load dynamics. This allows the propulsion motor to be easily and repeatedly loaded according to any desired operational profile for the propeller.

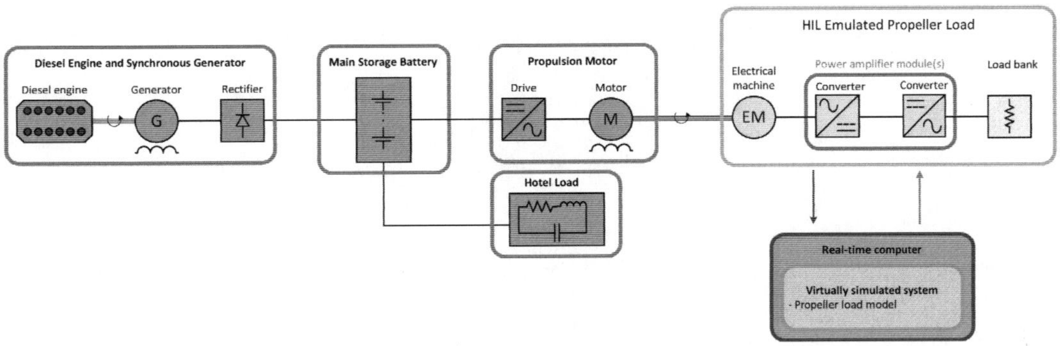

Figure 3: A simplified submarine power and energy system with emulated propeller load.

This HIL topology allows the test system to leverage the benefits of having a high quality propeller model into realistic stimulation of the propulsion system. Changes to the propeller model over time are also easily incorporated. An additional benefit of emulating the propeller via this topology is realised when used in conjunction with the following example.

Hardware-in-the-Loop for Mechanical Systems

Two key examples of how individual mechanical items can be tested with a HIL implementation are shown in Figure 4 and Figure 5. Figure 4 shows a HIL configuration which uses an electrical machine to draw power from a diesel engine. The mechanical load profile applied to the engine is controlled by the virtual simulation of the rest of the power system. The configuration in Figure 5 takes the concept one step further by interfacing a main propulsion motor to a virtual simulation at both its input and output. The simulation is configured to apply realistic supply and load stimuli such that the motor is tested as if it were connected to the actual power system. As shown in Figure 5, this topology also has the ability to recirculate the electrical power to minimise the overall power consumption.

Both topologies are useful for tests that do not require full fidelity, or when it is not desirable to operate the rest of the physical system if it exists. For example, an endurance test of an item of equipment over a long period of time would be expensive for a full test system in terms of fuel costs and maintenance overheads. Furthermore, it would prevent a full test system from being used to perform other tests. Subject to availability of hardware, the configurations outlined in Figure 4 and Figure 5 can be operated in parallel to reduce costs and maximise the utilisation of a test facility.

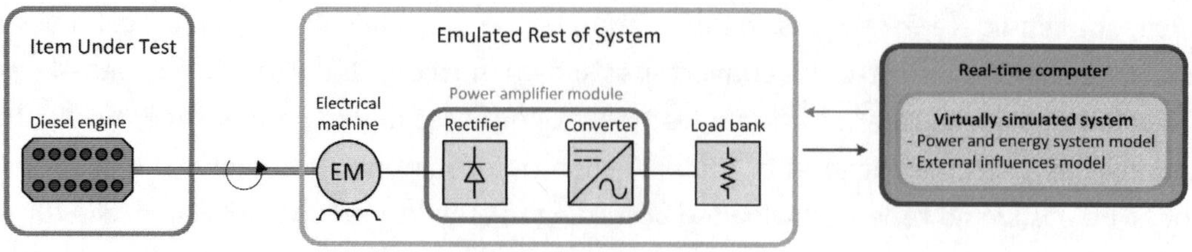

Figure 4: Hardware-in-the-Loop configuration for testing a diesel engine.

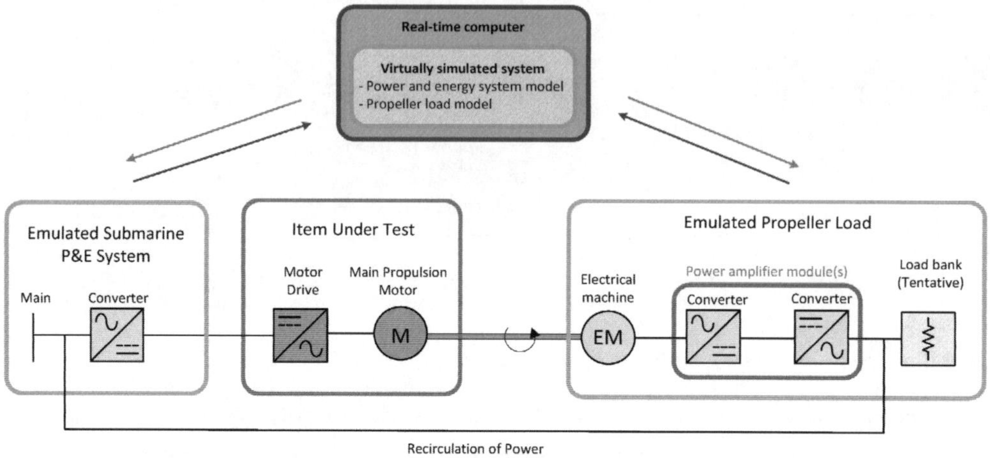

Figure 5: Hardware-in-the-Loop configuration for testing a main propulsion motor.

Hardware-in-the-Loop Emulation of a Main Storage Battery

A main storage battery strongly affects the performance, stability, and characteristics of the power and energy system of a maritime platform. Tests that involve the main storage battery can often be difficult to coordinate and time consuming, as significant efforts are required to condition the system to the required state.

Emulation of a battery using PHIL can be implemented as shown in Figure 6. The battery model is simulated and interfaced with the rest of the system via a power electronic converter. This topology has a number of advantages compared to a real battery. First, unlike a real battery, an emulated battery does not deteriorate with age. Second, the emulated battery does not require pre-conditioning prior to a test. This can eliminate or minimise significant long term running costs of such a system. Key parameters including age, state of charge, and state of health can be easily adjusted within the virtually simulated system and hence the emulated system, significantly reducing the preparation required for an experiment. An experimental proof of concept for emulating a battery via PHIL will be presented in the following section.

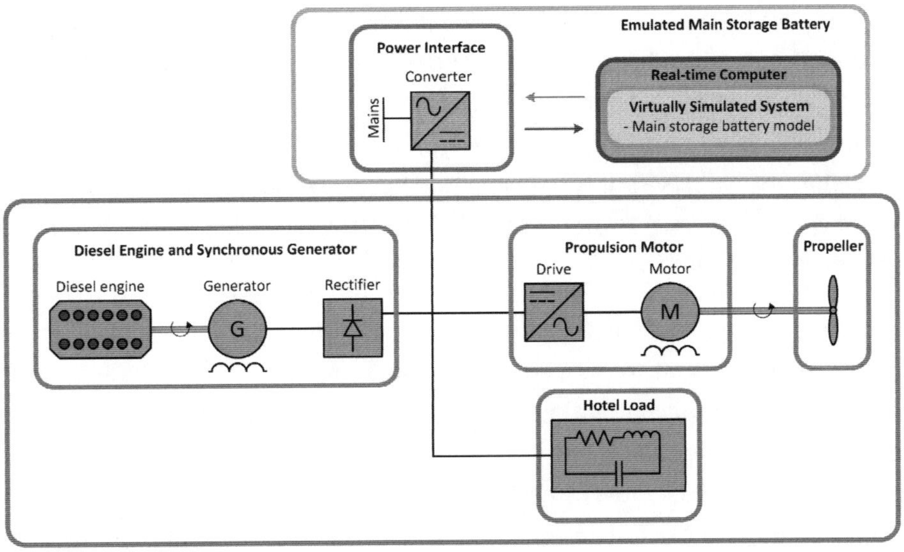

Figure 6: A simplified submarine power and energy system with an emulated main storage battery.

Discussion

Replicating the behaviour of a simulated system at various tear points using HIL provides an interesting means to scale hardware testing to satisfy the appetite for cost and risk in verifying a design. While it is expected that testing conducted in a large scale facility which fully replicates a platform should have the highest level of fidelity, it is also a very expensive solution. Examples in this section have demonstrated how HIL can be used to enable testing from a single component to the entire system. While subject to the accuracy of simulation models and the interface, a HIL system will not be able to fully replicate the exact behaviour of the original system. However, for most applications HIL based systems should provide sufficient fidelity to deliver the necessary level of assurance. Hardware-in-the-Loop ultimately gives the designer an extra tool for achieving the objectives of the test and evaluation program.

Broadening the range of tools available for testing provides options for test designers to satisfy various considerations such as: reducing overall program costs, improving scheduling and utilisation of test facilities, investigating faults, and early assessment of the impact of new technologies.

For the maritime platform considered in this paper, emulation of the main storage battery demonstrates an opportunity for a PHIL system to replace expensive consumables to achieve savings, while increasing overall flexibility. The implementation of this example will be further explored in a case study in the next section. Replacement of diesel generators with a power electronics based emulator powered by the main supply can further reduce fuel and maintenance costs. An emulated system can also work in parallel to partially replace a real system, providing trade-offs between cost and fidelity.

Schedule risk is a common consideration, as inevitably unforeseeable events such as damage to key components can cause delays in the test program. The impact of delays can be lessened by replacing unavailable components with an emulated system, enabling critical tests to continue with well understood penalties in overall fidelity. HIL implementations also enable a large test system to be broken into smaller test benches that can be operated in parallel, such as the mechanical system testing example. This provides additional scheduling options and maximises system utilisation.

By modifying models in the virtually simulated system, HIL implementations can be manipulated to replicate faults in a controlled and safe manner. This provides a valuable tool for troubleshooting and fault mitigation. For example, testing a main storage battery short circuit could cause significant damage to the battery and surrounding systems. A battery emulation based on a sufficiently rated power electronic interface allows the test to be conducted repeatedly in a controlled manner. As the understanding of a platform improves and the virtually simulated system is enhanced to more accurately reflect the behaviour of a real system, selected high risk tests can be conducted in the virtual system. Additionally, CHIL provides the option for testing software without risk to the real equipment, and can be conducted in parallel with other testing activities.

The ability to emulate behaviour of virtually simulated models using HIL also provides an interesting tool for research, development and testing of upgrades for a platform. The traditional process for upgrading a component is: simulation development, manufacture of a prototype, and verification in a complete system. HIL can provide an intermediate step between simulation and prototyping, by emulating the developmental component in a full test system. This enhances confidence in the design by allowing problems to be identified early, and potentially reduces the number of design cycles required. Initial integration testing can also be performed alongside the design verification. Both measures should reduce problems encountered with the eventual prototype. In the long term, emulated components with verified performance provide alternatives for testing upgraded components in a reference system for the platform. This reduces the need to maintain a full inventory of all versions of key components over the life of the platform. Additionally, the stress placed on the test system due to configuration changes, and requirements to work on parallel versions of the same platform are eased.

In summary, HIL provides a tool for designers and decision makers to scale their test regime according to the risk appetite throughout the life of a platform. Obviously the fidelity of a test is affected by how much of the real system is emulated, which directly affects the level of assurance a test can offer. However, the HIL implementation offers reduced setup effort, running cost, and risk, which can be an acceptable compromise.

The following section presents an experimental case study of PHIL emulation of a main storage battery to demonstrate the practical implementation and performance of the technique.

POWER HARDWARE-IN-THE-LOOP EMULATION OF A MAIN STORAGE BATTERY CASE STUDY

A low power, proof of concept PHIL experiment where a main storage battery is emulated in a physical test system has been performed at DST. The purpose of the experiment was to demonstrate the technique itself, as well as the performance.

A simplified equivalent electrical circuit for the submarine power train is shown at the top of Figure 7. It consists of the battery and the current flow to or from the rest of the system. To demonstrate this PHIL implementation, the experimental system needs to simulate the battery model and provide a power interface to the physical rest of system (ROS) as in Figure 6. However, in a low power system such as that at DST, the rest of the physical system does not exist. Therefore, the experimental proof of concept system also needs to emulate the ROS. The physical ROS was implemented as a current exchange with the emulated battery terminal as shown at the bottom of Figure 7. The exchange was configured to be representative of the power that the actual submarine systems would draw from or supply to the battery.

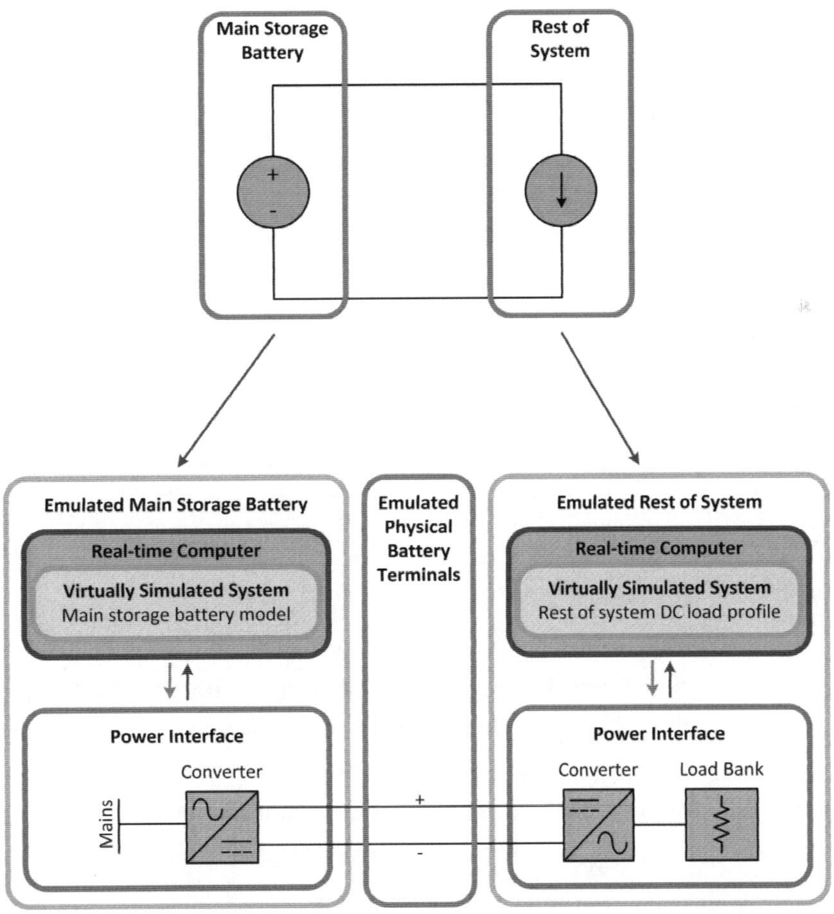

Figure 7: Reduced scale Power Hardware-in-the-Loop emulation of a main storage battery – simplified equivalent electrical circuit of a submarine power train (top); Power Hardware-in-the-Loop implementation (bottom).

Battery Simulation Model

A generic lead-acid battery model has been constructed from experimental data collected by DST from a single lead acid battery cell. The experimental data captured the response of the terminal voltage of the battery cell to a stimulation current profile. The model was built from the Ceraolo flooded lead-acid model [8], which is a network of electrical components tuned to represent the dynamics of a battery as shown in Figure 8. The experimental data was used to determine the parameters of the model via a parameter optimisation process. The resulting model provides good reproduction of the original experimental terminal voltages when stimulated with the same current profile. A larger battery string is obtained by simply scaling the output of the single cell model to the desired number of series cells.

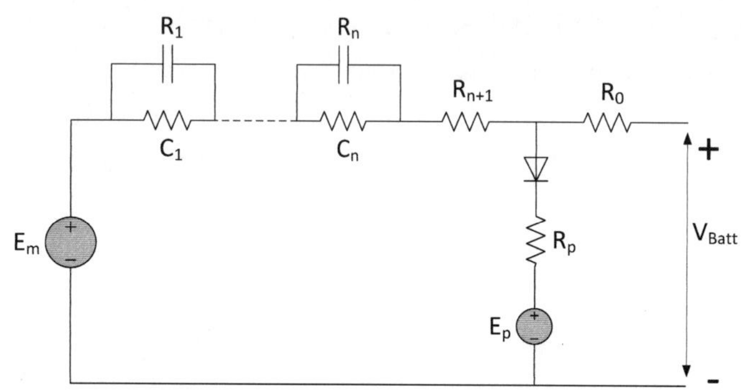

Figure 8: Ceraolo flooded lead-acid battery model.

Experimental Configuration

The configuration for the PHIL experiment is shown in Figure 9. It consists of three power amplifier modules (PAMs), two load banks, and the Real Time Digital Simulator (RTDS) Technologies' real-time computer. The PAMs are a power electronic building block approach to enable virtually any interface topology to be constructed, including the DC terminals necessary for emulation of a battery as shown in Figure 9. The PAMs were designed to maximise bandwidth for their power level to facilitate interfaces with as much transparency as possible while maintaining stable operation [9].

The control objectives of the interface for this configuration are the DC voltage and the DC power draw. The DC voltage is controlled by PAM2 which regulates the power exchange with PAM1 to obtain the target voltage provided by the terminal voltage of the simulated battery model. The DC power is regulated by PAM3 and the measured current is fed back to the simulated battery model. This creates the closed loop system where the battery simulation model is stimulated by the experimental current, i.e. Power Hardware-in-the-Loop.

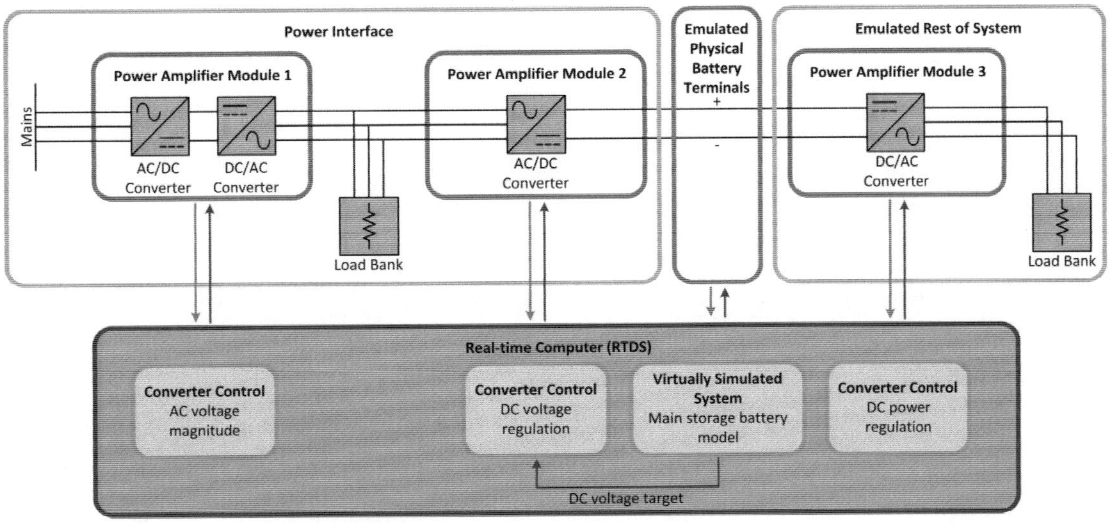

Figure 9: Experimental system configuration.

Power Hardware-in-the-Loop Results

The PHIL experiment was performed for 20 minutes, with 5 minute alternating cycles of charge and discharge. A constant current at the emulated battery terminal was used for both cycles as shown in Figure 10, and the transition between the modes was instantaneous.

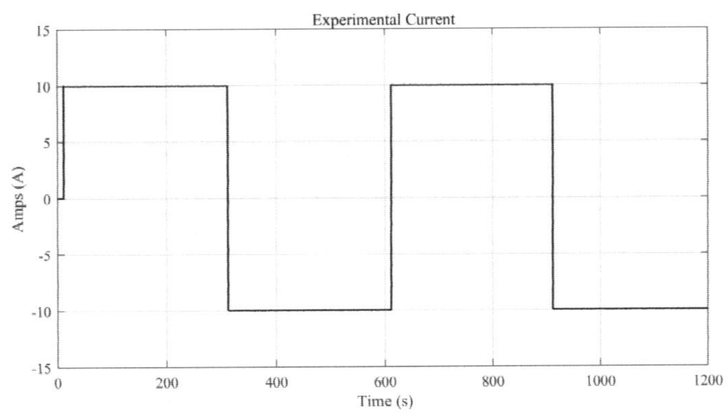

Figure 10: Experimental battery current during 20 minute PHIL experiment.

The simulated and experimental battery voltage waveforms obtained from the PHIL experiment are shown in Figure 11. A closer view of the charging cycle is shown in Figure 12. Both plots show good agreement between the simulated and experimentally emulated battery terminal voltages.

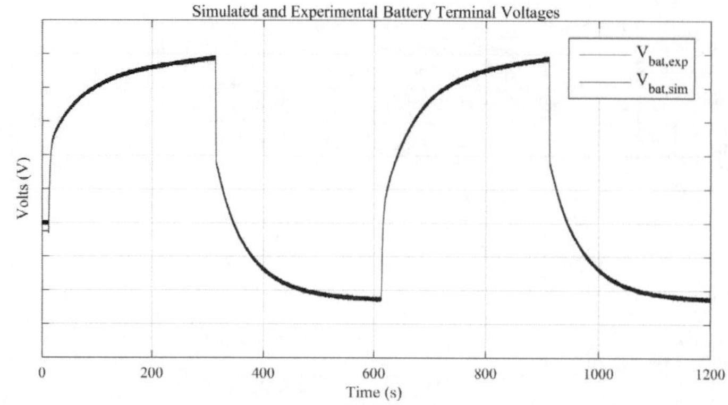

Figure 11: Simulated and experimental battery terminal voltages during 20 minute PHIL experiment.

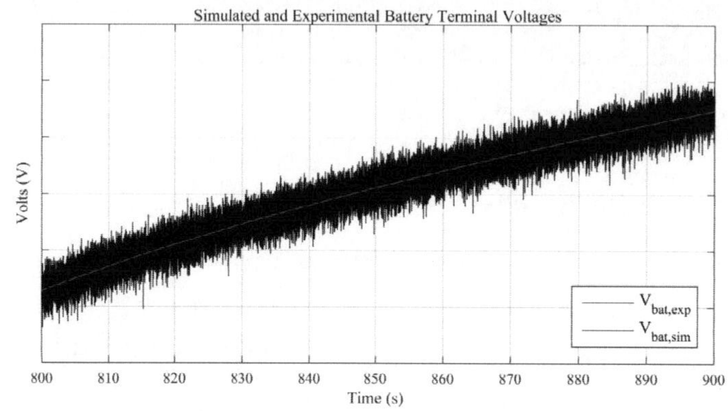

Figure 12: Simulated and experimental battery terminal voltages during 20 minute PHIL experiment – zoomed on peak of charge.

The dynamic performance of the emulation is demonstrated in Figure 13 where the change in voltage at the transition from charge to discharge is followed accurately at the emulated battery terminals. Both the steady state and dynamic performance demonstrate that the physical ROS is being presented with an accurate representation of the simulated battery terminal characteristic. Additionally, the shape of the voltage waveform is largely as expected indicating that the battery model is being stimulated appropriately.

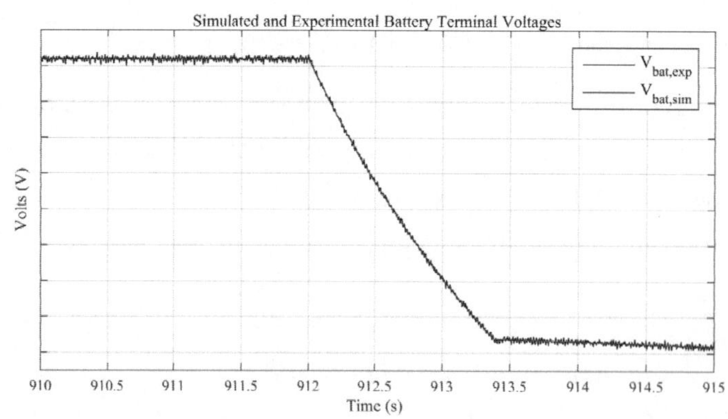

Figure 13: Simulated and experimental battery terminal voltages during 20 minute PHIL experiment – zoomed on charge to discharge mode change.

A battery is an excellent candidate for PHIL due to its slow dynamics. The real-time simulation model can achieve excellent fidelity under these conditions since the trade-offs required to achieve real-time operation are minimised. Furthermore, the experimental system can offer significantly faster dynamics than the battery, which minimises the impact of implementing the PHIL experiment.

While this experiment has been carried out at low power, the experimental control systems and the simulation model are fully transferable to higher power levels. The excellent performance is also expected to be maintained, as shifting both the experimental and simulation systems to higher power should allow the experimental system to remain significantly faster than the simulation.

CONCLUSION

Through life test and evaluation of the power and energy system for a naval platform is difficult due to the complex functional requirements, continuous upgrade cycles and advanced bespoke equipment. To meet these challenges, testing is increasingly being conducted in dedicated land based test facilities, prior to installation into a platform. However, this approach requires a great deal of effort and cost. Therefore there are strong imperatives to maximise such utilisation while minimising overall program budget.

Hardware-in-the-Loop (HIL) is a technique which enables a real item of equipment to seamlessly interact with a virtual real-time simulation. It provides a means to realistically emulate operation of a system at almost any point. This paper has discussed the basic concepts behind HIL. Through a number of examples and a case study, the application and opportunities that HIL can provide for test and evaluation of maritime platforms have been demonstrated. The strength of HIL is its ability to realistically emulate key components of a system, and realistically stimulate real items without the full complement of equipment. This provides a tool for designers to manage schedule and cost risks via more thorough testing. The concept can also benefit the through life support of a platform, from the design stage, to long term research and development of new technologies to maintain a capability edge. Ultimately, HIL provides a tool to scale and customise the testing regime according to the risk appetite, while trading off against the required fidelity and effort.

REFERENCES

1. C. S. Edrington, M. Steurer, J. Langston, T. El-Mezyani and K. Schoder, "Role of Power Hardware in the Loop in Modelling and Simulation for Experimentation in Power and Energy Systems," in *Proceedings of the IEEE,* vol. 103, no. 12, pp. 2401-2409, Dec. 2015.

2. M. Steurer, C. S. Edrington, M. Sloderbeck, W. Ren and J. Langston, "A Megawatt-Scale Power Hardware-in-the-Loop Simulation Setup for Motor Drives," in *IEEE Transactions on Industrial Electronics,* vol. 54, no. 4, pp. 1254-1260, Apr. 2010.
3. W. Ren, M. Steurer and T. L. Baldwin, "Improve the Stability and the Accuracy of Power Hardware-in-the-Loop Simulation by Selecting Appropriate Interface Algorithms," in *IEEE Transactions on Industry Applications*, vol. 44, no. 4, pp. 1286-1294, Jul.-Aug.. 2008.
4. G. F. Lauss, M. O. Faruque, K. Schoder, C. Dufour, A. Viehweider and J. Langston, "Characteristics and Design of Power Hardware-in-the-Loop Simulations for Electrical Power Systems," in *IEEE Transactions on Industrial Electronics*, vol. 63, no. 1, pp. 406-417, Jan. 2016.
5. P. C. Kotsampopoulos, F. Lehfuss, G. F. Lauss, B. Bletterie and N. D. Hatziargyriou, "The Limitations of Digital Simulation and the Advantages of PHIL Testing in Studying Distributed Generation Provision of Ancillary Services," in *IEEE Transactions on Industrial Electronics*, vol. 62, no. 9, pp. 5502-5515, Sep. 2015.
6. J. Langston et al., "Power hardware-in-the-loop testing of a 500 kW photovoltaic array inverter," in *IECON 2012 - 38th Annual Conference on IEEE Industrial Electronics Society*, pp. 4797-4802, Oct. 2012.
7. F. Huerta, R. L. Tello and M. Prodanovic, "Real-Time Power-Hardware-in-the-Loop Implementation of Variable-Speed Wind Turbines," in *IEEE Transactions on Industrial Electronics*, vol. 64, no. 3, pp. 1893-1904, Mar. 2017.
8. M. Ceraolo, "New Dynamical Models of Lead-Acid Batteries," in *IEEE Transactions on Power Systems*, vol. 15, no. 4, pp. 1184-1190, Nov. 2000.
9. N. D. Marks, W. Y. Kong and D. S. Birt, "Stability of a Switched Mode Power Amplifier Interface for Power Hardware-in-the-Loop," in *IEEE Transactions on Industrial Electronics*, vol. 65, no. 11, pp. 8445-8454, Nov. 2018.

AUTHOR INDEX

AUTHOR INDEX